14th Conference on Recent Developments in Gravity 2010

(NEB XIV)

Journal of Physics: Conference Series Volume 283

Ioannina, Greece
8-11 June 2010

ISBN: 978-1-61782-664-1
ISSN: 1742-6588

14th Conference on Recent Developments in Gravity 2010

(NEB XIV)

Journal of Physics: Conference Series Volume 283

Ioannina, Greece
8-11 June 2010

ISBN: 978-1-61782-664-1
ISSN: 1742-6588

IOP Institute of Physics

Printed from e-media with permission by:

Curran Associates, Inc.
57 Morehouse Lane
Red Hook, NY 12571

Some format issues inherent in the e-media version may also appear in this print version.

Printed by Curran Associates, Inc. (2011)

For permission requests, please contact the Institute of Physics
at the address below.

Institute of Physics
Dirac House, Temple Back
Bristol BS1 6BE UK

Phone: 44 1 17 929 7481
Fax: 44 1 17 920 0979

techtracking@iop.org

Additional copies of this publication are available from:

Curran Associates, Inc.
57 Morehouse Lane
Red Hook, NY 12571 USA
Phone: 845-758-0400
Fax: 845-758-2634
Email: curran@proceedings.com
Web: www.proceedings.com

TABLE OF CONTENTS

PREFACE

The 14th conference in the series "Recent Developments in Gravity" was organized by the Theoretical Physics group of the Department of Physics, University of Ioannina, Greece. NEB-XIV (NEB-14) took place in Ioannina, Greece, from 8–11 June, 2010 at the Grand Serai Hotel.

The abstracts of all the talks, along with other details of the meeting, are to be found on the website of the conference http://neb14.physics.uoi.gr/index.htm.

The NEB series (initials from the Greek: Νεώτερες Εξελίξεις στην Βαρύτητα) is a series of meetings initiated in 1984 in Crete by the late Professor Vassilis Xanthopoulos and a group of Greek relativists. Since then this meeting has been held every other year at various cities in Greece, organized by different groups of Greek relativists and cosmologists. This was the third time it was organized at Ioannina (the other two times were in 1988 and 2000).

It was our pleasure to see that the international participation in these meetings has been rising constantly during the last few years and NEB-14 continued this tradition. The Conference was attended by a total of 106 participants, with 62 of them from Institutions outside of Greece. We had 20 plenary session invited speakers (15 of whom were of international origin), 42 parallel session talks and 2 posters.

The topics that are traditionally covered by the NEB conference series are:

- Cosmology (Dark Energy, Dark Matter, CMB etc)
- Gravitational Waves
- Alternative Theories of Gravity
- Relativistic Astrophysics
- Mathematical Relativity
- Quantum Gravity

During the present conference, the following variety of topics were addressed through the participants' talks and posters:

- mainstream and alternative models of dark energy and dark matter,
- construction and testing of new cosmological models,
- alternative (either 4-dimensional or higher-dimensional) theories of gravity and their predictions,
- the interplay of gravity with gauge theories,
- properties of astrophysical and miniature black holes and other massive astrophysical objects,
- cosmological perturbation spectra,
- modeling of astrophysical and cosmological observables via alternative theories,
- attempts to build a consistent quantum gravity theory.

We chose to place the emphasis in this meeting on subjects related to Observational and Theoretical Cosmology due to the rapid recent development of these research areas.

Our goal was to provide a stimulating environment for the presentation and discussion of the most recent cutting-edge research results in cosmology and classical and quantum gravity. We believe that this goal was achieved and we hope that the next meeting that will take place at Chania (Crete) in 2012 (NEB-15) will further raise the standards of this series.

Special thanks are due to the invited and keynote speakers for providing exciting talks and to all the participants for presenting interesting contributions and initiating fruitful discussions. We would also like to thank the Organizing Committee and the Scientific Committee for their valuable contributions to the organization of the meeting.

Finally we would like to thank our main sponsors – the University of Ioannina, the Academy of Athens, the National Bank of Greece and the 'Prokos' Bookstore – for providing financial support and making this meeting possible. Last but not least, we are grateful to the European Research and Training Network "UniverseNet" (MRTN-CT-2006035863-1) for providing the funds for the publication of the proceedings of our meeting.

The Editors
Leandros Perivolaropoulos
Panagiota Kanti

The Organizing Committee:
L Perivolaropoulos (Ioannina) (Chair)
P Kanti (Ioannina) (Co-Chair)
C Kolasis (Ioannina)
N Stergioulas (Thessaloniki)
K Kokkotas (Thessaloniki, Tuebingen)
D Papadopoulos (Thessaloniki)
M Plionis (Athens)
S Basilakos (Athens)
E Vagenas (Athens)
S Nesseris (Niels Bohr)
N Pappas (Ioannina)

The Scientific Committee:
A Ashtekar (Penn-State)
N Batakis (Ioannina)
D Christodoulou (ETH)

G Contopoulos (Academy of Athens)
C Frenk (University of Durham)
V Frolov (Alberta)

The Plenary Invited Speakers:
Prof. J A Font (University of Valencia)
Prof. V Frolov (University of Alberta)
Prof. R Gregory (University of Durham)
Prof. O Lahav (UCL London)
Prof. D Psaltis (University of Arizona)
Prof. S Sarkar (University of Oxford)
Prof. J Sola (Barcelona University)
Prof. R Woodard (University of Florida)

Keynote speakers:
Prof. I Bakas (University of Patras)
Prof. K Dimopoulos (University of Lancaster)
Prof. D Giannios (Princeton University)
Prof. N Mavromatos (Kings College, London)
Prof. D Polarski (University Montpellier II)
Prof. M Sakellariadou (Kings College, London)
Prof. T Sotiriou (University of Cambridge)
Prof. N Tetradis (University of Athens)
Prof. C Tsagas (University of Thessaloniki)

Dates

Issue 1 (2011)

Recent Developments in Gravity (NEB XIV) IOP Publishing
Journal of Physics: Conference Series **283** (2011) 012001 doi:10.1088/1742-6596/283/1/012001

Horizons, singularities and causal structure of the generalized McVittie space-times

Malcolm Anderson
Mathematics, Faculty of Science, Universiti Brunei Darussalam
Jalan Tungku Link, Gadong BE 1410, Brunei Darussalam

Enail: malcolm.anderson@ubd.edu.bn

Abstract. The generalized McVittie space-times, which are the natural extensions of a class of metrics introduced by McVittie in 1933 to model the space-time outside a spherical object embedded in an expanding universe, have recently been proposed by Faraoni and Jacques as candidate space-times for cosmological black holes. In this paper I analyze the singularities and horizon structure of the generalized McVittie space-times, and show that any expanding space-time of this type that satisfies the null energy condition and develops apparent horizons must asymptote to a standard McVittie solution. Furthermore, if the scale factor is asymptotically exponential then the space-time is future-incomplete and can be joined smoothly to an eternally-inflating Kottler (Schwarzschild-de Sitter) solution. I argue that no generalized McVittie space-time, apart from the Schwarzschild solution, can adequately represent a black hole, because all singular points are surrounded by anti-trapped regions rather than trapped regions.

1. Introduction

The (spatially-flat) McVittie space-times are a family of spherically symmetric, non-vacuum space-times with the general line element

$$ds^2 = -[1-M/(2ra)]^2[1+M/(2ra)]^{-2}dt^2 + a^2[1+M/(2ra)]^4(dr^2 + r^2 d\Omega^2) \tag{1}$$

where $d\Omega^2$ is the metric on the unit 2-sphere, M is a positive constant, and $a \equiv a(t)$ is a positive function to be specified.[1]

McVittie [1] introduced the line element (1) in 1933, in an attempt to model the space-time outside a non-rotating black hole embedded in an asymptotically Friedmann-Lemaître-Robertson-Walker (FLRW) universe.[2] The McVittie line element (1) reduces to the Schwarzschild metric in isotropic coordinates if $a \equiv 1$, and to the spatially-flat FLRW line element if $M = 0$.

The physical interpretation of the McVittie space-times is controversial even today. Sussmann [2] and Ferraris, Francaviglia and Spallicci [3] have rejected the view that the McVittie line element describes a black hole in an otherwise homogeneous universe because in almost all cases there is a

[1] Geometrized units are used throughout, so that $G = c = 1$.

[2] McVittie [1] also considered line elements that asymptote to universes with constant positive or negative spatial curvature, but attention will be restricted to the case of zero asymptotic curvature here.

curvature singularity at $r = M/(2a)$, and not at $r = 0$ as McVittie [1] originally claimed.[3] Moreover, the pressure of the underlying matter source is unbounded near this singularity while the density remains finite there, and such behavior is difficult to incorporate into a realistic equation of state.

The one case in which the surface $r = M/(2a)$ is regular occurs when the scale factor a is an exponential function of t. If $a(t) = a_0 \exp[(\Lambda/3)^{1/2} t]$ with a_0 and Λ positive constants, a transformation from (t,r) to coordinates (T,R) defined by

$$R = a_0 r \exp[(\Lambda/3)^{1/2} t]\{1 + (2a_0 r)^{-1} M \exp[(\Lambda/3)^{1/2} t]\}^2 \tag{2}$$

and

$$dT = dt + (\Lambda/3)^{1/2}(1 - 2M/R)^{-1/2}(1 - 2M/R - \tfrac{1}{3}\Lambda R^2)^{1/2} R\, dR \tag{3}$$

casts the McVittie line element (1) in the static form

$$ds^2 = -(1 - 2M/R - \tfrac{1}{3}\Lambda R^2) dT^2 + (1 - 2M/R - \tfrac{1}{3}\Lambda R^2)^{-1} dR^2 + R^2 d\Omega^2, \tag{4}$$

which is commonly referred to as the Schwarzschild-de Sitter solution, although it was first discovered by Kottler [5] in 1918. The Kottler solution, and the Schwarzschild solution to which it reduces when $\Lambda = 0$, are unique amongst the McVittie space-times in having a regular surface at $r = M/(2a)$ (or equivalently at $R = 2M$), and finite underlying density and pressure.

In the first of a series of papers on the McVittie space-times, Nolan [6] initially defended the black hole interpretation by arguing that, in the spatially-flat case, the McVittie space-times are the unique spherically symmetric space-times with a shear-free fluid source and homogeneous energy density that asymptote to an FLRW metric at space-like infinity. In a second paper, Nolan [4] rewrote the line element (1) in terms of the geometric radius

$$R = ar[1 + M/(2ar)]^2 \tag{5}$$

and characterized the resulting space-time as a "point mass at $R = 0$ surrounded by a singularity [at $R = 2M$]".[4] Nonetheless, he also demonstrated that, if the universe is expanding (so that $\frac{d}{dt}a(t) > 0$), then the singular surface at $R = 2M$ is surrounded by an anti-trapped region – the signature of a white hole – rather than the trapped region expected of a black hole.

More recently, Kaloper, Kleban and Martin [8] have criticized Nolan's stipulation that the energy density near a black hole in a cosmological background be homogeneous as "very restrictive and not very well-motivated physically", and have pointed out that it would almost inevitably induce an unbounded pressure gradient to balance the strong gravitational frame-dragging effects present at the horizon of a black hole. Despite this, they have registered their support for the black hole interpretation by demonstrating that, for some choices of the scale factor a, ingoing radial null geodesics that asymptote to the surface $r = M/(2a)$ are future-incomplete, and arguing that in these cases the completion of the McVittie space-time could contain a black hole region as well as a white hole region.

A natural generalization of the McVittie space-time that allows for an inhomogeneous density profile is the line element

$$ds^2 \equiv g_{ab} dx^a dx^b = -[1 - m/(2r)]^2 [1 + m/(2r)]^{-2} dt^2 + a^2 [1 + m/(2r)]^4 (dr^2 + r^2 d\Omega^2), \tag{6}$$

---

[3] In fact, the effect of replacing r in the McVittie line element (1) with $r^* \equiv M^2/(4r)$ is to recover the same line element with a^{-1} in place of a [4], and so the limit $r \to 0$ corresponds to the asymptotically FLRW region of a second McVittie space-time. A closely analogous situation occurs in the Schwarzschild metric, where isotropic coordinates are known to provide a double cover of the region outside the event horizon.

[4] However, Nolan [7] in a later paper showed that it is not possible to extend a McVittie space-time down to $R = 0$ unless the source fluid is tachyonic in the region $R < 2M$.

where now $a \equiv a(t)$ and $m \equiv m(t)$ are positive functions to be specified. Clearly, the McVittie line element (1) is recovered from (6) if m is chosen to be M/a. Faraoni and Jacques [9] have claimed that replacing t with a proper time coordinate in (6) can eliminate any singularities in the fluid pressure or density at the boundary surface $r = m/2$, but Carrera and Giulini [10] have correctly dismissed this claim and shown that the Ricci scalar diverges at $r = m/2$ in all cases except the Kottler solution.[5]

In what follows I will refer to any line element of the form (6) as a generalized McVittie (or gMcV) space-time. The analog of the geometric radius (5) for the gMcV space-times is the quantity

$$R = r(g_{rr})^{1/2} = ar[1+m/(2r)]^2. \tag{7}$$

Although it is known that all gMcV space-times bar the Kottler solution are singular on a surface with non-zero geometric radius $R = 2am$, it might nonetheless be possible that some members of the family avoid other undesirable features of the McVittie space-times. It is conceivable, for example, that there exist expanding gMcV space-times in which the singular surface is surrounded by a trapped region rather than an anti-trapped region. Furthermore, the suggestion by Kaloper, K;eban and Martin [8] that those McVittie (or gMcV) space-times that are known to be future-incomplete might contain black hole regions once they are completed has not yet been fully explored.

It is the purpose of this note to report the results of an extensive analysis of the salient features of the gMcV space-times, particularly the apparent horizons and causal structure. The overall conclusion is that the behavior of the gMcV space-times does not differ markedly from that of the McVittie space-times. In all cases the singular surface at $r = m/2$ is surrounded by an anti-trapped region if the space-time is expanding, and any future-incomplete McVittie or gMcV space-time that has $\lim_{t\to\infty}(\dot{a}/a) > 0$ can be completed by joining it smoothly to a Kottler extension, which also contains no trapped regions.

Indeed, the connections between the gMcV and McVittie space-times are even more intimate than might first appear, as any gMcV space-time that forms an apparent horizon inevitably asymptotes to a McVittie solution as $t \to \infty$ if the null energy condition is imposed.

2. Overview of the generalized McVittie space-times

In what follows, the functions a and m appearing in the gMcV line element (6) are assumed to be smooth, positive functions of the time coordinate t for all $t > 0$. If the space-time supports a cosmological singularity – a moment of time at which $a = 0$ – then this is assumed to occur in the limit as $t \to 0^+$. Further restrictions on the behavior of a and m will be imposed by the null energy condition, which is discussed in some detail below.

The radial coordinate r is in all cases taken to satisfy the restriction $r > m(t)/2$, as the region inside $r = m/2$ is simply a copy of a second exterior gMcV region. This can be seen by replacing r with $r^* \equiv m_0^2/(4r)$ in (6), where m_0 is any constant mass scale. The line element that results then has the same form as (6), save with m_0^2/m in place of m and am^2/m_0^2 in place of a. The isotropic coordinates of (6) therefore break the range $(0,\infty)$ of r into the exterior regions of two distinct gMcV space-times, in much the same way as happens with the McVittie and Schwarzschild space-times.

[5] Carrera and Giulini [10] have refuted a further statement made by Faraoni and Jacques [9], and assumed extensively by Faraoni in [11], that the line element of the Sultana-Dyer solution [12] can be transformed into the generalized McVittie form (6). This error has also been noticed by Sun [13].

2.1 The Einstein tensor and curvature singularities

With $d\Omega^2 = d\theta^2 + \sin^2\theta\, d\phi^2$, the non-zero components of the Einstein tensor generated by the gMcV line element (6) are[6]:

$$G_t^t = 3a^{-2}\Delta^{-2}(\Delta\dot{a} + 2\xi)^2, \tag{8}$$

$$-\tfrac{1}{16} r^{-4}\Delta^{-2}a^2(2r+m)^6 G_t^r = G_r^t = 8a^{-1}\Delta^{-3}(2r+m)\xi \tag{9}$$

and

$$G_r^r = G_\theta^\theta = G_\phi^\phi = 3a^{-2}\Delta^{-2}(\Delta\dot{a}+2\xi)^2 + 2\Delta^{-1}(2r+m)\tfrac{\partial}{\partial t}[a^{-1}\Delta^{-1}(\Delta\dot{a}+2\xi)], \tag{10}$$

where an overdot denotes d/dt, while $\xi \equiv \frac{d}{dt}(am)$ and $\Delta \equiv 2r - m$. Carrera and Giulini [10] refer to the notable feature that $G_r^r = G_\theta^\theta$ as "spatial Ricci-isotropy".

An immediate consequence of (9) is that the space-time carries a non-zero radial energy flux unless $\xi \equiv 0$, or equivalently $m \propto a^{-1}$. It was with the objective of ruling out any transfer if energy between the ambient medium and the central object that McVittie [1] restricted his attention to the line element (1) that bears his name.

Also, the Ricci scalar of the gMcV space-times is:

$$R_a^a = -6a^{-2}\Delta^{-3}\{(a\ddot{a}+\dot{a}^2)\Delta^3 + 2[3\dot{a}\xi + a\dot{\xi} + m(a\ddot{a}-\dot{a}^2)\Delta^2 + 2(5\xi^2 - 3\dot{a}m\xi + 2am\dot{\xi})\Delta + 4am\dot{m}\xi]\}\,. \tag{11}$$

It is clear that the Ricci scalar diverges at $r = m/2$ unless $\dot{m}\xi \equiv 0$. If $\xi \equiv 0$ the standard McVittie space-time is recovered, but R_a^a still diverges at $r = m/2$ unless $a\ddot{a} \equiv \dot{a}^2$, in which case *a* is constant or an exponential function of *t*. On the other hand, if $\dot{m} \equiv 0$ then $\xi \equiv \dot{a}m$ and $\dot{\xi} \equiv \ddot{a}m$, and the Ricci scalar diverges at $r = m/2$ unless $a\dot{a} + \dot{a}^2 \equiv 0$, or equivalently $a \propto t^{1/2}$. If $a = a_0 t^{1/2}$ for some constant a_0, and *m* is constant, then in fact R_a^a is zero everywhere, but the square of the Ricci scalar,

$$R^{ab}R_{ab} = t^{-4}\Delta^{-4}[\tfrac{3}{4}(2r+m)^4 - 512a_0^{-2}(2r+m)^{-4}m^2r^4t], \tag{12}$$

remains divergent at $r = m/2$. Thus, the Schwarzschild and Kottler solutions are the only members of the gMcV family that are non-singular at $r = m/2$, as Carrera and Giulini [10] claim.

The question of whether the Ricci scalar diverges at a cosmological singularity, where $a(t) \to 0$, is more delicate. R_a^a will be divergent in this limit unless all the coefficients of the various powers of Δ in (11) go to zero at least as rapidly as a^2. As will be seen below, $\lim_{t\to 0^+} m(t) = 0$ is inconsistent with the null energy condition in an expanding universe, which requires that $\dot{m} \le 0$, while the cosmological singularity is inaccessible to past-directed null geodesics if $\lim_{t\to 0^+} m(t) = \infty$. So it is sufficient to consider only the case where $\lim_{t\to 0^+} m(t)$ is a positive constant. The right-hand side

---

[6] The convention for the Riemann tensor used here is that

$$R_{abcd} = \tfrac{1}{2}(\partial_a\partial_c g_{bd} + \partial_b\partial_d g_{ac} - \partial_a\partial_d g_{bc} - \partial_b\partial_c g_{ad}) + g_{mn}(\Gamma_{ac}^m\Gamma_{bd}^n - \Gamma_{ad}^m\Gamma_{bc}^n)$$

where Γ_{ac}^m is the Christoffel symbol. Equivalently, $\nabla_a\nabla_c V_b - \nabla_c\nabla_a V_b = R_{abcd}V^d$ for any smooth vector field V^d, and in geometrized units the Einstein equation reads $G_b^a = -8\pi T_b^a$.

of (11), if bounded, then imposes four independent constraints on $\dot{a}$, $\ddot{a}$, ξ and $\dot{\xi}$ which are satisfied iff $\dot{a}/a$, $\ddot{a}/a$, ξ/a and $\dot{\xi}/a$ (or equivalently $\dot{m}$ and $\ddot{m}$) remain finite as $a \to 0$. But a cannot go to zero in a finite coordinate time t if $\dot{a}/a$ is bounded, so the Ricci scalar as guaranteed to diverge at a cosmological singularity in an expanding universe if the null energy condition holds,

2.2 The null energy condition

In view of the wide range of choices available for the metric functions a and m, it is useful to isolate a subset of the gMcV space-times that are potentially of physical interest by imposing a suitable condition on the stress-energy tensor T^a_b that the space-time inherits from the Einstein equations. Given that almost all the standard McVittie space-times violate the dominant energy condition – the requirement that $-T^a_b w^b$ be a future-directed timelike or null vector for all futire-directed timelike vectors w^b – I will instead follow Kaloper, Kleban and Martin [8] in imposing the null energy condition, which specifies that $T_{ab}v^a v^b \geq 0$ for all (future-directed) null vectors v^a.

In the case of the gMcV space-times, the null energy condition reads simply

$$G^t_t - G^r_r \geq 8a^{-1}r^2\Delta(2r+m)^{-3}\,|\,G^t_r\,|. \tag{13}$$

In particular, it follows that $G^t_t - G^r_r \geq 0$, which in view of (8) and (10) is equivalent to:

$$2\dot{m}(\xi/a) + 2\Delta\tfrac{d}{dt}(\xi/a) + \Delta^2\tfrac{d}{dt}(\dot{a}/a) \leq 0 \tag{14}$$

This in turn is true only if $\dot{m}\xi \leq 0$ and $\frac{d}{dt}(\dot{a}/a) \leq 0$. Given that $\xi = a\dot{m} + \dot{a}m$, while a and m are positive by assumption and $\dot{a} \geq 0$ in an expanding universe, an immediate consequence is that $\dot{m} \leq 0$ and $\xi \geq 0$.

So the function m and the Hubble parameter $\dot{a}/a$ are non-increasing, and the function am is non-decreasing, in any expanding gMcV space-time that satisfies the null energy condition. In addition, since now $G^t_r \geq 0$ the null energy condition (13) imposes the residual constraint

$$G^t_t - G^r_r \geq 8a^{-1}r^2\Delta(2r+m)^{-3}G^t_r, \tag{15}$$

or equivalently

$$a^2(2r+m)^3[2\dot{m}(\xi/a) + 2\Delta\tfrac{d}{dt}(\xi/a) + \Delta^2\tfrac{d}{dt}(\dot{a}/a)] + 8r^2\Delta\xi \leq 0. \tag{16}$$

This inequality needs considerable work to analyze fully, and will be examined in more detail in Section 3.2.

2.3 Possible matter sources

Faraoni and Jacques [9] and Carrera and Giulini [10] have already considered a number of possible matter sources for the gMcV space-times, and I will add little that is new here.

The simplest realistic matter source is a perfect fluid, with stress-energy tensor

$$T^a_b = (\rho + p)u^a u_b + p\delta^a_b, \tag{17}$$

where

$$u^a = (-g_{tt})^{-1/2}\cosh\beta\,\delta^a_t + (g_{rr})^{1/2}\sinh\beta\,\delta^a_r \tag{18}$$

is a radial 4-velocity field with rapidity $\beta(t,r)$. In particular,

$$T^r_r = \rho\sinh^2\beta + p\cosh^2\beta \quad \text{and} \quad T^\theta_\theta = p. \tag{19}$$

Hence, the metric requirement $G^r_r = G^\theta_\theta$ can be satisfied by a perfect-fluid source only if $(\rho + p)\sinh\beta = 0$. But then $T^t_r \propto (\rho + p)\sinh\beta\cosh\beta$ is also zero, and so from (9) $\xi = 0$.

Those gMcV space-times that do not belong to the standard McVittie class therefore require a more elaborate source model. Carrera and Giulini [10] have suggested adding a term representing a radial heat flux vector

$$q^a = q[(-g_{tt})^{-1/2} \sinh\beta\, \delta_t^a + (g_{rr})^{1/2} \cosh\beta\, \delta_r^a] \tag{20}$$

orthogonal to u^a, so that the stress-energy tensor becomes[7]

$$T_b^a = (\rho + p) u^a u_b + p\, \delta_b^a + q^a u_b + u^a q_b . \tag{21}$$

Then

$$T_r^r = \rho \sinh^2\beta + p\cosh^2\beta + 2q \sinh\beta \cosh\beta \quad \text{and} \quad T_\theta^\theta = p , \tag{22}$$

and the metric condition $G_r^r = G_\theta^\theta$ is satisfied if

$$q = -\tfrac{1}{2}(\rho + p)\tanh\beta . \tag{23}$$

Combining the stress-energy tensor (21) with the components (8)-(10) of the Einstein tensor gives the following expressions for the phenomenological variables:

$$8\pi\rho = 3a^{-2}(\dot{a} + 2\Delta^{-1}\xi)^2 , \tag{24}$$

$$8\pi p = -3a^{-2}(\dot{a} + 2\Delta^{-1}\xi)^2 - 2\Delta^{-3}(2r+m)[2\dot{m}(\xi/a) + 2\Delta \tfrac{d}{dt}(\xi/a) + \Delta^2 \tfrac{d}{dt}(\dot{a}/a)] , \tag{25}$$

$$8\pi q = 32 r^2 a^{-2} (2r + m)^{-2} \Delta^{-2} \xi \tag{26}$$

and

$$\tanh\beta = 32 r^2 a^{-2} (2r+m)^{-3} [2\dot{m}(\xi/a) + 2\Delta \tfrac{d}{dt}(\xi/a) + \Delta^2 \tfrac{d}{dt}(\dot{a}/a)]^{-1} \Delta\xi . \tag{27}$$

Note that for generic choices of the metric functions a and m the right-hand side of (27) is not necessarily consistent with the constraint $|\tanh\beta| \le 1$. However, if source fluid satisfies the null energy condition (13) then

$$|\tanh\beta| \equiv -8a^{-1} r^2 \Delta (2r + m)^{-3} (T_t^t - T_r^r)^{-1} |T_r^t| \tag{28}$$

is automatically less than or equal to 1. So the heat flux model (21) is in some sense a natural choice for the source fluid underlying the subclass of gMcV space-times that satisfy the null energy condition.

From (24)-(27), the fluid density in the heat flux model is guaranteed to be positive or zero, and both ρ and q diverge as Δ^{-2} in the limit $r \to m/2$ unless $\xi \equiv 0$, in which case ρ remains bounded at $r = m/2$ and of course $q = 0$ everywhere. Also, the pressure p diverges as Δ^{-3} unless $\dot{m}\xi = 0$. If $\xi = 0$ it diverges as Δ^{-1} unless $a\ddot{a} - \dot{a}^2 \equiv 0$, in which case the solution is Kottler or Schwarzschild and the pressure remains finite at $r = m/2$; if $\dot{m} = 0$ the pressure diverges as Δ^{-2} unless $2a\ddot{a} + \dot{a}^2 \equiv 0$, or equivalently $a \propto t^{2/3}$, in which case p is everywhere zero. Finally, the rapidity β is undefined if $\xi = 0$, $\tfrac{d}{dt}(\xi/a) = 0$ and a is an exponential function of time, and is otherwise zero if $\xi = 0$. Leaving these special cases aside, β goes to zero linearly in Δ as $r \to m/2$ unless $\dot{m} = 0$, in which case

$$\lim_{r \to m/2} \tanh\beta = \tfrac{1}{2}(am)^{-1} [\tfrac{d}{dt} \ln(\dot{a}/a)] . \tag{29}$$

Furthermore, if the universe is expanding ($\dot{a} \ge 0$) and the null energy condition is satisfied then $\rho + p \ge 0$, the heat flux is outwards ($q \ge 0$) and the bulk flow is inwards ($\beta \le 0$).

---

[7] Faraoni and Jacques [9] were the first to consider the addition of a heat flux vector, but chose q^a to be a purely spatial vector $qa^{-1}[1 + m/(2r)]^{-2} \delta_r^a$.

2.4 The mass of the central object

A quasi-local mass function that is commonly used to measure the energy of spherically-symmetric space-times is the Misner-Sharp mass M_{MS}, which is defined by [14]

$$1 - 2M_{\text{MS}}/r = g^{ab}\partial_a R\,\partial_b R\,, \tag{30}$$

where R is the geometric radius given explicitly for the gMcV space-times in (7), and g^{ab} is of course the inverse metric tensor. In the case of the gMcV space-times,

$$M_{\text{MS}} = am + \tfrac{1}{128}ar^{-3}(2r+m)^6(\dot{a} + 2\Delta^{-1}\xi)\,. \tag{31}$$

It is evident from (31) that the Misner-Sharp mass is positive everywhere outside $r = m/2$ but is bounded in the vicinity of $r = m/2$ only if the space-time is McVittie (i.e. $\xi = 0$).

An alternative characterization of the Misner-Sharp mass is

$$M_{\text{MS}} = -\tfrac{1}{2}R^3 R^{\theta\phi}{}_{\theta\phi}\,, \tag{32}$$

where the Riemann tensor component $R^{\theta\phi}{}_{\theta\phi}$ can be interpreted as the extrinsic curvature of the surfaces of constant r and t. By decomposing $R^{\theta\phi}{}_{\theta\phi}$ in the form

$$R^{\theta\phi}{}_{\theta\phi} = \tfrac{1}{2}(R^{\theta}_{\theta} + R^{\phi}_{\phi}) - \tfrac{1}{6}R^{a}_{a} + C^{\theta\phi}{}_{\theta\phi}\,, \tag{33}$$

where $C^{ab}{}_{cd}$ is the Weyl tensor, it is possible to isolate the contributions to M_{MS} of the fluid source (the Ricci terms) and the gravitational field of the putative central object (the Weyl term), which Carrera and Giulini [10] denote by E_{R} and E_{W} respectively. A straightforward calculation gives:

$$E_{\text{W}} = am \quad \text{and} \quad E_{\text{R}} = \tfrac{1}{128}ar^{-3}(2r+m)^6(\dot{a} + 2\Delta^{-1}\xi)\,. \tag{34}$$

Thus, if the universe is expanding and the null energy condition holds, the mass E_{W} of the central object – which Carrera and Giulini [10] call the "Weyl energy" and Nolan [4] calls "Hawking's renormalized quasi-local mass" – is positive and non-decreasing, as $\xi \equiv \frac{d}{dt}(am) \geq 0$. Similarly, the energy E_{R} of the fluid source is everywhere positive or zero, and is zero everywhere only if the space-time is McVittie with $\dot{a} = 0$ (the Schwarzschild solution). In fact, in view of (24), $E_{\text{R}} = \frac{4\pi}{3}\rho R^3$ [10].

2.5 Radial null geodesics and apparent horizons

The causal structure of the gMcV space-times is largely determined by the behavior of the radial null geodesics. These are the integral curves of the equations

$$dr/dt = \pm(g_{rr})^{-1/2}(-g_{tt})^{1/2} = \pm 4a^{-1}r^2(2r+m)^{-3}(2r-m)\,, \tag{35}$$

where the upper and lower signs refer to outgoing and ingoing geodesics respectively. The corresponding null tangent vectors $u^a_{\pm} \equiv dx^a/d\lambda$ are parallel to the null vector fields

$$v^a_{\pm} = (-g_{tt})^{-1/2}\delta^a_t \pm (g_{rr})^{1/2}\delta^a_r\,. \tag{36}$$

That is, $u^a_{\pm} = \kappa_{\pm}v^a_{\pm}$ where, because λ is an affine parameter, the conformal factors $\kappa_{\pm}$ are fixed (up to a positive multiplicative factor) by the equations

$$u^b_{\pm}\nabla_b u^a_{\pm} = 0\,. \tag{37}$$

Taking the inner product of (37) with $v_{\mp a}$ and making use of (36) gives

$$\kappa^{-2}_{\pm}u^b_{\pm}\nabla_b\kappa_{\pm} = -(v_{\pm a}v^b_{\pm}\nabla_b v^a_{\pm})/(v_{\pm}\bullet v_{\mp})\,, \tag{38}$$

or equivalently

$$\tfrac{d}{d\lambda}\kappa_{\pm}^{-1} = (-g_{tt}\,g_{rr})^{-1/2}[\pm\partial_r(-g_{tt})^{1/2} + \partial_t(g_{rr})] = a^{-1}(2r-m)^{-1}[\pm 16mr^2(2r+m)^{-3} + (2r-m)\dot{a} + 2\xi]. \tag{39}$$

Also, the expansions $\theta_{\pm} \equiv \nabla_a u_{\pm}^a$ of the two sets of null geodesic congruences are given by

$$\theta_{\pm} = \kappa_{\pm}[\nabla_a v_{\pm}^a - (v_{\pm a} v_{\pm}^b \nabla_b v_{\pm}^a)/(v_{\pm}\bullet v_{\mp})] \tag{40}$$

or, in terms of the metric functions,

$$\begin{aligned}\theta_{\pm} &= 2\kappa_{\pm} r^{-1}(g_{rr})^{-1/2}[(-g_{tt})^{-1/2}\partial_t R \pm (g_{rr})^{-1/2}\partial_r R] \\ &= 2\kappa_{\pm} a^{-1}(2r-m)^{-1}[(2r-m)\dot{a} + 2\xi \pm 4r(2r+m)^{-3}(2r-m)^2]\end{aligned} \tag{41}$$

In particular, the expansions vanish whenever the term in square brackets in the first line of (41) is zero. Note that the expression $g^{ab}\partial_a R\partial_b R \equiv (g_{tt})^{-1}(\partial_t R)^2 + (g_{rr})^{-1}(\partial_r R)^2$ appearing in the definition (30) of the Misner-Sharp mass is proportional to $\theta_+\theta_-$, so the geometric radius R changes from space-like to time-like, or vice versa, whenever one of the expansions changes sign.

The apparent horizons of the gMcV space-times are defined to be the surfaces on which either θ_+ or θ_- changes sign. Inside a "normal" region of space-time $\theta_+ > 0$ and $\theta_- < 0$, while R is a space-like coordinate with $\frac{d}{dt}R > 0$ ($\frac{d}{dt}R < 0$) on outgoing (ingoing) radial null geodesics. I will call any region of space-time in which both θ_+ or θ_- are negative "trapped", and any region in which both expansions are positive "anti-trapped".[8] It is evident from (41) that $\theta_+ > 0$ at all points outside the surface $r = m/2$ in an expanding gMcV space-time that satisfies the null energy condition, and so there are no trapped regions in such space-times (or at least, not in the coordinate patch covered by the gMcV line element (6)). Nonetheless, expanding gMcV space-times do support anti-trapped regions, and the apparent horizons which form their boundaries are important for the causal structure of the space-times. Inside the anti-trapped regions, R is a time-like coordinate and $\frac{d}{dt}R < 0$ on all ingoing and outgoing radial null geodesics.

The behaviour of the radial null geodesics is fully described by Equations (35), (39) and (41), plus the relationship between the affine parameter λ and the coordinate time t on any particular geodesic, which reads

$$dt/d\lambda = (-g_{tt})^{-1/2}\kappa_{\pm} = (2r-m)^{-1}(2r+m)\kappa_{\pm}\,, \tag{42}$$

or equivalently

$$\lambda = \int (2r+m)^{-1}(2r-m)\kappa_{\pm}^{-1}\,dt\,. \tag{43}$$

A radial null geodesic is future-complete if, starting from a point (t_*, r_*) with $t_* > 0$ and $r_* > m(t_*)/2$, the total forward lapse in λ (the limiting value as $t \to t_s$, where $t_s > t_*$ is the time coordinate of the first future singularity on the geodesic, or else as $t \to \infty$) diverges. Otherwise, the geodesic is future-incomplete. Similarly, a radial null geodesic is past-complete if, starting from the same point (t_*, r_*), the total backwards lapse in λ (the limiting value as $t \to t_s$, where now $t_s < t_*$ is the time coordinate of the most recent past singularity on the geodesic, or else as $t \to -\infty$) diverges, and is past-incomplete otherwise.

3. A phase-space description of the generalized McVittie space-times

It turns out that the analysis of the long-term behaviour and null geodesic structure of the gMcV space-times can be greatly simplified by introducing a new radial coordinate

[8] This is in contrast to Hayward [15], who calls the two types of regions "future-trapped" and "past-trapped" respectively.

$$x \equiv (2r - m)/m\,, \tag{44}$$

which is positive at all points on the gMcV coordinate patch, and goes to zero at the singular surface $r = m/2$.

Furthermore, the time evolution of any gMcV space-time can be represented by a trajectory in a two-dimensional phase-space spanned by the functions $\xi(t) \equiv \frac{d}{dt}(am)$ and $\sigma(t) \equiv -a\frac{d}{dt}m$. Note that if the space-time is expanding and satisfies the null energy condition then – because $\xi \geq 0$ and $\dot{m} \leq 0$ – the trajectory lies at all times inside or on the boundary of the first quadrant in the σ-ξ plane.

Now, since $\Delta = mx$ and $\dot{a}/a = (\sigma + \xi)/(am)$, the residual null energy condition (16) can be rewritten in terms of σ, ξ and x as a polynomial constraint $Q(x,t) \leq 0$, where

$$Q(x,t) \equiv (am)x(2+x)^3[x\dot{\sigma} + (2+x)\dot{\xi}] - \{(2+x)^3[x(2+x)\xi + (2+2x+x^2)\sigma] - 8x(1+x)^2\}\xi\,, \tag{45}$$

or equivalently

$$(am)[x\dot{\sigma} + (2+x)\dot{\xi}] \leq x^{-1}(2+x)^{-3}\xi\, q(x,t)\,, \tag{46}$$

where

$$q(x,t) \equiv 16\sigma + 8(5\sigma + 2\xi - 1)x + 4(11\sigma + 8\xi - 4)x^2 + 2(13\sigma + 12\xi - 4)x^3 + 8(\sigma + \xi)x^4 + (\sigma + \xi)x^5\,. \tag{47}$$

It should be stressed that, at any given time $t > 0$, the inequality (46) must be satisfied for all values of $x \in (0, \infty)$. The interesting and non-trivial constraints this places on the evolution of the gMcV space-times will be examined shortly.

Similarly, since $dx/dt = m^{-1}[2dr/dt - (x+1)m]$ the radial null geodesic equation (35) becomes:

$$dx/dt = (am)^{-1}(2+x)^{-3}(1+x)n_{\pm}(x,t)\,, \tag{48}$$

where

$$n_{\pm}(x,t) \equiv 8\sigma + 2(6\sigma \pm 1)x + 2(3\sigma \pm 1)x^2 + \sigma x^3\,. \tag{49}$$

Also, the equation (39) for the conformal factor $\kappa_{\pm}$ – once (42) has been used to convert the λ-derivative to a t-derivative – reads:

$$\tfrac{d}{dt}(\ln \kappa_{\mp}^{-1}) = (am)^{-1}(2+x)^{-4}[\,p_{\pm}(x,t) \pm 2(1+x)(2+2x-x^2)]\,, \tag{50}$$

while the equation (41) for the expansions becomes:

$$\theta_{\pm}/\kappa_{\pm} = 2(am)^{-1}x^{-1}(2+x)^{-3}\,p_{\pm}(x,t)\,, \tag{51}$$

where

$$p_{\pm}(x,t) \equiv 16\xi + 8(4\xi + \sigma)x + 2(12\xi + 6\sigma \pm 1)x^2 + 2(4\xi + 3\sigma \pm 1)x^3 + (\xi + \sigma)x^4\,. \tag{52}$$

Finally, the equation (43) for the affine parameter can be recast as:

$$\lambda = \int (2+x)^{-1} x\, \kappa_{\pm}^{-1} dt\,. \tag{53}$$

3.1 Emergence and development of apparent horizons

It has already been remarked that expanding gMcV space-times do not contain trapped regions. This is evident from the fact that the polynomial $p_{+}(x,t)$, defined in (52), is positive definite for all $x > 0$ if σ and ξ are non-negative.

However, the polynomial $p_{-}(x,t)$ is not necessarily positive definite, as the coefficients $2(12\xi + 6\sigma - 1)$ and $2(4\xi + 3\sigma - 1)$ can be negative. In fact, if degeneracies are ignored, either the two coefficients are both positive or both negative, or $4\xi + 3\sigma - 1$ is negative and $12\xi + 6\sigma - 1 \geq 4\xi + 3\sigma - 1$ is positive. In the first case p_{-} has no positive roots, and in the remaining cases the coefficients of p_- change sign twice and so there are either no positive roots or two positive roots.

The boundary in the σ-ξ plane between the regimes where p_- has no positive roots and p_- has two positive roots can be found by solving the equations $p_- = 0$ and $\frac{\partial}{\partial x} p_- = 0$ simultaneously for σ and ξ. This gives the parametric relations

$$\sigma = (2 + x_h)^{-3} x_h (4 + 4x_h - x_h^2) \quad \text{and} \quad \xi = (2 + x_h)^{-4} x_h^2 (x_h^2 - 2x_h - 2), \tag{54}$$

where x_h is the value of x at which the double root of $p_-(x,t)$ occurs. A plot of ξ against σ on this boundary curve – which is labeled Γ_h – is shown in Figure 1. If (σ, ξ) lies inside Γ_h the space-time is "normal" over some compact, connected range of x on the corresponding constant-t hypersurface, while if (σ, ξ) lies outside Γ_h the space-time is anti-trapped at all points on the corresponding hypersurface.

Note that the value of x_h always lies between the positive root of $x^2 - 2x - 2$ (which corresponds to $\xi = 0$) and the positive root of $x^2 - 4x - 4$ (corresponding to $\sigma = 0$). That is, $1 + \sqrt{3} \le x \le 2(1 + \sqrt{2})$ at any point where a "normal" region of space-time first appears. Also, the maximum values of σ and ξ on the boundary curve Γ_h are $\frac{1}{8}$ and $\sqrt{3}/9 \approx 0.19245$ respectively.

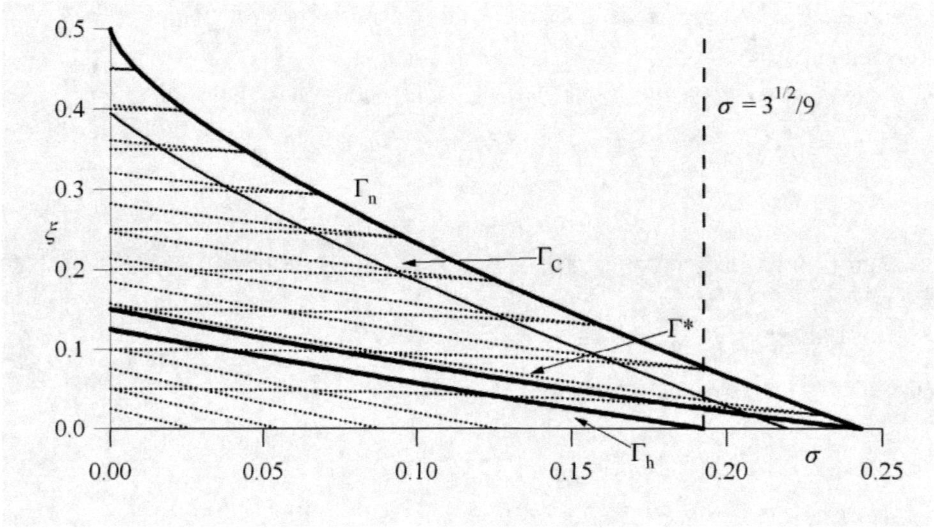

Figure 1. Regions in the σ-ξ plane that prove to be important for the long-term evolution of the gMcV space-times

In the McVittie case, $\xi = 0$ and $\dot{\sigma} = M \frac{d}{dt}(\dot{a}/a) \le 0$ according to the null energy condition, so all McVittie space-times can be represented by a point moving leftwards along the σ-axis in Figure 1. Thus, once a McVittie space-time enters the parametric regime where p_- has two positive roots (inside Γ_h) it will never subsequently leave it. The region of "normal" space-time in a McVittie space-time (if one forms at all) therefore appears with the bifurcation of two apparent horizons at $x = 1 + \sqrt{3}$ at a time $t = t_h$ for which $\sigma(t_h) = \sqrt{3}/9$. As will be seen shortly, the inner horizon in a McVittie space-time moves monotonically inwards (that is, with $\frac{d}{dt} x \le 0$) and the outer horizon monotonically outwards for all $t > t_h$.

The persistence and expansion of the "normal" region of space-time in the McVittie case is just an example of a well-known theorem, due to Hayward [15], which states that the area of an inner (outer) apparent horizon decreases (increases) monotonically if the null energy condition is satisfied. It should therefore come as no surprise that analogous results exist for the generalized McVittie space-times.

In fact, if the parametric equations (54) for Γ_h are substituted into the polynomial $q(x,t)$ defined by (47), it is easily checked that $q(x_h,t)=0$ and so the null energy condition (46) at $x=x_h$ becomes

$$\dot{\xi} \le -(2+x_h)^{-1} x_h \dot{\sigma} \tag{55}$$

on Γ_h. Furthermore, it can be seen by differentiating the two parametric equations in (54) that the direction tangent to Γ_h has $d\xi/d\sigma = -(2+x_h)^{-1} x_h$, and therefore that a gMcV space-time cannot cross Γ_h outwards if the null energy condition holds. Thus, if a "normal" region of space-time appears, it persists forever.

Similarly, the area of the inner (outer) apparent horizon is $4\pi R^2$, where

$$R = \tfrac{1}{2} am(1+x)^{-1}(2+x)^2 \tag{56}$$

is the geometric radius of the horizon, with x now the smaller (larger) of the two roots of $p_-(x,t)$. The rate of change of R is:

$$\tfrac{d}{dt}R = \tfrac{1}{2} am(1+x)^{-2} x(2+x) \tfrac{d}{dt}x + \tfrac{1}{2}\xi(1+x)^{-1}(2+x)^2 , \tag{57}$$

where $\frac{d}{dt}x$ is evaluated by taking the derivative of the equation $p_-(x,t)=0$:

$$\tfrac{d}{dt}x = -(2+x)^3[x\dot{\sigma} + (2+x)\dot{\xi}] \Big/ \tfrac{\partial}{\partial x} p_- . \tag{58}$$

Combining (57) and (58) gives:

$$\tfrac{d}{dt}R = -\tfrac{1}{2}(1+x)^{-2}(2+x)[Q(x,t) - (3x+4)\xi p_-(x,t)] \Big/ \tfrac{\partial}{\partial x} p_- , \tag{59}$$

where $Q(x,t) \le 0$ is defined in (45). Since $p_- = 0$ at an apparent horizon, and it is clear from the equation (52) for p_- that $\frac{\partial}{\partial x} p_- < 0$ at the inner horizon and $\frac{\partial}{\partial x} p_- > 0$ at the outer horizon, it follows that the area of the inner horizon must be decreasing ($\frac{d}{dt}R \le 0$) and the area of the outer horizon must be increasing ($\frac{d}{dt}R \ge 0$), as expected.

Equation (57) also gives some clues to the behavior of $\frac{d}{dt}x$ on the two horizons. Since $\frac{d}{dt}R \le 0$ on the inner horizon, $\frac{d}{dt}x$ must be strictly negative there in a gMcV space-time with $\xi > 0$. However, if $\xi > 0$ then $\frac{d}{dt}x$ can in principle be either positive or negative (or zero) on the outer horizon. In the McVittie case ($\xi = 0$) it is clear from (57) that $\frac{d}{dt}R$ and $\frac{d}{dt}x$ everywhere have the same sign, and so $\frac{d}{dt}x \ge 0$ (≤ 0) on the outer (inner) horizon, as mentioned above.

3.2 Long-term evolution of the generalized McVittie space-times

Consider now the constraints that the residual null energy condition (46) places on the time-development of the gMcV space-times. In the McVittie case (46) reduces to the inequality $\dot{\sigma} \le 0$ and so adds nothing new to the problem, but if $\xi > 0$ it does place a non-trivial restriction on the direction of the vector $(\dot{\sigma}, \dot{\xi})$ if the polynomial $q(x,t)$ is negative or zero for at least one positive value of x.

From (47), it is evident that $q(x,t)$ – like $p_-(x,t)$ – has either two or no positive roots if σ and ξ are non-negative. By solving the equations $q = 0$ and $\frac{\partial}{\partial x} q = 0$ simultaneously for σ and ξ, it can be seen that the boundary between the parametric regimes where q has no positive roots and q has two positive roots has the form

$$\sigma = 4(2+x_n)^{-3}x_n^3 \quad \text{and} \quad \xi = 4(2+x_n)^{-4}(2+4x_n-2x_n^3-x_n^4), \tag{60}$$

where x_n is the value of x at which the double root of q occurs.

This boundary is plotted as the curve labeled Γ_n in Figure 1. It cuts the ξ-axis at $\xi = \frac{1}{2}$, and the σ-axis at $\sigma = \frac{4\sqrt{3}}{9}(2\sqrt{3}-3)^{3/2} \approx 0.24339$. The polynomial $q(x,t)$ is positive definite outside Γ_n, but is negative for some range of positive values of x at all points inside Γ_n. Note that the first equation in (60) can be solved to give:

$$x_n(\sigma) = 2(4^{1/3}-\sigma^{1/3})^{-1}\sigma^{1/3}, \tag{61}$$

and if this is substituted into the second equation in (60) the boundary curve Γ_n can be represented explicitly as

$$\xi = \tfrac{1}{8}[4-12(2^{-1/6}\sigma^{1/3})^2-3(2^{-1/6}\sigma^{1/3})^4] \equiv \xi_n(\sigma). \tag{62}$$

[It is from this equation that the ordinate $\sigma = \frac{4\sqrt{3}}{9}(2\sqrt{3}-3)^{3/2}$ is taken.]

The actual form of the restrictions that apply inside Γ_n can be analyzed by returning to the inequality $Q(x,t) \le 0$ and solving the polynomial equations $Q = 0$ and $\frac{\partial}{\partial x}Q = 0$ for $\dot{\sigma}$ and $\dot{\xi}$. On defining $S \equiv (am)\dot{\sigma}/\xi$ and $T \equiv (am)\dot{\xi}/\xi - \xi$, this gives:

$$S = 8(2+x)^{-3}x(1+x) - 2x^{-2}(1+x)\sigma \tag{63}$$

and

$$T = -8(2+x)^{-4}(1+x)(1+x+x^2) + x^{-1}(2+x)\sigma\,. \tag{64}$$

For each fixed value of σ, equations (63) and (64) describe a curve in the plane of values of S and T. Since

$$\begin{aligned}\tfrac{1}{2}x^3(2+x)^4\tfrac{d}{dx}S &= -\tfrac{1}{2}x^2(2+x)^5\tfrac{d}{dx}T \\ &= 32\sigma + 80\sigma x + 80\sigma x^2 + 8(1+5\sigma)x^3 + 2(4+5\sigma)x^4 + (\sigma-4)x^5,\end{aligned} \tag{65}$$

each curve starts with $S \ll -1$ and $T \gg 1$ for small values of x (provided that $\sigma > 0$), approaches and crosses the T-axis with $\frac{d}{dx}S > 0$ and $\frac{d}{dx}T < 0$, until (if $\sigma < 4$) there is a simultaneous turning point in S and T at some point on a locus that connects $(S,T) = (\frac{4\sqrt{3}}{9}, -\frac{2}{3})$ (for $\sigma \to 0$) to $(S,T) = (0,4)$ (for $\sigma \to 4$), after which the curve continues with $\frac{d}{dx}S < 0$ and $\frac{d}{dx}T > 0$ to $(S,T) = (0,\sigma)$.

Now, unlike the polynomials p_- and q considered earlier, the number of positive roots of Q can increase or decrease by 1, so new positive roots need not appear in pairs. This is because the coefficient of the highest power (x^5) in Q, which is $am(\dot{\sigma}+\dot{\xi}) - \xi(\sigma+\xi)$, has no fixed sign, and single positive roots can appear or disappear (at infinity) whenever this coefficient is zero. The locus of points in the S-T plane on which this occurs is the straight line

$$S + T = \sigma\,, \tag{66}$$

which joins the curve (63) and (64) for the double roots of Q at $(S,T)=(0,\sigma)$, but also intersects the curve before the latter reaches its turning point.[9]

For $\sigma>0$, the union of the curve (63) and (64) for the double roots with the straight line $S+T=\sigma$ covers all the points in the S-T plane where positive roots of Q appear or disappear, and has a characteristic "swallow-tail" shape.[10] The boundary curves are plotted for the cases $\sigma=0$ and $\sigma=\sqrt{3}/9$ in Figure 2. For each fixed value of σ, the polynomial Q is negative definite (for $x>0$) only in the domain that lies entirely on or below this set of points, and so is bounded above by the curve for the double roots as far as the point C where it intersects the line $S+T=\sigma$, and then above by the straight line as it goes off to infinity in the fourth quadrant.

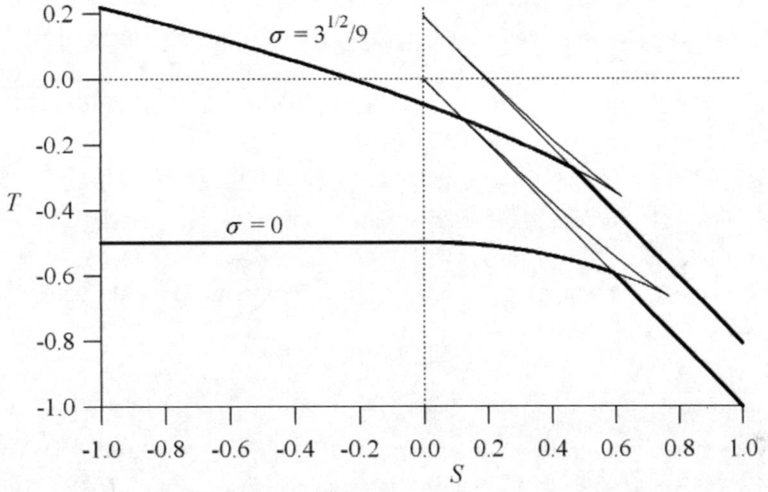

Figure 2. The boundary curves defined by (63) and (64), or (66), in the cases $\sigma=0$ and $\sigma=\sqrt{3}/9$. For each value of σ, the allowable region in the S-T plane lies on or below the thickened parts of the curve.

The constraints that the null energy condition places on the phase-space evolution of the gMcV space-times can now be calculated as follows. From the definitions of S and T, the zero point of $(\dot{\sigma},\dot{\xi})$ lies at $(S,T)=(0,-\xi)$, which for $\xi\geq 0$ is a point on the T-axis on or below the S-axis in Figure 2. If,

[9] This can be seen by substituting the expression (72) for $\sigma(x)$ at the intersection point C into the right-hand side of (65), and verifying that the polynomial that results, $4x^2(4x^2+x-2)$, is positive definite for $|x|\geq 1$ (the range of values for which $\sigma(x)\geq 0$).

[10] The limiting case when $\sigma=0$ is slightly more complex. Here, the curve for the double roots does not start at infinity in the second quadrant, but instead at $(S,T)=(0,-\frac{1}{2})$ with $dT/dS=0$. It initially descends to a simultaneous turning point at $(S,T)=(\frac{4\sqrt{3}}{9},-\frac{2}{3})$, then asymptotes to the origin. Single positive roots appear or disappear on the straight line $S+T=0$, which of course goes through the origin and cuts the first curve to form a "swallow-tail". However, because the constant term in Q is zero when $\sigma=0$, further single positive roots can appear or disappear when the coefficient of x, which is $8[2(am)\dot{\xi}-2\xi^2-\xi]$, vanishes. This corresponds to a second, horizontal, line $T=-\frac{1}{2}$, which completes the boundary of the region where Q is negative definite in the third quadrant.

Journal of Physics: Conference Series **283** (2011) 012001 doi:10.1088/1742-6596/283/1/012001

for given values of σ and ξ, the point $(0,-\xi)$ lies inside the region where Q is negative definite, all possible values of $d\xi/d\sigma$ are permitted by the null energy condition, or equivalently (σ,ξ) lies outside the curve Γ_n in Figure 1.

On the other hand, if $(0,-\xi)$ lies outside the region where Q is negative definite, the permissible values of $d\xi/d\sigma$ are the gradients of those lines in Figure 2 which start at $(0,-\xi)$ and subsequently pass through at least one point in the region where Q is negative definite. So if $\dot{\sigma}<0$ then $d\xi/d\sigma$ must be greater than or equal to the gradient of the tangent line from $(0,-\xi)$ to the boundary curve (63) and (64) in the half-plane $S<0$, while if $\dot{\sigma}>0$ then $d\xi/d\sigma$ must be less than or equal to the gradient of either the tangent line from $(0,-\xi)$ to the boundary curve in the half-plane $S>0$ or of the line segment from $(0,-\xi)$ to the intersection point *C*. That there is a tangent line from $(0,-\xi)$ to the boundary curve on the left, and that the line from $(0,-\xi)$ to *C* does not otherwise cut the boundary curve on the right, follows from the fact that the curve (63) and (64) is concave downwards and has a shallower gradient than the line $S+T=\sigma$ everywhere to the left of *C*. This can in turn be seen by noting that

$$dT/dS=-(2+x)^{-1}x>-1 \quad \text{and so} \quad dT^2/dS^2=-2(2+x)^{-2}(dS/dx)^{-1}<0 \tag{67}$$

on the curve (63) and (64) (before the turning point), while of course $dT/dS=-1$ on the line $S+T=\sigma$.

If it is assumed that $d\xi/d\sigma$ takes on either of its two extreme possible values, then from the definitions of *S* and *T*

$$d\xi/d\sigma=(T+\xi)/S, \tag{68}$$

with the right-hand side evaluated at the point where the line from $(0,-\xi)$ touches the boundary curve. If the point of contact is a tangent point then, from (67), $d\xi/d\sigma=-(2+x)^{-1}x$. This equation is equivalent to the constraint $q(x,t)=0$, as can be seen by combining it with (68) and the expressions (63) and (64) for *S* and *T*.

The equation $q(x)=0$ can now be used to solve for ξ and $\frac{d}{dx}\xi$ in terms of *x*, σ and $\frac{d}{dx}\sigma$, and eliminating ξ and $\frac{d}{dx}\xi$ from (68) results in a linear ODE for σ

$$\tfrac{d}{dx}\sigma-2x^{-1}(2+x)^{-1}\sigma=-8(2+x)^{-4}x^2, \tag{69}$$

whose general solution is

$$\sigma=(2+x)^{-3}x(1+x)+K(2+x)^{-1}x, \tag{70}$$

where *K* is a constant of integration. Substituting this equation back into $q(x,t)=0$ and solving for ξ then gives

$$\xi=-8(2+x)^{-4}(1+x)(1+x+x^2)-K(2+x)^{-2}(2+2x+x^2). \tag{71}$$

On the other hand, if the point of contact is the intersection point *C* then substituting the expressions (63) and (64) for *S* and *T* into the equation (66) for the straight-line boundary gives a polynomial equation which can be solved directly to give:

$$\sigma=4(2+x)^{-4}x^2(x^2-1). \tag{72}$$

If this equation is used to eliminate σ and $\frac{d}{dx}\sigma$ from (68) then a linear equation for ξ results:

$$(4x^2+x-2)^{-1}(1+2x)\tfrac{d}{dx}\xi-(1+x)^{-1}(2+x)^{-1}x\xi=(2+x)^{-5}x(1+x)(x^2-2x-2), \tag{73}$$

which solves to give:

$$\xi=-4(2+x)^{-4}x^2(x^2-1)+\tfrac{6561}{2\sqrt{3}}\bar{K}(2+x)^{-8}(1+x)(1+2x)^{1/2}e^{2(x-1)}, \tag{74}$$

Journal of Physics: Conference Series **283** (2011) 012001 doi:10.1088/1742-6596/283/1/012001

where $\bar{K}$ is a second constant of integration.

Together, equations (70) and (71) describe a family of curves that doubly foliate the region inside Γ_n in Figure 1. Each curve starts on the ξ-axis with $x=0$ and $\xi=-\frac{1}{2}(K+1)$. As x increases from 0, the point (σ,ξ) follows the shallower of the two sets of curves in Figure 1, with σ increasing and ξ decreasing. Given that $0\le\xi<\frac{1}{2}$ inside Γ_n, the only relevant values of K on the shallower curves are $-2<K<-1$.

When x is equal to the positive root of the equation $(K+2)(1+x)+(\frac{1}{4}K-1)x^2=0$, the point (σ,ξ) lies on Γ_n, and here crosses over to the steeper of the two sets of curves. It then follows these curves, with σ decreasing and ξ increasing, until it returns to the ξ-axis when x has reached the positive root of $(K+2)(1+x)+\frac{1}{4}Kx^2=0$.[11] The relevant values of K on the steeper set of curves are $-2<K<0$. Note that the curve Γ_h bounding the solutions that contain apparent horizons corresponds to the steep curve with $K=-1$.

Throughout much of the region inside Γ_n the family of curves described by (72) and (74) are almost indistinguishable from the steeper set of curves described by (70) and (71). They start on the ξ-axis with $x=1$ and $\xi=\bar{K}$, and – provided that $\bar{K}<\frac{16}{27}$ – initially descend towards the σ-axis, although those curves with $\bar{K}>0.1685$ pass through a horizontal turning point before ever reaching the σ-axis. Nonetheless, the turning points are of little consequence for the evolution of the gMcV space-times, as the curves described by (72) and (74) constrain the vector $(\dot{\sigma},\dot{\xi})$ only inside the curve marked Γ_C in Figure 1, and all the turning points on the curves (72) and (74) lie outside Γ_C. Between Γ_C and Γ_n the vector $(\dot{\sigma},\dot{\xi})$ is instead constrained (for phase curves with $\dot{\sigma}>0$) by the steeper set of curves described by (70) and (71)

The curve Γ_C itself is defined by the condition that the tangent line from the point $(0,-\xi)$ in Figure 2 touches the boundary curve in the right-half plane at the point where this curve intersects the straight line $S+T=\sigma$. Substituting (70) and (71) into (63) and (64) and imposing (66) gives an equation for λ as a function of x which reduces (70) and (71) to

$$\sigma=4(2+x)^{-4}x^2(x^2-1) \quad\text{and}\quad \xi=-4(2+x)^{-5}(1+x)(x^4+x^3-2x^2-8x-4). \tag{75}$$

The corresponding curve Γ_C cuts the ξ-axis in Figure 1 at $\xi=\frac{32}{81}\approx0.3951$, and the σ-axis at $\sigma\approx0.2196$. As a result, the only physically relevant values of $\bar{K}$ are $0<\bar{K}<\frac{32}{81}$.

To summarize the constraints that the null energy condition places on the evolution of gMcV space-times inside Γ_n in Figure 1, phase curves with $\dot{\sigma}<0$ must lie on or below the shallower set of curves described by (70) and (71), and phase curves with $\dot{\sigma}>0$ must either lie or on below the steeper set of curves described by (70) and (71) (between Γ_C and Γ_n) or lie or on below the curves described by (72) and (74) (inside Γ_C).

It is evident from the picture presented in Figure 1, therefore, that region of phase space from which the gMcV space-times cannot escape includes not only the region inside the curve Γ_h, but a slightly larger region bounded by the steep curve, labeled Γ^*, that passes through the point $(\sigma,\xi)\approx(0.24339,0)$ where Γ_n cuts the σ-axis. This is the steeper curve from the set defined by (70)

[11] However, these steep curves are shown in Figure 1 only as far as the curve marked Γ_C. Inside Γ_C the steeper set of constraining curves, shown in Figure 1, satisfy (72) and (74).

and (71) that has $K = 4\sqrt{3} - 8 \approx -1.072$, and its continuation inside Γ_C from the set defined by (72) and (74), which has $\bar{K} \approx 0.1501$.

By contrast, the evolution of gMcV space-times represented by phase points outside Γ_n is virtually unconstrained. The phase curves can arbitrary form in this region, although the null energy condition continues to place restrictions on the magnitudes of $\dot{\sigma}$ and $\dot{\xi}$. In particular, any space-time with σ and ξ constant satisfies the null energy condition, provided that the point (σ,ξ) lies on or outside Γ_n.

Other possible solutions of interest include space-times that cycle endlessly between the regions inside and outside Γ_n. As an example, consider a phase curve that starts at a point (σ_1,ξ_1) on Γ_n with $0 < \xi_1 < \frac{1}{2}$, traces the shallow curve left from (σ_1,ξ_1), then crosses over onto one of the steep curves before reaching the steep curve that passes through $(\sigma,\xi) \approx (0.2196, 0)$, and follows the steep curve down and to the right until it returns to Γ_n at a point (σ_2,ξ_2) with $0 < \xi_2 < \xi_1$. From (σ_2,ξ_2) it traces the vertical line $\sigma = \sigma_2$ through the region outside Γ_n until ξ increases to ξ_1, then returns to (σ_1,ξ_1) along the horizontal line $\xi = \xi_1$. In view of (46), the two-straight line segments will satisfy the null energy condition provided that

$$(am)\dot{\xi}/\xi \le \min_{x>0}[x^{-1}(2+x)^{-4}\,q(x,t)] \quad \text{and} \quad (am)\dot{\sigma}/\xi \le \min_{x>0}[x^{-2}(2+x)^{-3}\,q(x,t)] \qquad (76)$$

at each point on the respective segments. The phase loop described here can clearly be repeated indefinitely.

3.3 Further results

Two further results that can be established on the basis of the phase-plane analysis illustrated in Figures 1 and 2, but are too technical to be worthy of a detailed proof here, are:

1. For any expanding gMcV space-time that satisfies the null energy condition and whose phase curve enters the region inside the curve Γ^*, the Weyl energy *am* must tend to a finite positive limit $(am)_\infty$ as $t \to \infty$. This follows again from the null energy condition (46), which can be recast as the requirement that

$$\begin{aligned}(am)\xi^{-1}[(2+x)^{-1}x\dot{\sigma}+\dot{\xi}] &\le \xi + x^{-1}(2+x)^{-1}(2+2x+x^2)\sigma - 8(2+x)^{-4}(1+x)^2 \\ &\equiv x^{-1}(2+x)^{-4}q(x,t)\end{aligned} \qquad (77)$$

for all $x > 0$. It is easily verified that, if ξ and $\sigma < 4$ are fixed, the function on the right-hand side of (77) is minimized when $x = x_n(\sigma) = 2(4^{1/3} - \sigma^{1/3})^{-1}\sigma^{1/3}$, and so

$$(am)\xi^{-1}(4^{-1/3}\sigma^{1/3}\,\dot{\sigma} + \dot{\xi}) \le \xi - \xi_n(\sigma)\,, \qquad (78)$$

where $\xi_n(\sigma)$, defined in equation (62), is the value of ξ on the curve Γ_n. Since $\xi = \frac{d}{dt}(am)$, this inequality is equivalent to

$$\frac{d}{d\ln(am)}[\tfrac{3}{4}(2^{-1/6}\sigma^{1/3})^4 + \xi] \le \xi - \xi_n(\sigma)\,. \qquad (79)$$

Now, $\xi < \xi_n(\sigma)$ at all points inside Γ_n in Figure 1, and once the trajectory of a gMcV space-time is inside Γ^* it can never again touch or cross Γ^*. Hence, for such a space-time, $\xi - \xi_n(\sigma)$ has a strictly negative upper bound at sufficiently late times, and so if *am* increases without bound the function $\frac{3}{4}(2^{-1/6}\sigma^{1/3})^4 + \xi$ must eventually become negative, in violation of the constraints $\sigma \ge 0$ and $\xi \ge 0$.

2. Any expanding gMcV space-time that satisfies the null energy condition and whose phase curve enters the region inside the curve Γ^* must have $\lim_{t\to\infty} \xi(t) = 0$. The proof of this result is too long

and technical to be worth reproducing here, but hinges on the fact that if ξ does not tend to zero but is also not bounded away from zero (to avoid $am \to \infty$) then there must be a positive constant ξ^* and an infinite sequence of disjoint intervals $\{I_k = [t_k, t_k + \varepsilon_k]\}$ with finite total measure and $\xi(t_k) = \frac{1}{2}\xi^* \le \xi(t) \le \xi^* = \xi(t_k + \varepsilon_k)$ on I_k. What this means in effect is that $\dot{\xi}$ is unbounded above, and in view of the parametric equations (63) and (64) for the boundary of the allowable region in *S-T* space if (σ, ξ) is inside Γ_n, which imply that $\dot{\sigma} \propto -\dot{\xi}^2$ in the limit of large $\dot{\xi}$, it can be shown that $\dot{\sigma}$ diverges so rapidly that σ must eventually become negative.

3. For any expanding gMcV space-time that satisfies the null energy condition and whose phase curve enters the region inside the curve Γ^*, the function σ must tend to a finite limit σ_∞ as $t \to \infty$, with $0 \le \sigma_\infty < \sigma^*$ where $\sigma^* = \frac{4\sqrt{3}}{9}(2\sqrt{3}-3)^{3/2} \approx 0.24339$ is the value of σ where Γ^* meets the axis $\xi = 0$. To verify this result, note first that those steep curves described by equations (70) and (71) or (72) and (74) that cut the axis $\xi = 0$ all have gradients $|d\xi/d\sigma|$ bounded below by $L = \sqrt{2/\sqrt{3}-1} \approx 0.39332$, which is the gradient of curve Γ^* at the point where it touches the axis $\xi = 0$. If σ does not converge to a limit σ_∞ on a particular trajectory inside Γ^* then there must exist two constants $0 \le \sigma_1 < \sigma_2 \le \sigma^*$ with the property that, for all times $\bar{t} > 0$, $\inf_{t>\bar{t}} \sigma(t) \le \sigma_1$ and $\sup_{t>\bar{t}} \sigma(t) \ge \sigma_2$. Since $\xi \to 0$ from Result 2, for every $\varepsilon > 0$ there exists a time $t_\varepsilon > 0$ for which $\xi(t_\varepsilon) < \varepsilon$ and $\sigma(t_\varepsilon) - \sigma_1 < \varepsilon$. But whenever σ is increasing with *t*, the phase curve of the space-time must lie on or below the steeper family of curves in Figure 1, and so $\sigma(t) \le \sigma_1 + \varepsilon + L^{-1}\varepsilon$ for all $t > t_\varepsilon$. Fixing $\varepsilon < (1+2L)^{-1}L(\sigma_2 - \sigma_1)$ therefore guarantees that $\sigma(t) < \sigma_2 - \varepsilon$ for all $t > t_\varepsilon$, which contradicts the assumption that $\sup_{t>\bar{t}} \sigma(t) \ge \sigma_2$.

Together, these results imply that any gMcV space-time that develops an apparent horizon (and thereby enters the region inside Γ_h) inevitably asymptotes to a standard McVittie space-time with $\xi = 0$ and σ constant as $t \to \infty$.

4. Radial null geodesics

4.1 Existence and uniqueness

In contrast to the null energy condition, the equation (48) for radial null geodesics in the gMcV space-times, which can be written as

$$dx/dt = (am)^{-1}[(1+x)\sigma \pm 2(2+x)^{-3}(1+x)^2 x], \tag{80}$$

is relatively easy to analyze. Apart from the pre-factor $(am)^{-1}$ – which is positive and constant (in the McVittie case) or decreasing – the right-hand side of (80) depends on *t* only through σ. As a result, there is very little difference between the behavior of the radial null geodesics in a gMcV space-time and in the corresponding McVittie space-time (with $\xi = 0$ but σ unchanged).

Nolan [4] has shown that, with *a* assumed to have a power-law dependence on *t* (so that $m = M/a$ has the inverse dependence, and $\sigma \propto t^{-1}$), there are unique ingoing and outgoing radial null geodesics through each point $(t, r) = (t_*, r_*)$ with $t_* > 0$ and $r_* > m/2$, and that every point in this family has an initial point on the surface $r = m/2$ at some time $t > 0$. The analogs of these results for the gMcV space-times – with no specific functional forms assumed for *a* or *m* – can be established as follows.

Note first that the form of the term in (80) proportional to σ suggests replacing *x* as the dependent variable with $u \equiv \ln(1+x)$. The equation then becomes

$$du/dt = (am)^{-1} v_{\pm}(u,t) \quad \text{with} \quad v_{\pm}(u,t) \equiv \sigma \pm 2(1+e^u)^{-3}(e^u - 1)e^u \,. \tag{81}$$

Furthermore, at each fixed value of $t > 0$, both $v_{\pm}$ and $\partial v_{\pm}/\partial u$ are bounded:

$$|v_{\pm}(u,t)| \leq \sigma(t) + \sqrt{3}/9 \quad \text{and} \quad |\tfrac{\partial}{\partial u} v_{\pm}(u,t)| \leq \tfrac{1}{4} \quad \text{for all } u \in (-\infty, \infty)\,, \tag{82}$$

while the functions a and m are assumed to be C^∞ for all $t \in (0,\infty)$. Hence, by the Picard-Lindelöf Theorem, unique ingoing and outgoing solutions to the equation (81) through each point $(t,u) = (t_*, u_*)$ with $t_* > 0$ and $u_* \geq 0$, and each solution can be extended to all $t > t_*$. With u replaced by x and u_* by x_* – noting that $u_* = 0$ corresponds to $x_* = 0$ – the same statement is true of the original equation (80).

4.2 Outgoing radial null geodesics

In order to analyze the asymptotic properties of the radial null geodesics (both outgoing and ingoing) through a given point $(t,u) = (t_*, u_*)$, it is useful to introduce the following characteristic integrals:

$$\tau_\infty \equiv \int_{t_*}^{\infty} (am)^{-1} dt\,, \quad \mu_\infty \equiv \int_{t_*}^{\infty} (am)^{-1}\sigma\, dt = -\lim_{t\to\infty} \ln[m(t)/m_*]$$

$$\tau_0 \equiv \int_{t_*}^{0} (am)^{-1} dt \quad \text{and} \quad \mu_0 \equiv \int_{t_*}^{0} (am)^{-1}\sigma\, dt = -\lim_{t\to 0^+} \ln[m(t)/m_*]\,, \tag{83}$$

where $m_* \equiv m(t_*)$.

Because $\frac{d}{dt}(am) \geq 0$ and $\dot{m} \leq 0$, in the limit $t \to \infty$ the Weyl energy *am* remains finite and positive or goes to ∞, while m remains finite and positive or goes to 0; and in the limit $t \to 0^+$ the reverse is true: *am* remains finite and positive or goes to 0, while m remains finite and positive or goes to ∞. So the integrals τ_∞ and μ_∞ are either finite and positive or diverge to ∞, while τ_0 and μ_0 are either finite and negative or diverge to $-\infty$. Although all four integrals depend on the geodesic in question through the value of t_*, their convergence or divergence are characteristics of the space-time only. The McVittie space-times, for example, all have $\tau_\infty = \infty$ and τ_0 finite, while – in view of Result 1 in Section 3.3 – any gMcV space-time that enters the region inside Γ^* in Figure 1 must also have $\tau_\infty = \infty$.

It is now possible to determine the fate of the outgoing radial null geodesics at early and late times. From (81),

$$(am)^{-1}\sigma \leq du/dt \leq (am)^{-1}(\sigma + \tfrac{\sqrt{3}}{9}) \tag{84}$$

along the outgoing geodesics, and so integrating this inequality forwards and backwards from $t = t_*$ gives

$$\mu_\infty \leq \lim_{t\to\infty}(u - u_*) \leq \mu_\infty + \tfrac{\sqrt{3}}{9}\tau_\infty \quad \text{and} \quad \mu_0 \geq \lim_{t\to 0^+}(u - u_*) \geq \mu_0 + \tfrac{\sqrt{3}}{9}\tau_0 \tag{85}$$

in the respective cases.

It follows from the first inequality in (85) that if μ_∞ and τ_∞ are both finite then u (and so x) remains bounded as $t \to \infty$, while if μ_∞ diverges then u and x also go to ∞ as $t \to \infty$. The one indeterminate possibility, when μ_∞ is finite and τ_∞ diverges, can be dealt with by noting from (81) that

$$du/dt \geq 2(am)^{-1}(1+e^u)^{-3}(e^u - 1)e^u \tag{86}$$

on the outgoing geodesics, which after integrating forwards from $t = t_*$ gives:

$$[\cosh u + 4\ln(e^u - 1) - 2u]_{u_*}^{u} \geq \int_{t_*}^{t} (am)^{-1} dt\,. \tag{87}$$

If τ_∞ diverges this inequality can be satisfied only if u (and x) diverges. So $x \to \infty$ as $t \to \infty$ in all cases except when μ_∞ and τ_∞ are both finite.

When the geodesics are traced backwards towards $t \to 0$, there are two possible outcomes. Either $\lim_{t\to 0^+}(u - u_*) > -u_*$, in which case the geodesic begins at $t = 0$ – which may or may not mark the location of a cosmological singularity – at some value of $x > 0$, or $\lim_{t\to 0^+}(u - u_*) \le -u_*$ and the geodesic emerges from the singularity at $x = 0$ (the surface $r = m/2$) at some time $t \ge 0$. From the second inequality in (85) it follows that both possibilities remain open if μ_0 and τ_0 are finite, but that all outgoing geodesics emerge from $x = 0$ if $\mu_0 = -\infty$. The one remaining case, when μ_0 is finite and τ_0 diverges, can be treated by integrating (86) backwards from $t = t_*$ to give:

$$[\cosh u + 4\ln(e^u - 1) - 2u]_{u_*}^{u} \le \int_{t_*}^{t} (am)^{-1}\, dt\,, \tag{88}$$

from which it follows that u (and x) must go to 0 at some positive time if $\tau_0 = -\infty$.

The coordinate x has, of course, no intrinsic physical significance, so it is instructive also to examine the behavior on the outgoing radial geodesics of the geometric radius R, defined in equation (56). Because x is monotonically increasing along the outgoing geodesics, as is evident from (80), it is clear from the equation (57) for $\frac{d}{dt}R$ that the same is true of R. Furthermore, since am is a positive increasing function of t, it follows from (56) that if $x \to \infty$ then so does R. On the other hand, if x converges to a finite limit as $t \to \infty$ along an outgoing geodesic, then from the discussion above $\tau_\infty = \int_{t_*}^{\infty} (am)^{-1} dt$ must also converge. But this is possible only if am increases without bound, and hence R again diverges. In all possible cases, therefore, the geometric radius goes to infinity along the outgoing radial null geodesics.

If an outgoing geodesic is traced backwards towards $t = 0$, it is evident from (56) that $R \to 0$ along the geodesic in question if and only if $am \to 0$. Since $\dot m \le 0$, this in turn is possible if and only if $a \to 0$ along the geodesic, and so the geodesic emerges from a cosmological singularity. In all other situations, the limiting of R (as $t \to 0$ or $x \to 0$, whichever is encountered first) is positive.

4.3 Ingoing radial null geodesics

Ingoing radial null geodesics are solutions of the geodesic equation (48) with $n_-(x,t)$ on the right. Despite their name, ingoing geodesics have $\frac{d}{dt}R > 0$ at all points (t,x) except those on or inside the apparent horizons (if any form), as was explained in Section 2.5. In particular, in view of the equation (52) for $p_-(x,t)$, the geometric radius R is increasing along the ingoing geodesics for both small and large values of x, unless σ and ξ are identically zero (and so the metric is Schwarzschild).

In principle, the analysis of Section 4.2 can be repeated to explore the asymptotic behaviour of the ingoing radial null geodesics, but unfortunately the results are not as clear-cut as for the outgoing geodesics. The equivalent of the inequality (84) for the ingoing geodesics is

$$(am)^{-1}(\sigma - \tfrac{\sqrt{3}}{9}) \le du/dt \le (am)^{-1}\sigma\,, \tag{89}$$

which when integrated forwards and backwards from $(t,x) = (t_*, x_*)$ gives

$$\mu_\infty - \frac{\sqrt{3}}{9}\tau_\infty \le \lim_{t\to\infty}(u - u_*) \le \mu_\infty \quad \text{and} \quad \mu_0 - \frac{\sqrt{3}}{9}\tau_0 \ge \lim_{t\to 0^+}(u - u_*) \ge \mu_0\,. \tag{90}$$

All that can be inferred from the second inequality in (90) about the provenance of the ingoing geodesics is that they emerge from the singularity $x = 0$ at some time $t > 0$ if τ_0 is finite and μ_0 diverges. In connection with this, it should be noted that Nolan [4] has already shown that the singularity at $r = m/2$ effectively hides the cosmological singularity from view in the restricted class of McVittie space-times with $\sigma \propto t^{-1}$ (which do have τ_0 finite and $\mu_0 = -\infty$).

However, the problem of the fate of the ingoing geodesics can be simplified by appealing to a result due to Hayward [15], which states that if apparent horizons form then no ingoing geodesic can cross the horizons from the normal region into the surrounding anti-trapped region. This can be seen directly by differentiating the equation (52) for $p_-(x,t)$ along an ingoing geodesic. In view of (48) and (49), this gives:

$$\tfrac{d}{dt} p_-(x,t) = (2+x)^3[x\dot{\sigma} + (2+x)\dot{\xi}] + \tfrac{\partial}{\partial x} p_-(x,t)\tfrac{dx}{dt} = (am)^{-1} x^{-1}[Q(x,t) + (2+x)^{-3}\, \overline{p}(x,t) p_-(x,t)], \tag{91}$$

where $\overline{p}$ is a fourth-order polynomial in x whose exact form is unimportant. So, given that $p_- = 0$ on the apparent horizons and $Q \le 0$ if the null energy condition holds, it follows that $\tfrac{d}{dt} p_- \le 0$ whenever an ingoing geodesic crosses a horizon, as claimed.

Another important technical question is whether the ingoing geodesics can reach the singularity at $x = 0$ in a finite time t. It should be evident from (48) and (49) that this is not possible, as $dx/dt \to (am)^{-1}\sigma \ge 0$ in the limit as $x \to 0$, and in the marginal case $\sigma = 0$ equation (48) becomes $dx/dt \approx -\tfrac{1}{4}(am)^{-1} x$ for small values of x. It follows therefore that x can tend to zero on an ingoing geodesic only in the asymptotic limit $t \to \infty$, and this only if σ is not eventually bounded away from zero and the integral τ_∞ diverges.

The analysis of the asymptotic behavior of the ingoing geodesics can now be divided into a small number of manageable cases.

(i) *The phase curve remains on or outside* Γ^*. Substituting (48) into (57) gives

$$\tfrac{d}{dt} R = \tfrac{1}{2}(1+x)^{-1}(2+x)^{-2} p_\pm(x,t) \tag{92}$$

on each of the radial null geodesics. If the phase curve of the space-time remains in the region on or outside the contour Γ^* in Figure 1 then $p_-(x,t) > 0$ for all positive x and t, and R is monotonic increasing on each ingoing geodesic. If x is unbounded above on the geodesic, it follows from (56) that R itself is unbounded above and so $R \to \infty$ as $t \to \infty$. On the other hand, if x is bounded above by some number x_+ and $p_-(x,t)$ is eventually bounded below by some positive constant ε then eventually $\tfrac{d}{dt} R \ge \tfrac{1}{2}(1+x_+)^{-1}(2+x_+)^{-2}\varepsilon$ is bounded below, and again $R \to \infty$ as $t \to \infty$.

The only points in the region on or outside the contour Γ^* at which $p_-(x,t)$ is not bounded away from zero for $x \in (0,\infty)$ are those on the horizontal axis $\xi = 0$, as $\lim_{t\to 0} p_-(x,t) = 0$ if $\xi = 0$. It is easily seen from (49) that the function $n_-(x,t)$ is positive definite for positive x whenever $\sigma > \sqrt{3}/9$. Let ξ^* be the value of ξ at the point in Figure 1 where Γ^* intersects the line $\sigma = \sqrt{3}/9$. Then $dx/dt \ge 0$ on all ingoing geodesics whenever $\xi \le \xi^*$.

Consider now a particular geodesic that passes through the point $(t,x) = (t_*, x_*)$. If the time interval $[t_*, \infty)$ is written as the union $S_+ \cup S_-$ of two closed subsets, with $\xi \ge \xi^*$ on S_+ and $\xi \le \xi^*$ on S_-, then there are two possible cases. If S_+ has infinite measure then, since ξ is bounded away from zero on I_+, it again follows from (92) that $R \to \infty$. Alternatively, if S_+ has finite measure M_+ it can be further divided into two disjoint subsets T_+ and T_-, with $dx/dt \ge 0$ on T_+ and $dx/dt \le 0$ on T_-. T_- can then be written as the union $\bigcup_k I_k$ of disjoint intervals $\{I_k = [a_k, b_k]\}$ with $x(b_k) \le x(a_k)$ and $x(a_{k+1}) \ge x(b_k)$ for all k. The geodesic equation (48) can be recast in the form

$$dx/dt = (am)^{-1}[(1+x)\sigma - 2(2+x)^{-3}(1+x)^2 x] \geq -\tfrac{8}{27}(am)_*^{-1} x, \tag{93}$$

which when integrated over $\bigcup_{k=1}^{N} I_k$ for any $N \geq 1$ gives

$$\sum_{k=1}^{N} \ln[x(b_k)/x(a_k)] \geq -2(am)_*^{-1} M_+ . \tag{94}$$

But the left-hand side of (94) satisfies the inequality

$$\sum_{k=1}^{N} \ln[x(b_k)/x(a_k)] = \ln[x(b_N)/x(a_1)] + \sum_{k=1}^{N-1} \ln[x(b_k)/x(a_{k+1})] \leq \ln[x(b_N)/x(a_1)], \tag{95}$$

and so (94) implies that the sequence $\{x(b_k)\}$ is bounded away from zero. Hence, x itself and $p_-(x,t)$ are also bounded away from zero on the geodesic, and once again $R \to \infty$.

So in all possible cases $R \to \infty$ as $t \to \infty$ on the ingoing geodesics if the phase curve remains in the region on or outside Γ^*.

(ii) *The phase curve enters the region inside* Γ^*. If the phase curve of the space-time enters the region inside Γ^* in Figure 1 then according to the result 1 of Section 3.3 the function am tends to a finite positive limit $(am)_\infty$ as $t \to \infty$. Given that the geometric radius R is bounded below by $2am > 0$, that $\frac{d}{dt}R$ along an ingoing geodesic is positive in an anti-trapped region and negative in a normal region (*i.e.* between any apparent horizons), and that no ingoing geodesic once inside the normal region can ever cross the apparent horizons into the anti-trapped region, it follows that R is eventually a monotonic function of t, and so either $R \to \infty$ or R converges to a finite value $R_\infty \geq 2(am)_\infty$.

In the first case it is evident from (56) that $x \to \infty$, while in the second case x converges to a finite value $x_\infty \geq 0$. Since $\xi \to 0$ and $\sigma \to \sigma_\infty \geq 0$ in the limit as $t \to \infty$, it follows that the right-hand side of (92) also converges, and therefore must converge to zero because R converges. But $p_-(x,t) = 0$ only at an apparent horizon or (if $\xi = 0$) at $x = 0$. If $x_\infty = 0$ then $R_\infty = 2(am)_\infty$ and – as was mentioned above – σ cannot eventually be bounded away from zero, so it follows that $\sigma_\infty = 0$, in which case (52) indicates that the inner horizon asymptotes to $x \to 0$ and the outer horizon has $x \to \infty$.

In summary, therefore, if the phase curve enters the region inside Γ^* then every ingoing geodesic will either diverge to $R = \infty$ (if apparent horizons never form, or the geodesic is early enough to avoid the horizons) or asymptote to the inner or outer horizon. Furthermore, since $\frac{d}{dt}R < 0$ between the horizons and $\frac{d}{dt}R > 0$ outside the horizons, it is clear that any ingoing geodesics that asymptote to the outer horizon will be unstable to small perturbations, so geodesics of this type form a set of zero measure.

The fate of the ingoing geodesics as they are traced backwards from a point $(t,x) = (t_*, x_*)$ is a matter of considerably less importance, but it should be clear from the foregoing discussion what the main possibilities are. If the phase curve of the space-time lies outside Γ_h initially then all ingoing geodesics have $\frac{d}{dt}R \geq 0$ at sufficiently small values of t, and so the geodesic either emerges from the singularity at $x = 0$ at some $t > 0$, or tends in the limit $t \to 0^+$ to a finite positive value of R when $\lim_{t \to 0^+}(am) > 0$, or to $R = 0$ when $\lim_{t \to 0^+}(am) = 0$.

If the phase curve lies on or inside Γ_h initially then inner and outer horizons exist for all $t > 0$. If (t_*, x_*) lies on or inside the inner horizon then $\frac{d}{dt}R \geq 0$ on the geodesic at sufficiently small values of t, and the alternatives mentioned in the previous paragraph apply, with the geodesic inside the inner horizon for all $t < t_*$. If (t_*, x_*) lies between the horizons the same alternatives again apply, but the point of emergence of the geodesic in the limit $t \to 0^+$ can lie on or inside the inner horizon, between the horizons, or on or outside the outer horizon. Finally, if (t_*, x_*) lies on or outside the outer horizon then the geodesic remains outside the outer horizon for all $t < t_*$, and tends to a finite non-negative value of R as $t \to 0^+$.

4.4 Geodesic completeness

A radial null geodesic passing through a point $(t,x) = (t_*, x_*)$ with $t_* > 0$ and $x_* > 0$ is future-complete if the total forward lapse in the affine parameter

$$\Delta\lambda_\infty \equiv \int_{t_*}^{\infty} (2+x)^{-1} x\kappa_\pm^{-1} dt \tag{96}$$

diverges, and is future-incomplete if $\Delta\lambda_\infty$ is finite. Here, the conformal factor $\kappa_\pm$ can be chosen to have an arbitrary positive value κ_* at (t_*, x_*), but its subsequent evolution is governed by equation (50), which can be rewritten in the form

$$\tfrac{d}{dt}(\ln \kappa_\pm^{-1}) = (am)(2+x)^{-4} k_\pm(x,t), \tag{97}$$

where

$$k_\pm(x,t) \equiv 16\xi \pm 4 + 8(4\xi + \sigma \pm 1)x + 4(6\xi + 3\sigma \pm 1)x^2 + 2(4\xi + 3\sigma)x^3 + (\xi + \sigma)x^4 . \tag{98}$$

Similarly, the geodesic is past-complete if the total backwards lapse in the affine parameter

$$\Delta\lambda_0 \equiv \int_{t_*}^{t_0} (2+x)^{-1} x\kappa_\pm^{-1} dt \tag{99}$$

diverges – where the bounding value t_0 here is either 0 or the value of t at which the geodesic emerges from $x = 0$, whichever is larger – and is past-incomplete if $\Delta\lambda_0$ is finite.

Now, the polynomial $k_+(x,t)$ is clearly positive definite (for positive x) if $\xi \geq 0$ and $\sigma \geq 0$, so it is evident from (97) that κ_+^{-1} is monotonically increasing with t, and is therefore bounded below by its value κ_*^{-1} at (t_*, x_*). The weighting factor $(2+x)^{-1}x$ in (96) is also bounded below by $(2+x_*)^{-1}x_*$, so it follows that $\Delta\lambda_\infty$ diverges for all outgoing radial null geodesics, which are therefore guaranteed to be future-complete.

Conversely, for all $t \leq t_*$ the function κ_+^{-1} is bounded above by κ_*^{-1}, while the factor $(2+x)^{-1}x$ is bounded above by 1, so $\Delta\lambda_0$ is finite for all outgoing radial geodesics. The outgoing geodesics are therefore past-incomplete, and their terminal points on the surface $x = 0$ or – if there is a cosmological singularity – at $t = 0$ are true singular boundaries of the space-time.

In the case of the ingoing geodesics, a useful consequence of (98) is that

$$k_-(x,t) = p_-(x,t) + 2(x+1)(x^2 - x - 2) . \tag{100}$$

The function $x^2 - x - 2$ is positive for all $x > 1 + \sqrt{3}$, and so if $p_-(x,t)$ remains positive or zero and x is eventually bounded below by $1 + \sqrt{3}$ on an ingoing geodesic then, from (97), κ_-^{-1} is eventually monotonically increasing with t, and the geodesic is future-complete.

In particular, if the phase curve of space-time enters the region inside Γ^* in Figure 1 then any ingoing geodesic on which $R \to \infty$ is future-complete, because the geodesic has $\tfrac{d}{dt}R \geq 0$ and $R \to \infty$. Similarly, although it is not generally true that $\tfrac{d}{dt}x \geq 0$ on the outer horizon, it can be seen by solving the equation $p_- = 0$ for σ and thereby eliminating σ from $\tfrac{\partial}{\partial x}p_-$ that

$$\tfrac{\partial}{\partial x}p_-(x,t) = -2x^{-1}\xi(2+x)^3 + 2(2+x)^{-1}x(x^2 - 2x - 2) \tag{101}$$

on the apparent horizons, and so $\frac{\partial}{\partial x}p_-(x,t)<0$ on the horizon if $x<1+\sqrt{3}$. This in turn is possible only if the horizon is an inner horizon, and it follows that $x\geq 1+\sqrt{3}$ on any outer horizon. Furthermore, an ingoing geodesic that asymptotes to the outer horizon must approach the horizon from outside the horizon, as $\frac{d}{dt}R\geq 0$ on the horizon. Hence, any ingoing geodesic that asymptotes to the outer horizon is also future-complete.

If the phase curve remains on or outside the contour Γ^*, then any ingoing geodesic on which $x\to\infty$ is also future-complete, as $\frac{d}{dt}R\geq 0$ on all geodesics in this case. So the only ingoing geodesics that are possibly future-incomplete are those that asymptote to the inner horizon (if horizons form) and those on which $R\to\infty$ and $am\to\infty$ but x is not eventually bounded below by $1+\sqrt{3}$ (if the phase curve remains outside Γ^*).

To demonstrate that geodesics of the last type are always future-complete requires another long technical argument, and I will only sketch the proof here. Combining equations (48) and (50) gives

$$\frac{d}{dt}[\ln\{(2+x)^{-1}x\kappa_-^{-1}\}]=(am)^{-1}x^{-1}(2+x)^{-4}q(x,t) \tag{102}$$

along an ingoing geodesic, where the function $q(x,t)$ – defined in (47) – is positive definite (for $x>0$) if the phase curve lies outside $\Gamma_{\rm n}$. As was mentioned in Section 3.3, the minimum of the function on the right-hand side of (102) is occurs when $x=2(4^{1/3}-\sigma^{1/3})^{-1}\sigma^{1/3}$ if $\sigma<4$, and in the limit as $x\to\infty$ if $\sigma\geq 4$. So

$$\frac{d}{dt}[\ln\{(2+x)^{-1}x\kappa_-^{-1}\}]\geq \xi^{-1}\psi(t)\equiv\begin{cases}\xi^{-1}[\xi-\xi_{\rm n}(\sigma)] & \text{if } \sigma<4\\ \xi^{-1}(\xi+\sigma) & \text{if } \sigma\geq 4\end{cases}. \tag{103}$$

Now, on defining the function

$$\chi(t)\equiv\begin{cases}\frac{3}{4}(2^{-1/6}\sigma^{1/3})^4+\xi & \text{if } \sigma<4\\ \sigma+\xi+2 & \text{if } \sigma\geq 4\end{cases} \tag{104}$$

it is evident that (78) together with its extension to $\sigma\geq 4$ read simply $\frac{d}{dt}\chi\leq\psi$. It therefore follows that $\frac{d}{dt}[\ln\{(2+x)^{-1}x\kappa_-^{-1}\}]\geq 0$ outside $\Gamma_{\rm n}$, that $\frac{d}{dt}\chi\leq 0$ inside $\Gamma_{\rm n}$, and that

$$\frac{d}{dt}[\ln\{(2+x)^{-1}x\kappa_-^{-1}\}]\geq \xi^{-1}\frac{d}{dt}\chi \tag{105}$$

everywhere.

If the geodesic is future-incomplete, then according to (96) $(2+x)^{-1}x\kappa_-^{-1}$ cannot be bounded away from zero in the limit of late time, and so $\ln[(2+x)^{-1}x\kappa_-^{-1}]$ cannot be bounded below on any unbounded interval $[t_1,\infty)$. Furthermore, because $\frac{d}{dt}[\ln\{(2+x)^{-1}x\kappa_-^{-1}\}]\geq 0$ outside $\Gamma_{\rm n}$ the phase curve cannot lie entirely on or outside $\Gamma_{\rm n}$ on any interval $[t_2,\infty)$. Also, $\frac{1}{2}>\chi>7-4\sqrt{3}\approx 0.07180$ in the region between $\Gamma_{\rm n}$ and Γ^*, so the net change $\Delta\chi\leq 0$ in χ along any segment of the phase curve that lies entirely in this region is bounded below.

An immediate consequence of (105) is that the net change in $\ln[(2+x)^{-1}x\kappa_-^{-1}]$ on the time interval $[t_*,t]$ satisfies

$$\Delta[\ln\{(2+x)^{-1}x\kappa_-^{-1}\}]\geq \sum_k\int_{I_k}\xi^{-1}\frac{d}{dt}\chi\,dt\ , \tag{106}$$

where $\{I_k\}$ is the set of sub-intervals in $[t_*,t]$ on which the phase curve lies on or inside Γ_n. If χ is monotonically decreasing on $[t_*,\infty)$ then, because $\Delta\chi \le 0$ is bounded below, $\Delta[\ln\{(2+x)^{-1}x\kappa_-^{-1}\}]$ will also be bounded below unless ξ is arbitrarily close to zero on $\{I_k\}$.

In the latter case, the phase curve must eventually enter every neighborhood of the point (σ,ξ) on the axis $\xi=0$ where Γ_n and Γ^* meet. At this point, $\chi = 7-4\sqrt{3}$ and since χ is monotonically decreasing the phase curve will eventually be trapped on or below the curve[12] $\chi = 7-4\sqrt{3}+\varepsilon$ for any positive ε. However, if ε is less than approximately 0.0055569 the curve $\chi = 7-4\sqrt{3}+\varepsilon$ intersects Γ^* at a point (σ,ξ) with $\sigma > \sqrt{3}/9$ and the phase curve is then trapped in the region $\sigma > \sqrt{3}/9$ [see Figure 1]. As mentioned in the previous section, dx/dt is then positive and bounded away from zero at all later times, contradicting the assumption that x is not eventually bounded below by $1+\sqrt{3}$.

On the hand, if the supremum of χ on $[t,\infty)$ does not tend to its minimum possible value $7-4\sqrt{3}$ as $t\to\infty$ then for each sufficiently late time interval I_k it is possible to find a finite union $O_k = \bigcup_{j=1}^{n_k} s_k^{(j)}$ of later sub-intervals with the property that the phase curve lies outside Γ_n on O_k, $\frac{d}{dt}\chi > 0$ on each $s_k^{(j)}$, χ is continuous from $s_k^{(j)}$ to $s_k^{(j+1)}$ for all $j \le n_k - 1$, and $\int_{O_k} \frac{d}{dt}\chi\, dt = -\int_{I_k} \frac{d}{dt}\chi\, dt$. The monotonic variation of χ on I_k and O_k allows ξ to be written as functions of χ on the two sets, which I will refer to as ξ_k^- and ξ_k^+ respectively. Because each curve $\chi =$ constant has gradient $d\xi/d\sigma < 0$, and the point defined by $(\chi,\xi_k^-(\chi))$ lies inside Γ_n while $(\chi,\xi_k^+(\chi))$ lies outside Γ_n, it follows that $\xi_k^-(\chi) > \xi_k^+(\chi)$ for all χ.

Hence, the net change in $\ln[(2+x)^{-1}x\kappa_-^{-1}]$ on $I_k \cup O_k$ is

$$\begin{aligned}\Delta[\ln\{(2+x)^{-1}x\kappa_-^{-1}\}] &\ge \int_{I_k} (\xi_k^-)^{-1}\frac{d}{dt}\chi\, dt + \int_{O_k} (\xi_k^+)^{-1}\frac{d}{dt}\chi\, dt \\ &= \int_{\chi_{\min}}^{\chi_{\max}} [(\xi_k^+)^{-1} - (\xi_k^-)^{-1}]\, d\chi \ge 0\end{aligned}, \tag{107}$$

and the net variation of the integrand in (96) is eventually positive.

By contrast, it is easily verified that all ingoing radial null geodesics that asymptote to the inner horizon are future-incomplete. As was mentioned in Section 3.3, if apparent horizons appear and the null energy condition is satisfied then am tends to a finite positive limit $(am)_\infty$ as $t\to\infty$, while $\xi\to 0$ and $\sigma\to\sigma_\infty$, where $0 \le \sigma_\infty < \sqrt{3}/9$. In addition, R and x tend to finite non-negative limits on any ingoing geodesic that asymptotes to the inner horizon, with the limiting value x_∞ of x fixed by the value of σ_∞ and lying in the range $[0, 1+\sqrt{3})$.

The geodesic equation (48) for an ingoing geodesic in this class can be solved in the limit of large t to give

$$x(t) \approx x_\infty + C_1 e^{-\alpha t} \quad \text{with} \quad \alpha \equiv 2(am)_\infty^{-1}(2+x_\infty)^{-4}(1+x_\infty)(2+2x_\infty - x_\infty^2) > 0 \tag{108}$$

where C_1 is an arbitrary constant. Substituting (108) into the equation (102) for $\frac{d}{dt}\ln[(2+x)^{-1}x\kappa_-^{-1}]$ then gives, for large values of t,

[12] Note that, at each point in the region in Figure 1 inside Γ_n, the curves $\chi =$ constant have a gradient $d\xi/d\sigma < 0$ that lies between the gradients of the steeper and shallower dotted curves through the same point.

$$(2+x)^{-1}x\kappa_{-}^{-1} \approx C_2 e^{-\alpha t} \quad (109)$$

where C_2 is an arbitrary positive constant. It is evident now from (96) that $\Delta\lambda_\infty$ converges and the geodesic is future-complete, as claimed.

The causal structure of gMcV space-times in which apparent horizons appear is therefore no different from the causal structure of the standard McVittie space-times with horizons. As was mentioned in Section 1, the causal structure of the McVittie solutions has been explored by Kaloper, Kleban and Martin [8], who have demonstrated that, if the space-time has the property that $\lim_{t\to\infty}(\dot{a}/a)$ is a finite positive constant, all ingoing radial geodesics that emerge from the surface $x=0$ at sufficiently late times (as so cannot avoid the inner horizon) are future-incomplete.

Since $\dot{a}/a=(am)^{-1}(\xi+\sigma)$, the requirement that $\lim_{t\to\infty}(\dot{a}/a)$ be a finite positive constant is equivalent to stipulating that $(am)_\infty^{-1}\sigma_\infty$ be positive. Kaloper, Kleban and Martin [8] exclude the case $\sigma_\infty=0$ because it entails that $x_\infty=0$, and so the incomplete ingoing radial geodesics terminate at the singularity $x=0$.

Figure 3 shows the conformal diagram for a gMcV (or McVittie) space-time that initially has a cosmological singularity and no apparent horizons, but later develops horizons and has $\sigma_\infty>0$. The ingoing geodesics are represented by the 45° lines inclined up and to the left, and the outgoing geodesics by the lines inclined up and to the right.

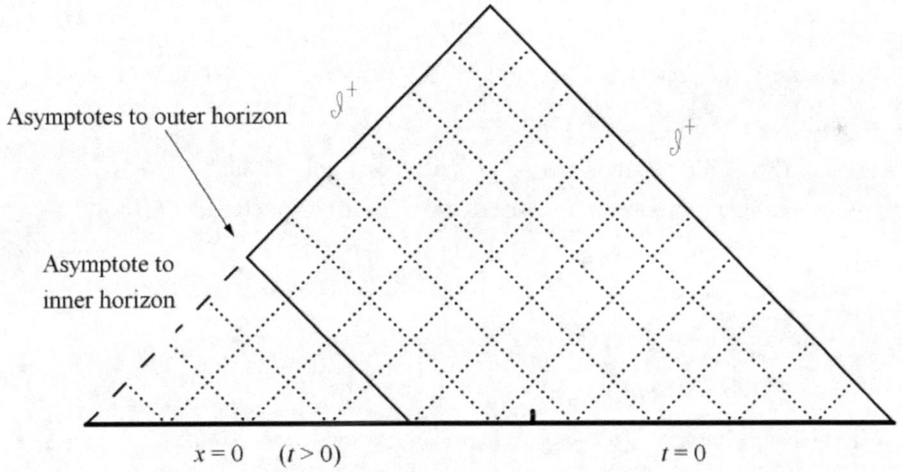

Figure 3. Conformal diagram for a gMcV space-time in which apparent horizons first appear at some time $t>0$ and σ tends to a non-zero limit. The future-incomplete ingoing radial null geodesics emerge from the singular surface $x=0$ and pass through the broken line at the left of the diagram.

5. Conclusions

The principal conclusion of this paper is that the long-term behavior and causal structure of the generalized McVittie space-times are no different from those of the standard McVittie space-times. In particular, if a gMcV space-time satisfies the null energy condition and develops apparent horizons then it inevitably asymptotes to a McVittie space-time, with $\xi\to 0$, $am\to(am)_\infty>0$ and $\sigma\to\sigma_\infty\geq 0$ as $t\to\infty$. Also, the only future-incomplete radial null geodesics in both the gMcV and McVittie cases are ingoing geodesics that asymptote to the inner apparent horizon.

It is tempting to conclude further that, if $\sigma_\infty > 0$, the space-time actually converges to a Kottler solution, which – as was mentioned in Section 1 – has $\xi \equiv 0$ and σ constant. This is not strictly speaking true, as the Kottler solution is regular at $x = 0$ while all other gMcV space-times except the Schwarzschild solution are singular at $x = 0$, and typically remain singular there in the limit as $t \to \infty$, as can be seen from (11).

Nonetheless, Kaloper, Kleban and Martin [8] have chosen to ignore this technicality, and argue that "in the case $H \to H_0 > 0$ it [the surface $x = x_\infty$, $t = \infty$] is a regular event horizon, and the McVittie solution really is a black hole, which asymptotes the Schwarzschild-de Sitter [Kottler] solution".[13] As evidence in support of this claim, they demonstrate that the ingoing radial null geodesics can be continued across the surface $x = x_\infty$, $t = \infty$, and that the limiting values of all the components of the Riemann tensor are finite at the surface, coinciding in fact with the corresponding Kottler values if $\lim_{t\to\infty} \dot{H} = 0$.

Indeed, there is a quicker way to show that a gMcV space-time that develops apparent horizons can always be matched onto the corresponding Kottler solution across the surface $x = x_\infty$, $t = \infty$, by calculating the first and second fundamental forms of the surfaces of constant *t* in both the gMcV and Kottler space-times, and verifying that they coincide in the limit as $t \to \infty$. If equation (44), which solves to give $r = \frac{1}{2}(1+x)m$, is used to replace *r* with *x* as the radial coordinate in (6), the gMcV line element becomes:

$$\begin{aligned} ds^2 &= \tfrac{1}{4}(am)^2(1+x)^{-4}(2+x)^4 \\ &\quad \times [dx - (am)^{-1}(2+x)^{-3}(1+x)n_+(x,t)dt][dx - (am)^{-1}(2+x)^{-3}(1+x)n_-(x,t)dt] \\ &\quad + \tfrac{1}{4}(am)^2(1+x)^{-2}(2+x)^4 d\Omega^2, \end{aligned} \tag{110}$$

where the functions $n_\pm(x,t)$ are defined in (49), and depend on *t* only through σ.

Now, the future-pointing unit vector orthogonal to a given surface of constant *t* is:

$$n^a = x^{-1}(2+x)[\delta_t^a + (am)^{-1}(1+x)\sigma\delta_x^a]. \tag{111}$$

Hence, the first fundamental form $h_{ab} = g_{ab} + n_a n_b$ of the surface has non-zero components

$$h_{tt} = \tfrac{1}{4}(1+x)^{-2}(2+x)^4\sigma^2, \quad h_{tx} = -\tfrac{1}{4}(1+x)^{-3}(2+x)^4 am\sigma,$$

$$h_{xx} = \tfrac{1}{4}(1+x)^{-4}(2+x)^4(am)^2 \quad \text{and} \quad h_{\theta\theta} = h_{\phi\phi}\csc^2\theta = \tfrac{1}{4}(1+x)^{-2}(2+x)^4(am)^2. \tag{112}$$

It is clear from (112) that in the limit as $t \to \infty$, when $\sigma \to \sigma_\infty$ and $am \to (am)_\infty$ in the gMcV space-time, the components of h_{ab} in the gMcV and the corresponding Kottler space-times do coincide.

Similarly, the second fundamental form $K_{ab} = h_a^c h_b^d D_{(c}n_{d)}$ of the surface has non-zero components

$$K_{tt} = \tfrac{1}{4}(am)^{-1}x^{-1}(1+x)^{-2}(2+x)^4\sigma^2[x\sigma + (2+x)\xi] \tag{113}$$

$$K_{tx} = -\tfrac{1}{4}x^{-1}(1+x)^{-3}(2+x)^4\sigma[x\sigma + (2+x)\xi] \tag{114}$$

$$K_{xx} = \tfrac{1}{4}(am)x^{-1}(1+x)^{-4}(2+x)^4[x\sigma + (2+x)\xi] \tag{115}$$

and

$$K_{\theta\theta} = K_{\phi\phi}\csc^2\theta = \tfrac{1}{4}(am)x^{-1}(1+x)^{-2}(2+x)^4[x\sigma + (2+x)\xi], \tag{116}$$

[13] Here, H is shorthand for the Hubble parameter $\dot{a}/a$.

amd it is again evident that the components of K_{ab} in the gMcV and Kottler space-times coincide in the limit $t \to \infty$.

What is questionable about the claims made by Kaloper, Kleban and Martin [8] is not that a McVittie space-time with $H \to H_0 > 0$ can be matched onto a Kottler solution, but that the Kottler solution can be interpreted as describing a "black hole". Although the gMcV line element covers only the coordinate patch $x > 0$ and the Kottler solution is regular at $x = 0$, the static form (4) of the Kottler line element has a curvature singularity at $R = 0$, and it is known that the region between $R = 0$ and the inner apparent horizon is everywhere anti-trapped.

Furthermore, the maximal analytic extension of the Kottler solution consists of an infinite chain of isometric copies of the original Kottler patch (4), as has been demonstrated by Bażański and Ferrari [16], on the basis of a more qualitative discussion of the problem by Gibbons and Hawking [17]. There is no cosmological singularity in the Kottler solution, and ingoing radial null geodesics are past-incomplete as well as future-incomplete, so each link in the chain of Kottler patches is formed by joining the future endpoints of the geodesics in one patch to the past endpoints of the geodesics in the next patch.

As a result, there is no prospect that the geodesic completion of any of the incomplete expanding gMcV space-times will contain a horizon surrounded by a trapped region, rather than an anti-trapped one. The search for a realistic line element describing a non-rotating black hole embedded in an expanding universe therefore needs to proceed in other directions.

References

[1] McVittie G C 1933 *Mon. Not. R. Astr. Soc.* **93** 32
[2] Sussmann R A 1988 *J. Math. Phys.* **29** 1177
[3] Ferraris M, Francaviglia M and Spallicci A 1996 *Nuovo Cimento* B **111** 1031
[4] Nolan B C 1999 Class. *Quant. Grav.* **16** 1227
[5] Kottler F 1918 *Ann. Phys. (Germany)* **56** 410
[6] Nolan B C 1998 *Phys. Rev.* D **58** 064006
[7] Nolan B C 1999 *Class. Quant. Grav.* **16** 3183
[8] Kaloper N, Kleban M and Martin D 2010 *Phys. Rev.* D **81** 104044
[9] Faraoni V and Jacques A 2007 *Phys. Rev.* D **76** 063510
[10] Carrera M and Giulini D 2010 *Phys. Rev.* D **81** 043521
[11] Faraoni V 2009 *Phys. Rev.* D **80** 044013
[12] Sultana J and Dyer C C 2005 *Gen. Relativ. Gravit.* **37** 1347
[13] Sun C-Y Trapping horizons in the Sultana-Dyer space-time *Preprint* gr-qc/1004055
[14] Misner C W and Sharp D H 1964 *Phys. Rev.* **136** B571
[15] Hayward S A 1994 *Phys. Rev.* D **49** 6467
[16] Bażański S L and Ferrari V 1986 *Nuovo Cimento* B **91** 126
[17] Gibbons G W and Hawking S W 1977 *Phys. Rev.* D **15** 2738

Searching for non-Kerr objects

Theocharis A. Apostolatos

Section of Astrophysics, Astronomy, and Mechanics, Department of Physics, University of Athens, Panepistimiopolis Zografos GR15783, Athens, Greece

Georgios Lukes-Gerakopoulos

Academy of Athens, Research Center for Astronomy, Soranou Efesiou 4, GR-11527, Athens, GREECE
Section of Astrophysics, Astronomy, and Mechanics, Department of Physics, University of Athens, Panepistimiopolis Zografos GR15783, Athens, Greece

George Contopoulos

Academy of Athens, Research Center for Astronomy, Soranou Efesiou 4, GR-11527, Athens, GREECE

E-mail: thapostol@phys.uoa.gr

Abstract. We suggest a method that could be used to discriminate a Kerr black hole from any other supermassive axisymmetric astrophysical object by analyzing the gravitational-wave signal from an extreme mass ratio inspiral (EMRI). The method is based on the quite distinct qualitative features that characterize a slightly nonintegrable system. According to the Poincaré-Birkhoff theorem, whenever a resonance of frequencies arise in an axisymmetric perturbed Kerr metric, instead of the anticipated KAM curves of the integrable Kerr case, a Birkhoff chain of islands appears on a surface of sections. The orbits of this chain of islands have a fixed ratio of frequencies. The idea is to exploit this feature to check if the inspiraling low-mass object spends a finite interval of time to cross this resonance, while its orbit evolves adiabatically due to the radiation of gravitational waves.

1. Introduction

One of the most promising sources of gravitational waves for LISA [1] (the future space interferometric detector that is planned to be launched by NASA and ESA) are solar mass ($1 - 10M_{\odot}$) compact objects (black holes or neutron stars) orbiting around supermassive black holes ($\simeq 10^5 - 10^8 M_{\odot}$). The extreme-mass ratio of such binaries render the analytic approximate Post-Newtonian formulae, that have been produced by a large number of people (see [2] and the cited references) during the last couple of decades, an extremely accurate tool to describe the adiabatic change of the orbit of the small object in the Kerr background of the larger one. The analysis of the signal of LISA will give us the opportunity to check if the waveforms according to general relativity are the right ones. Unfortunately the general relativistic waveforms depend on too many parameters of the binary; thus a confusion problem may arise while analyzing the data since a slight deviation of the waveforms due to a different background than Kerr, or a different theory of gravity than General Relativity may be mimicked by a pure relativistic waveform in a Kerr background with wrong parameters.

In order to make sure that whatever signal we monitor is exactly what we anticipate according to our conventional wisdom we have to think of "smoking guns", that is of qualitatively new features that could reveal a deviation of theory from what we expect (see e.g. [3]). Such a new feature that could reveal the deviation of the background metric from Kerr is any orbital effect that could discriminate a slightly non-integrable and an integrable system (like the Kerr one).

The KAM theorem of Kolmogorov, Arnold and Moser [4] states that most of the KAM tori on which the phase orbits of an integrable system evolve are not destroyed when the system becomes slightly perturbed. However, according to the Poincaré-Birkhoff theorem the resonant tori (the ones that are characterized by commensurate ratios of fundamental frequencies) of the integrable system disintegrate when the systems become slightly perturbed and they form the so-called Birkhoff chains of islands on a surface of sections. These islands have a finite thickness and all their interior corresponds to orbits that are characterized by the same rational ratio of frequencies.

This is exactly the idea that we have quantitatively explored in order to check if the background metric on which an EMRI evolves is the integrable case of Kerr or something else. We have used the Manko-Novikov metric [5] as an example of a non-Kerr metric. This is a solution of vacuum Einstein's equations that is axisymmetric and asymptotically flat. It is described by a different multipole moments from the Kerr metric starting from the quadrupole moment. Actually we have used the simplest family of the general Manko-Novikov solutions, which is characterized by a single parameter q, apart of its mass M and spin S. The q parameter measures the deviation of the quadrupole moment from the corresponding Kerr metric (the one with the same mass and spin).

We have shown that the geodesic orbits of this Manko-Novikov metric exhibit Birkhoff islands whenever a resonance is present, as expected for a slightly non-integrable system. The thickness of these islands has been measured as a function of the orbital energy, and the angular momentum, as well as the spin, and the quadrupole deviation of the metric. The size of the islands is expected to have a similar behavior for a generic type of a non-Kerr metric background.

Next we studied what will be the effect of the presence of these islands in the gravitational wave signal radiated by such an EMRI source. As the orbiting object loses energy and angular momentum the orbit changes adiabatically. As a consequence the phase orbit may pass through a chain of resonance islands. During the time interval that the orbit finds itself in such a chain of islands the corresponding frequencies encoded in the gravitational wave signal stay locked to a rational ratio, although both are changing. If such a signal behavior is observed we could be sure that the metric in which the small object orbits is not that of a Kerr black hole.

We end our analysis by presenting some indicative estimates of the time intervals spent by the signal with a locked ratio of frequencies.

2. A toy model for a non-Ker metric

2.1. The line element

In 1992 Manko and Novikov [5] constructed a multiparametric family of spacetimes, as a generalization of the Kerr metric, in order to describe the gravitational field of an arbitrary rotating and axially symmetric isolated object. It is actually an exact solution of the vacuum Einstein equations, built by a non-linear superposition of the Kerr spacetime with an arbitrary static vacuum Weyl field (see [6]). The Manko-Novikov solution that we have used (hereafter called MN metric) is a subfamily of the general multiparametric family of the Manko-Novikov solution. It depends on only three parameters: its mass monopole M, its spin S, and its dimensionless deviation from the corresponding Kerr's quadrupole q. The rest multipole moments are defined from these three parameters. Thus the first four non-zero mass and current-mass moments, M_l and S_l respectively, are

$$\begin{array}{ll} M_0 = M, & S_1 = S, \\ M_2 = -M\left[\left(\frac{S}{M}\right)^2 + qM^2\right], & S_3 = -M\left[\left(\frac{S}{M}\right)^3 + 2qM^2\left(\frac{S}{M}\right)\right]. \end{array} \tag{1}$$

By setting $q = 0$ the MN solution turns into an exact Kerr metric; thus all its multipole moments are then described by the formula

$$M_l + \imath S_l = M\left(\imath\frac{S}{M}\right)^l. \tag{2}$$

The particular MN metric in Weyl-Papapetrou coordinates is described by the following line element

$$ds^2 = -f(dt - \omega d\phi)^2 + f^{-1}\left[e^{2\gamma}(d\rho^2 + dz^2) + \rho^2 d\phi^2\right] \tag{3}$$

where all metric functions f, ω, γ should be considered as functions of the prolate spheroidal coordinates x, y. The coordinates ρ, z are the corresponding cylindrical coordinates which could be expressed as functions of x, y as well. Thus

$$\rho = k\sqrt{(x^2-1)(1-y^2)}, \quad z = kxy \tag{4}$$

and

$$\begin{aligned}
f &= e^{2\psi}\frac{A}{B}, && (5)\\
\omega &= 2ke^{-2\psi}\frac{C}{A} - 4k\frac{\alpha}{1-\alpha^2}, && (6)\\
e^{2\gamma} &= e^{2\gamma'}\frac{A}{(x^2-1)(1-\alpha^2)^2}, && (7)\\
A &= (x^2-1)(1+a\,b)^2 - (1-y^2)(b-a)^2, && (8)\\
B &= [(x+1)+(x-1)a\,b]^2 + [(1+y)a+(1-y)b]^2, && (9)\\
C &= (x^2-1)(1+a\,b)[(b-a)-y(a+b)] \\
&\quad +(1-y^2)(b-a)[(1+a\,b)+x(1-a\,b)], && (10)\\
\psi &= \beta\frac{P_2}{R^3}, && (11)\\
\gamma' &= \ln\sqrt{\frac{x^2-1}{x^2-y^2}} + \frac{3\beta^2}{2R^6}(P_3^2 - P_2^2) \\
&\quad + \beta\left(-2 + \sum_{\ell=0}^{2}\frac{x-y+(-1)^{2-\ell}(x+y)}{R^{\ell+1}}P_\ell\right), && (12)\\
a &= -\alpha\exp\left[-2\beta\left(-1+\sum_{\ell=0}^{2}\frac{(x-y)P_\ell}{R^{\ell+1}}\right)\right], && (13)\\
b &= \alpha\exp\left[2\beta\left(1+\sum_{\ell=0}^{2}\frac{(-1)^{3-\ell}(x+y)P_\ell}{R^{\ell+1}}\right)\right], && (14)\\
R &= \sqrt{x^2+y^2-1}, && (15)\\
P_\ell &= P_\ell(\frac{x\,y}{R}). && (16)
\end{aligned}$$

$P_\ell(z)$ denotes the Legendre polynomial of order l given by

$$P_\ell(z) = \frac{1}{2^\ell \ell!}\left(\frac{d}{dz}\right)^\ell (z^2-1)^\ell. \tag{17}$$

The three parameters k, α, β that appear in the formulae above characterize the metric and are related to the mass M, the spin S, and the quadrupole deviation q through the following expressions:

$$\alpha = \frac{-M+\sqrt{M^2-(S/M)^2}}{(S/M)}, \quad k = M\frac{1-\alpha^2}{1+\alpha^2}, \quad \beta = q\left(\frac{1+\alpha^2}{1-\alpha^2}\right)^3. \tag{18}$$

2.2. Description of the orbits

In order to study the geodesic orbits in MN we have relied on a Lagrangian formulation. Thus the geodesics are simply the equations of motion described by the Lagrangian

$$L = \frac{1}{2}\,\mu\, g_{\mu\nu}\,\frac{dx^\mu}{d\tau}\frac{dx^\nu}{d\tau} \tag{19}$$

where μ is the rest mass of the test body, and τ is the proper time along the orbit. Besides μ there are two integrals of motion $p_t = -\mu E$ and $p_\phi = \mu L_z$ due to stationarity and axisymmetry of the metric, respectively. From suitable combinations of E, L_z and the metric functions one can compute by simple integration the coordinates t and ϕ, once $\rho(\tau)$ and $z(\tau)$ are known. Thus the computation of the geodesic motion is constrained in the 4-dimensional phase space: $(\rho, \dot{\rho}, z, \dot{z})$ where the dots represent derivatives with respect to τ. An intuitive concise tool to obtain some basic characteristics of the orbits is the effective potential $V_{\text{eff}}(\rho, z)$ (see Eq. (14) of [7]) which connects the Newtonian analogue of kinetic energy to the position of the test particle on the $\rho - z$ plane. Thus

$$\frac{1}{2}(\dot{\rho}^2 + \dot{z}^2) + V_{\text{eff}}(\rho, z) = 0. \tag{20}$$

The locus of $V_{\text{eff}}(\rho, z) = 0$ defines the boundary of any orbit, and its shape is characterized by E, L_z, q, S and M.

While for $q < 0$ there is only a single region of non-plunging bound orbits ($E < 1$), for $q > 0$ a new small region of non-plunging bound orbits appears between the previous region and the central region for a range of the rest parameters. Depending on E, L_z, S the two regions are either distinct or they could be joined in a single one.

In order to numerically integrate the orbits we have applied a sixth order Runge Kutta integration scheme with a constant step of integration $\delta s = 10^{-1}$, except for the cases in which an orbit reaches the surface of section; then the integration step was gradually reduced to $\delta s = 10^{-8}$ in order to get a fine approximation of the surface of section at $z = 0$. The relative error of the integration, measured through the relative variation of the Lagrangian value $\Delta L/L = |(L_n - L)/L|$ after 10^3 crossings through the $z = 0$ surface of section, did not exceeded the order of 10^{-10}. (L_n is the value of the Lagrangian due to the integration). For the more demanding integration in the inner region of allowed non-plunging orbits the integration step was chosen even smaller to keep the errors of L at the same order of magnitude.

By changing the initial conditions of an orbit we obtained a series of KAM invariant curves that are foliated within each other on a surface of section according to the KAM theorem. The Poincaré-Birkhoff chains of islands are fine structures and one should vary the initial conditions through very small steps to find them. Fortunately there is a very useful tool to find the suitable initial conditions that lead to resonances; it is the rotation number, that is the average fraction of a full circle along the KAM curve that the series of subsequent intersecting points (Poincaré consequents) advance around a point that represents the central periodic orbit. Whenever this number is close to a ratio of integer numbers then we are approaching a resonance. (For more details about the use of rotation number see [7, 8].) Thus by using the rotation number as a tool to choose the initial conditions we managed to produce two chains of Birkhoff islands in the outer region of bound orbits; one corresponding to the resonance of $2/3$ and one corresponding to the resonance of $1/2$. If we choose initial conditions within a Birkhoff island the corresponding Poincaré consequents form a set of closed curves each lying inside the islands of that chain. In Fig. 1 we have plotted a series of KAM curves and the two aforementioned chains of islands on the surface of sections at $z = 0$ corresponding to orbits in the outer allowed region.

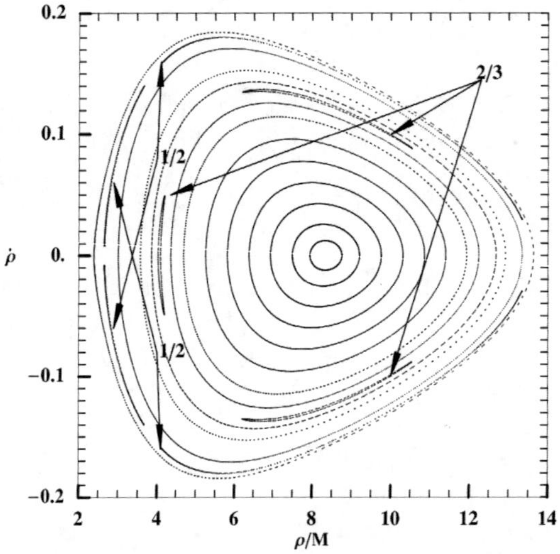

Figure 1. The surface of sections of the outer region on the $(\rho,\ \dot{\rho})$ plane for the parameter set $E = 0.95,\ L_z = 3M,\ \chi = 0.9,\ q = 0.95$. There is one triplet of islands corresponding to the resonance of $2/3$, and 2 doublets corresponding to the resonance of $1/2$.

The inner region of bound orbits, as long as it is distinct from the outer one, has a more complicated surface of section (see Fig. 2): Most of the orbits are chaotic and the intersecting points are scattered irregularly on a large region of the available phase space on the $(\rho - \dot{\rho})$ plane. In contrast to previous investigations [9] there are also regular quasi-periodic orbits represented by the invariant curves and at least one chain of 3 resonant islands (Fig. 2) Thus the general picture of a surface of section through the whole range of phase space $(\rho - \dot{\rho})$ is that of a metric that is "close" to the integrable Kerr one. The further we are located from the central point of the field, the less important are the effects of the higher multipole moments on the orbits; thus the MN metric behaves like a perturbed Kerr and the only effect on the phase space is the appearance of the Birkhoff chains. In the inner region the higher multipole moments are essential for a geodesic orbit. Since all these moments differ from the corresponding Kerr, the metric does not look like a slightly perturbed Kerr. Instead the metric is quite different from Kerr in this region and chaos is dominant. For a more thorough study of the orbits in the MN metric see [7].

3. Inspiraling orbits

Next we explored the effect of resonances on an adiabatically changing orbit, that represents an actual EMRI orbit in a MN background. We have used the best known analytic formula that describes the radiation losses of an EMRI orbit in a Kerr metric [10], suitably adjusted to allow for the difference of the quadrupole moments between the two metrics. Since the radiation

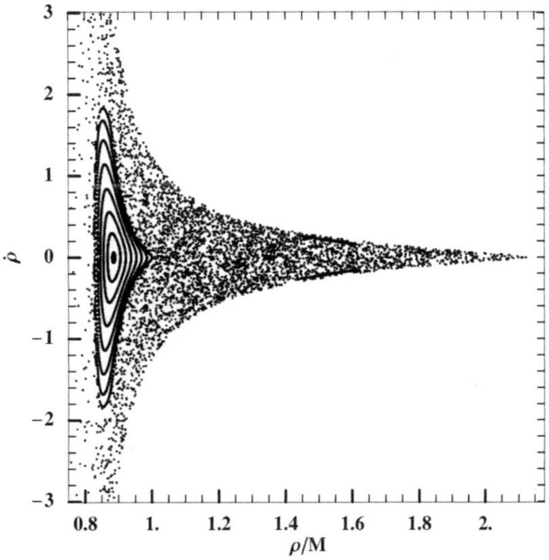

Figure 2. The surface of sections at $z = 0$ of the inner region on the ρ, $\dot{\rho}$ plane for the parameter set $E = 0.95$, $L_z = 3M$, $\chi = 0.9$, $q = 0.95$. The corresponding phase space is dominated by a chaotic sea, while there is also present a region of periodic orbits, and a very thin chain of resonant islands (thin dark regions).

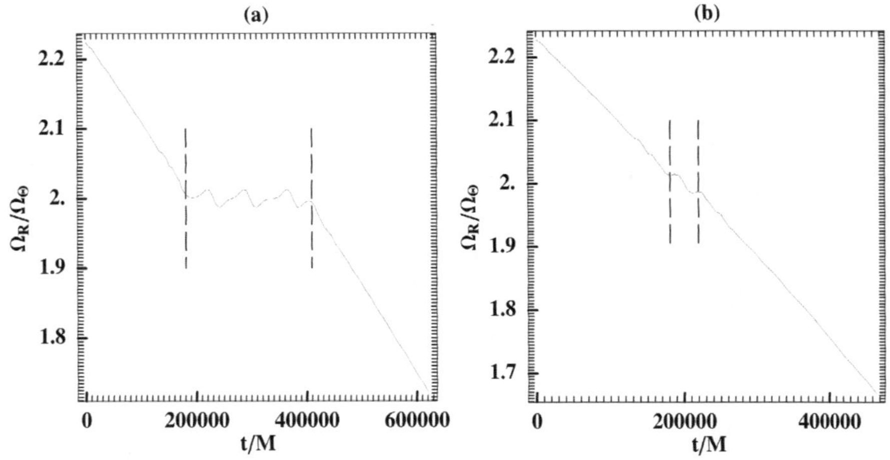

Figure 3. These two diagrams show how the ratio Ω_ρ/Ω_z varies with time while the phase orbit crosses the resonance of 2/3 for two slightly different initial conditions. Both cases correspond to a MN metric with $\chi = 0.9$, $q = 0.95$, and orbital parameters $E = 0.95$, $L_z = 3M$, while the ratio of masses is assumed to be $\mu/M = 8 \times 10^{-5}$.

losses of E and L_z are expressed as functions of the orbital parameters, we have used such initial conditions for an orbit that starts very close to a particular resonance. From the corresponding geodesic orbit we computed its orbital parameters and then the corresponding losses. Then we assumed a linear evolution of $E(t)$ and $L_z(t)$ and we evolved the orbit accordingly. Depending on the location of the entrance point of the phase orbit in a Birkhoff island the subsequent evolution of the orbit was either a longer or a shorter passage of the orbit through that resonant island. In any case the two fundamental frequencies of the orbit remained locked to each other at a constant ratio while they both evolved. Thus a plateau in the evolution of the ratio of frequencies was present in the Fourier snapshots of the signal. The frequencies of the corresponding gravitational waves should have then analogous behavior.

This analysis is useful in order to check if the background of a detected EMRI signal is that of a Kerr or not. Although it seems quite difficult technically to incorporate a plateau effect in the bank of templates used to detect and analyze a gravitational wave signal, since this would increase dramatically the parameter space of templates, we could simply focus our attention at that part of the signal that its frequencies are close to a ratio of simple integers and look for a possible freezing of the ratio of frequencies while they evolve.

In Fig. 3 we have plotted the evolution of the ratio of the two frequencies Ω_ρ/Ω_z as a function of time for two slightly different initial conditions. The two cases spend different time intervals inside the resonance of 2/3; thus the corresponding plateaus differ from each other. For a typical EMRI this time interval is of the order of a few hours to many days.

Acknowledgments

T. A. acknowledges the research funding program "Kapodistrias" of ELKE (Grant No 70/4/7672) and the I.K.Y. (IKYDA 2010). G. L.-G. was supported by the Research Committee of the Academy of Athens.

References

[1] Bender P, Danzmann P, and the LISA Study Team 1998 "Laser Interferometer Space Antenna for the Detection of Gravitational Waves, Pre-Phase A Report" **MPQ 233** (Garching: Max-Planck-Institüt für Quantenoptik)
[2] G. Schäfer to appear in "Mass and Motion in General Relativity" Proceedings of the CNRS School in Orleans/France, eds. Blanchet L, Spallicci L A, and Whiting B (*Preprint* gr-qc/09102857)
[3] Johannsen T and Psaltis D 2010 *Astrophys. J.*, **716** 187; and **718** 446
[4] Kolmogorov A N 1954 *Dokl. Akad. Nauk. SSSR* **98** 527; Arnold V I 1963 *Russian Math. Survey* **18** 13-40, 85-191; Moser J 1962 *Nachr. Akad. Wiss. Gttingen Math.-Phys. Kl. II* 1-20
[5] Manko V S and Novikov I D 1992 *Class. Quant. Grav.* **9** 2477
[6] Castejon-Amenedo J and Manko V S 1990 *Phys. Rev. D* **41** 2018
[7] Lukes-Gerakopoulos G, Apostolatos T A and Contopoulos G 2010 *Phys. Rev. D* **81** 124005
[8] Contopoulos G 2002 "Order and chaos in dynamical Astronomy" (Berlin: Springer)
[9] Gair J R, Li C and Mandel I 2008 *Phys. Rev. D* **77** 024035
[10] Gair J and Glampedakis K 2006 *Phys. Rev. D* **73** 064037

New Cosmological Constraints on Axions and Axion-Like Particles

Anastasios Avgoustidis

Centre for Theoretical Cosmology, DAMTP, CMS, Wilberforce Road, Cambridge CB3 0WA, England

E-mail: a.avgoustidis@damtp.cam.ac.uk

Abstract. We present constraints on the axion-photon coupling scale M for light axion-like particles by combining recent Supernova Type Ia data with the latest measurements of the Hubble expansion at redshifts between 0 and 2. Allowing for a coupling between axions and photons leads to a modification of the inferred luminosity distances for supernovae due to the conversion of photons to axions in the presence of intergalactic magnetic fields. We constrain such couplings by considering deviations from the luminosity-angular diameter distance relation $d_L = d_A(1+z)^2$. We find that for intergalactic magnetic fields of order 1 nG, current supernova and Hubble expansion data rule out a region in the coupling scale M between 10^{10} and 10^{11} GeV.

1. Introduction

Axions are light pseudo-scalar particles, first introduced by Peccei and Quinn (PQ) [1] as a solution to the strong-CP problem in Quantum ChromoDynamics (QCD). The essence of this problem is that the dimensionless parameter θ, controlling the CP-violating term allowed by QCD:

$$\mathcal{L}_{\not{C}\not{P}} = \frac{\alpha_s}{4\pi}\,\theta\,\mathrm{tr}G_{\mu\nu}\tilde{G}^{\mu\nu} \equiv \frac{\alpha_s}{4\pi}\,\theta\,\epsilon^{\mu\nu\kappa\lambda}\,\mathrm{tr}G_{\mu\nu}\tilde{G}^{\kappa\lambda}\,, \tag{1}$$

where $G_{\mu\nu}$ is the gluonic field strength (α_s being the strong coupling constant), is constrained experimentally [2] to be unnaturally small, $|\bar{\theta}| \lesssim 10^{-10}$. Since θ can a priori take values anywhere in the range between 0 and 2π, this is a fine-tuning problem with no explanation within the Standard Model. The PQ solution is then to promote θ to a dynamical field, the axion a, with a global shift symmetry $a \to a + \mathrm{const}$ that can only be broken by anomalous CP-violating terms. The axion thus acquires a potential with minimum at $a = 0$ and so its vacuum expectation value is dynamically driven to small values, thus solving the strong CP-problem. Due to the explicit breaking of the shift symmetry down to a discrete subgroup, the axion becomes a Pseudo-Nambu-Goldstone boson [3] with a parametrically small mass:

$$m_a \approx 0.6\ \mathrm{meV} \times \left(\frac{10^{10}\mathrm{GeV}}{f_a}\right)\,, \tag{2}$$

where, f_a is the axion decay constant determining the strength of interaction between the axion and the other Standard Model fields. In particular, the axion couples to the photon through a term:

$$\mathcal{L}_{a\gamma\gamma} = \frac{\alpha}{8\pi f_a} a F_{\mu\nu}\tilde{F}^{\mu\nu}\,, \tag{3}$$

where $F_{\mu\nu}$ is now the electromagnetic field strength and α the fine structure constant.

Even though axions were historically introduced to solve a concrete particle physics problem arising from a puzzling experimental result, as summarised above, it was soon realised that they could play an important role in cosmology and, in particular, provide a significant contribution to Cold Dark Matter (CDM). Indeed, an initial misalignment of the angle θ from zero, leads to oscillations of the zero-mode of the axion field as the universe cools to $T \ll f_a$. This corresponds to a zero-momentum condensate, behaving as CDM, with a potentially significant contribution to the matter density at the present epoch (for classic reviews see [4, 5]). Further, it is now understood that similar light (pseudo-)scalars, coupling to electromagnetism through two-photon terms like in equation (3), are ubiquitous in fundamental physics (see for example [6, 7, 8]). Notably, such Axion-Like Particles (ALPs) arise in both Heterotic and Type II string theories as zero-modes of certain antisymmetric form fields.

Of particular cosmological interest is the fact that ALPs occur generically in the very constructions used in Type IIB string theory to build models of cosmic inflation [9, 10, 11]. Indeed, the worldvolume action for a Dp-brane, on which these models are based, reads:

$$S_p = -T_p \left(\int d^{p+1}\xi \, e^{-\phi} \sqrt{\det(g + B + 2\pi\alpha' F)} + i \int \sum_q C_q \wedge e^{B+2\pi\alpha' F} \right), \tag{4}$$

where T_p is the brane tension, ϕ the dilaton, g the induced metric on the brane worldvolume, B the antisymmetric 2-form, F the field strength of $U(1)$ gauge fields living on the worldvolume, and C_q the Ramond-Ramond (RR) q-forms.

The inflationary sector is contained in the first term, known as the Dirac-Born-Infeld (DBI) piece, which generates kinetic terms for worldvolume scalars that play the role of inflaton fields. In the most well-studied class of brane inflation models [12, 13, 14, 15], the inflaton is identified with a certain combination of the embedding fields Y^M, worldvolume scalars describing the position of the brane in the bulk. This is captured by the $\sqrt{\det g}$ term, where the pullback $g_{\mu\nu} = G_{MN}\, \partial_\mu Y^M \partial_\nu Y^N$ of the bulk metric G_{MN} on the worldvolume generates kinetic terms[1] for the scalar fields Y^M. The gauge field term $F_{\mu\nu}$ generates, to lowest order, Maxwell electrodynamics $F_{\mu\nu}F^{\mu\nu}$. For $p > 3$, the Dp-brane can wrap one-cycles, in which case $F_{\mu\nu}F^{\mu\nu}$ contains kinetic terms for scalar fields that correspond to Wilson Lines along a combination of these one-cycles. This term can also be used for inflation in the context of Wilson line models [19, 20]. Finally, the term involving the (pullback) 2-form $B_{\mu\nu}$ has also been successfully used to generate (axion monodromy) inflation [21][2].

The second term in the action (4) is known as the Wess-Zumino (WZ) piece, generating all possible combinations of wedge products of the RR q-forms C_q and powers of B and F that integrate on the $(p+1)$-dimensional worldvolume. In most brane inflation models, its main role[3] is to cancel the constant (energy) term in the DBI expansion, but it is of central interest here because it contains axions. In the case of a $D3$-brane, for example, expanding the exponential to 2nd order in F yields a term:

$$\mathcal{S}_{\rm axion} \propto \int C_0\, F \wedge F \;\; = \;\; \int C_0\, F_{\mu\nu} F_{\kappa\lambda}\, dx^\mu \wedge dx^\nu \wedge dx^\kappa \wedge dx^\lambda =$$

[1] The potential $V(Y)$ comes from a Coulomb-like interaction between different branes and can be computed through a cylinder string amplitude [16]. It receives, however, important corrections from moduli stabilisation effects [14, 17, 18].

[2] The dilaton, ϕ, even though was historically the first field to be considered as an inflaton candidate in string theory, generically has too steep (exponential) potential and is unsuitable for inflation [22].

[3] Another important role of the WZ piece is in understanding the relation between cosmic strings produced at the end of brane inflation and $D1$-branes. Expanding the exponential to linear order yields a term $C_2 \wedge F$, which can be used to argue that topological defects produced by Higgsing the gauge field F couple to the RR 2-form C_2 and so can be identified with $D1$-branes [23].

$$= \int C_0\, F_{\mu\nu} F_{\kappa\lambda}\, \epsilon^{\mu\nu\kappa\lambda} d^4x \equiv \int C_0\, F_{\mu\nu} \tilde{F}^{\mu\nu}\, d^4x\,. \qquad (5)$$

This is the same term as in (3) with the axion identified with the RR scalar C_0, while the constant of proportionality (coupling strength) depends on a combination of fundamental parameters like the string scale and string coupling. Thus, even though axions are not necessarily part of the inflationary sector, they are generically present in the underlying theory and could be used to provide additional constraints on fundamental hi-energy physics parameters through their coupling (5) to photons. Indeed, such couplings can lead to the conversion of ordinary photons into axions (or ALPs) in the presence of a magnetic field (cf the Primakoff effect [24]).

Here, we present constraints on axion-photon couplings by allowing for the conversion of photons originating from Type Ia Supernovae (SN) into axions in the presence of intergalactic magnetic fields. This gives rise to an opacity effect at cosmological redshifts, which can be constrained by confronting luminosity and angular diameter distance measurements with the Etherington relation [25]

$$d_L(z) = (1+z)^2 d_A(z)\,, \qquad (6)$$

where $d_L(z)$ and $d_A(z)$ are respectively the luminosity and angular diameter distances at redshift z. This "distance-duality" relation is valid in any cosmological background where photons travel on null geodesics, provided that local Lorentz invariance is respected and, crucially, photon number is conserved. Since, for the optical frequencies we are interested in (SN observations), deviations from Lorentz invariance are strongly constrained [26], one can interpret any such opacity effect in terms of photon-axion conversions and use the data to constrain the fundamental coupling scale quantifying terms like (3) and (5). We find that, for intergalactic magnetic fields of the order of 1 nG, current SN and cosmic expansion data rule out a region in the axion-photon coupling between 10^{10} and 10^{11} GeV.

2. Constraining cosmic opacity

Let us first describe the general method and place constraints on cosmic opacity $\tau(z)$ (defined below) without specifying the details of the underlying theory [27]. In section 3 we will interpret these constraints as a photon-axion mixing effect, which actually imposes a particular parametrisation of τ [28]. We use Type Ia SN brightness data (namely the SCP Union 2008 compilation [29]) in combination with measurements of cosmic expansion $H(z)$ from differential ageing of luminous red galaxies (LRGs) [30, 31, 32] to constrain opacity at redshifts up to $z \approx 2$. The basic idea is to study possible deviations from the Etherington relation (6), the distance duality between luminosity and angular diameter distance. As discussed in section 1, systematic violations of this relation can be interpreted as an opacity effect in the observed luminosity distance, which we quantify in terms of an opacity function $\tau(z)$:

$$d^2_{L,obs} = d^2_{L,true} e^{\tau(z)}\,. \qquad (7)$$

Therefore, the inferred distance moduli for the observed SNe pick a term linear in $\tau(z)$:

$$DM_{obs}(z) = DM_{true}(z) + 2.5[\log e]\tau(z)\,. \qquad (8)$$

On the other hand, we can also use other observations independent of τ, to determine the right-hand-side (d_A) in equation (6). This approach was initiated in [33] (see also [34, 35, 36, 37]) where the authors used measurements [38] of the baryon acoustic oscillation (BAO) scale at two redshifts, $z = 0.20$ and $z = 0.35$, obtaining an upper-bound for the difference in opacity between these two redshifts, $\Delta\tau < 0.13$ at 95% confidence. In reference [27] we improved this constraint by one order of magnitude and extended it over a wider redshift range by using instead measurements of cosmic expansion $H(z)$ from differential ageing of LRGs at redshifts $z \lesssim 2$. Introducing a parameter ϵ, we studied deviations from the Etherington relation of the form:

$$d_L(z) = d_A(z)(1+z)^{2+\epsilon}, \tag{9}$$

and constrained this parameter to be $\epsilon = -0.01^{+0.08}_{-0.09}$ at 95% confidence. Using the latest $H(z)$ data [32] combined with the most recent determination of H_0 (Riess *et al* [39]) improves this constraint to $\epsilon = -0.04^{+0.08}_{-0.07}$, again at 95% confidence [28].

Fig. 1 shows our joint constraints combining the $H(z)$ and SN datasets described above. On the left, the dark blue contours correspond to the (two-parameter) 68% and 95% confidence levels obtained from SN data alone, while lighter blue contours are the corresponding confidence levels for $H(z)$ data. Solid-line transparent contours are for joint SN+$H(z)$ data. On the right we show one-parameter (marginalized over all other parameters) constraints on ϵ. Note on the left panel that even the SN contours alone close, excluding a flat universe without a cosmological constant by more than 2-σ. This is a known feature of high-redshift SN data. The statistical significance is dramatically improved by including the $H(z)$ data.

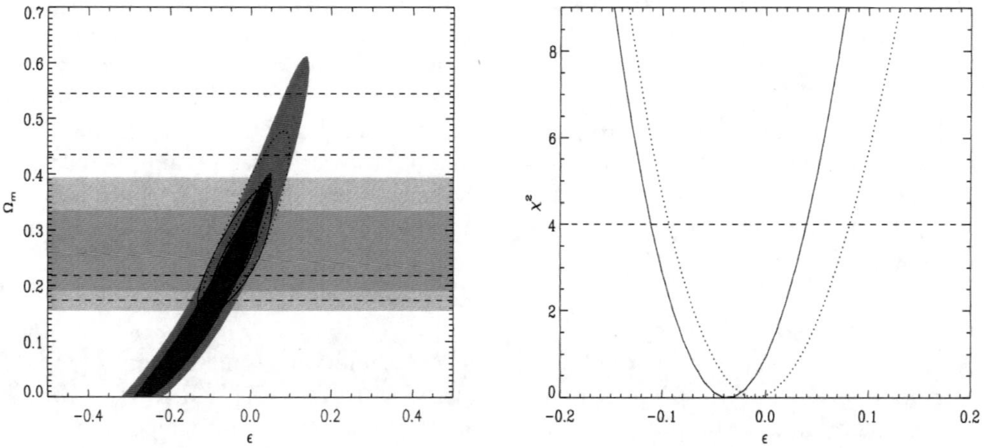

Figure 1. *Left:* Two-parameter constraints on the $\epsilon - \Omega_m$ plane. Dark blue contours correspond to 68% and 95% confidence levels obtained from SN data alone, light blue contours are for $H(z)$ data, and solid line transparent contours are for joint SN+$H(z)$. *Right:* One-parameter joint constraints on ϵ. The dashed line shows the 95% confidence level, $\Delta\chi^2 = 2$. The dashed (left) and dotted (right) lines show the corresponding constraints using the HST Key project determination of H_0 [40] instead of [39].

We now move on to study the implications of these opacity constraints on models of (pseudo-)scalar ALPs coupling to photons.

3. Opacity from Axions and Axion-Like-Particles

Scalar or pseudo-scalar particles from physics beyond the standard model, here denoted as ϕ, may couple to photons through

$$\mathcal{L}_{scalar} = \frac{1}{4M} F_{\mu\nu} F^{\mu\nu} \phi \tag{10}$$

and

$$\mathcal{L}_{pseudo-scalar} = \frac{1}{8M} \epsilon_{\mu\nu\lambda\rho} F^{\mu\nu} F^{\lambda\rho} \phi \tag{11}$$

where M is the energy scale of the coupling, $F_{\mu\nu}$ the electromagnetic field strength and $\epsilon_{\mu\nu\lambda\rho}$ the Levi-Civita symbol in four dimensions. As described in the introduction, a coupling of the form (11) arises for the axion, introduced by Peccei and Quinn (PQ) to solve the strong CP problem [1], but, more general (pseudo-)scalar ALPs are ubiquitous in fundamental theory and particular string theory [7]. Interestingly, they arise in the constructions used to build models of brane inflation, thus providing a new potential observational window to the physics of the early universe.

In the context of PQ theory, the global shift symmetry is explicitly broken and the axion mass is set by the coupling (11), see equation (2). For generic ALPs, however, the mass is in principle independent of the coupling strength, and can be zero if the related shift symmetry remains intact. Here, we will focus on very light ALPs, for which opacity constraints from $H(z)+SN$ data are strongest.

As briefly discussed in the intoduction, the mixing terms (10) and (11) imply that ALPs can affect photon propagation. Indeed, if photons traverse a magnetic field there is a finite probability that they will oscillate into ALPs [41]. Notice however that only photons polarized perpendicular (parallel) to the magnetic field mix with scalar (pseudo-scalar) particles. Therefore, the interactions between photons and ALPs in the presence of a magnetic field not only imply that photon number is not conserved, but can also alter the polarization of the light beam. Both effects have been exploited in searches for ALPs both in the laboratory and astronomical observations, see [8] for a recent review.

3.1. Effects of ALPs on SN brightness

ALPs can have an impact on SN observations if photons from SNae are observed through intergalactic magnetic fields. This would give rise to changes in the observed SN luminosities, in a redshift-dependent way.

Indeed, the probability of a suitably polarized photon converting into an ALP after traveling a distance L in a constant, coherent magnetic field is given by [41]:

$$P(L) = \sin^2(2\theta)\sin^2\left(\frac{\Delta(L)}{\cos 2\theta}\right), \tag{12}$$

where

$$\Delta(L) = \frac{m_{\text{eff}}^2 L}{4\omega} \tag{13}$$

$$\tan 2\theta = \frac{2B\omega}{M m_{\text{eff}}^2}. \tag{14}$$

Here, $\omega = 2\pi\nu$ is the photon energy, B is the strength of the magnetic field and $m_{\text{eff}}^2 = |m_\phi^2 - \omega_P^2|$, with m_ϕ the mass of the ALP and $\omega_P^2 = 4\pi^2\alpha n_e/m_e$ the plasma frequency of the medium acting as an effective mass for the photons (α is the fine structure constant, n_e the local number density of electrons, and m_e the mass of the electron). In what follows we restrict our attention to very light fields, $m_\phi^2 < \omega_P^2$, for which constraints on cosmic opacity have the most power for constraining the strength of the coupling between ALPs and photons.

The probability (12) depends strongly on the intergalactic magnetic field and its coherence length. Unfortunately, the astrophysical uncertainties for both of these quantities are large. Magnetic fields coherent over the whole Hubble volume are limited, by observations of the CMB and Faraday rotation, to $B \lesssim 10^{-9}$ G [42, 43], but fields with shorter coherence lengths are also constrained. In particular, fields coherent on scales $\sim$ 50 Mpc must satisfy $B \lesssim 6\times10^{-9}$ G, while fields coherent on scales $\sim$ Mpc are constrained to $B \lesssim 10^{-8}$ G [43]. These values are relevant because, to explain the origin of observed galactic magnetic fields, intergalactic magnetic fields with coherence lengths $\sim$ Mpc are needed [44].

When considering axion effects on SN brightness, one is interesting in the integrated effect of the probability (12) along the line of sight. The intergalactic magnetic field is not expected to be coherent all the way from the earth to supernovae sources and the standard approach for modelling its fluctuations over cosmological scales is to use the *cell magnetic field* model[4]. In this model, one assumes that the magnetic field is split up into a large number of equally sized domains of length L each. Within each domain the magnetic field is taken to be constant, its strength being the same in all domains, while its orientation is chosen randomly. The photon-axion conversion probability in each domain is then given by equation (12) and one must sum over all domains along the line of sight to each given supernova.

In addition, one also needs to know the plasma frequency of the intergalactic medium along the line of sight, a complicated issue given that no measurements of the electron density are available in the large voids of the interstellar medium. A large-scale average can be easily inferred from the total amount of electrons determined by the CMB estimation of the baryon to photon ratio, giving $\omega_P \simeq 1.8 \times 10^{-14}\,\text{eV}$ in the present era, see e.g. [47]. Note, however, that average values up to a factor of 15 smaller may be plausible [48]. The accepted approach is to assume that ω_P^2 is homogeneous given by a fixed value in this range. To check the dependence of our results on this assumption, we will allow a range of a couple of orders of magnitude around the average.

3.2. *The integrated effect*

As we are interested in opacity effects out to redshifts $z \gtrsim \mathcal{O}(1)$ we must also take into account the redshift evolution of the environment that causes mixing between photons and (pseudo-)scalars. Assuming the magnetic fields are frozen into the plasma, their strength scales as $B(z) = B_0(1+z)^2$ [44] while $\omega(z) = \omega_0(1+z)$ and $\omega_P^2(z) = \omega_{P0}^2(1+z)^3$ (since it is proportional to the electron density). Here, the subscript 0 indicates values in the present epoch. The physical length of a magnetic domain scales as $L(z) = L_0(1+z)^{-1}$ as long as it is smaller than the Hubble radius. Then, the two quantities appearing in the probability of conversion (12), θ and Δ, redshift as $\Delta(z) = \Delta_0(1+z)$ and $\tan 2\theta(z) = \tan 2\theta_0$.

Ignoring this redshift evolution, the photon survival probability for a source at redshift z was first calculated in [49]. In [28] we computed this probability including the above redshift evolution. The overall effect can be modelled by a product of N redshift-dependent matrices, which can be diagonalised by a redshift-independent transformation. If the number of domains N is large, this product can be approximated by an integral, yielding for the survival probability $\mathcal{P}(z)$ [28]:

$$\mathcal{P}(z) = A + (1-A)\exp\left(-\frac{P}{H_0 L}\frac{H(z)-H_0}{\Omega_m H_0}\right), \tag{15}$$

where we have allowed for an initial flux of axions $I_\phi(z_I)$ and defined:

$$\mathcal{P}(z) = \frac{I_\gamma(0)}{I_\gamma(z_I)} \quad ; \quad A = \frac{2}{3}\left(1 + \frac{I_\phi(z_I)}{I_\gamma(z_I)}\right). \tag{16}$$

Note that with this "averaging" procedure one looses sensitivity on whether the effect comes from scalar or pseudo-scalar particles, i.e. form terms like (10) or (11) respectively.

[4] A more accurate choice for a model of the magnetic field would be to assume a power spectrum of its fluctuations. However, at high frequencies, $\Delta \ll \pi/2$, the cell and power spectrum models give the same results, and at lower frequencies the cell model captures all the qualitative features of ALP-photon interactions, but underestimates the probability of conversion [45, 46]. Therefore, using the cell magnetic field model will give rise to conservative constraints.

Equation (15) is valid for small P; in the case where P is of order unity, one should replace $3P/2 \to -\ln(1-3P/2)$.

Also note that the redshift dependence in (15) is absent when $A=1$. When this is the case, the initial flux of photons and scalars is thermalised, $I_\gamma(z_I) = 2I_\phi(z_I)$, so that, on average, the effect of photons converting into scalars is compensated by that of scalars converting into photons, and thus there is no propagation effect.

As explained above, the finite probability of conversion gives rise to an apparent change in luminosity distance. Thus, if photons are converted to ALPs along the line of sight, then the inferred and true and luminosity distance squared (cf equation (7)) will differ by the (redshift-dependent) factor $\mathcal{P}(z)$, so:

$$d_L \to d_L/\sqrt{\mathcal{P}(z)}\,. \tag{17}$$

This allows us to search for systematic deviations from the Etherington relation (6) using the methodology of section 2, and interpret them in terms of ALPs, placing constraints on the axion-photon coupling strength M in equations (10) and (11). Therefore, interpreting opacity as coming from ALPs imposes a special parametrization for $\tau(z)$:

$$\tau(z) = -\ln \mathcal{P}(z)\,, \tag{18}$$

where $\mathcal{P}(z)$ is given by equation (15) above.

We now proceed to use this result and the methodology of section 2 to examine constraints on the ALP parameters A and P, that can in turn be translated into a constraint for the scalar-photon coupling strength M.

4. Constraints on Axion-Like-Particles

In Fig. 2 we show our results marginalized over Ω_m and H_0. The dark and light contours represent 68% and 95% joint confidence levels respectively, using the SN data only (left) and joint SN+$H(z)$ data (right). For very small conversion probability $P \lesssim 10^{-5}$ (weak mixing) our constraints become weak because of the lack of photon-ALP mixing. On the other extreme, when the probability is very strong ($P \gtrsim$ few $\times\, 10^{-2}$) the photons and axions mix until thermalisation and the redshift dependence of the opacity is lost, so, again, our constraints become weak. Finally we can also observe that a band around $A=1$ (yellow line) again cannot be constrained. $A=1$ corresponds to an equilibrated photon-ALP flux at the SNe such that photon$\to$ALP conversions are compensated by the reverse process, making the photon number redshift-independent.

The first notable feature is that our 2-σ bounds (white regions) are significantly improved when including the $H(z)$ data (right panels) compared to using SN data alone (left panels). This improvement is particularly visible in the weak mixing regime.

An interesting feature appears in the upper left plot, corresponding to a redshift-dependent background using SN data only. Here, the 68% C.L. contour shows not one, but three regions where photon-ALP mixing improves the fitting of the data compared to standard ΛCDM cosmology. The lower-left region corresponds to the parameters invoked by Csaki *et al.* [50] to explain SN dimming without cosmic acceleration. As we show below, the SN data alone, exclude a flat ΛCDM model with zero cosmological constant at > 3-σ. Most importantly, the joint SN+$H(z)$ analysis (upper right panel) completely rules out this region.

The preferred region at large A and small P/L also deserves some comments. A value of A greater than unity produces an increase of the SN luminosity with redshift because the SNe would shine more ALPs than photons. In Fig. 1 we have also seen a slight preference for this scenario for the SN data alone, as the 1-σ region lies in negative values of ϵ. In the joint SN+$H(z)$ analysis, this trend is softened but, still, small negative values (slight SN brightening) are slightly preferred, even though this is not statistically significant. In Fig. 2 this possibility cannot be excluded, but the statistical preference for this region decreases compared to the rest of the allowed parameter space.

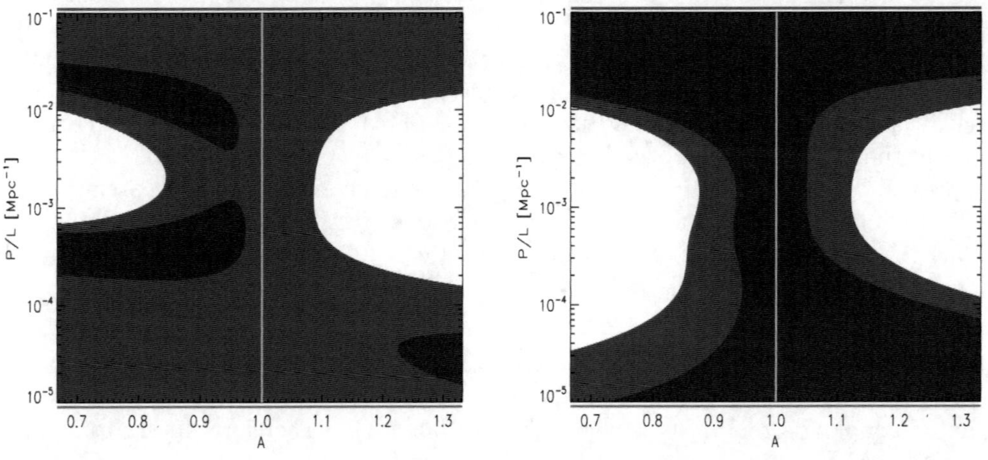

Figure 2. Two parameter constraints in the A-P/L plane for general ALPs. Contours represent the 68% (dark) and 95% (light) confidence levels. For the left panels we have used SN data only, while in the right panels we show the joint SN+$H(z)$ analysis. The contours are marginalised over cosmologies and H_0.

4.1. Cosmological Constraints

Let us now interpret our results by keeping dependence of our likelihood contours on cosmological parameters, in particular Ω_m. We will finally express our results in terms of a constraint on the axion-photon coupling strength M.

ALPs having no other interactions than the two-photon coupling were shown to contribute very little to the SN luminosity [49], corresponding to the case $A \simeq 2/3$. It is evident from Fig. 2 that only a range of conversion probabilities around $P/L \sim \mathcal{O}(10^{-3})$ can be excluded. Fig. 3 shows 1- and 2-σ, two-parameter likelihood contours on the P/L–Ω_m plane after marginalisation over H_0. Note that even the SN constraints alone (dark blue contours) rule out this model as an alternative to a cosmological constant, at greater than 3-σ significance. However, there is still a notable degeneracy in the weak mixing regime, and a value for Ω_m greater than 0.8 is still allowed at the 2-σ level, this constraint being slightly weaker when the background redshift dependence is taken into account (left). Including the $H(z)$ data (light blue contours) breaks this degeneracy, yielding strong joint constraints in $P/L - \Omega_m$ (solid line contours).

We can now translate our bounds on P/L into constraints on the strength of the ALP coupling to photons. Since the coupling always appears multiplied by the magnetic field (which is also unknown) we quote bounds on the combination B/M. Let us also define appropriately normalised values of the magnetic field strength B and the energy-scale of the the axion-photon coupling M, as

$$B_{\rm nG} = \frac{B}{1{\rm nG}} \quad ; \quad M_{10} = \frac{M}{10^{10}{\rm GeV}} . \tag{19}$$

In Fig. 4 we show our constraints for the case $L = 1$ Mpc as a function of the uncertain value of the average electron density, or, equivalently, the plasma frequency[5]. The exclusion limit is a horizontal band which bends upwards around $n_e \simeq 0.2 \times 10^{-7}$ cm^{-3}. In the same figure, we

[5] To get rid of the oscillations of equation (12), which not only will be averaged out by energy binning but also by small fluctuations in the sizes of the domains and the values of the plasma frequencies, we used the substitution $\sin^2 x \to (1 - \exp(-2x^2))/2$.

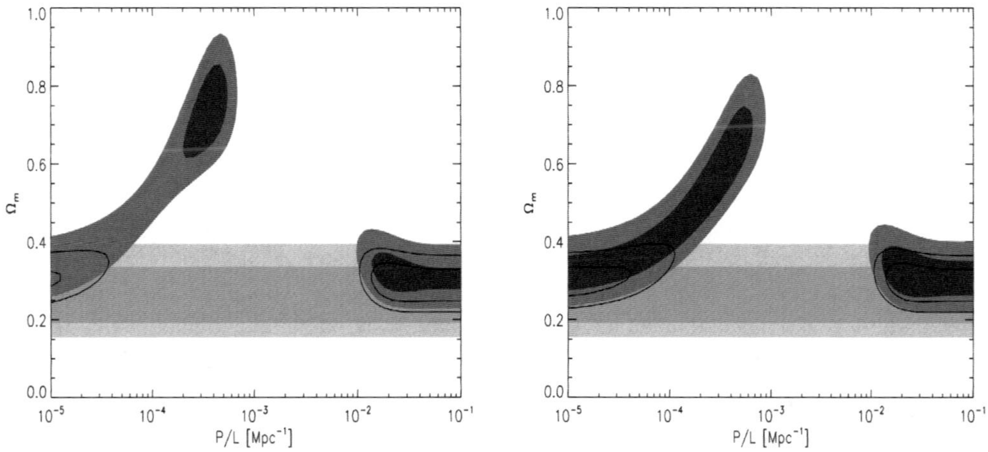

Figure 3. Confidence levels (68% and 95%) on the $P/L - \Omega_m$ plane for the simplest axion-like-particle model ($A = 2/3$). The small and large P regions correspond to the weak and strong mixing regimes respectively. Dark blue contours show constraints from SN data only, light blue from $H(z)$ data, and solid line contours from joint SN+$H(z)$. In the left panel, the redshift dependence of the background is taken into account, while in the right panel these affects are ignored.

have also reproduced the constraints of Mirizzi *et al* [51] from distortions of the CMB (region above the dashed line) and those of [52] (blue region) from QSO spectra. For $n_e \gtrsim 10^{-9}\text{cm}^{-3}$ our bounds are stronger than the CMB ones while still competitive with the QSO bounds. Our approach provides a complementary, independent way to obtain these constraints. Each of the three approaches reported in the figure is affected by different, unrelated, systematics: their agreement promotes one's confidence in these results. Significant improvement in these bounds is expected with future SN and BAO data, see reference [28] for forecasts.

5. Conclusions

If new particles from physics beyond the standard model couple to photons then the propagation of light may be altered. Here we have focused on Axion-Like-Particles coupling to electromagnetism through 2-photon interactions. Measurements of cosmic opacity provide a strong tool to constraining such scenarios, as interactions between photons and (pseudo-)scalar ALPs in the magnetic fields of the intergalactic medium leads to a new source of cosmic opacity.

Indeed, deviations from cosmic transparency can be constrained through their effects on distance duality, by considering possible deviations from the Etherington relation: the luminosity distance is $(1+z)^2$ times the angular diameter distance if photon number is conserved. Both luminosity distance and angular diameter distance depend on the Hubble parameter $H(z)$. We have thus combined direct measurements of cosmic expansion (from the latest determinations of the Hubble parameter) at redshifts $0 < z < 2$ and recent SN data yielding the luminosity distance. SN-inferred luminosity distances are affected by violation of photon conservation, but the $H(z)$ measurements we use are not. Assuming an underlying flat ΛCDM model, we have placed tight limits on possible deviations from photon-conservation. Photon-conservation can be violated by simple astrophysical effects which give uniform attenuation such as grey dust and we have reported constraints on this effect.

We have considered the implications of these constraints on more exotic sources of photon number violation involving a coupling of photons to ALPs. Remarkably, this allows us to use optical photons at energies as low as ~ 1 eV to constraint the axion-photon coupling strength, a fundamental parameter of the Lagrangian describing physics at much higher energies $\sim 10^{10}$ GeV. Indeed, assuming intergalactic magnetic fields of strength 1 nG and coherent over lengthscales of order 1Mpc, our constraints rule out a region of the axion-photon coupling M between 10^{10} and 10^{11}GeV (Fig. 4).

It is also possible to search for the effects of new physics on the propagation of photons in the laboratory, for a review see [8]. In regions of parameter space of these models where the transparency of the universe imposes constraints, the bounds from the analysis given here are at least two orders of magnitude better than what can currently be achieved with laboratory experiments. Future measurements of BAOs, and an increase in the number of observations of high redshift supernovae will lead to further improvements of constraints on ALPs.

Acknowledgements
I would like to thank the organisers of the NEB XIV Conference in Ioannina, Grecce, for hosting a successful conference and providing a stimulating environment for scientific interaction. The work presented here was done in collaboration with C. Burrage, J. Redondo, R. Jimenez and L. Verde. I am supported by a CTC Postdoctoral Fellowship.

References

[1] Peccei, R. D., & Quinn, H. R., Phys. Rev. Lett. **38** (1977) 1440.
[2] C. Amsler *et al.* [Particle Data Group], Phys. Lett. B **667** (2008) 1.
[3] S. Weinberg, Phys. Rev. Lett. **40** (1978) 223.

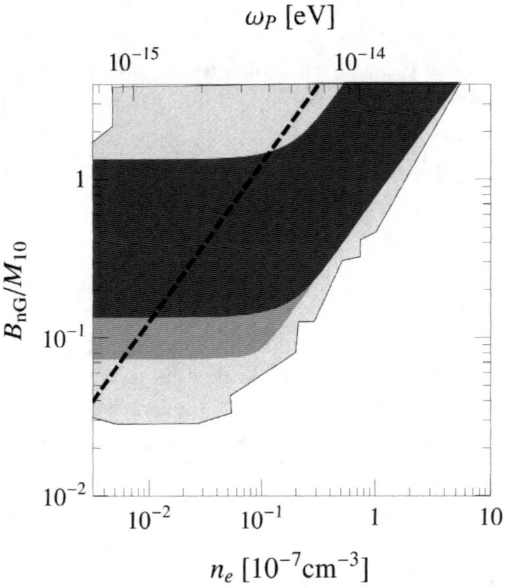

Figure 4. Bounds for the product of ALP coupling times magnetic field $B_{\rm nG}/M_{10}$ as a function of the average electron density or corresponding plasma frequency. The grey regions are excluded at 95% C.L. for $L = 1$ Mpc in the redshift-independent (dark) and dependent (light) cases. Also shown are the regions constrained from CMB [51] (above the dashed line), which dominate at low n_e, and QSO [52] (blue) spectral distortions.

[4] M. S. Turner, Phys. Rept. **197** (1990) 67.
[5] G. G. Raffelt, Phys. Rept. **198** (1990) 1.
[6] J. P. Conlon, JHEP **0605** (2006) 078 [arXiv:hep-th/0602233].
[7] Svrcek, P., & Witten, E., JHEP **0606** (2006) 051 [arXiv:hep-th/0605206]
[8] Jaeckel, J., & Ringwald, A., arXiv:1002.0329 [hep-ph]
[9] F. Quevedo, Class. Quant. Grav. **19** (2002) 5721 [arXiv:hep-th/0210292].
[10] C. P. Burgess, Class. Quant. Grav. **24** (2007) S795 [arXiv:0708.2865 [hep-th]].
[11] L. McAllister and E. Silverstein, Gen. Rel. Grav. **40** (2008) 565 [arXiv:0710.2951 [hep-th]].
[12] G. R. Dvali and S. H. H. Tye, Phys. Lett. B **450** (1999) 72 [arXiv:hep-ph/9812483].
[13] C. P. Burgess, M. Majumdar, D. Nolte, F. Quevedo, G. Rajesh and R. J. Zhang, JHEP **0107** (2001) 047 [arXiv:hep-th/0105204].
[14] S. Kachru, R. Kallosh, A. D. Linde, J. M. Maldacena, L. P. McAllister and S. P. Trivedi, JCAP **0310** (2003) 013 [arXiv:hep-th/0308055].
[15] D. Baumann, A. Dymarsky, I. R. Klebanov, L. McAllister and P. J. Steinhardt, Phys. Rev. Lett. **99** (2007) 141601 [arXiv:0705.3837 [hep-th]].
[16] J. Polchinski, *Cambridge, UK: Univ. Pr. (1998) 531 p*
[17] D. Baumann, A. Dymarsky, I. R. Klebanov and L. McAllister, JCAP **0801** (2008) 024 [arXiv:0706.0360 [hep-th]].
[18] D. Baumann, A. Dymarsky, S. Kachru, I. R. Klebanov and L. McAllister, Phys. Rev. Lett. **104** (2010) 251602 [arXiv:0912.4268 [hep-th]].
[19] A. Avgoustidis, D. Cremades and F. Quevedo, Gen. Rel. Grav. **39** (2007) 1203 [arXiv:hep-th/0606031].
[20] A. Avgoustidis and I. Zavala, JCAP **0901** (2009) 045 [arXiv:0810.5001 [hep-th]].
[21] L. McAllister, E. Silverstein and A. Westphal, Phys. Rev. D **82** (2010) 046003 [arXiv:0808.0706 [hep-th]].
[22] R. Brustein and P. J. Steinhardt, Phys. Lett. B **302** (1993) 196 [arXiv:hep-th/9212049].
[23] J. Polchinski, arXiv:hep-th/0412244.
[24] H. Primakoff, Phys. Rev. 81, 899 (1951).
[25] Etherington, J. M. H. 1933, Phil. Mag., 15, 761
[26] Kostelecky, V. A., & Russell, N. 2008, arXiv:0801.0287 [hep-ph]
[27] Avgoustidis, A., Verde, L., & Jimenez, R., JCAP **0906** (2009) 012 [arXiv:0902.2006 [astro-ph.CO]]
[28] A. Avgoustidis, C. Burrage, J. Redondo, L. Verde and R. Jimenez, JCAP **1010** (2010) 024 [arXiv:1004.2053 [astro-ph.CO]].
[29] Kowalski *et al.* 2008, ApJ, 686, 749
[30] Jimenez, R., Verde, L., Treu, T., & Stern, D., 2003, ApJ, 593, 622
[31] Simon J., *et al.*, Phys. Rev. D **71** (2005) 123001
[32] D. Stern, R. Jimenez, L. Verde, M. Kamionkowski and S. A. Stanford, JCAP **1002** (2010) 008 [arXiv:0907.3149 [astro-ph.CO]].
[33] More, S., Bovy, J., & Hogg, D. W. 2008, arXiv:0810.5553.
[34] Bassett B. A., & Kunz, M., Astrophys. J. **607**, 661 (2004).
[35] Bassett B. A., & Kunz, M., Phys. Rev. D **69**, 101305 (2004).
[36] Uzan, J. P. Gen. Rel. Grav. **39** (2007) 307 [arXiv:astro-ph/0605313]
[37] Lazkoz, R., Nesseris, S. & Perivolaropoulos, L. 2008, JCAP **0807** 012 [arXiv:0712.1232 [astro-ph]]
[38] Percival, W. J., Cole, S., Eisenstein, D. J., Nichol, R. C., Peacock, J. A., Pope, A. C., & Szalay, A. S. 2007, MNRAS, 381, 1053
[39] Riess, A.G., *et al.*, Astrophys. J. **699** (2009) 539 [arXiv:0905.0695 [astro-ph.CO]]
[40] Freedman, W. L., *et al.* 2001, ApJ, 553, 47
[41] Raffelt, G., & Stodolsky, L., Phys. Rev. D **37** (1988) 1237.
[42] Barrow, J. D., Ferreira, P. G., & Silk, J., Phys. Rev. Lett. **78** (1997) 3610 [arXiv:astro-ph/9701063]
[43] Blasi, P., Burles, S., & Olinto, A. V., Astrophys. J. **514** (1999) L79 [arXiv:astro-ph/9812487]
[44] Kronberg, P. P., Rept. Prog. Phys. **57** (1994) 325.
[45] Davis, A. C., Schelpe, C. A. O., & Shaw, D. J., Phys. Rev. D **80** (2009) 064016 [arXiv:0907.2672 [astro-ph.CO]]
[46] Schelpe, C. A. O., arXiv:1003.0232 [astro-ph.CO]
[47] Peebles, P. J. E., *Princeton, USA: Univ. Pr. (1993) 718 p*
[48] Csáki, C., Kaloper, N., & Terning, J. 2002, Physics Letters B, 535, 33
[49] Grossman, Y., Roy, S., & Zupan, J., Phys. Lett. B **543** (2002) 23 [arXiv:hep-ph/0204216]
[50] Csáki, C., Kaloper, N., & Terning, J., Phys. Rev. Lett. **88** (2002) 161302
[51] Mirizzi, A., Raffelt, G. G. & Serpico, P. D. 2005, Phys. Rev. D **72** 023501 [arXiv:astro-ph/0506078]
[52] Ostman, L., & Mortsell, E., JCAP **0502** (2005) 005 [arXiv:astro-ph/0410501]

Recent Developments in Gravity (NEB XIV) IOP Publishing
Journal of Physics: Conference Series **283** (2011) 012004 doi:10.1088/1742-6596/283/1/012004

Gradient flows and instantons at a Lifshitz point

Ioannis Bakas
Department of Physics, University of Patras, 26500 Patras, Greece
E-mail: bakas@ajax.physics.upatras.gr

Abstract. I provide a broad framework to embed gradient flow equations in non–relativistic field theory models that exhibit anisotropic scaling. The prime example is the heat equation arising from a Lifshitz scalar field theory; other examples include the Allen–Cahn equation that models the evolution of phase boundaries. Then, I review recent results reported in arXiv:1002.0062 describing instantons of Hořava–Lifshitz gravity as eternal solutions of certain geometric flow equations on 3–manifolds. These instanton solutions are in general chiral when the anisotropic scaling exponent is $z = 3$. Some general connections with the Onsager–Machlup theory of non–equilibrium processes are also briefly discussed in this context. Thus, theories of Lifshitz type in $d + 1$ dimensions can be used as off–shell toy models for dynamical vacuum selection of relativistic field theories in d dimensions.

1. Preliminaries: Lifshitz theories

Let us consider a point particle system with configuration space Q and a system of local coordinates q_I that describe its physical degrees of freedom. We further assume that Q is endowed with a metric $\mathcal{O}^{IJ}$ which is symmetric and non–degenerate so that the inverse metric $\mathcal{O}_{IJ}$ defined through $\mathcal{O}^{IK}\mathcal{O}_{KJ} = \delta^I_J$ exists everywhere. The action comprises of two terms taking the difference of the kinetic and potential energy

$$S = \frac{1}{2}\int dt \sum_{I,J}\left(\frac{dq_I}{dt}\mathcal{O}^{IJ}\frac{dq_J}{dt} - \frac{\partial W}{\partial q_I}\mathcal{O}_{IJ}\frac{\partial W}{\partial q_J}\right), \quad (1)$$

where we also assume that the potential is derived from a superpotential W that is a function on Q. This particular choice of potential is synonymous to having detailed balance condition in the system. Here, we will only consider systems with positive definite metric $\mathcal{O}^{IJ}$ so that the potential term is manifestly positive definite and its minima coincide with the critical points of W, i.e., $\partial W/\partial q_I = 0$, $\forall\ I$. Then, the ground states of the system describe a particle sitting still at the minima of the potential and there can be more than one degenerate vacua depending upon the choice of W. It is often interesting to consider systems with configuration space having indefinite metric $\mathcal{O}^{IJ}$, although some of the nice properties that are discussed in the following will be generally missing.

When the dimension of the configuration space Q is finite, we have an ordinary point particle system, which is not particularly interesting by itself but it is a useful guide for some of the constructions described below. Here, we focus entirely on infinite dimensional systems determined by the action (1) and consider only those cases that Q is the configuration space of a relativistic field theory in d Euclidean space–time dimensions. The superpotential W is a functional over the field space, which for all practical purposes is taken to be the classical action of the relativistic field theory in d dimensions. This construction may look odd at first sight, as the corresponding action (1) describes a non–relativistic field theory in $d+1$ space–time dimensions with t being the physical time coordinate, but, in fact, it provides our main framework. The resulting field theories are called models of Lifshitz type, [1], and they exhibit anisotropic scaling (typically it is $z=2$ for theories with W that contains up to two derivative terms, but there can be more general examples as will be seen later).

The prime example of a Lifshitz theory in $d+1$ space–time dimensions is provided by taking Q to be the configuration space of a relativistic free field theory in d Euclidean dimensions. For this it is appropriate to set $q_I(t)=\varphi(t,x)$, $q_J(t)=\varphi(t,x')$, $\mathcal{O}^{IJ}=\delta(x-x')=\mathcal{O}_{IJ}$ and choose

$$W[\varphi]=\frac{1}{2}\int d^dx\,(\nabla\varphi)^2 \tag{2}$$

so that $\partial W/\partial q_I=\delta W/\delta\varphi(x)=-\nabla^2\varphi(x)$. Then, by integration over x' (equivalently, summation over J), the point particle action (1) takes the form

$$S=\frac{1}{2}\int dt\;d^dx\Big[(\partial_t\varphi)^2-\left(\nabla^2\varphi\right)^2\Big] \tag{3}$$

and describes a non-relativistic field theory with anisotropic scaling $z=2$ known as Lifshitz scalar theory. The resulting equations of motion exhibit scale invariance under the transformation $x\to ax$, $t\to a^zt$ with exponent $z=2$; it should be contrasted to the relativistic scalar free field theory in $d+1$ space-time dimensions whose Lagrangian density is $(\partial_t\varphi)^2-(\nabla\varphi)^2$ and the corresponding scaling exponent is $z=1$.

A simple variant of the model is obtained by considering the action of a self–interacting relativistic scalar field in d Euclidean dimensions as superpotential functional,

$$W[\varphi]=\int d^dx\Big[\frac{1}{2}(\nabla\varphi)^2+V(\varphi)\Big]\,, \tag{4}$$

in which case the corresponding Lifshitz theory in $d+1$ space–time dimensions takes the form

$$S=\frac{1}{2}\int dt\;d^dx\Big[(\partial_t\varphi)^2-\left(\nabla^2\varphi-V'(\varphi)\right)^2\Big]\,, \tag{5}$$

where $V'(\varphi)=\partial V/\partial\varphi$. This particular field theory will be discussed later in more detail as it connects with some interesting mathematical problems of current interest. One may also consider more complicated examples using multi–component scalar field models.

Another characteristic class of Lifshitz models is provided by vector field theories with anisotropic scaling in space and time. As before, it is appropriate to use the action (1) as starting point and identify q_I with gauge fields $A_i(x)$ that live in d–dimensional Euclidean space, letting $i=1,2,\cdots,d$. The metric $\mathcal{O}^{IJ}$ is provided by the standard Riemannian metric $\mathrm{Tr}(T_aT_b)\delta(x-x')$ in the space of all d–dimensional gauge field configurations $A_i=A_i^aT_a$ with values in a Lie algebra $[T_a,\ T_b]=if_{ab}{}^cT_c$. Then, it is natural to define a Lifshitz gauge field theory in terms of the action, [2],

$$S=\frac{1}{2}\int dt\;d^dx\Big[\mathrm{Tr}(E_iE^i)-\frac{1}{g^2}\mathrm{Tr}\left((\partial_iF^{ik})(\partial_jF^j{}_k)\right)\Big]\,, \tag{6}$$

where

$$E_i=\partial_tA_i\,,\qquad F_{ij}=\partial_iA_j-\partial_jA_i-i[A_i,\ A_j] \tag{7}$$

provide the electric and magnetic fields in $d+1$ space–time dimensions, respectively, in the axial gauge $A_0(t,x)=0$. This action follows from (1) using the superpotential functional

$$W = \frac{1}{4g^2}\int d^d x \ \mathrm{Tr}(F_{ij}F^{ij}) \tag{8}$$

associated to relativistic Yang–Mills theory in d Euclidean dimensions, and, therefore, the resulting Lifshitz vector theory exhibits anisotropic scaling with exponent $z=2$.

Finally, we consider Lifshitz theories of geometric type associated to tensors of rank 2, which were introduced in the literature as alternative theories of gravity and they became known as Hořava–Lifshitz gravities, [3, 4]. They are defined in arbitrary dimensions by assuming that space–time is of the form $M_{d+1} = R \times \Sigma_d$ and the metric admits the ADM (Arnowitt–Deser–Misner) decomposition, [5]

$$\mathrm{d}s^2 = -N^2\mathrm{d}t^2 + g_{ij}\left(\mathrm{d}x^i + N^i\mathrm{d}t\right)\left(\mathrm{d}x^j + N^j\mathrm{d}t\right). \tag{9}$$

The metric on the spatial slices Σ_d is g_{ij}, whereas N and N^i are the lapse and shift functions, respectively, which depend on all space–time coordinates, in general. The infinite dimensional space of all Riemannian metrics g_{ij} is called superspace and it is endowed with a metric

$$\mathcal{G}^{ijk\ell} = \frac{1}{2}\left(g^{ik}g^{j\ell} + g^{i\ell}g^{jk}\right) - \lambda g^{ij}g^{k\ell} \tag{10}$$

that generalizes the standard DeWitt metric, [6], using an arbitrary parameter λ (other than 1). Here, for simplicity, we are suppressing the delta functions that are needed to integrate over space. The inverse metric is

$$\mathcal{G}_{ijk\ell} = \frac{1}{2}\left(g_{ik}g_{j\ell} + g_{i\ell}g_{jk}\right) - \frac{\lambda}{d\lambda - 1}g_{ij}g_{k\ell} \tag{11}$$

so that

$$\mathcal{G}^{ijk\ell}\mathcal{G}_{k\ell mn} = \frac{1}{2}(\delta^i_m\delta^j_n + \delta^i_n\delta^j_m) . \tag{12}$$

The metric in superspace is positive definite provided that $\lambda < 1/d$, but otherwise it is arbitrary.

The action of Hořava–Lifshitz gravity in $d+1$ dimensions is written as a sum of kinetic and potential terms. By identifying q_I with g_{ij} and $\mathcal{O}^{IJ}$ with $\mathcal{G}^{ijk\ell}$, the action (1) satisfying detailed balance takes the form, [3, 4],

$$S = \frac{2}{\kappa^2}\int dt d^d x\sqrt{g}\ N\ K_{ij}\mathcal{G}^{ijk\ell}K_{k\ell} - \frac{\kappa^2}{2}\int dt d^d x\sqrt{g}\ N\ E^{ij}\mathcal{G}_{ijk\ell}E^{k\ell} , \tag{13}$$

where K_{ij} is the second fundamental form measuring the extrinsic curvature of the spatial slices Σ_d at constant t (playing the role of momentum conjugate to the metric g_{ij}),

$$K_{ij} = \frac{1}{2N}\left(\partial_t g_{ij} - \nabla_i N_j - \nabla_j N_i\right) \tag{14}$$

and

$$E^{ij} = -\frac{1}{2\sqrt{g}}\frac{\delta W[g]}{\delta g_{ij}} . \tag{15}$$

The gravitational coupling of the theory in $d+1$ dimensions is κ.

Note that the kinetic term contains two time derivatives of the metric g_{ij}, and, as such, it is identical to ordinary general relativity in canonical form (though λ is taken arbitrary here).

The potential term is different, however, as it is derived from a superpotential functional $W[g]$, which is typically chosen to be the Euclidean action of a relativistic gravitational theory in d dimensions. If W is the Einstein–Hilbert action, the resulting Hořava–Lifshitz theory will have anisotropic scaling $z = 2$, but if W is the action of a covariant higher derivative gravitational theory in d dimensions the scaling exponent will be $z > 2$. Examples of this kind will be discussed later. Finally, note that the action (13) is not generally covariant, since by construction it is only invariant under the restricted set of foliation preserving diffeomorphisms of the space–time $R \times \Sigma_d$. In the following, we restrict attention to the so called projectable version of Hořava-Lifshitz theory, meaning that the lapse function N associated to the freedom of time reparametrization is restricted to be a function of t, whereas the shift functions N_i associated to diffeomorphisms of Σ_d can depend on all space-time coordinates. Also, in view of the applications that will be discussed next, we choose

$$N(t) = 1 \, , \qquad N_i(t, x) = 0 \tag{16}$$

without great loss of generality. We will indicate later how the lapse and shift functions can be reinstated into the equations of motion.

2. Gradient flows as Euclidean solutions

Let us now consider the Euclidean form of the action (1) obtained by Wick rotation $t \to it$ and construct solutions that are applicable to all Lifshitz type theories. We have

$$S_{\rm Eucl.} = \frac{1}{2} \int dt \sum_{I,J} \left(\frac{dq_I}{dt} \mathcal{O}^{IJ} \frac{dq_J}{dt} + \frac{\partial W}{\partial q_I} \mathcal{O}_{IJ} \frac{\partial W}{\partial q_J} \right) , \tag{17}$$

which can be alternatively written as follows

$$S_{\rm Eucl.} = \frac{1}{2} \int dt \left(\frac{dq_I}{dt} \mp \mathcal{O}_{IK} \frac{\partial W}{\partial q_K} \right) \mathcal{O}^{IJ} \left(\frac{dq_J}{dt} \mp \mathcal{O}_{JL} \frac{\partial W}{\partial q_L} \right) \pm \int dt \, \frac{dq_I}{dt} \frac{\partial W}{\partial q_I} \tag{18}$$

by completing the square. Here, summation is implicity assumed over repeated indices in order to simplify the presentation. For now and later use we also consider the action

$$S'_{\rm Eucl.} = \frac{1}{2} \int dt \left(\frac{dq_I}{dt} \mp \mathcal{O}_{IK} \frac{\partial W}{\partial q_K} \right) \mathcal{O}^{IJ} \left(\frac{dq_J}{dt} \mp \mathcal{O}_{JL} \frac{\partial W}{\partial q_L} \right) , \tag{19}$$

which differs from (18) by boundary terms as

$$S_{\rm Eucl.} = S'_{\rm Eucl.} \pm \int dt \, \frac{dW}{dt} \, . \tag{20}$$

Since we are only considering positive definite metrics $\mathcal{O}^{IJ}$, the action $S'_{\rm Eucl.}$ is bounded from below by zero. Thus, minima of the action are provided by special configurations that satisfy the system of first order equations

$$\frac{dq_I}{dt} = \pm \mathcal{O}_{IJ} \frac{\partial W}{\partial q_I} \, . \tag{21}$$

These are gradient flow equations for the variables $q_I(t)$ derived from the superpotential $W[q]$. They also yield solutions of the equations of motion following from the action $S_{\rm Eucl.}$, since the difference is a total derivative term (20) that certainly cannot affect the classical equations of motion. However, this boundary term is needed to make the variational problem well–posed and will be treated more carefully in the next section for, otherwise, the action $S_{\text{Eucl.}}$ may become infinite for solutions that become singular at a finite instant of time. In any case, the fixed points of the flow equations (21) represent static solutions, where the point particle is sitting still at the bottom of the potential. On the other hand, time dependent solutions of the flow equations are more interesting to consider in general, but they are complicated; we will present some examples that are worth studying in detail. It should also be noted that the superpotential W changes monotonically along such flow lines when $\mathcal{O}^{IJ}$ is positive definite, since

$$\frac{dW}{dt} = \frac{dq_I}{dt}\frac{\partial W}{\partial q_I} = \pm\frac{\partial W}{\partial q_I}\mathcal{O}_{IJ}\frac{\partial W}{\partial q_J} , \tag{22}$$

and, therefore, it provides an entropy functional for the corresponding evolution.

The simplest (but very instructive) example in the class of Lifshitz field theories is provided by the scalar field model (3) in the Euclidean domain. The corresponding gradient flow is linear and coincides with the forward or backward heat equation

$$\frac{\partial\varphi}{\partial t} = \pm\nabla^2\varphi , \tag{23}$$

depending on the choice of sign. Then, solutions of the heat equation in d dimensions are solutions of the Lifshitz scalar free field theory in $d+1$ Euclidean dimensions. Note, however, that the fundamental solution of the heat equation, which is a Gaussian function with time dependent height $1/t^{d/2}$ and width $t^{1/2}$, cannot exist for ever since $\varphi(t,x)$ becomes singular at some finite instant of time, say $t=0$. This simple example illustrates the difference between finite action solutions following from $S_{\text{Eucl.}}$ and $S'_{\text{Eucl.}}$ where the boundary terms (20) can in fact play important role. If we were only concerned with solutions of the flow equations that existed for a short time, the difference would have been irrelevant. The singularities are also preventing to uplift these solutions to Lifshitz theories that exist for all time.

These remarks are not only tailored for the Lifshitz scalar free field theory, but, in fact, they apply to all Lifshitz models. They also motivate the construction of instanton solutions of Lifshitz theories that will be discussed in the next section.

The next more complicated example is provided by a self–interacting Lifshitz scalar field as described by (5). According to the general framework, Euclidean solutions are provided by the non–linear flow equation

$$\frac{\partial\varphi}{\partial t} = \pm\left(\nabla^2\varphi - V'(\varphi)\right), \tag{24}$$

which is called Allen–Cahn equation, [7]. This equation arose first in the literature as phenomenological model for the motion of phase boundaries by surface tension, but it was subsequently studied in mathematics very extensively. In this context, which is deeply connected to motion by mean curvature flow and the problem of minimal surfaces, [8, 9], it seems more appropriate to consider the closely related equation

$$\frac{\partial\varphi^\epsilon}{\partial t} = \pm\left(\nabla^2\varphi^\epsilon - \frac{1}{\epsilon^2}\,V'(\varphi^\epsilon)\right), \tag{25}$$

which follows from above by implementing the anisotropic scaling transformation of the Lifshitz theory, $x \to x/\epsilon$ and $t \to t/\epsilon^2$, and investigate this in the limit of small ϵ when the potential has two equal wells, choosing, for example, $V(\varphi) = (\varphi^2 - \varphi_0^2)^2$. Solving the Allen–Cahn equation is not an easy task and it provides an active area of research in mathematics (see, for instance, [10], for some basic facts).

Euclidean solutions of gauge theories of Lifshitz type (6) are also interesting to consider as the corresponding gradient flow assumes the form

$$E_i \equiv \frac{\partial A_i}{\partial t} = \pm\frac{1}{g}\,\nabla_j F^j{}_i \tag{26}$$

and coincides with the so called Yang–Mills flow in d dimensions. This is another very interesting system of equations that has been studied for some time in mathematics (e.g., [11, 12]), but many of its aspects remain open problems to this day. Of course, one may also consider further generalizations by combining the action functionals (4) and (8) into a W that describes the action of a relativistic Yang–Mills–Higgs theory in d Euclidean dimensions. Then, the gradient flow equations of the corresponding Lifshitz theory will be mixture of the Allen–Cahn equation and the Yang–Mills flow whose general properties remain to be studied (see, however, [13]).

Finally, we come to the Euclidean solutions of geometric Lifshitz theories, such as Hořava–Lifshitz gravity described by the action (13). In this case, the gradient flow equation becomes, [3, 4], [14],

$$K_{ij} \equiv \frac{1}{2}\frac{\partial g_{ij}}{\partial t} = \pm\frac{\kappa^2}{2}\,\mathcal{G}_{ijk\ell}E^{k\ell} \tag{27}$$

and gives rise to geometric flows for the metric g_{ij} on Σ_d that depend on the choice of W. They can have second or higher order derivatives in space coordinates.

More precisely, we have the following evolution equation in the general case of Hořava–Lifshitz models

$$\frac{\partial g_{ij}}{\partial t} = \mp\frac{\kappa^2}{2\sqrt{g}}\,\mathcal{G}_{ijk\ell}\frac{\delta W}{\delta g_{k\ell}}\,, \tag{28}$$

which for W given by the Einstein–Hilbert action in $d > 2$ dimensions,

$$W[g] = \frac{2}{\kappa_{\rm w}^2}\int d^dx\sqrt{g}\,R\,, \tag{29}$$

it specializes to a variant of the celebrated Ricci flow equation (see, for instance, [15])

$$\frac{\partial g_{ij}}{\partial t} = \mp\frac{\kappa^2}{\kappa_{\rm w}^2}\left(R_{ij} - \frac{2\lambda-1}{2(d\lambda-1)}\,Rg_{ij}\right). \tag{30}$$

It runs forward or backward in time depending on the choice of the overall sign and the fixed points are Ricci flat metrics, $R_{ij} = 0$, independent of λ.

More general choices of $W[g]$ can also be made, depending on d, in which case the exponent z of anisotropic scaling can become bigger than 2. Although the corresponding flow equations are mathematically much more complicated for $z > 2$, as they involve higher derivative terms, they are physically better behaved for various reasons. The simplest choice (29) that leads to the Ricci flow shares many similarities with the heat equation in that the geometry on Σ_d becomes singular at some finite instant of time, [15]. This is actually a generic phenomenon that can render the action $S_{\rm Eucl.}$ (but not $S'_{\rm Eucl.}$) of $z = 2$ Hořava–Lifshitz gravity infinite along those flow lines (W becomes infinite at the ultra–violet fixed point of type I ancient solutions and vanishes at the singularity in $d > 2$ dimensions). More importantly, the corresponding space–time metric on $R \times \Sigma_d$, which is supposed to exist for all time, will be incomplete by the presence of singularities, thus making such solutions totally unacceptable. In the next section we will provide simple criteria that circumvent this problem in Lifshitz type theories and obtain non–singular solutions, where it is appropriate.

Concluding this section, we mention that the geometric flow equations can be modified by allowing arbitrary reparametrizations on Σ_d generated by a vector field $\xi_i(t, x)$ that may also depend on time. Then, the evolution (27) generalizes to

$$\frac{\partial g_{ij}}{\partial t} = \mp\frac{\kappa^2}{2\sqrt{g}}\,\mathcal{G}_{ijk\ell}\frac{\delta W}{\delta g_{k\ell}} + \nabla_i\xi_j + \nabla_j\xi_i\,. \tag{31}$$

The projectable version Hořava–Lifshitz theory with lapse and shift functions $N(t)$ and $N_i(t, x)$ accommodates nicely this modification by choosing $\xi_i = N_i/N$. Of course, $N(t)$ can always be set equal to 1 by appropriate redefinition of time, but $N_i(t,x)$ is left arbitrary, in general, since the theory is invariant under foliation preserving diffeomorphisms of the space–time manifold $R \times \Sigma_d$. The choice (16) with $N_i = 0$ amounts to taking $\xi_i = 0$ in the flow equations.

3. Instantons at a Lifshitz point

The definition of instantons in Lifshitz theories requires careful treatment of the boundary terms arising in the Euclidean action. Returning back to it we have

$$S_{\text{Eucl.}} = \frac{1}{2} \int dt \left(\frac{dq_I}{dt} \mp \mathcal{O}_{IK} \frac{\partial W}{\partial q_K} \right) \mathcal{O}^{IJ} \left(\frac{dq_J}{dt} \mp \mathcal{O}_{JL} \frac{\partial W}{\partial q_L} \right) \pm \int dt \, \frac{dW}{dt} \, , \tag{32}$$

which for theories with positive definite metric $\mathcal{O}^{IJ}$ yields immediately

$$S_{\text{Eucl.}} \geq \pm \int dt \, \frac{dW}{dt} \, . \tag{33}$$

The lower bound is saturated for special configurations satisfying the gradient flow equation

$$\frac{dq_I}{dt} = \pm \mathcal{O}_{IJ} \frac{\partial W}{\partial q_I} \, , \tag{34}$$

which thus provide extrema of the Euclidean action and hence solutions to the classical equations of motion as for $S'_{\text{Eucl.}}$ that was considered before.

Note, however, that the lower bound of the action (33) need not be finite on general grounds. All solutions of the gradient flow equations that are taken over a finite time interval $(t_1, \ t_2)$ will have finite action,

$$S_{\text{Eucl.}} = |W(t)|_{t_1}^{t_2} \, , \tag{35}$$

but this does not necessarily extend smoothly to the entire line $-\infty < t < +\infty$. What we really need is to have eternal solutions of the flow equations, which exist for all time, and interpolate smoothly between different critical points of the superpotential W. Then, the bound will have topological meaning and the corresponding trajectories will be bona fide instantons with finite action

$$S_{\text{instanton}} = |\Delta W| \equiv |W(t = +\infty) - W(t = -\infty)| \, . \tag{36}$$

This requirement defines the notion of instantons in Lifshitz theories, [14], and places attention to those models that have degenerate vacua. Also, to avoid unnecessary complications that may arise when considering Lifshitz field theories on $R \times \Sigma_d$, we implicitly assume that Σ_d is compact without boundaries (e.g., S^d) so that no additional space–boundary terms will come into play when putting lower bounds on the Euclidean action.

Let us elaborate more on the construction. First note that the lower bound of $S_{\text{Eucl.}}$ cannot be zero, since there are no periodic trajectories in Euclidean time; if that were the case, it would have been in contradiction with the monotonicity of $W(t)$ along the gradient flow lines. Thus, the lower bound is either positive and finite or infinite. Second, it is not at all guaranteed that the gradient flow equations will admit solutions that exist for sufficiently long time, as this is not easy to prove mathematically in general; if not, the Euclidean solutions we are considering will not be viable choices. Third, even if one can prove short time existence of the solutions, it is not clear whether these will extend to solutions for all time. Typically, the flow lines terminate at some finite instant of time by developing singularities and they cannot be continued further. Investigating the formation of singularities and their properties is a formidable task in general and special mathematical tools need to be developed in each case separately depending upon W. However, it is reasonable to expect that such singularities, if present, will be irremovable and that W can become infinite along the corresponding flow lines; the simple examples of the well studied heat equation and Ricci flow fully support this point. Thus, the only systematic way to avoid configurations with infinite Euclidean action is to consider eternal solutions of the flow equations. These are also natural from a physical point of view, since there is no a priori reason to have Lifshitz theories that make good sense only for finite time intervals.

In conclusion, we are only considering eternal solutions of the gradient flow equations as viable first order Euclidean time solutions of Lifshitz theories. Thus, to implement this program we do not need to prove the short time existence of the flow lines, in general, nor to investigate the possible formation of singularities in finite time. These may be interesting mathematical problems in their own right, but we can live without them when considering instantons. The only mathematical problems that have to be addressed in the present context is the existence, the construction and the classification of eternal solutions of gradient flows which are derived from a W with at least two critical points. There are many examples of Lifshitz theories that one can study in this context, but we will confine our attention to non–relativistic gravitational models, following [14], as there has been considerable activity in this area in recent times.

4. Instantons of Hořava–Lifshitz gravity

In this section we concentrate for definiteness to geometric flows as Euclidean solutions of Hořava–Lifshitz gravity theories and provide explicit examples of instanton configurations. We will consider models in $3+1$ space–time dimensions and choose the action of three–dimensional topologically massive gravity as our superpotential functional, [16],

$$W = \frac{2}{\kappa_{\rm w}^2} \int d^3x \sqrt{g}\, (R - 2\Lambda) + \frac{1}{\omega} W_{\rm CS} \, , \tag{37}$$

where

$$W_{\rm CS} = \int d^3x \, \sqrt{g} \, \varepsilon^{ijk} \Gamma^{\ell}{}_{im} \left(\partial_j \Gamma^{m}{}_{\ell k} + \frac{2}{3} \Gamma^{m}{}_{jn} \Gamma^{n}{}_{k\ell} \right) \tag{38}$$

is written in terms of the usual Levi–Civita connection of the three–dimensional metric g and ε^{ijk} is the fully anti-symmetric symbol with $\varepsilon^{123} = 1$. This choice yields a non–relativistic theory of gravity in $3+1$ dimensions with anisotropic scaling $z = 3$. As will be seen shortly, going beyond $z = 2$ is necessary in order to be able to produce examples of instanton configurations in Hořava–Lifshitz gravity.

Let us briefly describe some of the salient features of topologically massive gravity that will be needed in the following. The first term of W is the usual Einstein–Hilbert term, which is also augmented with a three–dimensional cosmological constant Λ, whereas the second term is the so called gravitational Chern–Simons action. The latter flips sign under orientation reversing transformations and, as a result, the theory of topologically massive gravity is not invariant under parity. The classical equations of motion that follow from W by varying the metric read as

$$R_{ij} - \frac{1}{2} R g_{ij} + \Lambda g_{ij} + \frac{\kappa_{\rm w}^2}{\omega} C_{ij} = 0 \, , \tag{39}$$

where C_{ij} is the Cotton tensor of the metric g, which is defined as follows

$$C_{ij} = \frac{\varepsilon_i{}^{k\ell}}{\sqrt{g}} \nabla_k (R_{j\ell} - \frac{1}{4} R g_{j\ell}) \tag{40}$$

and it is a traceless and covariantly conserved symmetric tensor. Thus, the classical equations of motion contain second and third order derivative terms, in general, explaining the scaling exponent $z = 3$ in the associated $(3+1)$–dimensional Hořava–Lifshitz gravity.

According to the general theory, Euclidean solutions of the Lifshitz theory on $R \times \Sigma_3$ are provided by trajectories of the corresponding gradient flow equation which takes the form, [14],

$$\partial_t g_{ij} = \mp \frac{\kappa^2}{\kappa_{\mathrm{w}}^2} \left(R_{ij} - \frac{2\lambda - 1}{2(3\lambda - 1)} R g_{ij} - \frac{\Lambda}{3\lambda - 1} g_{ij} \right) \mp \frac{\kappa^2}{\omega} C_{ij} \tag{41}$$

keeping both sign options for completeness. This is called Ricci–Cotton flow for the metric g on Σ_3. Its fixed points are independent of the parameter λ and coincide with the classical solutions of topologically massive gravity given by (39). Instanton solutions exist provided that there are more than one fixed points and they correspond to trajectories that interpolate smoothly between them as t varies from $-\infty$ to $+\infty$. It is certainly very difficult to derive general mathematical results about the properties of the Ricci–Cotton flow because it is a third order equation and the standard tools based on maximum principle do not apply. Nevertheless, it is possible to make appropriate ansatz that reduce the flow equation into a simpler system of ordinary differential equations, which in turn can be studied in detail and yield explicit configurations.

We are led to consider locally homogeneous geometries on Σ_3, which for simplicity it is assumed to have the topology of S^3, and make the Bianchi IX ansatz for the three–dimensional metrics

$$ds^2 = \gamma_1(t)\sigma_1^2 + \gamma_2(t)\sigma_2^2 + \gamma_3(t)\sigma_3^2 \tag{42}$$

so that the isometry group is $SU(2)$. Here, σ_i are the left–invariant one–forms of $SU(2)$ satisfying the defining relations $2d\sigma_i + \varepsilon_i{}^{jk}\sigma_j \wedge \sigma_k = 0$. Remarkably, this class of metrics provides consistent truncation of the Ricci–Cotton flow, as for all other homogeneous model three–geometries that arise in Bianchi classification. Sticking to this mini–superspace model, we will be able to describe all instanton solutions of $z = 3$ Hořava–Lifshitz gravity with $SU(2)$ isometry, following [14]. Since the instantons of Lifshitz theories correspond to eternal solutions of the gradient flow equations, their space–time interpretation is straightforward as they give rise to complete and regular metrics on $R \times S^3$. Here, due to space limitation, we only present in words the construction of these instantons and refer the reader to our published work for the details. However, we are permitted to give one simple example later to illustrate our general results on $SU(2)$ gravitational instantons.

First, we need to characterize all fixed points, i.e., homogeneous vacuum solutions of three–dimensional topologically massive gravity, [17, 18, 19] (but see also [20] for an overview). It turns out that apart from the maximally symmetric vacuum with $\gamma_1 = \gamma_2 = \gamma_3$, which always exists for all values of parameters, there can be additional fixed points that are formed by balancing the effect of the Einstein and Cotton terms to geometry. Clearly the sign of the Chern–Simons coupling ω plays important role in this problem as it distinguishes the case that the Einstein and Cotton tensors compete against each other from the case that they work together at the fixed points. Additional fixed points are present only in the first case and they correspond to axially symmetric metrics on S^3. The exact shape of the squashed or elongated spheres that can arise depends on the actual values of the couplings. When $\Lambda = 0$, it is also possible to have totally anisotropic fixed points beyond a critical value of ω. Then, the instanton solutions are configurations interpolating smoothly between these fixed points and describe eternal solutions of the Ricci–Cotton flow. Depending on the values of parameters, there are axially symmetric deformations of the metric that connect the fixed points, in which case the instantons have enhanced $SU(2) \times U(1)$ group of isometries, or there can be more general instantons with $SU(2)$ isometry alone, which are, however, difficult to describe in closed form. In any case, it is possible to classify all such instanton solutions, compute their action and find their moduli, as it is explained in the original work [14] when $\lambda < 1/3$ and $\Lambda \geq 0$.

An important property of these instantons is their chirality. Since the Cotton tensor is odd under parity, orientation reversing transformations on S^3 flip the sign of the coupling constant

ω. This has dramatic effect on the formation of fixed points since the Einstein and Cotton tensor will work together and not against each other. Thus, the instantons of $z = 3$ Hořava–Lifshitz gravity will cease to exist as they rely on the existence of more than one fixed point to support themselves, which is possible only for one sign of the coupling ω (assuming that $\kappa_{\rm w}$ is finite). Chiral instantons are not common to field theories, but here we have a model provided by Lifshitz theories that make them possible. Another important point is the absence of instantons in $z = 2$ Hořava–Lifshitz gravity, which explains the need to resort to models with higher anisotropy scaling exponent. In that case, the gradient flow equation is the Ricci flow, which admits only one fixed point on S^3, [15], as consequence of the Poincaré conjecture. Thus, there can be no gravitational instantons of the kind we are considering here. Finally, one may also consider instantons of Hořava–Lifshitz gravity with higher scaling exponent, say $z = 4$, by choosing W to be the action of three–dimensional new massive gravity (and generalizations thereof), [21]. Homogeneous vacua with $SU(2)$ symmetry have been systematically classified in this case, [22], and, therefore, gravitational instantons with $SU(2)$ isometry can be made available by considering eternal solutions of a certain fourth order geometric flow equation for Bianchi IX metrics. The details are lying beyond the scope of this presentation.

In the remaining part of this section, we illustrate the results given in [14], by presenting the simplest possible example of an instanton solution of $z = 3$ Hořava–Lifshitz gravity. We choose W to be the gravitational Chern–Simons action by dropping the Einstein–Hilbert term in the limit $\kappa_{\rm w} \to \infty$. Then, the underlying three–dimensional theory is the conformal rather than the topologically massive gravity, [23]. The fixed points obey $C_{ij} = 0$ and correspond to conformally flat metrics on Σ_3. The associated Lifshitz theory is pure Cotton and its Euclidean solutions are simply described by trajectories of the Cotton flow, which was also introduced in [24],

$$\partial_t g_{ij} = \mp \frac{\kappa^2}{\omega} C_{ij} \ , \tag{43}$$

and it is independent of λ. Since the Cotton tensor is traceless, this deformation of the metric preserves the volume of space and the calculations simplify a lot.

Within the mini–superspace model of Bianchi IX metrics on S^3, we are led to consider axially symmetric configurations with fixed volume $2\pi^2 L^3$ setting, in particular,

$$\gamma_1 = \gamma_2 \equiv x \ \frac{L^2}{4} \ , \qquad \gamma_3 = \frac{L^2}{4x^2} \ . \tag{44}$$

Then, the Cotton flow equation reduces consistently to a single equation for the variable $x(t)$ that parametrizes the shape of space (squashing) and reads as

$$\frac{dx}{dt} = \pm \frac{4\kappa^2}{\omega L^3} \ \frac{x^3 - 1}{x^5} \ . \tag{45}$$

Clearly, there are two fixed points, one with $x = 1$ and another with $x = \infty$. They both describe conformally flat metrics on S^3 as dx/dt vanishes there. The first one corresponds to the round metric on S^3 having $\gamma_1 = \gamma_2 = \gamma_3$, whereas the second one arises in the correlated limit $\gamma_1 = \gamma_2 = \infty$ and $\gamma_3 = 0$ with the volume held fixed. As such it looks degenerate, since S^3 is completely flattened out, but it is non–singular (it can be explicitly verified that it has no curvature singularities). These are the two bona fide fixed points that can and will support an instanton.

Simple integration of equation (45) yields a branch with $x \geq 1$ that exists for all time and it is explicitly given by

$$\pm \frac{t}{\tau} = x^3 - 1 + \log(x^3 - 1) \ , \tag{46}$$

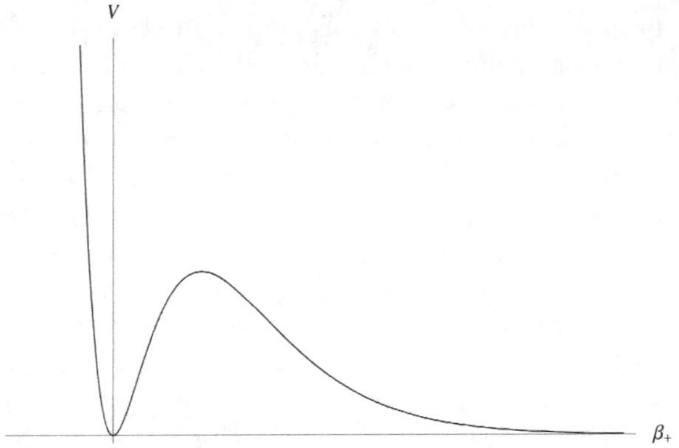

Figure 1. Plot of the potential $C_{ij}C^{ij}$ for metrics with $SU(2) \times U(1)$ isometry

where $\tau = |\omega| L^3/12\kappa^2$ is the characteristic time scale of the problem. As t ranges from $-\infty$ to $+\infty$, $x(t)$ interpolates smoothly between the values 1 and ∞ that describe the location of the two fixed points. As noted before, the choice of sign distinguishes instantons from anti–instantons. The instanton we have obtained in this fashion has finite Euclidean action, which turns out to be $S_{\text{inst.}} = 4\pi^2/|\omega|$, and it has no modulus other than L. It can also be seen without much effort that the solution (46) is the only instanton with $SU(2)$ isometry (up to permutations of the three principal axes of S^3) of the pure Cotton theory. In this case, the instanton and the anti–instanton are simply interrelated by parity transformations on S^3.

The figure above serves to illustrate the situation. It is a plot of the potential term $V = C_{ij}C^{ij}$ of pure Cotton $z = 3$ Hořava–Lifshitz gravity for axially symmetric Bianchi IX metrics on S^3, which turns out to be proportional to $(e^{-6\beta_+} - e^{-3\beta_+})^2$. The horizontal axis is the shape modulus of the sphere, which is conveniently described here by the variable $\beta_+ = \log x$; the notation is borrowed from mixmaster dynamics where this parametrization is widely used, [25], [26]. Thus, the two fixed points are located at the origin and at infinity and clearly they are degenerate since the potential vanishes there. There is also a small bump in between the two fixed points which is peaked at $\beta_+ = (\log 2)/3$, where $\gamma_1 = \gamma_2 = 2\gamma_3$.

The instanton interpolates between the two degenerate vacua, as in ordinary particle systems. This analogy is made exact by noting that the Euclidean theory of Hořava–Lifshitz gravity reduces in this case to the dynamics of a single mode $\beta_+(t)$ derived from the action

$$S_{\text{eff.}} = \frac{3\pi^2 L^3}{\kappa^2} \int dt \Big[\left(\frac{d\beta_+}{dt}\right)^2 + \frac{16\kappa^4}{\omega^2 L^6} \left(e^{-6\beta_+} - e^{-3\beta_+}\right)^2 \Big] . \qquad (47)$$

The change of variable $\beta_+ = \log x$ is necessary to cast the kinetic term in canonical form. The instanton solutions of this effective point particle model satisfy equation (45) written in terms of the variable β_+. Also, for these configurations, the effective action equals $4\pi^2/|\omega|$, as required for consistency of the interpretation. The calculation is easily done by noting that the potential term in (47) is itself derived from a superpotential, as

$$\left(e^{-6\beta_+} - e^{-3\beta_+}\right)^2 = \left(\frac{dW}{d\beta_+}\right)^2 ; \qquad W(\beta_+) = \frac{1}{3} e^{-3\beta_+} - \frac{1}{6} e^{-6\beta_+} \qquad (48)$$

so that

$$S_{\text{eff.}}^{\text{inst.}} = 2\, \frac{3\pi^2 L^3}{\kappa^2}\, \frac{4\kappa^2}{|\omega| L^3}\, |W(+\infty) - W(0)| = \frac{4\pi^2}{|\omega|} . \qquad (49)$$

Similar considerations apply to all other instanton solutions arising from consistent truncation of the Ricci–Cotton flow to the Bianchi IX mini–superspace sector. Such effective point particle systems, which are also commonly used in mixmaster dynamics, prove useful for comparing $SU(2)$ gravitational instantons of Hořava–Lifshitz gravity with those of general relativity. Recall that in (3+1)–dimensional Einstein gravity, the instantons are defined as regular geometries with self–dual Riemann (or more generally Weyl) curvature tensor. This property is not shared by the instantons of Hořava–Lifshitz gravity. Yet the $SU(2)$ instantons of Einstein gravity admit an alternative description as trajectories of a point particle moving under the influence of an effective potential, [27]. The difference is that the metric in the space of truncated degrees of freedom is indefinite (inherited by the choice $\lambda = 1$ in the Einsteinian DeWitt superspace metric) and the effective potential does not exhibit degenerate vacua. In that case, the $SU(2)$ instantons (such as the self–dual Taub–NUT solution) correspond to special trajectories of the particle extending from a local maximum to a local minimum of the effective potential. In terms of the four–dimensional geometry, they are supported by removable (nut) singularities at one end, leading to complete and everywhere regular space–time metrics. On the other hand, the singularities that the effective point particle may encounter in Hořava–Lifshitz gravity are not removable, which explains the need to have interpolating trajectories between degenerate vacua to account for instantons in non–relativistic gravitational theories. An important consequence of this difference is in the asymptotic structure of the corresponding space–time metrics: asymptotically locally flat metrics are only possible in Einstein gravity.

5. Concluding remarks

The framework we have presented here is quite broad and can be used to provide a toy model for dynamical vacuum selection in relativistic field theories associated to an action W in d Euclidean dimensions. The instanton solutions of the associated Lifshitz theory in $d+1$ space–time dimensions describe off-shell transitions among the many different vacua that populate the landscape of the d–dimensional theory. This alternative interpretation of Lifshitz models is in the spirit of the Onsager-Machlup theory for non-equilibrium processes in thermodynamics, [28], which is also based on the action (1) (or better to say $S'_{\rm Eucl.}$) and it is often called Onsager–Machlup functional in the literature. In this general context, W is the entropy function that changes monotonically in time and it is proportional to the logarithm of the probability of a given fluctuation. The gradient of W is the thermodynamic force measuring the tendency of a system to seek equilibrium. Linearization of the flow equations around the fixed points describe small fluctuations away from equilibrium states, whereas the instanton solutions incorporate non-linear effects for large transitions between different equilibrium states of the system. It will be interesting to strengthen the analogies between non-equilibrium processes and gradient flow equations in the future.

Some aspects of our work are also reminiscent of renormalization group equations, in particular for geometric theories, and we would like to understand them better. Recall that transitions among vacua of string theory are often described by running solutions of the beta–function equations. These off–shell processes are themselves gradient flows derived from an effective gravitational action coupled to a dilaton and possibly other massless fields of string theory. Thus, they can be alternatively viewed as Euclidean solutions of a Lifshitz theory in one dimension higher by identifying the renormalization group time with the extra Euclidean time dimension. In this context, the off–shell dilaton field is treated as a Lifshitz scalar field, and, likewise, all other off-shell massless fields are treated as Lifshitz tensor fields. However, there is an important technical difference that prevent us from taking immediate advantage of the results we described above. Here, we have been working under the assumption that the metric in field space is positive definite (the same also applies to Onsager–Machlup theory) in order to obtain instantons as extrema of the Euclidean action and interpret W as entropy functional.

On the other hand, the renormalization group equations used in the off–shell formulation of string theory follow from an indefinite metric by choosing, in particular, $\lambda = 1/2$ in DeWitt's superspace metric (it is the same choice that turns (30) into the ordinary Ricci flow equation for the pure metric sector of the theory). Of course, one can adjust the value of λ accordingly to make this metric positive definite, thus taking advantage of all goodies that come with it, and then use the resulting equations as toy model for dynamical vacuum selection in string theory. This modification does not affect the structure of the fixed points, whose defining relations are inert to λ, but the flow lines will not be trajectories of the renormalization group equations. It remains to be seen what else can be learned from this analogy.

Another aspect of the present work that is potentially interesting for mathematics is the unifying framework that Lifshitz theories provide to all gradient flow equations. Our results on Hořava–Lifshitz theories in $3+1$ dimensions also provide a good reason to study higher order geometric flows more systematically. The Cotton flow is the simplest example of this kind, being an equation of third order, but one should also study more systematically equations of mixed order such as the Ricci–Cotton flow. These systems possess an abundance of fixed points, and, hence, they have much richer structure than the ordinary second order Ricci flow allowing for eternal (instanton) solutions. The main technical problem is to develop methods to estimate curvature bounds and obtain general criteria for the formation of singularities along such flow lines. These issues remain largely unexplored at this time and they call for immediate attention by the experts.

Finally, we comment on the possible use of our instanton solutions to Hořava–Lifshitz theory when viewed as an alternative theory of gravitation at short distances. It will be interesting to obtain a Euclidean path integral formulation of such a non–relativistic theory of gravitation, where our instanton configurations will contribute substantially, since they are weighted by the exponential of (minus) their action. Also, mini–superspace models of quantum cosmology will be interesting to investigate in this context by focusing, in particular, to the mixmaster universe based on the homogeneous three–geometries, [26]. Comparison with ordinary gravity may provide some valuable lessons. Of course, it still remains to settle some open questions regarding the general validity of Hořava's theory, which appears to contain an unphysical scalar graviton mode, and its consistency with large scale gravitational physics. Also, the condition of detailed balance on which our instanton constructions were based appears to be rather restrictive on physical grounds, as it does not seem to allow for asymptotically flat solutions. A generalization of the original theory was recently made to circumvent some of these problems, [29], and it might still take time to settle the situation one way or another.

Acknowledgments

I am very grateful to the conference organizers for their kind invitation and financial support. I also thank Francois Bourliot, Dieter Lüst, Marios Petropoulos as well as Christos Sourdis for fruitful collaboration on the main aspects of the present work.

References

[1] Lifshitz E M 1941 On the theory of second order phase transitions *Zh. Eksp. Teor. Fiz.* **11** 255
[2] Hořava P 2008 Quantum criticality and Yang–Mills gauge theory *Preprint* arXiv:0811.2217 [hep-th]
[3] Hořava P 2009 Membranes at quantum criticality *J. High Energy Phys.* JHEP03(2009)020 (*Preprint* arXiv:0812.4287 [hep-th])
[4] Hořava P 2009 Quantum gravity at a Lifshitz point *Phys. Rev.* D **79** 084008 (*Preprint* arXiv:0901.3775 [hep-th])
[5] Misner C W, Thorne K S and Wheeler J A 1973 *Gravitation* (San Francisco: Freeman)
[6] DeWitt B S 1967 Quantum theory of gravity: 1. The canonical theory *Phys. Rev.* **160** 1113
[7] Allen A M and Cahn J W 1979 A microscopic theory for antiphase boundary motion and its application to antiphase domain coarsening *Acta Metallurgica* **27** 1085

[8] Ilmanen T 1993 Convergence of the Allen–Cahn equation to Brakke's motion by mean curvature *J. Diff. Geom.* **38** 417
[9] del Pino M, Kowalczyk M and Wei J 2009 Entire solutions of the Allen–Cahn equation and complete embedded minimal surfaces of finite total curvature in R^3 *Preprint* arXiv: 0902.2047 [math.AP]
[10] Alikakos N D 2009 Some basic facts on the system $\Delta u - W_u(u) = 0$ *Preprint* arXiv: 0909.4392 [math.AP]
[11] Struwe M 1994 The Yang–Mills flow in four dimensions *Calc. Var.* **2** 123
[12] Schlatter A 1997 Long time behaviour of the Yang–Mills flow in four dimensions *Ann. Glob. Anal. Geom.* **15** 1
[13] Hassel A 1993 The Yang–Mills–Higgs heat flow on R^3 *J. Funct. Anal.* **111** 431
[14] Bakas I, Bourliot F, Lüst D and Petropoulos M 2010 Geometric flows in Hořava–Lifshitz gravity *J. High Energy Phys.* JHEP04(2010)131 (*Preprint* arXiv:1002.0062 [hep-th])
[15] Cao H D, Chow B, Chu S C and Yau S T eds. 2003 *Collected Papers on Ricci Flow* (Somerville: International Press)
[16] Deser S, Jackiw R and Templeton S 1982 Topologically massive gauge theories *Ann. Phys.* **140** 372; Erratum-ibid. 1988 **185** 406
[17] Vuorio I 1985 Topologically massive planar universe *Phys. Lett.* B **163** 91
[18] Nutku Y and Bakler P 1989 Homogeneous, anisotropic three manifolds of topologically massive gravity *Ann. Phys.* **195** 16
[19] Ortiz M E 1990 Homogeneous solutions to topologically massive gravity *Ann. Phys.* **200** 345
[20] Chow D D K, Pope C N and Sezgin E 2010 Classification of solutions in topologically massive gravity *Class. Quant. Grav.* **27** 105001 (*Preprint* arXiv:0906.3559 [hep-th])
[21] Bergshoeff E A, Hohm O and Townsend P K 2009 Massive Gravity in Three Dimensions *Phys. Rev. Lett.* **102** 201301 (*Preprint* arXiv:0901.1766 [hep-th])
[22] Bakas I and Sourdis C 2010 Homogeneous vacua of (generalized) new massive gravity *Preprint* arXiv:1006.1871 [hep-th]
[23] Horne J H and Witten E 1989 Conformal gravity in three dimensions as a gauge theory *Phys. Rev. Lett.* **62** 501
[24] Kisisel A U O, Sarioglu O and Tekin B 2008 Cotton flow *Class. Quant. Grav.* **25** 165019 (*Preprint* arXiv:0803.1603 [hep-th])
[25] Misner C W 1969 Mixmaster universe *Phys. Rev. Lett.* **22** 1071; Misner C W 1969 Quantum cosmology 1 *Phys. Rev.* **186** 1319
[26] Bakas I, Bourliot F, Lüst D and Petropoulos M 2010 Mixmaster universe in Hořava–Lifshitz gravity *Class. Quant. Grav.* **27** 045013 (*Preprint* arXiv:0911.2665 [hep-th])
[27] Gibbons G W and Pope C N 1979 The positive action conjecture and asymptotically Euclidean metrics in quantum gravity *Commun. Math. Phys.* **66** 267
[28] Onsager L and Machlup S 1953 Fluctuations and irreversible processes *Phys. Rev.* **91** 1505; Tisza L and Manning I 1957 Fluctuations and irreversible thermodynamics *Phys. Rev.* **105** 1695
[29] Hořava P and Melby–Thompson C M 2010 General covariance in quantum gravity at a Lifshitz point *Phys. Rev.* D **82** 064027 (*Preprint* arXiv:1007.2410 [hep-th])

Violation of the Carter-Israel conjecture and its astrophysical implications

Cosimo Bambi

Institute for the Physics and Mathematics of the Universe, The University of Tokyo
5-1-5 Kashiwanoha, Kashiwa-shi, Chiba 277-8583, Japan

E-mail: cosimo.bambi@ipmu.jp

Abstract. On the basis of the Carter-Israel conjecture, today we believe that some compact and massive objects in the Galaxy and in the Universe are Kerr black holes. However, this idea cannot yet be confirmed by observations. We can currently obtain reliable estimates of the masses of these objects, but we do not know if the space-time around them is described by the Kerr metric and if they have an event horizon. A fundamental limit for a Kerr black hole is the Kerr bound $|a_*| \leq 1$. Here I discuss some astrophysical implications associated with the violation of this bound, which can thus be used to test the Carter-Israel conjecture.

1. Introduction

Today we believe that the final product of the gravitational collapse is a Kerr-Newman black hole. This is the Carter-Israel conjecture and it is based on the following argument. First, there are some singularity theorems showing that in general relativity the collapsing matter produces space-time singularities [1]. These theorems do not say anything about the nature of the singularities. The second step is thus to assume the Weak Cosmic Censorship conjecture [2], according to which space-time singularities must be hidden behind an event horizon. Surprisingly, in general relativity in 4 dimensions, the only possibility is the Kerr-Newman space-time [3, 4, 5, 6].

The Kerr-Newman metric has three parameters: the mass of the object, M, its spin, J, and its electric charge, Q. The spin J can be replaced by the spin parameter $a = J/M$, or by the dimensionless spin parameter $a_* = J/M^2$. For macroscopic bodies, the electric charge is usually very small and can be neglected[1]. Hereafter, we will thus restrict our attention to Kerr space-times with $Q = 0$. A fundamental property of Kerr black holes is the Kerr bound $|a_*| \leq 1$, which is the condition for the existence of the event horizon.

Observational evidences supporting the Carter-Israel conjecture are still definitively weak [8]. Astronomical observations have led to the discovery of at least two classes of astrophysical black hole candidates[2]: stellar-mass objects in X-ray binary systems ($M \sim 5 - 20\ M_\odot$) and super-massive objects at the center of many galaxies ($M \sim 10^5 - 10^{10}\ M_\odot$). All these objects

[1] For example, the electric charge is important for black holes with a mass $M \lesssim 10^{20}$ g in a ionized plasma [7]. Such small black holes cannot be produced today in the Universe, but could have been produced in the early Universe, when the matter density was much higher.

[2] The existence of a third class of astrophysical black holes, intermediate mass objects with $M \sim 10^2 - 10^4\ M_\odot$, is still quite controversial, because there are no reliable dynamical measurements of their masses.

are supposed to be Kerr black holes because they cannot be explained with something else without introducing new physics. For example, stellar-mass black hole candidates in X-ray binary systems are too heavy to be neutron/quark stars for any reasonable matter equation of state [9, 10]. The super-massive black hole candidate at the center of the Galaxy is too massive and compact to be a cluster of non-luminous bodies [11]. On the other hand, we do not have any observational evidence of the Kerr metric and of the existence of the event horizon.

New physics may instead be expected: it is indeed difficult to believe that space-time singularities can exist, even if behind an event horizon. So, the assumption of the Weak Cosmic Censorship conjecture seems to be a quite artificial trick, to exclude space-times in which new physics can be causally connected to distant observers. In other words, the conjecture would be motivated by our poor knowledge of the theory at high energy, but deviations from the Kerr space-times may be possible in Nature. Motivated by such a simple and general argument, one can consider the possibility that the Carter-Israel conjecture can be violated. Here, I will discuss the case of super-spinars, compact objects with $|a_*| > 1$. I will review the basic features of the accretion process onto super-spinars; more details on the subject can be found in the original papers [12, 13, 14]. Other observational properties of super-spinars were studied in [15, 16, 17].

2. Super-spinars

The Weak Cosmic Censorship conjecture was originally motivated by the fact that space-times with naked singularities present several kind of pathologies. One can however interpret such pathological features with the necessity of new physics. For example, there is a connection between the existence of naked singularities and regions with closed time-like curves [18]. The physical interpretation of space-times with causality violating regions has been recently investigated by some authors in Supergravity [19, 20, 21, 22]. Here, one finds space-times which apparently cannot be ruled out as unphysical and where causality can be violated. The solution of this puzzle seems to be in high energy corrections of the theory: at least in some cases, one can expect that the space-time goes to a new phase and a domain wall forms. Across the domain wall, the metric is non-differentiable and the expected region with closed time-like curves arises from the naive continuation of the metric ignoring the domain wall. The latter can be made of very exotic stuff, like super-tubes [19, 21, 22] or fundamental strings [20]. It is also remarkable that we know several counterexamples which look physically reasonable and in which the collapsing matter starts from regular initial data and evolves into a naked singularity, see e.g. Refs. [23, 24, 25, 26, 27, 28, 29].

The simplest object violating the Carter-Israel conjecture is probably the super-spinar [30], a compact body with $|a_*| > 1$. In absence of a uniqueness theorem similar to the one for Kerr black holes in 4 dimensions, super-spinars may be quite complex objects, characterized by many parameters. Nevertheless, at first approximation one can still expect to be able to describe their gravitational field with the Kerr metric. The first two terms in a multipole moment expansion of the space-time correspond to the mass and the spin of the massive object, while its deformation is encoded in higher order moments, which are typically much less important. In other words, deviations from the Kerr metric are usually very small, as one can see in [31]. If a compact object has $|a_*| > 1$, it cannot be a Kerr black hole. So, the Kerr bound can be used to test the Carter-Israel conjecture.

In absence of a complete theory of gravity, we have to take a phenomenological approach to study the astrophysical properties of super-spinars. They cannot be extremely compact, like a naked singularity, because otherwise they would be unstable due to the ergoregion instability [32]. It is probably reasonable to take their radius of order of their gravitational radius, which makes them more similar to a relativistic star made of very exotic matter than to a naked singularity.

Another issue is how super-spinars can be created. At present, it is not clear if it is possible to overspin an existent black hole [33]. If super-spinars can be the final product of the gravitational collapse of a star, then one should be able to obtain a super-spinning object from numerical simulations of a collapsing star. However this is not an easy job and there are so many unknown ingredients (and unknown physics) that it is unlikely to get an answer in a near future. We can however notice that recent attempts to measure the spin of stellar-mass black hole candidates suggest that these objects can rotate very fast [34][3]. Even if these measurements should be taken with great caution, it is intriguing the case of GSR 1915 + 105, whose spin parameter is estimated to be in the range 0.98 − 1. Since the evolution of the spin parameter due to accretion should be negligible in this system, the estimated a_* would reflect the initial spin of the object after its formation. One can thus think that the gravitational collapse of a star can produce a very fast rotating object. Since from the theory of stellar evolution we expect around 10^8 stellar-mass black holes in the Galaxy, even a low probability of violating the bound $|a_*| \leq 1$ may lead to a population of super-spinars in the Galaxy.

3. Numerical study of the accretion process

The study of the accretion process plays a fundamental role in the physics of compact objects, because it is the accretion process that determines how radiation is released by the accreting matter, and so what we can see from the compact object. It is thus not surprising that the accretion process onto Kerr black holes has been well studied in the literature and many research groups work on the subject [35, 36]. The quasi-steady-state configuration of adiabatic and spherically symmetric accretion onto a Schwarzschild black hole can be studied analytically [37]. However, in general, a numerical approach is necessary. The first numerical hydrodynamics simulations of the accretion process onto black holes can be traced back to the work of Wilson in 1972 [38], and were then extended in [39, 40]. After these works, the research was mainly devoted to the study of accretion from thick disks and tori, and to the study of the tori instabilities [41, 42, 43, 44].

In [12, 13, 14], I studied numerically the accretion process in Kerr space-time with arbitrary value of the spin parameter a_*. I neglected the back-reaction of the fluid to the geometry of the space-time, as well as the increase in mass and the variation in spin of the central object due to accretion. Such an approximation is surely reasonable if we want to study a stellar-mass compact object in a binary system, because in this case the matter captured from the stellar companion is typically small in comparison with the total mass of the compact object. The results of this simulations should not be applied to long-term accretion onto a super-massive object at the center of a galaxy, where accretion makes the mass of the compact object increase by a few orders of magnitude from its original value. The accreting matter was modeled as a perfect fluid.

The master formulas are the equations of conservation of baryon number and of the fluid energy-momentum tensor

$$\nabla_\mu J^\mu = 0 \, , \quad \nabla_\mu T^{\mu\nu} = 0 \, . \tag{1}$$

Here $J^\mu = \rho u^\mu$ and $T^{\mu\nu} = \rho h u^\mu u^\nu + p g^{\mu\nu}$, where ρ is the rest-mass energy density, p is the pressure, u^μ is the fluid 4-velocity, $h = 1 + \epsilon + p/\rho$ is the specific enthalpy, and ϵ is the specific internal energy density. In order to solve the system, an equation of state $p = p(\rho, \epsilon)$ must be specified.

The calculations were made with the relativistic hydrodynamics module of the public available code PLUTO [45, 46], properly modified for the case of curved space-time. I used the 3 + 1

[3] Let us notice that current estimates of black hole spin assume $|a_*| \leq 1$ and cannot be used to say that these objects are not super-spinars. If we allow for any value of the spin parameter, we would obtain two estimates, one with $|a_*| \leq 1$ and another with $|a_*| > 1$, because of the degeneracy discussed in [17].

Eulerian formalism of Ibanez and collaborators [47]. The line element of the space-time is written in the form

$$ds^2 = -\left(\alpha^2 - \beta_i\beta^i\right)dt^2 + 2\beta_i dt dx^i + \gamma_{ij}dx^i dx^j\,, \tag{2}$$

where α is the lapse function, β^i is the shift vector, and γ_{ij} is the 3-metric induced on each space-like slice. Here it is convenient to use two set of variables. The *primitive variables* are

$$\mathbf{V} = \left(\rho, v^i, p\right)^T \tag{3}$$

and are the quantities whose evolution and quasi-steady-state (if any) we want to determine. The hydrodynamical equations are instead solved in term of the *conserved variables*

$$\mathbf{U} = \left(D, S_i, \tau\right)^T\,, \tag{4}$$

which can be written in term of the primitive ones as $D = \rho W$, $S_i = \rho h W^2 v_i$, and $\tau = \rho h W^2 - D - p$, where W is the Lorentz factor. The equations of conservation (1) can now be written as

$$\frac{1}{\sqrt{-g}}\left[\frac{\partial}{\partial t}\left(\sqrt{\gamma}\,\mathbf{U}\right) + \frac{\partial}{\partial x^i}\left(\sqrt{-g}\,\mathbf{F}^i\right)\right] = \mathcal{S}\,, \tag{5}$$

where $\mathbf{F}^i$ and $\mathcal{S}$ are defined by

$$\mathbf{F}^i = \left(D\left(v^i - \beta^i/\alpha\right),\; S_j\left(v^i - \beta^i/\alpha\right) + p\delta^i_j,\; \tau\left(v^i - \beta^i/\alpha\right) + pv^i\right)^T, \tag{6}$$

$$\mathcal{S} = \left(0,\; T^{\mu\nu}\left(\partial_\mu g_{\nu j} - \Gamma^\lambda_{\mu\nu}g_{\lambda j}\right),\; \alpha\left(T^{\mu 0}\partial_\mu \ln\alpha - T^{\mu\nu}\Gamma^0_{\mu\nu}\right)\right)^T. \tag{7}$$

4. Results

Roughly speaking, naked singularities are not hidden behind an event horizon because their gravitational field is too weak to trap light rays. Close to the expected naked singularity, the gravitational field may be even repulsive [48, 49, 50, 51]. So, a quasi-steady-state accretion flow onto a naked singularity may be impossible: in some cases, the gas is accumulated around the massive object, forming a high density cloud that continues to grow [12, 52].

In [12, 13, 14], I considered adiabatic and spherically symmetric (at large radii) accretion process onto Kerr black holes and Kerr super-spinars. The initial configuration is a static cloud of gas around the massive object and then the system evolves to find a quasi-steady-state configuration. Gas is injected into the computational domain from the outer boundary at a constant rate and isotropically. With this set-up, the two parameters determining the accretion process are the spin parameter, a_*, and the radius of the compact object, R.

One can distinguish three kinds of accretion [13, 14]:

(i) *Black hole accretion.* For black holes and super-spinars with $|a_*|$ moderately larger than 1, one finds the usual accretion process onto a compact object. For a given R, the increase in $|a_*|$ makes the accretion process more difficult: in the quasi-steady-state configuration, the velocity of the gas around the compact object is lower, while the density and the temperature are higher. The gravitational field indeed becomes weaker for higher spin parameter, as one can easily understand by noticing that the radius of the event horizon of a black hole monotonically decreases with a_*. The difference, however, is very small and the exact value of the spin parameter does not affect significantly the process.

(ii) *Intermediate accretion.* As the spin parameter increases, the gravitational force around the super-spinar becomes weaker and even repulsive. Now the accretion process is significantly suppressed: the flow around the super-spinar becomes subsonic and the density and the temperature of the gas increase further.

(iii) *Super-spinar accretion.* For high value of the spin parameter, the process of accretion is very different: matter accretes from the poles, while the repulsive gravitational field produces outflows around the equatorial plane (see Fig. 1). 3-dimensional simulations show that the production of outflows is a quite chaotic phenomenon, without the formation of a stable structure [14].

In Figs. 2 and 3, I have plotted the quantity $v^{(r)} = e_i^{(r)} v^i$, where $e_b^{(a)}$ is the tetrad of a locally non-rotating observer [53]. Figs. 4 and 5 show the temperature profile around black holes and super-spinars. In these simulations, $R = 2.5\ M$ and the gas has a polytropic index $\Gamma = 5/3$ (non-relativistic particles). However, the qualitative behavior of the accretion process does not depend on the gas equation of state. For $a_* = 0$, 1, and 1.5, we find a black hole-like accretion; for $a_* = 2$, an intermediate accretion; for $a_* = 2.9$, the accretion is of super-spinar type. When $a_* = 2.5$ and 2.8, the accretion is essentially of the second kind, but there is some very weak ejection of matter near the equatorial plane.

Unlike jets and outflows produced around black holes, here the outflows are produced by the repulsive gravitational force at a small distance from the super-spinar and are ejected around the equatorial plane. These outflows become more energetic for higher value of the spin parameter. In some circumstances, the amount of matter in the outflow is considerable, which can indeed significantly reduce the mass accretion rate. On the other hand, for lower values of the spin parameter, the outflows may not be energetic enough to be ejected at large radii and escape from the gravitational field of the object. In these cases, one finds a convective region around the super-spinar, where the ejected gas is pushed back by the accreting fluid. This possibility is shown in Fig. 6, for the case of a super-spinar with $a_* = 2.8$ and a gas made of relativistic particles ($\Gamma = 4/3$).

Since the repulsive gravitational force around super-spinars seems to be able to create collimated jets with high Lorentz factor, in [13] I put forward the possibility that long gamma ray bursts might be explained with the formation of a super-spinar.

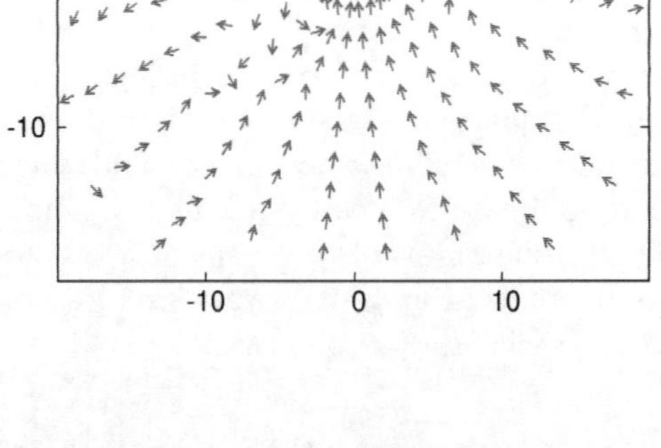

Figure 1. Snapshot of the direction of the gas velocity around a super-spinar with $a_* = 3.0$, on a plane containing the axis of symmetry z. Here matter accretes from the poles, while the repulsive gravitational field produces outflows around the equatorial plane. The unit of length along the axes is M.

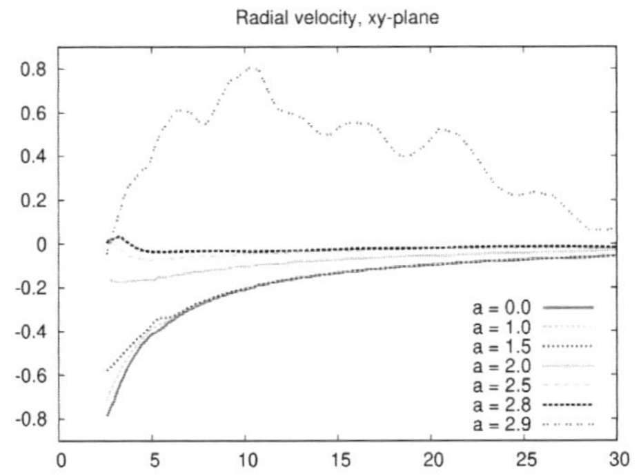

Figure 2. Snapshot at $t = 500\ M$ of the radial velocity as a function of the radial coordinate on the equatorial plane. $R = 2.5\ M$, radial coordinate in units $M = 1$.

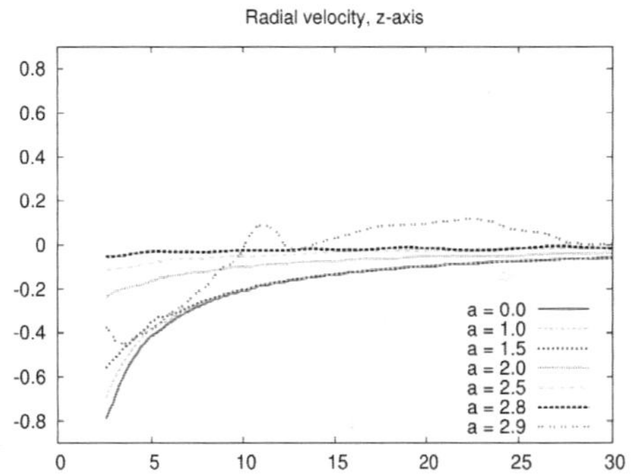

Figure 3. Snapshot at $t = 500\ M$ of the radial velocity as a function of the radial coordinate along the z-axis. $R = 2.5\ M$, radial coordinate in units $M = 1$.

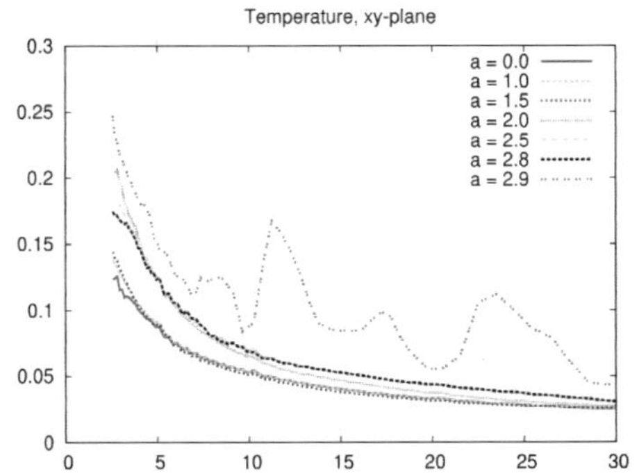

Figure 4. Snapshot at $t = 500\ M$ of the temperature as a function of the radial coordinate on the equatorial plane. $R = 2.5\ M$, temperature in GeV, radial coordinate in units $M = 1$.

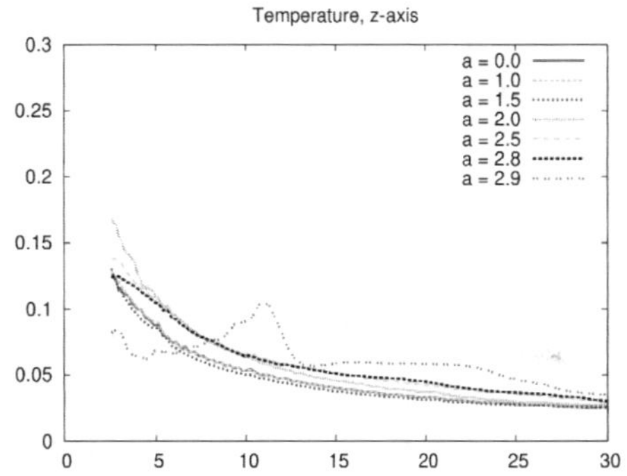

Figure 5. Snapshot at $t = 500\ M$ of the temperature as a function of the radial coordinate along the z-axis. $R = 2.5\ M$, temperature in GeV, radial coordinate in units $M = 1$.

Acknowledgments
I would like to thank Naoki Yoshida for careful reading of the first draft of this manuscript. This work was supported by World Premier International Research Center Initiative (WPI Initiative), MEXT, Japan, and by the JSPS Grant-in-Aid for Young Scientists (B) No. 22740147.

References
[1] S. W. Hawking and G. F. R. Ellis, *The Large scale structure of space-time*, (Cambridge University Press, Cambridge, UK, 1973).
[2] R. Penrose, Riv. Nuovo Cim. Numero Speciale **1**, 252 (1969) [Gen. Rel. Grav. **34**, 1141 (2002)].
[3] B. Carter, Phys. Rev. Lett. **26**, 331 (1971).
[4] D. C. Robinson, Phys. Rev. Lett. **34**, 905 (1975).
[5] P. O. Mazur, J. Phys. A **15**, 3173 (1982).
[6] P. T. Chrusciel and J. Lopes Costa, Astérisque **321**, 195 (2008) [arXiv:0806.0016 [gr-qc]].
[7] C. Bambi, A. D. Dolgov and A. A. Petrov, JCAP **0909**, 013 (2009) [arXiv:0806.3440 [astro-ph]].
[8] R. Narayan, New J. Phys. **7**, 199 (2005) [arXiv:gr-qc/0506078].
[9] C. E. . Rhoades and R. Ruffini, Phys. Rev. Lett. **32**, 324 (1974).

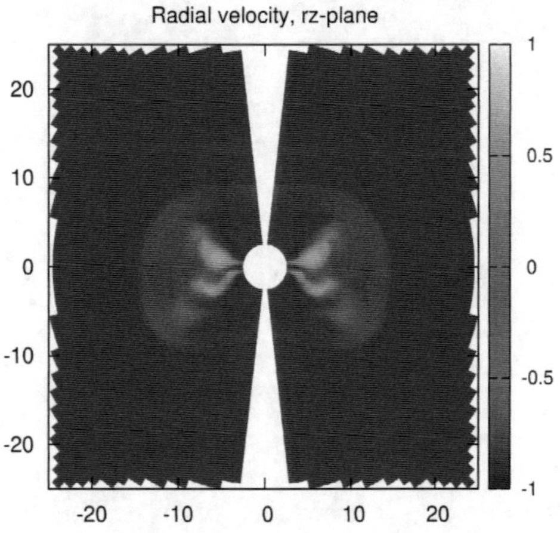

Figure 6. Radial velocity of the gas around a super-spinar with $a_* = 2.8$ and $R = 2.5\ M$.

[10] V. Kalogera and G. Baym, Astrophys. J. **470**, L61 (1996) [arXiv:astro-ph/9608059].
[11] E. Maoz, Astrophys. J. **494**, L181 (1998) [arXiv:astro-ph/9710309].
[12] C. Bambi, K. Freese, T. Harada, R. Takahashi and N. Yoshida, Phys. Rev. D **80**, 104023 (2009) [arXiv:0910.1634 [gr-qc]].
[13] C. Bambi, T. Harada, R. Takahashi and N. Yoshida, Phys. Rev. D **81**, 104004 (2010) [arXiv:1003.4821 [gr-qc]].
[14] C. Bambi and N. Yoshida, Phys. Rev. D (in press) [arXiv:1006.4296 [gr-qc]].
[15] C. Bambi and K. Freese, Phys. Rev. D **79**, 043002 (2009) [arXiv:0812.1328 [astro-ph]].
[16] C. Bambi, K. Freese and R. Takahashi, in *Windows on the Universe*, edited by L. Celnikier et al. (The Gioi Publishers, Ha Noi, Vietnam, 2010), pp. 575-578 [arXiv:0908.3238 [astro-ph.HE]].
[17] R. Takahashi and T. Harada, Class. Quant. Grav. **27**, 075003 (2010) [arXiv:1002.0421 [astro-ph.HE]].
[18] M. Kriele, Gen. Rel. Grav. **22**, 619 (1990).
[19] E. K. Boyda, S. Ganguli, P. Horava and U. Varadarajan, Phys. Rev. D **67**, 106003 (2003) [arXiv:hep-th/0212087].
[20] D. Israel, JHEP **0401**, 042 (2004) [arXiv:hep-th/0310158].
[21] N. Drukker, Phys. Rev. D **70**, 084031 (2004) [arXiv:hep-th/0404239].
[22] E. G. Gimon and P. Horava, arXiv:hep-th/0405019.
[23] D. Christodoulou, Commun. Math. Phys. **93**, 171 (1984).
[24] A. Ori and T. Piran, Phys. Rev. D **42**, 1068 (1990).
[25] P. S. Joshi and I. H. Dwivedi, Phys. Rev. D **47**, 5357 (1993) [arXiv:gr-qc/9303037].
[26] D. Christodoulou, Annals Math. **140**, 607 (1994).
[27] I. H. Dwivedi and P. S. Joshi, Commun. Math. Phys. **166**, 117 (1994) [arXiv:gr-qc/9405049].
[28] S. S. Deshingkar, I. H. Dwivedi and P. S. Joshi, Phys. Rev. D **59**, 044018 (1999) [arXiv:gr-qc/9805055].
[29] T. Harada, Pramana **63**, 741 (2004) [arXiv:gr-qc/0407109].
[30] E. G. Gimon and P. Horava, Phys. Lett. B **672**, 299 (2009) [arXiv:0706.2873 [hep-th]].
[31] C. Bambi and N. Yoshida, Class. Quant. Grav. (in press) [arXiv:1004.3149 [gr-qc]].
[32] P. Pani, E. Barausse, E. Berti and V. Cardoso, arXiv:1006.1863 [gr-qc].
[33] T. Jacobson and T. P. Sotiriou, Phys. Rev. Lett. **103**, 141101 (2009) [Erratum-ibid. **103**, 209903 (2009)] [arXiv:0907.4146 [gr-qc]].
[34] J. E. McClintock, R. Narayan, L. Gou, R. F. Penna and J. A. Steiner, arXiv:0911.5408 [astro-ph.HE].
[35] S. K. Chakrabarti, Phys. Rept. **266**, 229 (1996) [arXiv:astro-ph/9605015].
[36] J. A. Font, Living Rev. Rel. **11**, 7 (2007).
[37] F. C. Michel, Astrophys. Space Sci. **15**, 153 (1972).
[38] J. R. Wilson, Astrophys. J. **173**, 431 (1972).
[39] J. F. Hawley, L. L. Smarr and J. R. Wilson, Astrophys. J. **277**, 296 (1984).
[40] J. F. Hawley, L. L. Smarr and J. R. Wilson, Astrophys. J. Suppl. **55**, 211 (1984).
[41] J. F. Hawley, Astrophys. J. **381**, 496 (1991).
[42] M. Yokosawa, Publ. Astron. Soc. Japan **47**, 605 (1995).
[43] I. V. Igumenshchev and A. M. Beloborodov, Mon. Not. Roy. Astron. Soc. **284**, 767 (1997).
[44] J. P. De Villiers and J. F. Hawley, Astrophys. J. **577**, 866 (2002) [arXiv:astro-ph/0204163].

[45] A. Mignone, G. Bodo, S. Massaglia, T. Matsakos, O. Tesileanu and C. Zanni, Astrophys. J. Suppl. **170**, 228 (2007) [arXiv:astro-ph/0701854].
[46] `http://plutocode.to.astro.it/index.html`
[47] F. Banyuls, J. A. Font, J. M. Ibanez, J. M. Marti and J. A. Miralles, Astrophys. J. **476**, 221 (1997).
[48] M. Calvani and F. de Felice, Gen. Rel. Grav. **9**, 889 (1978).
[49] I. Gaida, H. R. Hollmann and J. M. Stewart, Class. Quant. Grav. **16**, 2231 (1999) [arXiv:hep-th/9811032].
[50] D. Batic, D. Chin and M. Nowakowski, arXiv:1005.1561 [gr-qc].
[51] O. Luongo and H. Quevedo, arXiv:1005.4532 [gr-qc].
[52] E. Babichev, S. Chernov, V. Dokuchaev and Yu. Eroshenko, arXiv:0806.0916 [gr-qc].
[53] J. M. Bardeen, W. H. Press and S. A. Teukolsky, Astrophys. J. **178**, 347 (1972).

Dark energy and matter perturbations in scalar-tensor theories of gravity

J C Bueno Sánchez[1,2] and L Perivolaropoulos[2]

[1] Departamento de Física Atómica, Molecular y Nuclear, Universidad Complutense de Madrid, 28040 Madrid, Spain

[2] Department of Physics, University of Ioannina, 541 10 Ioannina, Greece

E-mail: jcbueno@fis.ucm.es

Abstract. We solve analytically and numerically the generalized Einstein equations in scalar-tensor cosmologies to obtain the evolution of dark energy and matter linear perturbations. We compare our results with the corresponding results for minimally coupled quintessence perturbations. We find that Scalar-Tensor dark energy density perturbations are amplified by a factor of about 10^4 compared to minimally coupled quintessence perturbations on scales less than about $1000h^{-1}$Mpc (sub-Hubble scales). On these scales dark energy perturbations constitute a fraction of about 10% compared to matter density perturbations. Scalar-Tensor dark energy density perturbations are anti-correlated with matter linear perturbations on sub-Hubble scales. This anti-correlation of matter with negative pressure perturbations induces a mild amplification of matter perturbations by about 10% on sub-Hubble scales. The evolution of scalar field perturbations on sub-Hubble scales is scale independent and therefore corresponds to a vanishing effective speed of sound ($c_{s\Phi} = 0$). We briefly discuss the observational implications of our results which may include predictions for galaxy and cluster halo profiles which are modified compared to ΛCDM. The observed properties of these profiles are known to be in some tension with the predictions of ΛCDM.

1. Introduction

A wide range of cosmological observations indicate that the universe has entered a phase of accelerating expansion. These observations include both direct geometric probes of the expanding metric and dynamical probes of the growth rate of matter perturbations. This growth depends on both the expansion rate and the gravitational law on large scales.

Geometric probes of the cosmic expansion include (a) Type Ia supernovae (SnIa) standard candles [1, 2], (b) the angular location of the first peak in the CMB perturbations angular power spectrum, which probes the integrated cosmic expansion rate using the last scattering horizon as a standard ruler [3], (c) baryon acoustic oscillations of the matter density power spectrum. These oscillations also probe the integrated cosmic expansion rate on more recent redshifts using the last scattering horizon as a standard ruler [4], and (d) other less accurate standard candles (Gamma Ray Bursts [5], HII starburst galaxies[6]) and standard rulers (cluster gas mass fraction [7]) as well as probes of the age of the universe [8].

Dynamical probes of the cosmic expansion and the gravitational law on cosmological scales include X-Ray cluster growth data [9], large scale structure power spectrum at various redshift slices [10, 11], redshift distortion observed through the anisotropic pattern of galactic redshifts on cluster scales [11, 12] and weak lensing surveys [13, 14, 15].

These cosmological observations converge on the fact that the simplest Friedmann-Robertson-Walker (FRW) model describing well the cosmic expansion rate is the one corresponding to a cosmological constant [16] in a flat space, namely $H(z)^2 = H_0^2\left[\Omega_{0m}(1+z)^3 + (1-\Omega_{0m})\right]$ where $H(z)$ is the Hubble expansion rate at redshift z, $H_0 = H(z=0)$ and Ω_{0m} the present matter density normalized to the present critical density for flatness. This form of $H(z)$ and other similar, more complicated forms of it that are also consistent with cosmological data are predicted by broad classes of models: Dark energy models [17, 18, 19, 20, 21, 22, 23, 24, 25], modified gravity models [26, 27, 28, 29, 30, 31, 32] and local void models [33, 34].

Scalar-Tensor (ST) cosmological models [29, 30] (extended quintessence [35]) constitute a fairly generic representative of modified gravity models. They are based on the promotion of Newton's constant to a non-minimally coupled to curvature scalar field whose dynamics is determined by a potential $U(\Phi)$ and by the functional form of the non-minimal coupling $F(\Phi)$. The deviation of these models from GR is tightly constrained locally by solar system observations and by small scale gravitational experiments [36, 37]. These constraints, however, are significantly less stringent on cosmological scales [38, 39, 40] and may also be evaded by chameleon type arguments [41]. In the following we assume that the model considered passes the solar system tests in the context of such a mechanism.

The main focus of this work is to study the growth of perturbations in ST theories and its quantitative comparison with the corresponding growth in GR. This comparison can lead to the derivation of potential signatures of ST theories on the power spectrum and on other observables related to the growth of density perturbations. In ST theories, the non-minimal coupling of the scalar field to curvature perturbations (which in turn are driven by matter perturbations) leads to an amplification of the scalar field perturbations on sub-Hubble scales. Thus, it may be shown that on sub-Hubble scales the field perturbations $\delta\Phi$ are scale independent [29] and therefore the effective speed of sound $c_{s\Phi}$ for ST field perturbations vanishes. As discussed below, the corresponding scalar field density perturbations are also amplified but they are *anti-correlated* with respect to matter perturbations.

2. Perturbations in Scalar-Tensor Cosmologies

We consider the following ST action in the physical Jordan frame [29, 31]

$$S = \frac{1}{16\pi G}\int d^4x\sqrt{-g}\Big(F(\Phi)R - Z(\Phi)g^{\mu\nu}\partial_\mu\Phi\partial_\nu\Phi - 2U(\Phi)\Big) + S_m[\psi_m; g_{\mu\nu}]\,, \tag{2.1}$$

where G is the bare gravitational constant, R is the scalar curvature of the metric $g_{\mu\nu}$ and S_m is the action of matter fields. In what follows we use units such that $8\pi G = 1$. The variation of the dimensionless function $F(\Phi)$ describes the variation of the effective gravitational constant. This variation (spatial or temporal) is severely constrained by solar system experiments [36, 37]. The GR limit of ST theories is obtained either by fixing $F(\Phi) = \Phi_0 \simeq 1$ (Φ_0 is a constant) or by freezing the dynamics of Φ using the function $Z(\Phi)$ or the potential $U(\Phi)$.

Considering a flat FRW background where matter is described by a pressureless perfect fluid with density ρ_m we obtain the equations for the evolution of the background

$$3FH^2 = \rho_m + \frac{1}{2}\dot{\Phi}^2 - 3H\dot{F} + U \equiv \rho_{tot} \tag{2.2}$$

$$-2F\dot{H} = \rho_m + \dot{\Phi}^2 + \ddot{F} - H\dot{F} \equiv \rho_{tot} + p_{tot} \tag{2.3}$$

$$\ddot{\Phi} + 3H\dot{\Phi} = 3F_{,\Phi}\left(\dot{H} + 2H^2\right) - U_{,\Phi} \tag{2.4}$$

$$\dot{\rho}_m + 3H\,\rho_m = 0\,, \tag{2.5}$$

where we have rescaled Φ so that $Z = 1$ (assuming that $Z > 0$) and a subscript Φ stands for derivative with respect to Φ. To solve this system numerically we set initial conditions corresponding to the time of recombination ($z \simeq 1000$) and find the evolution of the background homogeneous field $\Phi(t)$, scale factor $a(t)$ and matter density $\rho_m(t) \propto a(t)^{-3}$. This scaling of the matter density would not be true in the Einstein frame due to the direct coupling of matter to the scalar field. However, here we assume that the physical frame is the Jordan frame and do not consider a direct coupling of the scalar field to matter.

In order to obtain the evolution of perturbations we consider the perturbed FRW metric which in the Newtonian gauge takes the form

$$ds^2 = -(1+2\phi)dt^2 + a^2(1-2\psi)\delta_{ij}dx^i dx^j \tag{2.6}$$

The linear gravitational potentials ϕ and ψ are related by $\phi = \psi - (F_{,\Phi}/F)\delta\Phi$, and along with the scalar field perturbations $\delta\Phi$ are sufficient to fully determine the cosmological perturbations in these theories. It is straightforward to obtain two additional differential equations for the perturbations ϕ, ψ and $\delta\Phi$ [17, 29, 42]

$$\dot{\xi} + 2H\xi + \left(3\dot{H} - \frac{k^2}{a^2}\right)\phi - \frac{1}{2}(\delta\mu + 3\delta q) = 0\,, \text{ with } \xi \equiv 3(H\phi + \dot{\psi}) \tag{2.7}$$

$$\ddot{\delta\Phi} + 3H\dot{\delta\Phi} + \left(\frac{k^2}{a^2} + U_{,\Phi\Phi} - \frac{R}{2}F_{,\Phi\Phi}\right)\delta\Phi - \dot{\Phi}\dot{\phi} - (2\ddot{\Phi} + 3H\dot{\Phi})\phi - \dot{\Phi}\xi - \frac{F_{,\Phi}}{2}\delta R = 0\,, \tag{2.8}$$

where R is the Ricci curvature scalar, δR is its perturbation, and $\delta\mu$ and δq are the effective energy density and pressure perturbations respectively. Also, to solve for $\delta\Phi$ and ψ we use of the generalized Poisson equation

$$\delta\mu = -2\left(H\xi + (k^2/a^2)\psi\right)\,. \tag{2.9}$$

Assuming specific forms for the field potentials F and U in Eqs. (2.7)-(2.8) we may numerically solve for the perturbations $\psi(k,t)$, $\delta\Phi(k,t)$. We solve the system with initial conditions at recombination ($z \simeq 1000$) corresponding to an initially smooth scalar field Φ and gravitational potential ψ in a background with small initial deviation from GR.

It is straightforward to obtain the density perturbations of both dark energy ($\delta_\Phi \equiv \delta\rho_\Phi/\rho_{tot}$) and matter ($\delta_m \equiv \delta\rho_m/\rho_{tot}$) in terms of the numerically derived perturbations $\phi(k,t)$, $\psi(k,t)$ and $\delta\Phi(k,t)$ and the corresponding background. Alternatively, the matter density perturbation, when normalized with respect to ρ_m, i.e. $\widehat{\delta_m} = \delta\rho_m/\rho_m$ is given by [17]

$$\ddot{\widehat{\delta}}_m + 2H\dot{\widehat{\delta}}_m + \frac{k^2}{a^2}\left(\psi - \frac{F_{,\Phi}}{F}\delta\Phi\right) - 3(\ddot{\psi} + 2H\dot{\psi}) = 0\,. \tag{2.10}$$

Even though the system (2.7), (2.8) for the evolution of metric and field perturbations on a scale k can only be solved numerically, there are several useful qualitative conclusions that can be obtained by considering appropriate limits of the scale k in these equations. There are four scales involved in Eqs. (2.7), (2.8): The physical scale $\frac{k}{a}$ of the perturbations, the Hubble expansion rate H, the mass scale of the potential $U_{,\Phi\Phi}^{1/2}$, and the shifted ST perturbation scale $F_{,\Phi}^{1/2}\frac{k}{a}$. In addition, in a cosmologically interesting setup, the expansion of the universe is driving the time evolution of every physical quantity f. Therefore, $|\dot{f}| \simeq H|f|$. For scalar fields that can play a role in the present accelerating expansion of the universe we require $U_{,\Phi\Phi}^{1/2} \simeq H$. Thus each term in Eqs. (2.7), (2.8) is determined by one of the three scales: $\frac{k}{a}$, H, $F_{,\Phi}^{1/2}\frac{k}{a}$. By identifying the terms that dominate in each range of perturbation scales $\frac{k}{a}$ we may simplify the perturbation equations and obtain approximate solutions for the corresponding range of scales. We consider the following ranges of perturbation scales:

- Sub-Hubble ST scales: $\frac{k}{a} \simeq F_{,\Phi}^{1/2}\frac{k}{a} \gg H$, $F_{,\Phi} \gtrsim 1$. In this case the perturbation scale and the shifted ST perturbation scale are of the same order. Ignoring subdominant terms we find that
$$\delta\Phi \simeq -\psi\frac{FF_{,\Phi}}{F+F_{,\Phi}^2}\,. \tag{2.11}$$
Thus, on sub-Hubble scales the field perturbations $\delta\Phi$ are independent of the scale k and can be a significant fraction of the total energy perturbations as demonstrated in the next section. The corresponding behavior in GR is very different. Setting $F=1$ and $\psi=\phi$ in (2.8) we obtain
$$\delta\Phi \simeq A\frac{a^2H^2}{k^2}(\Phi-\Phi_i)\,\psi \to 0\,, \tag{2.12}$$
where $\dot{\Phi} \simeq (\Phi-\Phi_i)\,H$ and A is a proportionality factor necessary to fit the numerical solution. As expected, the above implies that $\delta\Phi$ is negligible in GR on sub-Hubble scales. The field and matter density perturbations on these scales are respectively given by
$$\delta\rho_\Phi \simeq \frac{k^2}{a^2}\psi\frac{FF_{,\Phi}^2}{F+F_{,\Phi}^2} \quad \text{and} \quad \delta\rho_m \simeq -\frac{k^2}{a^2}\psi F\left(\frac{F_{,\Phi}^2}{F+F_{,\Phi}^2}+2\right)\,. \tag{2.13}$$
The sub-Hubble ratio $\delta\rho_\Phi/\delta\rho_m$ is therefore scale independent, indicating that the effective speed of sound for ST dark energy perturbations is $c_{s\Phi}=0$. Here we refer to the effective (rest frame) speed of sound as defined in Ref. [43]. This quantity determines the sound horizon, i.e. the scale below which perturbations can not grow. The fact that $\delta\rho_\Phi/\delta\rho_m$ is negative indicates an interesting anti-correlation between dark matter and dark energy perturbations, which is also confirmed numerically in the next section.
It is also straightforward to derive the equation for the evolution of matter density perturbations on sub-Hubble scales in ST theories. On such scales, where $\delta\Phi$ is given by Eq. (2.11) and $\delta\rho_m$ from Eq. (2.13), we obtain [44]
$$\ddot{\widehat{\delta}}_m + 2H\dot{\widehat{\delta}}_m - \frac{\rho_m\widehat{\delta}_m}{2}\frac{1}{F}\frac{2F+4F_{,\Phi}^2}{2F+3F_{,\Phi}^2} = 0\,. \tag{2.14}$$
It may be shown that $\frac{1}{F}\frac{2F+4F_{,\Phi}^2}{2F+3F_{,\Phi}^2} = \frac{G_{\rm eff}}{G}$, where $G_{\rm eff}$ is the effective gravitational constant in Cavendish-like experiments in the context of ST theories [29]. Therefore, Eq. (2.14) has the anticipated scale independent form (as in the case of GR) but Newton's constant G has been replaced by the effective
- Sub-Hubble GR scales: $\frac{k}{a} \gg H \gg F_{,\Phi}^{1/2}\frac{k}{a}$, $F_{,\Phi} \ll 1$. If $F_{,\Phi} \ll 1$ there is a range of sub-Hubble scales corresponding to $F_{,\Phi} \ll \frac{H^2a^2}{k^2} \ll 1$ for which the terms depending on the non-minimal coupling in Eq. (2.8) are negligible compared to all other terms. For this range of sub-Hubble scales the scalar field perturbations are negligible, scale dependent and behave as in GR (Eq. (2.12)). However, on small enough scales, i.e. when $\frac{H^2a^2}{k^2} \ll F_{,\Phi} \ll 1$, we recover the ST scale independent behavior of Eq. (2.11).
- Super-Hubble scales: $\frac{k}{a} \ll H$. In this case we may ignore the scale dependent terms in Eq. (2.8) to obtain $\delta\Phi \sim \psi$. Clearly, there is no scale dependence for the perturbations on super-Hubble scales. Similarly, for the matter component we find scale independent perturbations $\delta_m \sim \psi$.

The above qualitative features of the cosmological perturbations are confirmed and extended by the numerical derivation in the next section.

3. Numerical Solution

Firstly, we define specific forms for the potentials $F(\Phi)$ and $U(\Phi)$

$$F(\Phi) = 1 - \lambda_f \Phi^2 \quad , \quad U(\Phi) = 1 + \exp(-\lambda\Phi)\,. \tag{3.1}$$

This form of $F(\Phi)$ is consistent with solar system tests for $\Phi \simeq 0$. Indeed solar system constraints of ST theories imply that [36]

$$F_{,\Phi}^2/F\big|_{t=t_0} < 10^{-4}\,, \tag{3.2}$$

where t_0 refers to the present time.

We next solve the system (2.2),(2.4),(2.5) to determine the background evolution in the context of the above potentials. We set the initial conditions at t_i corresponding to recombination close to GR and check that our results are robust with respect to reasonable changes of the initial conditions. The parameters that need to be fixed for the solution of the background system (2.2),(2.4),(2.5) are λ, λ_f and $\Omega_{0\mathrm{m}}$. In most solutions discussed in this section we set $\Omega_{0\mathrm{m}} = 0.3$ and $\lambda = 2$, $\lambda_f = 5$ (ST cosmology) or $\lambda_f = 0$ (minimally coupled quintessence). We also check that deviations from the ΛCDM expansion rate are less than 3.5% for any value of the parameter λ with fixed value of the initial condition $\Phi(t_i) \simeq 0.1$. At late times, the constant part of $U(\Phi)$ prohibits a large deviation from the ΛCDM expansion rate despite of the somewhat increased kinetic energy of Φ.

In Fig. 1 we plot the background dynamics for a non-minimally coupled quintessence field Φ setting $\lambda_f = 5$ and $\lambda = 2$. In ST gravity the scalar field is moving under the influence of the effective potential $U_{\mathrm{eff}}(\Phi) = U(\Phi) - \frac{1}{2}RF(\Phi)$. Assuming $F \sim O(1)$ at all times implies that at early times $U_{\mathrm{eff}}(\Phi)$ is dominated by the second term. The effective mass of the field is then $m_{\mathrm{eff}} \sim (-RF_{,\Phi\Phi})^{1/2} \sim \lambda_f^{1/2} H$. As a result, for $\lambda_f \gtrsim 1$ and $\Phi_i \neq 0$ the scalar field is dynamically rapidly driven to its value corresponding to GR ($\Phi = 0$) and remains performing oscillations of decreasing amplitude around the attractor $\Phi = 0$ as the Universe expands. If $\Phi_i = 0$ then the field remains at $\Phi = 0$ ($F \simeq 1$) until $U_{\mathrm{eff}}(\Phi) \simeq U(\Phi)$. Therefore, Φ behaves as minimally coupled quintessence at early times and as non-minimally coupled quintessence at late times when the field is driven away from $\Phi = 0$. Deviations with respect to the ΛCDM expansion rate can be significant at early times if the field begins far from $\Phi = 0$. Therefore, large deviations can be avoided simply by tuning Φ_i so that $F(\Phi_i) \simeq 1$. With such initial conditions, the system approaches GR while the Hubble expansion rate becomes practically identical to ΛCDM at late times. When U starts dictating the field dynamics (late time evolution), we find that such deviations are always below 3% even for large values of λ. In particular, for steep potentials $U(\Phi\,;\lambda = 2)$ *the dynamical evolution of* Φ *leads to an amplified value* $F_{,\Phi}^2/F \simeq 10^{-1}$ *at redshifts* $z \lesssim O(1)$ violating solar system constraints (Eq. (3.2)) but not cosmological constraints $F_{,\Phi}^2/F|_{t=t_0} \lesssim O(1)$ [39]. Thus, in the context of the chameleon mechanism the increased value of $F_{,\Phi}^2$ has the potential of being consistent with observational constraints. Furthermore, and as discussed in the previous section (Eq. (2.13)), the value of $F_{,\Phi}^2$ determines the magnitude of the dark energy density perturbations compared to those of matter. We therefore anticipate *amplified dark energy perturbations* (compared to GR) when the dynamics of Φ is turned on by increasing the value of λ.

The non-minimal coupling of the field Φ to the curvature scalar results in a very characteristic behaviour of certain background quantities. In the lefthand panel of Fig. 2 we plot the evolution of the density parameter Ω_m as a function of the redshift. Owing to the negative effective energy contribution in Eq. (2.2) sourced by the field's motion, Ω_m features and oscillatory behaviour around $\Omega_m = 1$ at early times. Notice that it is possible to have $\Omega_\phi < 0$, $\Omega_m > 1$ since ρ_Φ is not positive definite (Eq. (3.3)). This effect is discussed in detail in Ref. [35]. Another remarkable feature distinguishing minimally coupled quintessence from the non-minimally coupled one is that the latter is able to cross the phantom divide line [45, 46, 47] corresponding to an effective dark energy equation of state $w_{\rm eff} = -1$. In general, the effective equation of state for the scalar field Φ is given by (see Eqs. (2.2), (2.3))

$$w_\Phi = \frac{p_\Phi}{\rho_\Phi} = \frac{\frac{1}{2}\dot{\Phi}^2 - U(\Phi) + \ddot{F} + 2H\dot{F}}{\frac{1}{2}\dot{\Phi}^2 + U(\Phi) - 3H\dot{F}}\,. \tag{3.3}$$

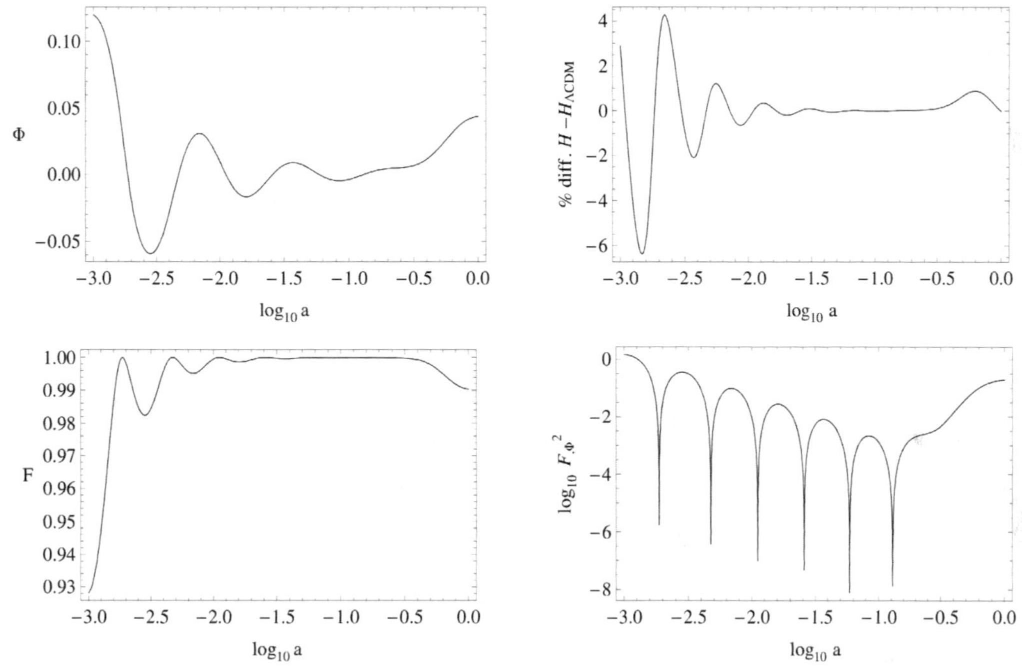

Figure 1. Background dynamics giving rise to the field contrasts depicted in Figs. 3 and 4. From left to right and from top to bottom: 1. Field oscillations about $\Phi = 0$; 2. % difference between H and $H_{\Lambda\rm CDM}$; 3. Evolution of $F(\Phi)$; 4. $F^2_{,\Phi}$ ($\simeq 10^{-1}$ at present). The parameters are $\lambda_f = 5$ and $\lambda = 2$.

In the righthand panel of Fig. 2 we show the evolution of the equation of state parameter for minimally ($\lambda_f = 0$, solid line) and non-minimally ($\lambda_f = 2$, dashed line) coupled quintessence, corresponding to GR and ST gravity respectively. At early times the background field oscillations give rise to divergences in w_Φ but such divergences do not reflect on the Hubble expansion rate. At late times, the scalar potential $U(\Phi)$ becomes relevant for the field dynamics, the field starts growing and w_Φ oscillates around the phantom divide line. Crossing of the phantom divide line is allowed by all current cosmological observations and is in fact favored by some of them [46]. This behavior is characteristic of ST gravities and cannot be achieved in minimally coupled quintessence [48].

In order to find the evolution of perturbations we solve Eqs. (2.7) and (2.8) using the numerical

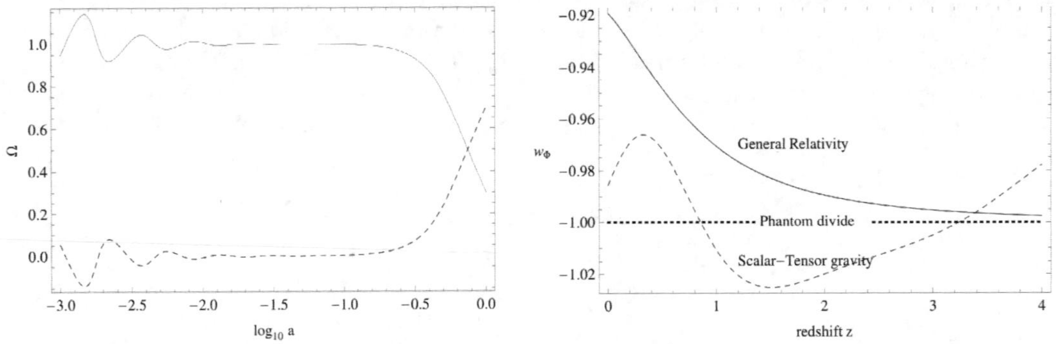

Figure 2. Plot of the density parameters Ω_m and $\Omega_\phi = 1 - \Omega_m$ corresponding to the background dynamics in Fig. 1 (lefthand panel). Equation of state parameter (EOS) w_Φ for the background field Φ as obtained in General Relativity (solid linc) with $\lambda = 2$ and Scalar-Tensor gravity (dashed line) with $\lambda_f = 5$ (righthand panel).

solution of the background and impose initial conditions

$$\delta\Phi(k, t_i) = \delta\dot{\Phi}(k, t_i) = 0 \tag{3.4}$$

$$\psi(k, t_i) = 1 \; ; \; \dot{\psi}(k, t_i) = 0 . \tag{3.5}$$

The evolution of the field $\delta\Phi$ vs the scale factor a (in logarithmic scale) for a perturbation of wavelength $\lambda_p = 30$ h^{-1}Mpc is shown in Fig. 3. In the lefthand panel we plot the numerical solution for the evolution of $\delta\Phi$ as obtained in GR (solid line) and as given by the approximation in Eq. (2.12) (dashed line) on subhorizon scales. We set $A \simeq 25$ in Eq. (2.12) in order to match the numerical solution. The righthand panel shows the numerical solution for the evolution of $\delta\Phi$ in ST gravity (solid line) and as given by Eq. (2.11) (dashed line) on subhorizon ST scales.

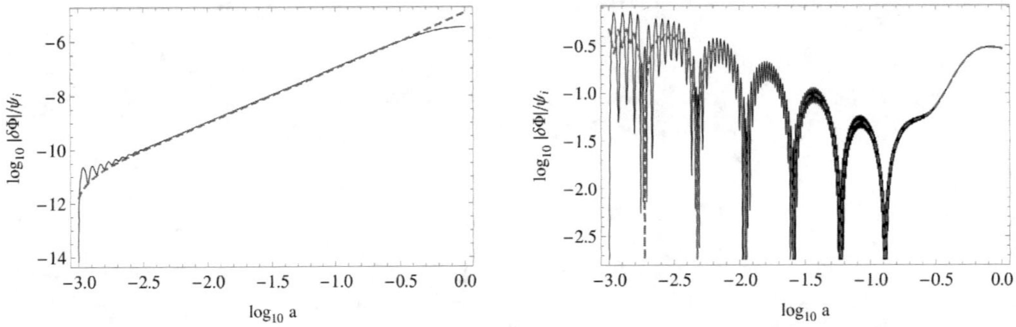

Figure 3. Evolution of the field perturbation $\delta\Phi$ as obtained in General Relativity (lefthand panel) and Scalar-Tensor gravity (righthand panel) for the scale $\lambda_p = 30$ h^{-1}Mpc. The numerical solution is the solid line and the analytical approximation on subhorizon ST and GR scales are the dashed lines. Oscillations in the background field Φ induce oscillations in $\delta\Phi$ through $F_{,\Phi} \propto \Phi$. We use $\lambda_f = 5$ for ST gravity and $\lambda = 2$ in both cases. The spikes in righthand panel correspond to changes of sign of $\delta\Phi$.

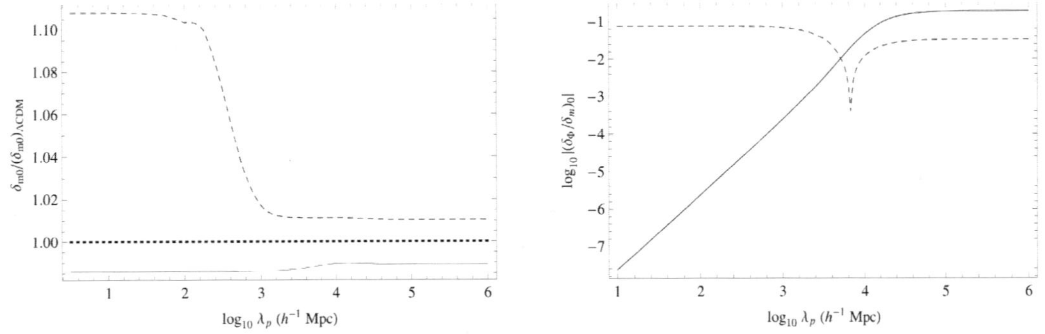

Figure 4. The lefthand panel shows the ratio $\frac{\delta_m(k,t_0)}{\delta_{m\Lambda CDM}(k,t_0)}$ for ST gravity (dashed line) and GR quintessence (solid line). The righthand panel shows the scale dependence of the ratio δ_Φ/δ_m at present as obtained in GR (solid line) and ST gravity (dashed line). The spike corresponds to a change of sign in δ_Φ/δ_m, thus revealing an anti-correlation between dark energy and matter perturbations. We use $\lambda_f = 5$ and $\lambda = 2$.

The evolution of matter overdensities is also affected by the introduction of a non-minimal coupling. This is demonstrated in the lefthand panel of Fig. 4 where we plot the present value of the ratio of δ_m, for minimally coupled quintessence (solid line) and ST gravity (dashed line), to the matter contrast in the ΛCDM cosmology for a range of scales.Clearly, the 10% amplification of matter perturbations in ST gravity is applicable on sub-Hubble scales while on larger scales the amplification is negligible. This mild amplification of matter perturbations may be attributed to the corresponding amplification of dark energy perturbations in these theories which also affects matter perturbations despite the predicted anti-correlation. Indeed, the dark energy void in a cluster of galaxies reduces the negative pressure inside the cluster and amplifies the gravitational collapse.

In the righthand panel of Fig. 4 we plot the scale dependence of the ratio δ_Φ/δ_m at present in GR (solid line) and in ST gravity (dashed line). Setting $\lambda_f = 5$, $\lambda = 2$, we find that the GR solution leads to negligible dark energy perturbations on sub-Hubble scales $\delta_\Phi \sim k^{-2}$ (Eq. (2.12)). In contrast, ST gravity produces amplified, anticorrelated with matter, dark energy perturbations on sub-Hubble scales. The ratio $\delta_\Phi/\delta_m = \delta\rho_\Phi/\delta\rho_m$ is scale independent on these scales as predicted by Eq. (2.13). In such models, where dark energy perturbations can grow on all scales, it may be shown that the speed of sound c_s vanishes [43]. The anti-correlation is evident by the spike of the dashed line, which corresponds to a change of sign of δ_Φ on sub-Hubble scales.

4. Conclusion

We have investigated in detail, analytically and numerically, the evolution of dark energy and matter linear density perturbations in Scalar-Tensor (ST) cosmologies. We have found that the evolution of dark energy perturbations in ST cosmologies is significantly different from the corresponding evolution in minimally coupled (GR) quintessence.

For natural values ($\mathcal{O}(1)$) of the ST Lagrangian parameters leading to a background expansion similar to ΛCDM (Fig. 1), dark energy density perturbations are amplified by a factor of about 10^6 compared to minimally coupled quintessence perturbations on scales less than about 100h^{-1}Mpc (Fig. 4). On sub-Hubble scales dark energy perturbations constitute a fixed fraction of about 10% compared to matter density perturbations (Fig. 4). The fixed scale independent fraction implies that the effective speed of sound for ST dark energy is

$c_{s\Phi} = 0$. The corresponding fraction for minimally coupled quintessence perturbations scales as k^{-2} and is about $\lesssim 10^{-4}\%$ (Fig. 4) corresponding to $c_{s\Phi} = 1$. Scalar-Tensor dark energy density perturbations are anti-correlated with matter linear perturbations on sub-Hubble scales (Eqs. (2.13) and Fig. 4). Thus clusters of galaxies overlap with voids of dark energy.

The evolution of scalar field perturbations on sub-Hubble scales, is scale independent and involves large oscillations (Fig. 3) induced by the amplified effective mass of the field. This mass amplification is due to the non-minimal coupling of the field to curvature, and therefore to matter (Eqs. (2.3), (2.4)). No such oscillations are present in minimally coupled quintessence perturbations which are suppressed on sub-Hubble scales and vary as k^{-2} (Eq. (2.12)). The evolution of matter density perturbations is affected by the introduction of non-minimal coupling (Fig. 4) and is amplified by about 10% in ST cosmology compared to minimally coupled quintessence and ΛCDM on sub-Hubble scales.

These results have interesting observational consequences. In particular

Dark Matter Halo Profiles: ΛCDM predicts shallow low concentration density dark matter halo profiles for clusters and galaxies in contrast to observations which indicate denser high concentration cluster haloes [49]. The amplified anti-correlated with matter dark energy perturbation profiles can lead to a modification of the predicted by ΛCDM dark matter halo profiles. In particular, the dark energy voids in clusters of galaxies can amplify locally dark matter clustering due to the local reduction of negative pressure in the region of the cluster.

Large Scale Structure Power Spectrum $P_m(k)$: For a non-minimal coupling $F_{,\Phi} = O(1)$ the ratio $(\delta_\Phi/\delta_m)^2 \sim P_\Phi(k)/P_m(k)$ is scale independent for practically all sub-Hubble scales (see Fig. 4). Thus it would be hard to identify a scale dependent signature of dark energy perturbations on the matter power spectrum for such values of $F_{,\Phi}$. For smaller values of the non-minimal coupling however, there exists a GR regime for large sub-Hubble scales where the dark energy perturbations are predicted to be scale dependent whereas on smaller scales we enter the ST regime where the ratio $P_\Phi(k)/P_m(k)$ becomes again scale independent. This transition from the GR regime on large sub-Hubble scales to the ST regime in small sub-Hubble scales may leave a trace (small glitch) on the matter power spectrum on a scale $k \simeq aHF_{,\Phi}^{-1/2}$.

In conclusion, the amplified and anti-correlated with matter, dark energy ST perturbations investigated in the present study provide a new direction of observational signatures for this class of modified gravity models.

Numerical Analysis Files: The mathematica files used for the numerical analysis and the production of the figures may be found at http://leandros.physics.uoi.gr/deperts/deperts.htm.

Acknowledgements
This work was supported by the European Research and Training Network MRTPN-CT-2006 035863-1 (UniverseNet).

References
[1] M. Hicken *et al.*, Astrophys. J. **700**, 1097 (2009) [arXiv:0901.4804 [astro-ph.CO]].
[2] M. Kowalski *et al.*, Astrophys. J. **686**, 749 (2008) [arXiv:0804.4142 [astro-ph]].
[3] E. Komatsu *et al.* [WMAP Collaboration], Astrophys. J. Suppl. **180**, 330 (2009) [arXiv:0803.0547 [astro-ph]].
[4] W. J. Percival *et al.*, arXiv:0907.1660 [astro-ph.CO].
[5] S. Basilakos and L. Perivolaropoulos, Mon. Not. Roy. Astron. Soc. **391**, 411 (2008) [arXiv:0805.0875 [astro-ph]].
[6] M. Plionis, R. Terlevich, S. Basilakos, F. Bresolin, E. Terlevich, J. Melnick and I. Georgantopoulos, J. Phys. Conf. Ser. **189**, 012032 (2009) [arXiv:0903.0131 [astro-ph.CO]].
[7] S. W. Allen, R. W. Schmidt, H. Ebeling, A. C. Fabian and L. van Speybroeck, Mon. Not. Roy. Astron. Soc. **353**, 457 (2004) [arXiv:astro-ph/0405340].
[8] L. M. Krauss and B. Chaboyer, Science **299**, 65 (2003).
[9] D. Rapetti, S. W. Allen, A. Mantz and H. Ebeling, arXiv:0911.1787 [Unknown].

[10] P. McDonald *et al.* [SDSS Collaboration], Astrophys. J. **635**, 761 (2005) [arXiv:astro-ph/0407377].
[11] S. Nesseris and L. Perivolaropoulos, Phys. Rev. D **77**, 023504 (2008) [arXiv:0710.1092 [astro-ph]].
[12] E. Hawkins *et al.*, Mon. Not. Roy. Astron. Soc. **346**, 78 (2003) [arXiv:astro-ph/0212375].
[13] L. Fu *et al.*, Astron. Astrophys. **479**, 9 (2008) [arXiv:0712.0884 [astro-ph]].
[14] J. Benjamin *et al.*, Mon. Not. Roy. Astron. Soc. **381**, 702 (2007) [arXiv:astro-ph/0703570].
[15] L. Amendola, M. Kunz and D. Sapone, JCAP **0804**, 013 (2008) [arXiv:0704.2421 [astro-ph]].
[16] P. J. E. Peebles and B. Ratra, Rev. Mod. Phys. **75**, 559 (2003) [arXiv:astro-ph/0207347];
T. Padmanabhan, Phys. Rept. **380**, 235 (2003) [arXiv:hep-th/0212290];
S. M. Carroll, Living Rev. Rel. **4**, 1 (2001) [arXiv:astro-ph/0004075];
V. Sahni, Class. Quant. Grav. **19**, 3435 (2002) [arXiv:astro-ph/0202076].
[17] E. J. Copeland, M. Sami and S. Tsujikawa, Int. J. Mod. Phys. D **15**, 1753 (2006) [arXiv:hep-th/0603057].
[18] J. Frieman, M. Turner and D. Huterer, Ann. Rev. Astron. Astrophys. **46**, 385 (2008) [arXiv:0803.0982 [astro-ph]].
[19] L. Perivolaropoulos, AIP Conf. Proc. **848**, 698 (2006) [arXiv:astro-ph/0601014].
[20] I. Zlatev, L. M. Wang and P. J. Steinhardt, Phys. Rev. Lett. **82**, 896 (1999) [arXiv:astro-ph/9807002].
[21] C. Armendariz-Picon, V. F. Mukhanov and P. J. Steinhardt, Phys. Rev. D **63**, 103510 (2001) [arXiv:astro-ph/0006373].
[22] D. Carturan and F. Finelli, Phys. Rev. D **68**, 103501 (2003) [arXiv:astro-ph/0211626].
[23] S. Capozziello, S. Nojiri and S. D. Odintsov, Phys. Lett. B **632**, 597 (2006) [arXiv:hep-th/0507182].
[24] N. Bilic, G. B. Tupper and R. D. Viollier, Phys. Lett. B **535**, 17 (2002) [arXiv:astro-ph/0111325].
[25] M. C. Bento, O. Bertolami and A. A. Sen, Phys. Rev. D **66**, 043507 (2002) [arXiv:gr-qc/0202064].
[26] S. Nojiri and S. D. Odintsov, eConf **C0602061**, 06 (2006) [Int. J. Geom. Meth. Mod. Phys. **4**, 115 (2007)] [arXiv:hep-th/0601213].
[27] L. Amendola, R. Gannouji, D. Polarski and S. Tsujikawa, Phys. Rev. D **75**, 083504 (2007) [arXiv:gr-qc/0612180].
[28] G. R. Dvali, G. Gabadadze and M. Porrati, Phys. Lett. B **485**, 208 (2000) [arXiv:hep-th/0005016].
[29] G. Esposito-Farese and D. Polarski, Phys. Rev. D **63**, 063504 (2001) [arXiv:gr-qc/0009034].
[30] J. P. Uzan, Phys. Rev. D **59**, 123510 (1999) [arXiv:gr-qc/9903004];
A. Coc, K. A. Olive, J. P. Uzan and E. Vangioni, Phys. Rev. D **73**, 083525 (2006) [arXiv:astro-ph/0601299];
A. Riazuelo and J. P. Uzan, Phys. Rev. D **66**, 023525 (2002) [arXiv:astro-ph/0107386];
C. Schimd, J. P. Uzan and A. Riazuelo, Phys. Rev. D **71**, 083512 (2005) [arXiv:astro-ph/0412120].
[31] P.G. Bergmann, Int. J. Theor. Phys. **1**, 25 (1968);
K. Nordtvedt, Astrophys. J. **161**, 1059 (1970);
R. Wagoner, Phys. Rev. **D1**, 3209 (1970).
[32] C. H. Brans, arXiv:gr-qc/0506063.
[33] H. Alnes, M. Amarzguioui and O. Gron, Phys. Rev. D **73**, 083519 (2006) [arXiv:astro-ph/0512006].
[34] S. Alexander, T. Biswas, A. Notari and D. Vaid, JCAP **0909**, 025 (2009) [arXiv:0712.0370 [astro-ph]].
[35] F. Perrotta, C. Baccigalupi and S. Matarrese, Phys. Rev. D **61**, 023507 (2000) [arXiv:astro-ph/9906066].
[36] T.M. Eubanks et al., Bull. Am. Phys. Soc., Abstract K 11.05 (1997);
J.O. Dickey et al., Science 265, 482 (1994);
J.G.Williams, X.X. Newhall, and J.O. Dickey, Phys. Rev. D 53, 6730 (1996);
C.M.Will, *Theory and Experiment in Gravitational Physics*, (Cambridge University Press, Cambridge, England, (1993).
[37] G. Esposito-Farese, AIP Conf. Proc. **736**, 35 (2004) [arXiv:gr-qc/0409081].
[38] J. P. Uzan, arXiv:0908.2243 [astro-ph.CO].
[39] K. i. Umezu, K. Ichiki and M. Yahiro, Phys. Rev. D **72**, 044010 (2005) [arXiv:astro-ph/0503578].
[40] T. Damour and B. Pichon, Phys. Rev. D **59**, 123502 (1999) [arXiv:astro-ph/9807176].
[41] J. Khoury and A. Weltman, Phys. Rev. D **69**, 044026 (2004) [arXiv:astro-ph/0309411].
[42] J. c. Hwang and H. Noh, Phys. Rev. D **71**, 063536 (2005) [arXiv:gr-qc/0412126].
[43] R. de Putter, D. Huterer and E. V. Linder, arXiv:1002.1311 [astro-ph.CO].
[44] B. Boisseau, G. Esposito-Farese, D. Polarski and A. A. Starobinsky, Phys. Rev. Lett. **85**, 2236 (2000) [arXiv:gr-qc/0001066].
[45] L. Perivolaropoulos, JCAP **0510**, 001 (2005) [arXiv:astro-ph/0504582].
[46] S. Nesseris and L. Perivolaropoulos, JCAP **0701**, 018 (2007) [arXiv:astro-ph/0610092].
[47] W. Hu, Phys. Rev. D **71**, 047301 (2005) [arXiv:astro-ph/0410680].
[48] A. Vikman, Phys. Rev. D **71**, 023515 (2005) [arXiv:astro-ph/0407107].
[49] T. J. Broadhurst, M. Takada, K. Umetsu, X. Kong, N. Arimoto, M. Chiba and T. Futamase, Astrophys. J. **619**, L143 (2005) [arXiv:astro-ph/0412192].

Galactically inertial space probes for the direct measurement of the metric expansion of the universe

Ivan Cagnani

Physics student, University of Bologna, Istituto di Fisica, via Irnerio 46, 40126, Bologna, Italy

Email: ivan@omgl.org

Abstract. Astrometric data from the future GAIA and OBSS missions will allow a more precise calculation of the local galactic circular speed, and better measurements of galactic movements relative to the CMB will be obtained by post-WMAP missions (ie Planck). Contemporary development of high specific impulse electric propulsion systems (ie VASIMIR) will enable the development of space probes able to properly compensate the galactic circular speed as well as the resulting attraction to the centre of our galaxy. The probes would appear immobile to an ideal observer fixed at the centre of the galaxy, in contrast of every other galactic object, which would appear moving according to their local galactic circular speed and their proper motions. Arranging at least three of these galactically static probes in an extended formation and measuring reciprocal distances of the probes over time with large angle laser ranges could allow a direct measurement of the metric expansion of the universe. Free-drifting laser-ranged targets released by the spacecrafts could also be used to measure and compensate solar system's induced local perturbations. For further reducing local effects and increase the accuracy of the results, the distance between the probes should be maximized and the location of the probes should be as far as possible from the Sun and any massive object (ie Jupiter, Saturn). Gravitational waves could also induce random errors but data from GW observatories like the planned LISA could be used to correct them.

1. Introduction

The interpretation of cosmic expansion depends on the interpretation of space. Do galaxies actually move physically as Mach would say[1], distancing themselves from each other at ever increasing speed, or rather, are they to be seen as pretty stationary objects embedded in a space whose metric is expanding at an accelerating pace?

Current astronomical observations cannot rule out either hypothesis and are used by proponents of both views to proof their hypothesis. New cosmological measurements are required to discriminate between these two views, and for determining the best cosmological description of the observable universe at all astronomical length scales. Particularly interesting, among the new experimental cosmology projects, are the space missions to measure directly, although necessarily locally, several cosmological parameters. Notable in this category are the LATOR[4], STEP[5], ASTROD I and Super-ASTROD[6] and SAGAS[16] missions. A common characteristic of most of these missions is the use of a triangular formation of deep space probes, often provided with powerful lasers and ion-optical clocks, and using, whenever possible, interferometric techniques, which cannot be applied to measurement of absolute proper distances but are very performing in measuring relative velocities between the probes. A similar experimental setup is applied to several gravitational wave detection

Published under licence by IOP Publishing Ltd

projects, like LISA[33], DECIGO[34] and the Big Bang Observer[35]. All of these missions can provide direct cosmological measurements[36]]37][39], and a possible dedicated future mission using the same technique and GW bandwidth (0.1Hz) could be able to measure the acceleration of expansion [38].

There have been attempts in recent years to recognize local effects of cosmological expansion in several high precision measurements within the solar system, in particular analyzing Lunar Laser Ranging data [3][7] and the Doppler radar tracking of several deep space probes, especially the Pioneer 11 probe [8][9][10][11][21]. Those conclusions have been contested by several authors, as there could be less exotic, non-cosmological explanations for both local phenomena [15][12][13][14]. In the case of the Earth-Moon system, the lunar recession rate is conventionally completely attributed to the energy released by tidal friction on both bodies. Tidal friction is difficult to measure accurately on Earth, but it is believed by most geophysicists to cause heating at the bottom of the oceans and in such way to contribute substantially to the energy budget of oceanic thermohaline circulation, also called Meridional Overturning Circulation. According to us, nothing should prevent the hypothesis that both cosmic expansion and tidal dissipation contribute to the recession motion of the Moon. According to [24], a lunar recession rate of 3.82 cm/year would implicate a total energy released by tidal friction of 2.5TW, assuming tidal friction as the only cause of lunar recession; anyway, the power dissipated by oceans as calculated from the surface is only 1.8TW. This energy difference should allow us to hypothesize that cosmic expansion could be a concurrent cause of lunar recession. A very interesting application of laser interferometric techniques has been proposed [3] as an enhancement of the current lunar laser ranging (LLR) project, in the prospective of placing active laser transponders on the Moon surface in addition to the current passive retroreflectors placed in the 1970s and used by LLR systems. The proposed active interferometric system would allow a direct very high precision determination of several cosmological parameters in the Moon-Earth-Sun system, including cosmic expansion and dark energy. According to us anyway the accuracy of such measurements could be debatable due to the complexity of Moon-Earth-Sun system. Current models contain already more than 170 parameters and still cannot fully explain millimeter-precision data currently obtained by the APOLLO LLR project [40][41].

For distances much greater than the Earth-Moon separation, laser interferometry cannot be applied because laser radiation would lose coherence before arriving to destination. Since the coherence length of the laser source is

$$L = \frac{2\ln(2)}{\pi n}\frac{\lambda^2}{\Delta\lambda} \tag{1.1}$$

regardless the choice of wavelength, λ and the spectral bandwidth Δλ obtainable from instrumentation, a different type of tracking must be used. While radio frequencies Doppler radar tracking was the technique of choice in the last century for deep space probes, this technique is impractical for probe to probe tacking at distances of 10AU or more. For example, the antenna installed on the probe cannot match a large radio-telescope on Earth. This leaves mission designers with the need of using a system more similar to current LLR, but using active laser transponders instead of retroreflectors. For very long distances, like Earth-Mars or more, it is convenient to use to transponders in asynchronous mode.

As illustrated in figure 1, Asynchronous Laser Transponders works emitting signals at a regular frequency instead of waiting for an answer signal to the original communication. This is justified by the time it takes light to cover distances of several AU or more.

The main requirements of this ranging and communication technique are sufficient laser power, a reflecting telescope with an aperture proportional to the distance to be ranged but never impractically large, and most importantly a very accurate clock for keeping the pulsing extremely regular in time.

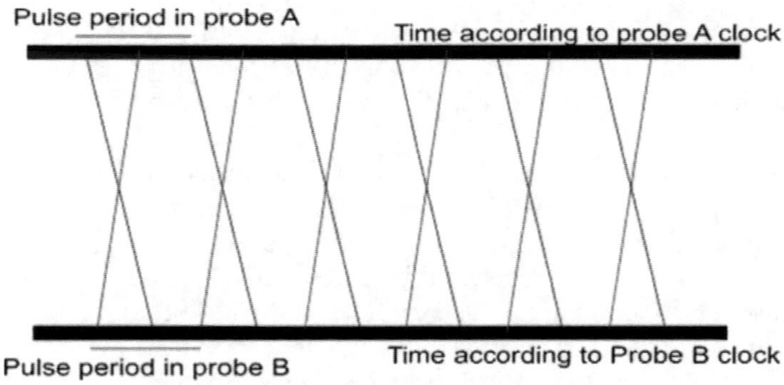

Figure 1 Asynchronous laser transponders emit periodic pulses. The emission time of the pulse from probe A is independent of the reception of a signal from probe B. The pulses can contain digital information so the transponders are both ranging and communication devices.

The telescope is used both for emission and reception of signals and is fundamental in transmission for collimating and directing the laser signal. Asynchronous Laser Transponders are the ranging and communication system adopted in our proposed mission.

Our proposed mission, a direct measurement of metric expansion by measuring the time variation of the area covered by 3 probes in a special dynamical state, would help to understand the fabric of spacetime in our solar system in a novel way, and could be determinant to falsify or verify Mach's hypothesis about the nature of space ad its relationship to matter and inertia.

The simple physics which describes the problem of the local expansion between two objects originally disconnected from the Hubble flow, called the problem of the "tethered galaxy", has been very well illustrated in [15], and we highly recommend the reading of that thesis as the best theoretical introduction to the issue. Another very detailed study of the problem [2], emphasizes the difference between dynamical and kinematic effects of cosmic expansion on a satellite or a probe in the solar system, and contains very noteworthy considerations about the Pioneer Anomaly and similar Doppler tracking revealed effects.

Theoretical work on the subject is far from being exhaustive anyway, and there are still disagreements, especially about the amount of possible effects on objects within the solar system.

In the case of Pioneer 11, the numerical similitude between the anomalous Sun-ward acceleration of the probe and the product of the Hubble constant with the speed of light got many people thinking:

$$8.6 \pm 1.34 \cdot 10^{-10}\, m\, s^{-2} \approx 7 \cdot 10^{-10}\, m\, s^{-2} \tag{1.2}$$

anyway the method itself used to determine the kinematics of the probe makes problematic the analysis of the violet-shift anomaly [2]. Nonetheless the Pioneer effect is interesting enough to warrant a space mission to study it. Several proposals have been made to space agencies, from dedicated missions [43][44] to the adaptation of planetary missions using the probe itself or a small dedicated probe released by the main probe [42].

Laser ranging is the best methodology for long range distance determination for general relativity tests and gravitational measurements, anyway a few authors expressed theoretical doubts about the invariance of the speed of light (for recent examples [25] [26] [27]). Any directional or temporal variation in the speed of light would be really complicate space experiments using lasers at long distances so any new investigation in that area would be welcome to remove any minor doubt before attempting very ambitious deep space laser measuring missions.

2. The mission proposal

After launch in a hyperbolic solar orbit, each probe's nuclear reactor is activated and the propulsion system powered up. The probes will reach a distance of 25AU from the Sun. They will create an equilateral triangle formation on a plane parallel to the mean ecliptic plane distanced by 7AU in the solar zenith direction and the center of the triangle will be vertically above the Sun. The projected Sun-probe distance on the mean ecliptic plane is 24AU. After this initial configuration has been achieved, a galactic stabilization procedure will start. The aim of the procedure is to cancel any galactic movement inherited by the probes as parts of the solar system. The stabilized probes will then vary their galactic position and their reciprocal distances only because of metric expansion (gravitational perturbations will still be possible from the Sun, the gaseous planets, comets passing nearby, as well as mechanical perturbations from solar wind, radiation pressure, but they will be assessed by releasing small laser retroreflecting test masses periodically from the from the main probes, and minor position corrections will be performed by specialized high accuracy ion thrusters).

The probe constellation ready for the data collection phase would appear like in figure 2:

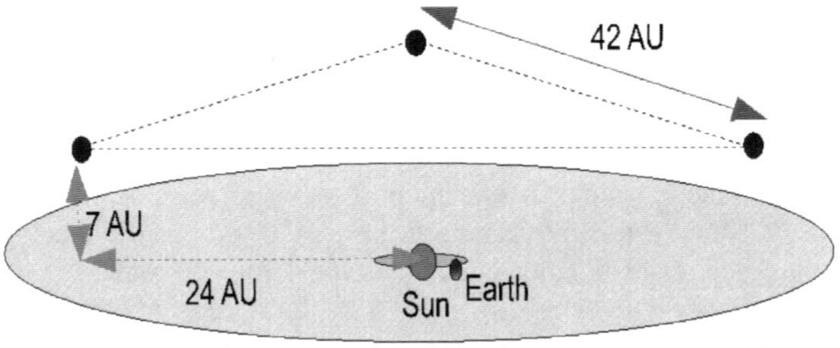

Figure 2: Not-to-scale illustration of the constellation of the three probes at the beginning of their scientific data acquisition phase. The Sun will not keep its position at the center of the triangle during the experiment as the probes are galactically stabilized but not stabilized with the Sun.

The reciprocal distance of each couple of probes is measured by the laser ranging system using asynchronous laser transponders repeating a 12s optical signals sequence transmitted continuously by each probe to the other.

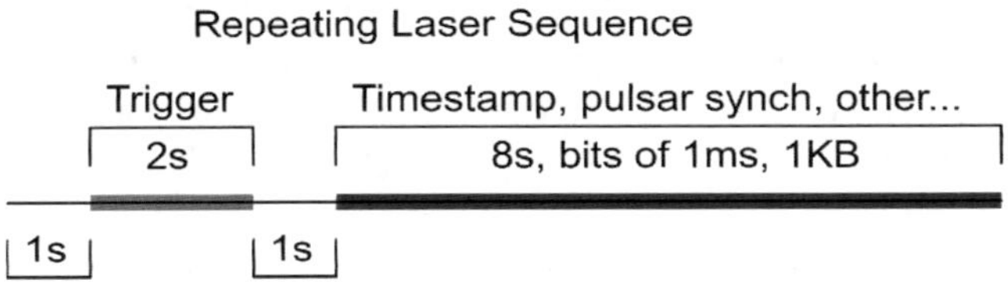

Figure 3: Sequence transmitted from each probe to the other two probes using its own asynchronous active laser transponders. The duration of the trigger must be as close to exactly 2s as feasible.

As illustrated in figure 3, the signal sequence is actually a mixed digital transmission containing a 2s continuous trigger and a 8s digital 1KB string (bits having a period of 1ms each) with the function of a timestamp. Both the trigger and the timestamp are followed by a 1s interval making the total duration of the sequence 12s. The trigger duration must be as close to exactly 2s as the probe clock and system allows, which ideally should mean a max error of $2 \cdot 10^{-18}$ s. The digital string following by 1s the trigger must contain the reading of the emitting probe's clock at the end of the trigger pulse. For redundancy the value of the clock reading is repeated 3 times in the 1KB string, and of course it is followed each time by an error correction code. Finally the string contains information about the pulsar used for clock synchronization by the emitter probe and the particular feature of the pulsar signal used for resetting the clock. The clock of each probe is reset and pulsar synchronized every 10^5 seconds. Of course the new synchronization information is immediately shared in the following pulse.

The measurement of the reciprocal speed between two probes is based on the difference from the expected time of arrival of the end of the next trigger pulse using the onboard clock to measure 12s with $2 \cdot 10^{-18}$ precision from the reception of the end of the last trigger. Any displacement between the two probes in the 12s separating two trigger emissions will generate a small delay (supposing metric expansion dominates) or a really minimal advance (supposing the Pioneer Anomaly dominates). Each probe's computer is aware that its clock can be affected by gravitational and mechanical perturbations slightly different than those affecting the clock of the emitting probe. Anyway, analyzing the timestamps, and arrival times of the end of the triggers, the receiver's computer can easily distinguish effects induced on the emitter's clock from the effect due to the variation of the reciprocal distances. In other words, it can distinguish a variation of the comoving positions or a perturbation of the spacetime fabric close to the emitter, from the effect that it must closely observe: the variation of the proper distance between the probes given unvarying comoving distance and no gravitational perturbations. It is possible to conceive the transmission of "I think you are under this disturbing effect..." messages from a probe to the other instead of the third redundant repetition of the trigger end emission time in the exchanged digital string. Of course, for error correction all internal delay times must be monitored by the emitter and their eventual variation must be communicated to the receiver too.

Given the standing lack of certain knowledge about the causes of the Pioneer Anomaly, the local effect of expansions, the nature of dark energy and its local effects too, a realistic simulation of the experiment at the high precision level of the measurement is not possible at the moment. Even if a perfect mathematical model considering all possible effects and their causes were developed, the uncertainty in cosmological and astrophysical parameters as actually known would make the model unable to simulate realistic results. This situation could change in the following years, as the precision in cosmological parameters determination is certainly going to increase significantly.

In an ideal universe with H_0 =70Km/s/Mpc the expected recession speed between the probes at fixed comoving distance of 42AU should be $1 \cdot 10^{-5}$ m/s, this means that the proper inter-probe distance will increase of 0.12 mm between two successive laser ranging pulse emissions (each transmission cycle lasting exactly 12s). A distance of 0.12 mm is covered in $4 \cdot 10^{-13}$ s at the speed of light in vacuum. With a nominal clock accuracy of $1 \cdot 10^{-18}$ and ideal measurement uncertainty of $2 \cdot 10^{-18}$, which could mean a real world measurement uncertainty of $1 \cdot 10^{-17}$ it seems realistic to be able to measure the 0.12 mm variation in length, and extract this simple displacement from the inevitable noise of unwanted effects.

Apart from the increase of the proper distance between the probes, and the following measurable increase in the timing of successive pulses, it must be considered the possibility of an effect of frequency shift in the laser signal. The Pioneer Anomaly has been measured, in the radio spectrum, as a $(5.99 \pm 0.01) \cdot 10^{-9}$ Hz s^{-1} effect. To avoid unnecessary complexity, let's suppose the probes are on the mean ecliptic plane. Since it takes the laser signal almost $2.1 \cdot 10^{4}$ s of time to cross the 42AU of distance between two probes, a frequency shift of $1.255 \cdot 10^{-4}$ Hz can be expected. Unfortunately a tenth of millihertz frequency shift in a laser is impossible to detect without an interferometric setup, and the coherence time of existing and conceivable lasers makes it impossible to use interferometry for distances of tens of AU. If such finely tunable laser could be built, the Pioneer effect could be measured directly in the proposed setup by recording the exact frequency of the trigger pulse at emission and communicating it to the receiver by adding it to the digital message following the trigger pulse, along with the timestamp and pulsar calibration information.

If the Pioneer Anomaly is a real acceleration and has a real effect on the probes, it should act very differently than the expansion. For two reasons: firstly expansion is not a force [15]. Unbound objects tend to follow naturally the Hubble flow, and previously tethered object getting untethered perceive no acceleration when joining the flow. Small acceleration effects may indeed be perceived but only because cosmic expansion is accelerating, possibly due to the presence dark energy. The Pioneer Anomaly appears instead to be a force but of course there is a need for more experimental data to be certain about that. The second difference is the direction, which is almost opposite in our setup. The Pioneer Anomaly appears an acceleration directed toward the Sun, the metric expansion should move the probes away from the an ideal object placed at the centre of the triangle and 7AU above the Sun, so while the two effects are not completely opposing each other, they can interact negatively. In this case, the proposed experimental setup should be able to separate the effects and measure them, anyway, as previously demonstrated, if the Pioneer Anomaly is not a real acceleration but only a cause of frequency shift of electromagnetic radiation, this will not be recorded by the proposed experimental setup and there will be not be a negative interaction to the proper recession measurement.

2.1. Galactic stabilization

Cosmic expansion should induce a subtle effect to the probes formation, so subtle it could be impossible to recognize in an environment full of disturbing interactions like that of an Earth orbit for example. As we can see from maps of the Cosmic Microwave Background radiation, the Earth motion with respect to the CMB rest frame (the closest we can have to an universal reference frame) is a combination of its rotation, its solar orbital motion, the solar motion in the galaxy (the Local Standard of Rest and the Sun's peculiar motion with respect to the LSR), then of course there are proper motions of the Milky Way in the local group and of motions of the local group.

The best kind of stabilization for each probe would be with respect to the CMB rest frame, as that would eliminate all disturbing motions and would put the probes in the same inertial system. There are several technological difficulties with this approach anyway (discussed later), which makes a less extreme stabilization worthy of consideration. In particular, galactic stabilization is the highest stabilization level that we are able to reach with technology already in development for other purposes.

Galactic stabilization means stabilization with respect to an ideal observer at the centre of the galaxy, non-corotating with the galaxy. This observer would see the constellation of 3 probes as stationary objects (shown in figure 4) in contrast of any another galactic object all orbiting around the galactic centre (including the hypervelocity stars, which while gravitationally unbounded to the Galaxy due to speeds largely greater than their local galactic escape speed, are still heavily gravitationally influenced by the Galaxy: they just follow hyperbolic orbits).

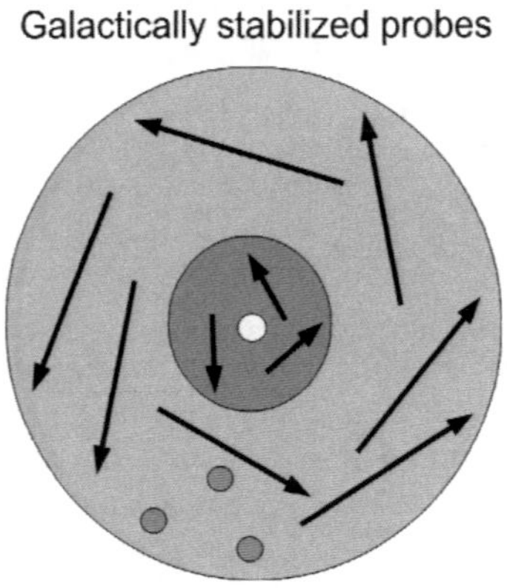

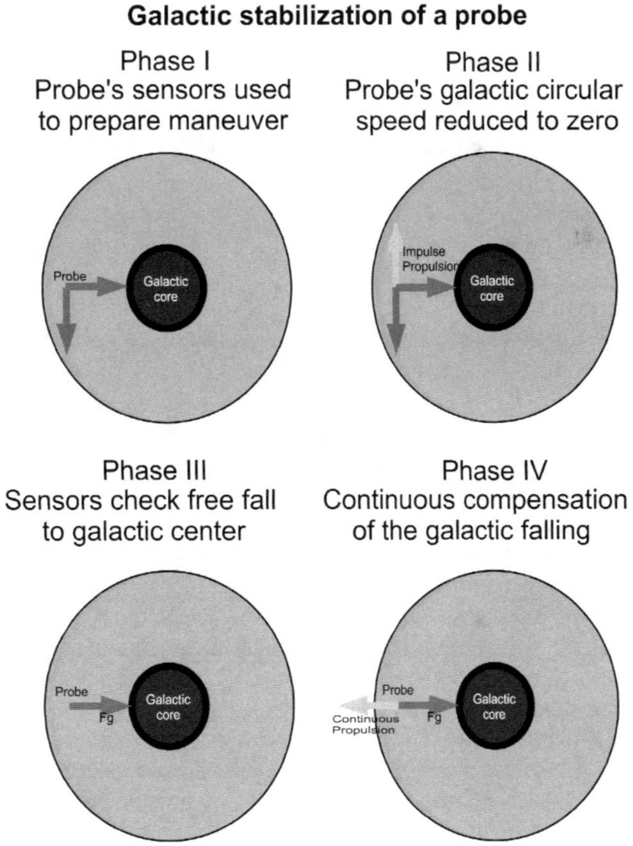

Figure 4 (up) : Galactically stabilized probes are seen from an observer non-corotating with the galaxy as the only "fixed" points with respect to the galactic center, all other objects would be moving according to their peculiar and radial velocity.

Figure 5 (right): The galactic stabilization maneuver is a 4 phase process. The first and 3rd phases use mostly the navigation sensors, the 2nd phase use high-thrust impulsive propulsion, the 4rd phase uses continuous propulsion.

Given the complexity of the Galaxy stellar orbits are not really circular, but the idealized circular orbit of the Sun, the Local Standard of Rest, is fundamental for any measurement in galactic coordinates, including those necessary to calculate the galactic stabilization manoeuvre. Knowing the deviations of the peculiar motion of the Sun from this ideal orbit is equally important. Any probe launched from the Earth is already orbiting the centre of the Galaxy as any other solar system object, and a probe launched at escape speed from the solar system will continue to follow this motion unless it can reach the same linear speed of the instantaneous tangential velocity of the Sun in its galactic orbit (estimated between 220Km/s and 260Km/s [47]) but in the exactly opposite direction. In that case the probe would find itself at zero speed in its own galactic orbit. The result of this manoeuvre of course would be the beginning of an accelerated motion towards the centre of the Galaxy (freefalling to it).

To keep the probe immobile with respect to the non-corotating galactic reference frame instead, a continuous equal and opposite acceleration would be required. This would require a very finely tunable continuous propulsion system. The galactic stabilization duration would then be limited by the amount of propellant in the engine, among other factors. After propellant depletion, the probes could attempt to continue measurements as long as possible to measure destabilized behavior and compare with stabilized data.

Both propulsion phases of the galactic stabilization maneuver (see figure 5) can be performed by a high power, variable specific impulse, electric ion engine like VASIMR [46], or better, an array of VASIMRs. Since the maximum velocity that a space vehicle can reach is bounded (although not limited) by the maximum exhaust speed of propellant, an engine like VASIMR, able to reach exhaust speeds of 300Km/s, is an ideal choice for this proposed application. Of course, using VASIMR at max power and max specific impulse mode requires a huge amount of electric power that only a space-rated nuclear generator is able to provide.

Our best LSR determinations so far come from elaborations with new methods of the Hipparcos database [19] and from the ongoing RAVE project [20] but they have uncertainties that would not allow the design of a galactic stabilization maneuver at this moment. High precision data about the Local Standard of Rest will probably only become available after the GAIA [48] astrometric mission has been launched, completed, and its results properly elaborated.

Each probe will also use its own pulsar positioning and navigational capability to test the level of galactic stabilization achieved before starting the cosmic expansion measurement along with the other probes. In particular, a precise determination of a probe's position and momentum relative to distant X-ray pulsars, to the Sun, and to a laser transponder satellite in Earth orbit (used for communication) will be required before and after the galactic circular velocity annulment maneuver, for calibration and correction.

2.2. Technology of the probes

The probes will be able to measure reciprocal distances with the maximum possible precision and accuracy. This will require extremely good clocks onboard the probes, like quantum-logic optical clocks, currently still in advanced development [17] but already used to detect subtle GR effects [18]. The clocks must be redundant. Each transponder in a probe should have 1 active optical clock and 2 inactive backup clocks, so at least 9 clocks are required in a probe. The clocks will be synchronized simultaneously to the same pulsar. Particular features of the pulsar signal will be used for synchronization as they can be recognized by all probes. These minor pulsar events will act as a trigger and all clocks in all probes will be synchronized at detection. The regular pulsar frequencies instead will be used by clocks to check internal accuracy and eventual delays.

Each transponder will use its own clock because calculating and measuring the internal delay of a circuit is crucial, so each circuit must be as simple as possible. Optical fibers should replace wires as much as possible in the circuits to reduce risks of interference and because light signals are faster. Synchronization with pulsars is common among all clocks but each transponder will compare its own active clock to the other transponders' active clocks, and if differences are significant, the clock will gets substituted by a backup unit within the transponder's circuit.

The first two transponders must be pointed at all times to the other probes and this prevents them to be used for communication with Earth. The third transponder, whose main function is laser tracking of the retroreflecting test masses released by each probe at regular intervals during the long measurement phase, will be charged if possible to look at Earth every fixed interval of time (for example every 24 Earth-hours) to send the latest measurement data to a satellite in Earth orbit and eventually receive laser messages). It could be reasonable to install a radio antenna as backup for data downloading, and for reception of emergency transmissions from Earth when the tracking transponder is not pointed to Earth. The probes anyway would be largely autonomous and regular communication with Earth will be important for the downloading of experimental data but never critical for the survival of the probes.

Position and momentum data will be determined by each probe using multiple sensors. The main instrument will be composed by redundant X-ray pulsar sensors (as in figure 6) combined with traditional startrackers and UV sensors. Pulsar sensors will also be fundamental for calibration and synchronization of the clocks of the probes, using asynchronous common view synchronization [17] between the probes.

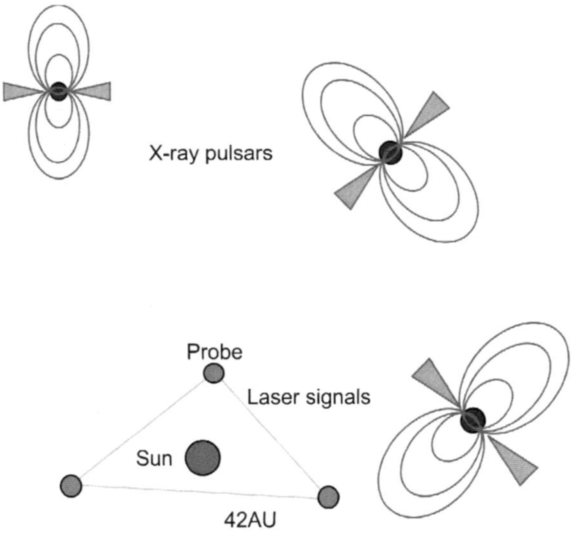

Figure 6: The probes have several types of navigation sensors but their most important instrument will be the X-ray pulsar sensor as the regular emission of millisecond pulsars is extremely useful also for clock synchronization between the probes. Especially useful for positioning and clock calibration will be events of detection of non-periodic imperfections in the pulse of those stars and the exchange of the arrival times of these events, as recorded by onboard clocks, between the probes.

The accuracy of current clocks has reached $8.6 \cdot 10^{-18}$ in the case of the quantum-logic ion clocks [18] but they need large improvements in reliability and autonomous operation. So while an improvement of accuracy to $1 \cdot 10^{-18}$ in the next few years should be expected, more years will be required to have very reliable test equipment in small space-rated packages.

The Variable Specific Impulse Magnetoplasma Rocket (VASIMR) [14] is currently the only existing electrical propulsion system which, coupled with developing space-rated nuclear power generators, is able to perform the impulsive and continuous phases of the galactic stabilization manoeuvre. A 200kW version of the engine is going to be installed soon at the International Space Station. More powerful versions and arrays of them are supposed to be used for fast human interplanetary travel. An unmanned mission like the one proposed here could be the best test of their most extreme performance and of their reliability and safety in a very demanding long duration mission.

Another important consideration is that while the propulsion system must be able to accelerate the probe to about 250Km/s it must also be able to be very accurate in doing so. Feedback from inertial accelerometers will certainly help the probe's computer to build a sophisticated model anticipating the engine behaviour and predicting its performance in many different situations and requirements. The models generated from the probes would be very valuable to designers working on VASIMR spacecrafts on Earth.

Each probe contains two main telescopes (transponders) at an angle of 60 degrees but with a modest freedom of pointing in all axes. The third telescope will necessarily be able to point laser retroreflecting targets in almost all directions.

All telescopes are collimating reflecting telescopes working in assembly with a laser, a fast specialized photosensor and a clock (the main components of an Asynchronous Laser Transponder). The required diameter for the telescopes is dependent on the power and characteristics of the laser emitter chosen (probably a fiber laser in the visible band) and of the photosensor but a diameter of 0.5 m should be considered the upper limit.

The navigation system will use X-ray pulsar based techniques that have been studied for decades. Their implementation has been anyway delayed by many difficulties, more conceptual than technical anyway. In particular, many problems in their implementation are associated with the need to add special and general relativistic corrections to a classical (Newtonian) reference frame. A new coordinate system called Coll relativistic positioning system has been recently proposed for this navigation technology with the aim of eliminating the need of complicated post-Newtonian and post-post-Newtonian corrections in a Newtonian reference frame [15].

For error correction of the galactic stabilization manoeuvre, several spherical, spin-stabilized retroreflecting test masses similar to the ones already tested in space by a Russia-NASA collaboration [45] will be released by the probe at a rate of 1 every three Earth-months. They will be followed by the dedicated third transponder, and their tracking will allow the onboard probe a mapping of the local gravitational perturbations. They will be released with a really small and well controlled acceleration and a spin of 50 to 100 Hz.

About the reactor, the only discriminating quality for a reactor in space is the power to mass ratio. Research on new reactors for human space travel applications has already started in America and Russia and important advancements should be expected in field in the next ten years.

2.3. Sources of errors

There are many possible sources of error, mostly related to the fact the Galaxy is so complex that we still do not have a model describing it (or any galaxy) with the necessary accuracy. The real Galaxy is not a disk, so all features (arms, bulge, satellite galaxies...) must be considered. Even considering an idealized circular model of the solar motion, recent literature still shows too much disagreement over the correct values of the solar circular speed.

Another error source is certainly the impossibility to predict exactly the local environment of the probe for simulating sufficiently well perturbations on its motion in the mission design phase. Gravitational waves must also be considered as a source of possible errors, but they will be better known and their induced errors eliminated thanks to the future space antennas for gravitational waves like LISA.

As already mentioned, the propulsion system itself, if not extremely well monitored and finely adjusted in an interactive way, can be a major source of error. The same consideration applies to the power generation and thermal control system. The high energy requirement of the engines can be covered only by an onboard nuclear reactor which can produce unwanted disturbances due to thermal emissions and its radiation pressure, but this induced disturbance can be calculated and greatly reduced adopting a counterbalancing symmetrical geometry for radiators, and actively balancing thermal loads of said radiators.

3. Enhanced mission with CMB rest frame stabilization and dark energy measurement

Galactic stabilization is not the finest imaginable simplification of the physical system with 3 laser ranging probes considered. It was just proposed because realizable with already existing, although not mature, technologies (like the VASIMR engines for propulsion and solid core fission reactors for power production).

In 1977 a dipole in the CMB measurement from a plane was discovered. This led to a wonderful hypothesis: the universe does have a sort of absolute reference frame and the velocity of any object in the universe is measurable with respect to that frame.

The galactocentric reference frame has a peculiar velocity with respect to the CMB, so that any cosmological measurement in galactocentric coordinates needs to be corrected using the best estimates available of that peculiar velocity. Unfortunately this adds further uncertainties to the uncertainties due to galactic stabilization accuracy. If a direct measurement of the deceleration parameter is to be attempted, an extremely high level of accuracy is required which simply cannot be achieved using galactic stabilization due to the intrinsic uncertainties.

The CMB rest frame is the ideal reference frame for direct cosmological measurement, anyway deciding to employ it for 3 laser ranging probes in deep space requires a deep modification to the hardware of the probes, adding CMB sensors and preferably a more powerful nuclear propulsion system, like a fission-fragment rocket [29]. To reach 0 m s^{-1} of relative motion with respect to the CMB frame, probes need a speed of 369.0 ± 0.9 km s^{-1} relative to LSR according to the 5-years WMAP CMB dipole measurement [49]. This latter delta-V requirement makes the mission challenging for a VASIMR propelled spacecraft but rather easy for a probe propelled by a fission-fragment rocket, a technology object of study for extreme high speed unmanned exploration missions.

3.1. CMB rest frame stabilized probes

If a probe velocity is such that the cosmic background radiation shows no dipole, it means the probe is immobile in the CMB rest frame, the only universal reference frame. This would ensure that 3 CMB stabilized probes are exactly in the same reference frame independently of their galactocentric position, and that they are immobile (as comoving distances) in that frame. Such configuration would make the distance measurement between the probes a simple to interpret and very significant measurement. It would be the most epistemologically correct way of performing the experiment. If all proper movements of each probe with respect to the CMB rest frame could be stabilized only local perturbations (measured with the free drifting retroreflecting targets) and expansion could move them.
While it is possible since 1977 to use the cosmic microwave radiation as a reference frame for astrophysical calculations, it's still impossible to measure the CMB dipole and anisotropies rapidly enough and precisely enough for real-time navigational purposes. Furthermore, a highly sensitive CMB sensor needs to be helium cooled. This would put another limit on the duration of the necessarily long measurement mission. Maybe, if only a crude dipole measurement can be sufficient (to check the continuous absence of a dipole in the CMB thanks to the work of the propulsion system) it could be possible to avoid using helium-cooling.

4. Conclusion

Studying local manifestations of expansion is studying the nature of space itself. Due to the uncertainties in the determination of the Hubble's constant value for this epoch (very recent results fluctuate from 70 to 74 Km/s/Mpc), a local, high precision determination could contribute to improvements in cosmological models too. The most interesting contemporary experimentation proposal is the use of Celestial Optical Interferometry to increase the relative uncertainty in determination of the Earth-Moon distance to 10^{-16}. Anyway, the Earth-Moon-Sun system is far too complex to obtain a clear and indisputable measurement of cosmological expansion on local scale. A much more simplified system is required, and the first requirement of this system is isolation. While a very high speed object within the Galaxy can be gravitationally unbounded (like hypervelocity stars) no physical system can be completely isolated from the galactic environment. A comparison with the CMB rest frame can give us an idea of its dynamical state. Galactic stabilization is a partial isolation condition that can be quite useful (if the stabilization manoeuvre is correctly calculated and performed), anyway the ultimate direct measurement of cosmic expansion and dark energy should involve a stabilization of the comoving coordinates of the probes with respect to the CMB rest frame.

References

[1] Rindler W, *Essential Relativity, Special, General and Cosmological*, Second Edition, Springer (1977). See in particular section 1.15

[2] Carrera, Giulini, Influence of global cosmological expansion on local dynamics and kinematics, *Rev. Mod. Phys.* **82**, 169–208 (2010)

[3] Unnikrishnan C S, Gillies G T, Celestial optical interferometry: a new tool for ultrahigh precision studies in gravity, *Int. J. Mod. Phys. D*, Vol. **17**, Nos. 3 & 4 (2008) 617–626

[4] Turyshev S G et Al., The Laser Astrometric Test of Relativity Mission, *Class. Quantum Grav.* 21 2773 (2004)

[5] Sumner T J, The STEP and GAUGE Missions, *Space Science Reviews* Vol. **148**, Num. 1-4, 475-487 (2009), Springer

[6] Ni W, Super-ASTROD: Probing primordial gravitational waves and mapping the outer solar system, *Class. Quantum Grav.* **26** 075021 (2009)

[7] Dumin Y V, On a probable manifestation of Hubble expansion at the local scales, as inferred from LLR data, *Preprint* astro-ph/0203151

[8] Rosales J, Sanchez-Gomez J, The "Pioneer effect" as a manifestation of the cosmic expansion in the solar system, *Preprint* gr-qc/9810085v3

[9] Nottale L, The Pioneer anomalous acceleration: a measurement of the cosmological constant at the scale of the solar system, *Preprint* gr-qc/0307042v1

[10] Fahr H, Siewert M, Does PIONEER measure local spacetime expansion?, *Preprint* gr-qc/0610034v2

[11] Oliveira F J 2006, Hubble law may explain the anomalous acceleration of the pioneer 10 & 11 spacecraft, *Preprint* gr-qc/0610029

[12] Williams J G, Turyshev S G et al. Reply to the Comment by Dumin on "Progress in Lunar Laser Ranging Tests of Relativistic Gravity", *Preprint* gr-qc/0612171

[13] Chodorowski M J, A direct consequence of the expansion of space?, *Monthly Notices of the Royal Astronomical Society*, **378**: 239–244

[14] Turyshev S G, Toth V T, The Pioneer Anomaly, *Living Rev. Relativity* **13**, 4 (2010)

[15] Davis T M, Fundamental Aspects of the Expansion of the Universe and Cosmic Horizons, *Preprint* astro-ph/0402278v1

[16] Wolf P et Al., Quantum physics exploring gravity in the outer solar system: the SAGAS project, *Experimental Astronomy* Volume **23**, Number 2, 651-687

[17] Chou C W et al., Frequency Comparison of Two High-Accuracy Al+ Optical Clocks, *Phys. Rev. Lett.* **104**, 070802 (2010)

[18] Chou C W et al., Optical Clocks and Relativity, *Science* 24 September 2010, Vol. **329**. no. 5999, pp. 1630 – 1633

[19] Schönrich R, Binney J, Dehnen W, Local Kinematics and the Local Standard of Rest (2010), Local kinematics and the local standard of rest. *Monthly Notices of the Royal Astronomical Society*, **403**: 1829–1833. doi: 10.1111/j.1365-2966.2010.16253.x

[20] Coskunoglu B et al., Local Stellar Kinematics from RAVE Data: I. Local Standard of Rest, *Preprint* 1011.1188v1 [astro-ph.GA]

[21] Nieto M M, Turyshev S G, Anderson J D, 2005, "Directly measured limit on the interplanetary matter density from Pioneer 10 and 11," *Physics Letters B* **613**, 11–19.

[22] Seto N et al., Possibility of Direct Measurement of the Acceleration of the Universe using 0.1Hz Band Laser Interferometer Gravitational Wave Antenna in Space, *Phys. Rev. Lett.* **87**, 221103 (2001)

[23] Clark R, Sheldon R, Dusty Plasma Based Fission Fragment Nuclear Reactor, American Institute of Aeronautics and Astronautics 15 April 2007

[24] Chan I, Parametric Subharmonic Instability and the β-effect, University of British Columbia, http://www.iam.ubc.ca/theses/IanChan/IanChan_MSc_Thesis.pdf

[25] Riofrio L, GM=tc3 Cosmology and the Moon, abstract: BAPS.2010.TSF.FA2.9, http://meetings.aps.org/link/BAPS.2010.TSF.FA2.9
[26] Cahill R T, Lunar Laser-Ranging Detection of Light-Speed Anisotropy and Gravitational Waves, *Progress in Physics* 2010:**2**.31-2.35, *Preprint* gen-ph/1001.2358v2 [physics.gen-ph]
[27] Bruchholz U E, Lunar Laser Ranging Test of the Invariance of c: a Correction, *Progress in Physics* 2010:**4**.30-4.31
[28] Clark R, Sheldon R, Dusty Plasma Based Fission Fragment Nuclear Reactor, American Institute of Aeronautics and Astronautics, 15 April 2007.
[29] Markwardt C B, 2002, "Independent Confirmation of the Pioneer 10 Anomalous Acceleration," , *Preprint* gr-qc/0208046.
[30] Naoki et al., Possibility of Direct Measurement of the Acceleration of the Universe Using 0.1 Hz Band Laser Interferometer Gravitational Wave Antenna in Space, *Phys. Rev. Lett.* **87**, 221103 (2001)
[31] Lämmerzahl C et al., OPTIS: a satellite-based test of special and general relativity, 2001 *Class. Quantum Grav.* **18** 2499
[32] Lämmerzahl et al., Is the Physics Within the Solar System Really Understood?, Lasers, Clocks and Drag-Free Control, *Astrophysics and Space Science Library*, 2008, Volume **349**, 1, 75-101, DOI: 10.1007/978-3-540-34377-6_3
[33] Danzmann K, Rüdiger A, LISA technology—concept, status, prospects, *Classical and Quantum Gravity*, 2003, **20** S1. DOI: 10.1088/0264-9381/20/10/301
[34] Sato S et al., DECIGO: The Japanese space gravitational wave antenna, 2009 *J. Phys.: Conf. Ser.* **154** 012040. doi: 10.1088/1742-6596/154/1/012040
[35] Harry G M et al., Laser interferometry for the Big Bang Observer, *Class. Quantum Grav.* **23** 4887, doi: 10.1088/0264-9381/23/15/008
[36] Nakayama K et al., Space laser interferometers can determine the thermal history of the early Universe, *Phys. Rev. D* **77**, 2008, DOI: 10.1103/PhysRevD.77.124001
[37] Krauss L M et al., Primordial Gravitational Waves and Cosmology, *Science* 21 May 2010: Vol. **328** no. 5981 pp. 989-992 DOI: 10.1126/science.1179541
[38] Seto N et al., Possibility of Direct Measurement of the Acceleration of the Universe Using 0.1 Hz Band Laser Interferometer Gravitational Wave Antenna in Space, *Phys. Rev. Lett.* **87**, 221103 (2001), DOI: 10.1103/PhysRevLett.87.221103
[39] Takahashi R, Nakamura T. Determination of the equation of the state of the Universe using ~ 0.1 Hz Gravitational Wave Detectors, *Prog. Theor. Phys.* **113** (2005) 63-71, DOI: 10.1143/PTP.113.63, arXiv:astro-ph/0408547v2
[40] Merkowitz S M, Tests of Gravity Using Lunar Laser Ranging, *Living Rev. Relativity*, **13**, (2010), 7
[41] Williams J G, Turyshev S G and Murphy Jr T W, "Improving LLR Tests Of Gravitational Theory", *Int. J. Mod. Phys. D*, **13**, 567–582, (2004)
[42] Rathke A, Izzo D, Options for a Nondedicated Mission to Test the Pioneer Anomaly, *Journal of Spacecraft and Rockets*, Vol. **43**, No. 4, July-August 2006
[43] Bertolami O, Páramos J, A mission to test the Pioneer anomaly: estimating the main scientific effects, *Int.J.Mod.Phys.D* **16**:1611-1623,2007, DOI: 10.1142/S0218271807011000, arXiv:gr-qc/0702149v2
[44] Dittus H, A mission to explore the Pioneer anomaly, *ESA Spec.Publ.* **588** (2005) 3-10, LA-UR-05-4907, *Preprint* gr-qc/0506139v1
[45] Burmistrov VB et al., Spherical Retroreflector with an Extremely Small Target Error: International Experiment in Space, http://cddis.nasa.gov/lw13/docs/papers/target_vasiliev_1m.pdf
[46] Díaz FRC, An overview of the VASIMR engine: High power space propulsion with RF plasma generation and heating, *AIP Conference Proceedings*, 2001
[47] McMillan P J, Binney J J, The uncertainty in Galactic parameters, *MNRAS*, **402**, 934, 2010,

(*Preprint* astro-ph.GA/0907.4685v2)

[48] Brown A G A, Learning about Galactic structure with Gaia astrometry, *Preprint* astro-ph.GA/0810.5437, 2008

[49] Hinshaw G et al., Five-Year Wilkinson Microwave Anisotropy Probe (WMAP1) Observations: Data Processing, Sky Maps, & Basic Results, Astrophysical Journal Supplement Series, **180**, 225-245 (*Preprint* astro-ph/0803.0570v2)

Journal of Physics: Conference Series **283** (2011) 012008 doi:10.1088/1742-6596/283/1/012008

Unlensing the CMB in real space: a new approach to extract the weak lensing convergence

C.S. Carvalho[1,2,3]

[1] Instituto de Plasmas e Fusão Nuclear, Instituto Superior Técnico Avenida Rovisco Pais, 1, 1049-001 Lisboa, Portugal

[2] Academy of Athens, Research Center for Astronomy and Applied Mathematics, Soranou Efessiou 4, 11-527, Athens, Greece

[3] Astrophysics and Cosmology Research Unit, School of Mathematical Sciences, University of KwaZulu-Natal, Durban, 4000, South Africa

E-mail: carvalho.c@gmail.com

Abstract. We present a new estimator defined in real space to extract the weak lensing convergence from temperature maps of the Cosmic microwave background radiation. This estimator is build upon the minimum variance estimator defined in harmonic space. Despite being by construction suboptimal, its implementation proves less sensitive to the experimental noise and to the bias of the mode coupling derived from the finite size of the map. [4]

1. Introduction

The Cosmic microwave background (CMB) radiation propagates almost unperturbed after recombination, so that what it contains are mainly the fluctuations at the last-scattering surface $z \sim 10^3$ of order 10^{-5} about the black body profile of temperature 2.725 K [1]. It also contains a polarization signal of order 10^{-6} as the imprint of a quadrupole produced during recombination [2]. Observations of the CMB temperature and polarization are thus a powerful probe of the physics of the early Universe and a robust discriminant of cosmological models.

At large angular scales ($\theta > 5'$), the observed fluctuations are mainly primary anisotropies described by small linear perturbations in the primordial plasma's density and velocity, which are reflected in the characteristic series of acoustic peaks. At small angular scales, the primary anisotropies become negligible due to Silk damping and non-linear effects dominate, which developed at more recent epochs from the primordial potential–matter coupling, forming secondary sources of anisotropies. Among the non-linear effects is weak gravitational lensing, which consists of the deflection of CMB photons by the matter fluctuations along their trajectory (see Ref. [3] for a review). Gravitational lensing generates small scale power at temperature gradients and distorts the temperature-temperature (TT) correlation power spectrum by smearing the acoustic peaks [4]. It also distorts and mixes the EE and BB polarisation power spectra [5]. Finally, weak lensing also correlates with other secondary effects, such as the integrated Sachs-Wolfe effect and the Sunyaev-Zel'dovich effect, thus inducing non-Gaussianities which are manifest in higher-order point correlation functions [6, 7].

[1] Present address

[4] This manuscript is based on a talk delivered by the author at the NEB 14, Ioannina, June 2010

The relevance of the CMB lensing reconstruction for cosmology is thus threefold. First, by changing the power spectra, CMB lensing affects the estimation of cosmological parameters. Second, by converting a fraction of the dominant E-polarisation mode into a B mode, CMB lensing introduces a contaminant in the measurement of the primordial B-polarisation mode and consequently of the inflation energy scale. Third, CMB lensing is a powerful probe of the gravitational potential, integrated from the last-scaterring surface to the presence, that governed the formation of the large scale structure. Thus it is also sensitive to the neutrino mass, the dark energy equation of state and the geometry of the University.

Current experiments, such as ACT [5], SPT [6] and PLANCK [7] are making robust measurements of unprecedented arcminute precision, so that weak lensing should soon become apparent both in the temperature and in the polarization observations. This calls for new statistical and modelling methods to be explored to extract the newly observable signals.

In this manuscript we report on a study where we explored the properties of a novel estimator of the CMB weak lensing, introduced in Ref. [8]. In section 2 we describe the lensing of the CMB temperature. In section 3 we present the conventional minimum variance estimator of the lensing field, upon which in section 4 we derive the newly proposed estimator. In section 5 we reconstruct the lensing field, implementing both estimators, and discuss the results. In section 6 we conclude.

2. Weak lensing of the CMB

Weak lensing of the CMB consists of the deflection of CMB photons from the original propagation direction $\boldsymbol{\theta}$ on the last scattering surface (the source plane) to an observed direction $\tilde{\boldsymbol{\theta}}$ on the sky today (the image plane). This deflection amounts to remapping the unlensed temperature anisotropies T to the lensed ones $\tilde{T}$ according to $\tilde{T}(\boldsymbol{\theta}) = T(\tilde{\boldsymbol{\theta}})$, where $\tilde{\boldsymbol{\theta}} = \boldsymbol{\alpha} + \boldsymbol{\theta}$. For small deflection angles $\boldsymbol{\alpha}$, we can expand

$$\tilde{T}(\boldsymbol{\theta}) = T(\boldsymbol{\theta} + \boldsymbol{\alpha}) = T(\boldsymbol{\theta}) + \boldsymbol{\alpha} \cdot \nabla T(\boldsymbol{\theta}) + O[\boldsymbol{\alpha}^2]. \tag{1}$$

The deflection angle is given by $\boldsymbol{\alpha} = \nabla\psi$, where the lensing potential ψ is the integral of the equation for a null geodesic from the source along the line of sight and ∇ is the covariant derivative on the image plane. Hence, for the line element

$$ds^2 = a^2(\eta)\left[-(1-2\Psi)d\eta^2 + (1+2\Psi)\left[dr^2 + f_k^2(r)\left(d\theta_x^2 + d\theta_y^2\right)\right]\right], \tag{2}$$

where the functional form of f_k depends on the spatial curvature k of the two-dimensional surface (θ_x, θ_y), integrating along the unperturbed path, $\theta_x =$ const and $\theta_y =$ const (the Born approximation), we find for the lensing potential ψ generated by the gravitational potential Ψ

$$\psi(\boldsymbol{\theta}) = -2\int_0^{r_{LS}} dr\, \frac{f_k(r_{LS}-r)}{f_k(r_{LS})f_k(r)}\Psi(r\boldsymbol{\theta}, \eta_0 - r), \quad f_k(r) = \begin{cases} \sin[r] : k = +1, \\ r : k = \ \ 0, \\ \sinh[r] : k = -1. \end{cases} \tag{3}$$

Here, η_0 is the conformal time at recombination, r is the comoving distance to the lens plane, r_{LS} is the comoving distance to the last-scattering surface and $\boldsymbol{\theta}$ is the two-dimensional position vector on the image plane.

It is useful to have an order of magnitude of the weak lensing effect on the CMB. On their way from the last-scattering surface, the CMB photons encounter gradients of matter densities, with

[5] `http://www.physics.princeton.edu/act/about.html`

[6] `http://pole.uchicago.edu/`

[7] `http://www.rssd.esa.int/index.php?project=planck`

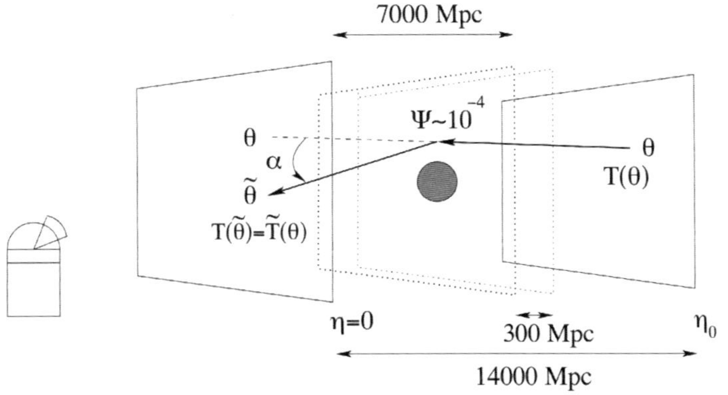

Figure 1. Weak lensing of the CMB photons. Schematic representation of the deflection of the CMB photons by the matter clustered along their trajectory from the last-scattering surface at conformal time η_0 to the present at $\eta = 0$.

the depth of the gravitational potential of order $\Psi \sim 10^{-4}$. The matter is clustered predominantly at a slice of thickness 300 Mpc, which corresponds to the characteristic size of the potential wells given by the peak of the matter power spectrum. Given that the distance to the last-scaterring surface is 14000 Mpc, the number of potentials passed through (hence of deflections suffered) is around 50, which for a random walk corresponds to a root mean square deflection of $\alpha(\theta) \simeq \sqrt{50} \times 10^{-4} \simeq 7 \times 10^{-4}\text{rad} = 2.4$ arcmin. Hence weak lensing is expected to become an important effect for $\ell > 3000$. The deflections will be correlated over the sky by the angular size of a characteristic potential halfway to the last-scattering surface, $300/7000 \simeq 2.5$ deg. Hence although the deflections are much smaller than the scale of the first acoustic peak, they are correlated over comparable scales. This effect is called weak because the deflections are small (~arcmin) and can only be detected as a statistical effect from coherent distortions (~ deg).

There are three ways to describe the lensing distortion of the CMB: a) the lensing potential ψ, b) the deflection vector $\boldsymbol{\alpha} = \nabla\psi$, and c) the convergence tensor $\boldsymbol{\kappa} = -\nabla\nabla\psi/2$ decomposed as

$$\boldsymbol{\kappa} = \begin{pmatrix} \kappa_0 + \kappa_+ & \kappa_\times \\ \kappa_\times & \kappa_0 - \kappa_+ \end{pmatrix}. \tag{4}$$

The descriptions of ψ and $\boldsymbol{\alpha}$ suffer from an ambiguity upon translation, since a patch of the sky and its translation have the same likelihood on account of isotropy. In contrast, the description of the convergence, which is a gradient of the deflection vector field, is locally well defined. Here we will develop the estimator to reconstruct the isotropic component of the tensor $\boldsymbol{\kappa}$, the convergence $\kappa_0(\boldsymbol{\theta}) = -(\partial_x^2 + \partial_y^2)\psi(\boldsymbol{\theta})/2$, which magnifies or demagnifies a feature on the last-scattering surface. We will treat the shear components, $\kappa_+(\boldsymbol{\theta}) = -(\partial_x^2 - \partial_y^2)\psi(\boldsymbol{\theta})/2$ and $\kappa_\times(\boldsymbol{\theta}) = -(\partial_x\partial_y)\psi(\boldsymbol{\theta})$, in a forthcoming study.

3. Minimum variance estimator of the weak lensing convergence

To construct an estimator for the lensing potential from the observed lensed CMB, we start from the observation that the weak lensing effect is apparent in the temperature power spectrum. For small angular scales, we can expand the lensing potential in plane waves

$$\psi(\boldsymbol{\theta}) = \int \frac{d^2\boldsymbol{\ell}}{(2\pi)^2}\ \exp[i\boldsymbol{\ell}\cdot\boldsymbol{\theta}]\ \psi(\boldsymbol{\ell}) \tag{5}$$

so that $(\nabla\psi)(\boldsymbol{\ell}) = i\boldsymbol{\ell}\psi(\boldsymbol{\ell})$. We can then write the linear order correction of the lensed temperature anisotropy in harmonic space as the convolution of the lensing potential with the unlensed temperature anisotropy

$$\tilde{T}(\boldsymbol{\ell}) = T(\boldsymbol{\ell}) - \int d^2\boldsymbol{\ell}'\, \boldsymbol{\ell}' \cdot (\boldsymbol{\ell} - \boldsymbol{\ell}')\, \psi(\boldsymbol{\ell}')\, T(\boldsymbol{\ell} - \boldsymbol{\ell}') + O[\psi^2(\boldsymbol{\ell})]. \tag{6}$$

Defining the temperature power spectrum by $\langle T(\boldsymbol{\ell})\, T(\boldsymbol{\ell}')\rangle = (2\pi)^2\, \delta(\boldsymbol{\ell} + \boldsymbol{\ell}')\, C_\ell$, we find that the power spectrum of the lensed temperature anisotropy is related to that of the unlensed as follows

$$\left\langle \tilde{T}(\boldsymbol{\ell}')\, \tilde{T}(\boldsymbol{\ell}' - \boldsymbol{\ell}) \right\rangle = (2\pi)^2\, \delta(\boldsymbol{\ell})\, C_{\ell'} + (2\pi)^2 \boldsymbol{\ell} \cdot \left[\boldsymbol{\ell}'\, C_{\ell'} + (\boldsymbol{\ell} - \boldsymbol{\ell}')\, C_{|\boldsymbol{\ell}-\boldsymbol{\ell}'|}\right] \psi(\boldsymbol{\ell}). \tag{7}$$

The first term in Eqn. (7) describes a statistically isotropic ensemble where modes with different $\boldsymbol{\ell}$ are uncorrelated. The anisotropies generated by the lensing potential introduce correlations among different $\boldsymbol{\ell}$ modes. Hence, an estimator for ψ is a weighted average of the $\boldsymbol{\ell} \neq 0$ term [9]

$$\hat{\psi}(\boldsymbol{\ell}, \boldsymbol{\ell}') = \frac{\tilde{T}(\boldsymbol{\ell}')\, \tilde{T}(\boldsymbol{\ell} - \boldsymbol{\ell}')}{\left[\boldsymbol{\ell} \cdot \boldsymbol{\ell}'\, C_{\ell'} + \boldsymbol{\ell} \cdot (\boldsymbol{\ell} - \boldsymbol{\ell}')\, C_{|\boldsymbol{\ell}-\boldsymbol{\ell}'|}\right]}. \tag{8}$$

The optimal estimator is the weighted combination of $\hat{\psi}(\boldsymbol{\ell}, \boldsymbol{\ell}')$ which makes the estimator unbiased and minimizes to leading order the variance. For the minimum variance estimator $\hat{\psi}(\boldsymbol{\ell})$ we find the convolution of the square of $\tilde{T}(\boldsymbol{\ell})$ by the weight function $Q^\psi(\boldsymbol{\ell}, \boldsymbol{\ell}')$

$$\hat{\psi}(\boldsymbol{\ell}) = \int \frac{d^2\boldsymbol{\ell}'}{(2\pi)^2}\, \tilde{T}(\boldsymbol{\ell}')\, \tilde{T}(\boldsymbol{\ell} - \boldsymbol{\ell}')\, Q^\psi(\boldsymbol{\ell}, \boldsymbol{\ell}'), \tag{9}$$

where

$$Q^\psi(\boldsymbol{\ell}, \boldsymbol{\ell}') = \mathcal{N}_\ell\, \frac{1}{2} \frac{\boldsymbol{\ell} \cdot \boldsymbol{\ell}' C_{\ell'} + \boldsymbol{\ell} \cdot (\boldsymbol{\ell} - \boldsymbol{\ell}') C_{|\boldsymbol{\ell}-\boldsymbol{\ell}'|}}{\tilde{C}_{\ell'} \tilde{C}_{|\boldsymbol{\ell}-\boldsymbol{\ell}'|}} \tag{10}$$

$$\mathcal{N}_\ell = \left[\int \frac{d^2\boldsymbol{\ell}'}{(2\pi)^2}\, \frac{1}{2} \frac{[\boldsymbol{\ell} \cdot \boldsymbol{\ell}'\, C_{\ell'} + \boldsymbol{\ell} \cdot (\boldsymbol{\ell} - \boldsymbol{\ell}')\, C_{|\boldsymbol{\ell}-\boldsymbol{\ell}'|}]^2}{\tilde{C}_{\ell'}\, \tilde{C}_{|\boldsymbol{\ell}-\boldsymbol{\ell}'|}}\right]^{-1} \equiv \mathrm{Var}[\hat{\psi}(\boldsymbol{\ell})] \tag{11}$$

(see Appendix A of Ref. [8] for a derivation). This reconstruction, based on the quadratic estimator in the lensed CMB map, will be noisy due to spurious correlations in the CMB map. The minimum variance estimator of the convergence $\kappa_0(\boldsymbol{\theta}) = -\nabla^2\psi(\boldsymbol{\theta})/2$ is then $\hat{\kappa}_0(\boldsymbol{\ell}) = \ell^2\, \hat{\psi}(\boldsymbol{\ell})$.

The expression of $Q^\psi(\boldsymbol{\ell}, \boldsymbol{\ell}')$ can be straightforwardly modified to include the detector noise and the finite beam width by replacing $\tilde{C}_{\ell'} \rightarrow \tilde{C}_{\ell'} + N_{\ell'}$. The noise power spectrum N_ℓ is the inverse-sum over the number of channels of the detector noise $n_i(\ell)$ of each channel i

$$N_\ell = \left(\sum_{i=0}^{\mathtt{num\_chann}} \frac{1}{n_i(\ell)}\right)^{-1}. \tag{12}$$

The detector noise is modelled by a Gaussian signal on the beam size and includes both the white noise amplitude and the beam profile attenuation factor

$$n_i(\ell) = (\theta_{fwhm_i}\sigma_{pix_i})^2 \exp[(\theta_{fwhm_i})^2 \ell(\ell+1)/(8\ln 2)], \tag{13}$$

where θ_{fwhm_i} is the beam full-width at half-maximum of the beam and σ_{pix_i} is the white noise amplitude per beam width.

4. Real space estimator of the weak lensing convergence

The optimal reconstruction in harmonic space discussed above implicitly assumes a full-sky coverage [10] and moreover without: galactic cuts, bad pixels due to excision of point sources, or a nonuniform weighting for an uneven sky coverage. To surpass the difficulties of the analysis of realistic maps, we consider a slightly less optimal estimator modified to have a kernel with compact support which acts in real space.

An estimator for the convergence in real space would be

$$\begin{aligned}\hat{\kappa}_0(\boldsymbol{\theta}) &= \int \frac{d^2\boldsymbol{\ell}}{(2\pi)^2}\, \exp[i\boldsymbol{\ell}\cdot\boldsymbol{\theta}] \int \frac{d^2\boldsymbol{\ell}'}{(2\pi)^2}\, \tilde{T}(\boldsymbol{\ell}')\, \tilde{T}(\boldsymbol{\ell}-\boldsymbol{\ell}')\, Q(\boldsymbol{\ell},\boldsymbol{\ell}') \\ &\equiv \int d^2\boldsymbol{\theta}'\, \tilde{T}(\boldsymbol{\theta}') \int d^2\boldsymbol{\theta}''\, \tilde{T}(\boldsymbol{\theta}'')\, Q(\boldsymbol{\theta},\boldsymbol{\theta}',\boldsymbol{\theta}''), \end{aligned} \tag{14}$$

where we define the corresponding weight function in real space by

$$Q(\boldsymbol{\theta},\boldsymbol{\theta}',\boldsymbol{\theta}'') = \int \frac{d^2\boldsymbol{\ell}}{(2\pi)^2}\, \exp[i\boldsymbol{\ell}\cdot\boldsymbol{\theta}] \int \frac{d^2\boldsymbol{\ell}'}{(2\pi)^2}\, \exp[-i\boldsymbol{\ell}'\cdot\boldsymbol{\theta}']\exp[-i(\boldsymbol{\ell}-\boldsymbol{\ell}')\cdot\boldsymbol{\theta}'']\, Q(\boldsymbol{\ell},\boldsymbol{\ell}'). \tag{15}$$

and $Q(\boldsymbol{\ell},\boldsymbol{\ell}') = \ell^2 Q^{\psi}(\boldsymbol{\ell},\boldsymbol{\ell}')$. The kernel $Q(\boldsymbol{\ell},\boldsymbol{\ell}')$ is a function of the lengths of $\boldsymbol{\ell}$ and $\boldsymbol{\ell}'$ as well as of the angle $\xi_{\boldsymbol{\ell}}$ between $\boldsymbol{\ell}$ and $\boldsymbol{\ell}'$, and can be expanded in eigenfunctions which factorize the radial and the angular dependence as follows

$$Q(\boldsymbol{\ell},\boldsymbol{\ell}') = \sum_{m=-\infty}^{+\infty} \exp[im\xi_{\boldsymbol{\ell}}]\, Q_m(\ell,\ell') \tag{16}$$

so that

$$Q_m(\ell,\ell') = \frac{1}{2\pi}\int d\xi_{\boldsymbol{\ell}}\, \exp[-im\xi_{\boldsymbol{\ell}}]\, Q(\boldsymbol{\ell},\boldsymbol{\ell}'). \tag{17}$$

In order to obtain the analogous factorization of the kernel in real space, which would make the implementation computationally less costly, we change to the variables $\boldsymbol{\ell}_+, \boldsymbol{\ell}_-$ such that $\boldsymbol{\ell} = \boldsymbol{\ell}_+$, $\boldsymbol{\ell}' = (\boldsymbol{\ell}_+ + \boldsymbol{\ell}_-)/2$. Upon this change of variables, Eqn. (9) yields

$$\hat{\kappa}_0(\boldsymbol{\ell}_+) = \int \frac{d^2\boldsymbol{\ell}_-}{(2\pi)^2}\, \tilde{T}\left(\frac{\boldsymbol{\ell}_+ + \boldsymbol{\ell}_-}{2}\right) \tilde{T}\left(\frac{\boldsymbol{\ell}_+ - \boldsymbol{\ell}_-}{2}\right) W(\boldsymbol{\ell}_+,\boldsymbol{\ell}_-), \tag{18}$$

where $W(\boldsymbol{\ell}_+,\boldsymbol{\ell}_-) = Q(\boldsymbol{\ell},\boldsymbol{\ell}')$. The corresponding estimator in real space becomes

$$\hat{\kappa}_0(\boldsymbol{\theta}) \equiv \int d^2\boldsymbol{\theta}_+ \int d^2\boldsymbol{\theta}_-\, \tilde{T}(\boldsymbol{\theta}_+ + \boldsymbol{\theta}_-)\, \tilde{T}(\boldsymbol{\theta}_+ - \boldsymbol{\theta}_-)\, W(\boldsymbol{\theta},\boldsymbol{\theta}_+,\boldsymbol{\theta}_-), \tag{19}$$

where we define the corresponding weight function in real space by

$$W(\boldsymbol{\theta},\boldsymbol{\theta}_+,\boldsymbol{\theta}_-) = \int \frac{d^2\boldsymbol{\ell}_+}{(2\pi)^2}\, \exp[i\boldsymbol{\ell}_+\cdot\boldsymbol{\theta}] \int \frac{d^2\boldsymbol{\ell}_-}{(2\pi)^2}\, \exp\left[-i(\boldsymbol{\ell}_+\cdot\boldsymbol{\theta}_+ + \boldsymbol{\ell}_-\cdot\boldsymbol{\theta}_-)\right] W(\boldsymbol{\ell}_+,\boldsymbol{\ell}_-). \tag{20}$$

We then expand the kernel $W(\boldsymbol{\ell}_+,\boldsymbol{\ell}_-)$ in terms of eigenfunctions that factorize the radial and the angular dependence so that

$$W_m(\ell_+,\ell_-) = \frac{1}{2\pi}\int d\chi_{\boldsymbol{\ell}}\, \exp[-im\chi_{\boldsymbol{\ell}}]\, W(\boldsymbol{\ell}_+,\boldsymbol{\ell}_-) \tag{21}$$

and $\chi_{\boldsymbol{\ell}}$ is the angle between $\boldsymbol{\ell}_+$ and $\boldsymbol{\ell}_-$. For the case that $\boldsymbol{\theta} = 0$, Eqn. (20) becomes

$$W(\theta_+, \theta_-, \chi_{\boldsymbol{\theta}}) \equiv \sum_{m=-\infty}^{+\infty} \exp[im\chi_{\boldsymbol{\theta}}]\, W_m(\theta_+, \theta_-), \tag{22}$$

where

$$W_m(\theta_+, \theta_-) = \frac{1}{(2\pi)^2} \int_0^\infty d\ell_+ \ell_+ \int_0^\infty d\ell_- \ell_-\; J_m(\ell_+\theta_+)\; J_m(\ell_-\theta_-)\; W_m(\ell_+, \ell_-) \tag{23}$$

and $\chi_{\boldsymbol{\theta}}$ is the angle between $\boldsymbol{\theta}_+$ and $\boldsymbol{\theta}_-$. Having achieved the factorization of the angular and the radial dependence of the kernel in real space, Eqn. (19) can be written as

$$\hat{\kappa}_0(\boldsymbol{\theta} = 0) = \sum_{m=-\infty}^{+\infty} \int d^2\boldsymbol{\theta}_+ \int d^2\boldsymbol{\theta}_- \exp[im\chi_{\boldsymbol{\theta}}]\, \tilde{T}(\boldsymbol{\theta}_+ + \boldsymbol{\theta}_-)\, \tilde{T}(\boldsymbol{\theta}_+ - \boldsymbol{\theta}_-)\, W_m(\theta_+, \theta_-). \tag{24}$$

The convergence map thus estimated contains two contributions: the convergence of the lensing potential $\kappa_0|_\psi$ and the convergence of the CMB (Gaussian noise) $\kappa_0|_{\psi=0}$

$$\hat{\kappa}_0 = \kappa_0|_\psi + \kappa_0|_{\psi=0}. \tag{25}$$

Hence, in the absence of lensing there is a non-vanishing contribution to the estimated convergence map and to the corresponding power spectrum. To remove the bias of the estimated map, i.e. $\langle \kappa_0|_{\psi=0} \rangle$, we take the pixel average of the convergence map obtained from a single realization of the lensed CMB and subtract from the convergence map. (This is equivalent to taking the average convergence map obtained from different realizations of the CMB lensed by the same realization of the lensing potential.) This causes the Gaussian noise to average out, yielding an unbiased estimate of the convergence map $\langle \hat{\kappa}_0 \rangle = \langle \kappa_0|_\psi \rangle$. To remove the bias of the estimated power spectrum, i.e. $\langle \kappa_0|_{\psi=0}\, \kappa_0^*|_{\psi=0} \rangle$, we compute the power spectrum of the convergence map obtained from different realizations of the unlensed CMB, take the average power spectrum and subtract from the power spectrum of the convergence map obtained from the single realization of the lensed CMB map, thus yielding an unbiased estimate of the convergence power spectrum $\langle \hat{\kappa}_0\, \hat{\kappa}_0^* \rangle = \langle \kappa_0|_\psi\, \kappa_0^*|_\psi \rangle$.

In this formulation, the kernel is the Bessel transform of the minimal variance weight function in harmonic space. Whereas in the harmonic space formulation the temperature map is transformed and then is weighted by the kernel in harmonic space, in the real space formulation the kernel is transformed and then weights the temperature map to yield the lensing field.

5. Implementation of the convergence estimator

We implement the real space estimator to reconstruct the convergence field. We will use two experiments based on the specification of the PLANCK experiment for the $\nu = 143$ GHz channel [11] which differ only in the detector noise. The experiments, denoted by *Designer* and *Planck*, both have $\theta_{fwhm} = 7.8$ arcmin for the beam full-width at half maximum, and $\sigma_{pix} = 0$ and $\sigma_{pix} = 6.8\, f_{sky}^{1/2}\, \mu K/\mathrm{rad}$ for the white noise amplitude per beam width respectively, where f_{sky} is the fraction area of the sky covered by the patch [12, 13]. We chose the experiments to have the same beam size in order not to entangle the effect of the noise with that of the beam.

The real space implementation is described as follows:

- First we synthesise the lensed CMB map by pixel remapping of the unlensed CMB map by the lensing map [14]. Then we convolve the lensed map with the beam profile $b_\ell = \sum_i \exp[-(\theta_{fwhm_i})^2 \ell(\ell+1)/(8\ln 2)/2]$ and finally add the map constructed from the detector noise power spectrum $1/\sum_i [1/\left(\theta_{fwhm_i}\sigma_{pix_i}\right)^2]$ (see Appendix B of Ref. [8]). Instead of deconvolving the lensed map with the beam profile, we deconvolve the kernel.

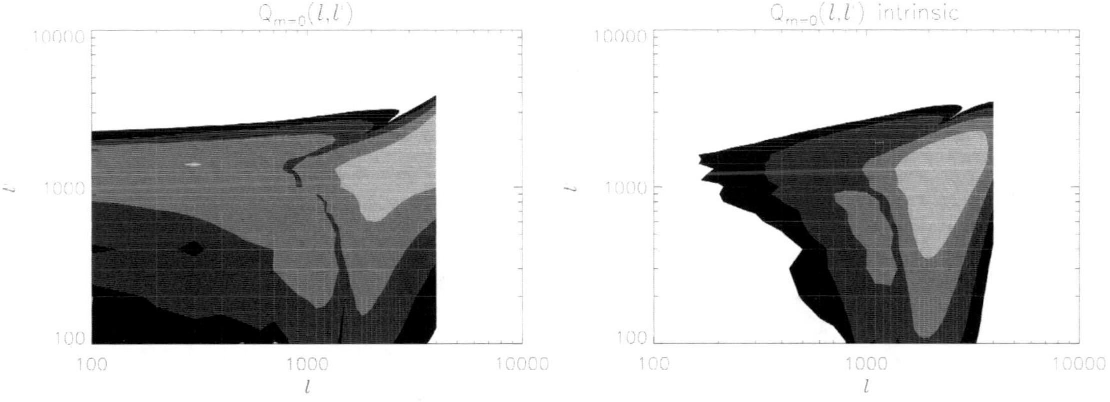

Figure 2. Contour plots of $Q_{m=0}(\ell,\ell')$ (left panel) and $Q_{m=0}^{intrinsic}(\ell,\ell')$ (right panel). The levels denote the orders of magnitude as fractions of the maximum value, delimited by $\{1,10^{-1},10^{-2},10^{-3},10^{-4}\}$ from the lightest to the darkest shade for PLANCK noise at $\nu=143$ GHz.

- Second we generate the kernel in harmonic space (Eqn. (10)), deconvolve with the beam profile (which is equivalent to deconvolving the lensed map with the beam profile) and finally transform to real space (Eqn. (23)).
- Third we convolve the square of the lensed map with the kernel (Eqn. (24)) and compute the unbiased convergence map. We repeat this step for different realizations of the unlensed CMB map and use the average power spectrum to compute the unbiased power spectrum of the convergence.

For comparison, we also implement the harmonic space estimator by applying the procedure described in Appendix C of Ref. [8]. In order to reproduce periodic boundary conditions, before convolving with the beam profile we apodize the CMB maps by smoothing to zero over a strip around the boundary. The requirement of periodic boundary conditions makes the implementation of the harmonic space estimator easier but is not necessary for the implementation of the real space estimator.

5.1. The kernel

For the real space implementation, we compute $W_m(\boldsymbol{\ell}_+,\boldsymbol{\ell}_-)$ using Eqn. (21) and the corresponding inverse Fourier transform $W_m(\boldsymbol{\theta}_+,\boldsymbol{\theta}_-)$ using Eqn. (23).

For the number of eigenfunctions m_{max} kept in the series in Eqn. (24), we took the order of the highest-order eigenfunction whose maximum amplitude was of order 1% the maximum amplitude of the dominant $m=0$ eigenfunction, which corresponds to $m_{max}=4$. For the kernel extent in harmonic space ℓ_{max}, we used the highest ℓ moment used in the synthesis of the lensed CMB, i.e. $\ell_{max}=4000$. This is also consistent with the value read off of $Q_{m=0}(\ell,\ell')$ which we plot in Fig. 2, left panel. Defining

$$Q(\boldsymbol{\ell},\boldsymbol{\ell}') = \mathcal{N}_\ell\, \ell^2\, \frac{1}{2}\frac{\boldsymbol{\ell}\cdot\boldsymbol{\ell}' C_{\ell'} + \boldsymbol{\ell}\cdot(\boldsymbol{\ell}-\boldsymbol{\ell}')C_{|\boldsymbol{\ell}-\boldsymbol{\ell}'|}}{\tilde{C}_{\ell'}\tilde{C}_{|\boldsymbol{\ell}-\boldsymbol{\ell}'|}} = \mathcal{N}_\ell\, Q^{intrinsic}(\boldsymbol{\ell},\boldsymbol{\ell}') \tag{26}$$

we also plot $Q_{m=0}^{intrinsic}$ in Fig. 2, right panel. Using higher values for ℓ_{max}, we observe that whereas $Q_m^{intrinsic}$ is compact, Q_m is largely ℓ_{max} invariant as a result of the presence of the estimator variance $\mathcal{N}_\ell$. We thus conclude that $\mathcal{N}_\ell$ weights the different ℓ modes in such a way as

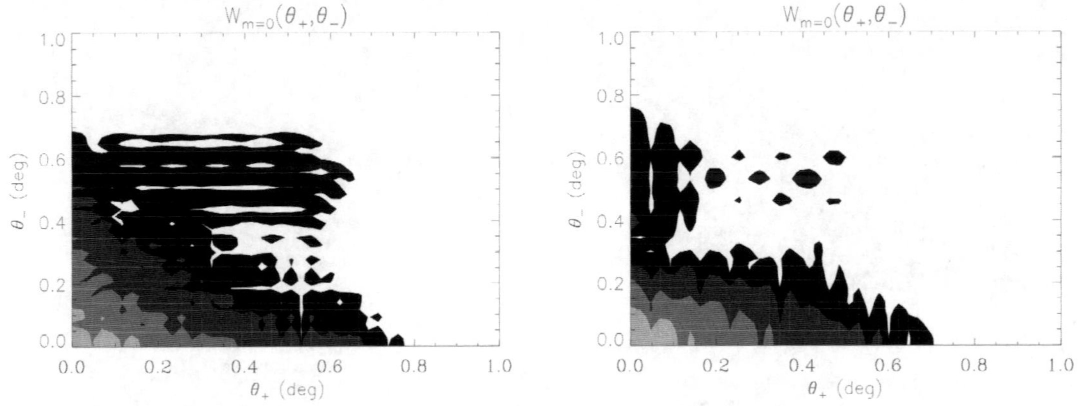

Figure 3. Contour plots of $W_{m=0}(\theta_+, \theta_-)$. The levels denote the orders of magnitude as fractions of the maximum value, delimited by $\{1, 10^{-1}, 10^{-2}, 10^{-3}, 10^{-4}\}$ from the lightest to the darkest shade. Here $m_{max} = 4$, $\ell_{max} = 4000$ for no detector noise (left panel) and for PLANCK noise at $\nu = 143$ GHz (right panel).

to enhance the contribution of the highest ℓ modes accessible. In the absence of detector noise, the kernel is displaced to higher values of ℓ and ℓ' up to the chosen ℓ_{max}.

In Fig. 3 we show the contour plot of the $m = 0$ eigenfunction of the kernel in real space for the *Designer* and the *Planck* experiment respectively. From these plots we read off the value for the spatial extent of the kernel θ_{max} used in the convolution to compute the convergence map. In harmonic space the kernel peaks at the scale of the beam but has support at smaller ℓ values. The spatial extent of the kernel is thus larger than the beam scale and scales roughly linearly with the beam size. We observe that the kernel in real space is compact and extends only up to $0.70°$ and $0.60°$ at the 1% level in the absence and in the presence of detector noise respectively. Since increasing the value of θ_{max} beyond $0.35°$ did not improve the reconstruction of the power spectrum, whereas reducing the value of θ_{max} below $0.35°$ degraded the reconstruction, we set $\theta_{max} = 0.35°$.

5.2. The convergence map

We compute the estimator by convolving the kernel with the lensed CMB map according to Eqn. (24). We implement the convolution using the discrete version of this equation as follows

$$\begin{aligned}\kappa_0(\boldsymbol{\theta} = 0) \quad &= \quad \sum_{m=-m_{max}}^{m_{max}} \sum_{\theta_{+x}} \sum_{\theta_{+y}} \sum_{\theta_{-x}} \sum_{\theta_{-y}} \exp\left[im \arctan\left(\frac{\theta_{+y} - \theta_{-y}}{\theta_{+x} - \theta_{-x}}\right)\right] \\ &\times \tilde{T}(\theta_{+x} + \theta_{-x}, \theta_{+y} + \theta_{-y})\, \tilde{T}(\theta_{+x} - \theta_{-x}, \theta_{+y} - \theta_{-y}) \\ &\times W_m\left(\sqrt{\theta_{+x}^2 + \theta_{+y}^2}, \sqrt{\theta_{-x}^2 + \theta_{-y}^2}\right). \qquad (27)\end{aligned}$$

The points in the temperature map which contribute to the convergence at a given point $\boldsymbol{\theta}$ span the region from $(\boldsymbol{\theta}_+ - \boldsymbol{\theta}_-)$ to $(\boldsymbol{\theta}_+ + \boldsymbol{\theta}_-)$, where $\boldsymbol{\theta}_+$ and $\boldsymbol{\theta}_-$ are in the kernel. Hence we need a region of the temperature map that is four times the length of the kernel to compute the convolution, while setting the summation upper limits equal to the spatial extent of the kernel θ_{max}.

We then average the reconstructed convergence map over all pixels to extract the Gaussian noise which is then subtracted from the map. Finally we compute the convergence power spectrum of a Monte Carlo realization of one hundred unlensed CMB maps and subtract the

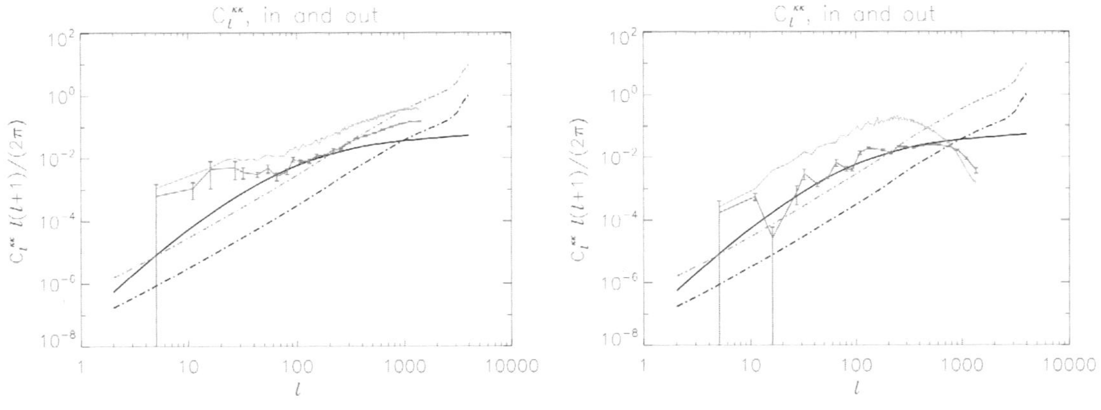

Figure 4. Mean square of the convergence map reconstructed with the harmonic space estimator (left panel) and the real space estimator (right panel). The smooth black curve is the input power spectrum. The light grey and dark grey solid lines are the output power spectra before and after the removal of the bias respectively. The error bars measure the total standard deviation binned over logarithmically spaced intervals in ℓ. Here $\theta_{fwhm} = 7.8'$, $FOV_{map} = 67°$ and $m_{max} = 4$, $\ell_{max} = 4000$ for no detector noise.

average variance of the Gaussian CMB noise from the convergence power spectrum of the lensed map, thus removing the bias from the estimated power spectrum.

5.3. The convergence power spectrum

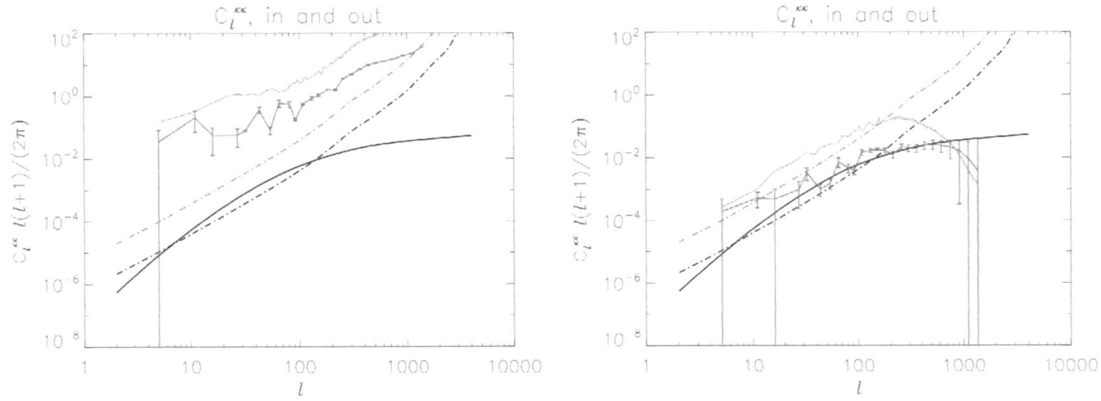

Figure 5. Mean square of the convergence map reconstructed with the harmonic space estimator (left panel) and the real space estimator (right panel). The smooth black curve is the input power spectrum. The light grey and dark grey solid lines are the output power spectra before and after the removal of the bias respectively. The error bars measure the total standard deviation binned over logarithmically spaced intervals in ℓ. Here $\theta_{fwhm} = 7.8'$, $FOV_{map} = 67°$, and $m_{max} = 4$, $\ell_{max} = 4000$ for PLANCK noise at $\nu = 143$ GHz.

In Figs. 4 and 5, for the *Designer* and the *Planck* experiment respectively, we plot the output power spectrum $C_\ell^{\kappa_{out}\kappa_{out}}$, both before (light grey solid line) and after (dark grey solid line) the removal of the bias, against the input power spectrum $C_\ell^{\kappa_{in}\kappa_{in}}$ (black solid line). We also plot the variance of the estimator $\text{Var}[\hat{\kappa}_0(\boldsymbol{\ell})]$ (black dot-dashed line) as well as $\text{Var}[\hat{\kappa}_0(\boldsymbol{\ell})]/f_{sky}$ (grey dot-dashed line) as an auxiliary tool for visualization. The error bars on the power spectra are computed from the standard deviation of the reconstructed power spectrum, binned over logarithmically spaced intervals in ℓ space and added in quadrature. The variance of the reconstructed power spectrum encompasses both the variance of the optimal estimator and the average value of the convergence fluctuation in the map, weighted by the fraction of the area of the sky covered by the patch, f_{sky}, as follows [13]

$$\text{Var}[C_\ell^{\kappa_0\kappa_0}] \approx \frac{1}{\ell\ f_{sky}} \left(\text{Var}[\hat{\kappa}_0(\boldsymbol{\ell})] + C_l^{\kappa_0\kappa_0}\right)^2. \tag{28}$$

We observe that at small ℓ the cosmic variance dominates the error of the reconstruction, whereas at large ℓ the variance of the estimator takes over. The region where $C_\ell^{\kappa_{in}\kappa_{in}} > \text{Var}[\hat{\kappa}_0(\boldsymbol{\ell})]$ indicates the interval in ℓ where the lensing potential can in theory be mapped. Accordingly, this is also the region where the reconstructed convergence power spectra track more closely the input power spectrum. Also visible are the fluctuations in the recovered power spectrum which arise from the subtraction of the noise bias term estimated from a finite number of realizations of the unlensed CMB map.

5.4. Discussion

Using the real space estimator, we achieve a reasonable reconstruction of the input power spectrum in the interval where the input signal is larger than the variance of the estimator, although on smaller scales, $\ell > 600$, there is a decrease in the recovered power. Comparing the reconstructed power spectra for the *Designer* and *Planck* experiment, we observe that the real space implementation is fairly insensitive to the detector noise and returns a reconstruction consistent with the input power spectrum. We can understand these results as follows.

The loss of power at small scales, beginning around the angular scale corresponding to the size of the kernel, is a consequence of averaging modes smaller than the finite extent of the kernel and hence is an intrinsic constraint of the real space implementation. This intrinsic constraint can be overcome by shrinking the extent of the kernel in real space, which can be achieved with a smaller beam and detector noise, thereby moving the support of the kernel to larger ℓ. The ℓ interval over which the power spectrum can be recovered by the real space estimator is determined as follows. The lower ℓ limit is determined by the size of the map, since it measures the longest wavelength mode that can be enclosed within the map. The upper ℓ limit is determined by the size of the kernel, since it measures the smallest wavelength mode that the kernel can probe, below which the averaging of modes smaller than the size of the kernel causes loss of power. In contrast, the ℓ range for a reasonable reconstruction with the harmonic space estimator is wider, being limited at higher ℓ by the angular scale corresponding to the beam size, since this is the scale which constrains the action of the kernel in harmonic space.

Within this ℓ range, the real space estimator seems to perform fairly well both in the absence and in the presence of detector noise. We explain the insensitivity of the real space estimator to the experimental noise as follows. Recalling that in the real space implementation the convergence in each pixel is given by the sum of the product of pairs of neighbouring pixels weighted by the kernel, the noise, being independent in each pixel, is averaged down. To test this result, we reconstructed the convergence map from a pure white noise input map using the real space estimator and the harmonic space estimator (Fig. 6). We observe that the power spectrum recovered by the real space estimator is about seven orders of magnitude smaller than the input power spectrum, whereas the power spectrum recovered by the harmonic space estimator is comparable to the input power spectrum.

Finally comparing the two implementations (Figs. 4 and 5), we observe that the removal of the bias was not effective in the harmonic space implementation. This remaining bias presumably

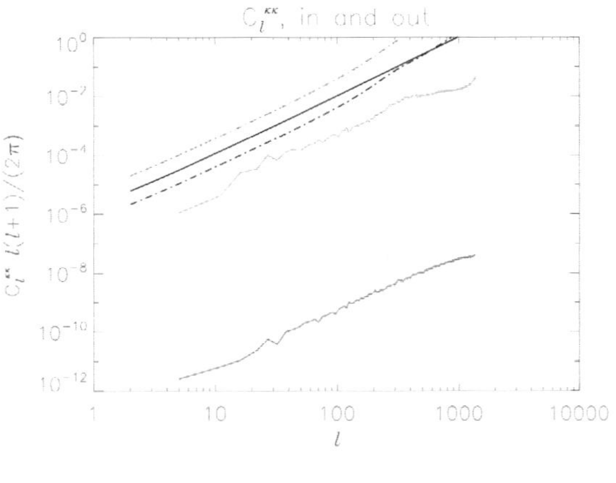

Figure 6. Mean square of the convergence map reconstructed with the harmonic space estimator (light grey line) and the real space estimator (dark grey line) from a pure white noise map. The grey solid lines are the output power spectra whereas the smooth black curve is the input power spectrum. Here $\theta_{fwhm} = 7.8'$, $FOV_{map} = 67°$, and $m_{max} = 4$, $\ell_{max} = 4000$ for PLANCK noise at $\nu = 143$ GHz.

arises from a coupling between unlensed temperature modes due to the finite size of the map, providing an additional source of lensing which we have not yet treated. This mode coupling is absent in the reconstruction over the full sky [15]. The real space implementation appears to be insensitive to this bias.

6. Concluding remarks

We report on the introduction of a new estimator of the weak lensing convergence of the CMB. The new estimator acts locally in real space and is thus able to treat the excision of pixels and nonuniform sky coverage in a flexible manner. The finite extent of the kernel is a desirable property of the proposed estimator, since in theory it allows the reconstruction of the lensing convergence, manifested mainly at very small scales, from a small map of the sky as long as the kernel probes sufficiently small angular scales.

We implemented the estimator on two experimental setups, one without and one with detector noise, based on the specifications of PLANCK for the $\nu = 143$ GHz channel. From the comparative implementation of the new estimator and the conventional estimator defined in harmonic space, we found that the real space implementation is less sensitive to the experimental noise, as well as to the bias of the mode coupling derived from the finite size of the map. Even though the new estimator shows a loss of power on scales smaller than the finite extent of the kernel, this effect could be studied using simulations to determine a form factor to be applied to correct the loss of power [16].

A preliminary study of the effect of the excision of points (e.g. resulting from the application of a mask for the subtraction of point sources) on the reconstruction of the lensing convergence seems to strengthen the advantages of a real space implementation of the estimator [17]. Follow-up studies include the formulation of the analogous estimators for the shear components of the convergence tensor, as well as for the CMB polarisation.

Acknowledgments

The author was supported by the South African National Research Foundation. The author is currently supported by the Fundação para a Ciência e a Tecnologia under the fellowship BPD/65993/2009.

References

[1] Bennett CL et al. 1996 4-Year COBE DMR Cosmic Microwave Background Observations: Maps and Basic Results *Astrophys.J.Lett.* **464** L1 (*Preprint* astro-ph/9601067)

[2] Hu W and White M 1997 A CMB Polarization Primer *New Astron.* **2** 323 (*Preprint* astro-ph/9706147)

[3] Lewis A and Challinor A 2006 Weak gravitational lensing of the CMB *Phys.Rep.* **429** 1 (*Preprint* astro-ph/0601594)

[4] Seljak U 1996 Gravitational lensing effect on cosmic microwave background anisotropies: a power spectrum approach *Astrophys.J.* **463** 1 (*Preprint* astro-ph/9505109)

[5] Seljak U and Zaldarriaga M 1997 Signature of gravity waves in polarization of the microwave background *Phys.Rev.Lett.* **78** 2054 (*Preprint* astro-ph/9609169); Zaldarriaga M and Seljak U 1998 Gravitational lensing effect on cosmic microwave background polarization *Phys.Rev.D* **58** 023003 (*Preprint* astro-ph/9803150)

[6] Zaldarriaga M 2000 Lensing of the CMB: Non-Gaussian Aspects *Phys.Rev.D* **62** 063510 (*Preprint* astro-ph/9910498)

[7] Cooray A 2002 Weak lensing of the CMB: power spectrum covariance *Phys.Rev.D* **65** 063512 (*Preprint* astro-ph/0110415); Cooray A and Kesden M 2003 Weak lensing of the CMB: extraction of lensing information from the trispectrum *New Astron.* **8** 231 (*Preprint* astro-ph/0204068)

[8] Carvalho CS and Moodley K 2010 Real space estimator for the weak lensing convergence from the CMB *Phys.Rev.D* **81** 123010 (*Preprint* 1005.4288 [astro-ph.CO])

[9] Hu W 2001 Mapping the dark matter through the CMB damping tail *Astrophys.J.* **557** 79 (*Preprint* astro-ph/0105424)

[10] Okamoto T and W Hu W 2003 Lensing reconstruction on the full sky *Phys.Rev.D* **67** 083002 (*Preprint* astro-ph/0301031)

[11] PLANCK Collaboration 2005 The Scientific Programme of PLANCK *ESA-SCI* **1** 5 (*Preprint* astro-ph/0604069)

[12] Knox L 1995 Determination of inflationary observables by cosmic microwave background anisotropiy experiment *Phys.Rev.D* **52** 4307 (*Preprint* astro-ph/9504054)

[13] Tegmark M 1997 CMB mapping experiments: a designer's guide *Phys.Rev.D* **56** 4514 (*Preprint* astro-ph/9705188)

[14] Lewis A 2005 Lensed CMB simulation and parameter estimation *Phys.Rev.D* **71** 083008 (*Preprint* astro-ph/0502469)

[15] Perotto L, Bobin J, Plaszczynski S, Starck J-L and Lavabre A 2009 Reconstruction of the CMB lensing for Planck, *Preprint* 0903.1308 (astro-ph.CO)

[16] Bucher M, Carvalho CS, Moodley K and Remazeilles M 2010 CMB Lensing Reconstruction in Real Space *Preprint* 1004.3285 [astro-ph.CO]

[17] Carvalho CS (in preparation)

Inflation in asymptotically safe $f(R)$ theory

Adriano Contillo
SISSA, Via Bonomea 265, Trieste, Italy
INFN, Sezione di Trieste, Italy

E-mail: contillo@sissa.it

Abstract. We discuss the existence of inflationary solutions in a class of renormalization group improved polynomial $f(R)$ theories, which have been studied recently in the context of the asymptotic safety scenario for quantum gravity. These theories seem to possess a nontrivial ultraviolet fixed point, where the dimensionful couplings scale according to their canonical dimensionality. Assuming that the cutoff is proportional to the Hubble parameter, we obtain modified Friedmann equations which admit both power law and exponential solutions. We establish that for sufficiently high order polynomial the solutions are reliable, in the sense that considering still higher order polynomials is very unlikely to change the solution.

1. Introduction

The idea that a perturbatively nonrenormalizable theory could be consistently defined in the UV limit at a nontrivial fixed point (FP), often called "asymptotic safety", is theoretically very attractive, especially when applied to quantum gravity [1]. Much progress in this direction has come from the direct application of Renormalization Group (RG) techniques to gravity. A particularly useful tool has been the non-perturbative Functional Renormalization Group Equation (FRGE) [2], defining an RG flow on a theory space which consists of all diffeomorphism invariant functionals of the metric $g_{\mu\nu}$. It defines a one parameter family of effective field theories with actions $\Gamma_k(g_{\mu\nu})$ depending on a coarse graining scale (or "cutoff") k, and interpolating between a "bare" action (for $k \to \infty$) and the ordinary effective action (for $k \to 0$).

The application of such concepts to a cosmological framework is based to the following logic [3]. If we want to study the quantum evolution of the cosmic scale factor, we should in principle use the full effective action $\Gamma(g_{\mu\nu})$, which, as we mentioned above, coincides with $\Gamma_k(g_{\mu\nu})$ in the limit $k \to 0$. However, our knowledge of this functional is rather poor. One way of gaining some traction on this issue is to observe that the Hubble parameter appears as a mass in propagators. Thus, the contributions of quantum fluctuations with wavelenghts greater than H^{-1} are suppressed. As a result, the functional Γ_k at k comparable to H should be a reasonable approximation for the same functional at $k = 0$. We do not know Γ_k much better than the full effective action, but we can easily calculate the dependence of some terms in Γ_k on k. By doing so we effectively take into account nonlocal terms that would be very hard to calculate otherwise.

Previous investigations along these lines, in particular [4], have been based on the Einstein-Hilbert (EH) truncation. It is important to establish that the results obtained there persist when further operators are included in the truncation. We know that at the FP the coefficients of these terms are not very small, but their presence does not seem to affect the values of the cosmological constant and Newton's constant too much. In other words, the FP that is is seen in the EH truncation seems to be robust. The question then is to see if this stability of the FP against the inclusion of new terms is reflected in the stability of the corresponding solutions. This question is important because the values of the couplings at the FP are fixed and as a consequence there are no free parameters to be varied. We will see that the (power-law or exponential) inflationary solutions are indeed stable against the inclusion of new terms, but establishing this fact requires including a rather large number of terms.

2. Asymptotic Safety

It is well known that general relativity can be treated as an effective field theory. This means that it is possible to compute quantum effects due to graviton loops, as long as the momenta of the particles in the loops are cut off at some scale. The results are independent of the structure of any ultraviolet completion, and therefore constitute genuine low energy predictions of any quantum theory of gravity.

Given for a field ϕ an effective action $\Gamma(\phi)$, it can always be written as a sum

$$\Gamma(\phi) = \sum g_i \mathcal{O}_i(\phi) \tag{1}$$

that runs over the infinitely many operators consistent with the undelying symmetries of the theory. Given an energy scale k, one can define *dimensionless* couplings $\tilde{g}_i = g_i k^{-d_i}$, where $[g_i] = [k]^{d_i}$. The resulting picture is an infinite dimensional space parametrized by the $\tilde{g}_i$'s, where a Renormalization Group flow can be seen as a vector field that shifts any action along a trajectory

$$\Gamma(\phi) \to \Gamma_k(\phi) = \sum \tilde{g}_i(k) k^{d_i} \mathcal{O}_i(\phi)\ . \tag{2}$$

The particular realization of RG that was used in this work is the one *à la* Wilson, based on the idea that the effective action describing physical phenomena at a momentum scale k can be thought of as the result of having integrated out all fluctuations of the field with momenta larger than k. This can be achieved adding to the bare action $\mathcal{S}$ a term

$$\Delta\mathcal{S}_k = \int dx\, \phi(x) R_k\, \phi(x) \tag{3}$$

where R_k is called the cutoff kernel (see [5] for details). This term acts as a mass for the fluctuations with momenta smaller than k. Following the usual procedure of background field method, we define a k-dependent generating functional of connected Green functions by

$$e^{-W_k(J)} = \int D\phi e^{-(S(\phi)+\Delta\mathcal{S}_k(\phi))-\int dx J\phi} \tag{4}$$

and the effective action as its Legendre transform

$$\Gamma_k(\phi) = W_k(J) - \int dx J\phi - \Delta\mathcal{S}_k(\phi)\ . \tag{5}$$

One can show that it obeys the so called functional renormalization group equation (FRGE)

$$k\frac{d\Gamma_k}{dk} = \frac{1}{2}\mathrm{Tr}\left(\Gamma_k^{(2)} + R_k\right)^{-1} k\frac{dR_k}{dk} \tag{6}$$

where $\Gamma_k^{(2)}$ is the inverse propagator of the graviton. From the r.h.s. of (6) one can then extract the β-functions of the couplings, as it is done in [5], and study the RG flow of the theory towards the UV limit $k \to \infty$.

A theory is then said to be asymptotically safe if

- for any coupling $\tilde{g}_i$ there exists a finite UV limit $\tilde{g}_i^*$ that is called a fixed point
- the stability matrix $M_{ij} = \left.\frac{\partial \beta_{\tilde{g}_i}}{\partial \tilde{g}_j}\right|_*$ evaluated in the fixed point has a finite number of negative eigenvalues m_i (in the following, it will come useful to define critical exponents $\theta_i = -m_i$).

If both the requests are satisfied one finds that the *critical surface*, the locus of the points that are attracted towards the FP, is finite dimensional. This means that only a finite number of experiments is needed to univocally identify a trajectory, and the theory is thereby predictive.

In [5] this procedure was applied to a (truncated) gravitational action consisting of a sum of powers of the curvature scalar R

$$\Gamma(g) = \int d^4x \sqrt{|g|} \sum_{i=0}^{n} g_i R^i \tag{7}$$

with $n = 2, \ldots, 8$, and it was further extended to $n = 9, 10$ in [6]. For every n a fixed point was found, with only three positive critical exponents. To understand if a finite truncation is a reliable approximation of the action (1), one can study the stability of the position of the fixed point $\tilde{g}_i^*$ and of the values of the critical exponents θ_i: these values should stay almost constant with respect to the inclusion of new operators in the action. In figure (2) it is clearly visible that this seems to be the case, so that our finite sum is a reliable truncation.

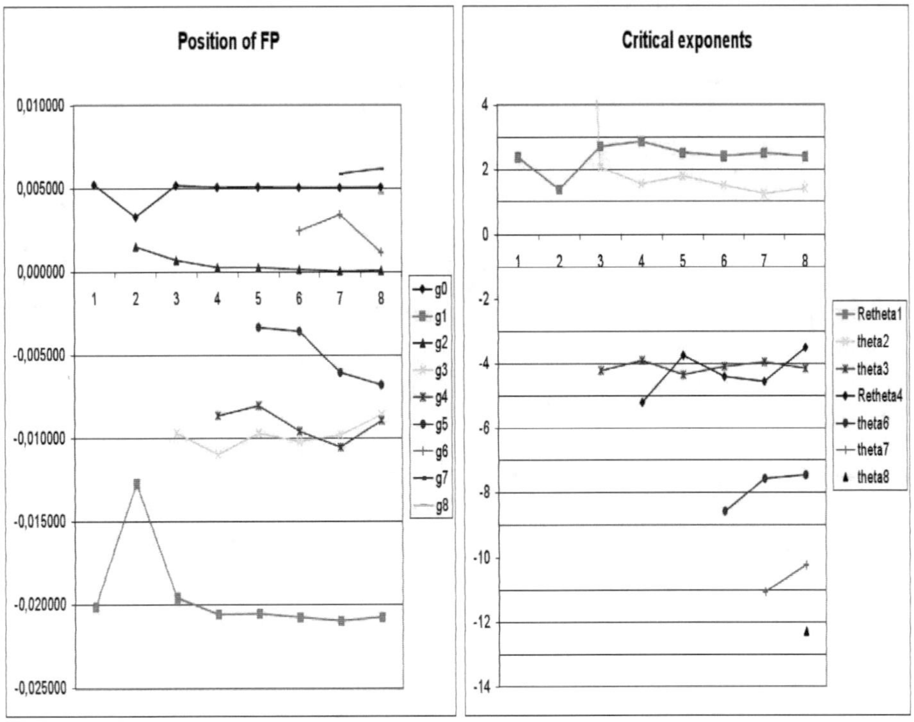

Figure 1. Position of the fixed point and values of the critical exponents as functions of the truncation order n.

Journal of Physics: Conference Series **283** (2011) 012009 doi:10.1088/1742-6596/283/1/012009

3. Asymptotically safe cosmology

It is now time to apply the above described formalism to a cosmological background. There are two main issues that are worth stressing, the first concerning the possibility to obtain any inflationary solution, and the second about the resulting stability of such solutions with respect to the inclusion of new terms in the action. We will address them after having defined a key ingedient of our analysis: the identidication of the RG scale k with a characteristic scale of the described system.

As it was said before, in the Wilsonian approach the cutoff scale is the momentum scale at which the process is taking place. This means that, in a Friedmann-Robertson-Walker metric, it should be set to the order of magnitude of the Hubble rate $H(t)$. Another argument can be found in the comparison with the framework of quantum field theory in curved spaces, where the Hubble rate appears as a mass term in the propagator of the field. We then set $k = \xi H$, where $\xi = \mathcal{O}(1)$ is a free parameter that in principle can be tuned to offset the scheme dependence carried by the cutoff scale.

3.1. *Inflationary cosmology*

We have now all the ingredients to extract the desired inflationary solutions. From the action

$$\Gamma(g) = \frac{1}{16\pi G}\int d^4x\sqrt{|g|}\,(f(R)-\Lambda) \tag{8}$$

where $f(R) = \sum f_i R^i$, we derive the modified Einstein equations that, after having imposed the FRW symmetry, appear as (modified) Friedmann equations

$$\mathcal{A}(H) = 8\pi G\rho + \Lambda \tag{9}$$

$$\mathcal{B}(H) = 8\pi G\rho(1-3w) + 4\Lambda \tag{10}$$

where w si the equation of state of the ordinary matter and $\mathcal{A}(H)$, $\mathcal{B}(H)$ are two functions of the Hubble rate whose shape depends on the chosen $f(R)$, and can be rearranged in the form

$$\mathcal{B}(H) = (1-3w)\mathcal{A}(H) + 3(1+w)\Lambda \tag{11}$$

$$\rho = \frac{1}{8\pi G}\left(\mathcal{A}(H)-\Lambda\right)\,. \tag{12}$$

Before solving the equations, we have to perform the above mentioned cutoff identification: the couplings are rewritten as

$$\Lambda = \tilde{\Lambda}k^2\ ;\quad G = \tilde{G}/k^2\ ;\quad f_i = \tilde{f}_i k^{2-2i} \tag{13}$$

and, because we are interested in describing the dynamics of the very early universe, are assumed to lie in the UV fixed point. With these substitutions, we look for solutions of the (11) in the form of a de Sitter ($H(t) = \bar{H}$ and $a(t) \propto e^{\bar{H}t}$)) and a power law ($H(t) = p/t$ and $a(t) \propto t^p$). It comes out that the system admits the former only for discrete values ξ_{dS} of the free parameter, while we found continuous solutions of the latter type, with $p = p(\tilde{\Lambda}^*, \tilde{G}^*, \tilde{f}_i^*, w, \xi)$. As a final remark, the functions $p(\xi)$ show vertical asymptotes, located exactly at the values ξ_{dS}: de Sitter is then recovered as a limit of power law.

In figure (3.1) the power law exponents p obtained for $n = 8$ are plotted as functions of ξ. There is in particular one growing solution that is greater than 1 (*i.e.* $\ddot{a}/a > 0$) for most values of the free parameter.

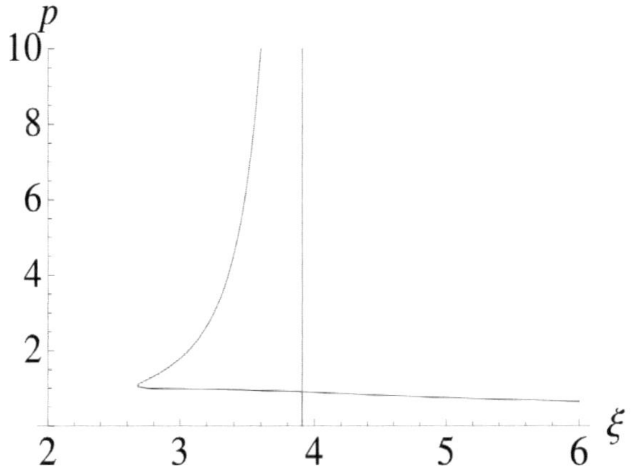

Figure 2. Power law exponents as functions of the parameter ξ. The asymptote is clearly visible for $\xi \simeq 3.9$.

3.2. Reliability analysis

As it was said before, the UV fixed point of the gravitational action (7) tends to stabilize as one widens the truncation. The issue is then whether the values of the solutions will show the same behaviour. We thus define a reliability criterion based on the following logic: taken the dimensionless curvature scalar $\tilde{R} = Rk^{-2}$, we write the dimensionless action with truncation n evaluated at the fixed point $\tilde{f}(\tilde{R})\Big|_n$ and we compare it with the same object in $n-1$. A truncation is then said to be reliable if

$$\left| \frac{\tilde{f}(\tilde{R})\Big|_n - \tilde{f}(\tilde{R})\Big|_{n-1}}{\tilde{f}(\tilde{R})\Big|_{n-1}} \right| < 5\% \qquad (14)$$

and in general this is only true for a limited range of values of $\tilde{R}$. One can schematically rewrite this as $\tilde{R} < c$, where

n	1	2	3	4	5	6	7	8	9	10
c	0.40	0.24	0.45	1.07	1.21	0.90	0.79	0.92	1.09	1.09

The table shows that the reliable range tends to increase, so that a larger truncation is more likely to correctly approximate the infinite truncation $n \to \infty$.

For a power law solution, the constraint on the dimensionless curvature scalar can be translated into a constraint on the values of the exponents, namely

$$c\xi^2 \gtrsim 6\left(2 + \dot{H}/H^2\right) = 6\,(2 - 1/p) \qquad (15)$$

that can be plotted together with the exponents themselves, as it is shown in figure (3.2) on the left. From the plot it comes clear that only a small portion of the growing solution is to be considered reliable, so we decided to extend the truncation to $n = 10$. The result is plotted in figure (3.2) on the right, that shows a great improvement in the reliability of the solutions.

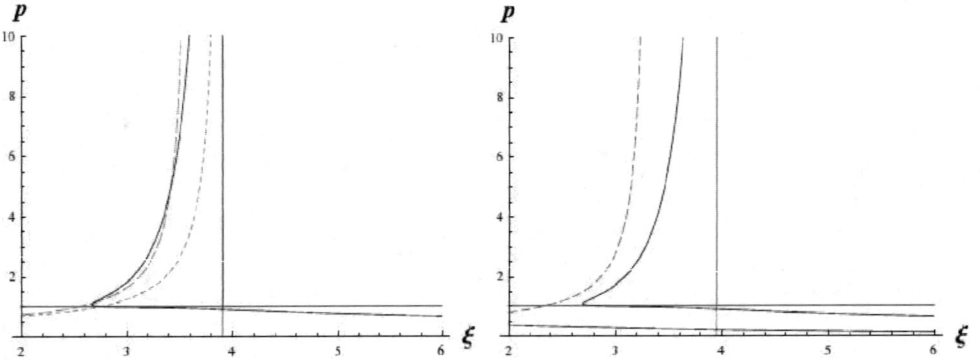

Figure 3. Power law exponents as functions of ξ for $n = 8$ (left) and $n = 10$ (right). Areas to the right of the dashed lines are the reliability regions, while the dotted line indicates the $n = 1$ solution described in [4].

4. Conclusions

The results obtained here can be summarized saying that, if we identify the cutoff with a multiple of the Hubble parameter, inflationary power law solutions (i.e. with exponent $p > 1$) exist for some range of values of ξ. The dependence on such parameter is strong, with the exponents diverging at some values ξ_{dS}. Exactly at those points, the theory admits also de Sitter solutions. Provided that ξ lies in the above range, that the starting point is close enough to the FP and that $p > 1$, it should always be possible to have a sufficient number of e-foldings.

Moreover, it has been shown that the increasing stability of the fixed point with respect to the widening of the truncation seems to be reflected in an increasing reliability of the cosmological solutions obtained using the truncated action. The overall conclusion that one can draw is that we found a strong hint of the convergence of the series in our fixed point action, so that any result coming from low order truncations should not be too misleading.

References

[1] S. Weinberg, "Ultraviolet divergences in quantum theories of gravitation", in "General Relativity, an Einstein Centenary Survey" ed. S. W. Hawking and W. Israel, Cambridge University Press, 1979.
[2] C. Wetterich, Phys. Lett. B**301** (1993) 90.
[3] M. Reuter and H. Weyer, Phys. Rev. D**69** (2004) 104022 and hep-th/0311196
[4] A. Bonanno and M. Reuter, JCAP **08** (2007) 024 and arxiv:hep-th/07060174.
[5] A. Codello, R. Percacci and C. Rahmede, Int. J. Mod. Phys. A **23** (2008) 143, arXiv:hep-th/07051769.
[6] A. Bonanno, A. Contillo and R. Percacci arXiv:gr-qc/10060192.

The quantum origin of cosmic structure: Theory and observations

Konstantinos Dimopoulos
Physics Department, Lancaster University, Lancaster LA1 4YB, UK
E-mail: k.dimopoulos1@lancaster.ac.uk

Abstract. The particle production process is reviewed, through which cosmic inflation can produce a scale invariant superhorizon spectrum of perturbations of suitable fields starting from their quantum fluctuations. Afterwards, in the context of the inflationary paradigm, a number of mechanisms (e.g. curvaton, inhomogeneous reheating etc.) through which such perturbations can source the curvature perturbation in the Universe and explain the formation of structures such as galaxies are briefly described. Finally, the possibility that cosmic vector fields also contribute to the curvature perturbation (e.g. through the vector curvaton mechanism) is considered and its distinct observational signatures are discussed, such as correlated statistical anisotropy in the spectrum and bispectrum of the curvature perturbation.

1. Introduction

The standard model of cosmology at present is comprised by Hot Big Bang Cosmology and Cosmic Inflation. The cosmology of the Hot big Bang accounts for the Hubble expansion of the Universe, the observed Cosmic Microwave Background (CMB) radiation, the primordial abundance of light elements (formed in the Early Universe through the process of Big Bang Nucleosynthesis) and the age of the Universe, which agrees well with the astrophysical estimates of the ages of the oldest globular clusters. In turn inflation overcomes or at least ameliorates certain fine-tuning problems regarding the initial conditions of the Hot Big Bang, namely the so-called horizon and flatness problems [1] [2].

What is cosmic inflation? In a nutshell, inflation is a brief period of superluminal expansion of space in the Early Universe. What inflation does is that it makes the observable Universe large and uniform. However, if the Universe were perfectly uniform there would be no structures like galaxies or galactic clusters, no stars with planets orbiting around them, no ... us. It is imperative, therefore, that there is a deviation from perfect uniformity, which can give rise to these structures. Indeed, we need a Primordial Density Perturbation (PDP) for structure formation to occur. It so happens that evidence for such a PDP exists as the latter reflects itself on the CMB through the so-called Sachs-Wolfe effect [3] which states that CMB light is redshifted when crossing growing overdensities. This effect directly connects the fractional amplitude of the PDP with the fractional perturbation of the temperature of the CMB:

$$\left.\frac{\delta T}{T}\right|_{\rm CMB} = \frac{1}{2}\left.\frac{\delta\rho}{\rho}\right|_{H} \approx 10^{-5}. \tag{1}$$

Even though the PDP appears to be very small, numerical simulations of structure formation

Journal of Physics: Conference Series **283** (2011) 012010 doi:10.1088/1742-6596/283/1/012010

have shown that it is enough to account for the observed structure in the Universe. What is the origin of this PDP? Well, it turns out that this too can be accounted for by Cosmic Inflation.

2. Particle production during cosmic inflation

To have an idea of how Cosmic Inflation is achieved consider the so-called Friedman equation, which is the temporal component of the Einstein equations for a homogeneous and isotropic spacetime. In flat space this equation reeds $H^2 = \frac{1}{3}\rho/m_P^2$, where $m_P^2 = (8\pi G)^{-1}$ is the reduced Planck mass[1] and $H \equiv \dot{a}/a$ is the Hubble parameter corresponding to the rate of the expansion of the Universe, with $a = a(t)$ being the scale factor of the Universe, parameterising the Universe expansion, and the dot denotes derivative with respect to the cosmic time t. Suppose now that, at some period in its early history, the Universe was dominated by an effective cosmological constant $\Lambda_{\rm eff}$. Then, $\rho \simeq m_P^2\Lambda_{\rm eff}$ = constant, which means that the Hubble parameter $H = \Lambda_{\rm eff}/3$ is constant and, therefore, $a \propto e^{Ht}$, i.e. space expands exponentially in time. Thus, inflation occurs when the Universe is dominated by an effective vacuum density. When inflation ends this density has to be transferred in the density of the thermal bath of the Hot Big Bang. This, in effect, amounts to a change of vacuum[2]. Therefore, vacuum states during inflation are not necessarily vacuum states after inflation, but instead they can become populated.

This process is called Particle Production and it is similar to the production of particles (in the form of Hawking radiation) on the event horizon of Black Holes [4]. Indeed, the cosmological horizon during inflation is an event horizon and can be viewed as an "inverted" (i.e. inside-out) black hole in the sense that nothing can escape being "sucked out" by the superluminal expansion. Virtual particle pairs, corresponding to quantum fluctuations, are broken up by the expansion and are pulled away to superhorizon distances, where they can no more find eachother and annihilate, becoming thereby real particles, giving rise to classical perturbations of the corresponding fields. The amplitude of these perturbations is determined by the Hawking temperature $\delta\phi \sim T_H$, which for de Sitter space is $T_H = H/2\pi$ [5]. Let us review now the particle production process in a more rigorous manner.

The standard paradigm considers a real, minimally coupled, scalar field ϕ of mass m with Lagrangian density

$$\mathcal{L} = \frac{1}{2}\partial_\mu\phi\,\partial^\mu\phi - \frac{1}{2}m^2\phi^2. \tag{2}$$

Using the above one obtains the equation of motion of the scalar field. The field is expected to become homogenised by the inflationary expansion as any inhomogeneities are inflated away. In this case the equation of motion becomes

$$\ddot{\phi} + 3H\dot{\phi} + m^2\phi = 0\,, \tag{3}$$

where $\phi = \phi(t)$. Deviations from homogeneity are introduced originally only from its vacuum fluctuations. To follow their evolution we perturb the field from its homogeneous value as $\phi = \phi(t) + \delta\phi(\vec{x}, t)$. Then we obtain the equation of motion of the Fourier components of the field $\delta\phi_k(\vec{k}) \equiv \int \delta\phi(\vec{x})e^{-i\vec{k}\cdot\vec{x}}d\vec{x}$. This equation reads

$$\left[\partial_t^2 + 3H\partial_t + m^2 + \left(\frac{k}{a}\right)^2\right]\delta\phi_k = 0\,. \tag{4}$$

The next step is to promote the perturbations of the field to quantum operators defined as

$$\delta\hat{\phi}(\vec{x}, t) = \int \frac{d^3k}{(2\pi)^3}\left[\hat{a}(\vec{k})\delta\varphi_k(k,t)e^{i\vec{k}\cdot\vec{x}} + \hat{a}^\dagger(\vec{k})\delta\varphi_k^*(k,t)e^{-i\vec{k}\cdot\vec{x}}\right], \tag{5}$$

[1] G is Newton's gravitational constant and we consider natural units where $c = \hbar = k_B = 1$.

[2] from the false vacuum of inflation, corresponding to $\Lambda_{\rm eff}$, to the true vacuum of the present, with $\Lambda \simeq 0$.

where $\hat{a}(\vec{k})$ and $\hat{a}^\dagger(\vec{k})$ are creation and annihilation operators respectively and we consider canonical quantisation with $[\hat{a}(\vec{k}), \hat{a}^\dagger(\vec{k}')] = (2\pi)^3\delta(\hat{k} - \hat{k}')$. The mode functions $\delta\varphi_k(k,t)$ satisfy the same equation of motion as the Fourier components of the field perturbations $\delta\phi_k(\vec{k},t)$ because this equation is linear. We can solve this equation considering that in the subhorizon limit ($k/aH \to +\infty$) the solution matches the so-called Bunch-Davies vacuum [6] which corresponds to the quantum fluctuations of a free scalar field in Minkowski spacetime[3]. Thus, the boundary condition reads

$$\delta\varphi_k \xrightarrow[k/aH\to+\infty]{} \frac{e^{-ik\tau}}{\sqrt{2k}} = \frac{e^{ik/aH}}{\sqrt{2k}}, \tag{6}$$

where $\tau = -1/aH$ is the conformal time, which factors out the expansion of the Universe. Using the above, the solution of Eq. (4) for the mode functions is

$$\delta\varphi_k = a^{-3/2}\sqrt{\frac{\pi}{H}}\frac{e^{i\frac{\pi}{2}(\nu-\frac{1}{2})}}{1-e^{i2\pi\nu}}[J_\nu(k/aH) - e^{i\pi\nu}J_{-\nu}(k/aH)], \tag{7}$$

where J_ν denotes Bessel functions of the first kind and $\nu \equiv \sqrt{\frac{9}{4} - (\frac{m}{H})^2}$. To investigate particle production we evaluate the above solution in the superhorizon limit ($k/aH \to 0$). We find

$$\delta\varphi_k \xrightarrow[k/aH\to 0]{} a^{-3/2}\sqrt{\frac{\pi}{H}}\frac{e^{i\frac{\pi}{2}(\nu-\frac{3}{2})}}{\sin(\pi\nu)}\frac{2^{\nu-1}}{\Gamma(1-\nu)}\left(\frac{aH}{k}\right)^\nu. \tag{8}$$

Using the above we can calculate the power spectrum $\mathcal{P}_{\delta\phi} \equiv \frac{k^3}{2\pi^2}|\delta\varphi|^2$ in the superhorizon limit. We find

$$\mathcal{P}_{\delta\phi} = \frac{8\pi|\Gamma(1-\nu)|^{-2}}{1-\cos(2\pi\nu)}\left(\frac{H}{2\pi}\right)^2\left(\frac{k}{2aH}\right)^{3-2\nu}. \tag{9}$$

Thus, we see that the scale dependence of the field perturbations is only included in the last term of the above. Considering a light field, i.e. $m \ll H \Rightarrow \nu \to \frac{3}{2}$, we see that the scale dependent term is eliminated and the spectrum becomes scale invariant. Indeed, for a light field the above spectrum can be written as

$$\mathcal{P}_{\delta\phi} = \left(\frac{H}{2\pi}\right)^2\left(\frac{k}{2aH}\right)^{2\eta}, \tag{10}$$

where $\eta = \frac{1}{3}(\frac{m}{H})^2$ and we considered $|\eta| \ll 1$. Thus, a light minimally coupled scalar field obtains a scale invariant spectrum of perturbations when they exit the horizon. The typical value of the field perturbations is $\delta\phi \approx \sqrt{\mathcal{P}_{\delta\phi}} \simeq H/2\pi$, i.e. it is determined by the Hawking temperature as expected.

What is the fate of these perturbations after horizon exit? The perturbations can now be treated classically and their evolution is determined by the equivalent of Eq. (3)

$$(\ddot{\delta\phi}) + 3H(\dot{\delta\phi}) + m^2(\delta\phi) = 0\,. \tag{11}$$

Considering that $H \approx$ constant the solution of the above for a light field is

$$\delta\phi = \frac{H}{2\pi}\left[e^{-\frac{1}{3}(\frac{m}{H})^2 H\Delta t} - \frac{1}{9}\left(\frac{m}{H}\right)^2 e^{-3H\Delta t}\right] \approx \frac{H}{2\pi}. \tag{12}$$

Thus, we see that the perturbations of a light scalar field freeze-out after horizon exit. This guarantees that scale invariance is maintained since, regardless of the time of horizon exit (i.e.

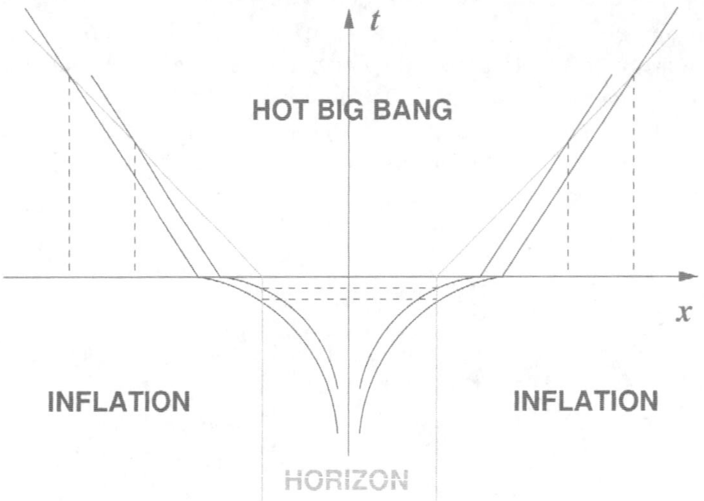

Figure 1. Schematic log-log plot of the evolution of the perturbations of a scalar field (enveloped by red solid lines) which follow the Universe expansion compared with the cosmological horizon (solid green lines). During inflation the cosmological horizon is an event horizon of constant physical radius. The perturbations of the scalar field start off as quantum fluctuations at subhorizon size, which are stretched to superhorizon scales by the (quasi)exponential expansion. After inflation, the size of the superhorizon perturbations continues to grow following the expansion of the Universe (i.e. proportional to $a(t)$) but the cosmological horizon, which is now a particle horizon) grows faster, with the speed of light. As a result, the perturbations reenter the horizon some time after the end of inflation. In the graph two different perturbations are depicted. They exit the horizon at different times, corresponding to the horizontal dashed lines. As a result they are inflated to different sizes so one is much larger that the other when they reenter the horizon. The size of the perturbations at horizon reentry is depicted by the vertical dashed lines. However, both the perturbations have the approximately the same amplitude, determined by the Hawking temperature during inflation $\delta\phi = \frac{H}{2\pi} = T_H$, which remains approximately constant. The fact that the scalar field perturbations retain the same amplitude even though they attain different sizes during their superhorizon evolution is the reason behind the scale-invariance of the perturbation spectrum.

regardless of how large the size of the perturbations becomes by inflation), the perturbations have the same amplitude because $H \approx$ constant until the end of inflation (see Fig. 1).

How are such field perturbations related with the PDP? The PDP arises because of the generation of a corresponding curvature perturbation ζ, as is discussed below. If the curvature perturbation is due to the perturbations of a light scalar field then their power spectra are proportional, i.e. $\mathcal{P}_\zeta \propto \mathcal{P}_{\delta\phi}$. This means that they have the same scale dependence. The latter can be parametrised in the form of a power-law: $\mathcal{P}_\zeta \propto k^{n_s-1}$, so that

$$n_s(k) - 1 \equiv \frac{d\ln\mathcal{P}_\zeta}{d\ln k}\,. \tag{13}$$

[3] Locally spacetime is flat. The subhorizon limit is well within the radius of curvature of spacetime during inflation so this curvature can be ignored. Similarly, in this limit the momentum of the virtual particles is much larger than their mass $k/a \gg m$ so that the field can be considered effectively massless.

Assuming that ζ is due to a light scalar field one obtains (cf. Eq. (10))

$$n_s = 1 + 2\eta + \mathcal{O}(\varepsilon)\,, \tag{14}$$

where $\varepsilon \equiv -\dot{H}/H^2 \ll 1$ quantifies the deviation from pure de Sitter expansion which we have ignored so far. Since, during inflation $|\eta| \ll 1$ for a light scalar field, we find that, if the curvature perturbation is due to this field then $n_s \simeq 1$ and the scale dependence of $\mathcal{P}_\zeta$ is eliminated. Thus, the PDP in this case would be approximately scale-invariant. Indeed, the latest WMAP observations suggest [7]

$$n_s = 0.963 \pm 0.012\,, \tag{15}$$

which agrees with the predictions of inflation with the quantum fluctuations of a light scalar field as the source of the PDP. Note that the observations deviate from exact scale invariance ($n_s = 1$) at 1-σ. This reveals some dynamics during inflation and agrees with the expectations of realistic inflation models.

3. The curvature perturbation

In general relativity the curvature of spacetime and the energy density of its content are interchangeable quantities, through the Einstein equations, depending on the choice of foliation of spacetime. Therefore, for the curvature perturbations it is useful to define a quantity ζ which is independent of such foliation (gauge invariant). This can be written as [8]

$$\zeta \equiv -\psi - H\frac{\delta\rho}{\dot{\rho}}, \tag{16}$$

where the first term on the right-hand-side is the curvature perturbation in uniform density slices of spacetime and the second term is the density perturbation in flat slices of spacetime. We call ζ the (gauge invariant) curvature perturbation from now on.

One can also define the power spectrum of the curvature perturbation $\langle\zeta^2(\vec{x})\rangle = \int_0^\infty d(\ln k)\mathcal{P}_\zeta(k)$ which is given by the two-point correlator as

$$\langle\zeta(\vec{k})\zeta(\vec{k}')\rangle = (2\pi)^2\delta^{(3)}(\vec{k}+\vec{k}')\frac{(2\pi)^3}{4\pi k^3}\mathcal{P}_\zeta(\vec{k}). \tag{17}$$

The latest observations of WMAP give [7]

$$\sqrt{\mathcal{P}_\zeta(k_0)} = (4.94 \pm 0.09) \times 10^{-5}, \tag{18}$$

where $k_0 = 0.002\,\mathrm{Mpc}^{-1}$ is the pivot scale. The corresponding density perturbation (at horizon re-entry) is given by

$$\left(\frac{\delta\rho}{\rho}\right)_H = \frac{2}{5}\zeta_{\rm LS} = (1.98 \pm 0.04) \times 10^{-5}, \tag{19}$$

which, as mentioned, is the observed measurement of the PDP.[4]

Another useful quantity is the so-called bispectrum B_ζ of the curvature perturbation, defined by the three-point correlator of ζ as follows

$$\langle\zeta(\vec{k})\zeta(\vec{k}')\zeta(\vec{k}'')\rangle = (2\pi)^2\delta^{(3)}(\vec{k}+\vec{k}'+\vec{k}'')B_\zeta(\vec{k},\vec{k}',\vec{k}'')\,. \tag{20}$$

The bispectrum can be related with the power spectrum is the following manner

$$B_\zeta(\vec{k}_1,\vec{k}_2,\vec{k}_3) = -\frac{6}{5}f_{\rm NL}[P_\zeta(k_1)P_\zeta(k_2) + P_\zeta(k_2)P_\zeta(k_3) + P_\zeta(k_3)P_\zeta(k_1)], \tag{21}$$

[4] The subscript 'LS' refers to the last-scattering surface, where the CMB is emitted.

where $P_\zeta(k) \equiv \frac{2\pi^2}{k^3}\mathcal{P}_\zeta(k)$.

The bispectrum is useful because it is exactly zero if the curvature perturbation is Gaussian, i.e. if it obeys Gaussian statistics. One expects that the perturbations of the fields which are generated during inflation from their quantum fluctuations are indeed Gaussian since the quantum fluctuations are random. Any non-Gaussian signal therefore would arise by the process which translates the perturbations of such fields, e.g. $\delta\phi$, into the curvature perturbation ζ. This process may be highly non-linear in which case ζ would feature non-Gaussian statistics. The level of such non-Gaussianity is quantified in the bispectrum of ζ by the so-called non-linearity parameter $f_{\rm NL}$. The latter also depends on the configuration of the three $\vec{k}$-vectors used to determine the bispectrum.

The two most popular configurations used to determine $f_{\rm NL}$ are 1) the equilateral configuration, where $k_1 = k_2 = k_3$ and 2) the squeezed configuration, where $k_1 = k_2 \gg k_3$. The WMAP findings for the values of $f_{\rm NL}$ in these configurations are [7]

$$f_{\rm NL}^{\rm eql} = 26 \pm 140 \quad \text{and} \quad f_{\rm NL}^{\rm sqz} = 32 \pm 21\,. \tag{22}$$

Note that there is a hint (at 1-σ) of non-Gaussianity in the squeezed configuration. It is likely that non-Gaussianity may be detected in the near future by the observations of the Planck satellite, which are expected to improve the precision of the measurement of $f_{\rm NL}$ by about an order of magnitude. It is important to stress here that the PDP is highly Gaussian. Indeed, remembering that $\mathcal{P}_\zeta(k_0) \sim 10^{-9}$ and also $B_\zeta \propto f_{\rm NL}\mathcal{P}_\zeta^2$ we see that, even if the observational upper bounds on $f_{\rm NL}$ are saturated, non-Gaussianity in the PDP is tiny. This also agrees with the conjecture that the PDP is due to perturbations of suitable fields (e.g. light scalar fields) arising from their quantum fluctuations during a period of inflation.

4. The inflationary paradigm

Before discussing the mechanisms through which the quantum fluctuations of suitable fields can generate the curvature perturbation from inflation, it is necessary to briefly present the so-called inflationary paradigm. This is the typical manner in which inflation is modelled in particle cosmology. According to the inflationary paradigm the Universe undergoes inflation when dominated by the potential density of a scalar field, which is called the inflaton field.

We return in the Lagrangian of Eq. (2) but this time instead of the mass term we consider a generic function $V(\phi)$, which corresponds to the potential density of the scalar field. In this case, Eq. (3) has the form

$$\ddot{\phi} + 3H\dot{\phi} + V'(\phi) = 0\,, \tag{23}$$

where the prime denotes derivative with respect to the scalar field ϕ. The above equation is similar to the equation of motion of a body sliding down the potential V and subject to a friction term determined by the rate of the Universe expansion H. Based on this analogy we will consider the field as rolling down the potential in field space.

Now, according to the inflationary paradigm, for inflation to take place the Universe must be dominated by the potential density of the inflation field, i.e. the kinetic density $\rho_{\rm kin}(\phi)$ of the field needs to be subdominant:

$$V(\phi) \gg \rho_{\rm kin}(\phi) \equiv \frac{1}{2}\dot{\phi}^2. \tag{24}$$

This means that the field hardly moves and, therefore, its potential density remains roughly constant, providing thereby the effective cosmological constant $\Lambda_{\rm eff} \equiv V/m_P^2$ needed for inflationary expansion. When the condition $\rho_{\rm kin} \ll V$ is valid it is said that the field undergoes slow-roll and its variation (motion in field space) is overdamped by the friction term in Eq. (23). In this case one can ignore the acceleration term in Eq. (23) which is then recast as

$$3H\dot{\phi} \simeq -V'(\phi)\,. \tag{25}$$

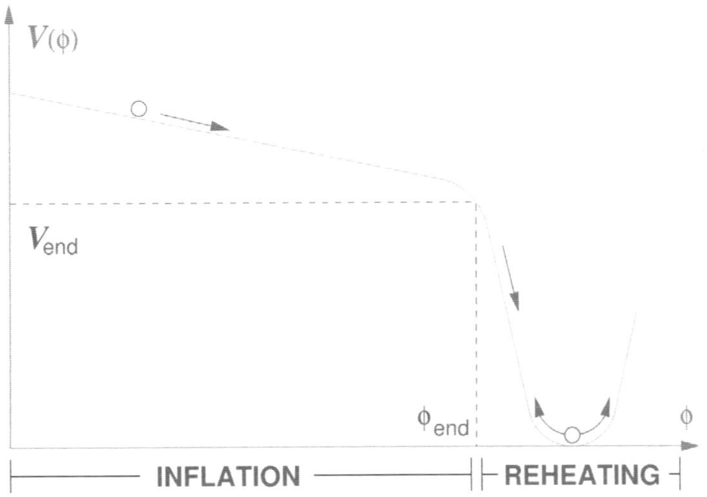

Figure 2. Schematic representation of the inflaton potential $V(\phi)$. During inflation the inflaton field ϕ (represented by a circle) rolls down its flat potential in field space until it reaches a critical value $\phi_{\rm end}$ (corresponding to $V_{\rm end} \equiv V(\phi_{\rm end})$) when slow-roll is terminated and inflation ends. Afterwards the field undergoes rapid oscillations around its VEV, corresponding to the minimum of the potential. The field decays while oscillating reheating thereby the Universe.

Because the roll of the field is overdamped, the value of $\dot{\phi}$ is very small, which means that the slope of the potential needs to be very small as well, according to the slow-roll equation above. Hence, the inflaton field corresponds to a flat direction in field space.

Inflation ends when the slow-roll condition is violated, i.e. at some critical value $\phi_{\rm end}$ of the inflaton field, the potential becomes steep and curved such that $\rho_{\rm kin} \sim V$. Afterwards, the inflaton condensate oscillates around its vacuum expectation value (VEV). Such coherent oscillations correspond to massive particles (inflatons)[5], which eventually decay into the standard model particles that comprise the thermal bath of the Hot big Bang (see Fig. 2). This process is called Reheating and, if it occurs in a perturbative manner, it is usually modelled by adding a phenomenological decay term in the field equation

$$\ddot{\phi} + 3H\dot{\phi} + \Gamma\dot{\phi} + V'(\phi) = 0\,, \tag{26}$$

where Γ stands for the decay rate of the inflaton field. Reheating occurs when the density of inflation is transferred to the newly formed thermal bath, whose temperature is $T_{\rm reh} \sim \sqrt{\Gamma m_P}$, called the reheating temperature.

5. Mechanisms for the formation of the curvature perturbation

We are ready now to discuss some ways that the perturbations of a light scalar field can give rise to the curvature perturbation in the Universe, which corresponds to the PDP that is responsible for structure formation.

5.1. The inflaton hypothesis

The traditional mechanism through which the curvature perturbation is generated employs the inflaton field itself for the job. This, so-called inflaton hypothesis, simply assumes that the field

[5] with zero momentum since the field is homogeneous

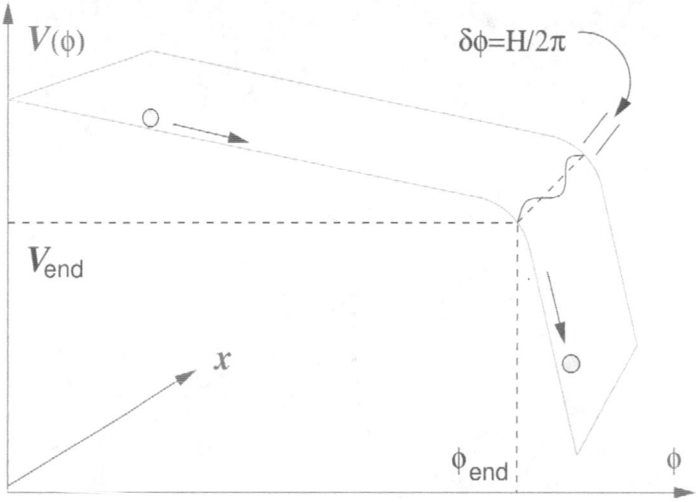

Figure 3. Schematic representation of the way that the curvature perturbation is generated in the inflaton hypothesis. The inflaton field perturbations vary its expectation value typically by $\delta\phi = H/2\pi$. This means that, while the inflaton field (represented by a circle) rolls down its potential in field space, its value is different throughout coordinate space, represented by the x-labelled axis. Thus, in some locations the rolling inflaton is ahead compared to others, which means that, in these locations, it will reach the critical value $\phi_{\rm end}$ that terminates inflation, somewhat earlier. At a given time the inflaton value is represented by the wiggly line which crosses the $\phi = \phi_{\rm end}$ line in several places. Therefore, the Universe inflates more in some places than in others, because, at the given time that corresponds to the wiggly line shown, in some locations inflation has ended while in others it still continues.

responsible for the formation of the curvature perturbation is the same field which determines the dynamics of inflation, i.e. the inflaton field.

As mentioned in the previous section, to undergo slow-roll the inflaton must correspond to a flat direction in field space and therefore is characterised by $V'' \ll H^2$, i.e. it is a light scalar field. This means that, through the particle production process, it obtains a scale-invariant superhorizon spectrum of perturbations. As a result, the perturbed inflaton field will reach the critical value $\phi_{\rm end}$ which terminates inflation at slightly different times at different points in space. Thus, inflation will continue a little bit more in some locations that in others (see Fig. 3). This is the cause of the generation of the curvature perturbation ζ. Indeed, it can be shown that the latter is determined by the difference in logarithmic expansion between the uniform density and the spatially flat slices, i.e. $\zeta = \delta(\ln a)$ [9].

The PDP in this case is given by

$$\left.\frac{\delta\rho}{\rho}\right|_H = \left.\frac{H^2}{5\pi\dot{\phi}}\right|_* , \tag{27}$$

where the asterisk denotes the epoch of horizon exit of the inflaton perturbations. Since in the inflaton hypothesis a single degree of freedom (the inflaton) determines both the dynamics of inflation and the PDP, the latter can be written in terms only of the characteristics of the inflaton potential. Indeed, Eq. (27) can be recast as [10]

$$\left.\frac{\delta\rho}{\rho}\right|_H = \left.\frac{1}{5\sqrt{3}\pi}\,\frac{V^{3/2}}{m_P^3|V'|}\right|_* , \tag{28}$$

which again is to be evaluated at horizon exit. Observational constraints on the amplitude of the PDP (cf. Eq. (19)) suggest that, if ε is not extremely small, the energy scale of inflation is comparable to that of grand unification, i.e. $V^{1/4} \sim 10^{15-16}\,\text{GeV}$. This seems a natural scale to introduce new physics but it turns out that, for single field models, it requires fine-tuning of model parameters.

The spectral index in the inflaton hypothesis is [10]

$$n_s = 1 + 2\eta_\phi - 6\varepsilon\,, \tag{29}$$

where $\eta_\phi \equiv m_P^2 V''/V$ corresponds to the curvature of the potential along the direction of the inflaton field. If $V(\phi) = \frac{1}{2}m_\phi^2\phi^2$ as in Eq. (2), then $\eta_\phi = \frac{1}{3}(\frac{m_\phi}{H})^2$ as discussed after Eq. (10)[6].

Under the inflaton hypothesis the generated non-Gaussianity is expected to be negligible with $f_{\rm NL} \ll 1$, i.e. below the limit of observability. This means that if non-Gaussianity is indeed observed in the PDP then all the single field inflation models are going to be falsified.

The beauty and the curse of the inflaton hypothesis is that it relies on a single degree of freedom, namely the inflaton field, to account for all the problems that inflation aims to address, namely provide the period of accelerated expansion that deals with the horizon and flatness problems and also generate the appropriate spectrum of curvature perturbations, which agrees with observational constrains on its amplitude and spectral index. As a result, models of single field inflation are typically overconstrained and suffer from substantial fine-tuning. This is why alternative hypotheses have been put forward for the generation of the PDP from inflation.

5.2. The curvaton hypothesis

This hypothesis assumes that the field responsible for the formation of the curvature perturbation has nothing to do with the dynamics of inflation, i.e. it is *other* than the inflaton field. This scalar field is called curvaton σ [11] (see also Refs. [12][13][14][15]). In order for the curvaton to play this role it needs to obtain a superhorizon spectrum of perturbations during inflation. Thus, it needs to be a light field during inflation so that it can undergo particle production.

It is evident that by introducing another degree of freedom, the fine-tuning problems of inflation model-building are substantially improved [16][17]. However, it must be stressed that the curvaton is not necessarily a new ad hoc addition to the theory, which is introduced by hand. Indeed, the fact that the curvaton is not connected to the inflaton sector allows it to correspond to physics at energy scales much smaller than that of inflation (e.g. the TeV scale which is accessible by LHC). Consequently, there exist a number of candidates for the curvaton which correspond to realistic fields already present in simple extensions of the Standard Model. Prominent examples are a right-handed neutrino [18], an MSSM flat direction [19][20] or the so-called orthogonal axion in supersymmetric realisations of the Peccei-Quinn symmetry [21][22].

Under the curvaton hypothesis the curvature perturbation is given by [23]

$$\zeta = \hat{\Omega}_\sigma \zeta_\sigma\,, \tag{30}$$

where $\hat{\Omega}_\sigma \equiv \frac{3\Omega_\sigma}{4-\Omega_\sigma} \simeq \Omega_\sigma$, where $\Omega_\sigma \equiv (\rho_\sigma/\rho)_{\rm dec}$ is the density parameter of the curvaton field at the time of its decay after inflation[7]. In the above ζ_σ is the curvature perturbation attributed to the curvaton field, which is determined by the fractional perturbation of the field itself:[8]

$$\zeta_\sigma \equiv -H\frac{\delta\rho_\sigma}{\dot{\rho}_\sigma} = \frac{1}{3}\frac{\delta\rho_\sigma}{\rho_\sigma} = \frac{2}{3}\frac{\delta\sigma}{\sigma} \simeq \frac{2}{3}\frac{\delta\sigma}{\sigma}\bigg|_* = \frac{H_*}{3\pi\sigma_*}\,, \tag{31}$$

[6] we considered also the Friedman equation with $\rho \simeq V$ since we have potential domination during inflation.

[7] $\hat{\Omega}_\sigma$ is also denoted as r in much of the literature.

[8] in spatially flat hypersurfaces

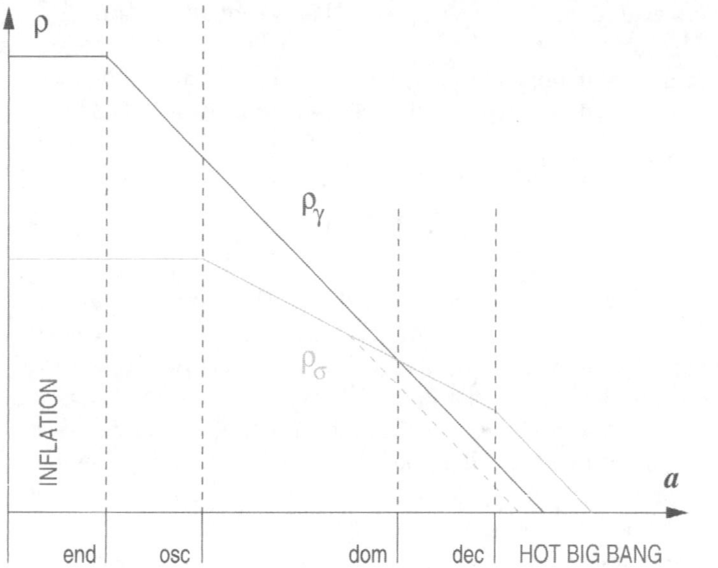

Figure 4. Log-log plot of the evolution of the inflaton energy density which decays into radiation ρ_γ (purple line) at the end of inflation (denoted by 'end') and the curvaton energy density ρ_σ (green line) (prompt reheating is assumed). During inflation, the curvaton density is negligible. After inflation $\rho_\gamma \propto a^{-4}$. In contrast, ρ_σ remains constant (the curvaton is frozen at some value σ_*) until $m \sim H(t)$, when the curvaton unfreezes and begins oscillating, after which time (denoted 'osc') $\rho_\sigma \propto a^{-3}$. At some moment (denoted 'dom') the curvaton density dominates the Universe ($\hat{\Omega}_\sigma = \Omega_\sigma = 1$) until, some time later (denoted 'dec') when it decays into the thermal bath of the Hot Big Bang. The dashed slanted line depicts the possibility that the curvaton decays before domination ($\hat{\Omega}_\sigma = \frac{3}{4}\Omega_\sigma \ll 1$), when substantial non-Gaussianity can be generated.

where we used that $\delta\sigma_* = H_*/2\pi$ and we assumed that near its decay the curvaton density is $\rho_\sigma \sim m_\sigma^2\sigma^2$, where $m_\sigma \ll H_*$ is the mass of the curvaton field. The spectral index of the PDP in this case is [23]

$$n_s = 1 + 2\eta_\sigma - 2\varepsilon\,, \tag{32}$$

where η_σ corresponds to the curvature of the potential along the curvaton direction.

Since by definition the curvaton should not affect the dynamics of inflation, during inflation we expect $\rho_\sigma \ll \rho$. Thus the density parameter of the curvaton is extremely small and its contribution to the overall curvature perturbation (cf. Eq. (30)) is also small. For the curvaton to significantly contribute to ζ we need to consider the period after inflation, when its contribution to the density can increase.

In the simplest case when $V(\sigma) \simeq \frac{1}{2}m_\sigma^2\sigma^2$ the equation of motion for the curvaton is of the same form as Eq. (3). Since during inflation the field is light and $m_\sigma \ll H$, the curvaton is overdamped and remains frozen in some value σ_*. After inflation, however, the Hubble parameter decreases in time $H(t) \propto 1/t$. As a result, there will be a moment when $m_\sigma \sim H(t)$, at which time the curvaton condensate will unfreeze and will begin coherent oscillations around its VEV. These oscillations correspond to massive particles (curvatons) whose density is diluted as $\rho_\sigma \propto a^{-3}$, which is less drastic that the density of the radiation background[9] $\rho_\gamma \propto a^{-4}$. Thus, the oscillating curvaton has a chance to dominate (or nearly dominate) the Universe before its decay

[9] This is the radiation that was generated by the decay of the inflaton field after the end of inflation.

(see Fig. 4). Thus, at curvaton decay $\hat{\Omega}_\sigma$ can be as large as unity and the curvaton contribution to ζ can be substantial. Consequently, the curvaton imposes its curvature perturbation onto the Universe at (or near) its domination.

Because of its spectrum of perturbations $\delta\sigma$ the amplitude of the curvaton oscillations is also perturbed, i.e. it is larger in some places than in others. This means that the density of the oscillating field is perturbed too and the field will dominate the Universe at different times at different locations. This is what generates the curvature perturbation (cf. Eq. (31)).

In contrast to the inflaton hypothesis, under the curvaton hypothesis non-Gaussianity can be substantial. Indeed, in this case [23]

$$f_{\rm NL} = \frac{5}{4\hat{\Omega}_\sigma}, \tag{33}$$

which can be large if the decay of the curvaton happens before domination when $\Omega_\sigma \ll 1$. In fact, the WMAP observations in Eq. (22) set the lower bound $\Omega_\sigma \gtrsim 0.01$.

5.3. *Other mechanisms*

There are numerous other ways to generate the PDP from a superhorizon spectrum of scalar field perturbations. This section briefly reviews two of them, namely the inhomogeneous reheating and the end of inflation mechanisms. As with the curvaton mechanism, these mechanisms assume that the contribution to the curvature perturbation from the inflaton field itself is negligible.

5.3.1. Inhomogeneous Reheating This mechanism generates the PDP by assuming that the inflaton decay rate Γ is modulated by another scalar field σ [24][25]. This scalar field is the one which undergoes particle production during inflation and which obtains thereby a superhorizon spectrum of perturbations.

In this scenario we can ignore the perturbations of the inflaton field. Thus, we can consider that inflation is terminated at the same time throughout space[10]. The inflaton then begins its coherent oscillations around its VEV until it decays when $\Gamma \simeq H(t)$. However, since in this case $\Gamma = \Gamma(\sigma)$ and σ is perturbed in space, the decay rate is different at different locations so that the Hot Big Bang begins at different times. This is what generates the curvature perturbation which corresponds to the PDP. Roughly speaking we have

$$\frac{\delta\rho}{\rho} \sim \frac{\delta\Gamma}{\Gamma} \sim \frac{\delta\sigma}{\sigma}. \tag{34}$$

In this scenario the spectral index is again given by Eq. (32). Non-Gaussianity in this model is quantified as [26]

$$f_{\rm NL} = 5\left(\frac{\Gamma''\Gamma}{\Gamma'^2} - 1\right), \tag{35}$$

where now the prime denotes derivative with respect to σ. If Γ has a power-law dependence on σ then $f_{\rm NL} = \mathcal{O}(1)$, which is marginally observable.

5.3.2. End of inflation This mechanism applies to a particular type of inflation model, the so-called hybrid inflation. Therefore, before discussing the mechanism a brief summary of hybrid inflation is necessary.

Hybrid inflation couples the inflaton field ϕ to another scalar field ψ, called the waterfall field, in such a way that inflation is terminated by a phase transition which sends the waterfall to its VEV [27]. Since inflation is likely to be at the scale of grand unification, in many cases

[10] at least in the observable Universe.

the waterfall field is assumed to be the Higgs field of a Grand Unified Theory (GUT). Thus, inflation is terminated by the breaking of grand unification, i.e. the GUT phase transition.

The scalar potential for hybrid inflation has the following form

$$V(\phi, \psi) = \frac{1}{4}\lambda(\psi^2 - M^2)^2 + \frac{1}{2}g\phi^2\psi^2 + V_\phi(\phi) \tag{36}$$

where M is the GUT energy scale, λ is the self-coupling of the waterfall field and g is the interaction coupling between the waterfall and the inflaton. The potential $V_\phi(\phi)$ is responsible for the slow-roll of the inflaton and its precise form is not relevant to this discussion[11]. From the above potential it is evident that the effective mass-squared of the waterfall is

$$m_\psi^2 = g\phi^2 - \lambda M^2 + \lambda\phi^2. \tag{37}$$

The above implies that there is a critical value of the inflaton

$$\phi_c = \sqrt{\frac{\lambda}{g}}M \tag{38}$$

such that if $\phi \gg \phi_c$ then $m_\psi^2 > 0$ and $\psi \to 0$. In this case, Eq. (36) becomes $V = \frac{1}{4}\lambda M^4 + V_\phi$. The constant contribution to the scalar potential provides the effective cosmological constant for (quasi)de Sitter inflation: $\Lambda_{\rm eff} \sim \lambda M^4/m_P^2$. If however, $\phi < \phi_c$ then $m_\psi^2 < 0$ and a phase transition occurs which results in $\psi \to M$, which gives the inflaton a large mass (through the interaction term) and so $\phi \to 0$. Assuming $V_\phi(0) = 0$, after the phase transition $V \to 0$ and inflation ends.

The end of inflation mechanism for the production of the PDP introduces an extra coupling between the waterfall field and another scalar field σ [30][31]. The scalar potential now becomes

$$V(\phi, \psi) = \frac{1}{4}\lambda(\psi^2 - M^2)^2 + \frac{1}{2}g\phi^2\psi^2 + V_\phi(\phi) + \frac{1}{2}h\sigma^2\psi^2 \tag{39}$$

where h parametrises the strength of the interaction between σ and the waterfall field ψ. With this addition the effective mass-squared of the waterfall becomes

$$m_\psi^2 = g\phi^2 - \lambda M^2 + \lambda\phi^2 + h\sigma^2 \tag{40}$$

and the critical value which triggers the phase transition that ends inflation is now

$$\phi_c(\sigma) = \left(\frac{\lambda}{g}M^2 - \frac{h}{g}\sigma^2\right)^{1/2}, \tag{41}$$

i.e. it is modulated by the value of σ. Therefore, if σ is light during inflation, then it undergoes particle production and obtains a superhorizon spectrum of perturbations with typical magnitude $\delta\sigma = H/2\pi$. This means that the value of ϕ_c is also perturbed. Hence, the phase transition which terminates inflation occurs earlier in some parts of the Universe than in others depending on when the inflaton reaches the critical value ϕ_c. Consequently, the Universe inflates more in some locations than in others and this generates the curvature perturbation. As in the curvaton scenario, the spectral index in this case is given by Eq. (32) [32].

[11] In supersymmetric versions of hybrid inflation the slow-roll potential is provided by radiative corrections and it is of the form $V_\phi \propto \ln\phi$ [28][29].

6. Cosmic vector fields and the curvature perturbation

Tantalising evidence exists of a preferred direction in the CMB temperature perturbations. In particular, the low multipoles in the CMB appear to be aligned [33][34] at a level which is statistically rather improbable [35][36]. A preferred direction in the CMB cannot be accounted for if inflation and the generation of the PDP is due to scalar fields only. Moreover, despite their abundance in theories beyond the Standard Model, scalar fields have not been observed as yet. If the Higgs field is not found in the LHC, the credibility of the ubiquitous usage of scalar fields in cosmology will be shaken.

Until recently only scalar fields have been considered both as responsible for the dynamics of inflation and also for the generation of the observed PDP. However, in the pioneering work of Ref. [37] the possibility that a vector field contributes in the generation of the PDP was first considered. Since then, a number of attempts have been made to investigate the role and the implications of cosmic vector fields in the generation of the PDP (see Ref. [38] and references therein).

Why vector fields were not considered originally for the generation of the PDP? After all, these are fields which are similar to the massive gauge bosons, which have indeed been observed in LEP. The main difficulty had to do with the inherent anisotropic nature of vector fields. Inflation would homogenise a vector field condensate, and a homogeneous vector field picks up a preferred direction in space. Thus, if this vector field condensate were to dominate the density of the Universe (so that it can affect the expansion and generate the PDP) it was felt that the anisotropic stress generated would lead to strongly anisotropic expansion which would be impossible to reconcile with the predominant isotropy of the CMB. However, as is discussed below, this difficulty can be circumvented in the vector field remains subdominant during inflation. Still, if (in analogy to scalar fields) light vector fields were needed for a scale invariant spectrum then a more subtle problem arises. Massless Abelian vector fields are conformally invariant and they cannot undergo particle production during inflation[12]. Thus, for light vector fields particle production is expected to be suppressed. Nevertheless, there are numerous mechanisms which break the conformality of vector fields so that particle production can be efficient. These mechanisms are model dependent which means that they can have distinct observational signatures as is discussed below.

7. Particle production of vector fields during inflation

Suppose that there is a suitable theory which breaks the vector field conformality during inflation. How are we to study particle production of the vector field? We follow the same recipe as in Sec. 2 assuming that the inflationary expansion remains isotropic.

Firstly we perturb the vector field around its homogenised value $A_\mu = A_\mu(t) + \delta A_\mu(\vec{x}, t)$. Then we Fourier transform the perturbations $\delta A_i(\vec{k}) = \int \delta A_i(\vec{x}) e^{-i\vec{k}\cdot\vec{x}} d\vec{x}$ and we obtain the equations of motion for the Fourier components in the theory which we are considering[13]. Then we promote the perturbations of the vector field into quantum operators

$$\delta \hat{A}_i(\vec{x}, t) = \int \frac{d^3k}{(2\pi)^3} \sum_\lambda \left[\hat{e}_i^\lambda \hat{a}_\lambda(\vec{k}) \delta \mathcal{A}_k(k,t) e^{i\vec{k}\cdot\vec{x}} + \hat{e}_i^{\lambda *} \hat{a}_\lambda^\dagger(\vec{k}) \delta \mathcal{A}_k^*(k,t) e^{-i\vec{k}\cdot\vec{x}} \right], \qquad (42)$$

where $\delta \mathcal{A}_k$ are the mode functions and we consider again canonical quantisation $[\hat{a}_\lambda(\vec{k}), \hat{a}_{\lambda'}^\dagger(\vec{k}')] = (2\pi)^3 \delta(\hat{k} - \hat{k}')\delta_{\lambda\lambda'}$. In the above we have introduced the polarisation vectors

$$\hat{e}^L \equiv \frac{1}{\sqrt{2}}(1, i, 0) \qquad \hat{e}^R \equiv \frac{1}{\sqrt{2}}(1, -i, 0) \qquad \hat{e}^{\|} \equiv (0, 0, 1), \qquad (43)$$

[12] They view the Universe expansion as a conformal transformation to which they are insensitive.

[13] We focus on the spatial components of the cosmic vector field, as they are the ones which would generate anisotropy. Also, the temporal component of a homogeneous massive Abelian vector field is zero.

where L, R denote the left and right transverse polarisations and $\|$ denotes the longitudinal polarisation[14].

As with the scalar field case, the mode functions of the above expansion are expected to satisfy the equations of motion for the Fourier components of the vector field perturbations. The form of these equations depends on the theory which breaks the conformality of the vector field. To solve them we again consider vacuum boundary conditions in the subhorizon limit $(k/aH \to \infty)$. They are [40]:

$$\delta\mathcal{A}_k^{L,R} \xrightarrow[k/aH\to+\infty]{} \frac{e^{ik/aH}}{\sqrt{2k}} \quad \text{and} \quad \delta\mathcal{A}_k^{\|} \xrightarrow[k/aH\to+\infty]{} \gamma\frac{e^{ik/aH}}{\sqrt{2k}}\,. \tag{44}$$

We see that the boundary condition is again the Bunch-Davies vacuum but the longitudinal component it is multiplied by the Lorentz boost factor $\gamma \equiv \frac{E}{m} = \sqrt{(\frac{k}{m})^2+1}$, which takes us from the frame with $\vec{k}=0$ where all components are equivalent, to the frame of momentum $\vec{k}$.[15]

After solving the equations of motion we evaluate the solutions in the superhorizon limit $(k/aH \to 0)$. Then we can obtain the power spectrum for each polarisation as $\mathcal{P}_\lambda(k) \equiv \frac{k^3}{2\pi^2}|\delta\mathcal{A}_k^\lambda|^2$, where $\delta\mathcal{A}_k^\lambda$ is evaluated in the superhorizon limit. Because we have three degrees of freedom there are three possibilities:

Case A: $\mathcal{P}_\| \neq \mathcal{P}_L \neq \mathcal{P}_R \neq \mathcal{P}_\|$
Case B: $\mathcal{P}_L = \mathcal{P}_R \neq \mathcal{P}_\|$
Case C: $\mathcal{P}_L = \mathcal{P}_R = \mathcal{P}_\|$

Even though Case A appears to be the most generic, in practice it is Case B that is the most common because usually the theories which break the vector field conformality are parity conserving. We have isotropic particle production only in the special Case C.

What would happen if the spectrum of the produced perturbations of the vector field affected the curvature perturbation? Since particle production is in general anisotropic we expect an anisotropic contribution to ζ. This contribution can be parametrised as follows. For the power spectrum we can write [39]

$$\mathcal{P}_\zeta(\vec{k}) = \mathcal{P}_\zeta^{\rm iso}(k)\left[1+g(\hat{A}\cdot\hat{k})^2\cdots\right], \tag{45}$$

where $\hat{A} \equiv \vec{A}/A$ and $\hat{k} \equiv \vec{k}/k$ with $A \equiv |\vec{A}|$ and the ellipsis denotes higher order contributions. The anisotropic part of the spectrum corresponds to a new observable, namely *Statistical Anisotropy*, which amounts to direction dependent patterns in the CMB [40] (see Fig. 5). Statistical anisotropy is quantified by the so-called anisotropy parameter g in Eq. (45). The WMAP observations set a surprisingly weak upper bound on the anisotropy parameter: $g \lesssim 0.3$ [41][16]. Thus we see that, at present, observations allow up to 30% statistical anisotropy in the spectrum of the PDP. The forthcoming data of the Planck satellite will improve precision by an order of magnitude and are likely to detect statistical anisotropy. If this will be so then a cosmic vector field will have to be involved in the generation of the PDP during inflation.

In a similar manner one can parametrise the statistically anisotropic contribution from a vector field in the bispectrum. Indeed, we can write [44][45]

$$f_{\rm NL} = f_{\rm NL}^{\rm iso}\left(1+\mathcal{G}\hat{A}_\perp^2\cdots\right), \tag{46}$$

[14] The index λ in Eq, (42) runs over these values.

[15] Note that for a massless vector field the longitudinal component is unphysical.

[16] In Ref. [42] statistical anisotropy with $g = 0.29 \pm 0.03$ is claimed to be detected at a level of 9-σ but the authors acknowledge that, because the direction is suspiciously near the ecliptic plane, this is probably due to an unknown systematic error. Hence we treat this number as an upper bound.

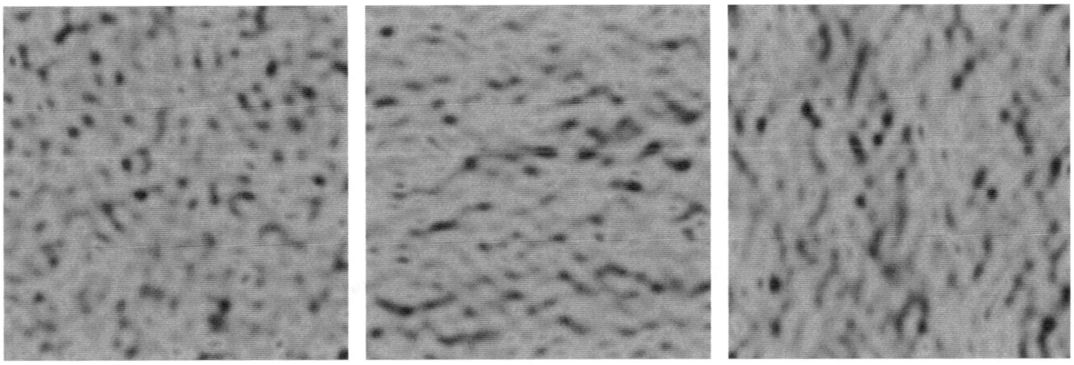

Figure 5. Patterns in the CMB temperature perturbations which may arise from statistical anisotropy in the CMB spectrum. The left panel shows an isotropic signal, while the middle and right panels show patterns due to statistical anisotropy along the vertical and horizontal direction respectively. The figure is taken from Ref. [43].

where $f_{\rm NL}^{\rm iso}$ denotes the isotropic part and $\hat{A}_\perp$ is the projection of $\hat{A}$ onto the plane of the three $\vec{k}$-vectors which are used to define the bispectrum. Note that, as non-Gaussianity has not been observed yet, there are no bounds on $\mathcal{G}$. If $\mathcal{G} > 1$ this means that non-Gaussianity is predominantly anisotropic (even though $\mathcal{P}_\zeta$ is not). If non-Gaussianity is indeed observed (e.g. by the Planck satellite) and no angular modulation of $f_{\rm NL}$ is found on the microwave sky then all models which predict $\mathcal{G} > 1$ will be falsified. It is important also to stress that the directions of statistical anisotropy in the spectrum and the bispectrum are correlated (determined by the direction of $\hat{A}$), which is a smoking gun for the contribution of a vector field to the PDP [44].

In the Cases A and B the spectra of the superhorizon perturbations of the vector field are predominantly anisotropic. Since $\mathcal{P}_\zeta$ is isotropic at least at the level of 70% the contribution of the vector field to the PDP in these two cases has to be subdominant and its significance is only the possibly observable statistical anisotropy in the spectrum and bispectrum. Thus, in the Cases A and B we still require some other, isotropic source (such as a scalar field) to provide the dominant contribution to the PDP. On the other hand, in Case C the perturbation spectra of the vector field are isotropic. In this case, the vector field alone can generate ζ and no input from any scalar field is necessary. Note that, in this case, the vector field may also produce statistical anisotropy if the $\mathcal{P}_\lambda$ are not exactly the same but differ slightly, albeit by no more that 30%.

8. Models for particle production of cosmic vector fields

In this section we discuss a couple of proposals for the generation of a flat superhorizon spectrum of vector field perturbations.

8.1. Non-minimal coupling to gravity

Consider the following theory

$$\mathcal{L} = -\frac{1}{4}F_{\mu\nu}F^{\mu\nu} + \frac{1}{2}(m^2 + \alpha R)A_\mu A^\mu, \tag{47}$$

where $F_{\mu\nu} = \partial_\mu A_\nu - \partial_\nu A_\mu$ is the field strength of the Abelian vector field A_μ, m is its bare mass, α is constant and R is the Ricci scalar. From the above we see that the vector field has effective mass-squared $m_A^2 = m^2 + \alpha R$.

Following the procedure outlined in the previous section, we obtain the following solution for the mode functions of the transverse perturbations of the physical vector field [46][17]

$$\delta\mathcal{A}_k^{L,R} = a^{-3/2}\sqrt{\frac{\pi}{H}}\frac{e^{i\frac{\pi}{2}(\nu-\frac{1}{2})}}{1-e^{i2\pi\nu}}[J_\nu(k/aH) - e^{i\pi\nu}J_{-\nu}(k/aH)]. \tag{48}$$

The above is identical to Eq. (7) with the crucial difference that $\nu \equiv \sqrt{\frac{1}{4} - (\frac{m_A}{H})^2}$. Using this solution we obtain the following expression for the power spectrum of the transverse perturbations in the superhorizon limit

$$\mathcal{P}_{L,R} = \frac{8\pi|\Gamma(1-\nu)|^{-2}}{1-\cos(2\pi\nu)}\left(\frac{H}{2\pi}\right)^2\left(\frac{k}{2aH}\right)^{3-2\nu}, \tag{49}$$

which again is identical to Eq. (9) but for the different value of ν. From the above it is evident that a scale invariant spectrum is attained for $\nu \approx \frac{3}{2}$. This translates into a requirement for both α and m as follows.

In a spatially flat, homogeneous and isotropic spacetime, the scalar curvature is $R = 3(3w-1)H^2$, where w is the barotropic parameter of the content of spacetime. During inflation, $w \simeq -1$ so that $R \simeq -12\,H^2$. This means that we can attain a scale invariant transverse spectrum if 1) $\alpha = \frac{1}{6}$ and 2) $m \ll H$, i.e. the vector field is a light field with a negative effective mass-squared [46][18]. Under these conditions the transverse spectra are indeed scale-invariant and are given by $\mathcal{P}_{R,L} = (\frac{H}{2\pi})^2$, exactly as a massless scalar field.

Employing the same conditions for α and m one obtains the following solution for the longitudinal component of the physical vector field [40]:

$$\delta\mathcal{A}_k^{\|} = \frac{1}{\sqrt{2}}\left[\left(\frac{k}{aH}\right) - 2\left(\frac{aH}{k}\right) + 2i\right]\frac{e^{ik/aH}}{\sqrt{2k}}. \tag{50}$$

Clearly the above is totally different from Eq. (48). In the superhorizon limit the power spectrum of the longitudinal component is $\mathcal{P}_{\|} = 2(\frac{H}{2\pi})^2 = 2\mathcal{P}_{R,L}$. This means that we are in Case B since $\mathcal{P}_{\|} > \mathcal{P}_{L,R}$ and $\mathcal{P}_L = \mathcal{P}_R$ because the theory is parity invariant. Thus, particle production in this theory is anisotropic at a level of 100%. Therefore, the vector field cannot alone be responsible for the PDP. Its contribution to ζ has to be subdominant and it can only be the source of statistical anisotropy. Before concluding, we should point out that this model has been criticised for suffering from instabilities (ghosts) [47][48] (see however Ref. [49]).

8.2. Varying kinetic function and mass

Now, consider the theory

$$\mathcal{L} = -\frac{1}{4}fF_{\mu\nu}F^{\mu\nu} + \frac{1}{2}m^2A_\mu A^\mu, \tag{51}$$

where $F_{\mu\nu} = \partial_\mu A_\nu - \partial_\nu A_\mu$, $f = f(t)$ is the kinetic function and we also consider $m = m(t) > 0$ during inflation. The Maxwell-type kinetic term in combination with the positive mass-squared guarantees the stability of the model [50] and therefore is motivated even if the vector field is

[17] In a flat homogeneous and isotropic Universe the spatial components of the physical (in contrast to comoving) vector field are A_i/a.

[18] The non-minimal coupling to gravity $\alpha = \frac{1}{6}$ has a special property. It turns the conformally invariant massless Abelian vector field equivalent (in the sense that its field equations are the same) to a (set of two) minimally coupled massless scalar field(s). If applied to a scalar field it renders it conformally invariant. In this respect it appears to reflect a deeper symmetry and therefore it may be a natural value for α.

not a gauge boson. Note also that a massive Abelian vector field is renormalisable even is it is not a gauge field [51].

The solutions for the mode functions of the field perturbations are too complicated to reproduce here. It suffices to say that scale invariance in the transverse components requires the kinetic function to scale with the expansion as $f \propto a^{-1\pm 3}$ and also the physical vector field to be light at horizon exit $M \ll H$, where $M \equiv m/\sqrt{f}$. Scale invariance for the longitudinal component additionally requires $m \propto a$ [38][52].

If the vector field is a gauge boson then f is the gauge kinetic function, which is related to the gauge coupling as $f \sim 1/e^2$. This means that only the case when $f \propto a^{-4}$ is acceptable because only then the gauge field remains weakly coupled during inflation. Note that the model is naturally realisable in supergravity theories where f is a holomorphic function of the scalar fields of the theory $f(\phi_i)$ [53]. In general, Kähler corrections to the scalar potential result in masses of order H for the scalar fields[19], which are therefore expected to be fast-rolling down the potential slopes during inflation, causing significant variation to f. Indeed, for a power-law dependence of the gauge kinetic function to the scalar fields, it is easy to show that $\dot{f}/f \sim H$ [38], i.e. f has a power-law dependence on a as assumed in this model. Now, if f is modulated by the inflaton field itself, then it can be shown that, under fairly general conditions, the backreaction to the inflaton's roll renders the scaling $f \propto a^{-4}$ an attractor solution [54]. Similarly, for a gauge boson, m can be modulated by a fast-rolling Higgs field with tachyonic mass $m_H = 2H$ [38].

Given the conditions $f \propto a^{-1\pm 3}$ and $m \propto a$ the power spectra for the transverse and longitudinal components depend on whether the vector field remains light until the end of inflation or not (note that $M = M(t)$). In particular we find [38][52]

$$M \ll H: \quad \mathcal{P}_{L,R} = \left(\frac{H}{2\pi}\right)^2 \quad \text{and} \quad \mathcal{P}_{\|} = \left(\frac{H}{2\pi}\right)^2 \left(\frac{3H}{M}\right)^2 \tag{52}$$

$$M \gg H: \quad \mathcal{P}_{L,R} = \mathcal{P}_{\|} = \frac{1}{2}\left(\frac{H}{2\pi}\right)^2 \left(\frac{3H}{M}\right)^2 . \tag{53}$$

From the above we see that, if the vector field remains light until the end of inflation, $\mathcal{P}_{\|} \gg \mathcal{P}_{L,R}$, i.e. we are in Case B. Therefore, particle production is strongly anisotropic and the vector field contribution to the PDP has to be subdominant but it can still give rise to substantial statistical anisotropy. In contrast, if the field becomes heavy by the end of inflation then particle production is isotropic and the vector field can be solely responsible for the generation of the PDP. Note that, because we need the field to be light when the cosmological scales exit the horizon, for it to become heavy we need $\dot{M} > 0$ during inflation, which is possible only in the case when $f \propto a^{-4}$. If $f \propto a^2$ then $M =$constant and we have $M \ll H$ throughout inflation. But if $f \propto a^{-4}$ then we have $M \propto a^3$ so that we may end up having $M \gg H$ at the end of inflation even though we started of with a light field at horizon exit. In the latter case the vector field begins coherent oscillations a few exponential expansions (e-folds) before the end of inflation. Assuming that at the end of inflation $f \to 1$ and the field becomes canonically normalised, we find that there is ample parameter space for Case C to be realised, namely: $1 < m/H < 10^6$ [38][52].

9. The vector curvaton paradigm

Through the examples in the previous section it is evident that a scale invariant spectrum of perturbations (isotropic or not) of a vector field can indeed be generated if one assumes some theory which appropriately breaks the conformality of the vector field. However, in order for these perturbations to affect or even generate the PDP the vector field needs to affect the Universe expansion, i.e. its density should become dominant (or nearly dominant) at some point.

[19] This is the source of the famous η-problem of inflation, since slow-roll requires $|\eta| = \frac{1}{3}(\frac{m}{H})^2 \ll 1$ for the inflaton.

But, even if particle production is isotropic the homogeneous zero-mode condensate is not. How can we avoid excessive anisotropic stress when the vector field dominates the expansion? It turns out that this is possible if the vector field plays the role of the curvaton.

Consider a minimally coupled massive Abelian vector field for which

$$\mathcal{L} = -\frac{1}{4}F_{\mu\nu}F^{\mu\nu} + \frac{1}{2}m^2 A_\mu A^\mu, \tag{54}$$

where $F_{\mu\nu} = \partial_\mu A_\nu - \partial_\nu A_\mu$ and $m = \text{constant} > 0$. It is clear that both models in the previous section eventually approach the above theory, when $m^2 \gg R \sim H^2$ and also after the end of inflation when $f \to 1$ and $m = \text{constant}$. The vector field condensate is homogenised by inflation so that $A_\mu = A_\mu(t)$. In this case it can be shown that the temporal component is zero and the spatial components satisfy the following equation [37]

$$\ddot{A}_i + H\dot{A}_i + m^2 A_i = 0\,, \tag{55}$$

which is similar to Eq. (3) for the scalar field. The energy-momentum tensor of this theory can be written in the form [37]

$$T_\mu^\nu = \text{diag}(\rho_A, -p_\perp, -p_\perp, +p_\perp)\,, \tag{56}$$

where

$$\begin{array}{lcl} \rho_A = \rho_{\text{kin}} + V & & \rho_{\text{kin}} \equiv -\frac{1}{4}F_{\mu\nu}F^{\mu\nu} \\ & \text{and} & \\ p_\perp = \rho_{\text{kin}} - V & & V \equiv -\frac{1}{2}m^2 A_\mu A^\mu. \end{array} \tag{57}$$

Eq. (56) is reminiscent of a perfect fluid with the crucial difference that the pressure in the longitudinal direction is of opposite sign compared to the transverse pressure. Thus, it seems that if the homogeneous vector field were to dominate the Universe it would indeed generate excessive anisotropic stress. Therefore, we assume that the vector field remains subdominant during inflation so that (quasi)de Sitter expansion is not spoilt.

After inflation the Hubble parameter decreases until $m > H(t)$. When this happens one can ignore the friction term in Eq. (55), which therefore suggests that the vector field condensate starts rapid (quasi)harmonic oscillations. It is easy to show that, during a Hubble time, on average $\overline{\rho_{\text{kin}}} \approx \overline{V}$, which means that, over many oscillations, the average pressure is zero: $\overline{p_\perp} = 0$ [37]. Thus the oscillating vector field condensate behaves as pressureless, *isotropic* matter. Its density, therefore, scales as $\rho_A \propto a^{-3}$ which not as drastic as the radiation background, which scales as $\rho_\gamma \propto a^{-4}$. Therefore, the oscillating vector field can gradually increase its density parameter and come to dominate (or nearly dominate) before its decay (see Fig. 6). Because it is isotropic, it dominates without causing any excessive anisotropic stress, so that the Universe expansion remains isotropic [37].

From the above we see that a massive Abelian vector field can follow the curvaton scenario and play the role of the curvaton without problem. Indeed, the perturbations of the vector field imply a perturbation in its local density $\rho_A(\vec{x})$, which means that in some locations it dominates the Universe earlier than in others. This is how it can generate the curvature perturbation according to the curvaton mechanism. It is important to note that, since ρ_A is a scalar quantity the curvature perturbation generated is scalar and not vector in nature. Also, note that the perturbations of the vector field δA_i follow a similar equation to Eq. (55), which means that they too are rapidly oscillating and do not introduce anisotropic stress. Thus, at domination the perturbations do not cause anisotropic expansion either, not even in a small scale. If particle

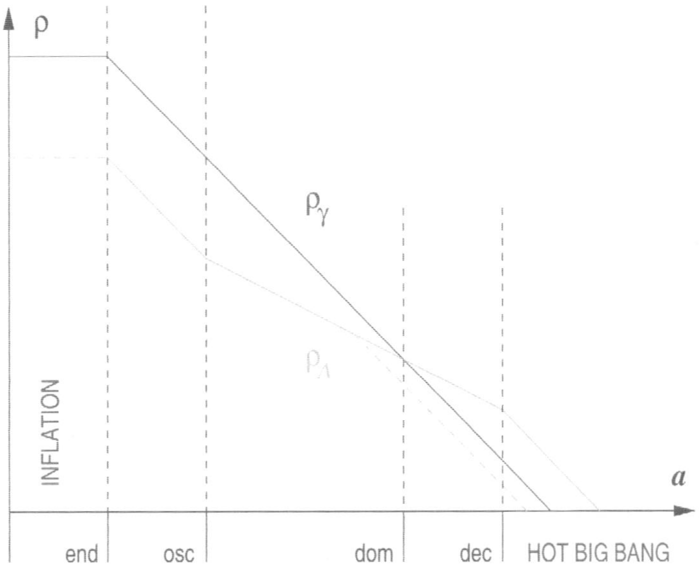

Figure 6. Log-log plot of the evolution of the inflaton energy density which decays into radiation ρ_γ (purple line) at the end of inflation (denoted by 'end') and the vector curvaton energy density ρ_A (green line) (prompt reheating is assumed). During inflation, the vector curvaton density is negligible (its evolution depends on the model which breaks its conformality). After inflation $\rho_\gamma \propto a^{-4}$. Similarly, ρ_A decreases also as radiation after inflation when the vector field is light (in contrast to the scalar curvaton case where ρ_σ remains constant when the curvaton is light, cf. Fig. 4). However, when $m \sim H(t)$, the vector curvaton becomes heavy and begins oscillating, after which time (denoted 'osc') $\rho_A \propto a^{-3}$. At some moment (denoted 'dom') the vector curvaton density dominates the Universe until, some time later (denoted 'dec') when it decays into the thermal bath of the Hot Big Bang. As the vector curvaton is rapidly oscillating it does not give rise to anisotropic stress at domination. The dashed slanted line depicts the possibility that the vector curvaton decays before domination ($\hat{\Omega}_A \ll 1$), when substantial non-Gaussianity can be generated.

production of the vector field is anisotropic then there are direction dependent patterns in the amplitude of the oscillating zero mode that lead to statistical anisotropy in the PDP [40].

In Ref. [44] statistical anisotropy in the bispectrum in the vector curvaton model was investigated. It was found that it manifests itself only quadratically, i.e. the expansion in Eq. (46) is truncated to $f_{\rm NL} = f_{\rm NL}^{\rm iso}\left(1 + \mathcal{G}\hat{A}_\perp^2\right)$. $\mathcal{G}$ was evaluated in the two models discussed in the previous section. In the non-minimally coupled to gravity model $\mathcal{G}^{\rm eql} = \frac{9}{8}$ in the equilateral configuration, whereas $\mathcal{G}^{\rm sqz} = 1$ in the squeezed configuration. This means that the angular modulation of $f_{\rm NL}$ is prominent and should be detected if non-Gaussianity is found. In the varying kinetic function and mass model $\mathcal{G}^{\rm eql} = \frac{1}{8}(\frac{3H}{M})^4 \gg \mathcal{G}^{\rm sqz} = (\frac{3H}{M})^2 \gg 1$ when $M \ll H$, i.e. if the field remains light until the end of inflation. Here we see that non-Gaussianity is predominantly anisotropic. Should no angular modulation of $f_{\rm NL}$ be observed this possibility will be excluded. However, if $M \gg H$, i.e. if the field becomes massive by the end of inflation, particle production is isotropic (Case C) and $f_{\rm NL} = f_{\rm NL}^{\rm iso} = 5/4\hat{\Omega}_A$, which is identical to the scalar curvaton case (cf. Eq. (33)).[20]

[20] $\hat{\Omega}_A$ is defined in the same way as $\hat{\Omega}_\sigma$.

The vector curvaton is an elegant mechanism for vector fields to contribute of even generate the PDP since it does not need to couple the fields to the inflaton sector. However, it is by no means the only way that a vector field can affect the curvature perturbation. For example, in Ref. [55], the end of inflation mechanism is employed (see Sec. 5.3.2), where, instead of coupling the waterfall field ψ of hybrid inflation to a scalar field σ, the authors considered introducing a coupling of the form $\Delta V = \frac{1}{2}hA_\mu A^\mu \psi^2$. This model too produces distinct observational signatures. For example, in this model $f_{\rm NL} = f_{\rm NL}^{\rm iso}(1 + \mathcal{G}\hat{A}_\perp^2 + \mathcal{G}'\hat{A}_\perp^4)$, where the maximum value for $\mathcal{G}'$ in the squeezed configuration is independent from the model parameters: $\mathcal{G}'^{\,\rm sqz}_{\rm max} = \frac{1}{4}$.

10. Conclusions

Cosmic structure originates from the growth of quantum fluctuations during a period of cosmic inflation in the Early Universe. The particle production process generates an almost scale invariant spectrum of superhorizon perturbations of suitable fields, e.g. light scalar fields. These perturbations give rise to the primordial density/curvature perturbation via a multitude of mechanisms (e.g. inflaton, curvaton, inhomogeneous reheating etc.). Observables such as the spectral index n_s or the non-linearity parameter $f_{\rm NL}$ will soon exclude whole classes of models. Indeed, the Planck satellite observations are expected to increase precision up to $f_{\rm NL} = \mathcal{O}(1)$.

Recently, the possibility that cosmic vector fields contribute or even generate the curvature perturbation ζ (e.g. through the vector curvaton mechanism) is being explored. If it is so then vector fields can produce distinct signatures such as correlated statistical anisotropy in the spectrum and bispectrum of ζ. WMAP observations allow up to 30% statistical anisotropy in the spectrum but the Planck satellite mission is expected to reduce this bound down to 2% [56], if statistical anisotropy is not observed. Anisotropy in the non-Gaussianity can be dominant, which means that $f_{\rm NL}$ may feature an angular modulation on the microwave sky.

The above suggest that cosmological observations allow for detailed modelling and open a window to fundamental physics complementary to Earth based experiments such as the LHC.

Acknowledgments

I would like to thank my collaborators Mindaugas Karčiauskas, David H. Lyth, Yeinzon Rodriguez-Garcia and Jacques M. Wagstaff. This work was supported by the European Union through the Marie Curie Research and Training Network "UniverseNet" (MRTN-CT-2006-035863). My participation to the NEB-14 conference was supported by a Royal Society Conference Grant.

[1] Guth A H 1981 The Inflationary Universe: A Possible Solution To The Horizon And Flatness Problems *Phys. Rev.* D **23** 347
[2] Starobinsky A A 1980 A new type of isotropic cosmological models without singularity, *Phys. Lett.* B **91** 99.
[3] Sachs R K and Wolfe A M 1967 Perturbations of a cosmological model and angular variations of the microwave background *Astrophys. J.* **147** 73; [2007 *Gen. Rel. Grav.* **39** 1929]
[4] Hawking S W 1975 Particle Creation By Black Holes *Commun. Math. Phys.* **43** 199 [1976 Erratum-ibid. **46** 206]
[5] Gibbons G W and Hawking S W 1977 Cosmological Event Horizons, Thermodynamics, And Particle Creation, *Phys. Rev.* D **15** 2738
[6] Bunch T S and Davies P C W 1978 Quantum Field Theory In De Sitter Space: Renormalization By Point Splitting *Proc. Roy. Soc. Lond.* A **360** 117
[7] Komatsu E *et al.* 2010 Seven-Year Wilkinson Microwave Anisotropy Probe (WMAP) Observations: Cosmological Interpretation *Preprint* arXiv:1001.4538 [astro-ph.CO]
[8] Lyth D H, Malik K A and Sasaki M 2005 A general proof of the conservation of the curvature perturbation *Journal of Cosmol. Astropart. Phys.* JCAP05(2005)004 (*Preprint* astro-ph/0411220)
[9] Sasaki M and Stewart E D 1996 A General Analytic Formula For The Spectral Index Of The Density Perturbations Produced During Inflation *Prog. Theor. Phys.* **95** (1996) 71 (*Preprint* astro-ph/9507001)
[10] Liddle A R and Lyth D H 2000 *Cosmological Inflation and Large Scale Structure* (Cambridge University Press).

[11] Lyth D H and Wands D 2002 Generating the curvature perturbation without an inflaton *Phys. Lett.* B **524** 5 (*Preprint* hep-ph/0110002)
[12] Enqvist K and Sloth M S 2002 Adiabatic CMB perturbations in pre big bang string cosmology *Nucl. Phys.* B **626** 395 (*Preprint* hep-ph/0109214)
[13] Moroi T and Takahashi T 2001 Effects of cosmological moduli fields on cosmic microwave background *Phys. Lett.* B **522** 215 [2002 Erratum-ibid. B **539** 303] (*Preprint* hep-ph/0110096)
[14] Mollerach S 1990 Isocurvature Baryon Perturbations And Inflation *Phys. Rev.* D **42** 313
[15] Linde A D and Mukhanov V F 1997 Nongaussian isocurvature perturbations from inflation *Phys. Rev.* D **56** 535 (*Preprint* astro-ph/9610219)
[16] Dimopoulos K and Lyth D H 2004 Models of inflation liberated by the curvaton hypothesis *Phys. Rev.* D **69** 123509 (*Preprint* hep-ph/0209180)
[17] Moroi T, Takahashi T and Toyoda Y 2005 Relaxing constraints on inflation models with curvaton *Phys. Rev.* D **72** 023502 (*Preprint* hep-ph/0501007)
[18] McDonald J 2003 Right-handed sneutrinos as curvatons *Phys. Rev.* D **68** 043505 (*Preprint* hep-ph/0302222)
[19] Enqvist K, Kasuya S and Mazumdar A 2003 Adiabatic density perturbations and matter generation from the MSSM *Phys. Rev. Lett.* **90** 091302 (*Preprint* hep-ph/0211147)
[20] Enqvist K, Jokinen A, Kasuya S and Mazumdar A 2003 MSSM flat direction as a curvaton *Phys. Rev.* D **68** 103507 (*Preprint* hep-ph/0303165)
[21] Chun E J, Dimopoulos K and Lyth D H 2004 Curvaton and QCD axion in supersymmetric theories *Phys. Rev.* D **70** 103510 (*Preprint* hep-ph/0402059)
[22] Dimopoulos K and Lazarides G 2006 Modular inflation and the orthogonal axion as curvaton *Phys. Rev.* D **73** 023525 (*Preprint* hep-ph/0511310)
[23] Lyth D H, Ungarelli C and Wands D 2003 The primordial density perturbation in the curvaton scenario *Phys. Rev.* D **67** 023503 (*Preprint* astro-ph/0208055)
[24] Dvali G, Gruzinov A and Zaldarriaga M 2004 A new mechanism for generating density perturbations from inflation *Phys. Rev.* D **69** 023505 (*Preprint* astro-ph/0303591)
[25] Dvali G, Gruzinov A and Zaldarriaga M 2004 Cosmological perturbations from inhomogeneous reheating, freezeout, and mass domination *Phys. Rev.* D **69** 083505 (*Preprint* astro-ph/0305548)
[26] Zaldarriaga M 2004 Non-Gaussianities in models with a varying inflaton decay rate *Phys. Rev.* D **69** 043508 (*Preprint* astro-ph/0306006)
[27] Linde A D Hybrid inflation 1994 *Phys. Rev.* D **49** 748 (*Preprint* astro-ph/9307002)
[28] Dvali G R, Shafi Q and Schaefer R K 1994 Large scale structure and supersymmetric inflation without fine tuning *Phys. Rev. Lett.* **73** 1886 (*Preprint* hep-ph/9406319)
[29] Lazarides G, Schaefer R K and Shafi Q 1997 Supersymmetric inflation at the grand unification scale *Phys. Rev.* D **56** 1324 (*Preprint* hep-ph/9608256)
[30] Lyth D H 2005 Generating the curvature perturbation at the end of inflation *Journal of Cosmol. Astropart. Phys.* JCAP11(2005)006 (*Preprint* astro-ph/0510443)
[31] Salem M P 2005 On the generation of density perturbations at the end of inflation," *Phys. Rev.* D **72** 123516 (*Preprint* astro-ph/0511146)
[32] Alabidi L and Lyth D H 2006 Curvature perturbation from symmetry breaking the end of inflation *Journal of Cosmol. Astropart. Phys.* JCAP08(2006)006 (*Preprint* astro-ph/0604569)
[33] Land K and Magueijo J 2005 The axis of evil *Phys. Rev. Lett.* **95** 071301 (it Preprint astro-ph/0502237)
[34] Land K and Magueijo J 2007 The Axis of Evil revisited *Mon. Not. Roy. Astron. Soc.* **378** 153 (*Preprint* astro-ph/0611518)
[35] Eriksen H K, Hansen F K, Banday A J, Gorski K M and Lilje P B 2004 Asymmetries in the CMB anisotropy field *Astrophys. J.* **605** 14 [2004 Erratum-ibid. **609** 1198] (*Preprint* astro-ph/0307507)
[36] Hansen F K, Banday A J and Gorski K M 2004 Testing the cosmological principle of isotropy: local power spectrum estimates of the WMAP data," *Mon. Not. Roy. Astron. Soc.* **354** 641 (*Preprint* astro-ph/0404206)
[37] Dimopoulos K 2006 Can a vector field be responsible for the curvature perturbation in the universe? *Phys. Rev.* D **74** 083502 (*Preprint* hep-ph/0607229)
[38] Dimopoulos K, Karčiauskas M and Wagstaff J M 2010 Vector Curvaton with varying Kinetic Function *Phys. Rev.* D **81** 023522 (*Preprint* 0907.1838 [hep-ph])
[39] Ackerman L, Carroll S M and Wise M B 2007 Imprints of a Primordial Preferred Direction on the Microwave Background *Phys. Rev.* D **75** 083502 [2009 Erratum-ibid. D **80** 069901] (*Preprint* astro-ph/0701357)
[40] Dimopoulos K, Karčiauskas M, Lyth D H and Rodriguez Y 2009 Statistical anisotropy of the curvature perturbation from vector field perturbations *Journal of Cosmol. Astropart. Phys.* JCAP05(2009)013 (*Preprint* 0809.1055 [astro-ph])
[41] Groeneboom N E and Eriksen H K 2009 Bayesian analysis of sparse anisotropic universe models and

application to the 5-yr WMAP data *Astrophys. J.* **690** 1807 (*Preprint* 0807.2242 [astro-ph])
[42] Groeneboom N E, Ackerman L, Wehus I K and Eriksen H K Bayesian analysis of an anisotropic universe model: systematics and polarization *Preprint* 0911.0150 [astro-ph.CO]
[43] Dimopoulos K 2009 Where galaxies really come from *Contemp. Phys.* **50** 633 (*Preprint* 0906.0903 [hep-ph])
[44] Karčiauskas M, Dimopoulos K and Lyth D H 2009 Anisotropic non-Gaussianity from vector field perturbations *Phys. Rev.* D **80** 023509 (*Preprint* 0812.0264 [astro-ph])
[45] Valenzuela-Toledo C A, Rodriguez Y and Lyth D H 2009 Non-gaussianity at tree- and one-loop levels from vector field perturbations *Phys. Rev.* D **80** 103519 (*Preprint* 0909.4064 [astro-ph.CO])
[46] Dimopoulos K and Karčiauskas M 2008 Non-minimally coupled vector curvaton *Journal of High Energy Phys.* JHEP 07(2008)119 (*Preprint* 0803.3041 [hep-th])
[47] Himmetoglu B, Contaldi C R and Peloso M 2009 Instability of anisotropic cosmological solutions supported by vector fields *Phys. Rev. Lett.* **102** 111301 (*Preprint* 0809.2779 [astro-ph])
[48] Himmetoglu B, Contaldi C R and Peloso M 2009 Instability of the ACW model, and problems with massive vectors during inflation *Phys. Rev.* D **79** 063517 (*Preprint* 0812.1231 [astro-ph])
[49] Karčiauskas M and Lyth D H 2010 On the health of a vector field with $(R\ A^2)/6$ coupling to gravity *Preprint* 1007.1426 [astro-ph.CO]
[50] Carroll S M, Dulaney T R, Gresham M I and Tam H 2009 Instabilities in the Aether *Phys. Rev.* D **79** 065011 (*Preprint* 0812.1049 [hep-th])
[51] Cornwall J M, Levin D N and Tiktopoulos G 1974 Derivation Of Gauge Invariance From High-Energy Unitarity Bounds On The S Matrix *Phys. Rev.* D **10** 1145 [1975 Erratum-ibid. D **11** 972]
[52] Dimopoulos K, Karčiauskas K and Wagstaff J M 2010 Vector Curvaton without Instabilities *Phys. Lett.* B **683** 298 (*Preprint* 0909.0475 [hep-ph])
[53] Dimopoulos K 2007 Supergravity inspired Vector Curvaton *Phys. Rev.* D **76** 063506 (*Preprint* 0705.3334 [hep-ph])
[54] Dimopoulos K and Wagstaff J M 2010 in preparation
[55] Yokoyama S and Soda J 2008 Primordial statistical anisotropy generated at the end of inflation *Journal of Cosmol. Astropart. Phys.* JCAP08(2008)005 (*Preprint* 0805.4265 [astro-ph])
[56] Pullen A R and Kamionkowski M 2007 Cosmic Microwave Background Statistics for a Direction-Dependent Primordial Power Spectrum *Phys. Rev.* D **76** 103529 (*Preprint* 0709.1144 [astro-ph])

Journal of Physics: Conference Series **283** (2011) 012011 doi:10.1088/1742-6596/283/1/012011

Relativistic MHD simulations of stellar core collapse and magnetars

José A. Font[1], Pablo Cerdá-Durán[2], Michael Gabler[1,2,3], Ewald Müller[2], Nikolaos Stergioulas[3]

[1]Departamento de Astronomía y Astrofísica, Universitat de València, 46100 Burjassot (Valencia), Spain
[2]Max-Planck-Institut für Astrophysik, Karl-Schwarzschild-Str. 1, 85741 Garching, Germany
[3]Department of Physics, Aristotle University of Thessaloniki, Thessaloniki 54124, Greece

E-mail: j.antonio.font@uv.es

Abstract. We present results from simulations of magneto-rotational stellar core collapse along with Alfvén oscillations in magnetars. These simulations are performed with the CoCoA/CoCoNuT code, which is able to handle ideal MHD flows in dynamical spacetimes in general relativity. Our core collapse simulations highlight the importance of genuine magnetic effects, like the magneto-rotational instability, for the dynamics of the flow. For the modelling of magnetars we use the anelastic approximation to general relativistic MHD, which allows for an effective suppression of fluid modes and an accurate description of Alfvén waves. We further compute Alfvén oscillation frequencies along individual magnetic field lines with a semi-analytic approach. Our work confirms previous results based on perturbative approaches regarding the existence of two families of quasi-periodic oscillations (QPOs), with harmonics at integer multiples of the fundamental frequency. Additional material is presented in the accompanying contribution by Gabler et al (2010b) in these proceedings.

1. Introduction

Understanding gravitational stellar core collapse is a long-standing problem in relativistic astrophysics (see e.g. [50] and references therein). Major progress has been accomplished through numerical modelling with increasing levels of complexity in the input physics and mathematics, regarding aspects as diverse as the treatment of the hydrodynamics, gravity, magnetic fields, nuclear matter equations of state, transport, etc. Numerical studies based upon Newtonian physics are vastly developed nowadays and state-of-the-art simulations are beginning to generate successful supernova explosions. On the other hand, relativistic approaches are becoming routine in recent years, aided by the development of conservative formulations of the general relativistic hydrodynamics equations and numerically-stable formulations of the Einstein equations (see e.g. [21] and references therein). While some of these developments also hold for the simulation of general relativistic MHD flows (GRMHD flows hereafter), their thorough numerical exploration is still ahead.

Only very recently the first GRMHD codes able to follow the evolution of matter flows in dynamical spacetimes have been developed [18, 40, 25, 1, 9]. Our numerical code, described in detail in [9], is able to handle ideal MHD flows in dynamical spacetimes in general relativity, and is particularly designed to investigate gravitational core collapse leading to neutron stars as well

Journal of Physics: Conference Series **283** (2011) 012011 doi:10.1088/1742-6596/283/1/012011

as the evolution of strongly magnetized compact objects such as magnetars. This code is based on the hydrodynamics `CoCoA/CoCoNuT` code (`www.mpa-garching.mpg.de/hydro/COCONUT`) described in [15, 16], and on its extensions discussed in [8], [11], and [12], to which interested readers are addressed for details. The Maxwell equations were already incorporated in the codes of [11] and [12], but only in the passive magnetic field approximation, where the contribution of the magnetic field to the energy-momentum tensor is neglected yielding no impact on the dynamics. In the current code, this assumption is relaxed and the effects of the magnetic field on the dynamics and the self-gravity of the fluid are incorporated following the approach laid out in [3].

In this paper we discuss two applications of our code, namely magneto-rotational collapse and Alfvén oscillations of magnetars. To date the weakest point of all existing magneto-rotational simulations is the unknown strength and geometry of the initial magnetic field in the core. If the magnetic field is initially weak, there exist several mechanisms, such as differential rotation (Ω-dynamo) and the magneto-rotational instability (MRI hereafter), which can amplify the field to a strength where it can influence the flow dynamics. Differential rotation transforms rotational energy into magnetic energy winding up any seed poloidal field into a toroidal field, while the MRI leads to an exponential growth of the field strength. The latter acts as long as the radial gradient of the angular velocity of the flow is negative, a situation which holds in core collapse simulations, as we discuss below.

The second topic considered in this paper concerns Alfvén oscillations of magnetars, neutron stars with magnetic fields larger than about 10^{14} G [19]. Such field strengths can be reached if efficient amplification mechanisms are operative during the first few seconds after gravitational collapse [19]. A class of these objects, the so-called Soft Gamma Repeaters (SGRs), are characterized by gamma-ray activity which sometimes intensifies as giant flares ($10^{44} - 10^{46}$ erg/s in ~ 0.2 s) followed by a decaying X-ray tail lasting for hundreds of seconds. Quasi-periodic oscillations have been observed in the X-ray tail of two such SGRs. The simulations reported here provide some first results in our long-term goal of understanding the link between the oscillation modes of magnetars and the observed QPO frequencies. For further results obtained with our code, the interested reader is refered to [10, 22] as well as the accompanying contribution of [23] in this volume.

2. Framework

2.1. Ideal GRMHD equations

The evolution of a magnetized fluid is determined by the conservation law of the energy-momentum, $\nabla_\mu T^{\mu\nu} = 0$, and by the continuity equation, $\nabla_\mu J^\mu = 0$, for the rest-mass current $J^\mu = \rho u^\mu$, where ρ is the rest-mass density and u^μ the 4-velocity. As usual, symbol ∇_μ is used to indicate the covariant derivative operator. In terms of the (Faraday) electromagnetic tensor $F^{\mu\nu}$, Maxwell's equations read

$$\nabla_\nu {}^*F^{\mu\nu} = 0, \qquad \nabla_\nu F^{\mu\nu} = \mathcal{J}^\mu, \tag{1}$$

where $F^{\mu\nu} = U^\mu E^\nu - U^\nu E^\mu - \eta^{\mu\nu\lambda\delta} U_\lambda B_\delta$, its dual ${}^*F^{\mu\nu} = \frac{1}{2}\eta^{\mu\nu\lambda\delta} F_{\lambda\delta}$, and $\eta^{\mu\nu\lambda\delta} = \frac{1}{\sqrt{-g}}[\mu\nu\lambda\delta]$, where $[\mu\nu\lambda\delta]$ is the completely antisymmetric Levi-Civita symbol. E^μ and B^μ stand for the electric and magnetic fields measured by an observer with 4-velocity U^μ, and $\mathcal{J}^\mu$ is the electric 4-current, $\mathcal{J}^\mu = \rho_q u^\mu + \sigma F^{\mu\nu} u_\nu$ where ρ_q is the proper charge density and σ is the electric conductivity.

Maxwell's equations can be simplified if the fluid is a perfect conductor. In this case the conductivity of the fluid is infinite and, to keep the current finite, the term $F^{\mu\nu} u_\nu$ must vanish, which results in $E^\mu = 0$ for a comoving observer. This case corresponds to the so-called ideal MHD condition. Under this assumption the electric field measured by the Eulerian observer has components

$$E^0 = 0, \qquad E^i = -\alpha\eta^{0ijk} v_j B_k, \tag{2}$$

and Maxwell's equations $\nabla^*_\nu F^{\mu\nu} = 0$ reduce to the divergence-free condition plus the induction equation for the evolution of the magnetic field

$$\frac{\partial(\sqrt{\gamma}B^i)}{\partial x^i} = 0, \tag{3}$$

$$\frac{1}{\sqrt{\gamma}}\frac{\partial}{\partial t}(\sqrt{\gamma}B^i) = \frac{1}{\sqrt{\gamma}}\frac{\partial}{\partial x^j}\{\sqrt{\gamma}[\alpha\tilde{v}^i B^j - \alpha\tilde{v}^j B^i]\}. \tag{4}$$

In the above expressions α is the spacetime lapse function, γ the determinant of the spatial metric, and $\tilde{v}^j = v^j - \beta^j/\alpha$, where v^j is the 3-velocity of the fluid and β^j the spacetime shift vector.

For a fluid endowed with a magnetic field the stress-energy tensor is the sum of that of the fluid and that of the electromagnetic field, $T^{\mu\nu} = T^{\mu\nu}_{\text{Fluid}} + T^{\mu\nu}_{\text{EM}}$, with

$$T^{\mu\nu}_{\text{Fluid}} = \rho h u^\mu u^\nu + p g^{\mu\nu} \qquad \textit{and} \qquad T^{\mu\nu}_{\text{EM}} = F^{\mu\lambda}F^\nu_\lambda - \frac{1}{4}g^{\mu\nu}F^{\lambda\delta}F_{\lambda\delta}, \tag{5}$$

where h is the enthalpy, p is the pressure, and $g^{\mu\nu}$ is the metric tensor. In ideal MHD, the electromagnetic part of the stress-energy tensor can be rewritten as

$$T^{\mu\nu}_{\text{EM}} = \left(u^\mu u^\nu + \frac{1}{2}g^{\mu\nu}\right)b^2 - b^\mu b^\nu, \tag{6}$$

where b^μ is the magnetic field measured by the observer comoving with the fluid and $b^2 = b^\nu b_\nu$. The total stress-energy tensor is thus given by

$$T^{\mu\nu} = \rho h^* u^\mu u^\nu + p^* g^{\mu\nu} - b^\mu b^\nu, \tag{7}$$

with $p^* = p + b^2/2$ and $h^* = h + b^2/\rho$, respectively.

Following [3] the conservation equations for the extended energy-momentum tensor, together with the continuity equation and the equation for the evolution of the magnetic field measured by the Eulerian observer, can be written as a first-order, flux-conservative, hyperbolic system.

$$\frac{1}{\sqrt{-g}}\left[\frac{\partial\sqrt{\gamma}\boldsymbol{U}}{\partial t} + \frac{\partial\sqrt{-g}\boldsymbol{F}^i}{\partial x^i}\right] = \boldsymbol{S}, \tag{8}$$

for a state vector $\mathbf{U}$, flux vector $\boldsymbol{F}^i$, and source vector $\boldsymbol{S}$. In this equation g is the determinant of the 4-dimensional metric. The state vector and the vector of fluxes of the fundamental GRMHD system of equations read:

$$\mathbf{U}(\mathbf{w}) = (D, S_j, \tau, B^k), \tag{9}$$

$$\begin{aligned}\mathbf{F}^i(\mathbf{w}) = {} & (D\tilde{v}^i, S_j\tilde{v}^i + p^*\delta^i_j - b_j B^i/W, \\ & \tau\tilde{v}^i + p^* v^i - \alpha b^0 B^i/W, \tilde{v}^i B^k - \tilde{v}^k B^i),\end{aligned} \tag{10}$$

where the definition of the conserved quantities is $D = \rho W$, $S_j = \rho h^* W^2 v_j - \alpha b^0 b_j$, and $\tau = \rho h^* W^2 - p^* - \alpha^2(b^0)^2 - D$, where $W = \alpha u^0$ is the Lorentz factor. The explicit expressions of the corresponding vector of sources can be found in [9].

The hyperbolic structure of these equations and the associated spectral decomposition of the flux-vector Jacobians, needed for their numerical solution with Riemann solvers, is given in [3, 4]. This information is required for numerically solving the system of equations using the class of high-resolution shock-capturing (HRSC) schemes implemented in our code.

2.2. Einstein equations (CFC)

In our work Einstein's field equations are formulated and solved using the conformally flat condition (CFC hereafter), introduced by [28]. In this approximation, the 3-metric in the ADM gauge is assumed to be conformally flat, $\gamma_{ij} = \phi^4\hat{\gamma}_{ij}$. Under the CFC assumption the gravitational field equations can be written as a system of five nonlinear elliptic equations for the set of variables $(\phi, \alpha\phi, \beta^i)$, namely

$$\hat{\Delta}\phi = -2\pi\phi^5\left(\rho h W^2 - p + \frac{K_{ij}K^{ij}}{16\pi}\right), \tag{11}$$

$$\hat{\Delta}(\alpha\phi) = 2\pi\alpha\phi^5\left(\rho h(3W^2-2) + 5p + \frac{7K_{ij}K^{ij}}{16\pi}\right), \tag{12}$$

$$\hat{\Delta}\beta^i = 16\pi\alpha\phi^4 S^i + 2\phi^{10}K^{ij}\hat{\nabla}_j\left(\frac{\alpha}{\phi^6}\right) - \frac{1}{3}\hat{\nabla}^i\hat{\nabla}_k\beta^k, \tag{13}$$

where $\hat{\nabla}_i$ and $\hat{\Delta}$ are the flat space Nabla and Laplace operators, respectively, K_{ij} is the extrinsic curvature, and S^i is the $i-$component of the momentum.

The simplified Einstein's equations introduced by the conformally-flat condition (CFC) are adequate in situations for which the deviations from spherical symmetry are not extreme. They have been used to simulate rotational core collapse in [15, 16, 17]. Modifications of the original CFC equations to account for additional post-Newtonian terms or to allow for black hole formation following core collapse are presented in [8, 14].

2.3. Numerical code

Our numerical code solves the coupled time evolution of the equations governing the dynamics of the spacetime, the fluid, and the magnetic field in general relativity. The equations are implemented using spherical polar coordinates $\{t, r, \theta, \varphi\}$ and axisymmetry and equatorial plane symmetry are assumed. The evolution of the matter fields is handled with a HRSC scheme which updates the variables (D, S_i, τ). Various cell-reconstruction procedures are available in the code, which are either second-order or third-order accurate in space, namely minmod, MC, and PHM (see [47] for definitions). The time update of the state vector $\boldsymbol{U}$ relies on the method of lines in combination with a second-order accurate Runge–Kutta scheme. The numerical fluxes at cell interfaces are obtained using either the HLL single-state solver of [26] or the symmetric scheme of [30]. The solenoidal condition of the magnetic field is ensured with the use of the flux constraint transport method, which has been adapted to the spherical polar coordinates used in the code, and uses cell interface-centered poloidal and (because of the assumption of axisymmetry) cell-centered toroidal magnetic field components. The time discretization of the induction equation is done in the same way as for the fluid equations. Correspondingly, we use a fix-point iteration scheme in combination with a linear Poisson solver to solve the CFC nonlinear elliptic equations (for further details see [8] and [15]). Furthermore, the code incorporates several equations of state, ranging from simple analytical expressions to tabulated microphysical equations of state.

The interested reader is addressed to [9] for details on the various tests passed by the code. We here simply point out that the test calculations performed demonstrate the ability of the code to properly handle all aspects appearing in the astrophysical scenarios the code is intended for, namely relativistic shocks, strongly magnetized fluids, and equilibrium configurations of magnetized neutron stars.

For the core collapse simulations we discuss below we consider two different initial models, namely a weakly magnetized model B10 with a central magnetic field of $|B|_\mathrm{c} = 10^{10}\sqrt{4\pi}$ Gauss, and a strongly magnetized model B12 with $|B|_\mathrm{c} = 10^{12}\sqrt{4\pi}$ Gauss. Correspondingly, for our study of magnetars we construct equilibrium magnetar models using the self-consistent solution

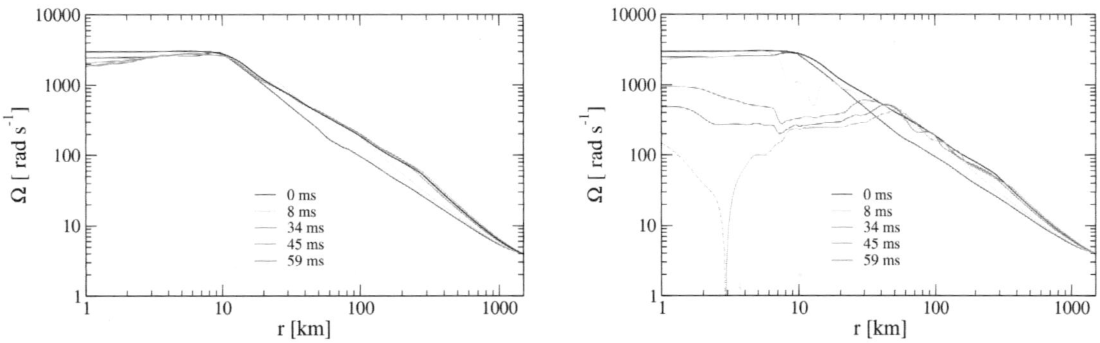

Figure 1. Radial profiles of the angular velocity Ω at the equator for model B10 (left panel) and B12 (right panel). The different lines correspond to different times after bounce: $t - t_b = 0$, 8, 34, 45 and 59 ms. The black line displays the profile of Ω at the end of the simulation for the corresponding unmagnetized model, and is shown for comparison.

of magnetized stars with a dipolar magnetic field presented by [6], where all the effects of the magnetic field on the matter and spacetime are taken into account. These equilibrium models are computed using the `LORENE` library (`www.lorene.obspm.fr`). We construct non-rotating polytropic equilibrium models with $\Gamma = 2$ and $K = 1.455 \times 10^5$ (cgs units). The specific central enthalpy is chosen to be $h_c = 1.256$ and the form of the current density which gives rise to the dipolar magnetic field is given in [6]. Here, we use the equilibrium model MNS2 of [10] as a reference model, with a mass of $1.4 M_\odot$ and a magnetic field strength at the pole equal to $B_{\rm pole} = 6.5 \times 10^{15}$ G. The interested reader is addressed to [10] for results with a larger sample of initial models.

3. Magneto-rotational core collapse simulations

The initial models used for the magneto-rotational core collapse simulations are evolved with the tabulated equation of state of [39] and an approximate deleptonization scheme [33] as described in [17] and [12]. Further discussion on the results reported next can be found in [9]. As the collapse proceeds both the density and the magnetic energy grow very similarly for both models, because even in the highly magnetized progenitor model B12 the strength of the magnetic field is not large enough to affect the collapse dynamics. After core bounce, however, both models behave differently. On the one hand, for model B10 the magnetic field is far from saturation and grows linearly with time at the end of the simulation, while model B12 shows a saturation of the magnetic field energy shortly after core bounce. Its central density continues to grow beyond bounce, and the model eventually approaches an equilibrium configuration with a central density about 10% larger than in the weakly magnetized model.

The behavior of the central density can be understood by analyzing the profiles of the angular velocity Ω, which are plotted in Figure 1. At the time of bounce those profiles are very similar for all models, since the magnetic field is still unimportant for the dynamics: the innermost 10 km of the core rotate rigidly, while further out Ω follows a power law with an exponent ~ -1.2. During the subsequent evolution the central region spins down, and the central density rises, the effect being more prominent in the strongest magnetized model B12 (right panel in Fig. 1). In the region $10\,\text{km} \le r \le 30\,\text{km}$ the magnetic field is strongest as differential rotation winds up the magnetic field more efficiently there [12]. On a time scale of about 50 ms, the angular velocity decreases by about a factor 10, and the innermost few kilometers of the core even acquire retrograde rotation, an effect already observed in the Newtonian simulations of [35] (see also [34]).

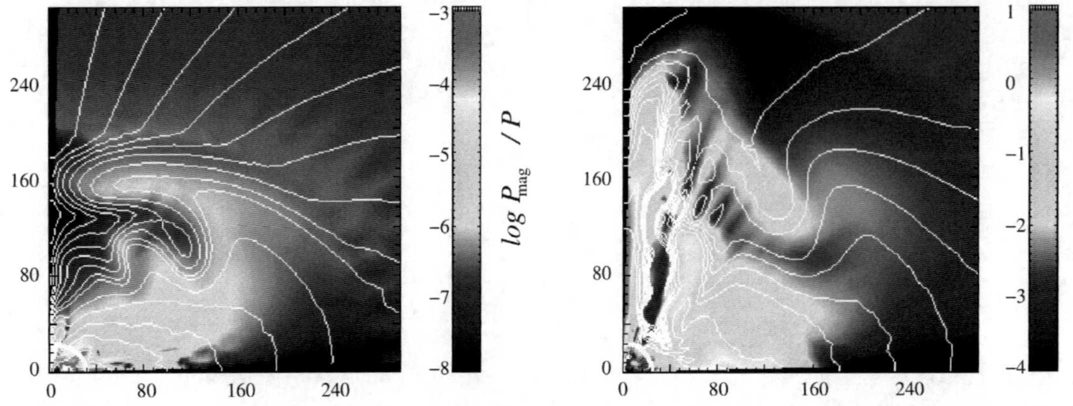

Figure 2. Structure of the magnetic field 51ms after core bounce for both models, weak field (B10; left panel) and strong field (B12; right panel).

The spin down of the core may be understood by means of the magneto-rotational instability (MRI hereafter; see [5]). In unstable regions the MRI grows exponentially for all length-scales larger than a critical length-scale $\lambda_{\rm crit} \sim 2\pi c_{\rm A}/\Omega$, where $c_{\rm A}$ is the Alfvén speed. The fastest growing MRI mode develops at length-scales near $\lambda_{\rm crit}$ on a typical time scale of $\tau_{\rm MRI} = 4\pi[\varpi\partial_\varpi\Omega]^{-1}$. Therefore, one needs to resolve length-scales of about the size of $\lambda_{\rm crit}$ in order to numerically capture the MRI. For model B12 such critical length-scale at bounce is $\lambda_{\rm crit} \sim 1 - 5\,{\rm km}$ inside the unstable region ($10\,{\rm km} \leq r \leq 30\,{\rm km}$). This region is covered with 60 radial and 30 angular zones, which corresponds to a resolution ($\Delta r, r\Delta\theta$) of $125\,{\rm m} \times 500\,{\rm m}$ at $r = 10\,{\rm km}$, and $900\,{\rm m} \times 1500\,{\rm m}$ at $r = 30\,{\rm km}$. This resolution is marginally sufficient to resolve the length-scale of the fastest growing mode of the MRI at bounce ($5 - 10$ radial zones, and $2 - 3$ angular zones). The strong redistribution of the angular momentum observed for model B12 might therefore be caused by the MRI. On the other hand, for model B10 the critical length-scale at bounce is about a factor of 100 smaller, ie $\lambda_{\rm crit} \sim 10 - 50\,{\rm m}$, and thus the fastest growing mode of the MRI cannot be resolved with our grid resolution.

In addition, we show in Fig. 2 the magnetic field topology for both models at the end of the simulation. In the low magnetized model B10 (left panel) the (transient) prompt convection developing after bounce twists the magnetic field outside the so-called neutrino-sphere, which is assumed to be located at about 30 km. In model B12 the magnetic field grows to values near equipartition, and a distinctive, strongly magnetized outflow propagates along the axis behind the shock front. Between $10\,{\rm km} < r < 30\,{\rm km}$, where the MRI is predominantly growing, axisymmetric channel flows form, which are morphologically similar to the flows found in the accretion disk simulations of [27].

4. Simulations of Alfvén QPOs in magnetars

4.1. Semi-analytic approach

Before discussing the results from our numerical simulations of Alfvén QPOs in magnetars, we consider a semi-analytic model which provides an alternative approach to this study. This model, described in full detail in [10], is based on the computation of Alfvén wave travel times along individual magnetic field lines (assuming a short-wavelength approximation). It only requires information on the initial equilibrium model and on the magnetic field configuration. Torsional Alfvén oscillations can be treated as axial perturbations traveling along individual magnetic field lines at the local Alfvén speed $\mathbf{v}_{\rm a}$. The path of an Alfvén wave inside the star can be obtained by integrating the equation

$$\frac{d\mathbf{x}}{dt} = \mathbf{v}_{\rm a}(\mathbf{x}), \tag{14}$$

starting from a given initial location $\mathbf{x}_0$ inside the star.

Instead of using polar coordinates (r, θ) we employ for convenience dimensionless coordinates adapted to the magnetic field lines (χ, ξ), which give the position of the magnetic field line and the location of points along an individual line, respectively. These are defined as $\chi \equiv r/r_c$ and $\xi \equiv t_a(r, \theta; \chi)/t_{\rm tot}(\chi) - 1/2$, where r_c labels the radial location in the equatorial plane where closed magnetic field lines have zero length, t_a is the arrival time of an Alfvén wave at the selected location and $t_{\rm tot}(\chi)$ is twice the total travel time of an Alfvén wave traveling along a magnetic field line. Let $Y(\xi, \chi; t)$ be the displacement describing traveling waves along a specific magnetic field line χ, which satisfies the wave equation

$$\frac{\partial^2 Y(\xi, t)}{\partial t^2} = \sigma^2 \frac{\partial^2 Y(\xi, t)}{\partial \xi^2}, \tag{15}$$

where $\sigma = 1/t_{\rm tot}$ is the wave speed. To find a correspondence with QPOs we are interested in solutions of this equation in the form of standing waves $Y(\xi, t) = A(\xi)\cos(2\pi f t + \varphi)$, where f is the oscillation frequency and φ is a phase. The spatial dependence can be written as $A(\xi) = a\,\sin(\kappa\xi + \phi)$, where κ is the wavenumber, ϕ a spatial phase, and a the amplitude of the oscillation. The boundary condition at the surface is chosen such that it mimics the presence of a solid crust. We therefore assume that the parts of the traction related to the crust have to vanish at the surface of the star. In practice, this condition leads to zero radial derivative of the displacement at this position. Such zero-traction boundary condition applied to a standing-wave yields the following equations:

$$\left(2\pi t \frac{\partial f}{\partial \chi} + \frac{\partial \varphi}{\partial \chi}\right)\frac{\partial \chi}{\partial r} = 0, \qquad \frac{\partial a}{\partial \chi}\frac{\partial \chi}{\partial r} = 0, \tag{16}$$

which are fulfilled only in two cases. First, if $\frac{\partial f}{\partial \chi} = 0$, $\frac{\partial \varphi}{\partial \chi} = 0$, $\frac{\partial a}{\partial \chi} = 0$, i.e. if $f(\chi)$, $\varphi(\chi)$ and $a(\chi)$ have a common turning point, or if they are constant. The second case is when $\frac{\partial \chi}{\partial r} = 0$, i.e. for those magnetic field lines that cross the stellar surface perpendicular to it, a condition which always holds in axisymmetry for the magnetic field line along the pole.

The time evolution of an initial perturbation will remain close to a standing wave solution as long as $2\pi \frac{\partial f}{\partial \chi} t \ll 1$, $\frac{\partial \varphi}{\partial \chi} \sim 0$ and $\frac{\partial a}{\partial \chi} \sim 0$, which provides the characteristic timescale on which phase mixing occurs and a global standing wave is damped, namely $\tau_{\rm d} = \frac{1}{2\pi}\left(\frac{\partial f}{\partial \chi}\right)^{-1}$. For timescales much shorter than the damping timescale a standing wave continuum of magnetic field lines throughout the star can be defined for each overtone with frequency $f(\chi) = \frac{n}{2t_{\rm tot}(\chi)}$, $n = 1, 2, 3, \cdots$. QPO overtones thus appear at exact integer multiples of the corresponding fundamental frequency.

Figure 3 displays the different branches of the standing-wave continuum of frequencies for our magnetar model, in a frequency vs. magnetic field line plot. For each overtone, the branch which corresponds to symmetric standing waves in the region of open field lines ($0 < \chi < 0.66$) smoothly joins to the two overlapping branches of symmetric and antisymmetric standing waves in the region of closed magnetic field lines ($0.66 < \chi < 1.0$). For the chosen magnetic field configuration, the standing-wave frequencies have a local maximum at $\chi = 0$ (i.e. at the pole) where the first set of conditions given by Eq. (16) are fulfilled. The predicted frequencies of these QPOs at the pole (the "upper" QPOs) are shown in Fig. 3 and labeled as $U_n^{(+)}$ and $U_n^{(-)}$, for symmetric and antisymmetric standing waves, respectively. Correspondingly, in the region of closed field lines, the standing-wave continuum for each overtone features a local minimum.

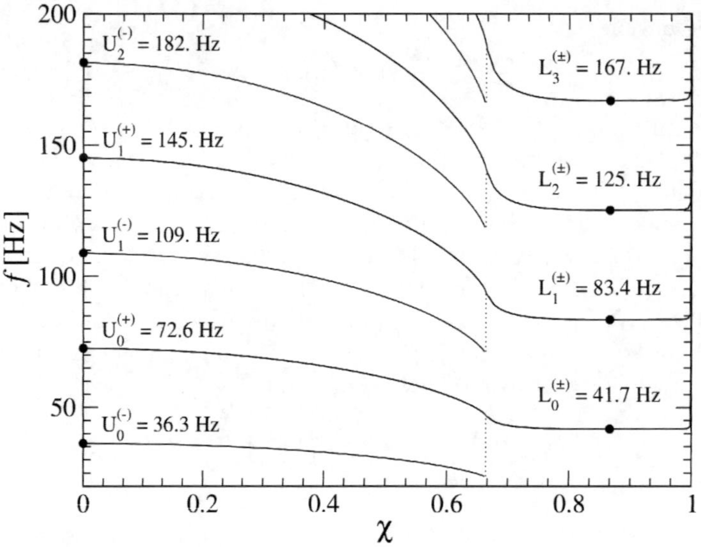

Figure 3. Standing-wave continuum and QPOs for the magnetar model considered. Filled circles denote the (upper and lower) QPOs.

These local minima also give rise to long-lived "lower" QPOs which are labeled as $L_n^{(\pm)}$ in Fig. 3 (the frequency of symmetric and antisymmetric standing waves coincide).

4.2. Alfvén QPO model

We have also performed numerical simulations of Alfvén QPOs which help to validate, and extend, the picture just described obtained with our semi-analytic approach. These simulations have been carried out using the same `CoCoA/CoCoNuT` code used for the work on magneto-rotational core collapse presented in the preceding section. For the simulations of Alfvén QPOs we adopt the Cowling approximation in which the dynamics of the spacetime metric is neglected and only consider small-amplitude torsional oscillations. Moreover, our numerical approach is based on the relativistic anelastic approximation [7] in which sound waves are neglected. This is a central approximation in our work as it reduces the time-step restrictions of numerical codes based on time-explicit schemes. These restrictions are imposed by the dependence of the eigenvalues of the hyperbolic system on the speed of sound (see [3]). We note that we can use the anelastic approximation since we deal with low-amplitude torsional oscillations of an equilibrium star. They have axial parity and couple weakly to density perturbations (density and pressure can be considered constant). As a result of our assumptions there only remain evolution equations for the azimuthal components of the momentum and of the magnetic field, S^φ and B^φ (see [10] for details).

The fundamental variable in our simulations is the 3-velocity component v^φ and the corresponding displacement Y, which we define as $\dot{Y} = \alpha v^\varphi$, where α is the spacetime lapse function and a dot denotes a time derivative. We choose an initial perturbation of the form

$$Y(t=0) = 0\,, \quad \dot{Y}(t=0) = f(r)\,b(\theta)\,, \tag{17}$$

with

$$f(r) = \sin\left(\frac{3\pi}{2}\frac{r}{r_*}\right)\,, \quad b(\theta) = b_2\frac{1}{3}\partial_\theta P_2(\cos\theta) + b_3\frac{1}{6}\partial_\theta P_3(\cos\theta)\,, \tag{18}$$

Table 1. QPO frequencies in the anelastic approximation. In parenthesis we give the relative difference with respect to the semi-analytic approach.

upper QPOs	f (Hz)	lower QPOs	f (Hz)
$U_0^{(-)}$	35.1 (4.%)	$L_0^{\pm}$	41.30 (1.0%)
$U_0^{(+)}$	72.6 (0.05%)	$L_1^{\pm}$	83.85 (0.6%)
$U_1^{(-)}$	110.2(1.1%)	$L_2^{\pm}$	126.4 (1.0%)
$U_1^{(+)}$	147.8(1.8%)	$L_3^{\pm}$	170.2 (2.0%)
$U_2^{(-)}$	182.8(0.7%)	$L_4^{\pm}$	210.3 (0.9%)
$U_2^{(+)}$	220.4(1.2%)	$L_5^{\pm}$	251.6 (0.6%)
$U_3^{(-)}$	256.7(1.0%)	$L_6^{\pm}$	295.4 (1.2%)
$U_3^{(+)}$	294.3(1.3%)	$L_7^{\pm}$	336.7 (0.9%)

where P_2 and P_3 are the corresponding Legendre polynomials and the coefficients b_2 and b_3 are chosen to fix the symmetry of the perturbation with respect to the equatorial plane.

Our simulations predict strong QPOs both, near the magnetic axis and within the region of closed magnetic field lines, and precisely at the frequencies obtained with our semi-analytic model. For overtones, the number of nodes also agrees with the predictions of the semi-analytic model. The corresponding frequencies of our magnetar model (model MNS2 of [10]) are summarized in Table 1, along with their relative difference (in parenthesis) when compared to the predictions of the semi-analytic model. The frequency of the fundamental QPO agrees to within 4% between the two approaches, while the agreement in all other frequencies is at the 1-2% level. The larger discrepancy for the fundamental upper QPO reflects the inaccuracy of the semi-analytic (short-wavelength) approach for this particular mode since the wavelength of the corresponding standing-wave is twice the size of the star itself.

The simulations performed in our study reported in [10] include models with either different central current density or central density with respect to our reference magnetar model. When varying the strength of the magnetic field we find that, as expected from the results of [43], the QPO frequencies change linearly with B. Also, the effect of changing the stellar mass is consistent (within a few percent) with the expansion in terms of the compactness, M/R, used in the empirical relations presented by [43]. This agreement allows us to construct empirical relations for all QPO frequencies, taking advantage of the M/R expansion in [43], which was constructed for a set of different EOSs (while here we only consider a polytropic EOS). These relations, which can be found in [10], allow us to provide constraints on the amplitude of the magnetic field for SGR 1806-20 and SGR 1900+14. In the former case the dipolar component of the magnetic field varies from 5×10^{15} G to 1.2×10^{16} G, while for the latter it is between 4.7×10^{15} G and 1.12×10^{16} G. These values give a mean surface magnetic field strength of about $3 - 8 \times 10^{15}$ G (as the mean value at the surface is $\sim 2/3$ of the value at the pole).

The interested reader is also addressed to the more recent results discussed by [23] in this volume (see also [22]) where the first numerical simulations of axisymmetric, torsional Alfvén oscillations in magnetars *including an extended crust* were presented. In the limit of very strong magnetic fields the results of [22] agree with previous studies of Alfvén oscillations in the absence of a solid crust [43, 10, 13], while for very weak magnetic fields crustal shear oscillations are recovered. In the intermediate regime, 5×10^{13} Gauss to 1×10^{15} Gauss, however, [22] find strong resonant absorption of crustal shear modes by the Alfvén continuum of the core, while the Alfvén oscillations are influenced by the shear modulus and become magneto-elastic oscillations.

5. Summary

Results from numerical simulations of magneto-rotational stellar core collapse and Alfvén oscillations in magnetars have been presented. These simulations have been performed with a numerical code, the `CoCoA/CoCoNuT` code, able to handle ideal MHD flows in dynamical spacetimes in general relativity. This code is based on high-resolution shock-capturing schemes to solve the flux-conservative hyperbolic GRMHD equations, and the flux constraint transport method to ensure the solenoidal condition of the magnetic field. The Einstein equations are formulated in the conformal flatness condition approximation, and the resulting elliptic equations are solved using a linear Poisson solver.

The first topic we have discussed has been the simulation of general relativistic magneto-rotational core collapse using a realistic stellar progenitor model and a microphysical equation of state. We have shown that the cores of collapse progenitors with unrealistically large values of the initial magnetic field, suffer a rapid spin down, an effect which may be understood resorting to the magneto-rotational instability

The second topic has been the study of Alfvén oscillations in magnetars using both a semi-analytic approach and nonlinear MHD simulations in the anelastic approximation. Our results show that the Alfvén oscillations form a continuum in the core of magnetars, whose turning points correspond to QPOs. When an extended crust is included in the magnetar model, strong resonant absorption of crustal shear modes by the Alfvén continuum of the core is found for intermediate values of the magnetic field (see [22, 23])

Acknowledgments

This work has been supported by a GSRT Greece-Spain bilateral grant, a DAAD Germany-Greece bilateral grant, the EU network ILIAS, the Spanish Ministry of Education and Science (AYA 2007-67626-C03-01), the Collaborative Research Center on *Gravitational Wave Astronomy* of the Deutsche Forschungsgesellschaft (DFG SFB/Transregio 7), and the European research community Compstar. One of the authors (MG) also acknowledges support through an individual DAAD exchange grant.

References

[1] Anderson, M., Hirschmann, E. W., Lehner, L., et al. 2008, ArXiv e-prints, 801
[2] Andersson N., Glampedakis K. , Samuelsson L., 2008, preprint (astro-ph/0812.2417)
[3] Antón, L., Zanotti, O., Miralles, J. A., et al. 2006, Astrophys. J., 637, 296
[4] Antón, L., Miralles, J. A., Martí, J.M., Ibáñez, J.M., Aloy, M.A., & Mimica, P., 2010, Astrophys. J. Suppl., 188, 1
[5] Balbus, S. A. & Hawley, J. F. 1991, Astrophys. J., 376, 214
[6] Bocquet M., Bonazolla S., Gourgoulhon E., Novak J., 1995, A&A, 301, 757
[7] Bonazzola, S., Villain, L., Bejger, M., 2007, Class. Quantum Grav., 24, 221
[8] Cerdá-Durán, P., Faye, G., Dimmelmeier, H., et al. 2005, A&A, 439, 1033
[9] Cerdá-Durán, P., Font, J.A., Antón, L., & Müller, E. 2008, A%A, 492, 937
[10] Cerdá-Durán, P., Stergioulas, N, & Font, J.A. 2009, MNRAS, 397, 1607
[11] Cerdá-Durán, P. & Font, J. A. 2007, Classical and Quantum Gravity, 24, 155
[12] Cerdá-Durán, P., Font, J. A., & Dimmelmeier, H. 2007, A&A, 474, 169
[13] Colaiuda A., Beyer, H., Kokkotas, K.D., 2009, MNRAS, 396, 1441
[14] Cordero-Carrión, I., Cerdá-Durán, P., Dimmelmeier, H., Jaramillo, J.L., Novak, J., & Gourgoulhon, E., 2009, Phys. Rev. D, 79, 024017
[15] Dimmelmeier, H., Font, J. A., & Müller, E. 2002, A&A, 388, 917
[16] Dimmelmeier, H., Font, J. A., & Müller, E. 2002, A&A, 393, 523
[17] Dimmelmeier, H., Ott, C. D., Janka, H.-T., Marek, A., & Müller, E. 2007, Phys. Rev. Lett., 98, 251101
[18] Duez, M. D., Liu, Y. T., Shapiro, S. L., & Stephens, B. C. 2005, Phys. Rev. D, 72, 024028

[19] Duncan R.C., Thompson C., 1992, ApJ, 392, L9
[20] Duncan R.C.,1998, ApJ, 1998, 498, L45
[21] Font, J. A. 2008, Living Rev. Relativity, 11, 7
[22] Gabler, M., Cerdá-Durán, P., Font, J. A., Müller, E., & Stergioulas, N., 2010, preprint [arXiv:1007.0856]
[23] Gabler, M., Cerdá-Durán, P., Font, J. A., Müller, E., & Stergioulas, N., 2010, these proceedings
[24] Glampedakis K., Samuelsson L., Andersson N., 2006, MNRAS, 371, L74
[25] Giacomazzo, B. & Rezzolla, L. 2007, Classical and Quantum Gravity, 24, 235
[26] Harten, A., Lax, P. D., & van Leer, B. 1983, SIAM Review, 25, 35
[27] Hawley, J. F. & Balbus, S. A. 1992, Astrophys. J., 400, 595
[28] Isenberg, J. A. 2008, Int. J. Mod. Phys. D, 17, 265
[29] Israel G. et al., 2005, ApJ, 628, L53
[30] Kurganov, A. & Tadmor, E. 2000, J. Comp. Phys., 160, 214
[31] Levin Y., 2006, MNRAS, 368, L35
[32] Levin Y., 2007, MNRAS, 377, 159
[33] Liebendörfer, M. 2005, Astrophys. J., 633, 1042
[34] Müller, E. 1979, A magnetohydrodynamic supernova model, PhD thesis, Darmstadt, Technische Hochschule
[35] Obergaulinger, M., Aloy, M. A., Dimmelmeier, H., & Müller, E. 2006, A&A, 457, 209
[36] Piro A.L., 2005, ApJ, 634, L153
[37] Samuelsson L., Andersson N., 2007, MNRAS, 374, 256
[38] Schumaker B.L., Thorne K.S., 1983, MNRAS, 203, 457
[39] Shen, H., Toki, H., Oyamatsu, K., & Sumiyoshi, K. 1998, Prog. Theor. Phys., 100, 1013
[40] Shibata, M. & Sekiguchi, Y. 2005, Phys. Rev. D, 71, 024014
[41] Sotani H., Kokkotas K.D., Stergioulas N., 2007a, MNRAS, 375, 261
[42] Sotani H., Colaiuda A., Kokkotas K.D., 2007c, preprint (gr-qc/0711.1518)
[43] Sotani H., Kokkotas K.D., Stergioulas N., 2008, MNRAS, 385, L5
[44] Sotani H., Kokkotas K.D., 2009, MNRAS, "Alfvén polar oscillations of relativistic stars", in preparation
[45] Strohmayer T.E., Watts A.L., 2005, ApJ, 632, L111
[46] Strohmayer T.E., Watts A.L., 2006, ApJ, 653, 593
[47] Toro, E. F. 1999, Riemann Solvers and Numerical Methods for Fluid Dynamics (Berlin: Springer Verlag)
[48] Watts A.L., Strohmayer T.E., 2006, ApJ, 637, L117
[49] Watts A.L., Strohmayer T.E., 2007, Advances in Space Research, 40, 1446
[50] Woosley, S. E., & Janka, H.-Th. 2005, Nature Physics, 1, 147

Journal of Physics: Conference Series **283** (2011) 012012 doi:10.1088/1742-6596/283/1/012012

Applications of hidden symmetries to black hole physics

Valeri Frolov

Institute of Theoretical Physics, Department of Physics University of Alberta, Edmonton, Alberta, T6G 2G7, CANADA

E-mail: vfrolov@ualberta.ca

Abstract. This work is a brief review of applications of hidden symmetries to black hole physics. Symmetry is one of the most important concepts of the science. In physics and mathematics the symmetry allows one to simplify a problem, and often to make it solvable. According to the Noether theorem symmetries are responsible for conservation laws. Besides evident (explicit) spacetime symmetries, responsible for conservation of energy, momentum, and angular momentum of a system, there also exist what is called hidden symmetries, which are connected with higher order in momentum integrals of motion. A remarkable fact is that black holes in four and higher dimensions always possess a set ('tower') of explicit and hidden symmetries which make the equations of motion of particles and light completely integrable. The paper gives a general review of the recently obtained results. The main focus is on understanding why at all black holes have something (symmetry) to hide.

1. Introduction

In this paper we discuss higher dimensional rotating black holes and their properties. We consider black holes with the spherical topology of the horizon in a spacetime which is either asymptotically flat (vacuum) or (A)deSitter (with cosmological constant). We demonstrate that such solutions of the Einstein equations always possess a "tower" of symmetries, which make the equations of geodesic motion completely integrable. This 'tower' is generated by a single object, which we call a *principal conformal Killing-Yano tensor*. This object is responcible for a set of Killing vectors reflecting a spacetime symmerty. It also generates a set of Killing tensors, connected with hidden symmetries of the spacetime. The corresponding conserved quantities, which are first and second order in the momentum, form a complete set of integrals of motion which make the geodesic equation completely integrable. The purpose of the paper is to describe a general scheme of this construction, and to discuss its application to concrete problems connected with particle motion and field propagation in the higher dimensional black holes. We also describe a class of metrics which admit the principal conformal Killing-Yano tensor and its generalizations.

2. Complete integrability

2.1. Phase space

Let us first remind three related notions which play an important role in study of dynamical systems: (i) complete integrability, (ii) separation of variables, and (iii) hidden symmetries (for more details see [1, 2]).

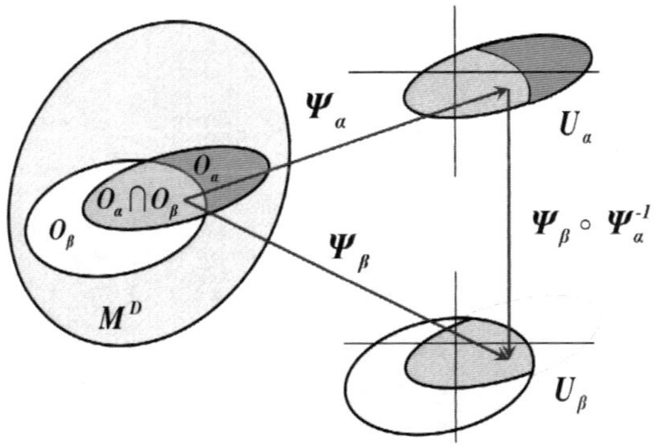

Figure 1. Coordinates on $n-$dimensional manifold are maps Ψ_α of open regions O_α on the regions U_α in the $n-$dimensional Euclidean space R^n. The map $\Psi_\beta \circ \Psi_\beta^{-1}$ determines a transformation between two coordinate systems.

A *phase space* is a set of three items $\{M^{2m}, \Omega, H\}$. M^{2m} is a $2m-$dimensional differential manifold. A symplectic form Ω is a closed, $d\Omega = 0$, non-degenerate 2-form. The non-degeneracy means that the corresponding matrix of its coefficient has the rank $2m$. Locally $\Omega = 0$ can be presented in the form

$$\Omega = d\alpha\,, \tag{1}$$

where α is a 1-form. H is a scalar function on M^{2m} called a *Hamiltonian*. We denote coordinates on M^{2m} by z^A $(A = 1, \ldots, 2m)$. A set of coordinates which covers all the manifold is called an *atlas*. Figure 1 illustrates these definitions.

Since the symplectic form is non-degenerate, the tensor Ω_{AB} has an inverse one Ω^{AB}, defined by the relation $\Omega^{AB}\Omega_{BC} = \delta^A_C$. The tensors Ω_{AB} and Ω^{AB} can be used to relate tensors with upper and lower position of indices. For example,

$$\eta^A = \Omega^{AB} H_{,B} \tag{2}$$

is a vector generating the Hamiltonian flow. The equation of motion of a particle in the phase space is

$$\dot{z}^A = \eta^A\,. \tag{3}$$

Poisson bracket for two functions on the phase space A and B is defined as

$$\{A, B\} = \Omega^{AB} A_{,A} B_{,B}\,. \tag{4}$$

One says that these function are in *involution* if their Poisson bracket vanishes. The Poisson brackets for any three functions A, B, and C on the symplectic manifold obey the *Jacobi identity*

$$\{\{A, B\}, C\} + \{\{B, C\}, A\} + \{\{C, A\}, B\} = 0\,. \tag{5}$$

If $F(z)$ is a function on the phase space, the equation of motion Eq.(3) determines its evolution $F_t = F(z(t))$

$$\dot{F}_t = \{H, F_t\}\,. \tag{6}$$

A function F for which $\{H, F\} = 0$ is an integral of motion of the Hamiltonian system. The Hamiltonian itself is a trivial integral of motion.

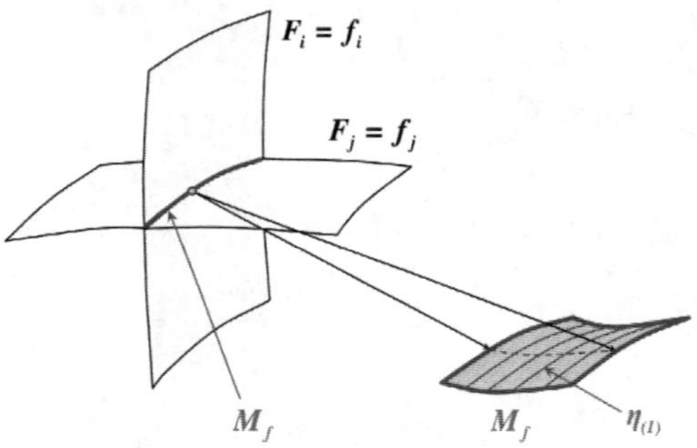

Figure 2. Illustration to the proof of Liouville theorem.

According to *Darboux theorem*, in the vicinity of any point of the phase space it is always possible to choose *canonical coordinates* $z^A = (q_1, q_2, \ldots, q_m, p_1, p_2, \ldots, p_m)$ in which the symplectic form is

$$\Omega = \sum_{i=1}^{m} dp_i \wedge dq_i \,. \tag{7}$$

In such canonical coordinates many relations are simplified and take a well known form. A set of canonical coordines which covers the phase space is called a *canonical atlas*.

2.2. Integrability

Integrability of equations of a dynamical system generically means that these equations can be solved by quadratures. Integrability is linked to 'existence of constants of motion'. For the integrability it is important to know: (i) How many constants of motion exist? (ii) How precisely are they related? and (iii) How the phase space is foliated by their level sets?

A system of differential equations is said to be integrable by quadratures if its solution can be found after a finite number of steps involving algebraic operations and integration. The following theorem (Bour, 1855; Liouville, 1855) establishes conditions required for the complete integrability of a Hamiltoniam system:

If a Hamiltonian system with m degrees of freedom has m integrals of motion $F_1 = H, F_2, \ldots, F_m$ in involution which are functionally independent on the intersection sets of the m functions, $F_i = f_i$, the solutions of the corresponding Hamiltonian equations can be found by quadratures.

The main idea behind Liouville's theorem is that the first integrals of motion F_i can be used as m coordinates. The involution condition implies that the m vector fields generated by gradients of F_i commute with each other and provide a choice of canonical coordinates. In these coordinates, the Hamiltonian is effectively reduced to a sum of m decoupled Hamiltonians that can be integrated.

We denote by M_f an intersection of the level sets $F_i = f_i$ for m integrals of motion (see Figure 2). Since the integrals of motion are independent, the tangent to M_f vectors $X_i^A = \Omega^{AB} F_{i,B}$ are linearly independent and M_f is a $m-$dimensional submanifold of the phase space. The condition that the integrals of motion are in involution implies

$$[X_i, X_j] = 0 \,. \tag{8}$$

Journal of Physics: Conference Series **283** (2011) 012012 doi:10.1088/1742-6596/283/1/012012

One also has

$$\Omega_{AB}X_i^A X_j^B = \{F_i, F_j\} = 0\,. \tag{9}$$

Let us use the equations $F_i(p,q) = f_i$ to get $p_i = p_i(f,q)$. Denote

$$S(F,q) = \int_{q^0}^{q} \sum_{i=1}^{m} p_i(f,q)dq_i\,. \tag{10}$$

The integral is taken over a path on M_f connecting two its points q^0 and q. The relation Eq.(9) implies that $S(F,q)$ does not depend on the choice of the path. Denote

$$\Psi_i = \frac{\partial S}{\partial F_i}\,, \tag{11}$$

then

$$dS = \sum_{i=1}^{m} (\Psi_i dF_i + p_i dq_i)\;. \tag{12}$$

Since the symplectic form Ω is closed, one has $d^2S = 0$ and hence

$$\Omega = \sum_{i=1}^{m} dp_i \wedge dq_i = \sum_{i=1}^{m} d\Psi_i \wedge dF_i\,. \tag{13}$$

This result means that there exists a canonical transformation (p_i, q_i) to (F_i, Ψ_i). To obtain 'new' canonical coordinates two operations are required: (1) finding of $p_i = p_i(f,q)$, and (2) calculation of some integrals. In the new variables the Hamiltonian equations take the form

$$\dot{F}_i = \{H, F_i\} = 0\,, \tag{14}$$

$$\dot{\Psi}_i = \{H, \Psi_i\} = \frac{\partial H}{\partial F_i} = \delta_i^1\,. \tag{15}$$

The solution of this system is trivial

$$F_i = \text{const}\,, \qquad \Psi_i = a_i + \delta_i^1 t\,. \tag{16}$$

Complete integrability and chaotic motion are at the two ends of 'properties' of a dynamical system. The integrability is exceptional, while the chaoticity is generic. In all cases, integrability seems to be deeply related to some symmetry, which might be partially hidden. The existence of integrals of motion reflects the symmetry.

Important known examples of completely integrable mechanical systems include:

(i) Motion in Euclidean space under a central potential;
(ii) Motion in the two Newtonian fixed centers;
(iii) Geodesics on an ellipsoid;
(iv) Motion of a rigid body about a fixed point;
(v) Neumann model[1].

[1] The Lagrangian of the Neumann model is

$$L = \frac{1}{2}\sum_{k=1}^{N}\left[\dot{x}_k^2 - a_k x_k^2 + \Lambda(x_k^2 - 1)\right]\,. \tag{17}$$

Journal of Physics: Conference Series **283** (2011) 012012 doi:10.1088/1742-6596/283/1/012012

3. Separation of variables and integrability

Complete integrability of the Hamiltonian systems is closely related with the separation of variables in the Hamilton-Jacobi equation. For a Hamiltonian $H(P,Q)$, $P = p_1, \dots, p_m$ and $Q = q_1, \dots, q_m$, the *Hamilton-Jacobi equation* is

$$H(\partial_Q S, Q) = 0\,, \qquad \partial_Q S = (\partial_{q_1} S, \dots, \partial_{q_m} S)\,. \tag{18}$$

If a variable q_1 and a derivative $\partial_{q_1} S$ enter this equation in a form of single expression $\Phi_1(\partial_{q_1} S, q_1)$, than one says that the variable q_1 is separated. In such a case one may try to search a solution in the form

$$S = S_1(q_1) + S'(q_2, \dots, q_m)\,. \tag{19}$$

Putting

$$\Phi_1(\partial_{q_1} S, q_1) = C_1\,, \tag{20}$$

one obtains an equation with a less number of variables

$$H_1(\partial_{q_2} S', \dots, \partial_{q_m} S', q_2, \dots, q_m; C_1) = 0\,. \tag{21}$$

Let $S'(q_2, \dots, q_m; C_1)$ be a solution of this equation depending on a parameter C_1, then Eq.(19) is a solution of Eq.(18) when S_1 obeys an ordinary differential equation Eq.(20), which is easily solvable by quadratures. If a variable q_2 can be separated in a new equation Eq.(21) one can repeat this procedure again. One says that the Hamilton-Jacobi equation Eq.(18) allows a *complete separation of variables* if after m steps we obtain a solution of the initial equation Eq.(18) which contains m constants C_i

$$S = S_1(q_1, C_1) + S_2(q_2; C_1, C_2) + \ldots + S_m(q_m; C_1, \dots, C_m)\,. \tag{22}$$

In this case one obtains a complete solution of the Hamilton-Jacobi equation which depends on m parameters and the corresponding Hamilton equations are integrable by quadratures (*Jacobi theorem*).

The constants $C_1, \dots, C_m$ for a completely separable Hamilton-Jacobi equation can be considered as functions on the phase space where they are integrals of motion. In a case, when these integrals on motion are independent and in involution, the system is completely integrable in the sense of Liouville.

4. Particle motion in General Relativity

4.1. Equation of motion in the Hamiltonian form

Consider a particle motion of mass m in the gravitational field. Its equation of motion is

$$m\,\frac{D^2 x^\mu}{d\tau^2} = 0\,. \tag{23}$$

Here, $D/d\tau$ is the covariant derivative with respect to the proper time τ. Introduce the affine parameter $\lambda = \tau/\mu$ and denote a derivative with respect to it by a dot. The Lagrangian for Eq.(23) is

$$L = \frac{1}{2} g_{\mu\nu} \dot{x}^\mu \dot{x}^\nu\,. \tag{24}$$

The momentum p_μ is defined as follows

$$p_\mu = \frac{\partial L}{\partial \dot{x}^\mu} = g_{\mu\nu} \dot{x}^\nu\,. \tag{25}$$

Denote by D the dimensionality of the spacetime. Coordinates (p_μ, x^μ) are canonical on the phase space M^{2D}. In these coordinates the symplectic form is

$$\Omega = \sum_{\mu=1}^{D} dp_\mu \wedge dx^\mu \, . \tag{26}$$

The Hamiltonian of the particle is

$$H = \frac{1}{2} g^{\mu\nu} p_\mu p_\nu \, . \tag{27}$$

It gives trivial integral of motion

$$H = \frac{1}{2} m^2 \, . \tag{28}$$

For null rays one must put $m = 0$. The Hamiltonian equations of motion

$$\dot{x}^\mu = \{H, x^\mu\} = g^{\mu\nu} p_\nu \, , \tag{29}$$

$$\dot{p}_\mu = \{H, p_\mu\} = -\frac{1}{2} g^{\nu\lambda}{}_{,\mu} p_\nu \, , \tag{30}$$

are equivalent to the geodesic equation Eq.(23), which can be written in the form

$$p^\nu p_{\mu;\nu} = 0 \, . \tag{31}$$

4.2. Integrals of motion and Killing tensors

Consider a special monomial in the momentum on the phase space of the relativistic particle of the form

$$\mathcal{K} = K^{\mu_1 \ldots \mu_s} p_{\mu_1} \ldots p_{\mu_s} \, . \tag{32}$$

A condition that this is an integral of motion implies

$$K^{(\mu_1 \ldots \mu_s ; \nu)} = 0 \, . \tag{33}$$

The symmetric tensor of the rank s, $K_{\mu_1 \ldots \mu_s}$, which obeys the equation Eq.(33), is called a *Killing tensor.* The metric $g_{\mu\nu}$ is a trivial example of the Killing tensor of rank 2.

Suppose we have two integrals of motion, $\mathcal{K}_{(s)}$ and $\mathcal{K}_{(t)}$, connected with the Killing tensors $K_{(s)}^{\mu_1 \ldots \mu_s}$ and $K_{(t)}^{\nu_1 \ldots \nu_t}$, respectively. Using the *Jacobi identity* Eq.(5) it is easy to check that $\{\mathcal{K}_{(s)}, \mathcal{K}_{(t)}\}$ also commutes with the Hamiltonian H and hence is an integral of motion. This commutator is a monomial of the power $s + t - 1$ in the momentum. The corresponding Killing tensor of the rank $s + t - 1$ is

$$[K_{(s)}, K_{(t)}] = K_{(s+t-1)} \, , \tag{34}$$

$$K_{(s+t-1)}^{\mu_1 \ldots \mu_{s-1} \lambda \nu_1 \ldots \nu_{t-1}} = s\, K_{(s)}^{\epsilon(\mu_1 \ldots \mu_{s-1}} \partial_\epsilon K_{(t)}^{\lambda \nu_1 \ldots \nu_{t-1})} - t\, K_{(t)}^{\epsilon(\nu_1 \ldots \nu_{t-1}} \partial_\epsilon K_{(s)}^{\lambda \nu_1 \ldots \mu_{s-1})} \tag{35}$$

The introduced operation for the Killing tensors is known as the *Schouten-Nijenhuis brackets.* Killing tensors form a Lie subalgebra of a Lie algebra of all totally symmetric contravariant tensor fields on the manifold with respect to these operations. The Killing tensors for which the Schouten-Nijenhuis bracket vanishes are said to be in involution.

In a simplest case when a monomial is of the first order in the momentum, the Killing tensor coincides with the Killing vector, and the *Schouten-Nijenhuis bracket* reduces to a usual commutator of two vector fields.

If there exist D non-degenerate functionally independent Killing tensors[2] in involution then the geodesic equations in $D-$dimensional spacetime are completely integrable.

Geodesic motion in the gravitational field of the most general solution describing a rotating black hole in 4 and higher dimensional spacetimes, which are asymptotically either flat or (A)dS, is a new class of physically interesting completely integrable systems.

[2] Some of them can be Killing vectors.

5. Killing-Yano tensors

5.1. Definitions

Killing tensors are natural symmetric generalizations of a Killing vector. Let us discuss another important generalization, known as a Killing-Yano tensor. We define first a *conformal Killing-Yano tensor* of rank p. It is an antisymmetric tensor $k_{\mu_1 \ldots \mu_p}$ which obeys the following equation

$$\nabla_{(\mu_1} k_{\mu_2)\mu_3 \ldots \mu_{p+1}} = g_{\mu_1\mu_2} \tilde{k}_{\mu_3 \ldots \mu_{p+1}} - (p-1) g_{[\mu_3(\mu_1} \tilde{k}_{\mu_2)\mu_4 \ldots \mu_{p+1}]} \,. \tag{36}$$

By tracing both sides of this equation one obtains

$$\tilde{k}_{\mu_2\mu_3 \ldots \mu_p} = \frac{1}{D-p+1} \nabla^{\mu_1} k_{\mu_1\mu_2 \ldots \mu_p} \,. \tag{37}$$

In the case when $\tilde{k}_{\mu_2 \ldots \mu_p} = 0$ one has a *Killing–Yano* (KY) tensor. For the KY tensor $\mathbf{k}$ the quantity

$$L_{\mu_1\mu_2 \ldots \mu_{p-1}} = k_{\mu_1\mu_2 \ldots \mu_p} p^{\mu_p} \,, \tag{38}$$

is parallel-propagated along the geodesic.

The 'square' $\mathbf{K}$ of KY tensor $\mathbf{k}$, defined by the relation

$$K_{\mu\nu} = (\mathbf{k} \circ \mathbf{k})_{\mu\nu} \equiv (p-1)! \, k_{\mu\mu_2 \ldots \mu_p} k_{\nu}{}^{\mu_2 \ldots \mu_p} \,, \tag{39}$$

is a Killing tensor. Notice that not an arbitrary Killing tensor can be written in the form .

5.2. Properties of conformal Killing-Yano tensors

The (conformal) Killing-Yano tensors have the following properties:

(i) Hodge dual of a conformal Killing-Yano tensor is a conformal Killing-Yano tensor;
(ii) Hodge dual of a closed conformal Killing-Yano tensor is a Killing-Yano tensor;
(iii) External product of two closed conformal Killing-Yano tensor is a closed conformal Killing-Yano tensor.

Figure 3 schematically illustrates these properties. The last of these statements was proved in [3, 4].

5.3. Principal conformal Killing-Yano tensor

Consider an antisymmetric tensor of rank 2 $h_{\mu\nu}$ which obeys the following equation

$$\nabla_\gamma h_{\mu\nu} = g_{\gamma\mu} \xi_\nu - g_{\gamma\nu} \xi_\mu \,. \tag{40}$$

This is a conformal Killing-Yano tensor. Equation (40) implies the following relations

$$\nabla_{[\gamma} h_{\mu\nu]} = 0 \,, \qquad \xi_\mu = \frac{1}{D} \nabla^\nu h_{\nu\mu} \,. \tag{41}$$

The first of these relations means that $\mathbf{h}$ is closed. Here $D = 2n + \varepsilon$ is the spacetime dimension. For even number of dimensions $\varepsilon = 0$, while for the odd number $\varepsilon = 1$. The tensor $\mathbf{h}$ is non-degenerate if its matrix rank is $2n$. A *principal conformal Killing-Yano tensor* is a non-degenerate closed conformal Killing-Yano tensor of rank 2. The existence of the principal conformal Killing-Yano tensor for the most general known solution [5] for higher dimensional rotating black holes with spherical topology of the horizon was proved in [6, 7].

It is possible to show that the vector ξ^μ defined by the second equation in (41) is a Killing vector. We call it a *primary Killing vector* [8].

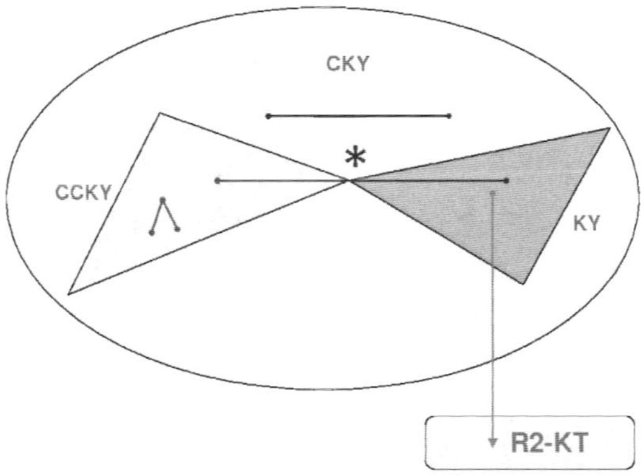

Figure 3. Schematical illustration of the properties of closed conformal Killing-Yano tensors (CCKY), Killing-Yano tensors (KY), and Killing tensors of rank 2 (R2-KT).

6. Killing-Yano 'tower'

In a spacetime with a principal conformal Killing-Yano tensor it is possible to construct a set of Killing vectors and tensors, which we call a *Killing-Yano 'tower'*. The idea of this constuction is following. For a given principal conformal Killing-Yano tensor $\mathbf{h}$ one can define the a set of $n-1$ objects

$$\mathbf{h}^{\wedge j} = \underbrace{\mathbf{h} \wedge \mathbf{h} \wedge \ldots \wedge \mathbf{h}}_{j \text{ times}}, \qquad j = 1, \ldots, n-1\,. \tag{42}$$

Each of these objects is a closed conformal Killing-Yano tensor, so that their Hodge dual are Killing-Yano tensors

$$\mathbf{k}_{(j)} = *\mathbf{h}^{\wedge j}\,. \tag{43}$$

By taking 'squares' of these tensors one obtains $n-1$ Killing tensors of the second rank

$$\mathbf{K}_{(j)} = \mathbf{k}_{(j)} \circ \mathbf{k}_{(j)}\,. \tag{44}$$

If ξ^ν is a primary Killing vector then it is possible to show that

$$\xi^\mu_{(j)} = K^\mu_{(j)\nu}\xi^\nu \tag{45}$$

are again Killing vectors. Thus one obtains n Killing vectors. In the odd dimensional spacetime there exists an additional Killing vector $\eta = *\mathbf{h}^{\wedge n}$. A set of constructed $n+\varepsilon$ Killing vectors, $n-1$ Killing tensors, and one trivial Killing tensor ($\mathbf{g}$) gives D conserved quantities for the geodesic motion. It is possible to show that the corresponding integrals of motion are independent and in involution. Thus geodesic motion equations are completely integrable in a spacetime with a principal conformal Killing-Yano tensor [9, 10, 11].

7. Canonical form of metric

7.1. Canonical coordinates

In a spacetime with a principal conformal Killing-Yano tensor $\mathbf{h}$ there exists a special convenient choice of coordinates. Consider the following eigen-value problem

$$h^\mu{}_\nu m^\nu_{\pm a} = \mp i x_a m^\mu_{\pm a}\,. \tag{46}$$

Complex eigen-vectors $m^{\mu}_{\pm a}$ can be written as

$$m^{\mu}_{\pm a} = e^{\mu}_{a} \pm i e^{\mu}_{\hat{a}} , \tag{47}$$

where $\mathbf{e}_a$ and $\mathbf{e}_{\hat{a}}$ are mutually orthogonal normalised real vectors. A non-degerate $\mathbf{h}$ has n different eigen-values x_a $(a = 1, \ldots, n)$, and the corresponding eigen-space for each of these eigen-values is two dimensional (see [12]). One can use x_a as n coordinates on the spacetime manifold. We call them *Darboux coordinates.* For each of the Killing vectors ξ from the Killing-Yano 'tower' one can introduce a Killing parameter so that the integral line of this vector field is a solution of the equation

$$\frac{dx^{\mu}}{d\psi} = \xi^{\mu} . \tag{48}$$

This gives us $n + \varepsilon$ Killing coordinates ψ_j $(j = 0, \ldots, n - \varepsilon)$. Total number of Darboux and Killing coordinates is D, which is sufficient for using as coordinate system in the D dimensional spacetime manifold. We call these coordinates *canonical.*

7.2. Off-shell canonical metric

In these coordinates the metric of the spacetime is of the form [13, 15, 14]

$$ds^2 = \sum_{a=1}^{n} \left[\frac{U_a}{X_a} (dx_a)^2 + \frac{X_a}{U_a} \left(\sum_{j=0}^{n-1} A^{(j)}_a d\psi_j \right)^2 \right] - \frac{\varepsilon c}{A^{(n)}} \left(\sum_{j=0}^{n} A^{(j)} d\psi_j \right)^2 . \tag{49}$$

Here

$$U_a = \prod_{b \neq a} (x_b^2 - x_a^2) , \qquad X_a = X_a(x_a) , \tag{50}$$

$$\prod_{a=1}^{n} (1 + \lambda x_a^2) = \sum_{j=0}^{n} \lambda^j A^{(j)} , \qquad (1 + \lambda x_b^2) \prod_{a=1}^{n} (1 + \lambda x_a^2) = \sum_{k=0}^{n-1} \lambda^k A^{(k)}_b . \tag{51}$$

A potential $\mathbf{b}$, which generates the principal conformal Killing-Yano tensor $\mathbf{h}$,

$$\mathbf{h} = d\mathbf{b} , \tag{52}$$

in the canonical coordinates is

$$\mathbf{b} = \frac{1}{2} \sum_{k=0}^{n-1} A^{(k+1)} d\psi_k . \tag{53}$$

The metric (49) is of the algebraical type D. As we indicated above, geodesic equations in this metric are completely integrable. Besides this, it also has the following nice properties. In the metric (49):

(i) Hamilton-Jacobi and Klein-Gordon equations allow complete separation of variables [16];
(ii) Massive Dirac equation is separable [17];
(iii) Stationary string equations are completely integrable [18];
(iv) Tensorial gravitational perturbation equations are separable [19];
(v) Equations of the parallel transport along timelike and null geodesics can be integrated [20, 21];
(vi) Equations for charged particle motion in such a spacetime in the presence of a test electromagnetic field are completely integrable provided this field is generated by the primary Killing vector [22].

7.3. On-shell metric

The metric (49) is valid for any geometry with a principal conformal Killing-Yano tensor. We call this metric *off-shell*, since it does not obey the Einstein equations. Let us consider now on-shell geometry, that is assume that the metric obeys the equations

$$R_{\mu\nu} = (D-1)\Lambda g_{\mu\nu} \,. \tag{54}$$

Λ is the cosmological constant. The Einstein equations restrict arbitrary functions $X_a(x_a)$, which enter (49), so that they take the form of polynomials [5, 23]

$$X_a = b_a x_a + \sum_{k=0}^{n} c_k x_a^{2k} \,. \tag{55}$$

As a result, the solution depends on $D-\varepsilon$ arbitrary parameters. This solution coincides with the most general solution for higher dimensional black holes in either asymptotically flat ($\Lambda = 0$), or asymptotically (A)dS spacetime, obtained in [5]. Arbiraty constants, which enter this solutions are: the cosmological constant Λ, the mass M, $(n-1+\varepsilon)$ rotation parameters, and $(n-1-\varepsilon)$ 'NUT' parameters. In the 4D case this is a Kerr-NUT-(A)dS metric.

7.4. Further developments

Let us mention two recent developments of the above described results.

(1) In our discussion we assumed that the closed conformal Killing-Yano tensor is non-degenerate. In particular, this implies that there exists a set of n different eigen-values of $\mathbf{h}$, which we used as Darboux coordinates. In a degenerate case, there may exist several eigen-values which are constants, and some of these constants can vanish. The general form of the canonical metric for such degenerate cases was constucted in [12].

(2) We discussed vacuum (with cosmological constant) solutions of the higher dimensional Einstein equations. An interesting generalization to a non-vacuum case was obtained recently in [24]. The authors consider a *five dimensional* minimally coupled gauged supergravity, which includes gravity and the Maxwell field with a Chern-Simons term. The corresponding Largrangian density is

$$L = *(R+\Lambda) - \frac{1}{2} F \wedge *F + \frac{1}{3\sqrt{3}} F \wedge F \wedge A \,. \tag{56}$$

The corresponding Einstein-Maxwell equations are

$$R_{\mu\nu} + \frac{1}{3}\Lambda = \frac{1}{2}(F_{\mu\lambda}F_{\nu}{}^{\lambda} - \frac{1}{6} g_{\mu\nu}F^2) \,, \tag{57}$$

$$dF = 0 \,, \qquad d * F - \frac{1}{\sqrt{3}} F \wedge F = 0 \,. \tag{58}$$

The main result of this work is the following. One can modify the covariant derivative by including a non-vanishing torsion $T = \frac{1}{\sqrt{3}} *F$, and generalize the equation (40), by subsituting the modified derivative insted of the covariant one. The authors demonstrated that the generalized principal conformal Killing-Yano tensor generates a 'tower' of integrals of motion which provides complete integrability of a charged particle motion in these spaces. An interesting example of a charged rotating black hole solution in this theory was obtained in [25]. It is interesting, that the corresponding metric is of a general algebraical type.

7.5. Acknowledgments

The author is grateful to the Natural Sciences and Engineering Research Council of Canada and the Killam Trust for their support.

References

[1] Arnold V I 1989 *Mathematical Methods of Classical Mechanics* Springer

[2] Babelon O, Bernard D, and Talon M 2003 Introduction to Classical Integrable Systems Cambridge Univ. Press.

[3] Krtouš P, Kubizňák D, Page D N, and Frolov V P 2007 *J. High Energy Phys.* **02** 004

[4] Frolov V P 2008 *Prog.Theor.Phys.Suppl.* **172** 210

[5] Chen W, Lü H, and Pope C N 2006 *Class. Quant. Grav.* **23** 5323

[6] Frolov V P and Kubizňák D 2007 *Phys.Rev.Lett* **98** 011101

[7] Kubizňák D and Frolov V P 2007 *Class. Quant. Grav.* **24** F1

[8] Frolov V P and Kubizňák D 2008 *Class.Quant.Grav.* **25** 154005

[9] Page D N, Kubizňák D, Vasudevan M, and Krtouš P 2007 *Phys. Rev. Lett.* **98** 061102

[10] Krtouš P, Kubizňák D, Page D N, and Vasudevan M 2007 *Phys. Rev.* D **76** 084034

[11] Houri T, Oota T and Yasui Y 2008 *J. Phys.* A **41** 025204

[12] Houri T, Takeshi Oota T and Yasui Y 2008 *Phys.Lett.* B **666** 391

[13] Houri T, Oota T and Yasui Y 2007 *Phys. Lett.* B **656** 214

[14] Houri T, Oota T and Yasui Y, 2009 *Class.Quant.Grav.* **26** 045015

[15] Krtous P, Frolov V P, and Kubiznak D 2008 *Phys.Rev.* D **78** 064022

[16] Frolov V P, Krtouš P, and Kubizňák D 2007 *J. High Energy Phys.* **02**) 005

[17] Oota T and Yasui Y 2008 *Phys.Lett.* B **659** 688

[18] Kubiznak D and Frolov V P 2008 *JHEP* **0802** 007

[19] Oota T and Yasui Y 2008 *Int.J.Mod.Phys.* A **25** 3055

[20] Connell P, Frolov V P and Kubiznak D 2008 *Phys.Rev.* D **78** 024042

[21] Kubiznak D, Frolov V P, Krtouš P, and Connell P 2009 *Phys.Rev.* D **79** 024018

[22] Frolov V and Krtouš P, *Charged particle in higher dimensional weakly charged rotating black hole spacetime* 2010 *arXiv:1010.2266 [hep-th]*

[23] Hamamoto N, Houri T, Oota T, and Yasui Y 2007 *J. Phys.* A **40**, F177

[24] Kubiznak D, Kunduri H K and Yasui Y 2009 *Phys.Lett.* B **678** 240

[25] Chong Z W, Cvetic M, Lu H and Pope C N 2005 *Phys. Rev. Lett.* **95** 161301

Recent Developments in Gravity (NEB XIV) IOP Publishing
Journal of Physics: Conference Series **283** (2011) 012013 doi:10.1088/1742-6596/283/1/012013

Magneto-elastic torsional oscillations of magnetars

Michael Gabler[1,2,3] , Pablo Cerdá-Durán[1], José A. Font[2] , Ewald Müller[1] and Nikolaos Stergioulas[3]

[1] Max-Planck-Institut für Astrophysik, Karl-Schwarzschild-Str. 1, 85741 Garching, Germany
[2] Departamento de Astronomía y Astrofísica, Universidad de Valencia, 46100 Burjassot (Valencia), Spain
[3] Department of Physics, Aristotle University of Thessaloniki, Thessaloniki 54124, Greece

E-mail: `miga@mpa-garching.mpg.de`

Abstract. We extend a general-relativistic ideal magneto-hydrodynamical code to include the effects of elasticity. Using this numerical tool we analyse the magneto-elastic oscillations of highly magnetised neutron stars (magnetars). In simulations without magnetic field we are able to recover the purely crustal shear oscillations within an accuracy of about a few per cent. For dipole magnetic fields between 5×10^{13} and 10^{15} G the Alfvén oscillations become modified substantially by the presence of the crust. Those quasi-periodic oscillations (QPOs) split into three families: Lower QPOs near the equator, Edge QPOs related to the last open field line and Upper QPOs at larger distance from the equator.

1. Introduction

The theoretical framework of relativistic elasticity has been developed recently in a series of papers by [1, 2, 3, 4]. One natural application of this theory is the crust of a neutron star, where the structure of the matter is crystalline and it is able to support shearing motions. When calculating the oscillatory modes supported by the crust, one has to include the effects of gravitation, which influence the theoretically obtained frequencies for isolated neutron stars significantly. Observational candidates to test theoretical models based upon relativistic elasticity can be found in the decaying tail of a giant burst of a soft-gamma repeater (SGR). In two such objects, SGR 1900+14 and SGR 1806-20, a number of long-lasting, quasi-periodic oscillations (QPOs) have been observed (see [5] and [6] for recent reviews).

The first attempts to explain those QPOs were based on models of purely crustal shear oscillations of an isolated neutron star (see e.g. [7, 8, 9, 10, 11, 12]). However, SGRs are believed to posses ultra strong magnetic fields $B \approx 10^{15}$ G, and it is necessary to construct self-consistent models including the interaction of the Alfvén oscillation with the crustal shear modes ([13, 14, 15, 16, 17]). A second approach to understand these QPOs is therefore based on purely Alfvén oscillations without a crust (see e.g. [18, 19, 20]). In those studies two families of QPOs were found. They are related to the open field lines near the pole and to the closed field lines near the equator. The Alfvén QPO model is very attractive, because it reproduces the near-integer-ratios of the observed 30, 92 and 150 Hz frequencies in SGR 1806-20. The results of the numerical simulation agree with a semi-analytic model based on standing waves in the short-wavelength limit [19].

Obviously, any realistic model has to include both contributions, the one from the crust and the one from the magnetised core. In first simplified models, it was shown that the effect of the coupling between crust and core may lead to an absorption of shear modes into a MHD continuum of Alfvén oscillations [15]. Additionally, long-lived QPOs may still appear at the turning points or edges of the continuum.

Here we extend a previous model of [19] to simulate coupled, magneto-elastic oscillations in a general-relativistic framework. We use a dipolar magnetic field and a tabulated equation of state (EOS) for dense matter. The numerical simulations are based on state-of-the-art Riemann solver methods for both the interior MHD fluid and the crust.

We use units where $c = G = 1$ with c and G being the speed of light and the gravitational constant, respectively. Latin (Greek) indices run from 1 to 3 (0 to 3).

2. Theoretical model

Following the analysis by [19], who considered purely Alfvén oscillations of magnetars with a two-dimensional, general-relativistic, ideal MHD code called MCoCoA [21], we upgrade this code by including the effects of the crust in general relativity. As in [22] this can be done by including an additional term $T^{\mu\nu}_{\text{elas}}$ in the stress-energy tensor

$$\begin{aligned} T^{\mu\nu} &= T^{\mu\nu}_{\text{fluid}} + T^{\mu\nu}_{\text{mag}} + T^{\mu\nu}_{\text{elas}} \\ &= \rho h u^\mu u^\nu + P g^{\mu\nu} + b^2 u^\mu u^\nu + \frac{1}{2} b^2 g^{\mu\nu} - b^\mu b^\nu - 2\mu_{\text{S}} \Sigma^{\mu\nu} , \end{aligned} \tag{1}$$

where ρ is the rest-mass density, h the specific enthalpy, P the isotropic fluid pressure, u^μ the 4-velocity of the fluid, b^μ the magnetic field measured by a co-moving observer (with $b^2 := b^\mu b_\mu$), $\Sigma^{\mu\nu}$ the shear tensor, and μ_{S} the shear modulus, respectively. The latter is obtained according to [10]. As in [19] we apply a number of simplifications: (i) a zero temperature EOS, (ii) axisymmetry, (iii) a purely poloidal magnetic field configuration, (iv) the Cowling approximation, (v) a spherically symmetric background and (vi) small amplitude oscillations. Assumptions (ii) and (iii) lead in the linear approximation to the decoupling of polar oscillations from axial ones. We further assume a conformally flat metric

$$ds^2 = -\alpha^2 dt^2 + \phi^4 \left(dr^2 + r^2 d\theta^2 + r^2 \sin\theta^2 d\varphi^2 \right) , \tag{2}$$

where α is the lapse function and ϕ the conformal factor. Employing the induction equation for the magnetic field, the equations for the conservation of energy and momentum can be cast into a conservation law of the following form

$$\frac{1}{\sqrt{-g}} \left(\frac{\partial \sqrt{\gamma} \mathbf{U}}{\partial t} + \frac{\partial \sqrt{-g} \mathbf{F}^i}{\partial x^i} \right) = 0 , \tag{3}$$

where g and γ are the determinants of the 4-metric and 3-metric, respectively. The two-component state and flux vectors are given by

$$\begin{aligned} \mathbf{U} &= [S_\varphi , B^\varphi] , & (4) \\ \mathbf{F}^r &= \left[-\frac{b_\varphi B^r}{W} - 2\mu_S \Sigma^r{}_\varphi , -v^\varphi B^r \right] , & (5) \\ \mathbf{F}^\theta &= \left[-\frac{b_\varphi B^\theta}{W} - 2\mu_S \Sigma^\theta{}_\varphi , -v^\varphi B^\theta \right] , & (6) \end{aligned}$$

where B^i are the magnetic field components as measured by an *Eulerian observer* [23], and $W = \alpha u^t$ is the Lorentz factor. The shear tensor $\Sigma^{i\varphi} = 1/2 g^{ii} \xi^\varphi{}_{,i}$ contains the spatial derivatives (denoted by a comma) of the fluid displacement ξ^φ due to the oscillations. The ξ^φ are related to the fluid 4-velocity according to ${\xi^{\varphi}}_{,t} = \alpha v^{\varphi} = u^{\varphi}/u^{t}$, where v^{φ} is the φ-component of the fluid 3-velocity.

The boundary conditions at the surface ${\xi^{\varphi}}_{,r} = 0$ is a consequence of the continuous traction condition and vanishing currents at the surface of the star. Moreover, at the crust-core interface the continuous traction condition together with the condition of continuous parallel electric field (and so continuous ξ_{φ}) imply $\xi^{\varphi}_{\text{core},r} = (1+\delta)\,\xi^{\varphi}_{\text{crust},r}$ with $\delta = \mu_S/(b_r b^r)$.

As an additional tool for the analysis of the coupled magneto-elastic oscillations of the neutron star, we extend the semi-analytic model of [19] by including a description of crust-core coupling. This provides a comparison aiding the interpretation of our numerical results. In the linear regime and in the limit of short wavelengths an Alfvén wave travels along the magnetic field line corresponding to

$$\frac{d\mathbf{x}}{dt} = \mathbf{v}_a(\mathbf{x})\,, \tag{7}$$

where $\mathbf{v}_a$ is the Alfvén velocity. Assuming standing waves along the magnetic field lines one can derive the following dispersion relation, see [19] for details,

$$f = \frac{\kappa}{2\pi t_{\text{tot}}}\,, \tag{8}$$

where κ is the wavenumber and t_{tot} is twice the total travel time of an Alfvén wave travelling along a magnetic field line starting from the equatorial plane and ending at the surface or another point at the equatorial plane. At this point the frequency of the oscillations is completely determined by the magnetic field topology and the boundary conditions. Each field line is oscillating independent from the others with its own frequency. Therefore, the collection of all lines forms a continuum of frequencies.

This model may be modified in the presence of the crust in two limiting cases. One may assume that the standing wave gets reflected at the crust-core interface, meaning that t_{tot} describes twice the total travel time up to the crust-core interface instead of the surface of the star. Simultaneously, the boundary condition at the crust-core interface changes to zero amplitude in the crust. This possibility should be valid for low magnetic fields.

In the second case one may integrate up to the surface of the neutron star, but additionally take the shear velocity in the crust $\mathbf{v}_s$ into account. The resulting travel time is then obtained from

$$\frac{d\mathbf{x}}{dt} = \sqrt{\mathbf{v}_a^2(\mathbf{x}) + \mathbf{v}_s^2(\mathbf{x})}\,, \tag{9}$$

where we assume that the wave is still travelling along the magneitc field lines. This is only an approximation because we are dealing with isotropic shear and so there is no preferred direction in contrast to the case with only magnetic field. Nevertheless it is expected to be valid approximately when the magnetic field dominates in the above equation.

3. Results

To test our method we first compare the frequencies of the purely shear oscillations obtained in our simulations with those available in the literature [10]. The equilibrium models for our simulations are constructed using the LORENE library (`www.lorene.obspm.fr`). These models are evolved in a numerical grid consisting of 60 points along the θ direction between 0 and $\pi/2$ and 120 points in the radial direction from 0 to the surface of the star, corresponding to about 20 points in the crust. The comparison can be found in Table 1. Despite the different methods, i.e. eigenmode calculation in [10] and Fourier analysis of the numerical evolution in this work, we are able to recover the frequencies of the crustal shear modes with a few per cent accuracy.

For the remainder of this section we will present results which refer to a specific stellar model with $1.4\,\mathrm{M}_{\odot}$ and circumferential radius of 12.26 km. The employed EOS for the core is APR [24]

Journal of Physics: Conference Series **283** (2011) 012013 doi:10.1088/1742-6596/283/1/012013

Table 1. Crustal shear frequencies of different equilibrium models described in [10]. The values in parenthesis are reference values given in [10].

Model	frequency in Hz for mode			
	n=0 (±1Hz) l=2	n=0 (±1Hz) l=3	n=0 (±1Hz) l=4	n=1 ±20Hz
APR+DH 1.6	23.5 (23.4)	37.1 (37.0)	49.8 (49.6)	880 (860)
APR+DH 2.0	21.9 (21.3)	35.1 (33.6)	46.8 (45.1)	1070 (1083)
L+DH 1.6	20.5 (20.6)	32.5 (32.5)	43.8 (43.7)	590 (586)
L+DH 2.0	19.0 (18.9)	30.2 (29.9)	40.5 (40.2)	720 (713)
WFF3+DH 1.6	25.2 (25.2)	39.8 (39.9)	53.4 (53.5)	1130 (1101)

and is matched to the DH EOS [25] for the crust. In any case we do not obtain qualitatively different results for different EOS or the same EOS but different masses of the model.

In the simulations we find three different regimes, depending on the relative strength between the magnetic field and the shear terms of the crust. For low magnetic fields $B < 5 \times 10^{13}$ G the purely crustal shear modes are recovered as shown above. With increasing field strength ($< 10^{15}$ G), the influence of the magnetic field increases and the crustal shear modes become damped [26]. For larger magnetic fields $B \gtrsim 10^{15}$ G the magnetic field dominates the evolution and our results approach those of [19].

Here we will focus on the intermediate regime, where the crust has a significant influence on the long-term oscillations occurring in the core of the neutron star. For magnetic fields between 5×10^{13} and 10^{15} G the coupled magneto-elastic oscillations hardly reach the surface of the star. Their largest oscillation amplitudes are located inside the liquid core of the star, see [26] for more details. Further differences with the case without crust [19], can be read off Figure 1, where we show the magnitude of the Fourier transform of the displacement averaged along individual magnetic field lines. The two panels show simulations with either odd (left) or even (right) symmetry with respect to the equatorial plane. The solid and dashed, red and green lines indicate the position of the Alfvén continuum obtained with the semi-analytic model, if reflection at the crust-core interface is assumed. First, we note that the structure with the lowest frequency, the fundamental QPO $U_0^{(+)}$, has even parity. This is in contrast to the simulations without crust, where this particular QPO did not even exist at all and the fundamental oscillation had odd parity. The local maxima in Figure 1, denoted by $U_n^{(\pm)}$, are directly related to the continuum (red lines). We therefore interpret them as similar to the *Upper* QPOs in [19] with the difference that in the presence of a crust they are located significantly away from the polar axis. The effects of the crust may interfere in the building of standing waves near the pole, such that no stable standing waves can be maintained there. We indicate the corresponding region by the dashed lines in Figure 1. The second family, called *Lower* QPOs $L_n^{(\pm)}$, found around the closed field lines, is reproduced qualitatively as in the case without crust. The only difference is that those QPOs are in the present case confined to the field lines which close inside the liquid core of the neutron star. This is not surprising, because the crust is not expected to influence the oscillations along those magnetic field lines. In contrast to [18, 19] we find a third family of QPOs related to the last magnetic field line, which just fails to close inside the core. QPOs at similar locations have been found by [20] in simulations without crust.

The different families can be interpreted according to [15] in the following way: the Lower QPOs are related to a turning point in the spectrum (green lines in Figure 1; obtained with the

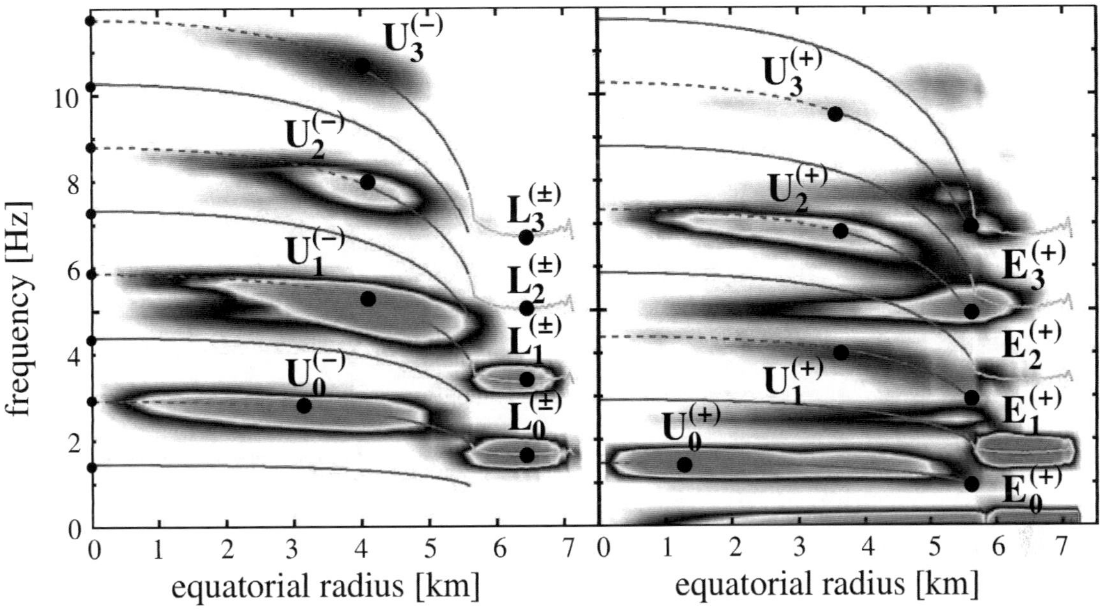

Figure 1. The magnitude of the Fourier transformation averaged along individual field lines denoted by their crossing point with the equatorial plane. *left* panel: simulation with odd parity; *right* panel: simulation with even parity. The solid and dashed lines represent the Alfvén continuum in the core, obtained with a semi-analytic model. Black points indicate the locations of Upper QPOs $U_n^{(\pm)}$, Lower QPOs $L_n^{(\pm)}$ and Edge QPOs $E_n^{(+)}$. The color scale reaches from white (minimum) to orange-red (maximum).

semi-analytic model), so they are turning-point QPOs; the Upper QPOs, which were related in [19] to the turning point at the polar axis change their character and become edge QPOs located at the edge of the continuum indicated by the end of the solid and the beginning of the dashed lines; the new family of even QPOs may be related to those parts of the continuum of the open field lines, which do not connect to that of the closed field lines. Since they are located at an edge of the continuum (solid red lines), we call this new family *Edge* QPO, $E_n^{(+)}$.

4. Conclusions

At magnetic field strength below 5×10^{13} G the torsional oscillations in magnetars are dominated by the crustal shear modes. For very large fields $B > 10^{15}$ G our results approach those of [19] without crust. In the intermediate regime the crustal modes disappear quickly, but the presence of the crust influences the magneto-elastic oscillations significantly in different ways: First, the structure of the obtained Upper QPOs changes as a consequence of a reflection at the crust-core boundary. Those QPOs are now located between the polar axis and the equator. Additionally, they have a minimum at the crust-core interface and vanishing amplitude inside the crust. Second, a new family of Edge QPOs appears due to the coupling of the field lines which just fail to close inside the liquid core. Finally, the Lower QPOs are limited to the field lines closing inside the liquid core.

To summarize, the shear oscillations of the crust are damped efficiently and the magneto-elastic QPOs have negligible amplitudes in the crust for magnetic field strengths in the range of 5×10^{13} to 10^{15} G. Therefore, our simulations favor magneto-elastic oscillations at field strengths $> 10^{15}$ G as a possible explanation for the observed magnetar QPOs.

References

[1] Karlovini M and Samuelsson L 2003 *Classical and Quantum Gravity* **20** 3613–3648
[2] Karlovini M, Samuelsson L and Zarroug M 2004 *Classical and Quantum Gravity* **21** 1559–1581 (*Preprint* arXiv:gr-qc/0309056)
[3] Karlovini M and Samuelsson L 2004 *Classical and Quantum Gravity* **21** 4531–4548 (*Preprint* arXiv:gr-qc/0401115)
[4] Karlovini M and Samuelsson L 2007 *Classical and Quantum Gravity* **24** 3171–3189 (*Preprint* arXiv:gr-qc/0703001)
[5] Israel G L, Belloni T, Stella L, Rephaeli Y, Gruber D E, Casella P, Dall'Osso S, Rea N, Persic M and Rothschild R E 2005 ApJ **628** L53–L56 (*Preprint* arXiv:astro-ph/0505255)
[6] Watts A L and Strohmayer T E 2007 *Advances in Space Research* **40** 1446–1452 (*Preprint* arXiv:astro-ph/0612252)
[7] Duncan R C 1998 ApJ **498** L45+ (*Preprint* arXiv:astro-ph/9803060)
[8] Strohmayer T E and Watts A L 2005 ApJ **632** L111–L114 (*Preprint* arXiv:astro-ph/0508206)
[9] Piro A L 2005 ApJ **634** L153–L156 (*Preprint* arXiv:astro-ph/0510578)
[10] Sotani H, Kokkotas K D and Stergioulas N 2007 MNRAS **375** 261–277 (*Preprint* arXiv:astro-ph/0608626)
[11] Samuelsson L and Andersson N 2007 MNRAS **374** 256–268 (*Preprint* arXiv:astro-ph/0609265)
[12] Steiner A W and Watts A L 2009 *Physical Review Letters* **103** 181101–+ (*Preprint* 0902.1683)
[13] Levin Y 2006 MNRAS **368** L35–L38 (*Preprint* arXiv:astro-ph/0601020)
[14] Glampedakis K and Andersson N 2006 MNRAS **371** 1311–1321 (*Preprint* arXiv:astro-ph/0607105)
[15] Levin Y 2007 MNRAS **377** 159–167 (*Preprint* arXiv:astro-ph/0612725)
[16] Lee U 2007 MNRAS **374** 1015–1029 (*Preprint* arXiv:astro-ph/0610182)
[17] Lee U 2008 MNRAS **385** 2069–2079 (*Preprint* 0710.4986)
[18] Sotani H, Kokkotas K D and Stergioulas N 2008 MNRAS **385** L5–L9 (*Preprint* 0710.1113)
[19] Cerdá-Durán P, Stergioulas N and Font J A 2009 MNRAS **397** 1607–1620 (*Preprint* 0902.1472)
[20] Colaiuda A, Beyer H and Kokkotas K D 2009 MNRAS **396** 1441–1448 (*Preprint* 0902.1401)
[21] Cerdá-Durán P, Font J A, Antón L and Müller E 2008 A&A **492** 937–953 (*Preprint* 0804.4572)
[22] Carter B and Samuelsson L 2006 *Classical and Quantum Gravity* **23** 5367–5388 (*Preprint* arXiv:gr-qc/0605024)
[23] Antón L, Zanotti O, Miralles J A, Martí J M, Ibáñez J M, Font J A and Pons J A 2006 ApJ **637** 296–312 (*Preprint* arXiv:astro-ph/0506063)
[24] Akmal A, Pandharipande V R and Ravenhall D G 1998 Phys. Rev. C **58** 1804–1828 (*Preprint* arXiv:hep-ph/9804388)
[25] Douchin F and Haensel P 2001 A&A **380** 151–167 (*Preprint* arXiv:astro-ph/0111092)
[26] Gabler M, Cerdá-Durán P, Font J, Müller E and Stergioulas N 2010 *submitted to* MNRAS (*Preprint* arXiv:astro-ph/1007.0856)

Black holes from generalized gauge field theories

J. Diaz-Alonso, D. Rubiera-Garcia

LUTH, Observatoire de Paris, CNRS, Université Paris Diderot. 5 Place Jules Janssen, 92190 Meudon, France

Departamento de Fisica, Universidad de Oviedo. Avda. Calvo Sotelo 18, E-33007 Oviedo, Asturias, Spain

E-mail: `joaquin.diaz@obspm.fr` ; `diego.rubiera-garcia@obspm.fr`

Abstract. We summarize the main results of a broad analysis on electrostatic, spherically symmetric (ESS) solutions of a class of non-linear electrodynamics models minimally coupled to gravitation. Such models are defined as arbitrary functions of the two quadratic field invariants, constrained by several physical admissibility requirements, and split into different families according to the behaviour of these lagrangian density functions in vacuum and on the boundary of their domains of definition. Depending on these behaviours the flat-space energy of the ESS field can be finite or divergent. For each model we qualitatively study the structure of its associated gravitational configurations, which can be asymptotically *Schwarzschild-like* or with an anomalous *non Schwarzschild-like* behaviour at $r \to \infty$ (but being asymptotically flat and well behaved anyhow). The extension of these results to the non-abelian case is also briefly considered.

1. Introduction

The non-linear Born-Infeld (BI) model [1], introduced in 1934 as an attempt to classically provide a finite energy for the electron field, has been the cornerstone of many developments in theoretical physics in the last three decades. The BI lagrangian density takes the form

$$L_{BI} = \beta^2 \left(1 - \sqrt{1 - \frac{X}{\beta^2} - \frac{Y^2}{4\beta^4}}\right), \tag{1}$$

where $X = -\frac{1}{2}F_{\mu\nu}F^{\mu\nu} = \vec{E}^2 - \vec{H}^2$ and $Y = -\frac{1}{2}F_{\mu\nu}F^{*\mu\nu} = 2\vec{E}\cdot\vec{H}$ are the two quadratic field invariants that can be constructed with the field strength tensor $F_{\mu\nu} = \partial_\mu A_\nu - \partial_\nu A_\mu$ and its dual $F^{*\mu\nu} = \frac{1}{2}\epsilon^{\mu\nu\alpha\beta}F_{\alpha\beta}$, and where $\vec{E}$ and $\vec{H}$ are the electric and magnetic fields, respectively.

When minimally coupled to gravity, its electrostatic spherically symmetric (ESS) solutions have been studied for a long time [2]. With no doubt, the finding that string theory leads to (abelian and non-abelian) BI-like lagrangians in its low-energy limit [3], has greatly contributed to the renewed interest in this kind of non-linear actions. As a result many non-linear electrodynamics (NED) models, aside from BI one, leading to asymptotically flat ESS structures, have been considered in the literature in the last decades [4]. Moreover, the characterization of the full class of physically admissible NED theories leading to asymptotically Schwarzschild-like configurations when coupled to gravity was performed in Ref.[5]. Other models have been found to lead to well behaved gravitational configurations but approaching asymptotic flatness slower than the Schwarzschild term [6, 7]. These lines of research have also lead to some promising attempts in the obtention of regular (singularity-free) ESS black hole solutions in NEDs minimally coupled to gravity [8]. Let us finally mention that black hole solutions generated by NEDs in other gravitational backgrounds have been also considered [9].

In this work we summarize the main results of a general procedure for the classification and proper characterization of asymptotically flat electrically charged black hole solutions of NEDs defined as arbitrary functions of the two quadratic field invariants *without explicitly fixing the lagrangian density function.* The key point in such an analysis is the existence of a first-integral of the field equations, which takes the same form with and without gravity. These models are constrained by a number of physical *admissibility* requirements and classified according to the asymptotic and central-field behaviours of their ESS solutions. Their associated flat-space energy can be finite or divergent, due to one or both of these field behaviours in the limits of the integral of energy ($r \to \infty$ and $r \sim 0$). For each one of these models we qualitatively characterize their associated (asymptotically flat and well behaved) gravitational configurations. Finally, the extension of these procedures to the non-abelian case is briefly discussed.

2. Gravitating ESS fields

Let us begin by introducing the action for NED models minimally coupled to the Einstein action as

$$S = \int d^4x\sqrt{-g}\left(\frac{R}{16\pi G} - \varphi(X,Y)\right), \tag{2}$$

where $\varphi(X,Y)$ is an arbitrary function of the two field invariants previously introduced. We consider an ansantz for the static spherically symmetric metric as

$$ds^2 = e^{\nu(r)}dt^2 - e^{\mu(r)}dr^2 - r^2 d\Omega^2, \tag{3}$$

where $d\Omega^2 = d\theta^2 + \sin^2\theta d\phi^2$ is the interval on the 2-sphere. However, due to the symmetry of the NED energy-momentum tensor $T_t^t = T_r^r$, only a single independent metric function remains as $\lambda(r) = e^{\nu(r)} = e^{-\mu(r)}$. The Einstein equations $G_{\mu\nu} = 8\pi T_{\mu\nu}$ (we are taking units $G = 1$) for this metric can be easily obtained to give [5]

$$\lambda(r) = 1 - \frac{2M}{r} + \frac{2\varepsilon_{ex}(r,q)}{r}, \tag{4}$$

where M is an integration constant, identified as the Arnowitt-Deser-Misner (ADM) mass, and we have introduced, by convenience, the *exterior integral of energy*, defined as

$$\varepsilon_{ex}(r,q) = \int_r^\infty R^2 T_0^0(R,q)dR. \tag{5}$$

However, the procedure leading to Eq.(4) needs the implicit assumption that the NED source is such that $\varepsilon_{ex}(r,q)$ converges in its upper limit ($r \to \infty$). As we shall see in the sequel there are NED models not satisfying this property, and in such cases the metric is integrated as

$$\lambda(r) = 1 + \frac{C}{r} - \frac{2\varepsilon_{in}(r,q)}{r}, \tag{6}$$

where now C is an integration constant not longer identified as the ADM mass (since, as we shall see, the large-r behaviour is governed in this case by the electrostatic field term, rather than the "Schwarzschild" one) and now

$$\varepsilon_{in}(r,q) = \int_0^r R^2 T_0^0(R,q)dR \tag{7}$$

is the *interior integral of energy*. In this case, the obtention of Eq.(6) needs the assumption that $\varepsilon_{in}(r,q)$ converges in the lower limit ($r \to 0$). But even if neither $\varepsilon_{ex}(r,q)$ nor $\varepsilon_{in}(r,q)$ can be defined, then the Einstein equations can be still integrated as

$$\lambda(r,q,D) = 1 + \frac{D}{r} - \frac{2\varepsilon(r,q,0)}{r}, \tag{8}$$

where

$$\varepsilon(r,q,B) = 4\pi \int r^2 T_0^0(r,q) + B, \tag{9}$$

B and $D = C - 2B$ being integration constants. Consequently, in order to properly characterize the metric function in the different cases, one is lead to analyze the conditions under which $\varepsilon_{ex}(r,q)$ and $\varepsilon_{in}(r,q)$ are well defined. This amounts to study the behaviour of $E(r)$ at $r \to \infty$ and around $r \sim 0$. To do this we take into account the fact that the NED field equations (we are defining $\varphi_X = \frac{\partial\varphi}{\partial X}$), obtained as $\nabla_\mu(\varphi_X F^{\mu\nu} + \varphi_Y F^{*\mu\nu}) = 0$ lead, for ESS fields ($\vec{E}(r) = E(r)\frac{\vec{r}}{r} \neq 0, \vec{H}(r) = 0$), to a first-integral

$$r^2 \varphi_X E(r) = q, \tag{10}$$

where q is an integration constant identified as the electric charge associated to the field $E(r)$. Remarkably, this first-integral takes the same form, in the Schwarzschild coordinate system, as in the absence of gravity in spherical coordinates. Consequently, we can perform the classification of the models in flat space according to the asymptotic and central-field behaviours of $E(r)$, and Eq.(10) will allow the translation of the results obtained there to the gravitational context.

2.1. Definition of the models

Let us now establish a set of constraints on the NED models, leading to what shall be called *admissible* models. We restrict the lagrangian density function $\varphi(X, Y = 0)$ to be *continuous, derivable and single-branched* in its domain of definition of the $X-Y$ plane, which must be open, connected and including the vacuum ($X = Y = 0$). The condition $\varphi(X,Y) = \varphi(X,-Y)$ must be also satisfied in order to implement parity invariance. In addition we require the positive definiteness of the flat-space energy density

$$\rho = T_0^0 = 2X\varphi_X - \varphi(X) \geq \left(\sqrt{X^2+Y^2} + X\right)\frac{\partial\varphi}{\partial X} + Y\frac{\partial\varphi}{\partial Y} - \varphi(X,Y) \geq 0, \tag{11}$$

whose fulfillment implies the conditions $\varphi(0,0) = 0$, $\frac{\partial\varphi}{\partial X}|_{(X>0,Y=0)} > 0$ and $\varphi(X > 0, 0) > 0$ (see [10]). These admissibility conditions endorse the definiteness and single-valued character of the ESS solutions and are essential in the analysis of the problem considered here. In particular, from the first-integral (10) it follows that $E(r)$ must be a monotonic function of r for a given q. Moreover, for admissible models $\varepsilon_{ex}(r,q)$ is a **monotonically decreasing and concave** function of r, while $\varepsilon_{in}(r,q)$ is a **monotonically increasing and convex** one.

For finite-energy ESS solutions, the energy associated to the ESS field is given by

$$\varepsilon(q) = \int_0^\infty r^2 T_0^0(r,q)dr = q^{3/2}\varepsilon(q=1), \tag{12}$$

where $\varepsilon(q = 1)$ is the energy of the solution of unit charge, a universal constant for a given model. We can now split the families of admissible models into two sets, according to the character (finite or divergent) of the energy (12). Such a character only depends, for admissible models, on the behaviour of the ESS fields as $r \to \infty$ and at $r = 0$. In this sense, we shall consider power-law expressions for the field behaviour around each of these limits. Let us then study each case separately.

I) Models with finite-energy ESS solutions
Asymptotically the field must behave as

$$E(r) \sim r^p, \tag{13}$$

with $p < -1$ for convergence of the energy there. The behaviour of the associated lagrangian density is obtained, by making use of the first-integral (10), as

$$\varphi(X) \sim X^{\frac{p-2}{2p}}. \tag{14}$$

In what follows we shall call this asymptotic behaviour, given by Eq.(13), as **B**-type fields. Such behaviour can be slower than Coulombian ($-2 < p < -1$, subcase **B1**), Coulombian ($p = -2$, **B2**) or faster than Coulombian ($p < -2$, **B3**).

As $r \sim 0$ there are two ESS field behaviours compatible with the finite energy requirement (**A**-type fields) for admissible models. In case **A1** the field behaves at the center as in Eq.(13) but now with $-1 < -p < 0$ for the convergence of the energy there (and, consequently, the field diverges in this case), and the lagrangian behaves also as in Eq.(14). On the other hand, in case **A2** the field becomes finite at the origin, behaving there as

$$E(r) \sim a - br^\sigma, \tag{15}$$

where a and $\sigma(> 0)$ are fixed parameters for a given model while b is a function of the charge [5]. Concerning the lagrangian density, its behaviour around the central values of the fields is given by (the case $\sigma = 2$ is singular, see Ref.[10])

$$\varphi(X, Y = 0) \sim \frac{2q\sigma b^{2/\sigma}}{2-\sigma}(a - \sqrt{X})^{\frac{\sigma-2}{\sigma}} + \Delta, \tag{16}$$

where $\Delta = \varphi(a^2, Y = 0)$ is an integration constant. Consequently we see that there are two different behaviours for $\varphi(X)$: if $\sigma \leq 2$ the lagrangian density diverges at $X = a^2$ while if $\sigma > 2$ it takes a finite value there given by Δ.

Consequently, for any admissible model $\varphi(X, Y = 0)$ to support finite-energy ESS solutions, these fields must belong to one of the **B**-cases at $r \to \infty$, and to one of the **A**-cases around $r \sim 0$. In between both domains the lagrangian density must be a strict monotonically increasing function of X.

II) Models with divergent-energy ESS solutions

ESS fields whose energy diverges at $r = 0$ are obtained from Eq.(13), but now imposing the restriction $p < -1$ there. We shall call this kind of behaviour as **UVD** (meaning "ultraviolet"-divergent). A similar behaviour is assumed in the asymptotic limit, with $-1 \leq p < 0$ for (asymptotically vanishing) ESS solutions with divergent energy there. The acronym **IRD** ("infrared"-divergent) will be employed for calling this family. Consequently the class of admissible NED models with divergent-energy ESS solutions is naturally classified into three families, resulting from the combination between i) divergent-energy fields at the center but not asymptotically (**UVD-B**), ii) asymptotic divergent-energy but not at the center (**A-IRD**) and iii) divergent energy in both limits (**UVD-IRD**). As a consequence of these behaviours $\varepsilon_{ex}(r, q)$ can only be defined in the first one of these cases, and $\varepsilon_{in}(r, q)$ only in the second one.

2.2. Metric structure. Horizons

With these ingredients we resume the previous discussion on the metric structure. First we look for the horizons r_h of the different configurations, which are the solutions of $\lambda(r) = 0$ in Eq.(4) (in such cases where $\varepsilon_{ex}(r, q)$ can be defined, see the previous section), implying

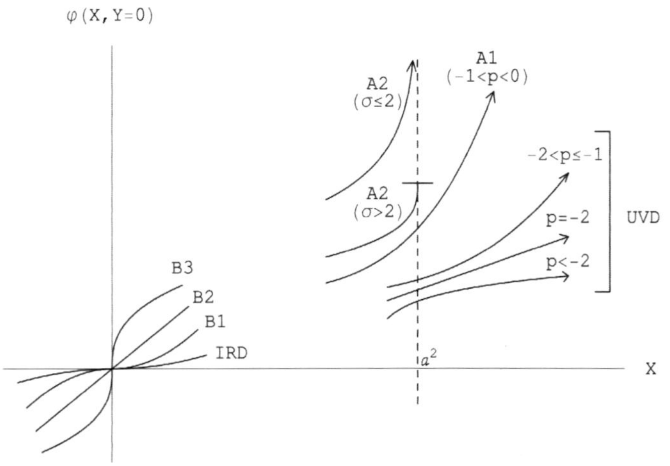

Figure 1. Qualitative behaviour of the families of admissible lagrangian density functions $\varphi(X, Y = 0)$. The behaviour at $X \sim 0$ ($r \to \infty$) is split into finite-energy cases (**B1**, **B2** and **B3**) and the energy-divergent one (**IRD**). At the center of the ESS solutions, there are two **A2** cases ($X \sim a^2$): for $\sigma \leq 2$, $\varphi(X)$ diverges there while if $\sigma > 2$, $\varphi(X)$ becomes finite. **A1** is a vertical parabolic branch with divergent slope for large X. **UVD** is the family of divergent-energy solutions there, with its different possible slopes plotted.

$$M - \frac{r}{2} = \varepsilon_{ex}(r,q), \tag{17}$$

and thus horizons are given by the cutting points between the monotonically decreasing and concave curve $\varepsilon_{ex}(r,q)$ and the beam of straight lines $M - \frac{r}{2}$, corresponding to different values of the ADM mass M (see Fig.2). Let us now study, following this procedure, each class of models, resulting from the combination of one of the possible field behaviours at the center (**UVD, A1** or **A2**) with one of the asymptotic cases (**B1, B2, B3** or **IRD**).

I) <u>A1-B</u>: At $r \to \infty$ the metric (4) vanishes as ($p < -1$)

$$\lambda(r) \to 1 - \frac{2M}{r} + \frac{32\pi q}{(p-2)(p+1)} r^p + \cdots \tag{18}$$

and, consequently, the large-r behaviour is governed by the Schwarzschild term, regardless of the **B**-subcase chosen (**B1**, **B2** or **B3**). At the center $\varepsilon_{ex}(r,q)$ takes the value $\varepsilon(q)$, with a negative divergent slope there. Consequently, the cutting points with the beam $M - \frac{r}{2}$ give the following configurations (see Fig.2): i) There is a family of extreme black holes (EBH), satisfying the conditions $\lambda(r) = \frac{d\lambda(r)}{dr}|_{r_h} = 0$, which give the EBH radius and mass as

$$8\pi r_{hextr}^2 T_0^0(r_{hextr}, q) = 1 \; ; \; M_{extr}(q) = \frac{r_{hextr}}{3} + \frac{16\pi q}{3} A_0(r_{hextr}, q), \tag{19}$$

where $A_0(r,q)$ is the electric potential; ii) $M < M_{hextr}(q)$: naked singularity (NS); iii) $M_{hextr}(q) < M < \varepsilon(q)$: two-horizons (Cauchy and event) BH; iv) $M > \varepsilon(q)$: single-horizon BH; v) $M = \varepsilon(q)$: for this critical configuration we must study the behaviour of the metric around $r \sim 0$ ($-1 < p < 0$):

$$\lambda(r) \sim 1 - \frac{2(M - \varepsilon(q))}{r} + \frac{32\pi q}{(p-2)(p+1)} r^p + \cdots, \tag{20}$$

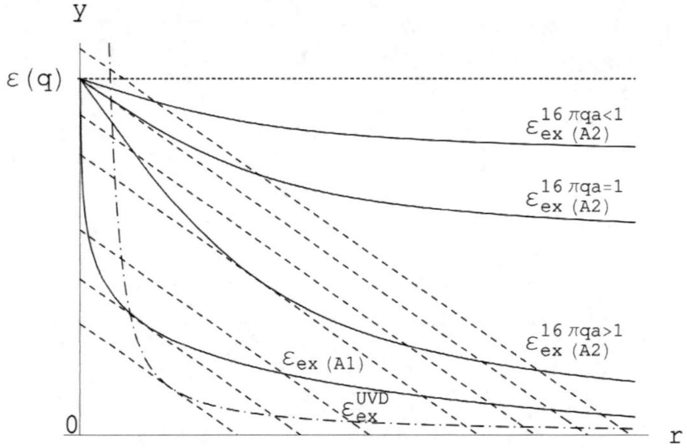

Figure 2. Qualitative behaviour of $\varepsilon_{ex}(r, q)$ for the families **A1**, **A2** (three subcases) and **UVD** at the center combined with **B**-fields at $r \to \infty$, cut by a set of dashed lines, corresponding to different values of M. The cut points give the horizons of the different configurations (see the discussion in the text). All the finite-energy cases have been chosen in order to begin at the same value of $\varepsilon(q)$ as $r \to 0$.

and we see that in this case $\lambda(r)$ diverges there to $-\infty$, as in subcase iv) and, consequently, this is a single-horizon BH as well. In subcases iv) and v) the central singularity is spacelike, as opposed to the timelike singularity of subcases i), ii), and iii). *Example of this family*: The Euler-Heisenberg lagrangian $\varphi(X, Y) = \frac{X}{2} + \mu(X^2 + \frac{7}{4}Y^2)$, μ being a constant (see Ref.[5] for details of this family in gravitation).

II) **A2-B:** Now there are three subcases, according to the (now finite) slope of $\varepsilon_{ex}(r, q)$ around $r \sim 0$. **A2a** ($16\pi qa > 1$): Here there are five classes of configurations: i) $M < M_{hextr}(q)$: NS; ii) $M = M_{hextr}(q)$: EBH; iii) $M_{hextr}(q) < M < \varepsilon(q)$: two-horizons BH; iv) $M > \varepsilon(q)$: single-horizon BH; v) $M = \varepsilon(q)$: for this configuration the metric at the center is finite, behaving there as ($\sigma > 0$)

$$\lambda(r) \sim 1 - 16\pi qa - \frac{2(M - \varepsilon(q))}{r} + \frac{32\pi bq}{(\sigma + 1)(2 - \sigma)} r^\sigma + \Delta r^2, \tag{21}$$

and, consequently, in this case $\lambda(0) = 1 - 16\pi qa < 0$, leading to a single-horizon BH as well. **A2b** ($16\pi qa < 1$): in this branch there are not EBH, and the available configurations are i) $M < \varepsilon(q)$: NS; ii) $M > \varepsilon(q)$: single-horizon BH; iii) $M = \varepsilon(q)$: in this case $\lambda(0) = 1 - 16\pi qa > 0$, as can be seen from (21), leading to a NS. **A2c** ($16\pi qa = 1$): the structures now are similar as in case **A1-B**, excepting for $M = \varepsilon(q)$, since $\lambda(0) = 0$ and this configuration is a kind of "black point" [5]. *Example of this family*: The BI lagrangian defined by Eq.(1).

III) **UVD-B:** In this case $\varepsilon_{ex}(r, q)$ diverges as $r \to 0$ (since now $\varepsilon(q)$ becomes divergent, as explained in subsection 2.1.) and there are three possible configurations: i) $M < M_{hextr}(q)$: NS; ii) $M = M_{hextr}(q)$: EBH; iii) $M > M_{hextr}(q)$: two-horizons BH. *Example of this family*: The Maxwell lagrangian $\varphi(X, Y = 0) = \alpha X$ (α a constant), leading to the Reissner-Nordström solution $\lambda(r) = 1 - \frac{2M}{r} + \frac{q^2}{\alpha r^2}$.

IV) **A1 (and A2)-IRD:** As already discussed, in this case $\varepsilon_{ex}(r, q)$ cannot be defined and the metric is integrated as in Eq.(6). Now horizons are determined by the equation

$$\frac{r+C}{2} = \varepsilon_{in}(r,q), \tag{22}$$

and by taking into account the monotonically increasing and convex character of $\varepsilon_{in}(r,q)$ the different configurations can be classified following the same procedure as in the previous cases, now using the cutting points with the beam of straight lines $\frac{r+C}{2}$ in Eq.(22), corresponding to different values of the integration constant C [7]. For the **A1** and **A2a** cases at the center we find: i) A family of EBHs, whose radius is given by that of Eq.(19), now with an associated value $C = C_{extr}(q)$; ii) $C > C_{extr}(q)$: NS; iii) $0 < C < C_{extr}(q)$: two-horizons BH; iv): $C < 0$: single-horizon BH; v) $C = 0$: single-horizon BH (now the metric diverges to $-\infty$ in case **A1**, while takes a finite value $\lambda(0) = 1 - 16\pi qa < 0$ in case **A2a**). In cases **A2b** and **A2c** the extreme black hole and two-horizons BH configurations disappear and we are lead to i) $C > 0$: NS; ii) $C < 0$: single-horizon BH; iii) $C = 0$ a NS in case **A2b** and an extreme black point [7] in case **A2c**. This class of gravitational configurations approaches asymptotic flatness slower than the Schwarzschild solution, since the asymptotic behaviour is governed by the ESS term (see Eq.(18)). *Example of this family*: The lagrangian $\varphi(X, Y = 0) = X^{\gamma}$ with $3/2 < \gamma < \infty$ (A1 family at the center).

V) UVD-IRD: In this case neither $\varepsilon_{ex}(r,q)$ nor $\varepsilon_{in}(r,q)$ exist, but the above procedure for classifying the different gravitational structures works again, employing now Eqs.(8) and (9), together with the value of the integration constant D. This leads to: i) as in the previous cases, EBH configurations are found for a value $D = D_{hextr}(q)$; ii) $D > D_{hextr}(q)$: NS; iii) $D < D_{hextr}(q)$: two-horizons BH. *Example of this family:* The lagrangian density $\varphi(X, Y = 0) = X^{3/2}$.

2.3. Extension to non-abelian fields

The above results can be extended to generalized non-abelian gauge groups. Let us summarize the main ideas. The whole analysis will appear in a forthcoming publication. In this case the field strength components in the Lie algebra are defined as $F_{a\mu\nu} = \partial_\mu A_{a\nu} - \partial_\nu A_{a\mu} - g\sum_{bc} C_{abc} A_{b\mu} A_{c\nu}$, where $1 \leq a \leq N$ (N the dimension of the group) and C_{abc} are the structure constants. The lagrangian density is again defined as $\varphi_{N-A}(X,Y)$ but now the field invariants are defined as (we are using the ordinary prescription in the calculation of the traces over the gauge group indices)

$$X = -\frac{1}{2}\sum_a (F_{a\mu\nu}F_a^{\mu\nu}) = \sum_a (\vec{E}_a^2 - \vec{H}_a^2)\ ;\ Y = -\frac{1}{2}\sum_a (F_{a\mu\nu}F_a^{*\mu\nu}) = 2\sum_a (\vec{E}_a \cdot \vec{H}_a). \tag{23}$$

The associated field equations in presence of gravitation take the form

$$\sum_c \nabla_{ac\mu}\left[\varphi_X F_c^{\mu\nu} + \varphi_Y F_c^{*\mu\nu}\right] = 0, \tag{24}$$

where $\nabla_{ac\mu} \equiv \delta_{ac}\nabla_\mu - g\sum_{bc} C_{abc} A_{b\mu}$ is the gauge-covariant derivative. Concerning the admissibility conditions, it can be shown that they take in the present case the same form as those obtained for NED models. For ESS solutions in this non-abelian case ($E_a \neq 0, H_a = 0, \forall a$), the N field equations arising from the $\nu = 0$ component of the field equations (24), in the metric (3), can be combined to give

$$r^2\sqrt{X}\varphi_X|_{Y=0} = Q, \tag{25}$$

where $Q = \sqrt{\sum_a^N Q_a^2}$ is the "mean-square charge", constructed with the integration constants Q_a arising from each one of the $\nu = 0$ components of the field equations and identified as "source color charges". As a consequence, there is a relation one-to-one between the ESS solutions of the generalized non-abelian gauge field theory and those of the NED models with the identification $\varphi_{N-A}(X, Y = 0) = \varphi_{NED}(X, Y = 0)$ and $q \leftrightarrow Q$. The associated energy $\varepsilon(Q)$ is the same as in the abelian case, but now there is a degeneration of energy in the sphere of radius Q. As a consequence of these equivalences, the integration of the metric from the Einstein equations, and the associated gravitational configurations for generalized non-abelian gauge groups can be shown to be the same as for the NED models analyzed above, once the previous identifications of lagrangian density functions and charges between abelian and non-abelian cases are performed.

3. Conclusions and perspectives

The gravitational configurations associated to general NED models minimally coupled to Einstein gravity can be qualitatively characterized without explicitly fixing the form of the lagrangian density function. In such a characterization the admissibility conditions play an essential role. A natural classification of the ESS solutions into several families is performed according to their asymptotic and central-field behaviours, which determine the character (finite or divergent) of the flat-space energy. Such NED models have been shown to lead to gravitating structures which are asymptotically Schwarzschild-like, or with an anomalous asymptotic non Schwarzschild-like behaviour (although still asymptotically flat and well behaved). In both cases the structure of horizons was studied, and explicit examples of lagrangian densities belonging to the different families of models were given. These procedures are not restricted to the asymptotically flat solutions of the Einstein gravity, but can be extended to (Anti-)de Sitter spaces and to higher-order curvature gravity theories (e.g. Gauss-Bonnet), a work in progress.

Acknowledgments

This work was partially supported by Spanish grants FC-08-IB08-154 and MICINN-09-FPA2009-11061. Hospitality of the organizers of the NEB14 meeting is acknowledged.

References

[1] Born M and Infeld L 1934 *Proc. R. Soc. London.* A **144** 425
[2] Hoffmann B 1935 *Phys. Rev.* **47** 877
Garcia A, Salazar H and Plebanski J F 1984 *Nuovo. Cim.* **84** 65
Demianski M 1986 *Found. of Phys.* **16** 187
Gibbons G W and Rasheed D A 1995 *Nucl. Phys.* B **454** 185
[3] Fradkin E S and Tseytlin A A 1985 *Phys. Lett.* B **163** 123
Leigh R G 1989 *Mod. Phys. Lett.* A **4** 2767
[4] de Oliveira H P 1994 *Class. Quant. Grav.* **11** 1469
Soleng H H 1995 *Phys. Rev.* D **52** 6178
Yajima H and Tamaki T 2001 *Phys. Rev.* D **63** 064007
Tamaki T 2004 *JCAP* **05** 004
Hassaine M and Martinez C 2007 *Phys. Rev.* D **75** 027502
[5] Diaz-Alonso J and Rubiera-Garcia D 2010 *Phys. Rev.* D **81** 064021
[6] Hassaine M and Martinez C 2008 *Class. Quant. Grav.* **25** 195023
[7] Diaz-Alonso J and Rubiera-Garcia D 2010 *Preprint* arXiv:1008.2710 [hep-th]
[8] Bronnikov K A 2001 *Phys. Rev.* D **63** 04405
Burinskii A and Hildebrandt S R 2002 *Phys. Rev.* D **65** 104017
Dymnikova I 2004 *Class. Quant. Grav.* **21** 4417
Ayón-Beato E and García A 1998 *Phys. Rev. Lett.* **80** 5056
[9] Wiltshire D L 1988 *Phys. Rev.* D **38** 2445
Fernando S and Krug D 2003 *Gen. Rel. Grav.* **35** 129
Dey T K 2004 *Phys. Lett.* B **95** 484
Cai R -G, Pang D -W and Wang A 2004 *Phys. Rev.* D **70** 124034
Aiello M, Ferraro R and Giribet G 2005 *Phys. Rev.* D **70** 104014
[10] Diaz-Alonso J and Rubiera-Garcia D 2009 *Ann. Phys.* **324** 827

Journal of Physics: Conference Series **283** (2011) 012015 doi:10.1088/1742-6596/283/1/012015

Acceleration and emission of MHD driven, relativistic jets

Dimitrios Giannios

Department of Astrophysical Sciences, Peyton Hall, Princeton University, Princeton, NJ 08544, USA

E-mail: `giannios@astro.princeton.edu`

Abstract. Strong magnetic fields can efficiently extract rotational energy from compact objects launching the relativistic jets observed in active galactic nuclei (AGN), microquasars and gamma-ray bursts (GRBs). Magnetohydrodynamical (MHD) acceleration of jets in the *ideal* MHD limit is not an ultra-efficient process. This makes internal collisions an unlikely mechanism for the jet emission. I argue that *non-ideal* MHD effects may both boost the acceleration efficiency and power the jet emission. I show that much of the GRB and blazar phenomenology can be understood as result of magnetic dissipation and that the early afterglow phases from GRBs appear to support the picture of MHD driving.

1. Introduction

Relativistic jets have been extensively observed from supermassive black holes in active galactic nuclei (AGN), solar-mass compact objects in binaries (microquasars) and gamma-ray bursts (GRBs). The theoretical paradigm for jet formation has been developed since the late 70s and can account in a unifying manner for all these sources. It envisions that jets come from strong magnetic fields that thread rotating objects (neutron stars, black hole ergospheres, or accretion disks) extracting their rotational energy (Bisnovatyi-Kogan & Ruzmaikin 1976; Blandford 1976; Blandford & Znajek 1977; Blandford & Payne 1982; for a review see Spruit 2010).

Despite substantial theoretical and observational progress in the field, we still lack a coherent connection between the jet dynamics and emission. Modeling of the acceleration of jets does not deal with the dissipative and emission processes. On the other hand, most of the theoretical work on the jet emission assumes emitting blobs of prescribed properties. Such studies, though informative for the properties of the emission region and the radiative mechanisms, cannot address which processes accelerate the particles. Often, internal collisions are invoked for the energy dissipation assuming (in contrast to the findings of MHD studies) that the acceleration of the jet is complete or almost complete.

I focus, here, on a different approach to the problem of jet emission that attempts to properly take into account the MHD nature of jets. Magnetic fields are considered to be dynamically important at the ejection region and likely remain important at the location of emission. I advocate that magnetic dissipation plays a critical role in both jet acceleration and emission. These considerations can be applied to both blazar and GRB jets providing common insights to the two, phenomenologically different, sources. Finally, I discuss how the residual magnetization of the jet at large distance affects the onset of the afterglow emission of GRBs.

2. Acceleration of MHD jets

For any flow to acquire relativistic speed, it must be clean, i.e., it must be launched with high energy to rest mass ratio. In MHD models[1] most of the energy is initially in the magnetic field: $\sigma_{in} \gg 1$, where σ_{in} is the magnetization as defined by Michel (1969) or, equivalently, the "baryon loading" of the flow (σ_{in} also stands for the maximum Lorentz factor to which the jet can potentially be accelerated). Further out the flow accelerates converting part of the magnetic energy into kinetic. During this process the bulk Lorentz factor Γ increases while the magnetization σ decreases. A large theoretical effort is devoted in calculating the efficiency with which the flow is accelerated. It turns out that, in relativistic flows, the efficiency of magnetic conversion is, at most, modest with the jet remaining strongly magnetized at larger distances where the emission takes place.

2.1. Ideal MHD acceleration

Acceleration of *non-relativistic* jets is rather efficient; the flow approaches its terminal speed close to the Alfvén radius (distance where the flow speed equals the Alfvén speed) where much of the magnetic energy is converted to kinetic (e.g. Blandford & Payne 1982). This is not the case for relativistic flows. Assuming an axisymmetric rotator (dipole field and rotation axis aligned), steady 1D relativistic MHD flows are only accelerated up to an asymptotic Lorentz factor $\Gamma \sim \sigma_{in}^{1/3}$ while the ratio of magnetic to kinetic energy remains high $\sim \sigma_{in}^{2/3}$ (Michel 1969; Goldreich & Julian 1970; Beskin, Kuznatova & Rafikov 1998). Rapid variations of the flow at the source can somewhat help to achieve further acceleration (Contopoulos 1995; Granot, Spitkovsky & Komissarov 2010; Lyutikov & Lister 2010).

2D *ideal* MHD calculations show that, depending on the shape of the jet, acceleration may be more efficient (Begelman & Li 1994; Vlahakis and Königl 2003, Komissarov et al. 2009, Lyubarsky 2009; Tchekhovskoy, Narayan & McKinney 2010). The acceleration rate is maximized for jet boundaries of parabolic shape. Such collimating boundaries, treated as a conducting wall by the simulations, may physically come from the confining pressure of surrounding gas (e.g. the envelop of the collapsing star, winds from the accretion disk) or external, large scale poloidal fields (Spruit, Foglizzo & Stehle 1997). In some cases, because of acceleration through confinement, the jet can reach a rough equipartition between kinetic and magnetic flux at large distance (the magnetization can approach $\sigma \sim 1$ there). Lyubarsky 2010 has shown that further acceleration to $\sigma \lesssim 1$ is extremely slow (and not relevant for astrophysical jets). Even for the most favourable conditions for ideal MHD acceleration, bulk Lorentz factors $\Gamma \sim 1000$ needed to explain bright bursts observed with Fermi (see, e.g., Abdo et al. 2009A) appear hard to achieve with high efficiency.

2.2. Non-Ideal MHD effects

More efficient acceleration can be achieved if one abandons the axisymmetry assumption and considers non-ideal MHD effects. If the magnetic field in the jet is non-axisymmetric, containing small-scale reversals, direct dissipation of magnetic energy through magnetic reconnection is possible. Dissipation steepens the pressure gradients in the flow resulting in additional acceleration. At the same time, about half of the magnetic energy is used for heating of the plasma and can power the jet emission. This counter-intuitive process of acceleration through destruction of Poynting flux is explained in Spruit & Drenkhahn (2004). It is a promising process to boost the acceleration and provide a natural mechanism for the jet emission (see next section).

[1] Alternatives to MHD acceleration can account for specific sources. For example, a viable model for GRB jets may be that of neutrino annihilation (Ruffert et al 1997; Popham, Woosley & Fryer 1999; Chen & Beloborodov 2007). However, the attraction of the MHD mechanism is that it can account for all observed relativistic flows in a unifying manner, negating the need to advocate a different mechanism for each type of source.

Non-axisymmetric fields may appear directly at the jet launching region or, alternatively, develop further out as result of MHD instabilities in the jet. The first situation can be realised if the polarity of the field that threads the inner disk reverses with time. The strong poloidal fields required for the jet launching may be produced locally because of the Parker instability (e.g., Tout & Pringle 1992) or advected by the environment (see, e.g., Spruit & Uzdensky 2005; Rothstein & Lovelace 2008). In both cases, separated regions of opposite polarity in the inner disk can be naturally produced. A characteristic configuration of non-axisymmetric jet at the launching region was studied by Drenkhahn (2002), Drenkhahn & Spruit (2002 see also Coroniti 1990; Lyubarky & Kirk 2001). Drenkhahn (2002) considered an oblique rotator (an object rotating with the magnetic dipole axis inclined to the rotational axis) [2] that leads to a flow with field reversals on a scale of the order of the light cylinder. In this configuration, magnetic energy is dissipated efficiently through reconnection within an (astrophysically) reasonable distance leading to fast jets. Furthermore, most of the energy is dissipated in regions of moderate or small opticall depths and can be efficiently radiated away.

The second possibility for the developments of small-scale fields in jets is connected to MHD instabilities of an initially axisymmetric jet. For the strongly magnetized flows discussed here, current driven (kink) instabilities are the most relevant ones (Eichler 1993; Begelman 1998; Nakamura & Meier 2004; Giannios & Spruit 2006). They are particularly relevant after the flow passes through the Alfvén radius where it acquires a strong toroidal field. Plasma with strong toroidal fields is notoriously unstable (Kadomtsev 1966). Whether the instability has enough time to grow and substantially affect the jet structure is not clear. Since kinks are intrinsically 3D, they can be appropriately studied with 3D simulations that cover a large range of lengthscales and timescales. Such simulations are, nowadays, starting to be feasible. Moll (2009; see also Moll, Spruit & Obergaulinger 2008) showed that, in Newtonian MHD, kink instabilities are very efficient in converting the large-scale toroidal fields in a small-scale configurations that favor magnetic dissipation. Initial investigation of instabilities in relativistic MHD jets finds weak effect of the current driven instabilities on the jet structure (McKinney & Blandford 2009). It is an exciting topic to the explored more in the future for different initial and boundary conditions.

3. Jet emission

I, now, turn my attention to processes responsible for the jet emission. The next two sections discuss internal-shock and magnetic dissipation models. The section (3.3) focuses on the implications from the observed variability of jets.

3.1. Internal shocks

A usual assumption for the jet emission is that it is result of internal collisions of different portions of the jet (see Rees & Meszaros 1994, Spada et al. 2001 for GRB and blazar jets respectively). The internal shock model typically considers collisions of unmagnetized fluids (see, however, Fan, Wei & Zhang 2004; Mimica & Aloy 2010). It can explain a good fraction of the observed phenomenology (e.g. Nakar & Piran 2002; Bošnjak, Daigne and Dubus 2009) *provided that* shocks can efficiently accelerate electrons into a nonthermal (power-law) distribution. The particles are accelerated on a very short timescale (through the first order Fermi mechanism at the shock front) and then cool radiatively through synchrotron and Synchrotron Self Compton (SSC) emission. In this picture, synchrotron emission is usually held responsible for the prompt GRB and synchrotron and SSC for the blazar emission.

One drawback of the internal shock model is that, since it only dissipates fraction of the relative kinetic energy, it generally suffers from low radiative efficiency. Realistic calculations

[2] This model is also directly applicable for the millisecond magnetar model for GRBs; Usov 1992; Uzdensky & MacFadyen 2006; Metzger, Thompson & Quataert 2007.

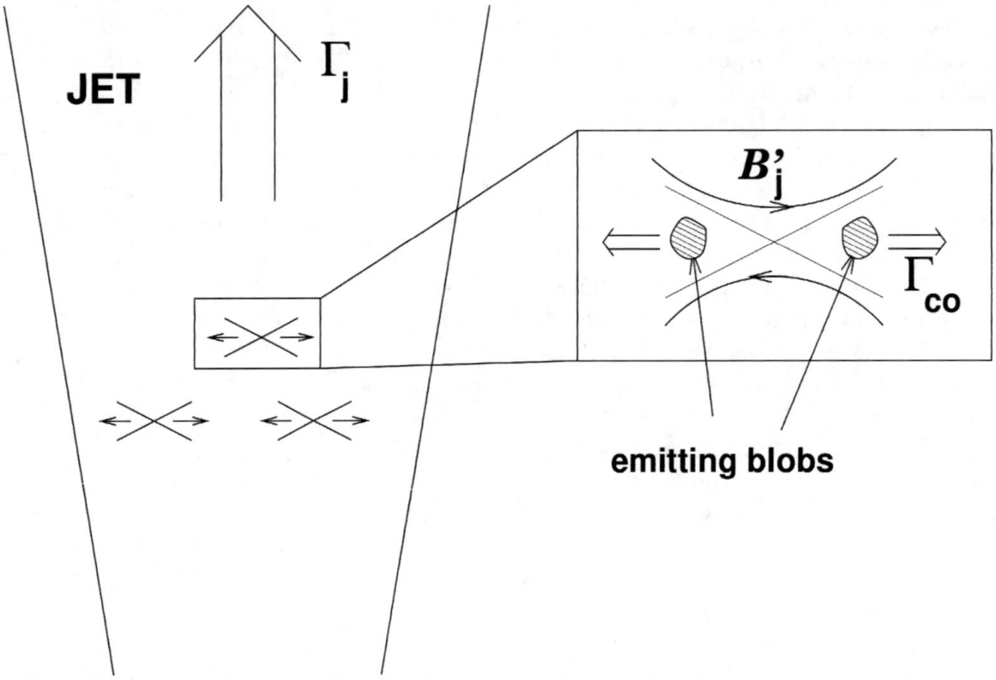

Figure 1. Schematic representation of the reconnection geometry shown in a frame comoving with the jet. Right: reconnection region enlarged. Heated and compressed plasma leaves the reconnection region at relativistic speed. Direct emission from the reconnection downstream can power ultra-fast evolving flares similar to those observed in the blazars PKS 2155–304 and Mrk 501. Furthermore, the ejected fast moving blobs can stir turbulent motions in the jet dissipating energy over a larger volume and powering to slower varying blazar and GRB emission.

bring the efficiency of such collisions down to the a $\sim 1\%$ level, maybe too low to account for GRB observations (Kumar 1999).

Internal shocks are not a favoured mechanism for jet emission if one takes into account the MHD nature of jets. As discussed above, ideal MHD models predict that magnetic fields are expected to maintain at least equipartition values $\sigma \gtrsim 1$ at large distances where the emission takes place. Any magnetization $\sigma \gtrsim 0.1$ greatly reduces the efficiency of shocks in heating the plasma by affecting the MHD shock conditions (Kennel & Coroniti 1984). To make things worse, the magnetic fields are expected to be close to parallel to the shock plane. Such configuration is highly unfavourable for particle acceleration in shocks. In these, so-called, "superluminal" shocks the particles are merely advected in the downstream by the field without entering the 1st order Fermi acceleration process (e.g. Kirk & Heavens 1989). The robust analytic arguments are verified by particle-in-cell simulations that show very weak nonthermal particle acceleration is expected even for low magnetization $\sigma \gtrsim 10^{-4}$ jets (Sironi & Spitkovsky 2010). No existing MHD model can achieve the required 99.99% conversion efficiency of magnetic energy into kinetic for particle acceleration at the shock to be feasible.

Non-ideal MHD processes may be relatively efficient in destroying magnetic energy and leading to a rather low-σ jet that could, in principle, be subject to efficient internal collisions. However, whenever this is the case, most of the magnetic energy has been already released in the dissipative MHD processes. Emission from these processes can easily outshine any shock signatures.

3.2. magnetic dissipation

I consider magnetic dissipation as a far more promising possibility for powering the jet emission (see, e.g., Lyutikov & Blandford 2003; Sikora et al. 2005). Magnetic energy can be efficiently dissipated for appropriate field arrangement via magnetic reconnection. Jets that contain small-scale magnetic field reversals are favorable configurations for reconnection. Magnetic reconnection has two generic outcomes: (1) particle heating/acceleration at the current layer or slow MHD shocks and (2) fast bulk fluid motions of plasma leaving the reconnection region (see Fig. 1). Strongly magnetized plasma (with magnetic energy density higher than its rest mass energy density as expected in relativistic jets) can enter the reconnection region rather cold and exit it relativistically hot and with relativistic bulk motions (Lyubarsky 2005; Zenitani, Hesse & Klimas 2010). When the accelerated particles emit efficiently close to the reconnection region, powerful, fast evolving emission can be produced (Giannios, Uzdensky & Begelman 2009). Furthermore, the energy in the bulk motions of the reconnection outflows can cascade to smaller scales by driving MHD turbulence in the jet. In a turbulent environment, particles can undergo stochastic (2nd order Fermi) acceleration with their typical energy determined by a balance of heating and cooling processes (Schlickeiser 1984; Thompson 1994; Stawarz & Petrosian 2008). Stochastic acceleration (slow heating) processes can play a very important role for the jet emission.

Ghisellini & Celotti (1999) have shown that balance of particle heating and cooling results in a mildly relativistic particle distribution that can go a long way in explaining the prompt GRB emission as result of Compton scattering. This idea has been further explored by Stern & Poutanen (2004) and Pe'er, Meszaros & Rees (2006) for various dissipation/heating prescriptions. We have carried out a radiative transfer calculation for the dissipation profile predicted by the reconnection model of Drenkhahn (2002) assuming homogeneously heated flow (Lazzati & Begelman 2010 relaxed this assumption by assuming localized heating and found qualitatively similar results). Our calculations for the emission expected from gradual dissipation models for GRB flows (Giannios 2006B; Giannios & Spruit 2007) showed that the model can naturally explain all the main features of the prompt emission spectrum in the energy range where it is typically observed (i.e. from X-rays to MeV γ-rays; see Fig. 2): the steady sub-MeV break and the spectral slopes below and above the break. Furthermore, comparison with multi-frequency observations of the the prompt emission from GRB 061121 that span from the optical to $\sim$ MeV range support the model (Giannios 2008). In general, gradual dissipation results in two main emission components: one that comes from layers with Thomson optical depths $\tau \sim 1$ and one from the Thomson thin region. The first component is the result of thermal seed photons upscattered by hot electrons close to the photosphere of the flow and appears in the X-ray and γ-ray bands. The second component is the combined result of (partially self absorbed) synchrotron emission and inverse Compton scattering and dominates the observed radiation in the optical and ultra violet and contributes to the $\sim$ GeV emission.

In its current form, the model predicts a exponential cutoff of the emission at $\sim$ 1 GeV and therefore cannot account for the multi-GeV emission observed in several *Fermi-LAT* bursts (e.g. Abdo et al. 2009A). However, there is strong evidence that the GeV emission has spectral and, most important, timing properties different from the sub-MeV one (e.g. Abdo et al. 2009B; Ghisellini et al. 2010). The mechanism responsible for the GeV emission is possibly not internal to the jet but related to the shock that the jet drives into the external medium (forward shock; see, e.g., Kumar & Barniol Duran 2009; Ghisellini et al. 2010).

3.3. Variability

Internal shock models rely on variability imprinted to the flow from the central engine (or from its collimation phases) for the emission (Nakar & Piran 2002). The central engine and therefore the jet emission can potentially vary on timescales as short as the light-crossing time (or dynamical

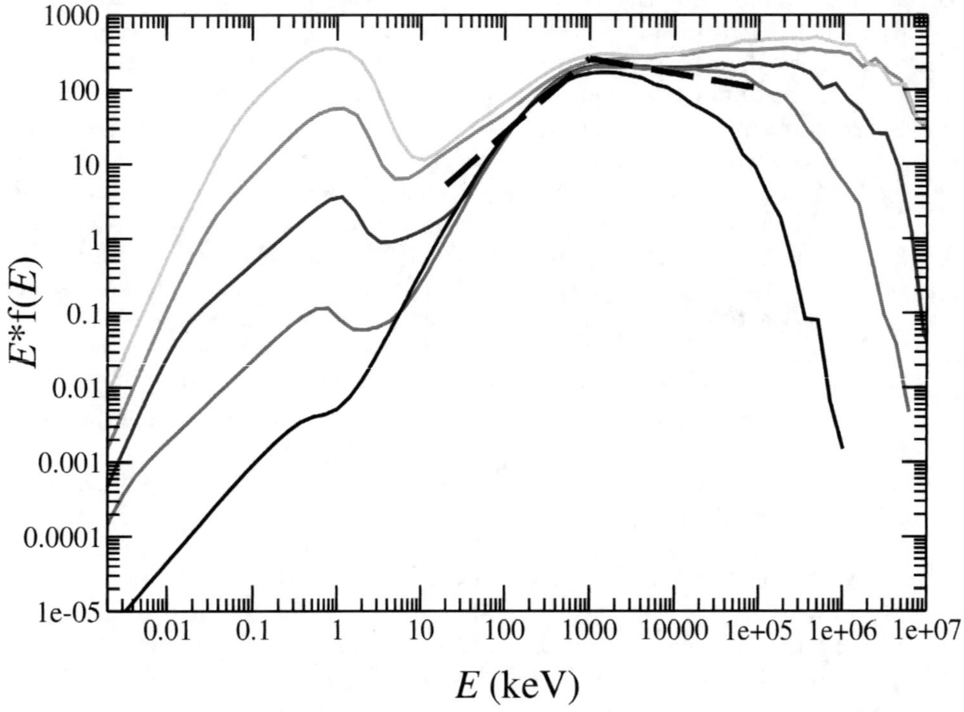

Figure 2. Resulting spectrum *(in the central engine frame)* from the gradual energy release model for GRB flows (Giannios 2008) compared to the typical GRB spectrum (schematically shown with the dashed line; e.g., Band et al. 1993). From bottom to top the curves correspond to initial magnetization (or baryon loading) σ_{in} =250, 350, 460, 590, 1000 respectively. The stable $\sim$ 1MeV break (shaped from the photospheric component) is followed by a flat $\nu \cdot f_\nu \propto \nu^0$ emission. The high σ_{in} models have bright UV emission (synchrotron component) and slope $\nu \cdot f_\nu \propto \nu^1$ below the MeV break (SSC emission).

time) of the compact object. The same is true in the reconnection model of Drenkhahn (2002) in which magnetic dissipation takes place close enough to the central engine. In this case there is a direct correspondence between the jet emission and the instantaneous luminosity of the flow (Giannios & Spruit 2007; See Fig. 3)[3]. These timescales are compatible with those observed in GRBs and, often (but not always; see following paragraph) in AGN jets.

Magnetic reconnection can potentially result to much faster variability provided that individual reconnection regions contribute substantially to the overall emission. This effect is pronounced when reconnection drives (as expected) fast motions within the jet (Lyutikov & Blandford 2003). The ultra-fast variability ($\sim$3-5 min) observed in Mrk 501 (Albert et al. 2007), PKS 2155–304 (Aharonian et al. 2007) may well be explained in the context of reconnection (mini)jets (Giannios, Uzdensky & Begelman 2009; 2010; Nalewajko et al. 2010). The minijets

[3] This is not the case for models where the magnetic dissipation is triggered by external interactions of the jet (Lyutikov & Blandford 2003; Thompson 2006). In these models the variability is generated in situ.

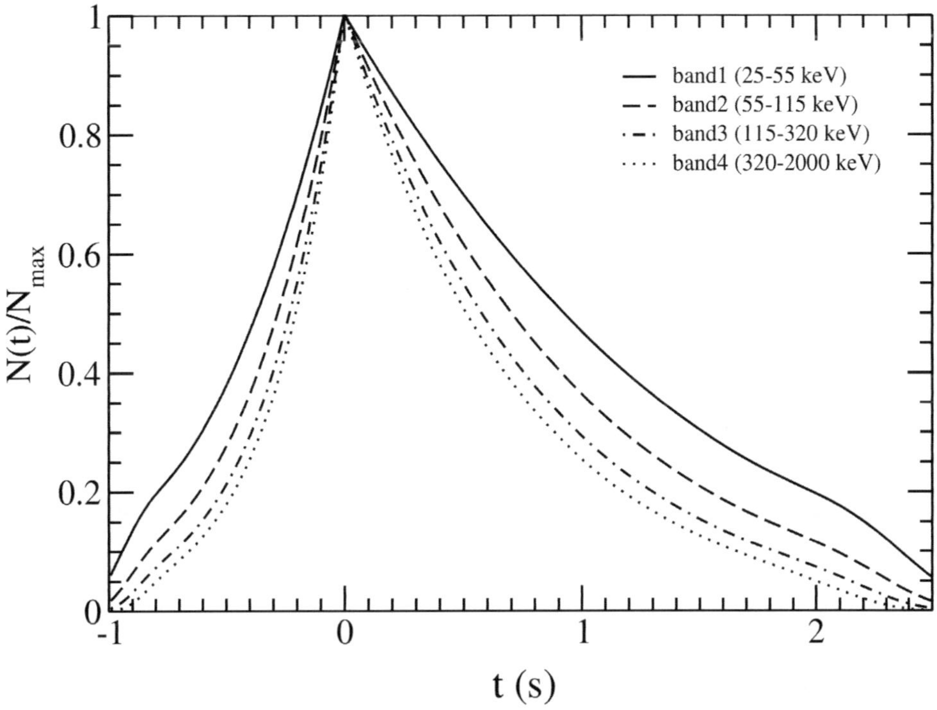

Figure 3. Narrowing of pulse width with photon energy, as predicted by the magnetic reconnection model for individual sub-pulses of a gamma-ray burst. Both the observed narrowing of the pulses (Norris et al. 1996) and the Amati relation (Amati et al. 2002; Amati 2006) can be explained under the assumption that the relation between luminosity and baryon loading $\sigma_{in} \propto L^{0.6}$, holds during the burst.

move faster than the bulk of the jet solving the Lorentz factor crisis (Begelman, Fabian & Rees 2008) that arises for these sources. It worths mentioning that equivalent timescales (shorter than the light-crossing time of the compact object) have not been observed in GRBs. However, they would correspond to sub-μsec variability which we probably do not have the (photon) statistics to detect. Summarising, ultra-fast flaring may be giving us unique probes to the very action of the magnetic reconnection while longer timescale variablity is result of variations of the bulk properties of the jet (e.g. luminosity) with time.

4. External interactions

After the jet acceleration and emission phases are complete, the flow interacts with the external medium and slows down. This interaction is believed to give rise to the GRB afterglow emission. The (hydro)dynamics of the deceleration of non-magnetized ejecta has been rather well understood (Sari & Piran 1995; Kobayashi, Piran & Sari 1999). The interaction with the external medium leads to the formation of a forward shock in the external medium and of a reverse shock (RS) into the ejecta. The RS crosses the ejecta on a short time scale leading to a brief emission episode (typically an optical flash). After the RS crossing, the forward shock

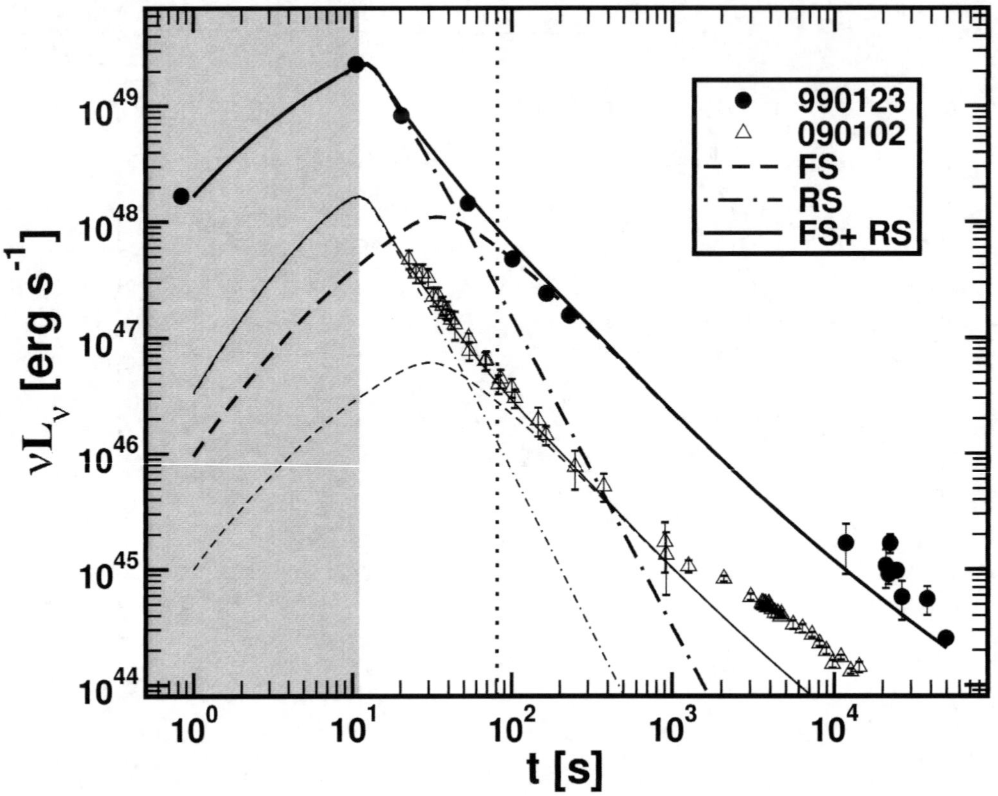

Figure 4. R-band light curve of GRB 990123 (Akerlof et al. 1999; Brigs et al. 1999) and GRB 090102 (Gendre et al. 2010; Steele et al. 2009) plotted in the rest frame of the central engine of the burst. Shown are modes which closely match the observed light curves. The σ=0.01 model (thick lines) has jet bulk Lorentz factor $\Gamma = 640$. The σ=0.1 model (thin lines) has jet bulk Lorentz factor $\Gamma = 940$. Full, dot-dashed and dashed lines denote the total (forward+reverse shock), reverse shock and forward shock emission respectively. The vertical dashed line approximately shows the time when the polarization of GRB 090102 was measured. The $\sim$ 10% polarization of GRB 090102 is naturally explained as the combination of highly polarized, dimmer reverse shock and weakly polarized, brighter forward shock emission.

dominates the afterglow emission.

MHD models for GRBs predict ejecta with dynamically important magnetic fields in the afterglow phases. The deceleration phase of strongly magnetized ejecta has only recently been understood. At a first approach to the problem, Zhang & Kobayashi (2005) solved for the shock conditions *assuming* that a pair of shock forms. Lyutikov (2006) and Genet, Daigne & Mochkovitch (2006) studied the deceleration of high σ ejecta. Giannios, Mimica & Aloy (2008) detive the condition for the formation of strong reverse shock and show that even moderate magnetization can suppress the shock, for typical GRB parameters. Relativistic MHD simulations studied the complete deceleration dynamics of ejecta with $\sigma \lesssim 1$ (Mimica, Giannios & Aloy 2009; 2010). Radiative transfer calculation coupled to the simulations gives the resulting synchrotron emission from the shock-accelerated particles. These simulations demonstrate that, for typical GRB parameters, the brightest reverse shock emission is achieved for moderate magnetization[4] $0.01 < \sigma < 0.1$ while for $\sigma \gtrsim 1$ the reverse shock emission is negligible.

Observationally, even after extensive campaigns (e.g. Klotz et al. 2009), only a handful of bursts with evidence for reverse shock emission have been found. GRB 990123 was the first and remains an standard example (Akerlof 1999). GRB 090102 showed evidence for reverse shock emission which was characterized by polarization of 10% (Steele et al. 2009). This high polarization degree is not expected from hydrodynamical models that assume small-scale field amplification at the shock.

The paucity of optical flashes-signatures of a RS in the large majority of GRBs can be understood if the bursts are typically characterized by $\sigma \sim 1$ or higher. The lightcurves of GRBs 990123, 090102 which have detected (what is thought to be) reverse shock emission, may come from jets with somewhat lower magnetization ($\sigma \sim 0.1$). Furthermore, the polarization of GRB 090102 is naturally explained as the combination of highly polarized reverse shock and weakly polarized forward shock emission (see Fig. 4). These findings agree well with the predictions of MHD models that strong fields survive at large distance.

Another finding of the Mimica et al. (2010) simulations is that while the bulk of the magnetic energy is transferred into the shocked external medium on a short timescale (comparable to the observed duration of the burst) some magnetic energy (a few % of the total) remains in the ejecta at least ten times longer that the burst duration. The deceleration of the ejecta may revive MHD instabilities resulting in delayed dissipative events (Giannios 2006A). The residual magnetic energy is sufficient to power some of the X-ray flaring observed in the early afterglow (Nousek et al. 2006; Chincarini et al. 2010). If this interpretation is correct, the X-ray flaring is powered by hot spots in the initial ejecta and does not necessarily imply late time revival of the GRB central engine.

5. (Instead of) Summary

Relativistic outflows have been observed or inferred in various astrophysical settings (galactic centers, binary systems, pulsars, collapsing massive stars). They are likely all emerging from compact objects (black holes or neutron stars). Our best bet for the origin of jets is that they are launched by magnetic fields that extract the rotational energy from the central engine.

Despite the theoretical expectation of a common physical driver of jets, the study of relativistic jets is fragmented both observationally and theoretically. Because of the different phenomenology and involved timescales, different communities observe jets from supermassive black holes, stellar mass compact objects or gamma-ray bursts. Furthermore, theoretical studies of the physics of jets in different scales are typically performed by separate groups. Other groups focus on (a) the jet launching from the compact object, (b) acceleration, (c) emission, or (d) propagation and interaction with the external medium. Despite the fact that in each sub-field there has been substantial theoretical and observational progress, we still lack a coherent and unifying description of the dissipative and emission processes in relativistic jets. Studies of MHD jet acceleration, rarely deal with the dissipative processes that lead to emission while studies for the emission typically neglect the intrinsically MHD structure of the jet.

I argue that one cannot understand the emission of jets without taking into account the main driver of the jet: the magnetic field. The presence of powerful magnetic fields makes internal collisions an unlikely mechanism for powering the emission. Instead, I advocate that similar mechanisms are responsible for the jet acceleration and emission in GRB and AGN jets, directly connected to dissipation of magnetic energy. In this picture, the ultra-fast flaring of blazars may come from plasma leaving from individual magnetic reconnection regions (minijets). Longer

[4] As a discussed in Sect. 3.1, such magnetisation is expected to suppress non-thermal particle acceleration in the shock (and emission over the relevant bands). However, according to the Sironi & Spitkovsky (2010) study, electrons are *heated* to high energies and can still result in an optical flash when the characteristic frequency of synchrotron emission is close to the optical band.

timescale variablity of jets likely tracks the variations of the lminosity of the flow. It can be connected to interactions of the reconnection minijets with the rest of the jet that dissipate bulk kinetic energy on larger volume.

Acknowledgments

I acknowledge support from the Lyman Spitzer, Jr. Fellowship awarded by the Department of Astrophysical Sciences at Princeton University.

References

[1] Abdo A A et al. 2009A *Science* **323** 1688
[2] Abdo A A et al. 2009B *Astroph. J.* **706** L138
[3] Aharonian F et al. 2007 *Astroph. J.* **664** L71
[4] Akerlof C et al 1999 *Nature* **398** 400
[5] Albert J et al. 2007 *Astroph. J.* **669** 862
[6] Amati L et al. 2002 *Astron. Astroph.* **390** 81
[7] Amati L 2006 *Mon. Not. Roy. Astron. Soc.* **372** 233
[8] Band D et al. 1993 *Astroph. J.* **413** 281
[9] Begelman M C 1998 *Astroph. J.* **493** 291
[10] Begelman M C and Li Z-Y 1994 *Astroph. J.* **426** 269
[11] Begelman M C, Fabian A C and Rees M J. 2008 *Mon. Not. Roy. Astron. Soc.* **384** L19
[12] Beskin V S, Kuznetova I V and Rafikov R R 1998 *Mon. Not. Roy. Astron. Soc.* **299** 341
[13] Bisnovatyi-Kogan G S and Ruzmaikin A A 1976 *Astroph. & Space Science* **42** 401
[14] Blandford R D 1976 *Mon. Not. Roy. Astron. Soc.* **176** 465
[15] Blandford R D and Znajek R L 1977 *Mon. Not. Roy. Astron. Soc.* **179** 433
[16] Blandford R D and Payne D G 1982 *Mon. Not. Roy. Astron. Soc.* **199** 883
[17] Bošnjak Ž, Daigne F and Dubus G 2009 *Astron. Astroph.* **498** 677
[18] Briggs M S et al. 1999 *Astroph. J.* **524** 82
[19] Camenzind M 1986 *Astron. Astroph.* **162** 32
[20] Chen W-X and Beloborodov A M 2007 *Astroph. J.* **657** 383
[21] Chincarini G et al. 2010 *Mon. Not. Roy. Astron. Soc.* **406** 2113
[22] Contopoulos J 1995 *Astroph. J.* **450** 616
[23] Coroniti F V 1990 *Astroph. J.* **349** 538
[24] Drenkhahn G 2002 *Astron. Astroph.* **387** 714
[25] Drenkhahn G and Spruit H C 2002 *Astron. Astroph.* **391** 1141
[26] Fan Y Z, Wei D M and Zhang B 2004 *Mon. Not. Roy. Astron. Soc.* **354** 1031
[27] Gendre B et al. 2010 *Mon. Not. Roy. Astron. Soc.* **405** 2372
[28] Genet F, Daigne F and Mochkovitch R 2006 *Astron. Astroph.* **457** 737
[29] Giannios D 2006A *Astron. Astroph.* **455** L5
[30] Giannios D 2006B *Astron. Astroph.* **457** 763
[31] Giannios D 2008 *Astron. Astroph.* **480** 305
[32] Giannios D and Spruit H C 2006 *Astron. Astroph.* **450** 887
[33] Giannios D and Spruit H C 2007 *Astron. Astroph.* **469** 1
[34] Giannios D, Mimica P and Aloy M A 2008 *Astron. Astroph.* **478** 747
[35] Giannios D, Uzdensky D. A and Begelman M C 2009 *Mon. Not. Roy. Astron. Soc.* **395** L29
[36] Giannios D, Uzdensky D A and Begelman M C 2010 *Mon. Not. Roy. Astron. Soc.* **402** 1649
[37] Ghisellini G and Celotti A 1999 *Astroph. J.* **511** L93
[38] Ghisellini G, Ghirlanda G, Nava L and Celotti A 2010 *Mon. Not. Roy. Astron. Soc.* **403** 926
[39] Granot J, Komissarov S and Spitkovsky A 2010, arXiv:1004.0959
[40] Goldreich P and Julian W H 1970 *Astroph. J.* **160** 971
[41] Eichler D 1993 *Astroph. J.* **419** 111
[42] Kadomtsev B B 1966 *Rev. Plasma Phys.* **2** 153
[43] Kennel C F and Coroniti F V 1984 *Astroph. J.* **283** 710
[44] Kirk J G and Heavens A F 1989 *Mon. Not. Roy. Astron. Soc.* **239** 995
[45] Klotz A, Boër M, Atteia J L and Gendre B 2009 *Astron. J.* **137** 4100
[46] Kobayashi S, Piran T and Sari R 1999 *Astroph. J.* **513** 669
[47] Komissarov S S, Vlahakis N, Königl A and Barkov M V 2009 *Mon. Not. Roy. Astron. Soc.* **394** 1182
[48] Kumar P 1999 *Astroph. J.* **523** L113

[49] Kumar P and Barniol Duran R. 2009 *Mon. Not. Roy. Astron. Soc.* **400** L75
[50] Lazzati D and Begelman M C 2010 *Preprint* arXiv:1005.4704
[51] Lyubarsky Y E 2005 *Mon. Not. Roy. Astron. Soc.* **358** 113
[52] Lyubarsky Y 2009 *Astroph. J.* **698** 1570
[53] Lyubarsky Y E 2010 *Mon. Not. Roy. Astron. Soc.* **402** 353
[54] Lyubarsky Y and Kirk J G 2001 *Astroph. J.* **547** 437
[55] Lyutikov M 2006 *New Journal of Physics* **8** 119
[56] Lyutikov M and Blandford R D 2003 *Preprint* astro-ph/0312347
[57] Lyutikov M and Lister M 2010, arXiv:1004.2430
[58] McKinney J C and Blandford R D 2009 *Mon. Not. Roy. Astron. Soc.* **394** L126
[59] Metzger B D, Thompson T A and Quataert E 2007 *Astroph. J.* **ApJ** 659 561
[60] Michel F C 1969 *Astroph. J.* **158** 727
[61] Mimica P and Aloy M A 2010 *Mon. Not. Roy. Astron. Soc.* **401** 525
[62] Mimica P, Giannios D and Aloy M A 2009 *Astron. Astroph.* **494** 879
[63] Mimica P, Giannios D and Aloy M A 2010 *Mon. Not. Roy. Astron. Soc.* **1005**
[64] Moll R 2009 *Astron. Astroph.* **507** 1203
[65] Moll R, Spruit H C and Obergaulinger M 2008 *Astron. Astroph.* **492** 621
[66] Nakamura M and Meier D L 2004 *Astroph. J.* **617** 123
[67] Nakar E and Piran T 2002 *Astroph. J.* **572** L139
[68] Nalewajko K, Giannios D, Begelman M C, Uzdensky D A and Sikora M 2010 *Mon. Not. Roy. Astron. Soc.* submitted arXiv:1007.3994
[69] Norris J P et al. 1996 *Astroph. J.* **459** 393
[70] Nousek J A et al. 2006 *Astroph. J.* **642** 389
[71] Pe'er A, Mészáros P and Rees M J 2005 *Astroph. J.* **635** 476
[72] Popham R, Woosley S E and Fryer, C. 1999 *Astroph. J.* **518** 356
[73] Rees M J and Meszaros P. 1994 *Astroph. J.* **430** L93
[74] Rothstein D M and Lovelace R V E 2008 *Astroph. J.* **677** 1221
[75] Ruffert M, Janka H-T, Takahashi K and Schaefer G 1997 *Astron. Astroph.* **319** 122
[76] Sari R and Piran T 1995 *Astroph. J.* **455** L143
[77] Schlickeiser R. 1984 *Astron. Astroph.* **136** 227
[78] Sikora M, Begelman M C, Madejski G M and Lasota J-P 2005 *Astroph. J.* **625** 72
[79] Sironi L and Spitkovsky A 2010 *Astroph. J.* submitted arXiv:1009.0024
[80] Spada M, Ghiselini G, Lazzati, D and Celloti A 2001 *Mon. Not. Roy. Astron. Soc.* **325** 1559
[81] Spruit H C, Foglizzo T and Stehle R 1997 *Mon. Not. Roy. Astron. Soc.* **288** 333
[82] Spruit H C and Drenkhahn G D 2004, Magnetically powered prompt radiation and flow acceleration in GRB, in Proceedings *Gamma Ray Bursts in the Afterglow Era, Third Workshop'* Rome Sept 2002. ASP Conference Proccedings 312 p 357
[83] Spruit H C and Uzdensky D A 2005 *Astroph. J.* **629** 960
[84] Spruit H C 2010 *Lecture Notes in Physics* Berlin Springer Verlag 794 pv233
[85] Stawarz Ł and Petrosian V 2008 *Astroph. J.* **681** 1725
[86] Steele I A, Mundell C G, Smith R J, Kobayashi S and Guidorzi C 2009 *Nature* **462** 767
[87] Stern B E and Poutanen J 2004 *Mon. Not. Roy. Astron. Soc.* **352** L35
[88] Tchekhovskoy A, Narayan R and McKinney J C 2010 *New Astron.* **15** 749
[89] Thompson C 1994 *Mon. Not. Roy. Astron. Soc.* **270** 480
[90] Thompson C 2006 *Astroph. J.* **651** 333
[91] Tout C A and Pringle J E 1992 *Mon. Not. Roy. Astron. Soc.* **259** 604
[92] Usov V V 1992 *Nature* **357** 472
[93] Uzdensky D A and MacFadyen A I 2006 *Astroph. J.* **647** 1192
[94] Vlahakis N and Königl A 2003 *Astroph. J.* **596** 1080
[95] Zhang B and Kobayashi S 2005 *Astroph. J.* **628** 315
[96] Zenitani S, Hesse M and Klimas A 2010, *Astroph. J.* **716** L214

DCPT-10/67

Holographic Superconductivity with Gauss-Bonnet gravity

Ruth Gregory
Centre for Particle Theory Durham University, South Road, Durham, DH1 3LE, UK

Abstract. I review recent work ([1, 2]) on holographic superconductivity with Einstein-Gauss-Bonnet gravity, and show how the critical temperature of the superconductor depends on both gravitational backreaction and the Gauss-Bonnet parameter, using both analytic and numerical arguments. I also review computations of the conductivity, finding the energy gap, and demonstrating that there is no universal gap ratio, ω_g/T_c, for these superconductors.

1. Introduction

Holographic superconductivity (see [3, 4, 5] for reviews) is a fascinating idea which proposes to describe strongly coupled high T_c superconductors via a classical gravitational system using the gauge gravity correspondence, [6]. The bulk theory typically has gauge and a charged scalar field, with a black hole providing a finite temperature. Choosing an appropriate scalar potential, such as a negative mass-squared above the Breitenlohner-Friedman bound, [7], allows the scalar to condense out of the vacuum near the black hole horizon, if the curvature is sufficiently large there [8]. This occurs for low temperature black holes, hence the analogy to superconductivity.

The superconducting phase corresponds to a "hairy" black hole, where the condensation of the scalar field out of its symmetric state screens the charge of the black hole. Typical no-hair theorems [9] do not apply in this case as the scalar potential is not positive definite. Near the boundary, the scalar has a power law fall-off, and the coefficient of this fall-off can be interpreted as a condensate in the boundary theory. Following the behaviour of this condensate as the temperature drops reveals the typical behaviour for an order parameter governing the superconducting phase transition, and has been observed in a wide variety of bulk thoeries, including varying the scalar mass and potential, varying the number of spacetime dimensions, as well as having magnetic fields present and the stability of the system [8, 10, 11, 12, 14, 15, 16]. Similar models have also been found from embeddings in string theory, [17], and it is therefore of interest to consider more general stringy aspects of these models.

In this presentation, I review the results of [1] and [2], which explored adding the higher curvature Gauss-Bonnet (GB) term, [18], to the gravitational action. I will also update on the talk where appropriate, and add a few new results and comments. In brief, many of the qualitative features of the holographic superconductors are stable under higher curvature corrections, however, there are quantitative differences, in particular, a universality relation for an energy gap in the superconductor was found not to be substantiated even in the Einstein limit. (See also [19] for other work on GB holographic superconductors.)

I first review the bulk superconductor, deriving some general analytical results and remarking on the probe superconductor. Then I summarize the numerical results, showing how the numerical results follow closely an analytical bound, revealing an interesting subtlety in the response of the superconductor to the Gauss-Bonnet parameter α. Finally, I review the results of [1] and [2] on conductivity.

2. The bulk superconductor

The general action is that of an Einstein-Gauss-Bonnet (EGB) gravitational action coupled to a massive charged complex scalar field and a U(1) gauge field:

$$S = \frac{1}{2\kappa^2}\int d^5x\sqrt{-g}\left[-R+\frac{12}{L^2}+\frac{\alpha}{2}\left(R^{abcd}R_{abcd}-4R^{ab}R_{ab}+R^2\right)\right] + \int d^5x\sqrt{-g}\left[-\frac{1}{4}F^{ab}F_{ab}+|\nabla_a\psi-iqA_a\psi|^2-m^2|\psi|^2\right] \quad (1)$$

where g is the determinant of the metric, and R_{abcd}, R_{ab} and R are the Riemann curvature tensor, Ricci tensor, and the Ricci scalar, respectively. The Gauss-Bonnet coupling constant is taken in the range $\alpha \in [0, L^2/4]$, and the cosmological constant is given in terms of a length scale, L, which for Einstein gravity ($\alpha = 0$) corresponds to the adS curvature. The Planck scale is set by $\kappa^2 = 8\pi G_5$, q is the charge, and m^2 the (squared) mass, of the scalar field.

The equations of motion satisfied by the bulk system are:

$$R_{ab} - \tfrac{1}{2}Rg_{ab} + \frac{6}{L^2}g_{ab} - \alpha\left[H_{ab} - \tfrac{1}{4}Hg_{ab}\right] = \kappa^2 T_{ab} \quad (2)$$

where

$$H_{ab} = R_a^{\ cde}R_{bcde} - 2R_{ac}R^c_b - 2R_{acbd}R^{cd} + RR_{ab}\,, \quad (3)$$

T_{ab} is the matter energy momentum tensor

$$T_{ab} = 2\mathcal{D}_{(a}\psi^\dagger\mathcal{D}_{b)}\psi - F_{ac}F_b^{\ c} - \left[|\mathcal{D}_c\psi|^2 - \tfrac{1}{4}F^2_{cd} - m^2|\psi|^2\right]g_{ab}\,, \quad (4)$$

and $\mathcal{D}_a = \nabla_a - iqA_a$ is the gauge covariant derivative.

To examine holographic superconductivity, we allow for a bulk black hole in order to have a system at finite temperature:

$$ds^2 = f(r)e^{2\nu(r)}dt^2 - \frac{dr^2}{f(r)} - \frac{r^2}{L_e^2}(dx^2+dy^2+dz^2) \quad (5)$$

where

$$f(r) = \frac{r^2}{2\alpha}\left[1-\sqrt{1-\frac{4\alpha}{L^2}\left(1-\frac{r_+^4}{r^4}\right)}\right] \quad (6)$$

is the black hole gravitational potential in the absence of backreaction, and

$$L_e^2 = \frac{L^2}{2}\left[1+\sqrt{1-\frac{4\alpha}{L^2}}\right] \quad (7)$$

is the actual curvature of the adS spacetime, which is renormalized away from the cosmological constant scale, L, once α is nonzero. The temperature of the background is:

$$T = \frac{1}{4\pi}f'(r_+)e^{\nu(r_+)}\,, \quad (8)$$

which is equal to $r_+/\pi L^2$ in the absence of backreaction.

Taking a static ansatz, $A_\mu = (\phi(r), 0...)$ and $\psi = \psi(r)$ (which can be taken to be real), the equations of motion become

$$\phi'' + \left(\frac{3}{r} - \nu'\right)\phi' - 2q^2\frac{\psi^2}{f}\phi = 0\,, \tag{9}$$

$$\psi'' + \left(\frac{3}{r} + \nu' + \frac{f'}{f}\right)\psi' + \left(\frac{q^2\phi^2}{f^2e^{2\nu}} - \frac{m^2}{f}\right)\psi = 0\,, \tag{10}$$

$$\left(1 - \frac{2\alpha f}{r^2}\right)\nu' = \frac{2\kappa^2}{3}r\left(\psi'^2 + \frac{q^2\phi^2\psi^2}{f^2e^{2\nu}}\right) \tag{11}$$

$$\left(1 - \frac{2\alpha f}{r^2}\right)f' + \frac{2}{r}f - \frac{4r}{L^2} = -\frac{2\kappa^2}{3}r\left[\frac{\phi'^2}{2e^{2\nu}} + m^2\psi^2 + f\psi'^2 + \frac{q^2\phi^2\psi^2}{fe^{2\nu}}\right] \tag{12}$$

where a prime denotes derivative with respect to r. These equations have several scaling symmetries, as noted in [2], which are used to set $L = Q = q = 1$. Note that by fixing $Q = 1$, the charge parameter is kept fixed in all numerical computations.

(i) $r \to ar$, $t, x^i \to at, ax^i$, $L \to aL$, $q \to q/a$, $\alpha \to a^2\alpha$, $A \to aA$.
(ii) $r \to br$, $t \to t/b$, $x^i \to x^i/b$, $f \to b^2 f$, $\phi \to b\phi$.
(iii) $\phi \to c\phi$, $\psi \to c\psi$, $q \to q/c$, $\kappa^2 \to \kappa^2/c^2$.

The horizon is defined in general by $f(r_+) = 0$, and demanding regularity of the solution at both the horizon and boundary gives the following boundary conditions:

• Horizon:

$$\phi(r_+) = 0, \qquad \psi'(r_+) = \frac{m^2}{f'(r_+)}\psi(r_+) \tag{13}$$

$$\nu'(r_+) = \frac{2\kappa^2}{3}r_+\left(\psi'(r_+)^2 + \frac{\phi'(r_+)^2\psi(r_+)^2}{f'(r_+)^2e^{2\nu(r_+)}}\right) \tag{14}$$

$$f'(r_+) = \frac{4}{L^2}r_+ - \frac{2\kappa^2}{3}r_+\left(\frac{\phi'(r_+)^2}{2e^{2\nu(r_+)}} + m^2\psi(r_+)^2\right) \tag{15}$$

• Boundary:

$$\nu \to 0\ , \quad f(r) \sim \frac{r^2}{L_e^2} \quad \Rightarrow \tag{16}$$

$$\phi(r) \sim P - \frac{Q}{r^2}\,, \qquad \psi(r) \sim \frac{C_-}{r^{\Delta_-}} + \frac{C_+}{r^{\Delta_+}}\,, \quad \text{as } r \to \infty\,, \tag{17}$$

where $\Delta_\pm = 2 \pm \sqrt{4 + m^2L_e^2}$ for a general mass m^2. We choose $C_- = 0$, then P and C_+ are fixed by consistency with the near horizon solution. According to the adS/CFT correspondence, we can interpret $\langle\mathcal{O}_{\Delta_+}\rangle \equiv C_+$, where $\mathcal{O}_{\Delta_+}$ is the operator with the conformal dimension Δ_+ dual to the scalar field.

In [1] we chose to set the mass of the scalar field as $m^2 = -3/L^2$, so that the mass remained the same as α was varied. However, the variation of the effective asymptotic AdS curvature, (7), with α relative to L means that this mass actually *increases* (i.e. becomes less negative) with respect to the asymptotic AdS scale, and therefore the dimension of the boundary operator corresponding to C_+ varies with α. Fixing $m^2 = -3/L^2_{\text{eff}}$ relative to the asymptotic AdS scale, avoids this problem, however the mass now varies with respect to the physical mass and temperature of the black hole as α varies. See [20] for a detailed examination of the effect of scalar mass on the holographic superconductor.

2.1. Analytic bounds

While the equations of motion (9-12) require a numerical solution in general, even in the absence of backreaction, near the critical temperature we can glean a measure of analytic information using the (uncondensed) charged black hole solution, [21]:

$$A = \phi_0(r)dt = \frac{Q}{r_+^2}\left(1 - \frac{r_+^2}{r^2}\right)dt \tag{18}$$

$$\nu_0 = 0 \tag{19}$$

$$f_0(r) = \frac{r^2}{2\alpha}\left[1 - \sqrt{1 - \frac{4\alpha}{L^2}\left(1 - \frac{r_+^4}{r^4}\right) + \frac{8\alpha\kappa^2 Q^2}{3r^4 r_+^2}\left(1 - \frac{r_+^2}{r^2}\right)}\right] \tag{20}$$

and linearizing in the scalar condensate. In the above, Q is the charge of the black hole (up to a geometrical factor of 4π), and r_+ is the event horizon, which determines the "ADM" mass of the black hole [22]. To avoid a naked singularity, we restrict the parameter range as $\alpha \leq L^2/4$.

Linearizing in the scalar condensate leaves the background solution for the metric and gauge field unchanged to leading order, and we can focus on (10) with f and ϕ taking their background values. Analytic bounds on the critical temperature can be obtained by looking for simple relations that must hold for the existence of a nontrivial scalar solution. First consider the variable $X_n = r^n\psi$, which satisfies to leading order:

$$X_n'' + \left(\frac{f_0'}{f_0} - \frac{3-2n}{r}\right)X_n' + \left(\frac{q^2\phi^2}{f_0^2} - \frac{m^2}{f_0} - \frac{nf_0'}{rf_0} + \frac{n(n-2)}{r^2}\right)X_n = 0\,. \tag{21}$$

To get an upper bound, let $n = 2$, and consider the general properties of a solution. At the horizon,

$$X_2'(r_+) = \frac{X_2(r_+)}{4\pi T_c}\left(\frac{8}{L^2} + m^2 - \frac{8\kappa^2 Q^2}{3r_+^6}\right) \tag{22}$$

which is positive for small $\kappa^2 Q^2$ (taking $X_2(r_+) > 0$ without loss of generality). Since $X_2 \sim 1/r^{2+\Delta_+}$ as $r \to \infty$, the solution must have a maximum for some r, which requires that

$$\left(\frac{q^2\phi^2}{f_0^2} - \frac{m^2}{f_0} - \frac{2f_0'}{rf_0}\right) > 0 \tag{23}$$

at this point. An examination of when this combination is never positive provides an upper bound for T_c. (A different n was used in [1], which sufficed for the probe limit although it led to a looser bound. For consistency and comparison between the scalar masses, this new bound is used here.)

We can also obtain a lower bound by considering $n = 3$. Manipulating (21) shows that *if* a solution exists, then the integral

$$\int_{r_+}^{\infty} \frac{1}{r^3}\left[\frac{\phi_0^2}{f_0} - m^2 + \frac{3f_0}{r^2} - \frac{3f_0'}{r}\right] = -\int_{r_+}^{\infty} \frac{f_0 X_3'^2}{r^3 X_3^2} \leq 0 \tag{24}$$

is negative. Note that negativity of this integral does not imply existence of a solution to the linearized equation near T_c, it is simply a necessary condition. Since this integral is always negative at large T, and positive as $T \to 0$ (for $\kappa^2 \lesssim 0.4$), observing where it changes sign provides a lower bound on T_c. This bound was found to give an extremely reliable indicator of T_c as computed numerically.

2.2. Analytic expansion

In instances where the exact solution is only known numerically, it can be useful to derive an analytic approximation to the exact solution using a technique of matching. The solution is Taylor expanded around the horizon and boundary, then matched at a midpoint. This technique was used in [1] to get an approximation to T_c, although the plot in [1] was inaccurate. In brief, we first transform the radial variable to $z = r_+/r$, and expand the near horizon solution of a generic field Y as:

$$Y \simeq Y_0 - Y_1(1-z) + \tfrac{1}{2}Y_2(1-z)^2 \tag{25}$$

The equations of motion give:

$$\psi_1 = \frac{m^2 r_+^2}{f_1}\psi_0 \tag{26}$$

$$\psi_2 = \frac{\psi_0}{2f_1^2}\left\{m^2 r_+^2 f_2 - m^2 r_+^2 f_1 \nu_1 + m^4 r_+^4 - q^2\phi_1^2 r_+^2 e^{-2\nu_0} - 3m^2 r_+^2 f_1\right\} \tag{27}$$

$$\phi_2 = \phi_1\left(1 + \nu_1 + \frac{2r_+^2 q^2 \psi_0^2}{f_1}\right) \tag{28}$$

where f_1, f_2, and ν_1 are given by (11,12), and depend on ϕ_0 and ψ_0. Since the asymptotic solutions for the gauge and scalar field given by (17) also have two unknowns, matching these fields and their derivatives at some intermediate point z_0 gives four equations which determine these four unknowns:

$$P - \frac{Qz_0^2}{r_+^2} = -\phi_1(1-z_0) + \frac{\phi_2}{2}(1-z_0)^2 \tag{29}$$

$$\frac{-2Qz_0}{r_+^2} = \phi_1 - \phi_2(1-z_0) \tag{30}$$

$$\frac{C_+ z_0^{\Delta_+}}{r_+^{\Delta_+}} = \psi_0 - \psi_1(1-z_0) + \frac{\psi_2}{2}(1-z_0)^2 \tag{31}$$

$$\Delta_+\frac{C_+ z_0^{\Delta_+ - 1}}{r_+^{\Delta_+}} = \psi_1 - \psi_2(1-z_0) \tag{32}$$

Clearly, given the complexity of the expressions (26)-(28), which themselves contain f_i, this is a rather involved algebraic process in the general backreacting and non-critical case. Nonetheless, the system is tractable analytically if we simply focus on deriving the critical temperature. In this case, we once again can take the background forms for ϕ, f and ν, which considerably reduce the complexity of the relations: (29) and (30) are automatically satisfied, $\nu \equiv 0$, (12) gives

$$f_1 = -4\frac{r_+^2}{L^2} + \frac{\kappa^2\phi_1^2}{3} \tag{33}$$

$$f_2 = -f_1 + \frac{2\alpha f_1^2}{r_+^2} + 2\kappa^2\phi_1^2 \tag{34}$$

and consistency of the ψ equations (31) and (32) implies

$$2\Delta_+\psi_0 - 2\left(\Delta_+ + (1-\Delta_+)z_0\right)\psi_1 + \left(\Delta_+ - 2(\Delta_+ - 1)z_0 + (\Delta_+ - 2)z_0^2\right)\psi_2 = 0 \tag{35}$$

Although this is a cumbersome expression for nonzero κ^2, it is just a quadratic for $\phi_1^2/r_+^2 = 4Q^2/r_+^6$, and a linear relation in the probe limit, which in either case can be straightforwardly

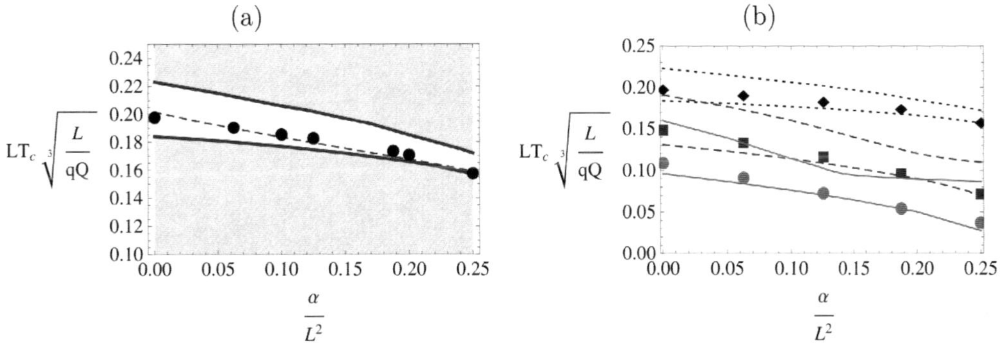

Figure 1. A plot of the critical temperature as a function of α for the scalar mass $m^2 = -3/L^2$. (a) shows a detailed plot of the probe limit, with upper and lower bounds shown as solid blue lines, the matching method value as a dotted line, and numerical data points. (b) shows a plot of the upper and lower bounds together with exact numerical results for the probe limit in black (dotted lines and diamond data points), $\kappa^2 = 0.05$ in blue (dashed lines and square data points), and $\kappa^2 = 0.1$ in red (solid lines and circular data points).

solved and the critical temperature obtained from

$$T_c = r_+ \left[\frac{1}{\pi L^2} - \frac{\kappa^2 Q^2}{3r_+^6} \right] \tag{36}$$

It only remains to choose a value for z_0. In [1] the value $z_0 = 1/2$ was chosen to be specific, however, this choice does not take into account the variation of the effective adS length scale with α. In essence, the matching point should be chosen to represent where the asymptotic expansion $f \simeq r_+^2/z^2 L_e^2$ turns over into the near horizon expansion. This obviously depends on the relative ratio of r_+ to L_e, although as $\alpha \to L^2/4$ this point moves rather significantly towards the boundary. A good approximation to the departure from the asymptotic regime is $z_0 \simeq \alpha/L^2 + 1/2$, which is the value used in figure 1(a). It must be noted however, that there is a degree of arbitrariness in this matching process. Although the choice of z_0 attempts to take into account the likely variation of the domains of validity of each expansion with α, this is a rather ad hoc process, and as backreaction is switched on, the matching procedure becomes less and less reliable. It is included here for completeness, however by far the best analytic guide to critical temperature is the lower bound.

2.3. Discussion of analytic results

By examining the behaviour of a putative condensed scalar in the bulk, it is possible to derive upper and lower bounds to the critical temperature. A matching method can also be used to get an approximate solution for the scalar, however, this method (as well as the upper bound) becomes unreliable as the backreaction is increased. On the other hand, the lower bound appears to be extremely representative of the behaviour of the actual critical temperature, and works well for a significant range of backreaction. Figures 1 and 2 show the analytical results, and will be discussed further together with numerical data in the next section.

3. Backreacting superconductors

In order to find the actual behaviour of the bulk superconductor, in [1] and [2] we integrated (9–12) numerically. As already stated, we took $L = Q = q = 1$, and varied r_+ to study how

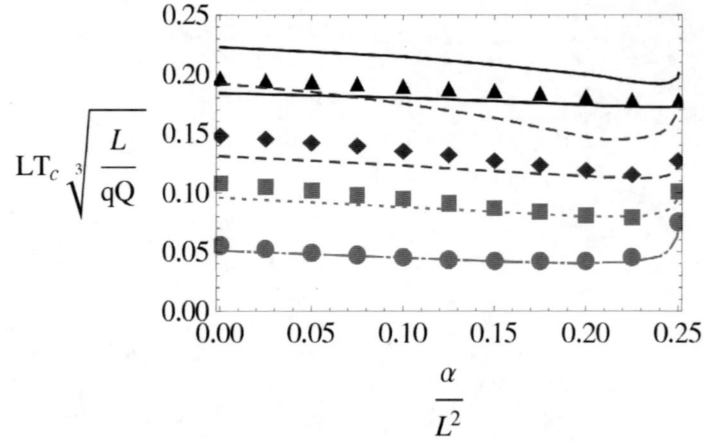

Figure 2. A plot of the critical temperature as a function of α for a selection of κ^2. The analytic bounds are shown as lines and the numerical data as points. Respectively: $\kappa^2 = 0$ is shown in black, with solid lines and triangular data points; $\kappa^2 = 0.05$ has blue dashed lines and diamonds; $\kappa^2 = 0.1$ has a grey dotted line and squares; $\kappa^2 = 0.2$ has a red dot-dash line with circular data points. The lower bound is shown for all κ^2 values, but the upper bound is shown only for the lowest two values of κ^2, as they overlap significantly with the other data and confuse the plot.

the system reacted to varying temperature. From these numerical solutions, we can deduce the critical temperature, determine the behaviour of the condensate with temperature, and compute the conductivity of the holographic superconductor as discussed in subsection 3.3.

3.1. The dependence of T_c on α and κ^2.
The analytic bounds for T_c have been plotted in figures 1 and 2 together with the exact values of T_c obtained by numerical computation for both masses of the scalar field. For $m^2 = -3/L^2$, both bounds are shown, as well as the matching method result for the probe limit. For $m^2 = -3/L_e^2$, the upper bound has only been shown for $\kappa^2 \leq 0.05$, as above this value it becomes less predictive and clutters the plot, and indeed beyond $\kappa^2 \sim 0.2$ (corresponding to $q \sim 2.25$ in the notation of [10]) it ceases to have quantitative value for any α. The lower bound on the other hand becomes successively more accurate as the values of κ^2 are stepped up, and gives a very good quantitative guide to the behaviour of T_c as we vary α and κ^2.

It is easy to see that in all cases, the effect of backreaction is to decrease T_c and thus make condensation harder. We can see this by using Gubser's rough argument, [8], that the effective scalar mass $m_{\text{eff}}^2 = m^2 - q^2\phi(r)^2/f(r)$ becomes more negative as backreaction is turned up. Essentially, the effect of backreaction is that the condensation of the scalar field not only screens the charge of the black hole, but also its mass, as the scalar and gauge fields now contribute to the ADM mass. This means that for a given charge and temperature, the radius of the black hole is increased, which makes it harder for the scalar to condense.

One very interesting feature clearly exhibited in the bounds is the turning point in T_c as a function of α for the mass $m^2 = -3/L_e^2$. In the probe case, this occurs very near the Chern-Simons limit $\alpha = L^2/4$, and is barely perceptible in the numerical data, however once backreaction is switched on, the minimum becomes much more pronounced, and indeed for large backreaction ($\kappa^2 = 0.2$) the Chern-Simons limit is showing a considerable enhancement of T_c over the typical values for lower α.

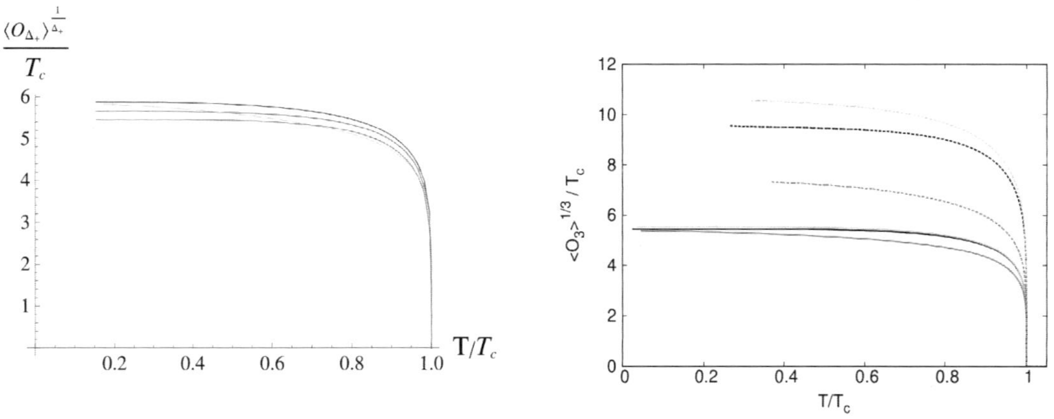

Figure 3. Two plots of the condensate as a function of temperature for $m^2 = -3/L^2$ (left) and $m^2 = -3/L_e^2$ (right). The plot on the left shows several values of α in the probe limit: $\alpha = 0$ in red, the lowest line, then $\alpha = 0.1$ in brown, 0.2 in blue and 0.25 in green. The plot on the right shows $\kappa^2 = 0$ in solid lines, and $\kappa^2 = 0.1$ as dotted lines. The black plot is $\alpha = 0$, green is $\alpha = 0.125$ and magenta is $\alpha = 0.25$.

3.2. The scalar condensate.

The numerical results allow a full exploration of the dependence of the condensate on temperature, and we expect to see the condensate switching on sharply at the critical temperature, then saturating at low temperatures, and this is indeed what is seen. Figure 3 shows $\langle \mathcal{O}_{\Delta_+}\rangle^{1/\Delta_+}$ as a function of temperature for both scalar masses.

Each line in the plot shows the characteristic curve of the condensate $\langle \mathcal{O}_{\Delta_+}\rangle^{1/\Delta_+}$ condensing at some critical temperature. For $m^2 = -3/L^2$ the curves are shown in the probe limit. While the critical temperature does not display the minimum in α, it can nonetheless be seen that the Chern-Simons limit does look rather different, and the scalar field saturates far more slowly as the temperature drops. For $m^2 = -3/L_e^2$, three different values of α $(0, 0.125, 0.25)$ and two different values of κ^2 $(0, 0.1)$ are chosen to display the features of the system. Both plots are shown with the curves normalized by T_c. In the latter plot, the effect of κ^2 is to increase the height of these graphs, in spite of the fact that the raw data tends to have lower values of $\langle \mathcal{O}_3\rangle$. This is clearly because the most significant impact of increasing gravitational backreaction is that the critical temperature of the system is lowered. In this case the effect of backreaction is extremely marked, with the condensate varying more widely with α.

3.3. Conductivity

In [11], Horowitz and Roberts observed an interesting phenomenon for the conductivity of the boundary theory. They considered the model (1) in the probe Einstein limit for a range of different bulk scalar masses, and on computing the conductivity found an apparent universal relation

$$\frac{\omega_g}{T_c} \simeq 8\,, \tag{37}$$

with deviations of less than 8 %.

Conductivity is conventionally expressed as the current density response to an applied electric field:

$$\sigma = \frac{\mathcal{J}}{\mathcal{E}}\,. \tag{38}$$

The boundary four-current J_μ is dual to the bulk field A_μ, thus we must consider perturbations of A_μ to compute the conductivity. The full perturbation problem must take into account perturbations of the metric (h_{ti} and h_{ri}) and the perturbation of the Gauss-Bonnet tensor H_{ti}, H_{ri}. After some algebra, these reduce to:

$$\dot{h}'_{ti} - \frac{2}{r}\dot{h}_{ti} - \ddot{h}_{ri} + \frac{L^2 f e^{2\nu}}{r^2 - 2\alpha f}\left(1 - \frac{\alpha(2\nu' f + f')}{r}\right)\Delta h_{ri} + \frac{2\kappa^2 r^2 \dot{A}_i \phi'}{r^2 - 2\alpha f} = 0 \quad (39)$$

$$\frac{e^{-\nu}}{rf}\left[rfe^{\nu}A_i'\right]' - \frac{\ddot{A}_i}{f^2 e^{2\nu}} + \frac{L^2}{r^2 f}\Delta A_i - \frac{2}{f}q^2\psi^2 A_i + \frac{\phi'}{fe^{2\nu}}\left(\dot{h}'_{ti} - \frac{2}{r}\dot{h}_{ti} - \dot{h}_{ri}\right) = 0 \quad (40)$$

where h_{ab} is the perturbation of the metric tensor, and A_i is the perturbation of the gauge field, which has only spatial components. Writing $A_i(t, r, x^i) = A(r)e^{i\mathbf{k}\cdot\mathbf{x} - i\omega t}e_i$, and setting $\mathbf{k} = \mathbf{0}$, (39) can be integrated and substituted in (40) to obtain:

$$A'' + \left(\frac{f'}{f} + \nu' + \frac{1}{r}\right)A' + \left[\frac{\omega^2}{f^2 e^{2\nu}} - \frac{2}{f}q^2\psi^2 - \frac{2\kappa^2 r^2 \phi'^2}{fe^{2\nu}(r^2 - 2\alpha f)}\right]A = 0\,. \quad (41)$$

This is solved under the physically imposed boundary condition of no outgoing radiation at the horizon:

$$A(r) \sim f(r)^{-i\frac{\omega}{4\pi T_+}}\,, \quad (42)$$

where T_+ is the temperature. In the asymptotic adS region ($r \to \infty$), the general solution takes the form

$$A = a_0 + \frac{a_2}{r^2} + \frac{a_0 L_e^4 \omega^2}{2r^2}\log\frac{r}{L} \quad (43)$$

where a_0 and a_2 are integration constants. Note there is an arbitrariness of scale in the logarithmic term, as pointed out in [11], however, this is related to an arbitrariness in the holographic renormalization process; see Appendix A of [2] for a full computation of the conductivity and discussion of this renormalization scale. After careful computation using the method of Skenderis [23], the conductivity of the EGB system is found to be [2]:

$$\sigma = \frac{2a_2}{i\omega L_e^4 a_0} + \frac{i\omega}{2} - i\omega\log\left(\frac{L_e}{L}\right)\,. \quad (44)$$

Note that the imaginary term linear in ω has an arbitrariness of scale from the counterterm subtraction, and this is capitalized in the presentation of numerical data, where a suitable renormalization scale is chosen to make the features of the plot clearest.

Figure 4 shows the real and imaginary parts of the conductivity as a function of ω/T_c, calculated for the scalar mass $m^2 = -3/L^2$ in the probe limit and for a selection of values of α/L^2. These plots clearly show that as α increases, the normalised gap frequency shifts to higher values. An additional feature is that the gap becomes softer with increasing α. These plots should be contrasted with figure 5, which shows similar conductivity plots, but now for scalar mass $m^2 = -3/L_e^2$, and with both no backreaction and a backreaction of $\kappa^2 = 0.05$. Three sample values of α are shown: 0, 0.125, and 0.25. In these plots, the gap if anything gets harder with increasing α, however once backreaction is included, the gap becomes more gentle and extended, and while the dip in $\mathrm{I{\scriptsize M}}(\sigma)$ is smoothed, it is still clearly apparent.

In all of these plots, the gap is clearly indicated by a rise in the real part of σ, which coincides with the global minimum of $\mathrm{I{\scriptsize M}}(\sigma)$. As already noted, the imaginary part of (44) is only valid up to a linear term in ω, the size of which is dependent on the renormalization scheme employed (and also on the charge Q). One can therefore tune this linear term in $\mathrm{I{\scriptsize M}}(\sigma)$ to create a finite global minimum if it is not initially present, and indeed the plots in figure 4 differ from those

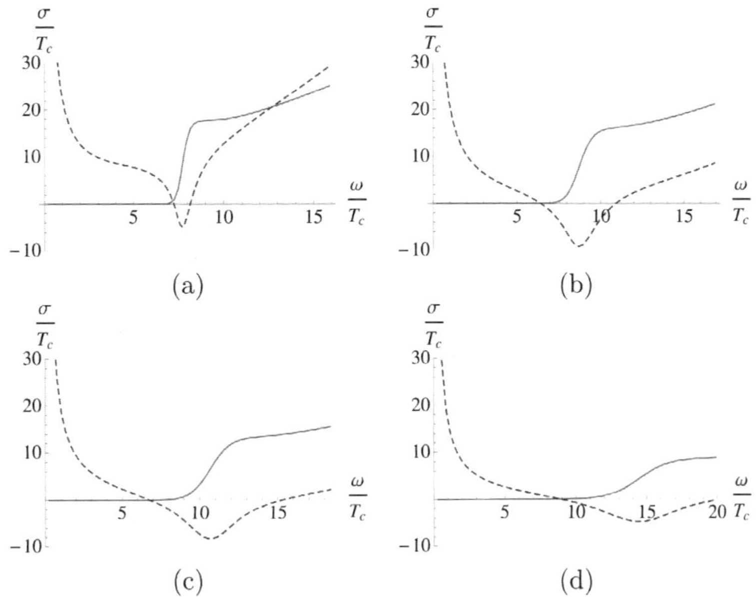

Figure 4. Conductivity: a range of plots showing the real (solid red line) and imaginary (blue dashed line) parts of the (normalized) conductivity as a function of (normalized) frequency for the scalar mass $m^2 = -3/L^2$. Plot (a) is $\alpha = 0$, (b) is $\alpha = 0.1$, and (c) is $\alpha = 0.2$, and (d) is $\alpha = 0.25$.

presented in [1] by precisely this sort of term so as to make the minimum apparent. This minimum can therefore be used to define ω_g, the value of the frequency gap.

The frequency gap is a distinct characteristic of a superconductor and in the BCS theory of superconductivity this frequency gap corresponds to the minimum energy required to break a Cooper pair. As mentioned above, in [11] it was claimed that for the holographic superconductor the relation $\omega_g/T_c \simeq 8$ had a certain universality, proving stable for a range of scalar masses and dimensions. In both [1] and [2], this relation was shown to be unstable to Gauss-Bonnet corrections. Figure 5(d) in particular gives a very clear indicator of how backreaction and higher curvature terms affect the gap. Increasing either α or κ^2 increases ω_g/T_c. For the case of increasing α, the effect occurs mainly because of a shift in the gap, rather than a significant alteration of T_c, which varies much more strongly with backreaction than α. On the other hand, varying κ^2 practically does not alter ω_g at all, whereas T_c drops dramatically, leading to a sharp rise in ω_g/T_c.

4. Summary

This presentation reviewed the work of [1] and [2] on exploring the implications of Gauss-Bonnet corrections to holographic superconductors. The results show that increasing backreaction lowers the critical temperature of the superconductor hence increasing ω_g/T_c. The effect of higher curvature terms is more subtle. Although these initially act in a similar fashion to backreaction in lowering the critical temperature, for significant GB coupling and larger scalar masses, the critical temperature eventually begins to increase. The conductivity gap is also modified, with both ω_g and T_c altering to increase the ratio ω_g/T_c. Clearly higher dimensional holographic superconductors have a rich structure, with or without higher curvature corrections.

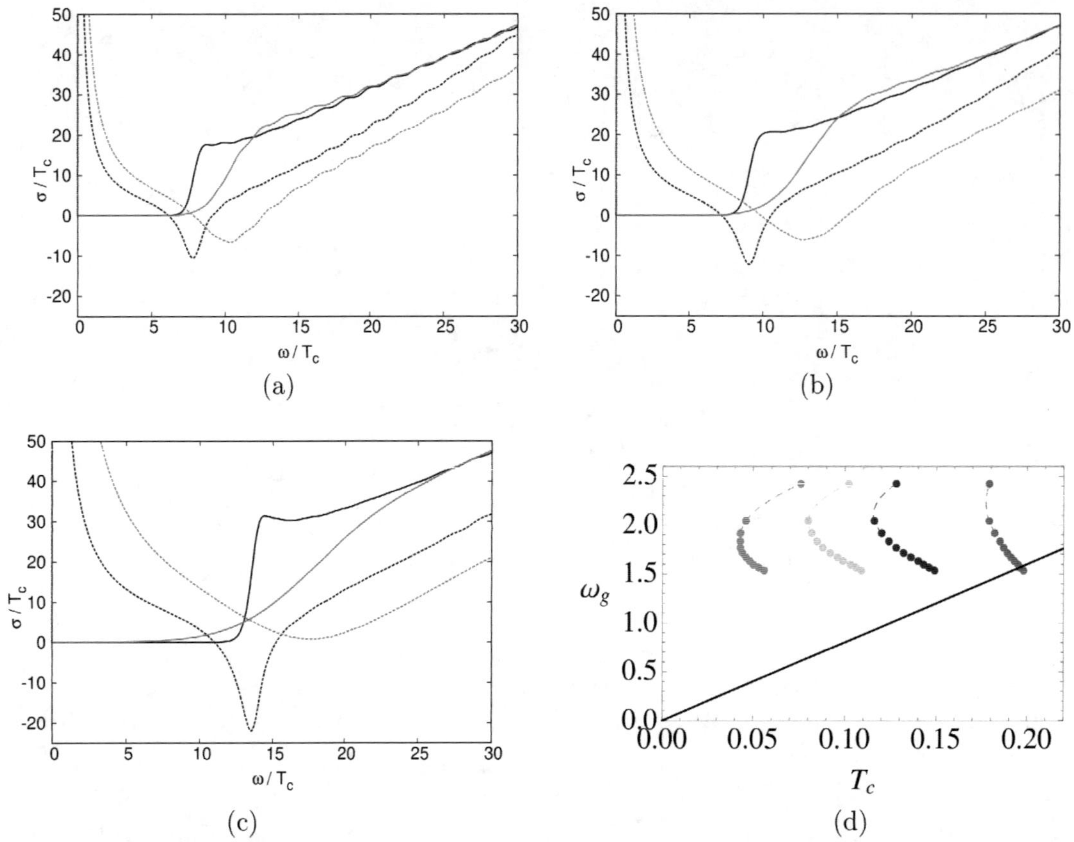

Figure 5. Conductivity: a range of plots showing the conductivity of the $m^2 = -3/L_e^2$ superconductor. Plots (a) – (c) show the real (solid line) and imaginary (dashed line) parts of the conductivity as a function of frequency for $\alpha = 0, 0.125$, and 0.25 respectively. In each case the conductivity is shown for no backreaction in black, and for a backreacting parameter $\kappa^2 = 0.05$ in magenta. The slight undulations in the plots at large ω is a numerical artefact. Plot (d) shows the gap frequency as a function of T_c with the line $\omega_g = 8T_c$ is shown in black. The different colours represent from right to left: Red is $\kappa^2 = 0$, Blue is $\kappa^2 = 0.05$, Green 0.1, and Orange $\kappa^2 = 0.2$. In each case α is incremented from 0 to 0.25. As the gap alters rapidly near the Chern-Simons limit, the dotted lines are added by hand to guide the eye.

Acknowledgments

I would like to thank Luke Barclay, Sugumi Kanno, Jiro Soda and Paul Sutcliffe for being such lively and pleasant collaborators. I would also like to thank Christos Charmousis, Rob Myers and Simon Ross for useful conversations. I acknowledge the support of STFC under the rolling grant ST/G000433/1.

References

[1] R. Gregory, S. Kanno and J. Soda, JHEP **0910**, 010 (2009) [arXiv:0907.3203 [hep-th]].
[2] L. Barclay, R. Gregory, S. Kanno and P. Sutcliffe, "Gauss-Bonnet Holographic Superconductors," arXiv:1009.1991 [hep-th].
[3] C. P. Herzog, J. Phys. A **42**, 343001 (2009) [arXiv:0904.1975 [hep-th]].
[4] S. A. Hartnoll, Class. Quant. Grav. **26**, 224002 (2009) [arXiv:0903.3246 [hep-th]].

[5] G. T. Horowitz, "Introduction to Holographic Superconductors," arXiv:1002.1722 [hep-th].
[6] J. M. Maldacena, Adv. Theor. Math. Phys. **2**, 231 (1998) [Int. J. Theor. Phys. **38**, 1113 (1999)] [arXiv:hep-th/9711200].
[7] P. Breitenlohner and D. Z. Freedman, Annals Phys. **144**, 249 (1982).
[8] S. S. Gubser, Class. Quant. Grav. **22**, 5121 (2005) [arXiv:hep-th/0505189].
Phys. Rev. D **78**, 065034 (2008) [arXiv:0801.2977 [hep-th]].
[9] J. D. Bekenstein, "Black hole hair: Twenty-five years after," arXiv:gr-qc/9605059.
T. Hertog, Phys. Rev. D **74**, 084008 (2006) [arXiv:gr-qc/0608075].
[10] S. A. Hartnoll, C. P. Herzog and G. T. Horowitz, Phys. Rev. Lett. **101**, 031601 (2008) [arXiv:0803.3295 [hep-th]].
S. A. Hartnoll, C. P. Herzog and G. T. Horowitz, JHEP **0812**, 015 (2008) [arXiv:0810.1563 [hep-th]].
[11] G. T. Horowitz and M. M. Roberts, Phys. Rev. D **78**, 126008 (2008) [arXiv:0810.1077 [hep-th]].
[12] S. S. Gubser and A. Nellore, Phys. Rev. D **80**, 105007 (2009) [arXiv:0908.1972 [hep-th]].
G. T. Horowitz and M. M. Roberts, JHEP **0911**, 015 (2009) [arXiv:0908.3677 [hep-th]].
R. A. Konoplya and A. Zhidenko, Phys. Lett. B **686**, 199 (2010) [arXiv:0909.2138 [hep-th]].
[13] S. S. Gubser and S. S. Pufu, JHEP **0811**, 033 (2008) [arXiv:0805.2960 [hep-th]].
K. Peeters, J. Powell and M. Zamaklar, JHEP **0909**, 101 (2009) [arXiv:0907.1508 [hep-th]].
[14] K. Maeda, M. Natsuume and T. Okamura, Phys. Rev. D **79**, 126004 (2009) [arXiv:0904.1914 [hep-th]].
[15] E. Nakano and W. Y. Wen, Phys. Rev. D **78**, 046004 (2008) [arXiv:0804.3180 [hep-th]].
T. Albash and C. V. Johnson, JHEP **0809**, 121 (2008) [arXiv:0804.3466 [hep-th]].
[16] T. Takahashi and J. Soda, Phys. Rev. D **79**, 104025 (2009) [arXiv:0902.2921 [gr-qc]].
S. Kanno and J. Soda, Phys. Rev. D **82**, 086003 (2010) [arXiv:1007.5002 [hep-th]].
[17] S. S. Gubser, C. P. Herzog, S. S. Pufu and T. Tesileanu, Phys. Rev. Lett. **103**, 141601 (2009) [[arXiv:0907.3510 [hep-th]].
J. P. Gauntlett, J. Sonner and T. Wiseman, Phys. Rev. Lett. **103**, 151601 (2009) [arXiv:0907.3796 [hep-th]].
[18] D. Lovelock, J. Math. Phys. **12**, 498 (1971).
[19] Q. Pan, B. Wang, E. Papantonopoulos, J. Oliveira and A. B. Pavan, Phys. Rev. D **81**, 106007 (2010) [arXiv:0912.2475 [hep-th]].
X. H. Ge, B. Wang, S. F. Wu and G. H. Yang, JHEP **1008**, 108 (2010) [arXiv:1002.4901 [hep-th]].
Y. Brihaye and B. Hartmann, Phys. Rev. D **81**, 126008 (2010) [arXiv:1003.5130 [hep-th]].
Q. Pan and B. Wang, Phys. Lett. B **693**, 159 (2010) [arXiv:1005.4743 [hep-th]].
R. G. Cai, Z. Y. Nie and H. Q. Zhang, Phys. Rev. D **82**, 066007 (2010) [arXiv:1007.3321 [hep-th]].
M. Siani, "Holographic Superconductors and Higher Curvature Corrections," arXiv:1010.0700 [hep-th].
[20] Luke Barclay, in preparation.
[21] D. G. Boulware and S. Deser, Phys. Rev. Lett. **55**, 2656 (1985).
R. G. Cai, Phys. Rev. D **65**, 084014 (2002) [arXiv:hep-th/0109133].
[22] L. F. Abbott and S. Deser, Nucl. Phys. B **195**, 76 (1982).
S. Deser and B. Tekin, Phys. Rev. D **67**, 084009 (2003) [arXiv:hep-th/0212292].
[23] K. Skenderis, Class. Quant. Grav. **19**, 5849 (2002) [arXiv:hep-th/0209067].

Scalar-tensor cosmologies with a potential in the general relativity limit

Laur Järv,[1] Piret Kuusk[1] and Margus Saal[2]

[1] Institute of Physics, University of Tartu, Riia 142, EE-51014 Tartu, Estonia
[2] Tartu Observatory, EE-61602 Tõravere, Tartumaa, Estonia

E-mail: laur.jarv@ut.ee

Abstract. We consider Friedmann-Lemaître-Robertson-Walker flat cosmological models in the framework of general Jordan frame scalar-tensor theories of gravity with arbitrary coupling functions, in the era when the energy density of the scalar potential dominates over the energy density of ordinary matter. To study the regime suggested by the local weak field tests (i.e. close to the so-called limit of general relativity) we propose a nonlinear approximation scheme, solve for the phase trajectories, and provide a complete classification of possible solutions. We argue that the topology of phase trajectories in the nonlinear approximation is representative of those of the full system, and thus can tell for which scalar-tensor models general relativity functions as an attractor. To the classes of models which asymptotically approach general relativity we give the solutions also in cosmological time and conclude with some observational implications.

1. Introduction

One of the most outstanding problems for contemporary physics is how to explain the dynamics behind the current accelerated expansion of the universe. A plethora of various approaches have been proposed, most of which can be subsumed under the categories of dark energy (extraordinary matter) or dark gravity (modifications to Einstein's general relativity). As new astrophysical and cosmological precision data is continuously becoming available, it remains a cumbersome task to systematically sift out the models which can meet all observational requirements. What one would like to have are reliable model independent tools for testing gravity at large scales, perhaps something akin to the parametrized post-Newtonian (PPN) formalism for weak fields.

While there has been some progress in designing such tools (the post-Friedmannian formalism), here we would like to introduce a bit more modest approach in scope, yet with a congruent aim. First we note that many proposed modified gravity theories can be cast in the form of scalar-tensor gravity (STG) – higher dimensional theories, braneworld models, $f(R)$ type gravities, varying speed of light theories, etc. Second, wide classes of STG cosmologies dynamically converge to the limit of general relativity favored by local weak field experiments, the so-called "attractor mechanism" [1]. Therefore we can use generic STG as a template or paradigm which implicitly incorporates many different theories, in order to confront them with various observations and see how good they are in comparison with each other. In principle this should provide an instrument of selection within the zoo of modified gravities.

In the present paper we review our recent work towards this direction [2, 3, 4, 5]. After setting the notation and describing the local weak field constraints in Section 2, we write the scalar-tensor cosmology as a dynamical system and briefly outline the features of the phase space in Section 3, followed by the discussion of the fixed points in Section 4. The fixed point corresponding to limit of general relativity needs a more careful consideration, facilitated by the approximation scheme of Section 5. This leads to approximate equations, which are used to find and classify the phase portraits in Section 6 and to find the solutions in time in Section 7. Finally, Section 8 provides a brief summary and outlook.

2. Scalar-tensor gravity and local weak field constraints

We start with a generic Jordan frame STG using the "Brans-Dicke" parametrization,

$$S = \frac{1}{2\kappa^2}\int d^4x\sqrt{-g}\left[\Psi R[g_{\mu\nu}] - \frac{\omega(\Psi)}{\Psi}\nabla^\rho\Psi\nabla_\rho\Psi - 2\kappa^2 V(\Psi)\right] + S_m\,. \tag{1}$$

In fact, the action above describes a family of theories, each pair of the functions $\omega(\Psi)$ and $V(\Psi)$ specifies a particular STG. The non-minimal coupling between curvature and the scalar field Ψ means variable gravitational "constant", $8\pi G = \frac{\kappa^2}{\Psi}$, and it makes sense to assume $0 < \Psi < \infty$. To ensure that the energy density of the scalar field is positive, one also needs $2\omega(\Psi) + 3 \geq 0$ and $V(\Psi) \geq 0$. The term S_m gives the usual matter part of the action.

Local weak field experiments (e.g. observations in the Solar System) reckoned in terms of the PPN formalism, impose rather severe constraints on the theory. If $V(\Psi)$ can be neglected, then

$$8\pi G = \frac{\kappa^2}{\Psi}\frac{2\omega+4}{2\omega+3}\,, \tag{2}$$

$$\beta - 1 = \frac{\kappa^2}{G}\frac{\frac{d\omega}{d\Psi}}{(2\omega+3)^2(2\omega+4)} \lesssim 10^{-4}\,, \tag{3}$$

$$\gamma - 1 = -\frac{1}{\omega+2} \lesssim 10^{-5}\,, \tag{4}$$

$$\frac{\dot G}{G} = -\dot\Psi\frac{2\omega+3}{2\omega+4}\left(G + \frac{2\frac{d\omega}{d\Psi}}{(2\omega+3)^2}\right) \lesssim 10^{-13}\ \mathrm{yr}^{-1}\,. \tag{5}$$

These constraints are satisfied by STG in the "Nordtvedt limit" [6]

$$\frac{1}{2\omega+3} \to 0\,, \qquad \frac{\frac{d\omega}{d\Psi}}{(2\omega+3)^3} \to 0\,, \tag{6}$$

in which case the predictions of STG and GR coincide and we may call (6) also "the limit of general relativity". (If $V(\Psi)$ gives a contribution, then the PPN parameters get a correction [7, 8].)

3. Scalar-tensor cosmology as a dynamical system

The field equations for flat ($k = 0$) Fridmann-Lemaître-Robertson-Walker spacetime filled with barotropic matter fluid $p = \mathrm{w}\rho$ read

$$H^2 = -H\frac{\dot\Psi}{\Psi} + \frac{1}{6}\frac{\dot\Psi^2}{\Psi^2}\,\omega(\Psi) + \frac{\kappa^2}{3}\frac{\rho}{\Psi} + \frac{\kappa^2}{3}\frac{V(\Psi)}{\Psi}\,, \tag{7}$$

$$2\dot H + 3H^2 = -2H\frac{\dot\Psi}{\Psi} - \frac{1}{2}\frac{\dot\Psi^2}{\Psi^2}\,\omega(\Psi) - \frac{\ddot\Psi}{\Psi} - \frac{\kappa^2}{\Psi}\mathrm{w}\rho + \frac{\kappa^2}{\Psi}\,V(\Psi)\,, \tag{8}$$

$$\ddot\Psi = -3H\dot\Psi - \frac{1}{2\omega(\Psi)+3}\,\frac{d\omega(\Psi)}{d\Psi}\,\dot\Psi^2 + \frac{\kappa^2}{2\omega(\Psi)+3}\,(1-3\mathrm{w})\,\rho + \frac{2\kappa^2}{2\omega(\Psi)+3}\left[2V(\Psi) - \Psi\,\frac{dV(\Psi)}{d\Psi}\right], \tag{9}$$

$$\dot\rho = -3H\,(\mathrm{w}+1)\,\rho\,, \tag{10}$$

where H is the Hubble parameter. Denoting $\Pi = \dot\Psi$ the equations (7)-(10) can be rewritten as a dynamical system

$$\dot\Psi = \Pi\,, \tag{11}$$

$$\dot\Pi = -\frac{1}{2\omega(\Psi)+3}\left[\frac{d\omega(\Psi)}{d\Psi}\Pi^2 - \kappa^2\,(1-3\mathrm{w})\,\rho + 2\kappa^2\left(\frac{dV(\Psi)}{d\Psi}\Psi - 2V(\Psi)\right)\right] - 3H\Pi\,, \tag{12}$$

$$\dot H = \frac{1}{2\Psi(2\omega(\Psi)+3)}\left[\frac{d\omega(\Psi)}{d\Psi}\Pi^2 - \kappa^2\,(1-3\mathrm{w})\,\rho + 2\kappa^2\left(\frac{dV(\Psi)}{d\Psi}\Psi - 2V(\Psi)\right)\right] - \frac{1}{2\Psi}\left[6H^2\Psi + 2H\Pi - \kappa^2(1-\mathrm{w})\rho - 2\kappa^2 V(\Psi)\right], \tag{13}$$

$$\dot\rho = -3H(1+\mathrm{w})\rho\,, \tag{14}$$

while the vector $(\dot\Psi, \dot\Pi, \dot H, \dot\rho)$ gives a tangent to the trajectories (solutions) in the phase space $\{\Psi, \Pi, H, \rho\}$. The trajectories lie on the 3-surface

$$H = -\frac{\Pi}{2\Psi} \pm \sqrt{(2\omega(\Psi)+3)\,\frac{\Pi^2}{12\Psi^2} + \frac{\kappa^2(\rho + V(\Psi))}{3\Psi}}\,. \tag{15}$$

By carefully inspecting Eq. (11)-(14) we can elicit the following generic information about the boundaries in the phase space [2, 3]:

- $|H| \to \infty$, $|\rho| \to \infty$, or $|\dot\Psi| \to \infty$ imply a spacetime curvature singularity;
- $\Psi \to 0$ generally also leads to a singularity (solutions can not slip from "attractive" to "repulsive" gravity);
- $\Psi \to \infty$ is not a singularity, but gravitational "constant" $\frac{\kappa^2}{\Psi}$ vanishes;
- $V \to \infty$ or $2\omega + 3 \to 0$ is again a singularity, solutions can not safely pass to the region where Ψ turns to a ghost ($2\omega + 3 < 0$);
- $\frac{1}{2\omega+3} \to 0$ turns out to be a singularity as well, unless $\dot\Psi = \Pi \to 0$.

4. Fixed points

If on cosmological scales the potential dominates over matter density ($V \not\equiv 0$, $\rho \equiv 0$) we can eliminate H using (15) and obtain a 2-dimensional system:

$$\dot{\Psi} = \Pi \tag{16}$$

$$\begin{aligned}\dot{\Pi} &= \left(\frac{3}{2\Psi} - \frac{1}{2\omega(\Psi)+3}\frac{d\omega}{d\Psi}\right)\Pi^2 + \frac{2\kappa^2}{2\omega(\Psi)+3}\left(2V(\Psi) - \Psi\frac{dV}{d\Psi}\right) \\ &\mp \frac{1}{2\Psi}\sqrt{3(2\omega(\Psi)+3)\Pi^2 + 12\kappa^2\Psi V(\Psi)}\,\Pi\,.\end{aligned} \tag{17}$$

The fixed points ($\dot{\Psi} = 0, \dot{\Pi} = 0$) are of two types, given by [2]:

$$\Psi_\bullet \; : \; \left.\frac{dV}{d\Psi}\right|_{\Psi_\bullet}\Psi_\bullet - 2V(\Psi_\bullet) = 0\,, \tag{18}$$

$$\Psi_\star \; : \; \frac{1}{2\omega(\Psi_\star)+3} = 0\,, \quad \frac{1}{(2\omega(\Psi_\star)+3)^2}\left.\frac{d\omega}{d\Psi}\right|_{\Psi=\Psi_\star} \neq 0. \tag{19}$$

(Strictly speaking, the latter case assumes $(2\omega(\Psi_\star)+3)\Pi_\star^2 = 0$, see below Section 5 for an improved approach.) The properties of the fixed points (stable, unstable) and the form of the solutions around the fixed points are determined by the eigenvalues, and these by the values $\omega(\Psi_{\bullet,\star})$, $\frac{d\omega}{d\Psi}|_{\Psi_{\bullet,\star}}$, $V(\Psi_{\bullet,\star})$, $\frac{dV}{d\Psi}|_{\Psi_{\bullet,\star}}$, $\frac{d^2V}{d\Psi^2}|_{\Psi_{\bullet,\star}}$. Notice that $\Psi_\star$ is compatible with the "limit of general relativity", i.e. the local weak field experiments (the matter density in the Solar System is still much higher than the energy density of the potential, even if on the cosmological scales the potential may dominate). On the other hand, the $\Psi_\bullet$ point is generally not in accord with the weak field observations, although in some special situations it may perhaps be possible salvage it by taking into account the contribution of the potential to the PPN paramaters.

If matter density dominates over potential ($\rho \not\equiv 0$, $V \equiv 0$) we can use a new time variable $dp \equiv \left|H + \frac{\dot{\Psi}}{2\Psi}\right| dt$ to eliminate H and obtain

$$\Psi' = \Upsilon \tag{20}$$

$$\Upsilon' = \pm\frac{2\omega(\Psi)+3}{8\Psi^2}\Upsilon^3 + \frac{6\omega(\Psi)+9-4\Psi\frac{d\omega(\Psi)}{d\Psi}}{4\Psi(2\omega(\Psi)+3)}\Upsilon^2 \mp \frac{3}{2}\Upsilon + \frac{3\Psi}{2\omega(\Psi)+3}\,. \tag{21}$$

The fixed point ($\Psi' = 0, \Upsilon' = 0$) in p-time corresponds to a fixed point ($\dot{\Psi} = 0, \dot{\Pi} = 0$) in t-time, and is given by [2]:

$$\Psi_\star \; : \; \frac{1}{2\omega(\Psi_\star)+3} = 0\,, \quad \frac{1}{(2\omega(\Psi_\star)+3)^2}\left.\frac{d\omega}{d\Psi}\right|_{\Psi=\Psi_\star} \neq 0. \tag{22}$$

Again it is good to see that the fixed point turns out to be compatible with the "limit of general relativity".

5. Approximation scheme for the "limit of general relativity"

In the GR limit $(\Psi_\star, \Pi_\star)$, i.e. (a) $\frac{1}{2\omega(\Psi)+3} \to 0$, (b) $\dot{\Psi} \equiv \Pi \to 0$ the equations contain a $\frac{0}{0}$ type indeterminacy (like $\frac{y}{x}$ at the origin), thus demanding a more careful analysis. Let us focus around the vicinity of this point, $\Psi = \Psi_\star + x$, $\Pi = \Pi_\star + y = y$, and expand in series (x and y being first order small):

$$\frac{1}{2\omega(\Psi)+3} = \frac{1}{2\omega(\Psi_\star)+3} + A_\star x + \ldots \approx A_\star x\,, \tag{23}$$

$$(2\omega(\Psi)+3)\Pi^2 = \frac{y^2}{0 + A_\star x + \ldots} = \frac{y^2}{A_\star x}(1 + O(x)) \approx \frac{y^2}{A_\star x}\,, \tag{24}$$

where (c) $A_\star \equiv \frac{d}{d\Psi}\left(\frac{1}{2\omega(\Psi)+3}\right)\Big|_{\Psi_\star} \neq 0$ ja (d) $\frac{1}{2\omega(\Psi)+3}$ is differentiable at $\Psi_\star$. In the following we will focus on the case when the matter density can be neglected in favor of the potential (i.e take $V \not\equiv 0$, $\rho \equiv 0$). Keeping only the terms which are of first order in x and y, the dynamical system (16), (17) becomes [4]

$$\dot{x} = y\,, \tag{25}$$

$$\dot{y} = \frac{y^2}{2x} - C_1\, y + C_2\, x\,, \tag{26}$$

where the constants

$$C_1 \equiv \pm\sqrt{\frac{3\kappa^2 V(\Psi_\star)}{\Psi_\star}}\,, \qquad C_2 \equiv 2\kappa^2 A_\star \left(2V(\Psi) - \frac{dV(\Psi)}{d\Psi}\Psi\right)\Big|_{\Psi_\star}\,, \tag{27}$$

encode the behavior of the functions ω and V near the GR point. In a similar manner the PPN parameters (3), (4) and $\frac{\dot{G}}{G}$ (5) are approximated by

$$\beta - 1 = -\frac{1}{2}A_\star^2 \Psi_\star x\,, \tag{28}$$

$$\gamma - 1 = -2A_\star x\,, \tag{29}$$

$$\frac{\dot{G}}{G} = -\frac{1 - A_\star\Psi_\star}{\Psi_\star}\,\dot{x}\,. \tag{30}$$

Via the Friedmann equation (15) we can also express $H(x(t))$, $\dot{H}(x(t))$ and get the equation of state parameter as

$$\mathrm{w} = -1 - \frac{2\dot{H}}{3H^2} = -1 + \frac{1}{C_1^2\Psi_\star}\left[\frac{3}{2}\left(1 + \frac{1}{\Psi_\star A_\star}\right)\frac{\dot{x}^2}{x} - 4C_1\dot{x} + 3C_2 x\right]\,. \tag{31}$$

6. Phase space trajectories

The phase trajectories for the nonlinear approximate system (25), (26) are determined by the equation

$$\frac{dy}{dx} = \frac{y}{2x} - C_1 + \frac{x}{y}\,C_2\,. \tag{32}$$

Its solutions depend on the sign of $C_1^2 + 2C_2 \equiv C$ [4]:

$$|x|K = \left|\frac{1}{2}y^2 + C_1 yx - C_2 x^2\right| \exp(-C_1 f(u))\,, \quad u \equiv \frac{y}{x}\,, \tag{33}$$

$$\begin{aligned} f(u) &= \frac{1}{\sqrt{C}}\ln\left|\frac{u + C_1 - \sqrt{C}}{u + C_1 + \sqrt{C}}\right| && \text{if } C > 0\,,\\ &= -\frac{2}{u + C_1} && \text{if } C = 0\,,\\ &= \frac{2}{\sqrt{|C|}}\left(\arctan\frac{u + C_1}{\sqrt{|C|}} + n\pi\right) && \text{if } C < 0\,. \end{aligned} \tag{34}$$

The phase portraits which emerge for the various values of the constants are classified in Table 1 and depicted on Figure 1. As the approximate system is non-linear the phase portraits split

Table 1. The topology of trajectories for the approximate nonlinear system (32).

	No.	Parameters		Topology of trajectories
$C>0$	1.a	$C_1>0$	$C_2>0$	2 hyperb., 2 st. & 2 unst. parab. sectors
	1.b	$C_1>0$	$C_2=0$	1 stable & 1 unstable parabolic sector, 2 stable sectors of degenerate fixed points
	1.c	$C_1>0$	$-\frac{C_1^2}{2}<C_2<0$	2 elliptic, 4 stable parabolic sectors
	1.d	$C_1=0$	$C_2>0$	2 hyperb., 2 st. & 2 unst. parab. sectors
	1.e	$C_1<0$	$C_2>0$	2 hyperb., 2 st. & 2 unst. parab. sectors
	1.f	$C_1<0$	$C_2=0$	1 stable & 1 unstable parabolic sector, 2 unst. sectors of degenerate fixed points
	1.g	$C_1<0$	$-\frac{C_1^2}{2}<C_2<0$	2 elliptic, 4 unstable parabolic sectors
$C=0$	2.a	$C_1>0$	$C_2=-\frac{C_1^2}{2}$	2 elliptic, 2 stable parabolic sectors
	2.b	$C_1=0$	$C_2=0$	2 stable & 2 unstable parabolic sectors
	2.c	$C_1<0$	$C_2=-\frac{C_1^2}{2}$	2 elliptic, 2 unstable parabolic sectors
$C<0$	3.a	$C_1>0$	$C_2<-\frac{C_1^2}{2}$	2 elliptic sectors
	3.b	$C_1=0$	$C_2<0$	2 elliptic sectors
	3.c	$C_1<0$	$C_2<-\frac{C_1^2}{2}$	2 elliptic sectors

into sectors, which can be hyperbolic, elliptic, or parabolic; stable or unstable. Typically there are many trajectories passing through the GR point either once or multiple times.

By comparing the tangents of the trajectories in full system (16)-(17) with the ones of the nonlinear approximate system (25)-(26), we can argue [4] that the topology of trajectories in the nonlinear approximation is representative of those of the full system. Therefore one should take the results of the nonlinear approximation seriously. In the end, the classification of the phase portraits reveals that only if $C_1>0$ and $C_2<0$, i.e.

$$\frac{d}{d\Psi}\left(\frac{1}{2\omega(\Psi)+3}\right)\Bigg|_{\Psi_\star}\left(2V(\Psi)-\frac{dV(\Psi)}{d\Psi}\Psi\right)\Bigg|_{\Psi_\star}<0 \tag{35}$$

does the GR point function as an asymptotic attractor for the flow of all trajectories in the vicinity.

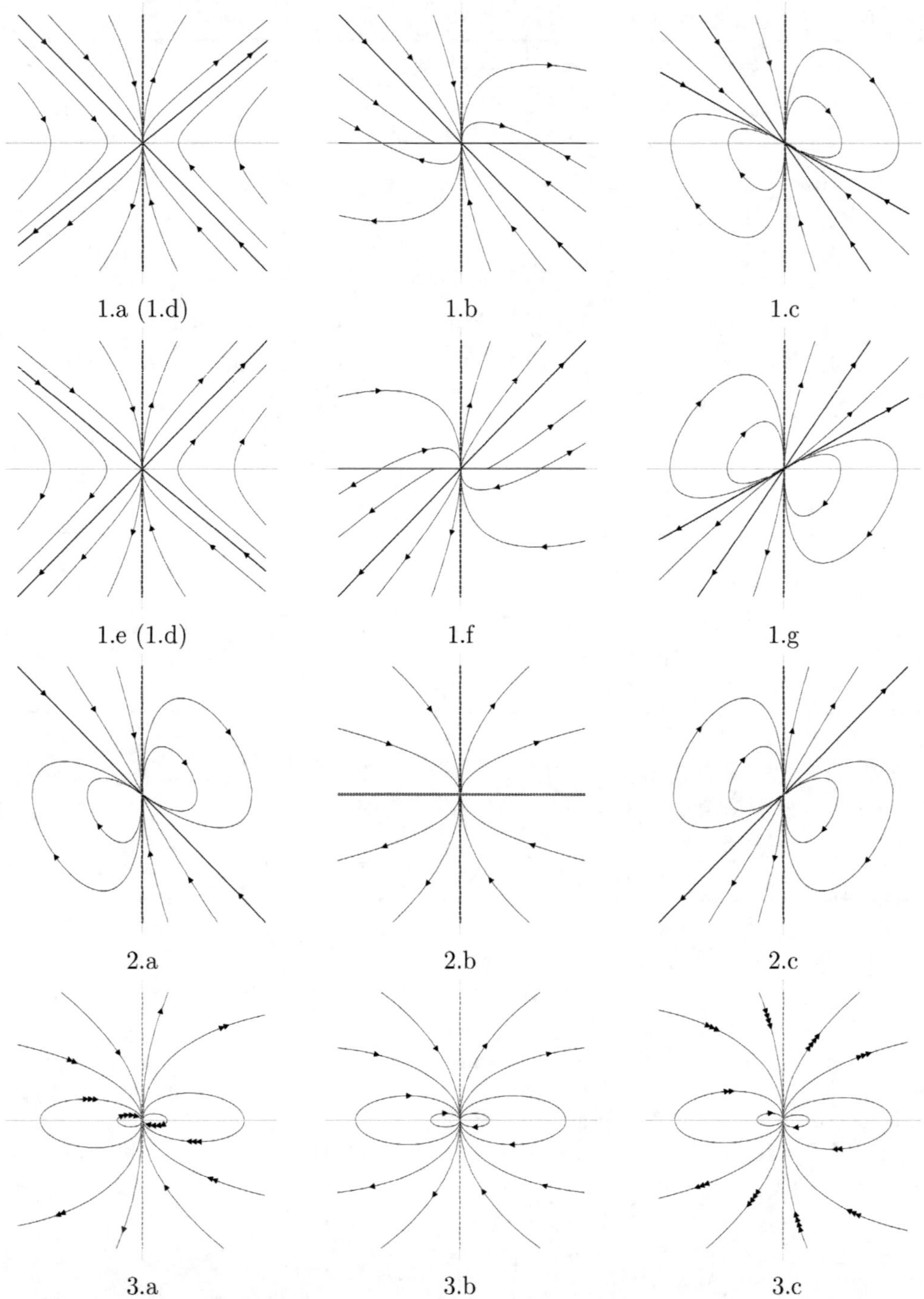

Figure 1. Phase portraits of the nonlinear approximation (32) near the GR point. (Axes: $x = \Psi - \Psi_\star$ horizontal and $y = \dot{\Psi}$ vertical.)

7. Time evolution

The approximate system (25)-(26) can be also combined into a second order equation

$$\ddot{x} + C_1 \dot{x} - C_2 x = \frac{\dot{x}^2}{2x}, \tag{36}$$

yielding solutions in terms of the cosmological time [5]:

$$\pm x = e^{-C_1 t}\left[M_1 e^{\frac{1}{2}t\sqrt{C}} - M_2 e^{-\frac{1}{2}t\sqrt{C}}\right]^2, \quad \text{if} \quad C > 0, \tag{37}$$

$$= e^{-C_1 t}\left[e^{\frac{1}{2}C_1 t_1} t - M_2\right]^2, \quad \text{if} \quad C = 0, \tag{38}$$

$$= e^{-C_1 t}\left[N_1 \sin(\frac{1}{2}t\sqrt{|C|}) - N_2 \cos(\frac{1}{2}t\sqrt{|C|})\right]^2, \quad \text{if} \quad C < 0. \tag{39}$$

where M_1, M_2, t_1, N_1, N_2 are constants of integration (determined by the initial conditions). In particular, the behavior of solutions which approach general relativity can be classified under two characteristic types: (i) exponential or linear exponential convergence, and (ii) damped oscillations around general relativity.

Given the solutions above, one can now plug $x(t)$ into the expressions for the PPN paramaters (28), (29), or $\frac{\dot{G}}{G}$ (30), or w (31) to find how they should evolve in time. The solutions for which GR is an attractor eventually converge to de Sitter, of course. But as they do it, some of them have oscillating, some monotonically evolving w. There are solutions which do not cross the phantom divide (w $= -1$), solutions which cross it only once, and solutions which cross it periodically. The particular behavior is determined by the constants C_1, C_2 and $A_\star$ (and to certain extent also by the initial conditions).

8. Summary and outlook

We have found and characterized the fixed points of STG cosmology in the case when potential dominates over cosmological matter density. In particular we have also found the general analytic form of solutions around the fixed points. This can be applied to cosmological expansion: our results inform whether the solutions of any particular theory have oscillating, phantom crossing etc behavior.

The analysis in the case of matter domination should be refined by carefully dealing with the indeterminacy in the equations, it's a work in progress. Next step, if possible, would be to study the cross-over phase from matter domination to potential domination.

The idea to be pursued is to rely upon the attractor mechanism – instead of scanning the full phase space range of all theories, it makes sense to focus upon the vicinity of certain points which are favored by the cosmological dynamics. Hence arises a selection principle: only those theories and models are viable which possess attractive fixed points around where the solutions satisfy observational constraints.

Acknowledgments

This work was supported by the Estonian Science Foundation Grant No. 7185 and by Estonian Ministry for Education and Science Support Grant No. SF0180013s07. M.S. also acknowledges the Estonian Science Foundation Postdoctoral research Grant No. JD131.

References

[1] Damour T and Nordtvedt K 1993 *Phys. Rev.* D **48** 3436
[2] Järv L, Kuusk P and Saal M 2008 *Phys. Rev.* D **78** 083530 (Preprint 0807.2159 [gr-qc])
[3] Järv L, Kuusk P and Saal M 2010 *Proc. Estonian Acad. Sci.* (in press) (Preprint 0903.4357 [gr-qc])

[4] Järv L, Kuusk P and Saal M 2010 *Phys. Rev.* D **81** 104007 (Preprint 1003.1686 [gr-qc])
[5] Järv L, Kuusk P and Saal M 2010 *Phys. Lett.* B doi:10.1016/j.physletb.2010.09.029 (in press) (Preprint 1006.1246 [gr-qc])
[6] Nordtvedt K 1970 *Astrophys. J.* **161** 1059
[7] Olmo G J 2005 *Phys. Rev.* D **72** 083505
[8] Perivolaropoulos L 2010 *Phys. Rev.* D **81** 047501 (Preprint 0911.3401 [gr-qc])

A conventional form of dark energy

K Kleidis[1,2] and N K Spyrou[1]

[1]Department of Astronomy, Aristoteleion University of Thessaloniki, 541.24 Thessaloniki, Greece

[2]Department of Mechanical Engineering, Technological Education Institute of Serres, 621.24 Serres, Greece

E-mail: kleidis@astro.auth.gr & spyrou@astro.auth.gr

Abstract. Motivated by recent results, indicating that the dark matter (DM) constituents can be *collisional*, we assume that the DM itself possesses also some sort of *thermodynamical properties*. In this case, the Universe matter-content can be treated as a gravitating fluid of positive pressure, and, therefore, together with all the other physical characteristics, *the energy of this fluid's internal motions* should be taken into account as a source of the universal gravitational field. In principle, this form of energy can compensate, also, the *extra (dark) energy*, needed to compromise spatial flatness, while, the post-recombination Universe remains ever-decelerating. What is more interesting, is that, at the same time (i.e., in the context of the collisional-DM approach), the theoretical curve, representing the distance modulus as a function of the cosmological redshift, fits the *Hubble diagram* of an extended sample of SN Ia events quite accurately. However, as we demonstrate, this is not the case for someone who, although living in a Universe filled with collisional DM, insists in adopting the traditional, *collisionless-DM* approach. From the point of view of such an observer, the distant light-emitting sources seem to lie farther (i.e., they appear to be dimmer) than expected, while, the Universe appears to be either accelerating or decelerating, depending on the value of the cosmological redshift.

1. Introduction

The current cosmological picture includes two major unresolved issues: (i) According to the observational data on the temperature-variations of the cosmic microwave background, the Universe can be described, adequately, by a spatially-flat Robertson-Walker (RW) model (see, e.g., [1]) and, therefore, it must contain a considerably-larger amount of energy, than the equivalent to the total rest-mass density of its matter content does (see, e.g., [2]). (ii) The various cosmologically-distant indicators appear to be dimmer than expected [3], [4], something that has been accommodated in the context of a recent phase of accelerated expansion [5], [6].

In order to compromise the above-mentioned observational results within a unified theoretical framework, an extra, dark energy (DE) component, of negative pressure, has been introduced [7], [8]. Other physically-motivated models have also appeared in the literature, including alternative-gravity theories [9], [10], braneworld scenarios [11], [12], Chaplygin gas [13], Cardassian Cosmology [14], theories of compactified internal dimensions [15], mass-varying neutrinos [16], [17], etc. However, each and everyone of these attempts suffers from the so-called *coincidence problem*, i.e., why is the Universe transitioning from deceleration to acceleration so recently; the *inflection point* is being (observationally) set at the cosmological redshift $z_t = 0.46 \pm 0.13$ (see, e.g., [18] and references therein).

The cosmological constant, Λ, could provide a phenomenological description of the DE [19]. In this case, the coincidence problem does not exist, since the *transition redshift*, z_t, depends on the value of Λ, which, consequently, can be determined observationally. The particle-physics vacuum does contribute an effective cosmological constant, but with an energy density 10^{123} times larger than what is observed [20]. This gross mismatch between theory and observation is one of the deepest physical enigmas.

Today, the need for an extra (dark) energy component has been confirmed, also, by other observational data, including galaxy-clusters' dynamics [21], the integrated Sachs-Wolfe effect [22] and baryon acoustic oscillations [23]. These results cry out for a unified theoretical explanation and, in the absence of a (so far) viable one, we should keep our options open, also, to alternative interpretations (see, e.g., [24] - [27]).

Nowadays, a lot of accumulated evidence suggests that, more than 80% (by mass) of the matter in the Universe consists of non-luminous and non-baryonic material [28]. Its name, dark matter (DM), reflects our ignorance on the exact nature of this constituent. Although we do not know for certain how the DM came to be formed, a sizeable relic abundance of weakly-interacting massive particles (WIMPs) is generally expected to have been produced as a by-product of the Universe's hot youth (see, e.g., [29] p. 369). Apart from their exact nature, the scientific community used to argue that, the WIMPs should be collisionless [30]. However, many recent results from high-energy-particle tracers, such as the ATIC [31] and PAMELA [32], combined with those of the Wilkinson microwave anisotropy probe (WMAP) survey [33], have revealed an unusually-high electron - positron production in the Universe, much more than what is anticipated by supernovae explosions or cosmic-ray collisions. These results have led many scientists to argue that, among the best candidate-sources of these high-energy events are the annihilations of WIMPs (see, e.g., [34] - [43], for an extensive, though incomplete list), i.e., that the DM constituents can be collisional [44] - [47]. If this is true, it could affect our perception on the nature of DE (in connection, see, e.g., [48] - [51]).

Indeed, if the (collisional-) DM constituents interact with each other frequently enough, so that their (kinetic) energy can be re-distributed, i.e., the DM itself possesses also some sort of thermodynamical properties, a conventional extra-energy component does exist in the Universe: *It is the energy of the internal motions of the collisional-DM fluid.*

On this basis, it is worth examining the evolution and the dynamical characteristics of a cosmological model, in which (in principle) there is no DE at all. The matter-energy content of this model consists only of two components: The DM (dominant) and the baryonic matter (subdominant), both having the abundances attributed to them by the WMAP survey [2].

Accordingly, we shall demonstrate that these two components are (by themselves) sufficient (i) to reproduce that (today) $\Omega = 1$, (ii) to account for the *observed* dimming of the distant light-emitting sources and (iii) to explain the *apparent* accelerated expansion of the Universe. All the above, simply provided that these two constituents (basically the dark one) form a gravitating perfect fluid of total rest-mass density ρ and a *positive* total pressure, p. In this case, together with all the other physical characteristics, the energy of this fluid's internal motions should (also) be taken into account as a source of the universal gravitational field.

This paper is organized as follows: In Section 2 we consider a cosmological model driven by an ideal fluid of thermodynamically-involved DM with positive pressure. As we find out, in this case, the extra energy (needed to compromise spatial flatness) can be compensated by the energy of the internal motions of this fluid, while, the post-recombination Universe remains ever-decelerating. However, as we demonstrate in Section 3, this is not the case for someone who (although living in a Universe filled with collisional DM) insists in adopting the traditional (collisionless-DM) approach. From the point of view of such an observer, besides the need of an extra-energy component (for confronting the CMB-based observational results), the distant light-emitting sources seem to lie farther (i.e., they appear to be dimmer) than expected, while the Universe appears to be either accelerating or decelerating, depending on the value of the cosmological redshift.

2. Collisional-DM Cosmology

It is generally accepted that, the study of the CMB has been proved a powerful tool in exploring the *post-recombination Universe*. According to the various CMB-oriented observational data, the Universe has emerged out of the radiation epoch as a spatially-flat RW model (see, e.g., [1])

$$ds^2 = S^2(\eta)\left[c^2 d\eta^2 - \left(dx^2 + dy^2 + dz^2\right)\right] , \tag{1}$$

where η is the conformal time and $S(\eta)$ is the scale factor. As a consequence, the value of the Hubble parameter at the present epoch is, by definition, given by

$$H_0^2 = \frac{8\pi G}{3}\rho_c \tag{2}$$

(see, e.g., [52] p. 77), where G is Newton's constant of gravitation and ρ_c is the *critical* rest-mass density for closing the Universe. The evolution of such a model depends on the nature of the source that drives the universal gravitational field, i.e., its matter-energy content.

In determining the Universe matter-energy content, within the context of a collisional-DM cosmology, we assume that, in principle, there is no DE at all. Instead, the DM, together with the small *baryonic contamination* (the latter being tightly bounded to the former), constitute a gravitating perfect fluid (i.e., a fluid that is, practically, homogeneous and isotropic at large scale), which possesses also some sort of thermodynamical properties. In this case, together with all the other physical characteristics, the energy of this fluid's internal motions should be taken into account as a source of the universal gravitational field and, therefore, the post-recombination Universe is no longer driven by *dust*, but, by a fluid of *positive pressure*

$$p = w\rho c^2 , \tag{3}$$

where ρ is the rest-mass density, c is the velocity of light and $0 \leq w = \left(\frac{c_s}{c}\right)^2 \leq 1$ is a dimensionless constant, which measures the square of the *speed of sound*, c_s, in terms of c^2.

The motions of the (luminous and dark) elements in the interior of such a fluid are governed by the equations

$$T^{\mu\nu}_{;\nu} = 0 , \tag{4}$$

where *Greek* indices refer to the four-dimensional space-time (in connection *Latin* indices refer to the three-dimensional spatial section), the *semicolon* denotes covariant derivative, and $T^{\mu\nu}$ is the energy-momentum tensor of the perfect fluid, which takes on the standard form

$$T^{\mu\nu} = (\varepsilon + p)u^\mu u^\nu - p g^{\mu\nu} , \tag{5}$$

where $u^\mu = dx^\mu/ds$ is the four-velocity ($u_\mu u^\mu = 1$) at the position of a fluid's volume element, $g^{\mu\nu}$ are the contravariant components of the Universe metric tensor and ε is the fluid's *total-energy* density, which, in this case, is decomposed as

$$\varepsilon = \rho c^2 + \rho\Pi \tag{6}$$

(see, e.g., [53] pp. 81 - 84 and 90 - 94), where $\rho\Pi$ is *the energy density of the internal motions* of the fluid.

In a maximally-symmetric (i.e., homogeneous and isotropic) cosmological setup, the geodesic motions and the hydrodynamic flows are, practically, *indistinguishable*. Hence, a comoving observer of the cosmic expansion also traces the hydrodynamic flow of the homogeneous cosmic fluid and the *Weyl's postulate* is valid (see, e.g., [54] p. 91). Accordingly, the dynamical evolution of the model (1) is governed by the Friedmann equation (with $\Lambda = 0$) of the classical Friedmann-Robertson-Walker (FRW) cosmology

$$H^2 = \frac{8\pi G}{3c^2}\varepsilon\,, \tag{7}$$

where

$$H = \frac{S'}{S^2} \tag{8}$$

is *the Hubble parameter* in terms of $S(\eta)$ and the *prime* denotes differentiation with respect to η. Nevertheless, inherently, there is an essential difference between our model and the rest of the classical FRW cosmologies: In our case, the basic matter-constituents (although they may look like particles receding from each other) are, in fact, the volume elements of a collisional-DM fluid, i.e., *they possess some sort of internal structure*; and, hence, *thermodynamical content*, as well, and so, the functional form of ε in Eq. (7), is no longer given by $\varepsilon = \rho c^2$, but, by Eq. (6).

On the other hand, in terms of the metric tensor (1), the conservation law $T^{0\nu}_{;\nu} = 0$, yields

$$\varepsilon' + 3\frac{S'}{S}(\varepsilon + p) = 0\,, \tag{9}$$

which, upon consideration of Eqs. (3) and (6), results in

$$\rho = \rho_0\left(\frac{S_0}{S}\right)^3\,, \tag{10}$$

where ρ_0 and S_0 can be considered as denoting the corresponding present-time values. Eq. (10) represents the conservation of the *total mass*, in a cosmological model where matter dominates over radiation, i.e., for every η within the *post-recombination* epoch.

Finally, in a cosmological model filled with collisional (i.e., thermodynamically-involved) DM, the *first law of thermodynamics* for *adiabatic flows*

$$d\Pi + p d\left(\frac{1}{\rho}\right) = 0 \tag{11}$$

(see, e.g., [55]), results in

$$\Pi = \Pi_0 + wc^2 \ln\left(\frac{\rho}{\rho_0}\right)\,, \tag{12}$$

where Π_0 is the energy per unit mass, Π, of this fluid's internal motions, at the present epoch. Accordingly, the total-energy density of the Universe matter-energy content is written in the form

$$\varepsilon = \rho c^2\left[1 + \frac{\Pi_0}{c^2} + w\ln\left(\frac{\rho}{\rho_0}\right)\right]\,. \tag{13}$$

Now, with the aid of Eqs. (2), (10) and (13), Eq. (7) reads

$$\left(\frac{H}{H_0}\right)^2 = \Omega_M\left(\frac{S_0}{S}\right)^3\left[1 + \frac{\Pi_0}{c^2} + 3w\ln\left(\frac{S_0}{S}\right)\right]\,, \tag{14}$$

where $\Omega_M = \frac{\rho}{\rho_c}$ is the rest-mass density parameter. According to Eq. (14), at the present epoch (where $S = S_0$ and $H = H_0$), the value of Π_0 is given by

$$\Pi_0 = \left(\frac{1}{\Omega_M} - 1\right)c^2\,. \tag{15}$$

In view of Eqs. (6) and (15), at the present epoch, the total-energy density parameter results in

$$\Omega = \frac{\varepsilon_0}{\varepsilon_c} = \frac{\rho_0 c^2}{\rho_c c^2} + \frac{\rho_0 \Pi_0}{\rho_c c^2} = \Omega_M + \Omega_M \frac{\Pi_0}{c^2} = 1 \, , \tag{16}$$

i.e., *the extra (dark) energy, needed to flatten the Universe, can be compensated by the energy of the internal motions of a collisional-DM fluid.*

Upon consideration of Eq. (15), Eq. (14) is written in the form

$$\left(\frac{H}{H_0} \right)^2 = \left(\frac{S_0}{S} \right)^3 \left[1 + 3w\Omega_M \ln \left(\frac{S_0}{S} \right) \right] \, . \tag{17}$$

Eq. (17) can become particularly transparent (and useful), if we take into account that, since $0 \leq w \leq 1$ and $\Omega_M \simeq 0.3$ (see, e.g., [2]), the combination $w\Omega_M$ can be quite small, i.e., $w\Omega_M \ll 1$. In this case, to terms linear in $w\Omega_M$, Eq. (17) results in

$$H \simeq H_0 \left(\frac{S_0}{S} \right)^{\frac{3}{2}(1+w\Omega_M)} \tag{18}$$

and can be solved, to determine the scale factor of the collisional-DM model (1), as follows

$$S = S_0 \left(\frac{\eta}{\eta_0} \right)^{\frac{2}{1+3w\Omega_M}} \, , \tag{19}$$

where we have defined the present-time value, η_0, of the conformal time, η, as

$$\eta_0 = \frac{2}{(1+3w\Omega_M) H_0 S_0} \, . \tag{20}$$

For $w \neq 0$, Eq. (19) is the *natural generalization* of the corresponding collisionless-DM model, the well-known Einstein-de Sitter (EdS) Universe ($S \sim \eta^2$) (see, e.g., [52] pp. 77, 83 and 142 - 144).

In the collisional-DM model (1), the *cosmological redshift* parameter is defined as

$$z + 1 = \frac{S_0}{S} \tag{21}$$

and, therefore, Eq. (18) is written in the form

$$H = H_0 (1+z)^{\frac{3}{2}(1+w\Omega_M)} \, . \tag{22}$$

We cannot help but noticing the surprising functional similarity of Eq. (22) to the corresponding result regarding *a dark-energy fluid* with equation of state in the form of Eq. (3) (cf. Eqs. (13) and (14) of [18]). In our case, however, $w \geq 0$ and, therefore, on the approach to $z = 0$, $H(z)$ decreases monotonically. In other words, a cosmological model filled with collisional DM necessarily decelerates its expansion.

This can be readily verified, by expressing the corresponding deceleration parameter, q, in terms of H and z, as

$$q(z) = \frac{dH/dz}{H(z)} (1+z) - 1 \tag{23}$$

(cf. Eq. (16) of [18]), which, in view of Eq. (22), yields

$$q(z) = \frac{1}{2} (1 + 3w\Omega_M) > 0 \, , \tag{24}$$

independently of z, even if $w = 0$. In other words, the model of the gravitating perfect fluid source seems to be inadequate for confronting the *apparent* accelerated expansion of the Universe. The actual reason is that, it does not have to. As we shall demonstrate in the next Section, in a Universe filled with collisional DM, both the observed dimming of the distant light-emitting sources and the apparent accelerating expansion of the Universe, can be due to the misinterpretation of several cosmologically-relevant parameters by those observers who, although living in a collisional-DM Universe, insist in adopting the collisionless-DM approach.

3. Mistreating dark matter as collisionless

At the time the (unexpected) dimming of the SNe Ia standard candles was first discovered, the common perception about the *cosmos* was that the DM constituents are collisionless. The *physical content* of a collisionless-DM Universe (in which both the pressure and the energy of the internal motions are assumed to be negligible and, therefore, disregarded) is entirely different than that of the thermodynamically-involved-DM model (where p, $\Pi \neq 0$). In other words, the dynamical properties of a pressureless Universe are no longer described by $g_{\mu\nu}$, i.e., Eq. (1), but, rather, in terms of *another* metric tensor, $\tilde{g}_{\mu\nu}$, for which, the corresponding (spatially-flat) line-element is written in the form

$$d\tilde{s}^2 = R^2(\eta) \left[c^2 d\eta^2 - \left(dx^2 + dy^2 + dz^2 \right) \right] . \tag{25}$$

Clearly, the evolution of such a model is given in terms of the scale factor $R(\eta)$ and, therefore, from the point of view of an observer who (mis)treats the DM as collisionless, $\tilde{g}_{\mu\nu}$ is the metric tensor upon which he/she should rely on, in interpreting observations.

However, we recall that, for such an observer, the accumulated evidence, in favor of spatial flatness, necessarily leads to the assumption of an extra (dark) energy component, in contrast to the collisional-DM case, where such an assumption is no longer necessary. Indeed, in the latter case, the appropriate candidate for the extra energy needed to flatten the Universe is already included in the model (the energy of the internal motions).

Furthermore, in the context of the collisionless-DM scenario, every effort to account for the (apparent) dimming of the SNe Ia standard candles, naturally, should also be based on $\tilde{g}_{\mu\nu}$ and the cosmologically-relevant parameters arising from it. Accordingly, a possible explanation could be that, recently, the Universe accelerated its expansion [3], [4]. Such an assumption, however, attributes unnecessarily-exotic properties to the extra amount of energy needed to compromise spatial flatness (e.g., it should be repulsive in nature, i.e., of negative pressure, etc.). Therefore, we cannot help but wondering, whether there is another (more conventional) explanation, that could be found (also) within the context of the collisional-DM model.

In fact, in what follows, we shall demonstrate that both the observed dimming of the distant light-emitting sources and the accelerated expansion of the Universe could be only apparent, based on the misinterpretation of several cosmologically-relevant parameters by someone who (although living in a Universe filled with collisional DM) insists in adopting the traditional (collisionless-DM) approach.

In order to explore such a possibility, we note that the collisional-DM treatment of the Universe (in terms of which $p \neq 0$ and the motions of its matter constituents are, in principle, hydrodynamic flows) can be related to the collisionless-DM approach (in terms of which $\tilde{p} = 0$ and the corresponding motions are, necessarily, geodesics) by means of a conformal transformation [56], [57]

$$\tilde{g}_{\mu\nu} = f^2(x^\kappa)\, g_{\mu\nu} \tag{26}$$

(in connection, see also [58], [59] pp. 24 - 29 and 54 - 61, [60]), where, upon consideration of *isentropic flows*, the conformal factor $f(x^\kappa)$ takes on the functional form [56]

$$f(x^\kappa) = \mathcal{C} \left(\frac{\varepsilon + p}{\rho c^2} \right) = \mathcal{C} \left[1 + \frac{1}{c^2} \left(\Pi + \frac{p}{\rho} \right) \right] , \tag{27}$$

where $\mathcal{C}$ is an arbitrary (integration) constant. According to Eq. (27), $f(x^\kappa)$ is essentially the *specific enthalpy* of the ideal fluid under consideration. Recently, Verozub [61] extrapolated these results to include every Riemannian space-time and not just the metric attributed to a bounded, perfect-fluid source. In particular, he showed that the adiabatic hydrodynamic motion of an ideal-fluid element in a space-time with metric tensor $g_{\mu\nu}$, takes place along the geodesic lines of a Riemannian manifold with metric tensor given by the combination of Eqs. (26) and (27).

With the aid of the technique developed by Kleidis and Spyrou [56], we can determine the scale factor of the spatially-flat cosmological model (25), i.e., the scale factor of the Universe as it is realized by someone who, although living in a collisional-DM Universe (where p, $\Pi > 0$ and $\frac{dp}{d\eta} \neq 0$), *mistreats* the DM as collisionless ($\tilde{p} = 0$).

In view of Eq. (26), the scale factor of the Universe as it is realized by a supporter of the collisionless-DM scenario, $R(\eta)$, is related to the corresponding quantity of the collisional-DM model, $S(\eta)$, as follows

$$R(\eta) = f(x^\kappa) S(\eta) \, , \tag{28}$$

where $f(x^\kappa)$ is given, in terms of z, by

$$f(z) = \frac{\mathcal{C}}{\Omega_M} \left(1 + w\Omega_M \left[1 + 3\ln(1+z)\right]\right) \, . \tag{29}$$

In Eq. (29), the arbitrary integration constant, $\mathcal{C}$, can be determined, by demanding that, *in the pressureless case, these two models should coincide*, i.e., $\tilde{g}_{\mu\nu} = g_{\mu\nu}$. In other words, for $w = 0 = p$, $R(\eta) = S(\eta, w = 0)$ - the EdS model; and, hence, $f(w = 0) = 1$.

For $p = 0$, the first law of thermodynamics results in $\Pi = constant = \Pi_0$, with Π_0 (in our case) being given by Eq. (15). Now, according to Eq. (27), the condition $f(w = 0) = 1$ leads to

$$\mathcal{C} = \Omega_M \tag{30}$$

and, therefore, Eq. (29) results in

$$f(z) = 1 + w\Omega_M \left[1 + 3\ln(1+z)\right] \, . \tag{31}$$

Next, using Eqs. (28) and (29), we shall express several cosmologically-relevant parameters of the collisional-DM model, in terms of their collisionless-DM counterparts.

3.1. The cosmological redshift

A supporter of the collisionless-DM scenario would define the corresponding cosmological redshift parameter, $\tilde{z}$, as

$$\tilde{z} + 1 = \frac{R(\eta_0)}{R(\eta)} \, , \tag{32}$$

which, upon consideration of Eq. (28), is written in the form

$$\tilde{z} + 1 = \frac{f(\eta_0)}{f(\eta)} (z+1) \, , \tag{33}$$

where

$$f(\eta_0) = \frac{R(\eta_0)}{S(\eta_0)} = 1 + w\Omega_M \, . \tag{34}$$

Taking into account Eq. (31), Eq. (33) yields

$$\tilde{z} + 1 = \frac{1 + w\Omega_M}{1 + w\Omega_M[1 + 3\ln(1+z)]} (z+1) \, , \tag{35}$$

which, to linear terms in $w\Omega_M$, results in

$$1 + \tilde{z} \simeq (1 + z)^{1-3w\Omega_M} . \tag{36}$$

Notice that, for every (fixed) value of the cosmological redshift z, i.e., as it is defined in the collisional-DM model, the corresponding collisionless-DM quantity $\tilde{z}$ is always a little bit *smaller* ($\tilde{z} < z$). In other words, on observing a light-emitting source of the collisional-DM model, an observer who adopts the collisionless-DM scenario will *"realize"* that this source lies farther (z) than he/she would expect ($\tilde{z}$).

3.2. The luminosity distance and the distance modulus

Nowadays, the most direct and reliable method to determine, observationally, the (relatively) recent history of the Universe expansion, is to measure the redshift and the apparent luminosity (equivalently, the apparent magnitude, m) of cosmologically-distant indicators (standard candles), whose absolute luminosity (equivalently, the absolute magnitude, M) is assumed to be known.

SN Ia events appear to be one of the most suitable cosmological standard candles. With the aid of these events, a number of scientific groups have attempted to find evidence in support of a recently-accelerating stage of the Universe [62] - [66]. In each and everyone of these surveys, the SN Ia events, at *peak luminosity*, appear to be *dimmer* (i.e., they seem to lie farther) than expected. This result was, eventually, accomodated within the context of the *concordance model*, by a DE fluid of negative pressure, with $\Omega_X \sim 0.7$ [67]. However, in view of Eq. (36), there may be another, more conventional, interpretation.

Photons travel along *null geodesics*, $d\tilde{s}^2 = 0 = ds^2$, which remain unaffected by conformal transformations. Accordingly, both in the collisional-DM treatment and in the collisionless-DM approach, the *radial distance* of a light-emitting source (in comoving coordinates) is the same, i.e.,

$$\tilde{r} = c\,(\eta_r - \eta_e) = r , \tag{37}$$

where η_r and η_e are the conformal times of *reception* and *emission* of light, respectively (usually, $\eta_r = \eta_0$).

In this case, with the aid of Eq. (36), the formula determining the luminosity distance in a spatially-flat collisional-DM model,

$$d_L(z) = rS(\eta_0)(1 + z) , \tag{38}$$

can be expressed in terms of the corresponding collisionless-DM quantity,

$$\tilde{d}_L(\tilde{z}) = \tilde{r}R(\eta_0)(1 + \tilde{z}) , \tag{39}$$

as follows

$$\frac{d_L}{\tilde{d}_L} = \frac{1}{1 + w\Omega_M}(1 + z)^{3w\Omega_M} . \tag{40}$$

This relation is very interesting. It suggests that, in a Universe filled with collisional DM (i.e., as long as $w \neq 0$), there exists a *characteristic (transition) value* of the cosmological redshift

$$z_c = (1 + w\Omega_M)^{\frac{1}{3w\Omega_M}} - 1 , \tag{41}$$

such that, the luminosity distance of the various light-emitting sources located at $z > z_c$ *is always larger* than what is realized by a supporter of the collisionless-DM scenario. Therefore, an observer who treats the DM as collisionless, necessarily admits that, any standard candle located at $z > z_c$, lies *farther* than expected ($d_L > \tilde{d}_L$).

The same thing happens, also, in the case of the distance moduli corresponding to d_L and $\tilde{d}_L$. Let us denote by

$$\mu(z) = 5 \log \left(\frac{d_L}{Mpc} \right) + 25 \tag{42}$$

the *K-corrected* distance modulus $(m - M)$ of a light-emitting source (see, e.g., [54], Eqs. (13.10) and (13.12), p. 359) in the collisional-DM model, where d_L is measured in *megaparsecs* (Mpc). In a similar fashion,

$$\tilde{\mu}(\tilde{z}) = 5 \log \left(\frac{\tilde{d}_L}{Mpc} \right) + 25 \tag{43}$$

is the distance modulus of the same source, as it is defined by someone who, although living in the collisional-DM model, insists in adopting the (traditional) collisionless-DM approach. Subtracting Eqs. (42) and (43) by parts, and using Eq. (40), we obtain

$$\Delta\mu = \mu - \tilde{\mu} = 15 w \Omega_M \log(1 + z) - 5 \log (1 + w\Omega_M) \,. \tag{44}$$

According to Eq. (44), any light-emitting source of the collisional-DM Universe located at $z > z_c$, from the point of view of an observer who treats the DM as collisionless, appears to be dimmer than expected, i.e., $\tilde{\mu} < \mu$.

We can not help but noticing, the prominent similarity between the characteristic value z_c and the *transition redshift*, z_t, that signals *the onset of dimming* of the SNe Ia standard candles, which, according to the supporters of the collisionless-DM approach, is interpreted as an entry into a phase of accelerated expansion. For $w\Omega_M = 0.1$, i.e., $w = \frac{1}{3}$ (the DM consists of relativistic particles), the characteristic transition value under consideration is set at $z_c = 0.37$, while, for lower values of w, z_c reaches up to 0.39. Each of these results lies within the observationally-determined range of values of the cosmological redshift, concerning the transition point, which distinguishes between accelerated and decelerated expansion of the Universe [6]. In other words, in a collisional-DM model, an *inflection point* (in the d_L versus z diagram) arises naturally (namely, z_c), without the need to assume a transition from deceleration to acceleration. Certainly, such a model no longer suffers from the *coincidence problem*.

Therefore, if the DM possesses some sort of thermodynamical properties, then, it is possible that: (i) The discrepancy between the *expected* value of the distance modulus ($\tilde{\mu}$) of a SN Ia standard candle and the corresponding *observed* one (μ), as well as (ii) the accompanying inflection point, z_t, that signals the transition from deceleration to acceleration, arise only because many cosmologists (although living in a collisional-DM model) rather insist in adopting the (traditional) collisionless-DM approach. We can readily demonstrate this, by overplotting Eq. (44) in the *Hubble (μ versus z) diagram* of a SN Ia dataset.

There is an extended sample of 192 SN Ia events, which has been used by Davis et al. [68] to scrutinize the viability of various DE scenarios. It is available, either at `http://www.ctio.noao.edu/essence` or at `http://braeburn.pha.jhu.edu/~ariess/R06`.

In order to overplot Eq. (44) on the μ versus z diagram of this dataset, first of all, we need to determine the luminosity distance $\tilde{d}_L(\tilde{z})$ (and, through that, the corresponding modulus, $\tilde{\mu}(\tilde{z})$), that is realized by someone who, although living in a collisional-DM model, insists in adopting the collisionless-DM approach. Such an observer, performs calculations in the (traditional) framework of a pressureless Universe, adopting the corresponding formula of the luminosity distance. In a spatially-flat model, such a formula is given by (see, e.g., [69])

$$\tilde{d}_L(\tilde{z}) = \frac{2c}{\tilde{H}_0} \, (1 + \tilde{z})^{1/2} \left[(1 + \tilde{z})^{1/2} - 1 \right] \,, \tag{45}$$

representing *the luminosity distance in the EdS Universe*, the (conformally) pressureless counterpart of the collisional-DM model (19).

However, we need to stress that, in a collisional-DM Universe, the measured quantity (corresponding to the cosmological redshift) is z and not $\tilde{z}$. Therefore, in order to include, also, the function $\tilde{\mu}(\tilde{z})$ in the Hubble diagram of the SN Ia dataset used by Davis et al. [68], we have to express $\tilde{d}_L(\tilde{z})$ in terms of the *truly measured* quantity, z. It can be done (appropriately) by inserting Eq. (36) into Eq. (45), to obtain

$$\tilde{d}_L(z) = \frac{2c}{(1 - 4w\Omega_M)\, H_0} (1+z)^{\frac{1}{2}(1-3w\Omega_M)} \left[(1+z)^{\frac{1}{2}(1-3w\Omega_M)} - 1\right] . \tag{46}$$

However, this is not the case for a supporter of the collisionless-DM scenario. Indeed, in depicting Eq. (43) - with $\tilde{d}_L(\tilde{z})$ being given by Eq. (45) - on the μ versus z diagram of a sample of SN events, such an observer (admitting that $w = 0$), unavoidably, misinterprets the measured quantity z as $\tilde{z}$ (also, the quantity H_0 is misinterpreted as $\tilde{H}_0$). In other words, the theoretical formula of the luminosity distance that is used by someone who, although living in a (spatially-flat) collisional-DM model, insists in adopting the collisionless-DM approach, rather is (falsely) written in the form

$$\tilde{d}_L(z) = \frac{2c}{H_0} (1+z)^{1/2} \left[(1+z)^{1/2} - 1\right] , \tag{47}$$

instead of that given by Eq. (46). In what follows, we admit that $H_0 = 70.5\ Km/sec/Mpc$ (Komatsu et al. 2009) and hence $2c/H_0 = 8509.8\ Mpc$.

Next, the theoretical curves, corresponding to the distance moduli $\mu(z)$ - for $w\Omega_M = 0.16$ - (green solid line), $\tilde{\mu}(z)$ - (also for $w\Omega_M = 0.16$) with $\tilde{d}_L(z)$ being given by Eq. (46) - (orange solid line) and $\tilde{\mu}(z)$ - with $\tilde{d}_L(z)$ being given by Eq. (47) - (dashed line), are overplotted in the Hubble diagram of the SN Ia dataset used by Davis et al. [68] (Fig. 1). We observe that, the (appropriately translated in terms of z) collisionless-DM quantity $\tilde{\mu}(z)$ - with $\tilde{d}_L(z)$ being given by Eq. (46) - (orange solid line) is quite far from fitting this sample of data, although, for $z \leq 1.75$, it is much closer to the μ versus z distribution of the SN Ia data available, than the falsely used quantity $\tilde{\mu}(z)$ - with $\tilde{d}_L(z)$ being given by Eq. (47) - (dashed line).

The situation changes, completely, when someone takes into account the thermodynamical content of a collisional-DM fluid with $w\Omega_M = 0.16$, thus using Eq. (44), instead of Eq. (43) alone. In this case, the function $\mu(z)$ (green solid line) fits the entire dataset under consideration quite accurately.

Taken together, these results suggest that, if the DM constituents interact with each other frequently enough (so that their energy is re-distributed and, hence, the DM fluid aqcuires some sort of thermodynamical properties), then, what is realized as "dimming" of the SNe Ia standard candles could be only "apparent", provided that the cosmologists no longer insist in adopting the collisionless-DM approach. In other words, in a Universe filled with collisional DM, the unexpected dimming of the distant light-emitting sources can be explained in a more conventional way, than that implemented within the context of the accelerated expansion.

3.3. The Hubble and the deceleration parameters

By virtue of Eq. (28), the Hubble parameter that is realized by a supporter of the collisionless-DM scenario, $\tilde{H}$, is written in terms of H as

$$\tilde{H} = H \frac{d}{dz}\left(\frac{1+z}{f}\right), \tag{48}$$

which, in view of Eqs. (22) and (31), results in

$$\tilde{H} = H_0(1+z)^{\frac{3}{2}(1+w\Omega_M)} \frac{1 - 2w\Omega_M + 3w\Omega_M \ln(1+z)}{(1 + w\Omega_M[1 + 3\ln(1+z)])^2} . \tag{49}$$

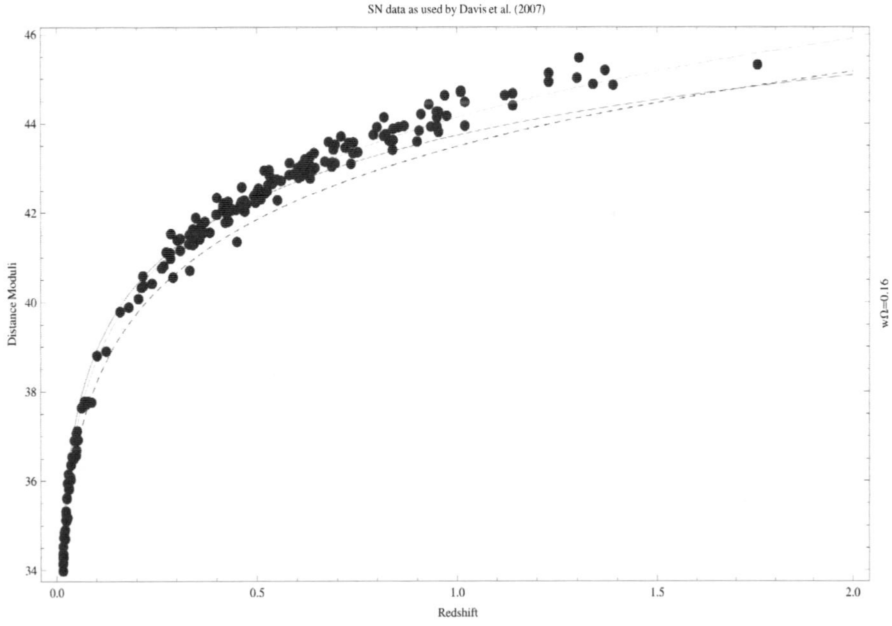

Figure 1. The SNe Ia Hubble diagram of the sample used by Davis et al [68]. Overplotted are the theoretical curves, corresponding to the distance moduli $\mu(z)$ - for $w\Omega_M = 0.16$ - (green solid line), $\tilde{\mu}(z)$ - (also for $w\Omega_M = 0.16$) with $\tilde{d}_L(z)$ being given by Eq. (46) - (orange solid line) and $\tilde{\mu}(z)$ - with $\tilde{d}_L(z)$ being given by Eq. (47) - (dashed line). We observe that, after the thermodynamical content of a collisional-DM fluid is taken into account, the theoretical curve representing the distance modulus, $\mu(z)$ (Eq. (44)), fits the entire dataset quite accurately (green line).

We note that, to terms linear in $w\Omega_M$,

$$\tilde{H}_0 = H_0(1 - 4w\Omega_M) \,, \tag{50}$$

i.e., within the context of the collisionless-DM approach, at the present epoch (where $z = 0$), the Universe expands only as long as $w\Omega_M < 0.25$, and, in any case, at a lower rate than what the collisional-DM treatment implies.

In view of Eq. (36), i.e., at relatively-low values of z, Eq. (49) can be written in terms of $\tilde{z}$, as

$$\begin{aligned} \tilde{H} &= H_0 \, (1+\tilde{z})^{\frac{3(1+w\Omega_M)}{2(1-3w\Omega_M)}} \, (1 - 3w\Omega_M) \\ &\times \frac{1 - 5w\Omega_M + 3w\Omega_M \ln(1+\tilde{z}) + O(w\Omega_M)^2}{[1 - 2w\Omega_M + 3w\Omega_M \ln(1+\tilde{z}) + O(w\Omega_M)^2]^2} \,. \end{aligned} \tag{51}$$

By analogy to Eq. (23), a supporter of the collisionless-DM scenario would define the corresponding deceleration parameter, $\tilde{q}$, as

$$\tilde{q}(\tilde{z}) = \frac{d\tilde{H}/d\tilde{z}}{\tilde{H}(\tilde{z})}(1+\tilde{z}) - 1, \tag{52}$$

which, by virtue of Eq. (51), yields

$$\tilde{q}(\tilde{z}) = \frac{1}{2} \cdot \left[\frac{1 - 4w\Omega_M + 6w\Omega_M \ln(1+\tilde{z}) + O(w\Omega_M)^2}{1 - 10w\Omega_M + 6w\Omega_M \ln(1+\tilde{z}) + O(w\Omega_M)^2} \right] . \tag{53}$$

Now, the condition for accelerated expansion ($\tilde{q} < 0$) is translated to

$$[1 - 4w\Omega_M + 6w\Omega_M \ln(1+\tilde{z})] \cdot [1 - 10w\Omega_M + 6w\Omega_M \ln(1+\tilde{z})] < 0 \, , \tag{54}$$

from which, to terms linear in $w\Omega_M$, we obtain

$$\tilde{q}(\tilde{z}) < 0 \Leftrightarrow 1 - 14w\Omega_M + 12w\Omega_M \ln(1+\tilde{z}) < 0 \, . \tag{55}$$

From Eq. (55) it becomes evident that, from the point of view of someone who insists in adopting the collisionless-DM approach, $\tilde{q}(\tilde{z}) < 0$ at cosmological redshifts

$$\tilde{z} < \tilde{z}_t = e^{\frac{14w\Omega_M - 1}{12w\Omega_M}} - 1. \tag{56}$$

This relation is very interesting: It suggests that, if the Universe matter-content is treated as a collisional-DM fluid with w being larger than a *critical* value, w_c, such that

$$w\Omega_M > w_c\Omega_M = \frac{1}{14} \approx 0.0714 \tag{57}$$

(i.e., $w > w_c \approx 0.238$), then, from the point of view of someone who treats the DM as collisionless, there exists a *transition* value, $\tilde{z}_t$, of $\tilde{z}$, below which, the post-recombination Universe (as being realized by such an observer) is *accelerating*.

In other words, if the Universe evolution is dominated by a collisional-DM fluid with $w > w_c$, then, the apparent acceleration of the cosmic expansion could (very well) be due to a misinterpretation of several cosmologically-relevant parameters, by an observer who (although living in a cosmological model filled with collisional DM) insists in adopting the collisionless-DM approach. At the same time, for such an observer, the cosmologically-distant indicators would appear to be dimmer than expected (cf. Eq. (44)).

We recall here that, the recent observational data concerning the SNe Ia standard candles set the *transition redshift* between accelerated and decelerated expansion at $z_t = 0.46 \pm 0.13$ [6]. In this case, the combination of Eqs. (36) and (56) results in the following non-linear *algebraic* equation, involving the transition value, z_t, of the *truly measured* quantity z,

$$(1 + z_t) \; e^{0.25/3w\Omega_M} = 3.2114 \, (1 + z_t)^{3w\Omega_M} \; . \tag{58}$$

Eq. (58) can be solved, numerically, with respect to the combination $w\Omega_M$. Accordingly, we verify that the value

$$(w\Omega_M)_t = 0.0932 \pm 0.0060 \tag{59}$$

reproduces (exactly) the above observational result for z_t. By virtue of Eq. (59), $w \simeq \frac{1}{3}$, i.e., compatibility of the collisional-DM approach with the (currently available) observational data, suggests that, the DM itself consists of *relativistic particles*. This means that, on the basis of the collisional-DM approach under study, the matter-content of the *dark sector* consists of *"hot"* DM. For the time being, the theory of hot dark matter does not appear to be compatible with the large-scale structure of the Universe (see, e.g., [30]), although there are recent studies that could debate this result (see, e.g., [70], [71]. In any case, a hot-DM model looks much less *exotic* than most of the (currently investigated) DE scenarios.

4. Discussion

In the present article, we have examined the possibility that, the extra (dark) energy, needed to flatten the Universe, can be compensated by the energy of the internal motions of a collisional-DM fluid. Accordingly, we ended up with the evolution of a cosmological model which is driven by a perfect fluid of *positive pressure*, and, consequently, the energy of this fluid's internal motions has also been taken into account as a source of the universal gravitational field. Then, we have asked ourselves, whether such a model can accomodate, also, the apparent dimming of the cosmologically-distant indicators and the associated phase of accelerated expansion.

In particular, based on recent observational data, indicating that the DM can be slightly collisional, we have assumed that, the matter-energy content which drives the evolution of the Universe (being modeled by a spatially-flat RW space-time) at every post-recombination epoch, is in the form of a perfect fluid with *positive pressure*. In this way, we have been able to determine the *"correct"* form of the scale factor, which (under the assumption that the DM is collisional) governs the evolution of the Universe, together with a series of cosmologically-relevant parameters.

The outcome is quite promising: In principle, the energy of the internal motions of the collisional-DM fluid can account for the (extra) DE, so that, at the present epoch, $\Omega = 1$ (cf. Eq. (16)), while the post-recombination Universe remains *ever-decelerated* (cf. Eq. (24)).

Accordingly, we have attempted to determine what is realized by someone who, although living in a collisional-DM model, insists in adopting the (traditional) collisionless-DM approach.

To do so, we have applied the technique developed by Kleidis and Spyrou [56]. With the aid of this technique, we have found the (conformal) transformation (cf. Eqs. (28) and (31)), which relates the collisional-DM description of a cosmological model (in terms of which p, $\Pi > 0$ and $\frac{dp}{d\eta} \neq 0$) to the corresponding collisionless-DM (pressureless) approach. Accordingly, we have explored the way that, a supporter of the collisionless-DM scenario interprets the observations carried out in a collisional-DM Universe.

In the collisional-DM Universe there is a *characteristic value*, z_c, of the cosmological redshift (cf. Eq. (41)), above which, the luminosity distance of the various light-emitting sources becomes larger than what is realized by an observer who treats the DM as a pressureless fluid (cf. Eq. (40)). In other words, from the point of view of someone who (although living in a collisional-DM model) insists in adopting the (traditional) collisionless-DM approach, the cosmologically-distant indicators, located at $z > z_c$, seem to lie farther (i.e., they appear to be dimmer) than expected (cf., also, Eq. (44)). The similarity between the characteristic value z_c and the (observationally-determined) *transition redshift*, z_t, that signals the onset of dimming of the SNe Ia standard candles, is obvious.

On the other hand, after the thermodynamical content of a collisional-DM fluid is taken into account, the theoretical curve representing the distance modulus, $\mu(z)$ (now given by Eq. (44)), fits the *Hubble diagram* of an extended sample of SN Ia standard candles quite accurately (green solid line in Fig. 1), in contrast to the corresponding collisionless-DM quantity, $\tilde{\mu}(\tilde{z})$, given either (appropriately) by the combination of Eqs. (43) and (46) (orange solid line in Fig. 1) or (falsely) by the combination of Eqs. (43) and (47) (dashed line in Fig. 1).

At the same time, as far as a supporter of the collisionless-DM scenario is concerned, the Universe appears to be *either accelerating or decelerating*, depending on the value of the cosmological redshift (cf. Eqs. (55) and (56)).

In this case, the quantity w, which, in the collisional-DM approach, parameterizes the various *isothermal flows*, plays also another (more interesting) role: As we have found, for $w\Omega_M \geq 0.0714$, there exists a (theoretically-determined) *transition value*, $\tilde{z}_t$, of the (collisionless-DM-oriented) cosmological redshift, $\tilde{z}$, such that, for $\tilde{z} < \tilde{z}_t$, we have $\tilde{q} < 0$, i.e., from the point of view of someone who adopts the (traditional) collisionless-DM approach, the Universe is *accelerating*. Accordingly, taking into account the observational result that, the transition

Journal of Physics: Conference Series **283** (2011) 012018 doi:10.1088/1742-6596/283/1/012018

redshift between accelerated and decelerated expansion is set at the value $z_t = 0.46 \pm 0.13$ of the *truly* measured quantity z, we have determined the exact value of the combination $w\Omega_M$, for which the collisional-DM approach to the post-recombination Universe is compatible with observations, namely $(w\Omega_M)_t = 0.0932 \pm 0.0060$. According to this result, compatibility of the collisional-DM approach with the observational data currently available, suggests that, the DM itself consists of relativistic particles ($w \simeq \frac{1}{3}$).

In any case, the assumption that the Universe matter-content (basically its DM component) can be slightly collisional, is to be seen as a natural effort to take into account all the (so far neglected) internal physical characteristics of a classical cosmological fluid as sources of the universal gravitational field. As we have shown, under this assumption, one can compensate for the majority of the recent observational data, regarding $\Omega \simeq 1$, as well as the *unexpected dimming* of the SNe Ia standard candles and the *apparent* accelerated expansion of the Universe, without the need of any DE or the cosmological constant, and (certainly) without suffering from the *coincidence problem*, since, in a collisional-DM Universe, a potential *inflection point* (in the d_L versus z diagram) arises naturally. Although speculative, the idea that the DE (needed to flatten the Universe) could be attributed to the internal motions of a collisional-DM fluid, is (at least) intriguing and should be further explored and scrutinized, in the search of conventional alternatives to the DE concept.

References

[1] de Bernardis P et al. [BOOMERanG Collaboration] 2000 *Nature* **404** 955
[2] Komatsu E et al. [WMAP Collaboration] 2009 *Astrophys. J. Suppl. Series* **180** 330
[3] Riess A G et al. [Supernova Search Team Collaboration] 1998 *Astron. J.* **116** 1009
[4] Perlmutter S et al. [Supernova Cosmology Project Collaboration] 1999 *Astrophys. J.* **517** 565
[5] Tonry J L et al. [High-z Supernova Search Team] 2003 *Astrophys. J.* **594** 1
[6] Riess A G et al. [High-z Supernova Search Team] 2004 *Astrophys. J.* **607** 665
[7] Caldwell R R, Dave R and Steinhardt P J 1998 *Phys. Rev. Lett.* **80** 1528
[8] Carroll S M 1998 *Phys. Rev. Lett.* **81** 3067
[9] Esposito-Farese G and Polarski D 2001 *Phys. Rev.* **D 63** 063504
[10] Capozziello S, Carloni S and Troisi A 2003 *Recent Res. Dev. Astron. Astrophys.* **1** 625
[11] Dvali G R, Gabadadze G and Porratti M 2000 *Phys. Lett.* **B 485** 208
[12] Bousso R and Polchinski J 2000 *JHEP* **0006** 1 - 26
[13] Kamenshchik A, Moschella U and Pasquier V 2001 *Phys. Lett.* **B 487** 7
[14] Freese K and Lewis M 2002 *Phys. Lett.* **B 540**, 1
[15] Perivolaropoulos L 2003 *Phys. Rev.* **D 67** 123516
[16] Fardon R, Nelson A E and Weiner N 2004 *JCAP* **10** 005
[17] Peccei R D 2005 *Phys. Rev.* **D 71** 023527
[18] Perivolaropoulos L 2007 *Accelerating Universe: Observational status and theoretical implications* in The Invisible Universe: Dark Matter and Dark Energy, ed. L. Papantonopoulos, *Lecture Notes in Physics* **720** 257 (New York: Springer Verlag)
[19] Sahni V and Starobinsky A 2000 *IJMP* **D 9** 373
[20] Sahni V 2004 *Dark matter and dark energy* in The Physics of the Early Universe, ed. L. Papantonopoulos, *Lecture Notes in Physics* **653** 141 (Berlin: Springer Verlag)
[21] Allen S W, Schmidt R W, Ebeling H, Fabian A C and van Speybroeck L 2004 *MNRAS* **353**, 457
[22] Boughn S and Crittenden R 2004 *Nature* **427** 45
[23] Eisenstein D J et al. [SDSS Collaboration] 2005 *Astrophys. J.* **633** 560
[24] Buchert T 2000 *Gen. Relativ. Gravit.* **32** 105
[25] Kolb E W, Matarrese S and Riotto A 2006 *New J. Phys.* **8** 322
[26] Celerier M-N 2007 *New Adv. Phys.* **1** 29
[27] Ellis G F R 2009 *JPhCS* **189** 012011
[28] Olive K A 2003 *TASI Lectures on Dark Matter* arXiv: 0301505 [astro-ph]
[29] Kolb E W and Turner M S 1990 *The Early Universe* (Menlo Park: Addison-Wesley)
[30] Hooper D 2009 *TASI 2008 Lectures on Dark Matter* arXiv: 0901.4090 [hep-ph]
[31] Chang J et al. [ATIC Collaboration] 2008 *Nature* **456** 362
[32] Adriani O et al. [PAMELA Collaboration] 2009 *Nature* **458** 607
[33] Hooper D, Finkbeiner D P and Dobler G 2007 *Phys. Rev.* **D 76** 083012

[34] Barger V, Keung W Y, Marfatia D and Shaughnessy G 2008 *Phys. Lett.* **B 672** 141
[35] Baushev A 2008 *Dark matter annihilation at cosmological redshifts: Possible relic signal from weakly interacting massive particles annihilation* arXiv: 0806.3108 [astro-ph]
[36] Bergstrom L, Bringmann T and Edsjo J 2008 *Phys. Rev.* **D 78** 103520
[37] Cholis I, Goodenough L, Hooper D, Simet M and Weiner N 2008a *High-energy positrons from annihilating dark matter* arXiv: 0809.1683 [hep-ph]
[38] Cholis I, Dobler G, Finkbeiner D P, Goodenough L and Weiner N 2008b, *The case for a 700+ GeV WIMP: Cosmic-ray spectra from ATIC and PAMELA* arXiv: 0811.3641 [astro-ph]
[39] Cirelli M and Strumia A 2008 *Minimal dark-matter predictions and the PAMELA positron excess* arXiv: 0808.3867 [astro-ph]
[40] Regis M and Ullio P 2008 *Phys. Rev.* **D 78** 103505
[41] Fornasa M, Pieri L, Bertone G and Branchini E 2009 *Anisotropy probe of galactic and extra-galactic dark matter annihilations* arXiv: 0901.2921 [astro-ph]
[42] Fox P J and Poppitz E 2009 *Phys. Rev.* **D 79** 083528
[43] Zurek K M 2009 *Phys. Rev.* **D 79** 115002
[44] Spergel D N and Steinhardt P J 2000 *Phys. Rev. Lett.* **84** 3760
[45] Arkani-Hamed N, Finkbeiner D P, Slatyer T R and Weiner N 2009 *Phys. Rev.* **D 79** 015014
[46] Cirelli M, Kadastik M, Raidal M and Strumia A 2009 *Nucl. Phys.* **B 813** 1
[47] Cohen T and Zurek K 2010 *Phys. Rev Lett.* **104** 101301
[48] Zimdahl W, Schwarz D J, Balakin A B and Pavon D 2001 *Phys. Rev.* **D 64** 3501
[49] Balakin A B, Pavon D, Schwarz D J and Zimdahl W 2003 *New J. Phys.* **5** 85
[50] Lima J A S, Silva F E and Santos R C 2008 *Class. Quantum Grav.* **25** 205006
[51] Basilakos S and Plionis M 2009 *A&A* **507** 47
[52] Peacock J A 1999 *Cosmological Physics* (Cambridge: Cambridge University Press)
[53] Fock V 1959 *The theory of space, time and gravitation* (London: Pergamon Press)
[54] Narlikar J V 1983 *Introduction to Cosmology* (Boston: Jones and Bartlett Publishers Inc.)
[55] Chandrasekhar S 1965 *Astrophys. J.* **142** 1488
[56] Kleidis K and Spyrou N K 2000 *Class. Quantum Grav.* **17** 2965
[57] Spyrou N K and Tsagas C G 2004 *Class. Quantum Grav.* **21** 2435
[58] Synge J L 1937 *Proc. London Math. Soc.* **43** 376 [reprinted 2002 in *Gen. Relat. Grav.* **34** 2177]
[59] Lichnerowicz A 1967 *Relativistic Hydrodynamics and Magneto-hydrodynamics* (New York: W. A. Benjamin Inc.)
[60] Carter B 1979 *Perfect fluid and magnetic field conservation laws in the theory of black-hole accretion rings* in Active Galactic Nuclei, ed. C Hazard and S Mitton (Cambridge: Cambridge University Press) p. 273
[61] Verozub L 2008 *IJMP* **D 17** 337
[62] Riess A G et al. [High-z Supernova Search Team] 2001 *Astrophys. J.* **560** 49
[63] Riess A G et al. [High-z Supernova Search Team] 2007 *Astrophys. J.* **659** 98
[64] Astier P et al. [SNLS Collaboration] 2006 *A&A* **447** 31
[65] Wood-Vasey W M et al. [ESSENCE Supernova Survey Team] 2007 *Astrophys. J.* **666** 694
[66] Kowalski M et al. [Supernova Cosmology Project Group] 2008 *Astrophys. J.* **686** 749
[67] Spergel D N et al. [WMAP Collaboration] 2003 *Astrophys. J. Suppl. Series* **148** 175
[68] Davis T M et al. [ESSENCE Supernova Survey Team] 2007 *Astrophys. J.* **666** 716
[69] Carroll S M, Press W H and Turner M S 1992 *ARA&A* **30** 499
[70] Farrar G R and Peebles P J E 2004 *Astrophys. J.* **604** 1
[71] Gubser S S and Peebles P J E 2004 *Phys. Rev.* **D 70** 123510

Journal of Physics: Conference Series **283** (2011) 012019 doi:10.1088/1742-6596/283/1/012019

Black Holes and Neutron Stars in Tensor-Vector-Scalar Theory

Paul D. Lasky

Theoretical Astrophysics, Eberhard Karls University of Tübingen, Tübingen 72076, Germany

E-mail: lasky@tat.physik.uni-tuebingen.de

Abstract. Bekenstein's Tensor-Vector-Scalar (TeVeS) theory has had considerable success as a relativistic theory of Modified Newtonian Dynamics. In the strong-field regime it makes many predictions that differ from General Relativity implying this regime provides an excellent testing ground for the theory using both current and future observations of neutron stars and black holes in both the electromagnetic and gravitational wave spectrum. I look at the current status of black hole and neutron star solutions in both the original TeVeS theory and its generalisation, including some key observational properties that differ from General Relativity.

1. Introduction

Modified Newtonian Dynamics (MoND) [1] provides an enticing alternative to the Cold Dark Matter paradigm by phenomenologically modifying Newton's law of gravity on galactic scales (for a review see Ref. [2]). MoND is nothing more than a toy model of gravity though as it is not a covariant theory. A fully relativistic theory generalising MoND has recently been proposed by Bekenstein [3], which includes the original tensor field of General Relativity coupled to an extra vector and scalar field, hence the theory is known by its acronym TeVeS.

In the weak acceleration limit, TeVeS has been shown to reproduce MoND implying the theory carries forth the various successes of that theory on galactic scales. In the Newtonian limit, the theory reproduces the parametrized post-Newtonian (PPN) coefficients consistently within limits imposed by solar system experiments (for a comprehensive review see Ref. [4]). More recently, TeVeS, when combined with a single sterile neutrino species of mass around 11 eV, can match the angular power spectrum of the cosmic microwave background [5] as well as the straight gravitational lensing arc of Abell 2390 [6] – two previous chinks in the TeVeS armour.

Meanwhile, the strong-field study of TeVeS is relatively under-developed. Giannios [7] first solved the TeVeS field equations for a spherically symmetric, static, vacuum spacetime. Sagi & Bekenstein [8] subsequently solved the field equations for charged spacetimes, thus deriving the Reissner-Nordström solution. Subsequently, in Lasky, Sotani & Giannios [9], we solved the spherically symmetric field equations in the presence of a perfect fluid, thus finding the equivalent TOV solution in order to study neutron star structure. The properties of these various solutions are reviewed in considerable detail below in sections 3 and 4.

The original formulation of TeVeS is, however, not without its shortcomings. In particular, Bekenstein's formulation [3] has been shown to suffer from dynamical problems. Seifert [10] showed that Giannios' Schwarzschild-TeVeS solution is unstable to linear perturbations for various relevant values of the theory's coupling parameters. Contaldi *et al.* [11] then showed that the vector field is prone to forming caustics under a variety of dynamical situations. Finally, Sagi [12] showed that the vector field is constrained by the cosmological value of the scalar field in such a way that it prevents the scalar field from evolving. Given these inconsistencies, Contaldi *et al.* [11] and Skordis [13] have provided a generalisation of TeVeS whereby the vector field action is not given by the Maxwellian form of the original TeVeS, but a more general Einstein-Aether form (the tensor and scalar fields are described equivalently to the original formulation).

Very little has been studied in the generalised version of TeVeS. Skordis [13] studied the cosmological equations of the theory and showed that they are identical to the original TeVeS equations up to a rescaling of Hubble's constant. Sagi [8] looked at the PPN parameters and found them not to conflict with solar system experiments for suitable values of the coupling parameters. In [14], I looked at the strong-field regime of this theory and found that the vacuum and charged black hole solutions and also the TOV solution are identical in this theory as the original theory up to a rescaling of various vector field coupling parameters.

In the present article I review the non-charged vacuum and neutron star solutions in the generalised and original versions of TeVeS. The black hole solutions are those found for the original version of the theory by Giannios [7], and subsequently derived for the generalised version in Lasky [14]. Meanwhile, the results for the neutron star spacetimes are reviewed from Lasky, Sotani & Giannios [9]. I first provide a brief description of the relevant field equations for the generalised TeVeS theory.

2. Generalised TeVeS

As the name suggests, TeVeS is a theory built upon three background fields – a tensor, a vector and a scalar field. The action describing the tensor field is simply the Einstein-Hilbert action from General Relativity. This implies that when the vector field and scalar field coupling parameters tend to zero, the theory reduces exactly to General Relativity. Herein I provide only a brief introduction to the relevant TeVeS equations. For more details the reader is referred to Bekenstein [3] for the equations of original TeVeS and to Contaldi *et al.* [11] and Skordis [13] for the generalised version of the theory.

Given the three background fields, one requires a metric in which physical measurements are made. This so-called *physical* metric, denoted $\tilde{g}_{\mu\nu}$, is related to the tensor field, $g_{\mu\nu}$ (also known as the *Einstein* metric), the vector field, A^μ, and the scalar field, φ, through the following relation

$$\tilde{g}_{\mu\nu} = \mathrm{e}^{-2\varphi}\left(g_{\mu\nu} + A_\mu A_\nu\right) - \mathrm{e}^{2\varphi} A_\mu A_\nu. \tag{1}$$

The modified Einstein field equations are found by varying the Einstein-Hilbert action where the Riemann tensor is that associated with the Einstein metric. These are

$$G_{\mu\nu} = 8\pi G\left[\tilde{T}_{\mu\nu} + \left(1 - \mathrm{e}^{-4\varphi}\right) A^\alpha \tilde{T}_{\alpha(\mu} A_{\nu)} + \tau_{\mu\nu}\right] + \Theta_{\mu\nu}, \tag{2}$$

where $G_{\mu\nu}$ is the Einstein tensor associated with the Einstein frame, $\tilde{T}_{\mu\nu}$ is the physical stress-energy tensor and $\tau_{\mu\nu}$ & $\Theta_{\mu\nu}$ are the effective stress-energy terms associated with the scalar and vector fields respectively. In particular

$$\tau_{\mu\nu} := \frac{\mu}{kG}\left[\nabla_\mu\varphi\nabla_\nu\varphi - \frac{1}{2}g^{\alpha\beta}\nabla_\alpha\varphi\nabla_\beta\varphi g_{\mu\nu} - A^\alpha\nabla_\alpha\varphi\left(A_{(\mu}\nabla_{\nu)}\varphi - \frac{1}{2}A^\beta\nabla_\beta\varphi g_{\mu\nu}\right)\right] - \frac{\mathcal{F}(\mu)}{2k^2\ell^2 G}g_{\mu\nu}, \tag{3}$$

where k is the scalar field coupling constant, μ is a function associated with the MoND acceleration scale, $\mathcal{F}(\mu)$ is related to the interpolation function in MoND and hence is not *a priori* predicted by the theory and ℓ is a fixed length scale associated with the free function. For the remainder of this article we only consider the strong-field regime of the theory whereby

Journal of Physics: Conference Series **283** (2011) 012019 doi:10.1088/1742-6596/283/1/012019

$\mu = 1$ is an excellent approximation (for more details see [3, 7, 8, 11]). In this case the function $\mathcal{F}$ does not contribute to the field equations [11], and is therefore ignored for the remainder of the article.

The generalised version of TeVeS has four vector field coupling components which describe the relative strengths of the individual terms in the vector field action. These are denoted by K, K_+, K_2 and K_4, where Bekenstein's original version of the theory is regained by setting the last three parameters to zero. The vector field contribution to the effective stress-energy in the modified Einstein equations, $\Theta_{\mu\nu}$, can be expressed as a sum of terms associated with each of the vector field coupling constants, plus a term associated with a Lagrange multiplier, λ;

$$\Theta_{\mu\nu} := \Theta^{K}_{\mu\nu} + \Theta^{K_+}_{\mu\nu} + \Theta^{K_2}_{\mu\nu} + \Theta^{K_4}_{\mu\nu} + \Theta^{\lambda}_{\mu\nu}, \tag{4}$$

where

$$\Theta^{K}_{\mu\nu} := K\left(F_{\alpha\mu}F^{\alpha}{}_{\nu} - \frac{1}{4}g_{\mu\nu}F_{\alpha\beta}F^{\alpha\beta}\right), \tag{5}$$

$$\Theta^{K_+}_{\mu\nu} := K_+\Big[S_{\mu\alpha}S_{\nu}{}^{\alpha} - \frac{1}{4}g_{\mu\nu}S_{\alpha\beta}S^{\alpha\beta} + \nabla_\alpha\left(A^\alpha S_{\mu\nu} - S^{\alpha}{}_{(\mu}A_{\nu)}\right)\Big], \tag{6}$$

$$\Theta^{K_2}_{\mu\nu} := K_2\Big[g_{\mu\nu}\nabla_\alpha\left(A^\alpha\nabla_\beta A^\beta\right) - A_{(\mu}\nabla_{\nu)}\left(\nabla_\alpha A^\alpha\right) - \frac{1}{2}g_{\mu\nu}\nabla_\alpha A^\alpha\nabla_\beta A^\beta\Big], \tag{7}$$

$$\Theta^{K_4}_{\mu\nu} := K_4\Big[\dot{A}_\mu\dot{A}_\nu + \dot{A}_\alpha A_{(\mu}\nabla_{\nu)}A^\alpha - \nabla_\alpha\left(\dot{A}^\alpha A_\mu A_\nu\right) - \frac{1}{2}g_{\mu\nu}\dot{A}_\alpha\dot{A}^\alpha\Big], \tag{8}$$

$$\Theta^{\lambda}_{\mu\nu} := -\lambda A_\mu A_\nu. \tag{9}$$

Here, $F_{\mu\nu} := \nabla_{[\nu}A_{\mu]}$, $S := \nabla_{(\nu}A_{\mu)}$ and $\dot{A}_\mu := A^\alpha\nabla_\alpha A_\mu$.

Dynamics of the vector field are governed by vector field equation

$$\begin{aligned} K\nabla_\alpha F^{\mu\alpha} + K_+\nabla_\alpha S^{\alpha\mu} + K_2\nabla^\mu\left(\nabla_\alpha A^\alpha\right) + \lambda A^\mu + K_4\left[\nabla_\alpha\left(\dot{A}^\mu A^\alpha\right) - \dot{A}^\alpha\nabla^\mu A_\alpha\right] \\ + \tfrac{8\pi\mu}{k}A^\alpha\nabla_\alpha\varphi g^{\mu\beta}\nabla_\beta\varphi = 8\pi G\left(1 - \mathrm{e}^{-4\varphi}\right)g^{\mu\alpha}\tilde{T}_{\alpha\beta}A^\beta. \end{aligned} \tag{10}$$

Contraction of this equation with the vector field isolates the Lagrange multiplier. This subsequent equation can then be used in the modified Einstein equations, in particular the term expressed in equation (9), such that the system is fully determined.

The final field equation is that of the scalar field

$$\nabla_\beta\Big[\mu\left(g^{\alpha\beta} - A^\alpha A^\beta\right)\nabla_\alpha\varphi\Big] = kG\Big[g^{\alpha\beta} + \left(1 + \mathrm{e}^{-4\varphi}\right)A^\alpha A^\beta\Big]\tilde{T}_{\alpha\beta}. \tag{11}$$

3. Black Holes

Giannios [7] first solved the original TeVeS field equations [i.e. equations (2-11) with $K_+ = K_2 = K_4 = 0$] for a static, spherically symmetric vacuum spacetime. He found two families of solutions dependent on the form of the vector field. When the vector field was constrained to be purely timelike, Giannios showed that the only black hole solution is exactly the Schwarzschild spacetime. Sagi & Bekenstein [8] subsequently generalised Giannios' result to include charge, showing that the Reissner-Nordström spacetime is also a solution of the TeVeS field equations. In Lasky [14], I showed that the solutions derived by Giannios and Sagi & Bekenstein for the original TeVeS theory are also valid in the generalised version of the theory, where the vector field coupling parameter K of TeVeS gets replaced by the combination of parameters $K + K_+ - K_4$.

As mentioned, the only black hole solution when the vector field is temporally aligned is the Schwarzschild solution. However, there exists a larger class of solutions than this which exhibit exotic phenomena such as naked singularities. The solution found by Giannios [7], and subsequently in Lasky [14], is usually represented in isotropic coordinates, which is for the physical metric

$$d\tilde{s}^2 = -\left(\frac{1 - r_c/r}{1 + r_c/r}\right)^a dt^2 + \left(1 + \frac{r_c}{r}\right)^{2+a}\left(1 - \frac{r_c}{r}\right)^{2-a}\left(dr^2 + r^2 d\Omega^2\right), \tag{12}$$

where $d\tilde{s}^2 := \tilde{g}_{\alpha\beta}dx^\alpha dx^\beta$ and $d\Omega^2 := d\theta^2 + \sin^2\theta d\phi^2$. The two constants in the physical metric, r_c and a, are related to the various coupling parameters of the theory, including the "scalar mass", m_φ, as well as the characteristic gravitational radius, r_g, according to

$$r_c = \frac{r_g}{4}\sqrt{1 + \frac{k}{\pi}\left(\frac{Gm_\varphi}{r_g}\right)^2 - \frac{\mathcal{K}}{2}}, \tag{13}$$

$$a = \frac{r_g}{2r_c} + \frac{kGm_\varphi}{4\pi r_c}, \tag{14}$$

where $\mathcal{K} := K + K_+ - K_4$.

Giannios [7] showed that r_c is a black hole event horizon if and only if $a = 2$. This is based on the requirement that the surface at $r = r_c$ must have a finite surface area, and also the singularity must be removable. Evaluating the physical Ricci scalar shows that it diverges for values of $a < 2$ and also $2 < a < 4$. Meanwhile, as the surface area is proportional to the $\tilde{g}_{rr}(r_c)$ component of the metric, one finds that this diverges for $a > 2$. One therefore has a black hole solution only when $a = 2$. When $0 < a < 2$ and $2 < a < 4$ one has a naked singularity at $r = r_c$, and when $a \geq 4$, r_c represents a removable singularity with a divergent surface area.

It is interesting to note that the metric (12) is exactly equivalent to that of Brans 'type I' metric [15] in Brans-Dicke scalar-tensor theory [16] with the assosciation $a = 2Q$ and $\chi = 0$ from the notation of Ref. [17]. Note that in the Brans I solution, setting $\chi = 0$ necessarily implies that the parameter Q is unity, and hence the Schwarzschild solution is recovered. This is because the Brans I solution has an extra algebraic equation which links Q and χ. The association with a Brans-Dicke solution is particularly interesting given the vast complexity of the field equations in TeVeS as compared with Brans-Dicke theory. However, the association is due to the simplifying assumption that the vector field is purely timelike. It was shown for the original version of TeVeS that black hole solutions with a non-zero radial component for the vector field produce significantly more complicated geometries than the case where the radial component vanishes [7]. This complexity will only increase when one analyses the black hole solutions with non-zero radial vector fields in the fully general theory.

Giannios [7] showed initial trepidation towards the above solution as the scalar field becomes negative close to the horizon, which therefore allows for superluminal propagation of scalar waves [3]. Sagi & Bekenstein [8] overcame this issue by showing that Giannios overlooked a branch of solutions where the scalar field was everywhere positive, although they did this only for the case where $a = 2$. By rearranging equations (13) and (14) we can re-derive the extra solution of Ref. [8] with the generalisation that a remains arbitrary. In this case, one can show that the scalar field throughout the spacetime is given by

$$\varphi(r) = \varphi_c + \delta_\pm \ln\left(\frac{1 - r_c/r}{1 + r_c/r}\right), \tag{15}$$

where φ_c is the cosmological value of the scalar field and $\delta_\pm$ can be found from equations (13) and (14) to be

$$\delta_\pm := \frac{a\left(2 - \mathcal{K}\right)k/2 \pm \sqrt{2k\left[\left(k - a^2\pi\right)\left(2 - \mathcal{K}\right) + 8\pi\right]}}{\left(2 - \mathcal{K}\right)k + 8\pi}. \tag{16}$$

Note here that when $a = 2$, equation (16) reduces to exactly equation (67) of Ref. [8] (where $\mathcal{K} \equiv K$ as they were working in the original TeVeS theory which has $K_+ = K_2 = K_4 = 0$).

Given that $0 < \mathcal{K} < 2$ and $k > 0$, one can trivially show that $\delta_+ > 0$. Moreover, after some extra work a condition is derived for δ_- that is

$$\delta_- < 0 \qquad \Longleftrightarrow \qquad a^2 < \frac{8}{2-\mathcal{K}}, \tag{17}$$

which further implies that $\delta_- < 0$ for all $a \leq 2$, whilst the opposite is not necessarily true.

The generic behaviour of the scalar field throughout the spacetime is of great importance for a forthcoming article [18] looking at the stability and quasinormal mode spectrum of black holes in generalised TeVeS and the relevance of superluminal propagation of scalar waves. Indeed, Bekenstein [3] showed that, in the eikonal approximation, scalar perturbations in the Newtonian limit of the theory travel with velocity $v = \exp(-2\varphi)/\sqrt{2}$. This implies that scalar perturbations travel superluminally if and only if $\varphi < 0$. Equation (15) shows that the scalar field necessarily diverges as $r \to r_c$. When $\delta_\pm > 0$, the scalar field diverges to positive infinity, and indeed the scalar field is everywhere positive (given $\varphi_c > 0$). However, when $\delta_\pm < 0$, the scalar field diverges to negative infinity as one approaches the critical radius, r_c, implying for some values of the radial coordinate the scalar field is negative. In Fig. 1 the scalar field is shown as a function of the radial coordinate for various values of the coupling parameters and also the spacetime parameter a.

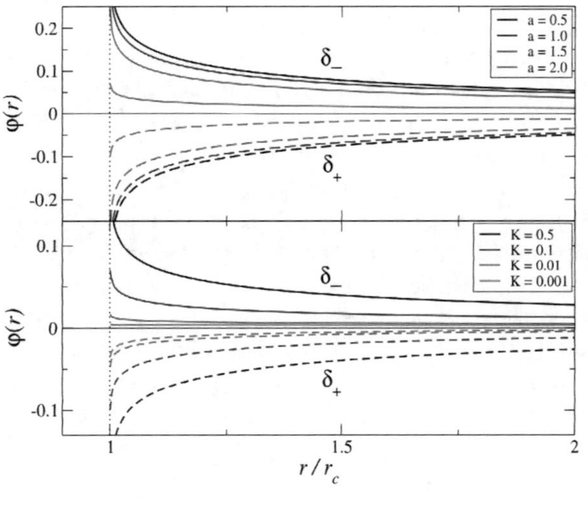

Figure 1. Scalar field, φ, as a function of the radial coordinate with $k = 0.003$ and $\varphi = 0.003$ for the black hole spacetimes. The top panel shows various values of a with $\mathcal{K} = 0.1$. The lower panel varies the vector field coupling $\mathcal{K} := K + K_+ - K_4$ for $a = 2$. For both plots the thick (dashed) lines represent δ_- (δ_+). In the eikonal approximation the velocity of scalar field perturbations is proportional to $\exp(-2\varphi_c)$, implying superluminal and subluminal propagation of perturbations for $\varphi < 0$ and $\varphi > 0$ respectively.

4. Neutron Stars

Neutron stars in TeVeS were first studied in Lasky, Giannios & Sotani [9] by solving the TOV equations of hydrodynamic equilibrium for a perfect fluid in spherical symmetry. In Lasky [14], I further showed that the neutron star solutions for the original TeVeS theory are also solutions of the generalised TeVeS field equations, with the same association that the coupling parameter K of TeVeS gets replaced by the combination of parameters $K + K_+ - K_4$. The following is a review of the results from [9, 14] on neutron star structure.

We now express the physical metric in Schwarzschild coordinates where the scalar field has been taken into account from equation (1)

$$d\tilde{s}^2 = -\mathrm{e}^{\nu+2\varphi}dt^2 + \mathrm{e}^{-2\varphi}\left(\frac{dr^2}{1-2m/r} + r^2 d\Omega^2\right). \tag{18}$$

Here, the metric components, ν and m, and the scalar field are only functions of the radial coordinate. Conservation of stress-energy, prescribed in the physical frame, reduces simply to the radial Euler equation

$$-\frac{d\tilde{P}}{dr} = -\left(\frac{1}{2}\frac{d\nu}{dr} + \frac{d\varphi}{dr}\right)\left(\tilde{\rho} + \tilde{P}\right). \tag{19}$$

The scalar field equation (11) can be once integrated to give

$$\frac{d\varphi}{dr} = \frac{kGM_\varphi \mathrm{e}^{-\nu/2}}{4\pi r^2\sqrt{1-2m/r}}, \qquad M_\varphi(r) := 4\pi\int_0^r \frac{\left(\tilde{\rho}+3\tilde{P}\right)\mathrm{e}^{\nu/2-2\varphi}r^2 dr}{\sqrt{1-2m/r}}. \tag{20}$$

Here, the scalar mass function, M_φ, has been defined according to [3]. After much algebra (for details see [9]), the remaining field equations can be reduced to two coupled equations which are analogous to the TOV equations in General Relativity

$$\frac{dm}{dr}\left(1-\frac{\mathcal{K}}{2}\right) - \frac{\mathcal{K}m}{2r} = 4\pi G\mathrm{e}^{-2\varphi}r^2\left(\tilde{\rho}+2\mathcal{K}\tilde{P}\right) + \left(1-\frac{2m}{r}\right)\left[\frac{2\pi r^2}{k}\left(\frac{d\varphi}{dr}\right)^2 - \frac{\mathcal{K}r}{4}\frac{d\nu}{dr}\left(1+\frac{r}{4}\frac{d\nu}{dr}\right)\right], \tag{21}$$

$$\frac{d\nu}{dr}\left(1+\frac{\mathcal{K}r}{8}\frac{d\nu}{dr}\right) = \frac{8\pi G\mathrm{e}^{-2\varphi}\tilde{P}r^2 + 2m/r}{r\left(1-2m/r\right)} + \frac{4\pi r}{k}\left(\frac{d\varphi}{dr}\right)^2. \tag{22}$$

It can be shown [9] that taking $\mathcal{K} = k = 0$ leads to the familiar TOV equations of General Relativity.

To build neutron star models from the above equations one must assume an equation of state relating the density and pressure. For these purposes we used equation of state A from Arnett & Bowers [19]. In Figs. 2 and 3, the ADM mass as a function of the central density and radius for various values of the vector field coupling constant $\mathcal{K}$ is shown. One can immediately see that strong variations of the vector coupling result in large deviations from the General Relativistic predictions. In Lasky, Sotani & Giannios [9] we also showed plots with the effect of varying the scalar field coupling constant. This was shown to have significantly smaller effect on the structure of neutron stars than the vector field parameter.

In general, Figs. 2 and 3 show that the mass and radii of neutron stars in TeVeS and generalised TeVeS are generically smaller than in General Relativity. The most accurate measurements of neutron star masses are from timing observations of binary pulsars, although significant progress is also being made using binaries with white dwarf companions (see Lattimer & Prakash [20] and references therein). A conservative estimate puts neutron star masses as $\gtrsim 1.5M_\odot$, although the true value is likely considerably higher. Taking this into account, together with variations in possibilities for the equation of state, in Ref. [9] we ruled out the parameter space $\mathcal{K} \gtrsim 1$, although we noted that this was a conservative estimate.

5. Conlclusions and Future Directions

The study of the strong-field regime of TeVeS and its generalisation is in its infancy. Whilst Sotani [21, 22, 23] has looked at oscillations and slow rotations of neutron stars in the original formulation of TeVeS, these results do not carry forth to the generalised version of the theory. Moreover, neutron stars do not provide the best testing ground for theories of gravity in the strong-field regime due to degeneracies associated with uncertainties in the equation of state and more detailed internal structure.

Black holes are, however, clean systems. The "No Hair" theorem of General Relativity implies that a black hole is characterized by only its mass and angular momentum. However,

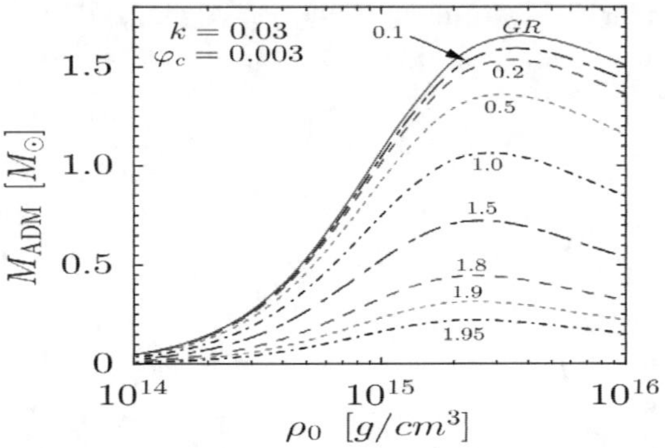

Figure 2. Central density as a function of the ADM mass for varying values of the vector field coupling constant $\mathcal{K}$. Here, the scalar field coupling constant is held constant at $k = 0.03$ and the cosmological values of the scalar field is $\varphi_c = 0.003$.

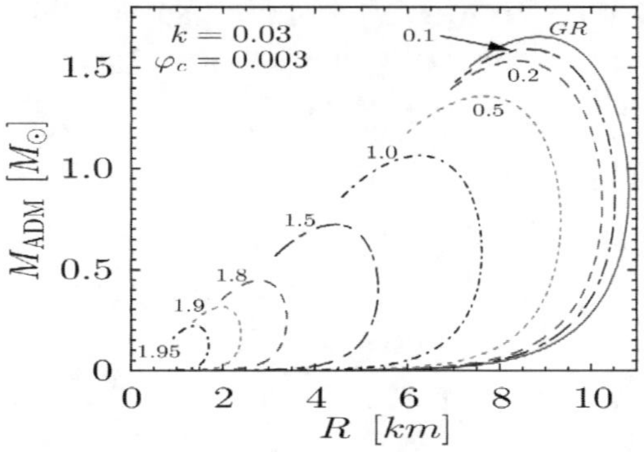

Figure 3. Mass-radius relation for various values of the vector coupling parameter $\mathcal{K}$. Again, $k = 0.03$ and $\varphi_c = 0.003$. For small values of the vector coupling there is little difference in the mass radius relation from the predictions of General Relativity.

any violation of this implies a violation of General Relativity. The solutions exhibiting naked singularities discussed above would have observational differences to black hole spacetimes in General Relativity. A relevant future study is therefore to establish the end state of gravitational collapse to determine if such solutions are physically meaningful. If the end state of gravitational collapse in TeVeS is a black hole, then, under the assumption that the vector field is aligned only in the temporal direction, such a black hole would be indistinguishable from a General Relativistic Schwarzschild black hole. However, the perturbations of such spacetimes would differ to those in General Relativity due to the different field equations as well as the additional degrees of freedom associated with the extra fields. For example, a spherically symmetric black hole in TeVeS would be allowed to emit gravitational waves by way of breathing modes of radiation, a phenomena not possible in General Relaltivity due to Birkhoff's theorem. In a forthcoming article [18], we look at perturbations of black holes in generalised TeVeS to determine both stability properties and also differences one would expect in gravitational wave observations of black holes in TeVeS as opposed to General Relativity.

Acknowledgments

This work is supported by the Alexander von Humboldt Foundation and the Transregio 7 "Gravitational Wave Astronomy," financed by the Deutsche Forschungsgemeinschaft DFG (German Research Foundation).

References

[1] Milgrom M 1983 *Astrophys. J.* **270** 365
[2] Sanders R H and McGaugh S S 2002 *Ann. Rev. Astron. Astrophys.* **40** 263
[3] Bekenstein J D 2004 *Phys. Rev. D* **70** 083509
[4] Skordis C 2009 *Class. Q. Grav.* **26** 143001
[5] Angus G W 2009 *Mon. Not. R. Astron. Soc.* **394** 527
[6] Feix M, Zhao H, Fedeli C, na J L G P and Hoekstra H 2010 Substructure lensing in galaxy clusters as a constraint on low-mass sterile neutrinaos in tensor-vector-scalar theory: The straight arc of Abell 2390 arXiv:1008.1963
[7] Giannios D 2005 *Phys. Rev. D* **71** 103511
[8] Sagi E and Bekenstein J D 2008 *Phys. Rev. D* **77** 024010
[9] Lasky P D, Sotani H and Giannios D 2008 *Phys. Rev. D* **78** 104019

[10] Seifert M D 2007 *Phys. Rev. D* **76** 064002
[11] Contaldi C R, Wiseman T and Withers B 2008 *Phys. Rev. D* **78** 044034
[12] Sagi E 2009 *Phys. Rev. D* **80** 044032
[13] Skordis C 2008 *Phys. Rev. D* **77** 123502
[14] Lasky P D 2009 *Phys. Rev. D* **80** 081501 (R)
[15] Brans C H 1962 *Phys. Rev.* **125** 2194
[16] Brans C and Dicke R H 1961 *Phys. Rev.* **124** 925
[17] Scheel M A, Shapiro S L and Teukolsky S A 1995 *Phys. Rev. D* **51** 4236
[18] Lasky P D and Doneva D D 2010 Stability and quasinormal modes of black holes in Tensor-Vector-Scalar theory in Preparation
[19] Arnett W D and Bowers R L 1977 *Astrophys. J.* **33** S415
[20] Lattimer J M and Prakash M 2007 *Phys. Rep.* **442** 109
[21] Sotani H 2009 *Phys. Rev. D* **79** 064033
[22] Sotani H 2009 *Phys. Rev. D* **80** 064035
[23] Sotani H 2010 *Phys. Rev. D* **81** 084006

CMB Polarization and Dark Energy Induced Cosmological Birefringence

Guo-Chin Liu[1], Seokcheon Lee[2] and Kin-Wang Ng[2,3,4]

[1]Department of Physics, Tamkang University, 251-37 Tamsui, Taipei County, Taiwan 251, R.O.C.
[2] Institute of Physics, Academia Sinica, Taipei, Taiwan 11529, R.O.C.
[3] Institute of Astronomy and Astrophysics, Academia Sinica, Taipei, Taiwan 11529, R.O.C.
[4] Leung Center for Cosmology and Particle Astrophysics, National Taiwan University, Taipei, Taiwan 10617, R.O.C.

E-mail: liugc@mail.tku.edu.tw

Abstract. The coupling between dark energy and the pseudoscalar of electromagnetism, if there is any, would induce a rotation of the polarization plane of the cosmic microwave background (CMB). This results in a non-vanishing B-mode and parity-violating TB and EB correlations. Taking into account this effect, we calculate the full set of power spectra of cosmic microwave background (CMB) temperature and polarization anisotropies. We also give the constraint on the coupling strength from WMAP seven year data.

1. Introduction

The existence of a dark component with an effective negative pressure, supported by several observations especially the Hubble diagram for the type-Ia supernovae (see, for example, [1, 2]), is still one of the puzzles in cosmology. Cosmological constant is the simplest possibility for such dark component. However, the observed value of the cosmological constant is completely different from theoretical expectation [3]. An alternative candidate, the so-called quintessence described by a dynamical scalar field ϕ, is naturally considered. The dynamics of ϕ in general quintessence models is governed by a scalar potential $V(\phi)$ which makes the dark energy dominant in recent epoch. There are many different kinds of proposed potentials, for example, pseudo Nambu-Goldston boson, inverse power law, exponential, hyperbolic cosine, and tracking oscillating [4]. To differentiate between the models and finally reconstruct $V(\phi)$ would likely require next-generation observations.

One particular way to study the quintessence is to consider the interaction of ϕ to the pseudoscalar of electromagnetism. Pseudoscalar couplings usually arise from the spontaneous breaking of a compact symmetry group, say, U(1) (see Frieman et al. 1995 at reference [4]). Carroll has argued that this coupling leads to the rotation of the polarization vector of propagating photons if ϕ is varying with time. This effect is called as the "cosmological birefringence" [5]. This rotation results in a non-vanishing B-mode and parity-violating TB and EB correlations. Here, we present the CMB power spectra in the presence of the cosmological birefringence by the use of the full Boltzmann code. We briefly describe the the power spectra of CMB polarization in the presence of the cosmological birefringence in Sec. 2. In Sec. 3, we apply the power spectra to give a constraint on the coupling strength using the CMB data

Published under licence by IOP Publishing Ltd

made by Wilkinson Microwave Background Probe (WMAP) [6]. Sec. 4 is our discussion and conclusion.

2. Power spectra of CMB polarization with the cosmological birefringence

Thomson scatterings of anisotropic radiation by free electrons give rise to the linear polarization, which is usually described by the Stokes parameters Q and U [7]. When the polarization plane is rotated by an angle α, the Stokes parameters are transformed into

$$\begin{pmatrix} Q' \\ U' \end{pmatrix} = \begin{pmatrix} \cos 2\alpha & \sin 2\alpha \\ -\sin 2\alpha & \cos 2\alpha \end{pmatrix} \begin{pmatrix} Q \\ U \end{pmatrix}. \tag{1}$$

In standard cosmology, the time evolution of the polarization perturbation is governed by the Boltzmann equation [8], which states that when one follows a light ray the polarization of the radiation can change only due to the Thomson scatterings. With the birefringence effect, the Q and U further change due to the rotation of the polarization plane. Thus, we can then write,

$$\frac{dX}{d\eta} = \dot{X}_{thomson} + \dot{X}_{rotation}, \tag{2}$$

$X = Q \pm iU$ and the derivatives are taken with respect to the conformal time η.

Using eq. (1), the angular velocity due to the cosmological birefringence of dark energy would lead to the temporal rate of change of the Stokes parameters:

$$\dot{Q} \pm i\dot{U} = \mp i2\omega \left(Q \pm iU\right), \tag{3}$$

and the evolution equations for the Fourier modes of the Stokes parameters are modified to

$$\begin{aligned} \dot{\Delta}_{Q\pm iU}(\mathbf{k},\eta) \quad + \quad ik\mu\Delta_{Q\pm iU}(\mathbf{k},\eta) = n_e\sigma_T a(\eta)\Big[-\Delta_{Q\pm iU}(\mathbf{k},\eta) \\ + \sum_m \sqrt{\frac{6\pi}{5}}\ {}_{\pm2}Y_2^m(\hat{\mathbf{n}})S_P^{(m)}(\mathbf{k},\eta)\Big] \mp i2\omega\Delta_{Q\mp iU}(\mathbf{k},\eta), \end{aligned} \tag{4}$$

where $\mu = \hat{\mathbf{n}} \cdot \hat{\mathbf{k}}$ is the cosine of the angle between the CMB photon direction and the Fourier wave vector, n_e is the number density of free electrons, σ_T is the Thomson cross section, and a is the scale factor. ${}_sY_l^m$ is spherical harmonics with spin-weight s with $m = 0, \pm1, \pm2$ corresponding to scalar, vector, and tensor perturbations, respectively, if the axis of ${}_sY_l^m$ is aligned with the wave vector $\mathbf{k}$. $S_P^{(m)}$ is the source term of generating polarization, which is the composition of the quadrupole components of the temperature and polarization perturbations $S_P^{(m)}(\mathbf{k},\eta) \equiv \Delta_{T2}^{(m)}(\mathbf{k},\eta) + 12\sqrt{6}\Delta_{+,2}^{(m)}(\mathbf{k},\eta) + 12\sqrt{6}\Delta_{-,2}^{(m)}(\mathbf{k},\eta)$. We have followed the notation in Ref. [9] and expanded the perturbations in the spin-0 and spin-2 spherical harmonics [10] according to the scalar and tensor properties of the temperature anisotropy and polarization respectively. Thus, $\Delta_{T2}^{(m)}$ and $\Delta_{\pm,2}^{(m)}$ are the respective expansion coefficients for Δ_T on basis $Y_{l,m}$ and $\Delta_{Q\pm iU}$ on basis ${}_{\pm2}Y_{lm}$.

We are used to decomposing the polarization on the sky into a divergence free component, the so-called E mode, and a curl component, the so-called B mode because the values for Q and U depend on the choice of a coordinate system. Whether B mode is generated depends on the existence of U for the local mode whose wave vector $\mathbf{k}$ parallels to the $\hat{\mathbf{z}}$ of the coordinates whereas Q is defined as the difference in intensity polarized in the $\hat{\theta}$ and $\hat{\psi}$ directions [11]. Mathematically, for $m = 0$, only Q is generated in the local mode. The axisymmetry of the radiation field about the mode axis guarantees that no B mode can be generated by scalar mode perturbations.

In the presence of cosmological birefringence, we can find two important features in Eq. (4). Firstly, the rotation of the polarization plane generates U contributions to the local mode polarization. This converts the power from the E mode to B mode. The conversion depends on how much rotation there is from the epoch when the polarization is generated to today. Secondly, TB and EB cross correlations are expected to vanish due to the parity (T and E have parity $(-1)^\ell$ while B has $(-1)^{\ell+1}$). The cosmological birefringence violates the parity and thus generates the TB and EB power spectra whose magnitudes depend on the integrated rotation of polarization too. Substituting the angular velocity of the polarization plane in Eq. (7) into Eq. (4), we calculate the power spectra T, E, B, TE, TB, and EB modes.

These six power spectra form a complete two-point statistics of CMB temperature and polarization anisotropies. To simplify the calculation, we only include the scalar perturbations. That is by setting $m = 0$ in Eq. (4). Without showing the details, we just write down the power spectra which are obtained from the solutions for the line-of-sight integration as

$$\begin{aligned}
C_\ell^{(E,B)} &= (4\pi)^2 \frac{9}{16}\frac{(\ell+2)!}{(\ell-2)!} \int k^2 \mathrm{d}k \left[\Delta_{(E,B)\ell}(k,\eta_0)\right]^2, \\
C_\ell^{EB} &= (4\pi)^2 \frac{9}{16}\frac{(\ell+2)!}{(\ell-2)!} \int k^2 \mathrm{d}k \Delta_{E\ell}(k,\eta_0)\Delta_{B\ell}(k,\eta_0), \\
C_\ell^{TE} &= (4\pi)^2 \sqrt{\frac{9}{16}\frac{(\ell+2)!}{(\ell-2)!}} \int k^2 \mathrm{d}k \Delta_{T\ell}(k,\eta_0)\Delta_{E\ell}(k,\eta_0), \\
C_\ell^{TB} &= (4\pi)^2 \sqrt{\frac{9}{16}\frac{(\ell+2)!}{(\ell-2)!}} \int k^2 \mathrm{d}k \Delta_{T\ell}(k,\eta_0)\Delta_{B\ell}(k,\eta_0), \qquad (5)
\end{aligned}$$

where

$$\begin{aligned}
\Delta_{T\ell}(k,\eta_0) &= \int_0^{\eta_0} \mathrm{d}\eta g(\eta) S_T(k,\eta) j_\ell(kr), \\
\Delta_{E\ell}(k,\eta_0) + i\Delta_{B\ell}(k,\eta_0) &= \int_0^{\eta_0} \mathrm{d}\eta g(\eta) S_P(k,\eta) \frac{j_\ell(kr)}{(kr)^2} e^{i2\alpha(\eta)}, \qquad (6)
\end{aligned}$$

where the visibility function $g(\eta)$ describes the probability that a photon scattered at epoch η reaches the observer at the present time, η_0. Similar to $S_P \equiv S_P^{(0)}$, S_T is the source term generating the temperature anisotropy. j_ℓ is the spherical Bessel function and $r = \eta_0 - \eta$. The rotation angle $\alpha(\eta) = \int_\eta^{\eta_0} d\eta' \omega(\eta')$. We do not present the formula for the temperature anisotropy because it is unchanged under the rotation of the polarization plane.

The last term in Eq. (4) appears due to the rotation of the polarization plane. This rotation can occur, as proposed by Carroll [5], due to the coupling $\beta_{F\tilde{F}}\phi/\bar{M} F_{\mu\nu}\tilde{F}^{\mu\nu}$, if ϕ is varying with time. Here $\beta_{F\tilde{F}}$ is the coupling strength, $\bar{M}$ is the reduced Planck mass, and $\tilde{F}^{\mu\nu}$ is the dual of the electromagnetic tensor. The dispersion relation for electromagnetic radiation coupling to the time varying quintessence field ϕ is given by $E^2 = k^2 \pm k\beta_{F\tilde{F}}\dot{\phi}/(a\bar{M})$, where $\pm$ refer to the right and left handed circular polarization, respectively. Therefore, the net angular velocity of the polarization plane is [5]

$$\omega = 2\beta_{F\tilde{F}} \frac{\dot{\phi}}{a\bar{M}}. \qquad (7)$$

In Figure 1, we show the integrated rotation angle of CMB polarization plane for different coupling strength. We consider the potential $V(\phi) = V_0 \exp(\lambda\phi^2/2\bar{M}^2)$ for our quintessence model, where λ is a parameter determining how shallow the potential is. Hereafter we fix $\lambda = 5$ and we will obtain similar results by choosing other values of λ.

We show the power spectra T, E, B, TE, TB, and EB modes with the coupling strength $\beta_{F\tilde{F}}$ ranging from 10^{-5} to 10^{-3} from Figure 2 to Figure 5. On small scales, increasing coupling

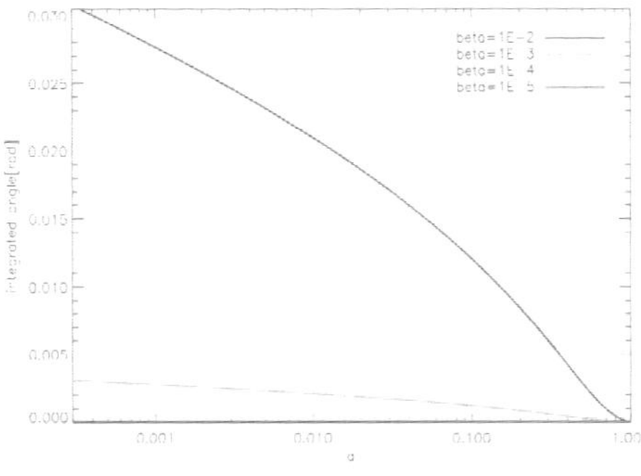

Figure 1. Integrated rotation angle of CMB polarization plane for different coupling strength

strength results in a suppression of the E mode in the standard model and non-vanishing B and EB modes. Furthermore, the shapes of the B and EB mode power spectra basically follow the standard E mode except the reionization bump on large scales.

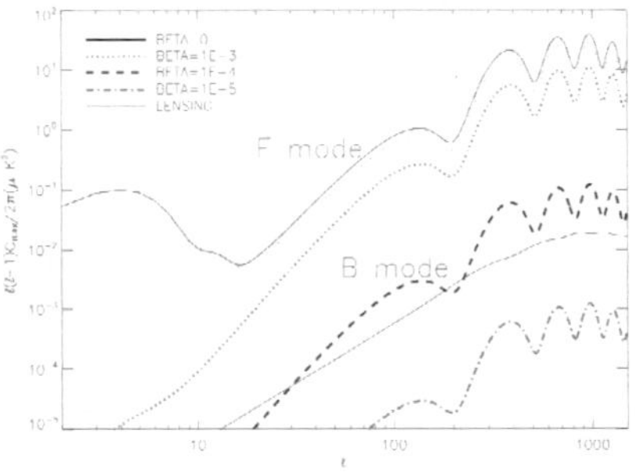

Figure 2. E, B mode power spectra from the cosmological birefringence with different coupling strength.

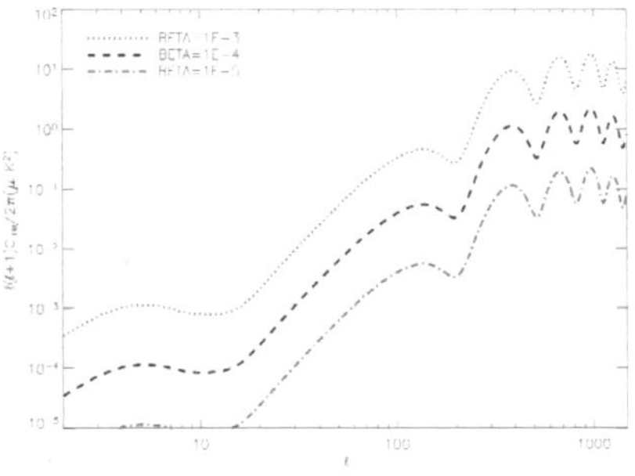

Figure 3. EB mode power spectra from the cosmological birefringence with different coupling strength.

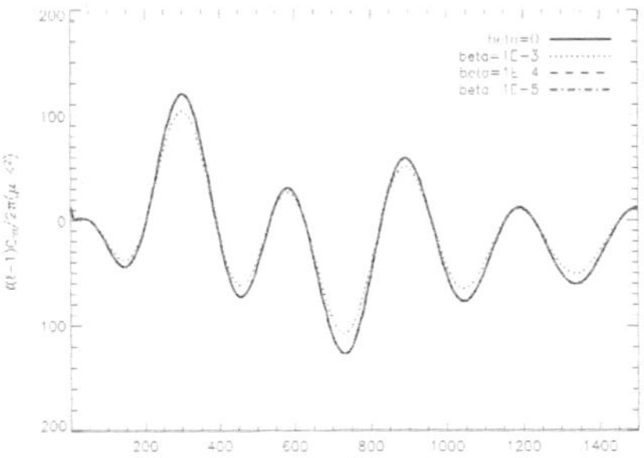

Figure 4. TE mode power spectra from the cosmological birefringence with different coupling strength.

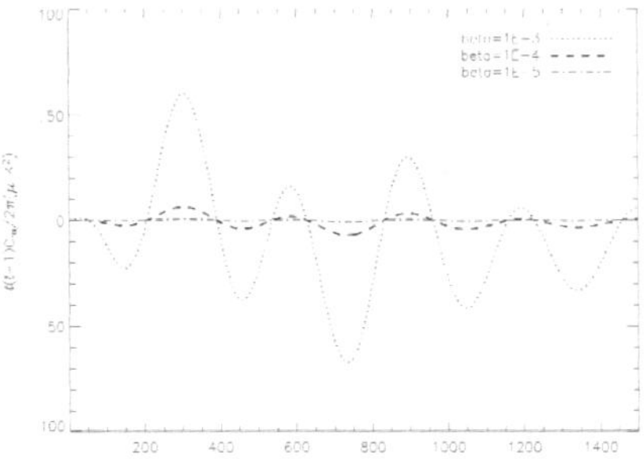

Figure 5. TB mode power spectra from the cosmological birefringence with different coupling strength.

If the coupling strength is large enough, the B mode induced by the cosmological birefringence will mix up with the gravitational lensing-induced B mode. Gravitational lensing by large scale

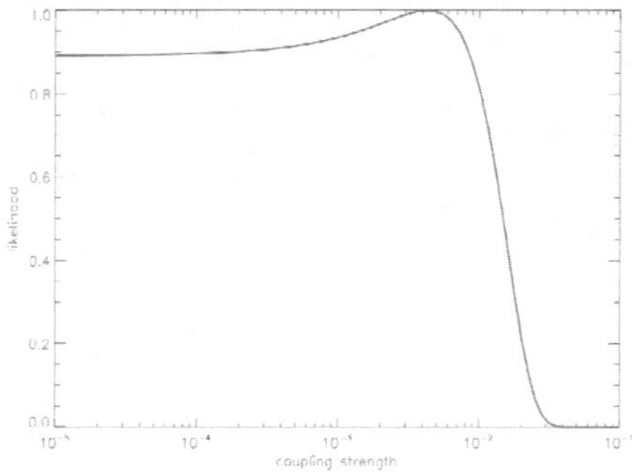

Figure 6. Likelihood function of the coupling strength.

structures modifies slightly the primary E mode power spectrum. Most noticeably it generates, through mode coupling, B mode polarization out of pure E mode signal [12]. The lensing-induced B power spectrum, which peaks around $\ell \sim 1000$, has the roughly similar shape with that from the birefringence. We also show the power spectrum of the lensing-induced B mode in Fig. 1 by a thin solid curve for comparison. The birefringence-induced B mode is indeed compatible with the lensing-induced B mode for $\beta_{F\tilde{F}} \sim 10^{-4}$.

3. Constraints from CMB data

The coupling strength can be constrained by the measured polarization data. It must be very sensitive, especially, to the TB and EB power spectra. Recently, more and more CMB experiments produce TB and EB data for this issue, for example, Boomerang 2003 [13], QUad 2006-2006 data [14] and WMAP 7-year data. Here we use a public likelihood software, which is used by the WMAP team to compute Fishre and Master matrices, and compute the likelihoods of various mode, to fit the coupling strength. The CMB power spectra is computed with the cosmological birefringence effect while fixing other cosmological parameters based on the 7-year WMAP data. We turn on the TE and low-ℓ TB, EB and high-ℓ master TB power spectra for our purpose. Then we convert χ^2 to the likelihood by $\mathcal{L} = e^{-\chi^2/2}$ and normalize the maximum likelihood value to unity. The result is shown in Figure 6. There is an "unclear" peak of the likelihood located at $\beta_{F\tilde{F}} = 4.6 \times 10^{-3}$, which corresponds to the integrated rotation angle from recombination epoch to today is about -0.9 degree. This result is consistent with Komatsu et al. 2009 [15], in which rotation angle is $-1.1^o \pm 1.3^o(\text{stat.}) \pm 1.5^o(\text{syst.})$ This small value of the coupling strength gives an insignificant change on TE and E modes and thus will not affect the determination of the cosmological parameters.

4. Discussion and conclusion

We have computed the power spectra of CMB polarization with the effect of cosmological birefringence by the use of the full Boltzmann code. Comparing with a rough estimation in Eq. (5): $C_{B\ell} \sim C_{E\ell} \sin^2 2\alpha_\ell$ and $C_{EB\ell} \sim 0.5 C_{E\ell} \sin 4\alpha_\ell$ used by some authors, the full Boltzmann code is more accurate. In the rough estimation, the total rotated angle, α_ℓ, for certain angular scales $\theta \sim \pi/\ell$ from last scattering epoch to today is assumed to be constant for all scales. This in general is not correct. The E mode power on small scales mainly comes from the recombination epoch at $z \sim 1100$. On the other hand, the boosting power on large scales comes from reionization epoch when the CMB photons are rescattered by free electrons at $z \sim 10$ [16]. From Eq. (7) and the evolution of ϕ, we find that the integrated rotation angle from the reionization epoch is much smaller than that from the recombination epoch. Therefore, there is much less power converted from E mode to B mode on large scales than small scales.

We have also constrained the coupling strength $\beta_{F\tilde{F}}$ by the use of WMAP 7-year data. We found an peak of the likelihood located at $\beta_{F\tilde{F}} = 4.6 \times 10^{-3}$. However, it is not a conclusive value because the peak is not clear. Furthermore, it is remarkable that the rotation-induced B mode with this value of the coupling strength exceeds the lensing-induced B mode. Therefore, more careful measurements of TB and EB are necessary for separating the two effects.

5. References

[1] Bahcall N A, Ostriker J P, Perlmutter S and Steinhardt P J 1999 *Science* **284** 1481
[2] Riess A G *et al* 2004 *Astrophys. J.* **607** 665
[3] Weinberg S 1989 *Rev. Mod. Phys.* **61** 1
[4] Ratra B and Peebles P J E 1988 *Phys. Rev. D.* **37** 3406
Frieman J A, Hill C T, Stebbins A and Waga I 1995 *Phys. Rev. Lett.* **75** 2077
Sahni V and Wang L 2000 *Phys. Rev. D* **62** 103517
Lee S, Olive K A and Pospelov M 2004 *Phys. Rev. D* **70** 083503
Olivares G, Atrio-Barandela F and Pavon D 2005 *Phys. Rev. D* **71** 063523
[5] Carroll S M 1998 *Phys. Rev. Lett.* **81** 3067
[6] Larson D *et al* 2010 *ApJS* **192** 16
[7] Chandrasekar S 1960 *Radiative Transfer* (Dover New York)
[8] Peebles P J E and Yu J T 1970 *Astrophys. J.* **162** 815
Bond J R and Efstathiou G 1984 *Astrophys. J.* **285** L45
Ma C P and Bertschinger E 1995 *Astrophys. J.* **455** 7
[9] Liu G C, Sugiyama N, Benson A J, Lacey C G and Nusser A 2001 *Astrophys. J.* **561** 504
[10] Newman E and Penrose R 1966 *J. Math. Phys.* **7** 863
[11] Hu W 1997 *New Astronomy* **2** 323
[12] Zaldarriaga M and Seljak U 1998 *Phys. Rev. D* **58** 023003
Benabed K, Bernardeau F and van Waerbeke L 2001 *Phys. Rev. D* **63** 043501
[13] Piacentini F *et al* 2006 *Astrophys. J.* **647** 833
Montroy T E *et al* 2006 *Astrophys. J.* **647** 813
[14] Brown M L *et al* 2009 *Astrophys. J.* **705** 978
[15] Komatsu E *et al* 2011 *ApJS* **192** 18
[16] Page L *et al* 2007 *ApJS* **170** 335

Recent Developments in Gravity (NEB XIV) IOP Publishing
Journal of Physics: Conference Series **283** (2011) 012021 doi:10.1088/1742-6596/283/1/012021

Quasi-equilibrium models of magnetized compact objects

Charalampos Markakis[1], **Kōji Uryū**[2] **and Eric Gourgoulhon**[3]

[1]Department of Physics, University of Wisconsin-Milwaukee, P.O. Box 413, Milwaukee, WI 53201, USA

[2]Department of Physics, University of the Ryukyus, Senbaru, Nishihara, Okinawa 903-0213, Japan

[3]Laboratoire Univers et Théories, UMR 8102 du CNRS, Observatoire de Paris, Université Paris Diderot, F-92190 Meudon, France

E-mail: markakis@uwm.edu[1], uryu@sci.u-ryukyu.ac.jp[2], eric.gourgoulhon@obspm.fr[3]

Abstract. We report work towards a relativistic formulation for modeling strongly magnetized neutron stars, rotating or in a close circular orbit around another neutron star or black hole, under the approximations of helical symmetry and ideal MHD. The quasi-stationary evolution is governed by the first law of thermodynamics for helically symmetric systems, which is generalized to include magnetic fields. The formulation involves an iterative scheme for solving the Einstein-Maxwell and relativistic MHD-Euler equations numerically. The resulting configurations for binary systems could be used as self-consistent initial data for studying their inspiral and merger.

1. Introduction

A uniformly rotating neutron star is significantly deformed when the ratio of kinetic energy T to gravitational energy W becomes $T/|W| \sim 0.1$. When the magnetic field energy,

$$\mathcal{M} := \frac{1}{8\pi}\int B^2 d^3x, \tag{1}$$

becomes such that $\mathcal{M}/|W| \sim 0.01$, the contribution of the magnetic field to the structure of the neutron srar may not be neglected. This is estimated to occur for

$$B \sim 4.4 \times 10^{16} \left(\frac{M\,[M_\odot]}{1.4\,[M_\odot]}\right)\left(\frac{10\,[\mathrm{km}]}{R\,[\mathrm{km}]}\right)^2 [\mathrm{G}]. \tag{2}$$

Recent observations of anomalous X-ray pulsars, or soft γ-ray repeaters suggest that the neutron stars in these systems may be associated with strong magnetic fields around $10^{14} - 10^{15}$ G at the surface (see, e.g. [1]). We may expect that the interior magnetic field of such a strongly magnetized neutron star, a magnetar, is a few orders of magnitudes stronger.

Such strong magnetic fields have not been found in binary neutron star systems, and may not survive until merger. Hypothetically, however, strongly magnetized neutron stars or black holes may form binary neutron star or black hole - neutron star systems. For instance, the poloidal

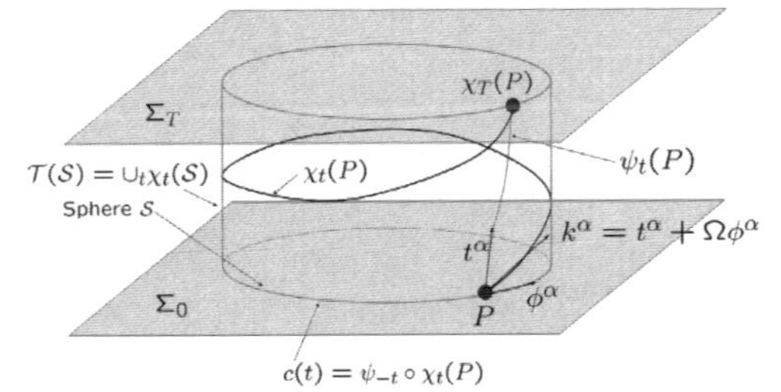

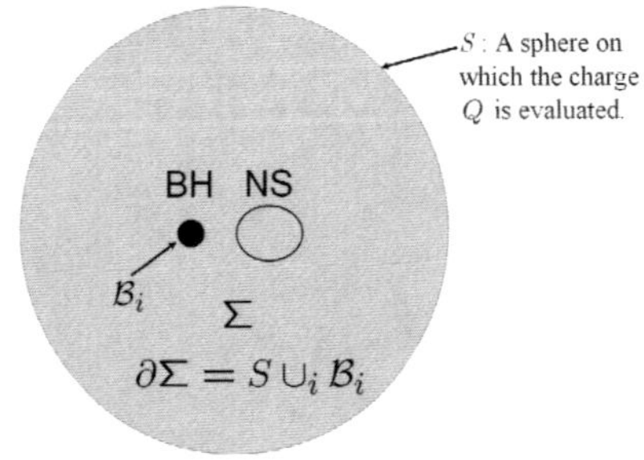

Figure 1. Left panel: the history of a sphere $\mathcal{S}$ generated by the helical vector k^α. Right panel: a domain of integration Σ and its boundaries.

field seen as the surface magnetic field decays but the toroidal field stays strong enough to affect the structure of the compact objects.

In several numerical relativity simulations of magnetized binary neutron stars, it has been shown that, after binary inspiral and merger, the magnetic fields are amplified around 10 times by magnetic winding and the magnetorotational instability in post-merger, pre-collapse objects [2]. Such an object is likely to form a black hole and a magnetized toroid system, which becomes the source of a short γ-ray burst. It is desirable to prepare realistic initial data sets for such merger simulations calculated by solving the Einstein-Maxwell equations and a first integral of the MHD-Euler equation, assuming (quasi-)equilibrium. In this article, we model such magnetized binary compact objects in close circular orbits, assuming that the spacetime and magnetic fields satisfy helical symmetry and that the stars are in equilibrium [3–5].

2. Thermodynamic laws for helically symmetric Einstein-Maxwell spacetimes with charged and magnetized perfect fluids

2.1. Zeroth law

We consider a globally hyperbolic spacetime $(\mathcal{M}, g_{\alpha\beta})$. The field k^α is transverse to each Cauchy surface, not necessarily timelike everywhere, and generates a one-parameter family of diffeomorphisms χ_t. The action of χ_t on a spacelike sphere $\mathcal{S}$ on a Cauchy surface generates a timelike surface, $\mathcal{T}(\mathcal{S}) = \cup_t \chi_t(\mathcal{S})$, called the *history of* $\mathcal{S}$. Then, k^α is a *helical vector* field if there is a smallest $T > 0$ for which P and $\chi_T(P)$ are timelike separated for every point P outside of the history $\mathcal{T}(\mathcal{S})$. A vector k^α written as

$$k^\alpha = t^\alpha + \Omega\phi^\alpha \tag{3}$$

is the helical vector, where t^α is a timelike vector and ϕ^α a spacelike vector that has circular orbits with a parameter length 2π (see, FUS). In Fig. 1, a schematic figure of the history $\mathcal{T}(\mathcal{S})$ generated by the diffeo χ_t is presented.

Each Cauchy surface of a helically symmetric spacetime does not admit flat asymptotics. Therefore, the future (past) horizon $\mathcal{H}^\pm$ is defined as the boundary of the future(past) domain of outer communication $\mathcal{D}^\pm$ of a history $\mathcal{T}(\mathcal{S})$ of each spacelike sphere $\mathcal{S}$. If the history $\mathcal{T}(\mathcal{S})$ of a sphere $\mathcal{S}$ is in $\mathcal{D}^\pm$, the future (past) horizon agrees with the chronological past (future) of the history $\mathcal{T}$, $\mathcal{H}^\pm = \partial I^\mp(\mathcal{T})$.

The existence of a global helical symmetry assures that the horizon is a Killing horizon. With the null energy condition $R_{\alpha\beta}l^\alpha l^\beta \geq 0$ for any null vector l^α, the surface gravity κ, defined on each connected component of the horizon $\mathcal{H}^\pm$ by

$$k^\beta \nabla_\beta k^\alpha = \kappa k^\alpha, \tag{4}$$

is constant. The proof is given in FUS, in which the conditions of theorems by Friedrich, Rácz, and Wald [6] are modified to make them suitable for helically symmetric spacetimes. Also the electric potential $\Phi^{\rm E}$ in the rotating frame, defined by

$$E_\alpha = F_{\alpha\beta}k^\beta = -\nabla_\alpha\Phi^{\rm E}, \qquad \Phi^{\rm E} = A_\alpha k^\alpha + \text{const}, \tag{5}$$

is constant on $\mathcal{H}^\pm$, assuming the symmetry $\pounds_k A_\alpha = 0$. Since E_α is null on $\mathcal{H}^\pm$ and $E_\alpha k^\alpha = 0$, E_α is parallel to the null generator on $\mathcal{H}^\pm$. Hence for any vector η^α tangent to $\mathcal{H}^\pm$, $\eta^\alpha E_\alpha = -\eta^\alpha\nabla_\alpha\Phi^{\rm E} = 0$. Therefore, as for the stationary and axisymmetric spacetimes shown by Carter [7, 8], the surface gravity κ and the electric potential $\Phi^{\rm E}$ in the rotating frame are constant on the horizon $\mathcal{H}^\pm$.

2.2. First law

Consider a family of spacetimes,

$$\mathcal{Q}(\lambda) := [g_{\alpha\beta}(\lambda), u^\alpha(\lambda), \rho(\lambda), s(\lambda), A_\alpha(\lambda), j^\alpha(\lambda)], \tag{6}$$

whose Lagrangian density is written

$$\mathcal{L} = \left(\frac{1}{16\pi}R - \epsilon - \frac{1}{16\pi}F_{\alpha\beta}F^{\alpha\beta} + A_\alpha j^\alpha\right)\sqrt{-g}, \tag{7}$$

where u^α, ρ, s, ϵ, and j^α are, respectively, the 4-velocity, baryon rest mass density, entropy per baryon mass, energy density, and electric 4-current. A generalized first law for the spacetime $\mathcal{Q}(\lambda)$ associated with the helical symmetry is derived as a variation formula of the Noether charge associated with the helical Killing vector defined by [9, 10]

$$Q(\lambda) = \oint_S Q^{\alpha\beta} dS_{\alpha\beta}, \tag{8}$$

$$\mathfrak{B}^\alpha(\lambda) = \frac{1}{16\pi}(g^{\alpha\gamma}g^{\beta\delta} - g^{\alpha\beta}g^{\gamma\delta})|_{\lambda=0}\, \mathring{\nabla}_\beta\, g_{\gamma\delta}(\lambda) + \frac{1}{4\pi}F^{\beta\alpha}|_{\lambda=0}\Big[A_\beta(\lambda) - \frac{1}{2}A_\beta(0)\Big] + \mathrm{O}(\lambda^2), \tag{9}$$

With this choice of $\mathfrak{B}^\alpha$, the charge $Q(\lambda)$ becomes finite and independent of the sphere S used to evaluate the charge $Q(\lambda)$ as long as S encloses all black holes and neutron stars.

To see this, we first check that $Q(0)$ is independent of S. The surface integral of (8) is rewritten in terms of integrals over a spacelike hypersurface Σ transverse to k^α and over the black hole boundary $\mathcal{B}_i$, where $\mathcal{B}_i$ is the i-th connected component of $\Sigma \cap \mathcal{H}^+$. In Fig.1, the region of integration is shown. Since, the boundary of Σ is the union of the sphere S and $\mathcal{B}_i$, $\partial\Sigma = S \cup_i \mathcal{B}_i$, a difference $Q - \sum_i Q_i$ is calculated at $\lambda = 0$:

$$\begin{aligned} Q - \sum_i Q_i =& -\frac{1}{8\pi}\int_\Sigma (G^\alpha{}_\beta k^\beta - 8\pi T_{\rm F}{}^\alpha{}_\beta) dS_\alpha - \frac{1}{16\pi}\int_\Sigma R k^\alpha dS_\alpha. \\ &+ \int_\Sigma \left(\frac{1}{8\pi}\nabla_\gamma F^{\beta\gamma}A_\beta k^\alpha - \frac{1}{4\pi}k^\gamma A_\gamma \nabla_\beta F^{\alpha\beta}\right) dS_\alpha - \sum_i \frac{1}{4\pi}\int_{\mathcal{B}i} k^\gamma A_\gamma F^{\alpha\beta} dS_{\alpha\beta}. \end{aligned} \tag{10}$$

where $Q_i(\lambda)$ is defined on $\mathcal{B}_i$ by

$$Q_i(\lambda) := \oint_{\mathcal{B}_i} Q^{\alpha\beta} dS_{\alpha\beta}, \tag{11}$$

$T_{\rm F}{}^{\alpha}{}_{\beta}$, the stress-energy tensor of the electromagnetic field, is defined by,

$$T_{\rm F}^{\alpha\beta} = \frac{1}{4\pi}\left(F^{\alpha\gamma}F^{\beta}{}_{\gamma} - \frac{1}{4}g^{\alpha\beta}F_{\gamma\delta}F^{\gamma\delta}\right). \tag{12}$$

To derive Eq. (10), we have also used a condition

$$\frac{1}{4\pi}\oint_S k^\gamma A_\gamma F^{\alpha\beta}dS_{\alpha\beta} = 0 \tag{13}$$

on the boundary sphere S to determine the constant of the electric potential $\Phi^{\rm E}$ in Eq. (5). Eq. (10) implies that $Q(0)$ does not depend on the sphere S as long as it encloses all black holes and neutron stars, because in the region where the sphere S is located, the integrand of the volume integrals in Eq. (10) vanishes when the Einstein and Maxwell equations are satisfied.

Next, the variation $\delta Q := dQ/d\lambda$ in the Noether charge is evaluated in terms of perturbations of the baryon mass, entropy, circulation and electric current of each fluid element, and the surface areas and charges of the black holes. The detailed calculation is described in [11]: the variation δQ becomes

$$\begin{aligned}\delta Q \;=\; &\int_\Sigma \left\{ \frac{T}{u^t}\Delta(s\,\rho u^\alpha\, dS_\alpha) + \frac{h-Ts}{u^t}\Delta(\rho u^\alpha\, dS_\alpha) + v^\beta\Delta(hu_\beta\,\rho u^\alpha\, dS_\alpha) - A_\beta k^\beta\,\Delta(j^\alpha dS_\alpha)\right.\\ &\left. - (j^\alpha k^\beta - j^\beta k^\alpha)\Delta A_\beta\, dS_\alpha\right\} + \sum_i \left(\frac{1}{8\pi}\kappa_i\delta\mathcal{A}_i + \Phi_i^{\rm E}\,\delta Q_i^{\rm E}\right),\end{aligned} \tag{14}$$

where T is the temperature, h is the relativistic specific enthalpy, and v^α is the spatial velocity defined by the following decomposition of the 4-velocity u^α with respect to the helical vector k^α:

$$u^\alpha = u^t(k^\alpha + v^\alpha) \quad \text{with} \quad v^\alpha\nabla_\alpha t = 0 \quad \text{and} \quad u^t = u^\alpha\nabla_\alpha t. \tag{15}$$

The electric charge of each black hole $Q_i^{\rm E}$ is defined by

$$Q_i^{\rm E} := \frac{1}{4\pi}\oint_{\mathcal{B}i} F^{\alpha\beta}dS_{\alpha\beta}, \tag{16}$$

which is related to the total electric charge of the system $Q^{\rm E}$ by Stokes' theorem:

$$Q^{\rm E} := \frac{1}{4\pi}\oint_S F^{\alpha\beta}dS_{\alpha\beta}. = \int_\Sigma j^\alpha dS_\alpha + \sum_i Q_i^{\rm E}. \tag{17}$$

Note that the $\Phi_i^{\rm E}$, defined on each $\mathcal{B}_i$ by

$$\Phi_i^{\rm E} = -A^\alpha k_\alpha = \Phi^{\rm E} + C, \tag{18}$$

is constant. In the case of stationary and axisymmetric spacetimes, the mass variation formula derived by Carter [7, 8] can be derived from Eq. (14).

We can now verify that $Q(\lambda)$ is independent of the location of the 2-surface S, because the charge $Q(\lambda)$ at $\lambda = 0$ is shown to be independent of S, and the variation formula (14) implies that $dQ/d\lambda = \delta Q$ is independent of S as long as it encloses the fluid and black holes. Moreover, in [11] we have shown that the difference,

$$\delta\Big(Q - \sum_i Q_i - \frac{1}{4\pi}\int_{\partial\Sigma} k^\gamma A_\gamma F^{\alpha\beta}dS_{\alpha\beta}\Big) = \delta Q - \sum_i \Big(\frac{1}{8\pi}\kappa_i\delta\mathcal{A}_i + \Phi_i^{\rm E}\,\delta Q_i^{\rm E}\Big), \tag{19}$$

is invariant under a gauge transformation that respects the symmetry, and hence we verify that so is δQ.

2.3. Application of the first law to solution sequences in equilibrium

When inspiraling binary systems, or isolated neutron stars are evolving adiabatically – in a timescale much longer than the dynamical timescale – they may be modeled by a sequence of solutions in equilibrium. When the first law for a stationary and axisymmetric perfect fluid spacetime is applied to a sequence of rotating neutron star solutions whose local changes of rest mass and entropy are constant, the first law becomes $\delta M = \Omega\delta J$. This is a condition for applying the turning point theorem [12] to determine the stability of solutions. In this section, we consider an application of the first law (14) to a sequence of helically symmetric solutions that model inspiraling binary black holes and/or neutron stars, assuming that the neutron star matter is a perfect conductor.

For a perfectly conducting medium, the ideal MHD condition,

$$F_{\alpha\beta}u^\beta \;=\; 0, \tag{20}$$

is satisfied, and hence its curl is written

$$\pounds_u F_{\alpha\beta} \;=\; 0, \tag{21}$$

which is the magnetic flux conservation law, Alfven's law. For each equilibrium solution, rest mass and entropy are conserved:

$$\pounds_u(\rho\sqrt{-g}) = 0, \quad \pounds_u s = 0. \tag{22}$$

When we consider a sequence of solutions along which the rest mass, entropy, and magnetic flux are all conserved, the perturbed conservation laws corresponding to the above Eqs. (21) and (22) have first integrals:

$$\Delta(\rho\sqrt{-g}) = 0, \quad \Delta s = 0, \quad \text{and} \quad \Delta F_{\alpha\beta} = 0. \tag{23}$$

For black holes, we may assume that areas and charges are constant. With these assumptions, the first law (14) is rewritten

$$\delta Q \;=\; \int_\Sigma \Big\{ v^\beta \Delta(hu_\beta\,\rho u^\alpha\, dS_\alpha) \;-\; A_\beta k^\beta\,\Delta(j^\alpha dS_\alpha) \Big\}. \tag{24}$$

Here $\Delta F_{\alpha\beta} = 0$ with $F_{\alpha\beta} = (dA)_{\alpha\beta}$, $\Delta(dA)_{\alpha\beta} = (d\Delta A)_{\alpha\beta}$, whence the Poincaré lemma implies that there exists a function Ψ such that $\Delta A_\alpha = \nabla_\alpha\Psi$. From a substitution of this to the last term in the volume integral of Eq. (14), the latter is shown to vanish. The remaining terms in Eq. (24) are related to the circulation of the magnetized flow.

For a perfect fluid without magnetic fields, the conservation of circulation is written as $\pounds_u\hat{\omega}_{\alpha\beta} = 0$ where $\hat{\omega}_{\alpha\beta} := (d(hu))_{\alpha\beta}$ is the vorticity tensor. When circulation is conserved along a sequence of solutions, it implies that $\Delta\hat{\omega}_{\alpha\beta} = 0$, and hence the first term in the integral of Eq. (24) vanishes [13]. In general, there is no such conservation law for the circulation of magnetized flow. However, Bekenstein and Oron [14] (Tarapov and Gorskii [15] for Newtonian MHD) have developed a formulation of ideal MHD, in which a generalized circulation of magnetized flow is conserved. Their theory is based on a Lagrangian in which the interaction term $A_\alpha j^\alpha$ in Eq. (7) is replaced by the term $F_{\alpha\beta}\rho u^\alpha q^\beta$, where the vector q^α is a Lagrange multiplier enforcing the perfect conductivity condition $F_{\alpha\beta}u^\alpha = 0$. Variation with respect to A_α yields the Maxwell equations with a current of the form

$$j^\alpha \;=\; \nabla_\beta(\rho u^\alpha q^\beta - \rho u^\beta q^\alpha). \tag{25}$$

When a vector q^α is found to give the current (25), the Lorenz force term in the MHD-Euler equation

$$u^\beta (d(hu))_{\beta\alpha} = \frac{1}{\rho} F_{\alpha\beta} j^\beta \tag{26}$$

can be absorbed into the left hand side, bringing eq. (26) in the canonical form

$$u^\beta \omega_{\beta\alpha} = 0, \tag{27}$$

where

$$\omega_{\alpha\beta} := (dw)_{\alpha\beta}, \quad w_\alpha := hu_\alpha + \eta_\alpha, \quad \eta_\alpha := F_{\alpha\beta} q^\beta, \tag{28}$$

because of a relation,

$$\frac{1}{\rho} F_{\alpha\beta} j^\beta = \frac{1}{\rho} F_{\alpha\beta} \left[\pounds_q (\rho u^\beta) + \rho u^\beta \nabla_\gamma q^\gamma \right] = (d\eta)_{\alpha\beta} u^\beta. \tag{29}$$

The one-from w_α and its exterior derivative $\omega_{\alpha\beta}$ can be respectively regarded as a canonical momentum of a magneto-fluid element and a generalized vorticity. The MHD-Euler equation (27) in this case implies conservation of circulation for the magnetized flow, because its curl, with $d\omega = 0$, yields

$$\pounds_u \omega_{\alpha\beta} = 0, \tag{30}$$

and the Lagrangian perturbation of this conservation law has a first integral

$$\Delta \omega_{\alpha\beta} = 0. \tag{31}$$

Substituting the current (25) to the first law (14), we have

$$\begin{aligned} \delta Q \;=\; & \int_\Sigma \left\{ \frac{T}{u^t} \Delta dS + \frac{h - Ts}{u^t} \Delta dM_{\rm B} + v^\alpha \Delta dC_\alpha - v^\beta q^\gamma \Delta F_{\beta\gamma}\, dM_{\rm B} - (j^\alpha k^\beta - j^\beta k^\alpha) \Delta A_\beta\, dS_\alpha \right\} \\ & + \sum_i \left(\frac{1}{8\pi} \kappa_i \delta A_i + \Phi_i^{\rm E}\, \delta Q_i^{\rm E} \right), \end{aligned} \tag{32}$$

where we introduced the notation

$$dM_{\rm B} := \rho u^\alpha dS_\alpha, \quad dS := s\, dM_{\rm B}, \quad dC_\alpha := (hu_\alpha + \eta_\alpha) dM_{\rm B}. \tag{33}$$

Applying the first law (32) to a sequence of solutions along which the quantities are conserved as in Eq. (23) and the circulation of magnetized flow is conserved as in Eq. (31), the first law becomes

$$\delta Q = 0, \tag{34}$$

or for asymptotically flat systems, such as in the post-Newtonian approximation, $\delta Q = \delta M - \Omega \delta J = 0$.

3. A formulation for computing magnetized neutron star equilibria

To compute solutions of single or binary compact objects in equilibrium, we apply a finite difference scheme, or a pseudo-spectral method, to a system of basic equations and numerically solve the system. Those equations include the Einstein equation, the Maxwell equations, the MHD-Euler equation, and the baryon mass conservation equation. Here, we assume that the flow is homentropic, and hence the neutron-star matter is described by a one-parameter EOS.

For the gravitational field, the Isenberg - Wilson - Mathews formulation, the waveless formulation, or a full set of Einstein's equations assuming helical symmetry may be used. These formulations are based on a 3+1 decomposition of spacetime, and reliable numerical methods to solve these equations have been developed [16–18]. We expect that the Maxwell equations can be solved using analogous formulations and applying one of the above numerical methods.

However, for the MHD-Euler equation, when stationarity or helical symmetry is imposed, it is no longer an evolution equation, and as a result it is difficult to integrate numerically. In self-consistent field methods [16–19], an equilibrium solution is computed using a first integral of the (MHD-) Euler equation which exists when the flow is assumed to be either corotational or irrotational. Therefore, finding the first integral of the MHD-Euler equation is a key, and also a restriction, for computing equilibrium solutions considered in the previous section successfully.

If we assume that the orbit of the stars is closed and quasi-circular, then the system appears stationary in a frame rotating with frequency equal to the orbital frequency Ω, and there exists an approximate helical Killing vector k^α, given by eq. (3), that Lie-derives all variables in (6). The Cartan identity implies

$$k^\beta (dw)_{\beta\alpha} = \pounds_k w_\alpha - \nabla_\alpha (k^\gamma w_\gamma) \tag{35}$$

It should be noticed that $\pounds_k w_\alpha \neq 0$ in general, since q^α is not an observable quantity. One has

$$\pounds_k w_\alpha \;=\; \pounds_k \eta_\alpha \;=\; F_{\alpha\beta} \pounds_k q^\beta. \tag{36}$$

(In fact, if one assumes $\pounds_k w_\alpha = 0$ for a corotational flow, no Lorenz force is exerted on the matter). As shown below (see also [20]), the Cartan identity (35) leads quickly to conserved quantities when certain assumptions for the flow are made.

3.1. Irrotational magneto-flow

If one assumes that the magneto-flow is irrotational, i.e. described by a potential Φ:

$$w_\alpha = h u_\alpha + \eta_\alpha = \nabla_\alpha \Phi, \tag{37}$$

then the vorticity $\omega = dw$ vanishes and the MHD-Euler equation (27) is automatically satisfied while the left hand side of eq. (35) vanishes as well, so that

$$\pounds_k w_\alpha - \nabla_\alpha (k^\gamma w_\gamma) = 0 \tag{38}$$

The above equation has a first integral iff the term $\pounds_k w_\alpha$ is the gradient of some scalar function. Because of eq. (36), this integrability condition is written as $\pounds_k q^\beta F_{\beta\alpha} = \nabla_\alpha f$, or, using the Cartan identity,

$$\pounds_{[k,q]} F_{\alpha\beta} \;=\; 0, \tag{39}$$

where $[k,q]^\alpha := \pounds_k q^\alpha$. Then, eq. (35) implies that the quantity

$$k^\gamma w_\gamma + f \;=\; \mathcal{E} \tag{40}$$

is constant throughout the magneto-fluid.

In this case, we have four variables for the matter, two thermodynamic variables and two variables for the velocity fields, $\{h, p, \Phi, u^t\}$. The set of equations to solve for these four variables is supplemented by a one-parameter EOS, $p = p(h)$, and the normalization of the 4-velocity $u_\alpha u^\alpha = -1$. The velocity potential Φ is obtained from the rest mass conservation equation which is an elliptic equation for Φ with Neumann boundary conditions on the stellar surface. An additional degree of freedom may be obtained from the first integral (40), if it exists.

It is not trivial to find a vector q^α that satisfies both the integrability condition (39) and the ideal MHD condition (20); q^α is not freely specifiable. Instead of solving for a q^α that satisfies both conditions, let us assume that

$$q^\alpha = q^t k^\alpha \tag{41}$$

with $\pounds_k q^t$ regarded a function of $A_\alpha k^\alpha$. If we select q^α as above, at least instantaneously on an initial hypersurface Σ, then the first integral is written

$$k^\alpha w_\alpha + \int \pounds_k q^t \, d(A_\alpha k^\alpha) = \mathcal{E} \tag{42}$$

It is likely that solutions with neutron stars and strong magnetic fields computed from Eq. (42) may not be strictly in equilibrium. Nonetheless, such solutions with strong magnetic fields that satisfy a set of hydrostationary equations will serve as interesting initial data sets for numerical relativity merger simulations and for further studies of such strongly magnetized compact binaries.

3.2. Corotational magneto-flows

If the flow is corotational, as is the case for a tidally locked binary or a rigidly rotating non-axisymmetric magnetar, then the velocity field can be written it terms of the helical Killing vector (3) as

$$u^\alpha = u^t k^\alpha \tag{43}$$

In this case, the MHD-Euler equation (27) implies that the left-hand side of the Cartan identity (35) vanishes, so that eq. (38) again holds. The integrability condition of the latter equation then is formally identical to eq. (39) and the first integral of eq. (38) again has the form of eq. (40):

$$k^\gamma w_\gamma + f = \mathcal{E} \tag{44}$$

In the corotational case, the perfect MHD condition (20) also admits a first integral:

$$k^\gamma A_\gamma = 0 \tag{45}$$

This follows from the Cartan identity,

$$k^\beta (dA)_{\beta\alpha} = \pounds_k A_\alpha - \nabla_\alpha (k^\gamma A_\gamma), \tag{46}$$

along with the facts that the term $k^\beta (dA)_{\beta\alpha}$ vanishes by virtue of eqs. (43) and (20), and that the term $\pounds_k A_\alpha$ can be made to vanish by a gauge transformation in A_α (c.f. Appendix C in [22]). One can then incorporate the first integrals (44) and (45) in a self-consistent iteration scheme analogous to that outlined for the irrotational case above, coupled to the Einstein-Maxwell equations, to construct equilibrium models of rigidly rotating configurations. A scheme based on eq. (44), with $f = 0$, has been implemented in [23] for constructing non-magnetized triaxial (Jacobi) spheroids in general relativity and can be extended to model magnetars that are triaxially deformed by their magnetic field.

4. Discussion

The thermodynamic laws derived for helically symmetric Einstein-Maxwell spacetimes with magnetized matter can be applied to a non-axisymmetirc single rotating star, or an axisymmetric rotating star beyond the stationary, axisymmetric *and circular* spacetime. Our project involves computing quasi-equilibrium configurations of magnetized compact objects with numerical codes based on self-consistent field methods. Such methods can be used to model not only binary compact objects, but also magnetars, as well as proto neutron stars which are likely to have strong magnetic fields.

Acknowledgments

We thank Brandon Carter, Yoshiharu Eriguchi, John Friedman and Ichiro Oda for enlightening discussions and suggestions. This work was supported by JSPS Grant-in-Aid for Scientific Research(C) 20540275, MEXT Grant-in-Aid for Scientific Research on Innovative Area 20105004, NSF Grant PHY100155 and ANR grant 06-2-134423 *Méthodes mathématiques pour la relativité générale.* CM thanks the Greek State Scholarships Foundation for support during the early stages of this work and the Paris Observatory and the University of Wisconsin-Milwaukee for travel support. KU and EG acknowledge support from the JSPS Invitation Fellowship for Research in Japan (Short-term) and the invitation program of foreign researchers at the Paris Observatory.

References

[1] Woods P M, & Thompson C 2006, In *Compact stellar X-ray sources* ed W Lewin and M van der Klis, Cambridge Astrophysics Series, No. 39, 547, Cambridge University Press.
[2] Liu Y T, Shapiro S L, Etienne Z B and Taniguchi K 2008, Phys. Rev. D **78**, 024012; Anderson M, Hirschmann E W, Lehner L, Liebling S L, Motl P M, Neilsen D, Palenzuela C, Tohline J E 2008, Phys. Rev. Lett. **100**, 191101. Giacomazzo B, Rezzolla L and Baiotti L 2009, *Preprint* gr-qc/0901.2722.
[3] Blackburn J K and Detweiler S 1992, Phys. Rev. D **46**, 2318; Detweiler S 1994, Phys. Rev. D **50**, 4929.
[4] Schild A 1963, Phys. Rev. 131, 2762
[5] Friedman J L and Uryu K 2006, Phys. Rev. D **73**, 104039; Glenz M M and Uryu K 2007, Phys. Rev. D **76**, 027501.
[6] Friedrich H, Rácz I and Wald R M 1999, Comm. Math. Phys. **204**, 691.
[7] Carter B 1973, *Black hole equilibrium states*, in *Black Holes (Les Astres Occlus)*, p. 57-214.
[8] Carter B 1979, *The general theory of the mechanical, electromagnetic and thermodynamic properties of black holes*, in *General Relativity: An Einstein centenary survey*, ed. S W Hawking & W Israel, p. 294-369; B Carter, *Generalised mass variation formula for a stationary axisymmetric star or black hole with surrounding accretion disc*, in *Anneau d'accrétion sur les trous noires*, (Faculté des Sciences, Angers), ed. I Moret-Bailly & C Latremolière, p. 166-182; B. Carter, *Perfect fluid and magnetic field conservation laws in the theory of black hole accretion rings*, in *Active galactic nuclei.* (A79-50785 22-90) Cambridge, Cambridge University Press, ed. C Hazard & S Mitton, p. 273-300.
[9] Wald R M 1993, Phys. Rev. D. **48**, R3427
[10] Iyer V and Wald R M 1995, Phys. Rev. D. **52**, 4430
[11] Uryu K, Gourgoulhon E and Markakis C 2010, Phys. Rev. D **82**, 104054
[12] Sorkin R 1981, Astrophys. J. **249**, 254.
[13] Friedman J L, Uryu K and Shibata M (FUS) 2002, Phys. Rev. D **65**, 064035 [Erratum-ibid. 2004, D **70**, 129904]
[14] Bekenstein J D and Oron A 2000, Phys. Rev. E **62**, 5594; Bekenstein J D and Oron A 2001, Found. Phys. **31**, 895
[15] Tarapov I E 1984, PMM U.S.S.R., 48, 275; Gorskii V B 1986, PMM U.S.S.R., 50, 388.
[16] Bonazzola S, Gourgoulhon E and Marck J-A 1999, Phys. Rev. Lett. 82, 892; Gourgoulhon E, Grandclement P, Taniguchi K, Marck J-A, Bonazzola S 2001, Phys. Rev. D 63, 064029; Taniguchi K and Gourgoulhon E 2002, Phys. Rev. D 66, 104019; ibid. 2003, 68, 124025; Bejger M, Gondek-Rosinska D, Gourgoulhon E, Haensel P, Taniguchi K and Zdunik J L 2005, Astron. Astrophys. **431**, 297; Taniguchi K and Shibata M 2010, Astrophys. J. Suppl. **188**, 187
[17] Uryu K and Eriguchi Y 2000, Phys. Rev. D **61**, 124023; Uryu K, Shibata M and Eriguchi Y 2000, Phys. Rev. D **62**, 104015; Uryu K, Limousin F, Friedman J L, Gourgoulhon E and Shibata M 2006, Phys. Rev. Lett. **97**, 171101; Uryu K, Limousin F, Friedman J L, Gourgoulhon E and Shibata M 2009, Phys. Rev. D **80**, 124004.
[18] Taniguchi K, Baumgarte T W, Faber J A and Shapiro S L 2005, Phys. Rev. D **72**, 044008; Taniguchi K, Baumgarte T W, Faber J A and Shapiro S L 2006, Phys. Rev. D **74**, 041502(R); Taniguchi K, Baumgarte T W, Faber J A and Shapiro S L 2007, Phys. Rev. D **75**, 084005 (2007); Taniguchi K, Baumgarte T W, Faber J A and Shapiro S L 2008, Phys. Rev. D **77**, 044003; Kyutoku K, Shibata M and Taniguchi K 2009, Phys. Rev. D **79**, 124018; Foucart F, Kidder L E, Pfeiffer H P and Teukolsky S A 2008, Phys. Rev. D **77**, 124051.
[19] Price R H, Markakis C and Friedman J L 2009, J. Math. Phys. **50**, 073505 (*Preprint* astro-ph.SR/0903.3074)
[20] Gourgoulhon E 2006, *An Introduction to Relativistic Hydrodynamics*, in *Stellar fluid dynamics and numerical simulations*, ed M Rieutord & B Dubrulle, EAS Publications Series **21** p. 43-79 (*Preprint* gr-qc/0603009)

[21] Ostriker J P and Mark J W-K 1968, Astrophys. J. **151**, 1075; Hachisu I 1986, Astrophys. J. Suppl. **62**, 461; ibid. 61, 479 (1986); Komatsu H, Eriguchi Y and Hachisu I 1989, Mon. Not. Roy. Astron. Soc. **237**, 355
[22] Gourgoulhon E, Markakis C, Uryu K and Eriguchi Y 2011, *Magnetohydrodynamics in stationary and axisymmetric spacetimes: a fully covariant approach*, *Preprint* gr-qc/1101.3497
[23] Huang X, Markakis C, Sugiyama N and Uryu K 2008, Phys. Rev. D **78** 124023 (*Preprint* gr-qc/0809.0673)

Journal of Physics: Conference Series **283** (2011) 012022 doi:10.1088/1742-6596/283/1/012022

Stringy Space-Time Foam and High-Energy Cosmic Photons

Nick E. Mavromatos

CERN, Theory Division, CH-1211 Geneva 23, Switzerland;
On leave from : King's College London, Physics Department, Strand, London WC2R 2LS, UK

E-mail: nikolaos.mavromatos@kcl.ac.uk

Abstract.
In this review, I discuss briefly stringent tests of Lorentz-violating quantum space-time foam models inspired from String/Brane theories, provided by studies of high energy Photons from intense celestial sources, such as Active Galactic Nuclei or Gamma Ray Bursts. The theoretical models predict modifications to the radiation dispersion relations, which are quadratically suppressed by the string mass scale, and time delays in the arrival times of photons (assumed to be emitted more or less simultaneously from the source), which are proportional to the photon energy, so that the more energetic photons arrive later. Although the astrophysics at the source of these energetic photons is still not understood, and such non simultaneous arrivals, that have been observed recently, might well be due to non simultaneous emission as a result of conventional physics effects, nevertheless, rather surprisingly, the observed time delays can also fit excellently the stringy space-time foam scenarios, provided the space-time defect foam is inhomogeneous. The key features of the model, that allow it to evade a plethora of astrophysical constraints on Lorentz violation, in sharp contrast to other field-theoretic Lorentz-violating models of quantum gravity, are: (i) transparency of the foam to electrons and in general charged matter, (ii) absence of birefringence effects and (iii) a breakdown of the local effective lagrangian formalism.

1. Introduction: Mysterious Results in High Energy Gamma-Ray Astronomy

On July 9th 2005, the MAGIC (***M***ajor ***A***tmospheric ***G***amma-ray ***I***maging ***C***herenkov) Telescope, located in the Canary Islands, observed [1] very high energy gamma rays from the active galactic nucleus Markarian 501 (Mkn 501), which lies at red-shift $z = 0.034$ (*i.e.* about half a million light years away) from Earth, with energies up to the order of 10 TeV (1 TeV $=$ 10^3 GeV $= 10^{12}$ eV), which were delayed up to four minutes as compared with their lower-energy counterparts (in the 0.6 TeV or lower range). It was the first observation of such a distinct delay.

Three years after the MAGIC observations, in September 2008, the FERMI (formerly known as GLAST) Satellite Telescope [2], also observed time delays of the higher-energy photons, from the distant Gamma Ray Burst (GRB) 080916c [3], at red-shifts $z = 4.35$, and later on from GRB 090510 [4], at red-shift $z = 0.9$ and from GRB 09092B, at redhifts $z = 1.822$ [5].

The delay effects may be due to the conventional astrophysics of the active galactic nucleus or Gamma-Ray Burst (*source* effect), which, however, is not well understood at present [6, 7]. In fact, currently there seem to be no consensus among the astrophysicists on the appropriate mechanism for the production of such high-energy photons at the source.

These uncertainties prompted more ambitious, although admittedly far-fetched, explanations for the MAGIC effect [8] and more general for the other observed delayed arrivals of photons, which pertain to new fundamental physics, affecting the photon *propagation*. This may be due, for instance, to space-time *foamy* vacuum structures due to quantum gravitational fluctuations [9] that lead to modified dispersion relations for photons [10, 11, 12]. If true, this would be a clear departure from the Lorentz invariant energy (E)-momentum ($\vec{p}$) relations of Special Relativity, $E = |\vec{p}|c$.

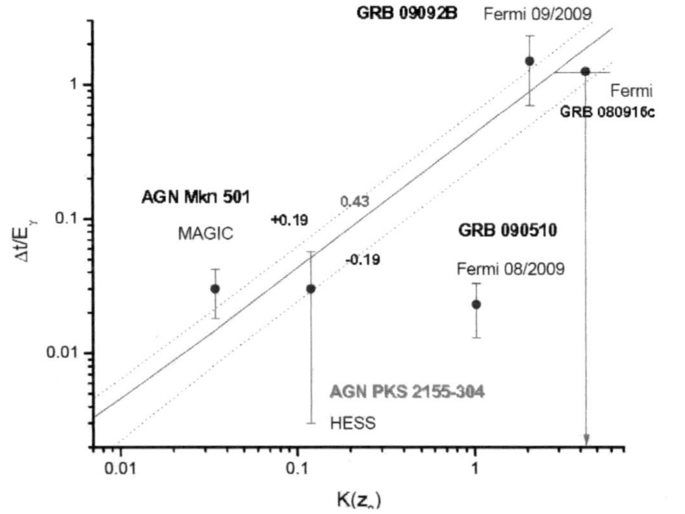

Figure 1. Comparison of data on delays Δt in the arrival times of energetic gamma rays from various astrophysical sources with models in which the velocity of light is reduced by an amount linear in the photon energy. The graph plots on a logarithmic scale the quantity $\Delta t/E$ and a function of the red-shift, $K(z)$, which is essentially the distance of the source from the observation point. The data include two Active Galactic Nuclei (AGN), Mkn 501 [8] and PKS 2155-304 [16], and three Gamma Ray Bursts (GRB) observed by the Fermi satellite, 090510 [4], 09092B [5] and 080916c [3].

If we plot the observed time delays Δt versus the energy E of the photon, including experimental errors, we obtain [13] the diagram of fig. 1. We observe from the figure that with the exception of GRB 090510 , observed by the FERMI satellite [4], the other four points can be fitted by a straight line:

$$\Delta t/E_\gamma = (0.43 \pm 0.19) \times K(z)\text{s/GeV}, \qquad K(z) \equiv \int_0^z \frac{(1+z)dz}{\sqrt{\Omega_\Lambda + \Omega_m (1+z)^3}}, \tag{1}$$

assuming an expanding Universe within the framework of the standard Cosmological-constant-Cold-Dark-Matter (ΛCDM) model. The function $K(z)$ expresses the effect of an expanding Universe onto the linear relation $\Delta t \propto E$, and includes (i) a time dilation factor [12] $(1+z)$ and (ii) the redshifting [11] of the photon energy which implies that the observed energy of a photon with initial energy E is reduced to $E_{\rm obs} = E_0/(1+z)$. Taking into account that in a Robertson-Walker expanding Universe, we assume for our analysis, the infinitesimal time interval dt is related to the Hubble rate $H(z)$, where z is the red-shift, via: $dt = -dz/[(1+z)H(z)]$, we obtain a total delay in the arrival times of photons with energy difference ΔE:

$$(\Delta t)_{\rm obs} = \frac{\Delta E}{M_{\rm QG1}} {\rm H}_0^{-1} \int_0^z \frac{(1+z)dz}{\sqrt{\Omega_\Lambda + \Omega_m (1+z)^3}} \tag{2}$$

where H_0 is the (current-era) Hubble expansion rate, which is fitted to the data, (1). As already mentioned, we have assumed for concreteness the ΛCDM standard model of cosmology, with $\Omega_i \equiv \frac{\rho_{i(0)}}{\rho_c}$ representing the present-epoch energy densities (ρ_i) of matter (including dark matter), Ω_m, and dark vacuum energy, Ω_Λ, in units of the critical density $\rho_c \equiv \frac{3H_0^2}{8\pi G_N}$ of the Universe (G_N is the Newton's gravitational constant). The current astrophysical measurements of the acceleration of the Universe are all consistent with a non zero Cosmological-Constant Universe with Cold-Dark-Matter (ΛCDM Model), with $\Omega_\Lambda \sim 73\%$ and $\Omega_m \sim 27\%$. The quantity $M_{\rm QG1}$ appearing in (2) may be viewed as a phenomenological parameter at this stage, with units of energy. In concrete models of space-time foam, where propagation effects due to QG are assumed to be the dominant cause of the delay, this scale will be identified with the Quantum Gravity scale, where space-time foam effects are expected to set in.

The best fit [13], based on (1), leads to the following result for the scale, $M_{QG1} = (0.98^{+0.77}_{-0.30}) \times 10^{18}$ GeV. This result is remarkable, because it provides the first hint that such delays might be related with string theory effects, given that the order of the scale is that of the conventional string mass scale [14].

However, the GRB 090510 data, provided that the latter can be trusted [1], do not fit to this linear energy scheme. In fact the time delays and energies pertaining to the GRB 090510 [4] are such that the pertinent energy scale $M_{\rm QG1}$ in (1), that would reproduce this point in the graph of fig. 1, is found to be at least $1.2M_P$, with $M_P \sim 1.2. \times 10^{19}$ GeV the Planck mass.

The immediate reaction to this result, of course, is [4] to *exclude* the QG as being responsible for the induced time delays, since one expects, on accounts of naturalness, that any QG effect should not appear at scales higher than the Planck scale. And, indeed, this may well be the case, since, as we have already mentioned, astrophysical mechanisms at the source, which are in general different between GRBs and AGNs, may be responsible for the observed delayed arrivals of the more energetic photons. Nevertheless, as we shall discuss below, the pattern of the observed photon delays fits [13] a string model of quantum-gravity-induced refractive index, with the pertinent quantum gravity energy scale being essentially the same as that inferred from the MAGIC observations (of order 10^{18} GeV). Viewed as a lower bound, this scale is also compatible with that obtained from other Gamma-Ray data of the H.E.S.S. Collaboration [15, 16]. In particular, we shall argue that a stringy model of space-time foam ("D-foam"), proposed in [17, 18, 19], which is based on scenarios involving brane worlds punctured by localized space-time brany defects (termed D-"particles"), can provide consistent fits to the current data on delayed arrival times of photons from MAGIC and FERMI Telescopes.

However, in order to accommodate the findings of MAGIC with the recently observed time delays of photons from the extremely short Gamma-Ray burst GRB 090510 [4], one needs inhomogeneous densities of D-particle defects in the foam [20]. In this sense, astrophysical observations of such intense, short Bursts, are quite essential in falsifying models. The important feature of the string model, which allows for a consistent fit to all available data, including the GRB 090510, is that the foam there consists of real and not virtual space-time defects. Hence, their density may be *non-uniform*, varying with the redshift, *i.e.* be different at various epochs of the Universe. As we shall discuss below [20], the effective QG scale that enters (2) is inversely proportional to this density. A depletion or reduction, therefore, of the number of defects encountered by the photons at the redshift region of the GRB 090510, $z = 0.9$ (which by the way is the regime where one expects a cosmic deceleration to acceleration transition for the Universe) may well be responsible for a reconciliation of the observed delays for the GRB 090510 with those of MAGIC, in the sense that they are both due to the (inhomogeneous) stringy

[1] Indeed, there are uncertainties in this measurement concerning, for instance, the precise emission time of photons due to the yet uncertain duration of pre-cursors to burst activities and other such issues, which need to be confirmed by other measurements of similarly short bursts, that are presently lacking.

foam. From this point of view, the very few available data on observed time delays available today (fig. 1) are not sufficient to exclude the model. One needs a significant improvement on statistics of relevant measurements at various redshifts and various directions (in order to exclude anisotropic foam situations), before definite conclusions on the falsification of the model are reached. Complementary tests from the Cosmology of these models may be helpful, although for inhomogeneous foam situations, studies in the early universe epoch cannot be used to constrain the present-era density of foam, as the two could be different.

2. Lorentz-Violating Stringy Space-time Foam model: D-foam

One of the cornerstones of Modern Physics is Einstein's theory of Special Relativity (SR), which is based on the assumption that the speed of light in vacuo c is an invariant under all observers. In fact, this implies invariance of the physical laws under the Lorentz transformations in flat space times, and the rôle of c as a universal limiting velocity for *all* particle species. The generalization (by Einstein) of SR to include curved space times, that is the theory of General Relativity (GR), encompasses SR locally in the sense of the *strong form of the equivalence principle.* According to it, *at every space-time point, in an arbitrary gravitational field, it is possible to choose a locally inertial ('free-float') coordinate frame, such that within a sufficiently small region of space and time around the point in question, the laws of Nature are described by special relativity,* i.e. *are of the same form as in unaccelerated Cartesian coordinate frames in the absence of Gravitation.* In other words, locally one can always make a coordinate transformation such that the space time looks *flat.* This is not true globally, of course, and this is why GR is a more general theory to describe gravitation. The equivalence principle relies on another fundamental invariance of GR, that of general coordinate, that is the invariance of the gravitational action under arbitrary changes of coordinates. This allows GR to be expressed in a generally covariant form.

In such a locally *Lorentz-invariant* vacuum, the photon dispersion relation, that is a local in space-time relation between the photon's four-wavevector components $k^\mu = (\omega, \vec{k})$ (where ω denotes the frequency, and $\vec{k}$ the momentum) reads in a covariant notation:

$$k^\mu k^\nu \eta_{\mu\nu} = 0 \tag{3}$$

where repeated indices $\mu, \nu = 0, 1, \ldots 3$, with 0 referring to temporal components, denote summation and $\eta_{\mu\nu}$ denotes the Minkowski space-time metric, with components $\eta_{00} = -1$, $\eta_{0i} = \eta_{i0} = 0$, $\eta_{ij} = \eta_{ji} = \delta_{ij}$, $i = 1, 2, 3$ with δ_{ij} the Kronecker delta symbol.

The above relation (3) implies the equality of all three kinds of photon velocities *in vacuo* that stem from its wave nature (due to the particle-wave duality relation):

$$\begin{aligned}
&\text{phase}: && v_{\rm ph} = \tfrac{\omega}{|\vec{k}|} \equiv \tfrac{c}{n(\omega)} = c \\
&\text{group}: && v_{\rm gr} = \tfrac{\partial\omega}{\partial|\vec{k}|} \equiv \tfrac{c}{n_{\rm gr}(\omega)} = c\ , \qquad n_{\rm gr}(\omega) = n(\omega) + \omega\frac{\partial n(\omega)}{\partial\omega} \\
&\text{front}: && v_{\rm front} = c/n(\infty) = c
\end{aligned} \tag{4}$$

since the phase and group *refractive indices* of the trivial *vacuum* equal unity $n(\omega) = n_{\rm gr}(\omega) = 1$. For brevity we shall work from now on in units where $\hbar = c = 1$.

GR is a classical field theory. In a quantum theory of space-time, i.e in a Quantum-Gravity (QG) model, there is a priori no fundamental reason for the strong equivalence principle to hold, and in this sense, local Lorentz invariance might not be respected. This is the situation, in fact, that characterizes certain QG models, which predict a space-time "*foamy*" structure at microscopic (Planck) scales [9]. The latter may not be Lorentz invariant. The foamy structures could include a variety of topologically non-trivial configurations at such microscopic scales, ranging from virtual black holes to space-time *non-commutativity* and (real) stringy defects, as is the case of our stringy D-particle foam models [17, 18, 19], which we now proceed to review.

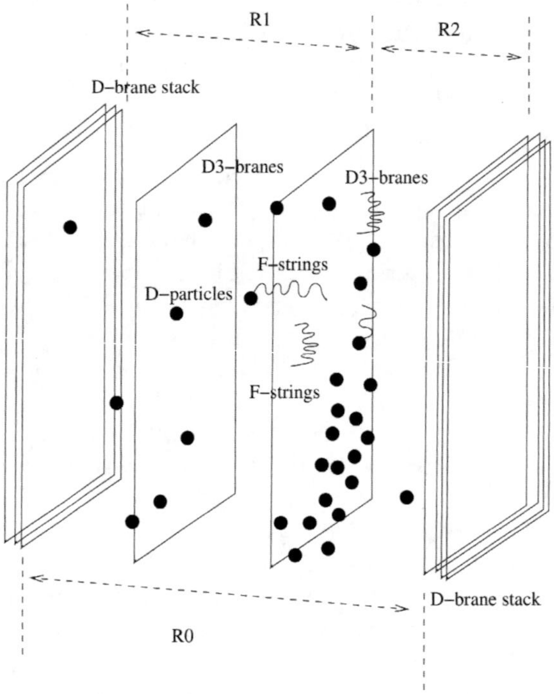

Figure 2. A string theory model of D-particle "foam". The model consists of appropriate stucks of parallel D(irichlet)branes, some of which are moving in a higher-dimensional Bulk space time, punctured by point-like D-brane defects (D-particles). The observable Universe is represented by one of such moving branes, compactified to three spatial dimensions (D3-brane). As the D3-brane world moves, D-particles from the bulk cross it and, thus, to an observer on the brane, they appear as "flashing on and off" space-time foam defects ("D-particle foam"). Photons, represented by open strings with their ends attached on the D3 brane, interact with these defects via capture/recoil, and this leads to non-trivial refractive indices. The effect is therefore "classical" from the bulk space time point of view, but appears as an effective "quantum foam" from the D3-brane observer effective viewpoint.

The model of [17] constitutes an attempt to construct a brane/string-inspired model of space time foam which could have realistic cosmological properties. For this purpose we exploited the modern approach to string theory [14], involving membrane hypersurfaces (D(irichlet)-branes). Such structures are responsible for reconciliating (often via duality symmetries) certain string theories (like type I), which before were discarded as physically uninteresting, with Standard-Model phenomenology in the low-energy limit.

In particular, we considered (*c.f.* figure 2) a ten-dimensional bulk bounded by two eight-dimensional orientifold planes, which contains two stacks of eight-dimensional branes, compactified to three spatial dimensions. Owing to special reflective properties, the orbifolds act as boundaries of the ninth-dimension. The bulk space is punctured by point-like D0-branes (D-particles), which are allowed in type IA string theory (a T-dual of type I strings [21]) we

Journal of Physics: Conference Series **283** (2011) 012022 doi:10.1088/1742-6596/283/1/012022

consider in [17] and here [2]. These are massive objects in string theory [14], with masses M_s/g_s, where M_s is the string mass scale (playing the rôle of the quantum gravity scale in string theory), and $g_s < 1$ is the string coupling, assumed weak for our purposes. These objects are viewed as space-time *defects*, analogous, *e.g.* to cosmic strings, but these are point-like and electrically neutral. I have to stress at this point that, according to modern ideas in string theory [14], the scale M_s is in general different from the four-dimensional Planck-mass scale $M_P = 1.2 \times 10^{19}$ GeV/c^2, and in fact it is a free parameter in string theory to be constrained by experiment. The energy scale $M_s c^2$ can be as low as a few TeV; it cannot be lower than this, though, since if this were the case we should have already seen fundamental string structures experimentally.

Supersymmetry dictates the number of D8-branes in each stack in the model, but does not restrict the number of D0-branes in the bulk. For definiteness, we restrict our attention for now to the type-IA model, in which the bulk space is restricted to a finite range by two appropriate stacks of D8-branes, each stack being supplemented by an appropriate orientifold eight-plane with specific reflecting properties, so that the bulk space-time is effectively compactified to a finite region, as illustrated in Fig. 2. We then postulate that two of the D8-branes have been detached from their respective stacks, and are propagating in the bulk. As discussed above, the bulk region is punctured by D0-branes (D-particles), whose density may be *inhomogeneous*. When there are no relative motions of the D3-branes or D-particles, it was shown in [17] that the ground-state energy vanishes, as decreed by the supersymmetries of the configuration. Thus, such static configurations constitute appropriate ground states of string/brane theory. On the other hand, relative motions of the branes break target-space supersymmetry explicitly, contributing to the dark energy density.

Our model assumes a collision between two branes from the original stack of branes (*c.f.* figure 2) at an early epoch of the Universe, resulting in an initial cosmically catastrophic Big-Bang type event in such non-equilibrium cosmologies [22]. After the collision, the branes bounce back. It is natural to assume that, during the current (late) era of the Universe, the D3-brane representing our Universe is moving slowly towards the stack of branes from which they emanated and the configuration is evolving adiabatically. Hence populations of bulk D-particles cross the brane worlds and interact with the stringy matter on them. To an observer on the brane the space-time defects will appear to be *"flashing" on* and *off*. The model we are using involves eight-dimensional branes and so requires an appropriate *compactification* scheme to three spatial dimensions e.g. by using manifolds with non-trivial *fluxes* (unrelated to real magnetic fields). Different coupling of fermions and bosons to such external fields breaks target space supersymmetry in a way independent of that induced by brane motion, which could be the dominant one in phenomenologically realistic models. The consequent induced mass splitting [23, 24] between fermionic and bosonic excitations on the brane world is proportional to the intensity of the flux field (a string generalization of the well-known Zeeman effect of ordinary quantum mechanics, whereby the presence of an external field leads to energy splittings, which are however different between (charged) fermions and bosons). In this way one may obtain phenomenologically realistic mass splittings in the excitation spectrum (at TeV or higher energy scales) owing to *supersymmetry obstruction* rather than spontaneous breaking (this terminology, which is due to E. Witten [25], means that, although the ground state could still be characterised by zero vacuum energy, the masses of fermion and boson excitations differ and thus supersymmetry is broken at the level of the excitation-spectrum). We also mention that the plausible assumption of a population concentration of massive D-particle defects in the haloes of galaxies can lead to modified galactic dynamics [26].

One can calculate the vacuum energy induced on the brane world in such an adiabatic situation by considering its interaction with the D-particles as well as the other branes in the

[2] One can extend the construction to phenomenologically realistic models of type IIB strings [19].

construction. This calculation was presented in [17, 20], where we refer the interested reader for further details. Here we mention only the results relevant for the present discussion. The important point to notice here is the fact that the forces exerted by the various structures, such as D-particles and other branes, on the brane world, representing the observable Universe in this kind of models, do not have a fixed sign.

We concentrate first on D0-particle/D8-brane interactions in the type-IA model of [17]. During the late era of the Universe when the approximation of adiabatic motion is valid, we use a weak-string-coupling approximation $g_s \ll 1$. In such a case, the D-particle masses $\sim M_s/g_s$ are large, *i.e.*, these masses could be of the Planck size: $M_s/g_s \sim M_P = 1.22 \cdot 10^{19}$ GeV or higher. In the adiabatic approximation for the relative motion, these interactions may be represented by a string stretched between the D0-particle and the D8-brane, as shown in Fig. 2. The world-sheet amplitude of such a string yields the appropriate potential energy between the D-particle and the D-brane, which in turn determines the relevant contribution to the vacuum energy of the brane. As is well known [14], parallel relative motion does not generate any potential, and the only non-trivial contributions to the brane vacuum energy come from motion transverse to the D-brane. Neglecting a velocity-independent term in the D0-particle/D8-brane potential that is cancelled for a D8-brane in the presence of orientifold O_8 planes [14] [3], we find [17]:

$$\mathcal{V}^{short}_{D0-D8} = -\frac{\pi\alpha'}{12}\frac{v^2}{r^3} \quad \text{for} \quad r \ll \sqrt{\alpha'} \,, \tag{5}$$

$$\mathcal{V}^{long}_{D0-D8} = +\frac{r\,v^2}{8\,\pi\,\alpha'} \quad \text{for} \quad r \gg \sqrt{\alpha'} \,. \tag{6}$$

where $v \ll 1$ is the relative velocity between the D-particle and the brane world, which is assumed to be non-relativistic. We note that the sign of the effective potential changes between short distances (5) and long distances (6). We also note that there is a minimum distance given by:

$$r_{\min} \simeq \sqrt{v\,\alpha'} \,, \qquad v \ll 1 \,, \tag{7}$$

which guarantees that (5) is less than r/α', rendering the effective low-energy field theory well-defined. Below this minimum distance, the D0-particle/D8-brane string amplitude diverges when expanded in powers of $(\alpha')^2 v^2/r^4$. When they are separated from a D-brane by a distance smaller than $r_{\min}$, D-particles should be considered as lying on the D-brane world, and two D-branes separated by less than $r_{\min}$ should be considered as coincident.

We now consider a configuration with a moving D8-brane located at distances $R_i(t)$ from the orientifold end-planes, where $R_1(t) + R_2(t) = R_0$ the fixed extent of the ninth bulk dimension, and the 9-density of the D-particles in the bulk is denoted by $n^\star(r)$: see Fig. 2. The total D8-vacuum-energy density ρ^8 due to the relative motions is [17]:

$$\rho^{D8-D0}_{\text{total}} = -\int_{r_{\min}}^{\ell_s} n^\star(r)\,\frac{\pi\alpha'}{12}\frac{v^2}{r^3}\,dr - \int_{-r_{\min}}^{-\ell_s} n^\star(r)\,\frac{\pi\alpha'}{12}\frac{v^2}{r^3}\,dr +$$

$$\int_{-\ell_s}^{-R_1(t)} n^\star(r)\frac{r\,v^2}{8\,\pi\,\alpha'}\,dr + \int_{\ell_s}^{R_2(t)} n^\star(r)\frac{r\,v^2}{8\,\pi\,\alpha'}\,dr + \rho_0 \tag{8}$$

where the origin of the r coordinate is placed on the 8-brane world and ρ_0 combines the contributions to the vacuum energy density from inside the band $-r_{\min} \le r \le r_{\min}$, which include the brane tension. When the D8-brane is moving in a uniform bulk distribution of D-particles, we may set $n^* \star (r) = n_0$, a constant, and the dark energy density $\rho^{D8-D0}_{\text{total}}$ on the D8-brane is also (approximately) constant for a long period of time:

$$\rho^{D8-D0}_{\text{total}} = -n_0\frac{\pi}{12}v(1-v) + n_0 v^2\frac{1}{16\pi\alpha'}(R_1(t)^2 + R_2(t)^2 - 2\alpha') + \rho_0 \,. \tag{9}$$

[3] This cancellation is crucial for obtaining an appropriate supersymmetric string ground state with zero ground-state energy.

Because of the adiabatic motion of the D-brane, the time dependence of $R_i(t)$ is weak, so that there is only a weak time dependence of the D-brane vacuum energy density: it is positive if $\rho_0 > 0$, which can be arranged by considering branes with positive tension.

However, one can also consider the possibility that the D-particle density $n^\star(r)$ is inhomogeneous, perhaps because of some prior catastrophic cosmic collision, or some subsequent disturbance. If there is a region depleted by D-particles [20] - a *D-void* - the relative importance of the terms in (9) may be changed. In such a case, the first term on the right-hand side of (9) may become significantly smaller than the term proportional to $R_i^2(t)/\alpha'$. As an illustration, consider for simplicity and concreteness a situation in which there are different densities of D-particles close to the D8-brane ($n_{\rm local}$) and at long distances to the left and right ($n_{\rm left}, n_{\rm right}$). In this case, one obtains from (8):

$$\rho_{\rm total}^{D8-D0} \simeq -n_{\rm local}\tfrac{\pi}{12}v(1-v) + n_{\rm left}v^2\tfrac{1}{16\pi\alpha'}\left(R_1(t)^2 - \alpha'\right) + \\ n_{\rm right}v^2\tfrac{1}{16\pi\alpha'}\left(R_2(t)^2 - \alpha'\right) + \rho_0 \qquad (10)$$

for the induced energy density on the D8-brane. The first term can be significantly smaller in magnitude than the corresponding term in the uniform case, if the local density of D-particles is suppressed. Overall, the D-particle-induced energy density on the D-brane world *increases* as the brane enters a region where the D-particle density is *depleted*. This could cause the onset of an *accelerating phase* in the expansion of the Universe. It is intriguing that the red-shift of GRB 090510 [4] is in the ballpark of the redshift range where the expansion of the Universe apparently made a transition from deceleration to acceleration [27]. According to the above discussion, then, within our string foam model this *may not* be a coincidence [20]. The result (10) was derived in an oversimplified case, where the possible effects of other branes and orientifolds were not taken into account. However, as we argued in [20], the ideas emerging from this simple example persist in more realistic structures.

The important point for our purposes in such models of space-time foam, is the fact that there is an induced refractive index *in vacuo*, as a consequence of the photon interactions with the D-particles that cross the moving brane Universe. It is the *linear density* of the D-particle defects $n(z)$ encountered by a propagating photon that determines the amount of refraction. The density of D-particles crossing the D-brane world cannot be determined from first principles, and so may be regarded as a parameter in phenomenological models. The flux of D-particles is proportional also to the velocity v of the D8-brane in the bulk, if the relative motion of the population of D-particles is ignored.

In order to make some phenomenological headway, we adopt some simplifying assumptions. For example, we may assume that between a redshift $z < 1$ and today ($z = 0$), the energy density has remained approximately constant, as suggested by the available cosmological data. As shown in [20], this assumption implies certain restrictions for the bulk density $n_{\rm short}(z)$ of D-particle defects near the brane world

$$n_{\rm short}(z) = n_{\rm short}(0) - \tfrac{12}{2^7\pi^{10}\alpha'^5}\tfrac{1}{\ell_s^9}\tfrac{c}{\ell_s}\tfrac{(v/c)^4}{(1-\frac{v}{c})}\int_0^z \tfrac{dz'}{H(z')(1+z')} \quad \text{where } \ell_s \equiv \sqrt{\alpha'}\,. \qquad (11)$$

We can make use of this result when we fit the available data from MAGIC and FERMI telescopes to this model. For instance, by requiring that this density falls from O(1) at the redshifts $z = 0.03$, relevant to MAGIC observations, by at least two orders of magnitude as we approach the red-shift $z = 0.9$ of GRB 090510, we may constrain the relative motion of the D3-brane world. In fact, as discussed in [20], this can be achieved for relatively small velocities of the D3-brane world in the bulk, $v < 10^{-4}c$, consistent with the Early Universe Cosmology of the model [22].

Before closing this section we would like to make a comment regarding the modification of (thermal) Dark Matter (DM) relic abundances in the foam model. As discussed in [28], the quantum fluctuations of the D-particles act as sources of particle production and thus affect the respective Boltzmann equation determining the relic abundance of DM particles. Additional modifications to this equation are due to the Finsler-like [29] character of the induced space-time metric during the interaction of neutral particles (like the DM ones) with the foam, due to its dependence on D-particle recoil velocities (and hence momentum transfer) (*c.f.* (24) below). We assume foam types in which the Lorentz symmetry is conserved on the average, and is violated only through fluctuations, namely we assume statistical ensembles of D-particles such that $\ll u_i \gg = 0, \quad \ll u_i u_j \gg \equiv \frac{g_s^2}{M_s^2}\Delta_i^2 \overline{p_i}\, \overline{p}_j \delta_{ij}$, where $\overline{p}$ is some average momentum scale. One finds [28]:

$$\frac{\Omega'_\chi h_0^2}{(\Omega_\chi h_0^2)_{\rm no\ source}} \simeq \left[1 + 207.38 g_s^2 \frac{m^2}{M_s^2} x_f^{-2}\left(\sum_{i=1}^3 \Delta_i^2\right)\right]^{1/2}\left[1 + g_s^2\frac{m^2}{M_s^2}\left(\sum_{i=1}^3 \Delta_i^2\right)\left(1 + 6x_0^{-1}\right)\right] \quad (12)$$

where $x = m/T$, T is the temperature, the suffix f denotes the freeze-out point of the DM species, the suffix 0 denotes present-day quantities, $\Omega'_\chi h_0^2$ denote the Hubble-constant-free present-day thermal relic abundance of a weakly interacting heavy Dark Matter particle of mass m, in the presence of the foam, and the suffix "no source" indicates the corresponding quantity in the Standard (foam-free) Cosmology. Clearly the dominant correction terms are of order $g_s^2 \frac{m^2}{M_s^2}\left(\sum_{i=1}^3 \Delta_i^2\right) > 0$.

The modifications are suppressed by the square of the string scale (actually the D-particle mass M_s/g_s) and hence for relatively high string scales (much higher than TeV, of interest to us here) are small and do not lead to significant constraints on the density of foam. Indeed, in our case, from the MAGIC experiment, if we explain the observed delays as being due to the stringy foam alone, we obtain $M_s/g_s \sim 10^{19}$ GeV, while phenomenologically realistic DM candidates have masses m in the range of at most a few TeV (usually a few hundreds of GeV) [30].

It is important in this latter respect to make a comment regarding the nature of our foam: the D-particles in our approach [17, 18, 19] are viewed as *background configurations* and *not* as *excitations* of the string vacuum. In other string/brane models, some authors have viewed the D-particles as localized excitations of the vacuum [31]. In those cases, the D-particles may be considered as dark matter candidates themselves and their density would be constrained by the cosmological observations on the DM sector to avoid overclosure of the Universe. As such they could not contribute to the refractive index. Indeed, for the latter property to occur, light must interact coherently with the D-particles, rather than scattering on them individually. Otherwise, there would not just be a time delay and thus an index of refraction, but the light would be incoherently deflected at arbitrarily large angles. Coherent scattering can only occur if the wavelength of light is much greater than the mean separation of the scatterers. In case the D0 particles behave like dark matter [31], with masses near the Planck scale (to account for the MAGIC delays [8]), their number density must be less than 10^{-20} m^{-3} to avoid overclosure of the Universe. Thus their mean separation would be greater than 100 m. The gamma rays that are of interest to us here have a much shorter wavelength (smaller than cm) and therefore cannot experience refraction due to these D-particles [4]. In our D-foam model [17, 18], where the D0-branes (or the compactified D3 branes around 3-cycles in the model of [19]) are viewed only as background defects, such an issue does not arise, and the density of the foam cannot be constrained by overclosure of the Universe issues. As we have discussed above, in this case the D-particles contribute to the dark energy sector (when in motion) consistently with current cosmological data. The cancellation between attractive (gravitational) and repulsive flux forces guarantees a supersymmetric vacuum, with zero vacuum energy, if no relative motion of foam occurs.

[4] At any rate, such super-heavy DM would have been washed out by inflation in any realistic cosmology, so the scenario of [31] for D-particle excitations to play the rôle of DM pertains to much lighter D-particles in theories with low string mass scales.

In the above picture, the presence of a ground-state D-particle defect would be reflected in the world-sheet boundary conditions for the open strings representing radiation or matter excitations. Thus, the only effects a recoiling D-particle would have can be summarized through its induced Finsler-like metric distortions and the associated time delays of the interacting neutral matter or radiation, which we now proceed to discuss.

3. Time Delays in D-particle foam and Stringy Uncertainties

In this section we discuss a possible origin of time delays induced in the arrival time of photons, emitted simultaneously from an astrophysical object, as a result of their propagation in the above-described D-particle space-time foam model. This comes by considering local interactions of photons, viewed as open string states in the model, with the D-particle defects. As we shall argue, the above model is in principle capable of reproducing photon arrival time delays proportional to the photon energies, of the kind observed in the MAGIC experiment [1, 8]. This microscopic phenomenon, which is essentially stringy and does not *characterize* local field theories, contributes to a sub-luminal non-trivial refractive index *in vacuo*, induced by the capture of photons or electrically neutral probes by the D-particle foam [18]. The capture process is described schematically in figure 3. An important feature of the model is that, on

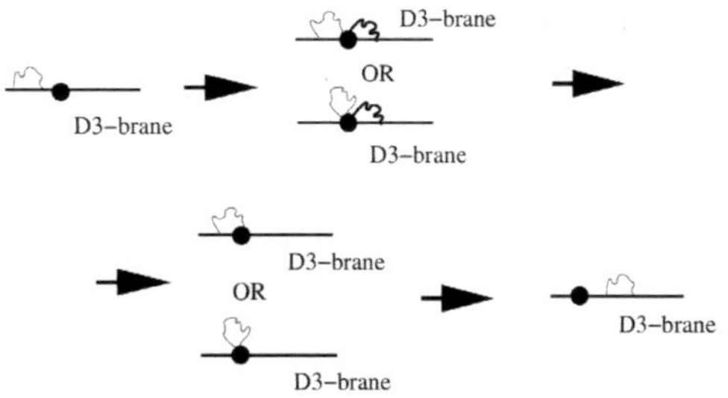

Figure 3. Schematic view of the capture process of an open string state, representing a photon propagating on a D3-brane world, by a D-particle on this world. The intermediate string state, indicated by thick wave lines, which is created on capture of the end(s) of the photon by the D-particle, stretches between the D-particle and the brane world, oscillates in size between 0 and a maximal length of order $\alpha' p^0$, where p^0 is the energy of the incident photon, and thus produces a series of outgoing photon waves, with attenuating amplitudes, constituting the re-emission process. The intermediate string state provides also the restoring force, necessary for keeping the D-particle roughly in its position after scattering.

account of electric charge conservation, only electrically neutral excitations are subjected to capture by the D-particles. To the charged matter the D-particle foam looks *transparent*. This is because the capture process of fig. 3 entails a splitting of the open string matter state. Charged excitations are characterized by an electric flux flowing across the string, and when the latter is cut in two pieces as a result of its capture by the D-particle defect, the flux should go somewhere because charge is conserved. The D-particle, being neutral, cannot support this conservation, and hence only *electrically neutral* excitations, such as photons, are subjected to this splitting and the associated delays. The reader should bear in mind of course that the D-particles carry other kinds of fluxes, unrelated to electromagnetism [14]. These are conserved separately, and it is for this reason that isolated D-particles cannot exist, but there must always be in the company of other D-branes, as in our model above, so that the relevant fluxes are carried by the stretched strings between the latter and the D-particles.

For our purposes in this work it is also important to remark that the D-particles are treated as static when compared to photons. This is because the ends of the open string representing the photon move on the D3 brane world with the speed of light in (normal) vacuo, while the relative velocities of the D-particles with respect to the brane world (which propagates in the bulk space) are much lower than this. For instance, as discussed in [22], to reproduce cosmological observations in this model, in particular the spectrum of primordial density fluctuations, which are affected by the relative motion of the brane world, the speed of propagation of the D3-brane Universe should be smaller than $10^{-4}c$.

When the end(s) of the open-string photon state are attached to the D-particle, there is an intermediate string state formed, *stretched* between the D-particle and the D3-brane, which absorbs the incident energy p^0 of the photon state to grow in size from zero to a maximum length L that is determined by the requirement of energy minimization as follows: one assumes [32] that the intermediate string state needs N oscillations to achieve its maximal length L, as in the standard string-string scattering case. Following the same logic, we first observe that energy conservation during the capture process of fig. 3 leads to relations of the form

$$p^0 = \tfrac{L}{\alpha'} + \tfrac{N}{L} \tag{13}$$

Minimizing the right-hand-side with respect to L and taking into account that the one end of the intermediate string state attached to the brane world moves with the *speed of light in vacuo*, $c = 1$ (in our units), one arrives at time delays for the outgoing photon waves of order [18]:

$$\Delta t \sim \alpha' p^0 \ . \tag{14}$$

This delay is *causal* and can also be obtained by considering scattering of open string states off D-particle backgrounds [5]. One assumes that the incident open string state splits into two open string states upon interaction with the D-particle defect, which, after their scattering in the D-particle background, the presence of which is implemented through the appropriate Dirichlet world-sheet boundary conditions of the open strings, re-join to produce the outgoing state. The intermediate stage involves two open strings that scatter off each other. The (quantum) oscillations of the intermediate string state will produce a series of outgoing wave-packets, with attenuating amplitudes, which will correspond to the re-emission process of the photon after capture. The presence of the stretched string state, which carries the characteristic flux of the D-brane interactions, provides the *restoring force*, necessary to keep the D-particle in its position after scattering with the photon.

The situation may be thought of as the stringy/brany analogue of the restoring force in situations in local field theories of photons propagating in media with non trivial refractive indices, as discussed by Feynman [33]. The rôle of the electrons in that case (represented as harmonic oscillators) is played here by the D-particle defects of the space-time. The stringy situation, however, is more complicated, since the D-particles have an infinite number of oscillatory excitations, represented by the various modes of open strings with their ends attached to them. Moreover, contrary to the conventional medium case, in the string model the refractive index is found proportional to the photon frequency, while the effective mass scale that suppresses the effect (14) is the quantum gravity (string) scale M_s. The latter property can be understood qualitatively by the fact that the mass of these D-particle defects is of order [14] $1/(g_s\sqrt{\alpha'}) = M_s/g_s$, where g_s is the string coupling (in units of $\hbar = c = 1$).

[5] For type IA strings, which admit point-like D-particles, the situation is a straightforward extension of techniques applied to the open string-string scattering case of [32]. For type IIB strings, the situation is technically more involved, due to the compactified D3-branes, wrapped up around three cycles, that play the rôle of "D-particles" in the model [19], but the results are qualitatively the same as in (14).

We remark at this stage that the above time delays are a direct consequence of the *stringy uncertainty principles*. Indeed, strings are characterized by two kinds of uncertainty relations: the phase-space Heisenberg uncertainty, modified by higher order terms in α' [34] as a result of the existence of the minimal string length $\ell_s \equiv \sqrt{\alpha'}$ in target space-time,

$$\Delta X \Delta P \geq \hbar + \alpha'(\Delta P)^2 + \ldots \tag{15}$$

and the time-space uncertainty relation [35]

$$\Delta X \Delta t \geq \alpha' \ . \tag{16}$$

Since the momentum uncertainty $\Delta P < p^0$, we have from (15), to leading order in α' (in units $\hbar = 1$):

$$\Delta X \geq \tfrac{1}{\Delta P} > \tfrac{1}{p^0} \ . \tag{17}$$

In view of (14), we then arrive at consistency with the space-time uncertainty (16),

$$\Delta X \geq \tfrac{\alpha'}{\alpha' p^0} \sim \tfrac{\alpha'}{\Delta t} \tag{18}$$

As in the conventional string theory photon-photon scattering [32], these delays are *causal, i.e.* consistent with the fact that signals never arrive before they occur. Hence they are *additive* for multiple scatterings of photons by the foam defects from emission till observation. As we shall discuss in section 4, this provides the necessary *amplification*, so that the total delay of the more energetic photons can be [18] of the observed order in MAGIC and FERMI experiments.

In addition to this leading refractive index effect (14), there are corrections induced by the recoil of the D-particle itself, which contribute to space-time distortions that we now proceed to discuss. From a world-sheet view point, the presence of D-particle recoil may be represented by adding to a fixed-point (conformal) σ-model action, the following deformation [36, 37]:

$$\mathcal{V}_{\rm D}^{imp} = \tfrac{1}{2\pi\alpha'} \sum_{i=1}^{D} \int_{\partial D} d\tau \, u_i X^0 \Theta\left(X^0\right) \partial_n X^i. \tag{19}$$

where D in the sum denotes the appropriate number of spatial target-space dimensions. For a recoiling D-particle confined on a D3 brane, $D = 3$.

There is a specific type of conformal algebra, termed logarithmic conformal algebra [38], that the recoil operators satisfy [36, 37]. This algebra is the limiting case of world-sheet algebras that can still be classified by conformal blocks. The impulse operator $\Theta(X^0)$ is regularized so that the logarithmic conformal field theory algebra is respected [6]. The conformal algebra is consistent with momentum conservation during recoil [36, 37], which allows for the expression of the recoil velocity u_i in terms of momentum transfer during the scattering

$$u_i = g_s \tfrac{p_1 - p_2}{M_s} \ , \tag{20}$$

with $\frac{M_s}{g_s}$ being the D-particle "mass" and $\Delta p \equiv p_1 - p_2$ the associated momentum transfer of a string state during its scattering with the D-particle.

We next note that one can write the boundary recoil/capture operator (19) as a total derivative over the bulk of the world-sheet, by means of the two-dimensional version of Stokes

[6] This can be done by using the world-sheet scale, $\varepsilon^{-2} \equiv \ln{(L/a)^2}$, with a an Ultra-Violet scale and L the world-sheet area, as a regulator [36, 37]: $\Theta_\varepsilon(X^0) = -i \int_{-\infty}^{\infty} \frac{d\omega}{\omega - i\varepsilon} e^{i\omega X^0}$. The quantity $\varepsilon \to 0^+$ at the end of the calculations.

theorem. Omitting from now on the explicit summation over repeated i-index, which is understood to be over the spatial indices of the D3-brane world, we write then:

$$\mathcal{V}_{\rm D}^{imp} = \frac{1}{2\pi\alpha'}\int_D d^2z\,\epsilon_{\alpha\beta}\partial^\beta\left(\left[u_iX^0\right]\Theta_\varepsilon\left(X^0\right)\partial^\alpha X^i\right) =$$
$$\frac{1}{4\pi\alpha'}\int_D d^2z\,(2u_i)\,\epsilon_{\alpha\beta}\partial^\beta X^0\left[\Theta_\varepsilon\left(X^0\right)+X^0\delta_\varepsilon\left(X^0\right)\right]\partial^\alpha X^i \tag{21}$$

where $\delta_\varepsilon(X^0)$ is an ε-regularized δ-function. For relatively large times after the impact at $X^0=0$ (which we assume for our phenomenological purposes in this work), this is equivalent to a deformation describing an open string propagating in an antisymmetric $B_{\mu\nu}$-background corresponding to an external constant in target-space "electric" field,

$$B_{0i}\sim u_i\ ,\quad B_{ij}=0\ ,\quad (X^0>0) \tag{22}$$

where the $X^0\delta(X^0)$ terms in the argument of the electric field yield vanishing contributions in the large time limit, and hence are ignored from now on.

To discuss the space time effects of a recoiling D-particle on an open string state propagating on a D3 brane world, we should consider a σ-model in the presence of the B-field (22), which leads to mixed-type boundary conditions for open strings on the boundary $\partial\mathcal{D}$ of world-sheet surfaces with the topology of a disc. Absence of a recoil-velocity u_i-field leads to the usual Neumann boundary conditions, while the limit where $g_{\mu\nu}\to 0$, with $u_i\neq 0$, leads to Dirichlet boundary conditions.

In analogy with the standard string case in background electric fields [39, 32], one obtains a *non-commutative space-time* if recoil of the D-particle is taken into account. This can be seen upon considering commutation relations among the coordinates of the first quantised σ-model in the background (22). As in the standard case of a constant electric background field, in the presence of a recoiling D-particle, the pertinent non commutativity is between time and the spatial coordinate along the direction of the recoil velocity field (for large times t after the impact at $t=0$):

$$[X^1,t]=i\theta^{10}\ ,\qquad \theta^{01}(=-\theta^{10})\equiv\theta=\frac{1}{u_c}\frac{\tilde{u}}{1-\tilde{u}^2} \tag{23}$$

where, for simplicity and concreteness, we assume recoil along the spatial X^1 direction. Thus, the induced non commutativity is consistent with the breaking of the Lorentz symmetry of the ground state by the D-particle recoil. The quantity $\tilde{u}_i\equiv\frac{u_i}{u_c}$ and $u_c=\frac{1}{2\pi\alpha'}$ is the Born-Infeld *critical* field. The space-time uncertainty relations (23) are consistent with the corresponding space-time string uncertainty principle (16).

Of crucial interest in our case is the form of the induced open-string *effective target-space-time metric*. The situation parallels that of open strings in external electric field backgrounds, discussed in refs. [32, 39]. Hence, the effective open-string metric, $g_{\mu\nu}^{\rm open,electric}$, which is due to the presence of the recoil-velocity field $\vec{u}$, whose direction breaks target-space Lorentz invariance, is obtained by extending appropriately the results of strings in constant electric field backgrounds [39, 32] to the background (22):

$$\begin{aligned} g_{\mu\nu}^{\rm open,electric} &= \left(1-\tilde{u}_i^2\right)\eta_{\mu\nu}\ , \qquad \mu,\nu=0,1 \\ g_{\mu\nu}^{\rm open,electric} &= \eta_{\mu\nu}\ ,\mu,\nu=\text{all other values}\ . \end{aligned} \tag{24}$$

For concreteness and simplicity, we considered a frame of reference where the matter particle has momentum only across the spatial direction X^1, *i.e.* $0\neq p_1\equiv p\parallel u_1\ ,p_2=p_3=0$. Moreover, as in the standard case of strings in an electric field background, there is a modified effective string coupling [39, 32]:

$$g_s^{\rm eff}=g_s\left(1-\tilde{u}^2\right)^{1/2} \tag{25}$$

The fact that the metric in our recoil case depends on momentum transfer variables implies that D-particle recoil induces Finsler-type metrics [29], *i.e.* metric functions that depend on phase-space coordinates of the matter (photon) state.

We now mention that the presence of the critical "background field" u_c, at which both the metric (24) and the effective coupling (25) vanish, is associated with the *destabilization of the vacuum* [40] when the field intensity approaches the *critical value*. Since in our D-particle foam case, the rôle of the 'electric' field is played by the recoil velocity of the D-particle defect, the critical field corresponds to the relativistic speed of light, in accordance with special relativistic kinematics, which is respected in string theory by construction. On account of (20), then, this implies an upper bound on the induced momentum transfer, and hence on the available momenta, for the effective field theory limit to be valid. Indeed, if we represent Δp in (20) as a fraction of the incident momentum p_1, $\Delta p = rp_1$, $r < 1$, then the condition that the recoil velocity of the D-particle is below the speed of light in vacuo, as required by the underlying consistency of strings with the relativity principle, implies [41]

$$\Delta p \equiv rp_1 < \frac{M_s}{g_s} \Rightarrow p_1 < \frac{M_s}{r\,g_s} \ . \tag{26}$$

When the incident momentum approaches the order of this cutoff, the effective string coupling (25) vanishes, while above that value the coupling becomes imaginary, indicating complete absorption of the string state by the D-particle. The space-time distortion due to recoil is so strong in such a case that there is no possibility of re-emergence of the string state, the defect behaves like a black hole, capturing permanently the string state.

It is important to notice that in modern string theory the quantities M_s, g_s are completely phenomenological. In fact, it is possible to construct phenomenologically realistic string/brane-Universe models, in the sense of being capable to incorporate the Standard Model particles at low energies, with string scales in such a way that M_s/g_s is significantly lower than the Planck scale. For instance, there are constructions [42] in the large extra dimension framework, for which M_s/g_s is of the order of 10^{19} eV, *i.e.* of order of the conventional GZK cutoff in Lorentz invariant particle physics models. Thus, by embedding the D-particle foam to such models, one may have the appearance of a *novel type* Gretisen Zatsepin Kusmin (GZK) cutoff [43], of similar order to the one obtained from conventional Lorentz invariance arguments, but of quite different origin: here it is the *subluminal* nature of the *recoil velocity* of the foam that sets the new upper bound in momentum transfer. This is not fine-tuning in our opinion, but indicates the appearance of a new upper bound in momenta, related implicity to the underlying Lorentz invariance of the string theory, which is broken spontaneously by the recoiling D-particle background. This point will become of importance later on, when we discuss constraints of our model coming from ultra-high-energy/infrared photon/photon scattering [44].

Before closing this section we make an important remark. The induced metric (24) will affect the dispersion relations of the photon state:

$$p^\mu p^\nu g_{\mu\nu}^{\text{open,electric}} = 0 \ . \tag{27}$$

However, because the corrections on the recoil velocity u_i are quadratic, such modifications will be suppressed by the square of the D-particle mass scale. On the other hand, the presence of the D-particle recoil velocity will affect the induced time delays (14) by higher-order corrections of the form, as follows by direct analogy of our case with that of open strings in a constant electric field [32]:

$$\Delta t_{\text{with D-foam recoil velocity}} = \alpha' \frac{p^0}{1-\tilde{u}_i^2} \ . \tag{28}$$

Thus, the D-particle recoil effects are quadratically suppressed by the D-particle mass scales, since (*c.f.* (20)) $\tilde{u}_i \propto g_s \Delta p_i / M_s$, with Δp_i the relevant string-state momentum transfer.

However, the leading order delay effect (28), obtained formally by considering the limit of vanishing recoil velocity, is linearly suppressed by the string scale, and thus the induced time delays are disentangled in the string foam model from the modified Finsler dispersion relations (27). This is important for the phenomenology of the model, as we shall discuss below.

Lack of a Local Effective Field Theory Formalism in recoiling D-particle Foam: An important comment arises at this stage concerning the construction of a possible local effective field theory action in the case of D-particle foam, when recoil of the D-particle defect is considered. In view of the formal analogy of the problem with that of an external electric background field, one is tempted to apply the same considerations as those leading to the non-commutative (NC) effective actions in field theory, for instance for case of NC quantum electrodynamics of ref. [45]. Naively, one might think of writing down a Standard-Model-Extension type Lagrangian [46], including non-commutativity (and thus Lorentz-violating but CPT conserving) terms in the presence of D-foam background, with the $\theta^{\alpha\beta}$ parameter being replaced by $\theta^{0i} = u_i = g_s \frac{\Delta p_i}{M_s}$, the D-particle recoil velocity.

However, this is not correct. As a result of the momentum-transfer dependence of the non-commutativity parameter in the D-foam case, it is not possible to write down the effective action in target space as a power series of local quantum operators. The momentum transfer is not represented by such operators when taking the Fourier transform.

Thus writing down a local effective action for the recoiling D-particle in interaction with, say, a photon, is not possible. This has important consequences for phenomenology. Indeed, the time delays (28), which are associated with the stringy uncertainty, are found proportional to the incident energy of the (split) photon state and are thus linearly suppressed by the string scale. The associated refractive index is therefore linearly suppressed, but this cannot be interpreted as an average propagation of photons in the context of some local effective action. As we have seen above, the associated anomalous photon dispersion terms in this case, induced by the Finsler metric (24), are *quadratically* suppressed by the string scale.

Therefore, any analysis, such as those involving high energy cosmic rays [44] using an effective linear dispersion relation, obtained from a local effective action, does not apply to our problem. Moreover, it is a generic feature of any local effective theory, that is a theory on flat space times involving higher derivatives operators, to yield birefringence, since the corresponding modified dispersion relations for photons reduce at the end of the day to a solution of the photon frequency as a function of the wave number k which is obtained from a quadratic equation. The latter admits two solutions with different propagation for the two photon polarizations. This is *not* the case of the uncertainty-related time delay (28), which, as we have seen, is independent of the photon polarization [18].

4. Multiple Photon-D-particle Scatterings in the Foam and Total (observed) Time Delays

The above-discussed time delays (14) pertain to a single encounter of a photon with a D-particle. In case of a foam, with a linear density of defects $n^*/\sqrt{\alpha'}$, *i.e.* n^* defects per string length, the overall delay encountered in the propagation of the photon from the source to observation, corresponding to a traversed distance D, is:

$$\Delta t_{\rm total} = \alpha' p^0 n^* \frac{D}{\sqrt{\alpha'}} = \frac{p^0}{M_s} n^* D \ , \qquad (29)$$

where p^0 denotes an average photon energy. When the Universe's expansion is taken into account, one has to consider the appropriate red-shift-z dependent stretching factors, which affect the measured delay in the propagation of two photons with different energies, as well as the Hubble expansion rate $H(z)$. Specifically, in a Roberston-Walker cosmology the delay due to any single scattering event is affected by: (i) a time dilation factor [12] $(1+z)$ and (ii) the

Journal of Physics: Conference Series **283** (2011) 012022 doi:10.1088/1742-6596/283/1/012022

redshifting [11] of the photon energy which implies that the observed energy of a photon with initial energy E is reduced to $E_{\rm obs} = E_0/(1+z)$. Thus, the observed delay associated with (14) is [11, 12]:

$$\delta t_{\rm obs} = (1+z)\delta t_0 = (1+z)^2 \sqrt{\alpha'} E_{\rm obs}. \tag{30}$$

For a line density of D-particles $n(z)$ at redshift z, we have $n(z)d\ell = n(z)dt$ defects per co-moving length, where dt denotes the infinitesimal Robertson-Walker time interval of a co-moving observer. Hence, the total delay of an energetic photon in a co-moving time interval dt is given by $n(z)(1+z)^2 C\sqrt{\alpha'}E_{\rm obs}\, dt$. The time interval dt is related to the Hubble rate $H(z)$ in the standard way: $dt = -dz/[(1+z)H(z)]$. Thus, from (30) we obtain a total delay in the arrival times of photons with energy difference ΔE, which has the form considered in [10, 11], namely it is proportional to ΔE and is suppressed linearly by the quantum gravity (string) scale, M_s:

$$(\Delta t)_{\rm obs} = \frac{\Delta E}{M_s} {\rm H}_0^{-1} \int_0^z n^*(z) \frac{(1+z)dz}{\sqrt{\Omega_\Lambda + \Omega_m (1+z)^3}} \tag{31}$$

where z is the red-shift, ${\rm H}_0$ is the (current-era) Hubble expansion rate, and we have assumed for concreteness the ΛCDM standard model of cosmology. The reader should compare this relation with (2), based on naive considerations. The presence in (31) of the linear density of defects $n^*(z)$, which is in general red-shift dependent, is a crucial difference, and it is this feature that allows the string model to fit simultaneously the MAGIC result and the data from the short burst GRB 090510, as we shall discuss below.

Notice in (31) that the essentially stringy nature of the delay implies that the characteristic suppression scale is the string scale M_s, which plays the rôle of the quantum gravity scale in this case. The scale M_s is a free parameter in the modern version of string theory, and thus it can be constrained by experiment. As we have discussed in this article, the observations of delays of energetic (TeV) photons from AGN by the MAGIC telescope [1] can provide such an experimental way of constraining n^*/M_s in (2). For $\Delta E \sim 10$ TeV, for instance, the delay (2) can lead to the observed one of order of minutes, provided $M_s/n^* \sim 10^{18}$ GeV (in natural units with $c = 1$) [8]. This implies natural values for both n^* and M_s, although it must be noted that n^* is another free parameter of the bulk string cosmology model of [17], considered here. In general, $n^*(z)$ is affected by the expansion of the Universe, as it is diluted by it, but also depends on the bulk model and the interactions among the D-particles themselves. For inhomogeneous foam situations, the dependence of n^* on the redshift can be quite complicated [20]. For redshifts of relevance to the MAGIC experiments, $z = 0.034 \ll 1$, one may ignore the z-dependence of n^* to a good approximation.

The total delay (31) may be thought of as implying [33] an effective *subluminal* refractive index $n(E)$ of light propagating in this space time, since one may assume that the delay is equivalent to light being slowed down due to the medium effects. On account of the theoretical uncertainties in the source mechanism, however, the result of the AGN Mkn 501 observations of the MAGIC Telescope translate to *upper* bounds for the quantity n^*/M_s in (2), which determines the strength of the anomalous photon dispersion in the string/D-particle foam model.

In view of the above discussion, if the time delays observed by MAGIC can finally be attributed partly or wholly to this type of stringy space-time foam, and therefore to the stringy uncertainties, then the AGN Mkn 501, and other such celestial sources of very high energy photons, may be viewed as playing the rôle of Heisenberg microscopes and amplifiers for the stringy space-time foam effects.

But how, then, can we incorporate the data from GRB 0905210 [4] in this model? It is clear that a fit with a linearly modified dispersion relation does not work in this case for the values of the Quantum Gravity scale $M_{\rm QG\,1}^{\rm MAGIC} \sim 10^{18}$ GeV that fit the MAGIC, H.E.S.S. and the other FERMI data. Indeed, the observed short delays of this burst can be explained on the basis of linearly modified dispersion relations only if the quantum gravity scale is *larger* than $M_{\rm QG\,1} \simeq 1.2 M_P$.

As we have already mentioned, this argument has been used in [4] in order to exclude all models of QG involving linearly suppressed time delays for photons, on account of naturalness. We would not agree with this statement. Leaving aside the fact that from a single measurement, with uncertainties on the precursor of the GRB, one cannot draw safe conclusions, we mention that what one calls a *natural scale of QG* is highly model dependent. For instance, as we have seen above, in the string foam model, the relevant "QG scale", that dictates the order of the foam-induced time delays of photons, is a complex function of the model's parameters and is not simply given by the string scale M_s [20]. Indeed, as becomes clear from (29) (or (2)) the relevant scale is not simply the D-particle mass, M_s/g_s, but a combination

$$M_{\rm QG-D-foam} \sim \frac{M_s}{g_s\, n^*(z)} \tag{32}$$

involving the linear number density $n^*(z)$ of the foam defect, encountered by the photon during its propagation from the source till observation. This quantity depends on the bulk density of the D-particles, which in the model of [17] is a free parameter.

Inhomogeneous bulks are perfectly consistent background configurations for our brane world scenario [20]. In order to match the photon delay data of Fig. 1 with the D-foam model, we need a reduction of the *linear* density of defects encountered by the photon by about two orders of magnitude in the region $0.2 < z < 1$, whereas for $z < 0.2$ there must be, on average, one D-particle defect per unit string length ℓ_s. We therefore assume that our D-brane encountered a D-void when $0.2 < z < 1$, in which there was a significant reduction in the bulk nine-dimensional density of defects. Such assumptions can be consistent with cosmological considerations on the dark sector of the model, as explained in [20]. In particular, an important parameter for the cosmology of the model is the propagation velocity v of the brane world in the bulk (*c.f.* fig. 2). As we have discussed previously (11) this is related [20] to the density of the defects near the brane world, $n^{\rm short}(z)$, and hence to the linear density of defects encountered by the photon on the brane world.

To have a reduced density by two order of magnitude at redshifts $z = 0.9$, while having a density of defects of $O(1)$ per string length at redshifts $z < 0.1$, one must consider the magnitude of v as well as the string scale ℓ_s. For instance, it follows from (11) that for string energy scales of the order of TeV, *i.e.* string time scales $\ell_s/c = 10^{-27}\ s$, one must consider a brane velocity $v \le \sqrt{10} \times 10^{-11}\, c$, which is not implausible for a slowly moving D-brane at a late era of the Universe [7]. This is compatible with the constraint on v obtained from inflation in [22], namely $v^2 \le 1.48 \times 10^{-5}\, g_s^{-1}$, where $g_s < 1$ for the weak string coupling we assume here. On the other hand, if we assume a 9-volume $V_9 = (K\ell_s)^9$: $K \sim 10^3$ and $\ell_s \sim 10^{-17}/{\rm GeV}$, then the brane velocity $v \le 10^{-4}c$. In our model, due to the friction induced on the D-brane by the bulk D-particles, one would expect that the late-epoch brane velocity should be much smaller than that during the inflationary era immediately following a D-brane collision [22].

We mention for completion, that the above considerations pertain strictly speaking to point-like D-particles, which are allowed only in type IA string constructions. However, as discussed in [20], similar conclusions can be applied to type-IIB constructions of D-foam [19].

5. Other (astrophysical) constraints on quantum-gravity foam

The sensitivity of the MAGIC (and FERMI) observations to Planck scale physics, at least for linearly suppressed modified dispersion relations, calls for an immediate comparison with other sensitive probes of non-trivial optical properties of QG medium.

[7] Much smaller velocities are required for small string scales that are comparable to the four-dimensional Planck length.

Indeed, from the analysis of [8], there was no microscopic model dependence of the induced modifications of the photon dispersion relations, other than the sub-luminal nature of the induced refractive index and the associated absence of birefringence, that is the independence of the refractive index on the photon polarization. The latter feature avoids the otherwise very stringent constraints on the photon dispersion relation imposed by astrophysical observations, as we now come to discuss.

We shall be very brief in our description of the complementary astrophysical tests on Lorentz invariance and quantum-gravity modified dispersion relations, to avoid large diversion from our main point of this review article which is string theory.

There are three major classes of complementary astrophysical constraints, to be considered in any attempt to interpret the MAGIC, FERMI or more general γ-ray Astrophysics results in terms of quantum-gravity induced anomalies in photon dispersion.

- ***Birefringence and strong constraints on QG-induced photon dispersion***
  In certain models of quantum gravity, with modified dispersion relations, for instance the so-called loop-quantum gravity [47], the ground state breaks reflexion symmetry (parity) and this is one of the pre-requisites for a dependence of the induced refractive index on the photon polarization, *i.e.* birefringence. We remind the reader that in birefringent materials this is caused precisely by the existence of some kind of anisotropies in the material. The velocities of the two photon polarizations (denoted by $\pm$) in such QG models may be parametrised by:
  $$v_{\pm} = c\,(1 \pm \xi(\hbar\omega/M_P)^n) \tag{33}$$
  where $M_P = 1.22 \times 10^{19}$ GeV is the Planck energy scale, and ξ is a parameter following from the underlying theoretical model, which is related with the modifications of the pertinent dispersion relations for photons. The order of suppression of these effects is described by n which in the models of [47] assumed the value $n = 1$, but in general one could have higher order suppression.
  Vacuum QG birefringence should have showed up in optical measurements from remote astrophysical sources, in particular GRBs. Ultraviolet (UV) radiation measurements from distant galaxies [48] and UV/optical polarization measurements of light from Gamma Ray Bursters [49] rule out birefringence unless it is induced at a scale (way) beyond the Planck mass (for linear models, the lower bound on the QG scale in such models can exceed the Planck scale ($\sim 10^{19}$ GeV) by as much as *seven orders of magnitude*). Indeed, in terms of the parameter η introduced above (*c.f.* (33)), for the case $n = 1$ of [47], one finds from optical polarization observations that the absence of detectable birefringence effects imply the upper bound $|\xi| < 2 \times 10^{-7}$, which is incompatible with the MAGIC observed delays, saturating from below the bounds imposed by the MAGIC experiment.
  At this point, we wish to mention that, using recent polarimetric observations of the Crab Nebula in the hard X-ray band by INTEGRAL [50], the authors of [51] have demonstrated that the absence of vacuum birefringence effects constrains linearly suppressed Lorentz violation in quantum electrodynamics to the level $|\xi| < 6 \times 10^{-10}$ at 95% C.L., thereby tightening by about three orders of magnitude the above-mentioned constraint.
- ***Synchrotron Radiation and further stringent constraints for electronic QG-induced anomalous dispersion in vacuo***
  Another important experimental constraint on models with QG-induced anomalous dispersion relations comes from observations of synchrotron radiation from distant galaxies [52, 53, 54], such as Crab Nebula (*c.f.* fig. 4). As well known, the magnetic fields at the core regions of galaxies curve the paths of (and thus accelerate) charged particles, in particularly electrons (which are stable and therefore appropriate for astrophysical

observations), and thus, on account of energy conservation this results in synchrotron radiation.

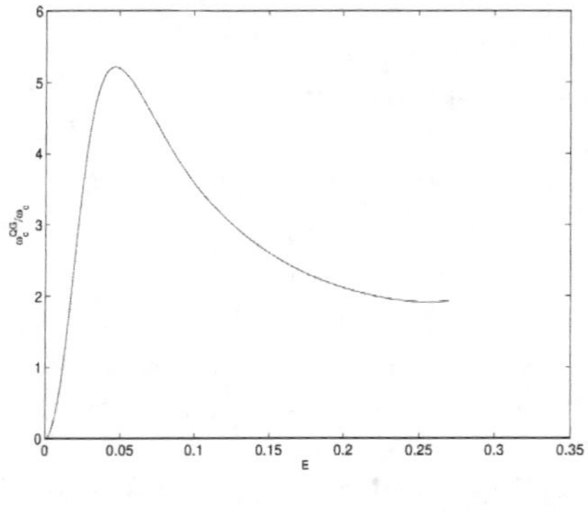

Figure 4. The Crab Nebula is a supernova remnant, with a rotating neutron star (the Crab Pulsar) at its centre. Observations of synchrotron radiation from such celestial objects place very stringent constraints on quantum gravity models with anomalous dispersion relations for electrons (from the first of Ref. [53]).

In standard electrodynamics [55], electrons in an external magnetic field H, follow helical orbits transverse to the direction of H. The so-accelerated electrons in a magnetic field emit synchrotron radiation with a spectrum that cuts off sharply at a frequency ω_c (*c.f.* fig. 4):

$$\omega_c^{LI} = \frac{3}{2}\frac{eH}{m_0}\frac{1}{1-\beta^2}, \tag{34}$$

where e is the electron charge, m_0 its mass, and $\beta_\perp \equiv v_\perp$ is the component of the velocity of the electron perpendicular to the direction of the magnetic field. The superfix LI in (34) stresses that this formula is based on a LI approach, in which one calculates the electron trajectory in a given magnetic field H and the radiation produced by a given current, using the relativistic relation between energy and velocity.

All these assumptions are affected by violations of Lorentz symmetry, such as those encountered in quantum-gravity space-time foam models, leading to modified dispersion relations of the form:

$$\omega^2(k) = k^2 + \xi_\gamma \frac{k^{2+\alpha}}{M_P^\alpha}, \tag{35}$$

$$E^2(p) = m_0^2 + p^2 + \xi_e \frac{p^{2+\alpha}}{M_P^\alpha}, \tag{36}$$

for photons (35) and electrons (36), where ω and k are the photon frequency and wave number, and E and p are the electron energy and momentum, with m_0 the electron (rest) mass. In the spirit of the MAGIC observation analysis above, we assume here linear ($\alpha = 1$) or quadratic QG ($\alpha = 2$) effects, characterized by parameters ξ_γ and ξ_e, extracting the Planck mass scale $M_P = 1.22 \times 10^{19}$ GeV. In fact one can do the analysis [53] for a general α (single power) and attempt to extract limits on this parameter by matching with observations.

A detailed analysis [52, 53], including the modifications in the electron's trajectories due to space-time foam [53], yields:

$$\omega_c^{QG} \propto \omega_c^{LI} \frac{1}{(1+\sqrt{2-1/\eta^2})^{1/2}\left(\frac{m_0^2}{E^2} + (\alpha+1)\left(\frac{E}{\mathcal{M}}\right)^\alpha\right)},$$

$$\mathcal{M} \equiv M_P/|\xi_e| \ , \quad \eta \equiv 1 - (E/\mathcal{M})^\alpha \ , \tag{37}$$

where ω_c^{LI} is given in (34) and the superscript "QG" indicates that the QG-modified dispersion relations (36) are used. This function is plotted schematically (for $\alpha = 1$) in fig. 4.

In [52, 53], the above QG-modified dispersion relations have been tested using observations from Crab Nebula. It should be emphasized that the estimate of the end-point energy of the Crab synchrotron spectrum and of the magnetic field used above are indirect values based on the predictions of the Synchrotron Self-Compton (SSC) model of very-high-energy emission from Crab Nebula [56]. In [53] the choice of parameters used was the one that gives good agreement between the experimental data on high-energy emission and the predictions of the SSC model [56, 57]. Estimating the magnetic field of Crab Nebula in the region 160×10^{-6} Gauss $< H < 260 \times 10^{-6}$ Gauss, and requiring $|\xi_e| \le 1$ (which thus sets the quantum gravity scale as at least M_P) one obtains the following bounds on the exponent α of the dispersion relations (36) [53]:

$$\alpha \ge \alpha_c \ : \qquad 1.72 < \alpha_c < 1.74 \tag{38}$$

These results imply already a sensitivity to quadratic QG corrections with Planck mass suppression M_P.

However, for photons there are no strong constraints on ξ_γ coming from synchrotron radiation studies, unless in cases where QG models entail birefringence [54], where, as we discussed above, strong constraints on the photon dispersion are expected at any rate from optical measurements on GRBs. In this sense, the result (38) *excludes the possibility* that the MAGIC observations leading to a four-minute delay of the most energetic photons are due to a quantum foam that acts *universally* among photons and electrons. However, the synchrotron radiation measurements cannot exclude anomalous photon dispersion with linear Planck-mass suppression, leading to a saturation of the lower bound imposed by the MAGIC experiment [8], in models where the foam is transparent to electrons, as in the string foam case.

- ***Strong constraints from Ultra-high-energy Cosmic photon annihilation***
Further strong constraints on generic modified dispersion relations for photons, like the ones used in the aforementioned QG-interpretation of the MAGIC results [8], comes from processes of scattering of ultra-high-energy photons, with energies above 10^{19} eV, off very-low energy cosmic photons, such as the ones of the cosmic microwave background (CMB) radiation that populates the Universe today, as a remnant from the Big-Bang epoch. In [44] it has been argued that the non-observation of such ultra-high energy (UHE) photons places very strong constraints on the parameters governing the modification of the photon dispersion relations, that are several orders of magnitude smaller than the values required to reproduce the MAGIC time delays, should the effect be attributed predominantly to photon propagation in a QG dispersive medium.
The main argument relies on the fact that an ultra-high-energy photon would interact with a low-energy ("infrared") photon of the CMB background (with energies in the eV range) to produce electron prositron pairs, according to the reaction:

$$\gamma_{\rm UHE} \quad + \quad \gamma_{\rm CMB} \quad \Rightarrow \quad e^+ \, e^- \ . \tag{39}$$

The basic assumption in the analysis is the strict energy and momentum conservation in the above reaction, despite the modified dispersion relations for the photons. Such an assumption stems from the validity of a *local-effective-lagrangian description* of QG foam effects on particles with energies much lower than the QG energy scale (assumed close to

Planck scale $M_{\rm Pl} = 10^{19}$ GeV). In this formalism, one can represent effectively the foam *dispersive effects* by higher-derivative *local operators* in a *flat-space-time* Lagrangian. The upshot of this is the modification of the pertinent equations of motion for the photon field (which in a Lorentz-invariant theory would be the ordinary Maxwell equations) by higher-derivative terms, suppressed by some power of the QG mass scale.
One considers the modified dispersion relations (35), (36), which in the notation of [44], taking explicit account of the various polarizations and helicities, can be written as:

$$\omega_\pm^2 = k^2 + \xi_n^\pm k^2 \left(\frac{k}{M_P}\right)^n , \qquad \omega_b^2 = k_b^2 ,$$
$$E_{e,\pm}^2 = p_e^2 + m_e^2 + \eta_n^{e,\pm} p_e^2 \left(\frac{p_e}{M_P}\right)^n \tag{40}$$

with $(\omega, \vec{k})$ the four-momenta for photons, and $(E, \vec{p}_e)$ the corresponding four-momentum vectors for electrons; the suffix b indicates a low-energy CMB photon, whose dispersion relations are assumed approximately the normal ones, as any QG correction is negligible due to the low values of energy and momenta. The +(-) signs indicate left(right) polarizations (photons) or helicities (electrons). Positive (negative) ξ indicate subluminal (superluminal) refractive indices. Upon the assumption of energy-momentum conservation in the process, one arrives at kinematic equations for the threshold of the reaction (39), that is the minimum energy of the high-energy photon required to produce the electron-positron pairs.
For the linear- or quadratic- suppression case, for which $n = 1, 2$ respectively (40), one finds that, for the relevant subluminal photon refractive indices corresponding to the saturation of the lower bound on the QG scale $M_{QG1} \sim 10^{18}$ GeV, the threshold for pair production disappears for ultra-high-energy photons, and hence such photons should have been observed. The non-observation of such photons implies constraints for the relevant parameters ξ, η which are stronger by *several orders of magnitude* than the bounds on the QG scale inferred from the MAGIC observations.
From the analysis of [44] one concludes that in the case of linear Planck-mass suppression of the sub-luminal QG-induced modified dispersion relations for photons, of interest for the QG-foam interpretation of the MAGIC results [8], parameters with size $\xi_1 > 10^{14}$ are ruled out. This exceeds the sensitivity of the MAGIC experiment to such Lorentz-symmetry violating effects by fifteen orders of magnitude ! Similar strong constraints are also obtained from the non observations of *photon decay* ($\gamma \to e^+e^-$), a process which, if there are modified dispersion relations, is in general allowed [44].

6. The String Foam Models Evade the above Constraints

From the above discussion it becomes clear that any model of refraction in space-time foam that exhibits effects at the level of the MAGIC experiment sensitivity [8] should be characterised by the following specific properties:

- (i) photons are *stable* (*i.e.* do *not* decay) but should exhibit a modified *subluminal* refractive index with Lorentz-violating corrections that grow linearly with $E/(M_{\rm QG\gamma}c^2)$, where $M_{\rm QG\gamma}$ is close to the Planck scale,
- (ii) the medium should not refract electrons, so as to avoid the synchrotron-radiation constraints [52, 53], and
- (iii) the coupling of the photons to the medium must be independent of photon polarization, so as not to have birefringence, thus avoiding the pertinent stringent constraints [48, 49, 54, 51].
- (iv) The formalism of local effective lagrangians should break down, in the sense that there are quantum fluctuations in the total energy in particle interactions, due to the presence

of a quantum gravitational 'environment', such that stringent constraints, which otherwise would have been imposed from the non-observation of ultra-high energy photons ($\hbar\omega > 10^{19}$ eV), are evaded.

The string-foam models (both type IA and type IIB) are characterized by all these properties [17, 53, 18], and thus avoid the stringent constraints. The absence of birefringence and the transparency of foam to electrons or charged probes, for reasons of charge conservation, make the models surviving the stringent constraints from synchrotron radiation from distant Nebulae [52, 54].

Moreover, the analysis in [44] is based on exact energy momentum conservation in the process (39), stemming from the assumption of the local-effective lagrangian formalism for QG-foam. As we discussed in [58, 41], and mentioned in section 3, however, such a formalism is not applicable in the case of the recoiling D-particle space-time foam model, where the fluctuations of space-time or other defects of gravitational nature paly the rôle of an external environment, resulting in *energy fluctuations* in the reaction (39). The presence of such fluctuations does affect the relevant energy-threshold equations, for the reaction to occur, which stem from kinematics, in such a way that the above stringent limits are no longer valid. In particular, we have seen that the pertinent anomalous dispersion terms are induced by the Finsler-like metric (24), which in turn arises by the distortion of space time due to the recoil of the D-particle space-time defect during its interaction with the photon. This metric is quadratically suppressed by the string scale, and hence the so-induced modified dispersion relations (27) contain anomalous terms quadratically suppressed by the QG scale, a case for which the constraints coming from ultra high energy cosmic rays are currently weak, but of course the situation can change in future experiments [44].

It should be stressed again that in the string model the modified dispersion relations are disentangled from the time delays (28) of the more energetic photons, which are linearly suppressed by the string scale. These delays are associated with the string uncertainty principle [35] and thus are not represented within the local effective field theory framework. We have also seen that this disentanglement allows for a consistent interpretation of the MAGIC delays with those of the short GRB 090510 observed by FERMI within the string D-foam framework, provided of course the foam is inhomogeneous. In addition, as discussed above, the form of the effective string coupling (25), and the induced metric (24), imply an upper bound in the photon momentum transfer (26), which characterizes the D-foam model as a result of the sub-luminal nature of the D-particle recoil velocity. These considerations imply that very high energy cosmic rays, when interacting with the D-foam, will have an extremely suppressed interaction rate (the string amplitudes are vanishing when the recoil velocity approaches the speed of light), thus providing additional reasons [41] for evading the strong constraints of [44]. There are also phenomenologically realistic brane models, with large extra dimensions [42], for which the ratio M_s/g_s (*i.e.* in our context the D-particle mass) may be of the order of the conventional GZK cutoff [43], 10^{20} eV. In these models, in view of (26), there would be no photons with energies higher than this that could not be completely absorbed when interacting with the D-particle foam. This, therefore, provides an explanation for the absence of such high-energy photons, in accordance with observations, and enables the model to evade the stringent constraints coming from ultra-high-energy cosmic rays.

7. Conclusions and Outlook

In this review we have discussed a stringy version of a (Lorentz-Invariance-Violating) space-time foam model, which appears as a candidate theory for an explanation of the delays of the more energetic photons from celestial sources, as observed by MAGIC and FERMI Telescopes, in agreement with all the other current astrophysical tests of Lorentz Violation. Although such

Journal of Physics: Conference Series **283** (2011) 012022 doi:10.1088/1742-6596/283/1/012022

delays could be due to conventional astrophysics at the source, nevertheless the latter is not well understood at present, and in fact there is no consensus among the relevant communities.

The fact that string theory (or better its modern version involving D-brane defects) is capable of explaining potentially the observed photon delays, in agreement with all the other astrophysical data currently available, is at least amusing. This means, if nothing else, that at least such tests provide a framework for experimentally testing/falsifying some models of string theory (entailing Lorentz Violation) at present or in the foreseeable future. The key point in the approach is the existence of space-time defects in the ground state of the model, whose topologically non-trivial interactions with the string states, via string-stretching during the capture process (*c.f.* fig. 3), are mainly responsible for the observed delays. The latter are found proportional to the incident photon energy. The situation may be graphically represented by the case of two boats, of different lengths, in rough seas: the waves in the sea represent the foam, the "medium" over which the two boats propagate; the longer boat, representing the longer wave-length (lower energy) photons, will cross the waves faster than the smaller boat, corresponding to the higher energy (shorter wavelength) photons, which will be relatively delayed by having to climb up and down the waves.

The peculiarity of the D-particle foam in being *transparent* to charged particles (as a result of electric charge conservation requirements), evades the stringent constraints on linear Planck scale suppression refractive indices that would otherwise have been induced by electron synchrotron radiation studies from Crab Nebula [52, 54]. Moreover, the absence of birefringence avoids the similarly stringent constraints on such models that would have been imposed by galactic [54] or extra-galactic measurements [48, 49]. Finally, it worths mentioning that the linear in energy time delays (14),(2), when applied to neutrinos, can be flavour (*i.e.* neutrino-species) independent, thus avoiding [59] the stringent constraints that would be obtained from models of quantum gravity with flavour-dependent modifications of neutrino propagation and thus modifications in their oscillations [60].

We must stress once more, however, that despite the apparent "success" of the D-foam model in dealing with stringent astrophysical constraints so far, one cannot as yet draw any safe conclusions on the validity of the model in Nature. Many more high-energy astrophysical photon measurements are needed in order to disentangle source from possible propagation effects due to fundamental physics. If a statistically significant population of data on photons from cosmic sources is collected, exhibiting refractive indices varying linearly with the distance of the source [10], as well as the photon energy, then this would be a very strong confirmation of the D-particle foam model, for reasons explained above. However, it must be noted that GRB's, which are expected to lead to statistically significant data in the next few years, will produce photons much lower in energies than the flares observed in AGN, and this could be a drawback. At any rate there are attempts to claim that observations from FERMI will have sensitivity close to the Planck scale [61] for such linear-suppression models soon. The case of GRB090510 is a perfect example of how a single measurement of a short, intense high energy burst, can place stringent limits on Lorentz Violation at the Planck scale. However, caution should be exercised here as to what one means by "sensitivity at the Planck scale". As demonstrated above (*c.f.* (32)), the effective quantum gravity scale is actually a complicated function of many microscopic parameters in the model. Nevertheless, if many observations at various redshift regimes on delayed arrivals of cosmic photons become available in the future, then we shall be able to make some definite conclusions regarding the order of magnitude of possible quantum gravity effects and thus falsify models, such as the D-particle space-time foam.

We believe we are entering exciting times for tests of fundamental physics concepts and symmetries by means of high-energy astrophysics. Therefore it seems that this branch of physics, together with collider and particle physics, will provide a platform for exploring our world at increasingly smaller microscopic length scales, thereby offering experimental guidance in our quest for the elusive theory of Quantum Gravity.

Acknowledgements

I would like to thank the organizers of the International Workshop *Recent Developments in Gravity (NEB 14)*, June 8-11 2010, Ioannina (Greece), for their invitation to present this review as a keynote lecture. This work is partially supported by the European Union through the Marie Curie Research and Training Network *UniverseNet* (MRTN-2006-035863).

References

[1] J. Albert *et al.* [MAGIC Collaboration], Astrophys. J. **669**, 862 (2007).
[2] See, for instance: A. Morselli [GLAST Collaboration], J. Phys. Conf. Ser. **110**, 062017 (2008), and references therein.
[3] A. A. Abdo *et al.* [The Fermi LAT and Fermi GBM Collaborations], DOI10.1126/science.1169101 (Science Express Research Articles), published online 19 February 2009.
[4] A. A. Abdo *et al.* [The Fermi LAT and Fermi GBM Collaborations], DOI:10.1038/nature08574, published online 19 November 2009.
[5] A. A. Abdo *et al.* [The Fermi/GBM collaboration and The Fermi/LAT Collaborations and The S Collaboration], Astrophys. J. **706**, L138 (2009) [arXiv:0909.2470 [astro-ph.HE]].
[6] For a review see: A. De Angelis, O. Mansutti and M. Persic, Riv. Nuovo Cim. **31**, 187 (2008).
[7] See, for instance: N. E. Mavromatos, J. Phys. Conf. Ser. **174**, 012016 (2009) [arXiv:0903.0318 [astro-ph.HE]], and references therein.
[8] J. Albert *et al.* [MAGIC Collaboration] and J. R. Ellis, N. E. Mavromatos, D. V. Nanopoulos, A. S. Sakharov and E. K. G. Sarkisyan, Phys. Lett. B **668**, 253 (2008).
[9] For a comprehensive account see: J. A. Wheeler and K. Ford, *Geons, black holes, and quantum foam: A life in physics, New York, USA: Norton (1998)*, and references therein.
[10] G. Amelino-Camelia, J. R. Ellis, N. E. Mavromatos, D. V. Nanopoulos and S. Sarkar, Nature **393**, 763 (1998); J. R. Ellis, K. Farakos, N. E. Mavromatos, V. A. Mitsou and D. V. Nanopoulos, Astrophys. J. **535**, 139 (2000).
[11] J. R. Ellis, N. E. Mavromatos, D. V. Nanopoulos, A. S. Sakharov and E. K. G. Sarkisyan, Astropart. Phys. **25**, 402 (2006) [Astropart. Phys. **29**, 158 (2008)];
[12] U. Jacob and T. Piran, JCAP **0801**, 031 (2008).
[13] J. Ellis, N. E. Mavromatos and D. V. Nanopoulos, Phys. Lett. B **674**, 83 (2009) [arXiv:0901.4052 [astro-ph.HE]].
[14] J. Polchinski, *String theory.* Vols. 1 and 2, *Cambridge, UK: Univ. Press (1998).*
[15] F. Aharonian, *et al.* [H.E.S.S. Collaboration], Astrophys. J. 664 (2007) L71.
[16] F. Aharonian *et al.* [H.E.S.S. Collaboration], Phys. Rev. Lett. **101**, 170402 (2008).
[17] J. R. Ellis, N. E. Mavromatos, M. Westmuckett, Phys. Rev. D **70**, 044036 (2004); *idid.* **71**, 106006 (2005); J. R. Ellis, N. E. Mavromatos, D. V. Nanopoulos and M. Westmuckett, Int. J. Mod. Phys. A **21**, 1379 (2006).
[18] J. Ellis, N. E. Mavromatos and D. V. Nanopoulos, Phys. Lett. B **665**, 412 (2008).
[19] T. Li, N. E. Mavromatos, D. V. Nanopoulos and D. Xie, Phys. Lett. B **679**, 407 (2009) [arXiv:0903.1303 [hep-th]].
[20] J. Ellis, N. E. Mavromatos and D. V. Nanopoulos, arXiv:0912.3428 [astro-ph.CO].
[21] J. H. Schwarz, arXiv:hep-th/9907061.
[22] J. R. Ellis, N. E. Mavromatos, D . V. Nanopoulos, A. Sakharov, New J. Phys. **6**, 171 (2004).
[23] C. Bachas, *A Way to break supersymmetry,* arXiv:hep-th/9503030.
[24] E. Gravanis, N. E. Mavromatos, Phys. Lett. B **547**, 117 (2002); N. E. Mavromatos in *Beyond the Desert, Oulu 2002 (Finland)* (ed. H.V. Klapdor-Kleingrothaus, IoP 2003), 3.
[25] E. Witten, Int. J. Mod. Phys. A **10** (1995) 1247.
[26] N.E. Mavromatos, M. Sakellariadou, Phys. Lett. B **652**, 97 (2007).
[27] See, for instance: A. Melchiorri, L. Pagano and S. Pandolfi, Phys. Rev. D **76**, 041301 (2007) [arXiv:0706.1314 [astro-ph]]; R. A. Daly, S. G. Djorgovski, K. A. Freeman, M. P. Mory, C. P. O'Dea, P. Kharb and S. Baum, arXiv:0710.5345 [astro-ph].
[28] N. E. Mavromatos, Sarben Sarkar and A. Vergou, arXiv:1009.2880 [hep-th].
[29] See, for instance: D. Bao, S. S. Chern and Z. Shen, *An introduction to Finsler Geometry* (Springer-Verlag (NY, 2000)). In the context of D-partidle foam, such Finsler-type metrics have been first derived in: J. R. Ellis, N. E. Mavromatos and D. V. Nanopoulos, Int. J. Mod. Phys. A **13**, 1059 (1998)

[arXiv:hep-th/9609238]. For a short recent review on this topic see: N. E. Mavromatos, PoS **QG-PH**, 027 (2007) [arXiv:0708.2250 [hep-th]] and references therein. Also Finsler metrics in string theory, but in a different context, have been previously suggested in : S. I. Vacaru, arXiv:hep-th/0211068; S. I. Vacaru, arXiv:hep-th/0310132; In a field theory context, such metrics have been discussed, among other works, in: G. Y. Bogoslovsky, arXiv:0706.2621 [gr-qc]; arXiv:0712.1718 [hep-th]. G. W. Gibbons, J. Gomis and C. N. Pope, Phys. Rev. D **76**, 081701 (2007) [arXiv:0707.2174 [hep-th]]; A. P. Kouretsis, M. Stathakopoulos and P. C. Stavrinos, arXiv:0810.3267 [gr-qc]; L. Sindoni, Phys. Rev. D **77**, 124009 (2008) [arXiv:0712.3518 [gr-qc]]; M. Anastasiei and S. I. Vacaru, J. Math. Phys. **50**, 013510 (2009) [arXiv:0710.3079 [math-ph]]. In the context of generic phenomenological models of non-standard dispersion relations in quantum gravity see: F. Girelli, S. Liberati and L. Sindoni, Phys. Rev. D **75**, 064015 (2007) [arXiv:gr-qc/0611024]; J. Magueijo and L. Smolin, Class. Quant. Grav. **21**, 1725 (2004) [arXiv:gr-qc/0305055]. J. Skakala and M. Visser, arXiv:0810.4376 [gr-qc].

[30] For a review, see: A. B. Lahanas, N. E. Mavromatos and D. V. Nanopoulos, Int. J. Mod. Phys. D **12**, 1529 (2003), and references therein.

[31] See for instance: G. Shiu, L. -T. Wang, Phys. Rev. **D69**, 126007 (2004), and references therein.

[32] N. Seiberg, L. Susskind and N. Toumbas, JHEP **0006**, 044 (2000).

[33] R.P. Feynman, R.B. Leighton and M. Sands, *The Feynman Lectures on Physics* Vol. 2, (Addison-Wesley, Reading Mass. 1977).

[34] G. Veneziano, Europhys. Lett. **2**, 199 (1986); D. Amati, M. Ciafaloni and G. Veneziano, Int. J. Mod. Phys. A **3**, 1615 (1988); D. J. Gross and P. F. Mende, Phys. Lett. B **197**, 129 (1987); Nucl. Phys. B **303**, 407 (1988).

[35] T. Yoneya, Prog. Theor. Phys. **103**, 1081 (2000).

[36] I. I. Kogan, N. E. Mavromatos and J. F. Wheater, Phys. Lett. B **387**, 483 (1996); J. R. Ellis, N. E. Mavromatos and D. V. Nanopoulos, Int. J. Mod. Phys. A **13**, 1059 (1998).

[37] N. E. Mavromatos and R. J. Szabo, Phys. Rev. D **59**, 104018 (1999) [arXiv:hep-th/9808124]; JHEP **0110**, 027 (2001) [arXiv:hep-th/0106259].

[38] For a very partial but somewhat relevant works to our discussion here see: V. Gurarie, Nucl. Phys. B **410**, 535 (1993) [arXiv:hep-th/9303160]; J. S. Caux, I. I. Kogan and A. M. Tsvelik, Nucl. Phys. B **466**, 444 (1996) [arXiv:hep-th/9511134]; I. I. Kogan and N. E. Mavromatos, Phys. Lett. B **375**, 111 (1996) [arXiv:hep-th/9512210]. M. A. I. Flohr, arXiv:hep-th/0407003. J. Fuchs, S. Hwang, A. M. Semikhatov and I. Y. Tipunin, Commun. Math. Phys. **247**, 713 (2004) [arXiv:hep-th/0306274]. M. S. Movahed, M. Saadat and M. Reza Rahimi Tabar, Nucl. Phys. B **707**, 405 (2005) [arXiv:cond-mat/0409486], and references therein.

[39] N. Seiberg and E. Witten, JHEP **9909**, 032 (1999) [arXiv:hep-th/9908142].

[40] C. P. Burgess, Nucl. Phys. B **294**, 427 (1987).

[41] J. Ellis, N. E. Mavromatos and D. V. Nanopoulos, arXiv:1004.4167 [astro-ph.HE], Phys. Lett. B in press (DOI information: 10.1016/j.physletb.2010.09.035).

[42] I. Antoniadis and B. Pioline, Nucl. Phys. B **550**, 41 (1999) [arXiv:hep-th/9902055].

[43] K. Greisen, Phys. Rev. Lett. **16**, 748 (1966); G. T. Zatsepin and V. A. Kuzmin, JETP Lett. **4**, 78 (1966) [Pisma Zh. Eksp. Teor. Fiz. **4**, 114 (1966)].

[44] M. Galaverni and G. Sigl, Phys. Rev. D **78**, 063003 (2008); Phys. Rev. Lett. **100**, 021102 (2008). L. Maccione, S. Liberati and G. Sigl, Phys. Rev. Lett. **105**, 021101 (2010) [arXiv:1003.5468 [astro-ph.HE]].

[45] S. M. Carroll, J. A. Harvey, V. A. Kostelecky, C. D. Lane and T. Okamoto, Phys. Rev. Lett. **87**, 141601 (2001) [arXiv:hep-th/0105082].

[46] For a relatively recent review on Standard Model Extension with Lorentz and/or CPT-Violating interactions see: V. A. (. Kostelecky, *Prepared for 3rd Meeting on CPT and Lorentz Symmetry (CPT 04), Bloomington, Indiana, 8-11 Aug 2007* and references therein.

[47] R. Gambini and J. Pullin, Phys. Rev. D **59**, 124021 (1999).

[48] R. J. Gleiser and C. N. Kozameh, Phys. Rev. D **64**, 083007 (2001).

[49] Y. Z. Fan, D. M. Wei and D. Xu, Mon. Not. Roy. Astron. Soc. **376**, 1857 (2006).

[50] A. J. Dean *et al.*, Science **321**, 1183 (2008).

[51] L. Maccione, S. Liberati, A. Celotti, J. G. Kirk and P. Ubertini, Phys. Rev. D **78**, 103003 (2008).

[52] T. Jacobson, S. Liberati and D. Mattingly, Nature **424**, 1019 (2003).

[53] J. R. Ellis, N. E. Mavromatos and A. S. Sakharov, Astropart. Phys. **20**, 669 (2004); J. R. Ellis, N. E. Mavromatos, D. V. Nanopoulos and A. S. Sakharov, Int. J. Mod. Phys. A **19**, 4413 (2004); Nature **428**, 386 (2004).

[54] L. Maccione, S. Liberati, A. Celotti and J. G. Kirk, JCAP **0710**, 013 (2007).

[55] See, for instance: J.D. Jackson, *Classical Electrodynamics* (John Wiley & Sons, Inc., 1975).

[56] R. A. Ong, Phys. Rept. **305**, 93 (1998).

[57] A. M. Hillas *et al.*, Astrophys. J. **503**, 744 (1998).
[58] J. R. Ellis, N. E. Mavromatos and D. V. Nanopoulos, Phys. Rev. D **63**, 124025 (2001).
[59] J. R. Ellis, N. Harries, A. Meregaglia, A. Rubbia and A. Sakharov, Phys. Rev. D **78** (2008) 033013 [arXiv:0805.0253 [hep-ph]].
[60] R. Brustein, D. Eichler and S. Foffa, Phys. Rev. D **65**, 105006 (2002).
[61] R. Lamon, JCAP **0808** (2008) 022.

Journal of Physics: Conference Series **283** (2011) 012023 doi:10.1088/1742-6596/283/1/012023

Open problems and results in the group theoretic approach to quantum gravity via the BMS group and its generalizations

Evangelos Melas

Technological Educational Institution of Lamia ,
Department of Informatics, GR 35–100, Lamia

E-mail: evangelosmelas@yahoo.co.uk

Abstract. The Bondi-Metzner-Sachs group B is the common asymptotic group of all asymptotically flat (lorentzian) space-times, and is the best candidate for the universal symmetry group of General Relativity. However, in quantum gravity, complexified or euclidean versions of General Relativity are frequently considered. McCarthy has shown that there are forty−two generalizations of B for these versions of the theory and a variety of further ones, either real in any signature, or complex. A firm foundation for quantum gravity can be laid by following through the analogue of Wigner's programme for special relativity with B replacing the Poincare group P. Here the main results which have been obtained so far in this research programme are reported and the more important open problems are stated.

1. History generalizations, other approaches and open problems

In 1939 Wigner, having followed a suggestion of Dirac, published a remarkable paper [1] which laid the foundations of quantum theory of fields . This paper greatly influenced also mathematics since it was the first explicit treatment of infinite dimensional representations of a Lie group, namely, of the Poincare group P.

It is interesting that G.W.Mackey developed his pioneering work on group representations of semi−direct products [2, 3, 4, 5, 6, 7, 8, 9, 10, 11, 12, 13] by extending Wigner's work and by using previous results derived by Clifford [14]. Remarkably, Mackey was able to show [2, 15] that Wigner's pioneering construction, as well as the Stone−von Neumann theorem [16, 17] on the uniqueness of operators satisfying the Heisenberg commutation relations could be derived as consequences of a principal theorem generalizing to weakly (and hence strongly) continuous unitary representations of separable locally compact groups of a well−known theorem for finite groups.

It is noteworthy that the entirely novel discrete results on the spectra of area and volume derived in loop quantum gravity stem from the fact that the Stone−von Neumann theorem cannot be employed any more: The Stone−von Neumann theorem tells us that there is (up to equivalence) precisely one irreducible regular representation of the Weyl algebra generated by the exponentiated position and momentum operators together with their Poisson relations. In loop quantum gravity one of the assumptions of the Stone−von Neumann theorem is not valid and as a result an inequivalent (unique!) representation of the corresponding Weyl algebra is obtained (see e.g. [23]).

Wigner's paper was also deeply satisfying philosophically, being based only on the most firmly established experimental facts and theoretical principles so that his work was free of any tacit or ad hoc assumptions. In particular he based his study on the following premises:

- The set of pure states of an isolated quantum mechanical system was described by the projective space $\mathcal{P}(H)$ of complex straight lines, through the origin, of a complex Hilbert space H.
- The relativity principle was expressed as the numerical invariance of all transition probabilities under Poincare transformations of the affine Minkowski space.

With these premises Wigner obtained:

- An exhaustive list of the strongly continuous unitary irreducible representations (IRs) of P.
- A complete classification of all possible solutions of all possible P-invariant wave equations without having to find or solve the equations!. These solution sets are precisely the IRs of the universal cover of P.
- The only known plausible theoretical definition of what a relativistic 'elementary particle' is: An 'elementary particle' is that entity which is described by one of the IRs of P. Quite unexpectedly, he showed that these IRs are fully parameterized by the mass and the spin of elementary particles. By doing so, he demonstrated in the most dramatic way 'The unreasonable effectiveness of mathematics in the natural sciences'.

In 1962, an intensive study of asymptotically flat axially symmetric space–times, representing bounded gravitational sources emitting gravitational radiation, was undertaken by Bondi, Metzner and Van der Burg [19, 20]. Soon after Sachs [21] showed that their results hold intact if the simplifying assumption of axial symmetry is dropped. This study was preceded by pioneering work of Trautman [22] on boundary conditions appropriate for gravitational radiation. Bondi, van der Burg, Metzner and Sachs analyzed the Einstein equations under the assumption that the spacetime metric admits a power series expansion in a distance from sources (measured along a system of null geodesies) with the leading term being the Minkowski metric. The main result was a simple definition of the total gravitational energy (the Bondi mass), which, due to the Einstein equations, is a nonincreasing function of a retarded time (a similar result was obtained earlier in [22], see [23]). Besides, there was a totally unexpected spin–off: it was found that the set of coordinate transformations which preserve the imposed, gravitational radiation–dictated, boundary conditions form a group, much larger than the Poincare group. the so called Bondi–Metzner–Sachs group B. Consequently, B is an unavoidable consequence of the presence of gravitational waves in General Relativity. B is the same for the whole class of asymptotically flat space–times which share the aformentioned boundary conditions – hence its universality and importance. In this sense, B is the common asymptotic group of all asymptotically flat (lorentzian) space–times, and is the best candidate for the universal symmetry group of General Relativity. As such, one may expect B to play a role in General Relativity similar to the role of the Poincare group P of Special Relativity. In fact, B aroused considerable interest as a possible candidate for replacing the Poincare group [24]. A study of unitary representations of the group was initiated by Sachs [24], and taken further by Cantoni [25, 26]. A little later, Komar [27] and Newman [28] suggested the possible relevance of the group to the problem of quantizing the gravitational field. Much later Awada, Gibbons and Shaw found [29] the appropriate generalization of this group for supergravity. Their analysis is valid for *any* space–time dimension.

B was first discovered [20, 21] *not* as a transformation group of (exact) global diffeomorphisms of a fixed manifold, but as a pseudo–group of local diffeomorphisms ('asymptotic isometries') of the asymptotic region of (Lorentzian) space–times which are asymptotically flat in lightlike future directions. However, Penrose showed with his conformal technique [30] that, by 'going to infinity', B *could* be interpreted as an (exact) global transformation group $B \times \Im^+ \to \Im^+$ of the 'future null boundary' $\Im^+$ of the space–times concerned. The importance of groups in Physics stems from the fact that they are automorphism groups of some underlying structure. Penrose showed that this is the case for B: He gave a geometric structure to $\Im^+$, the 'strong conformal geometry', such that the classical action $B \times \Im^+ \to \Im^+$ is the group of automorphisms of this geometry.

The key observation behind Penrose's conformal technique is that 'infinity' is far away with respect to the physical space–time metric. This means that one needs infinitely many 'meter sticks' in succession to 'get to infinity'. But what if by a conformal rescaling of the metric we replaced these meter sticks by ones that grow in length the farther out we go? Then it might be possible that only a finite number of them suffices to cover an infinite range, provided the growth rate is just right. This is not a new idea. In the theory of functions of one complex variable is common practice to bring infinity to a finite distance by using a conformal transformation. Penrose realized that he could use this type of transformation in General Relativity to map infinity to the boundary of an unphysical space–time which lies at finite distance with respect to the unphysical metric which is conformally related to the metric of the physical space–time; the effectiveness of the conformal technique in General Relativity partially stems from the fact that null geodesics are invariant under conformal rescalings of the metric. The conformal technique was in accord with Penrose's novel approach to General Relativity whose main ingredient was to take the conformal structure of space-time as fundamental . With Penrose's conformal technique, roughly speaking, points at infinity, which can be thought of as endpoints of null geodesics, are adjoined to the space–time by means of a conformal rescaling of the metric. The existence of a conformal factor which compresses the points at infinity into a pair of finite null hypersurfaces $\Im^\pm$ containing the endpoints of all null geodesics, together with a causality assumption and the vacuum (Einstein or Einstein–Maxwell) field equations in the asymptotic region, guarantee [31] that are non–shearing null hypersurfaces with the topology $R \times S^2$. Penrose's analysis provided a basis for later work due to Newman [32], Newman and Penrose [33], Schmidt [34], Winicour [35], and others. These investigations have been crucial to the study of radiation – especially gravitational radiation – and also to the development of several ideas concerning global issues in general relativity. In particular, a great deal of the analysis of black holes, singularities [36], and, more recently, of H–spaces [37] and of asymptotic quantization of zero rest mass fields [31, 38] relies heavily on concepts and techniques – the conformal technique playing a prominent role – introduced originally in the investigation of null infinity .

$B = C(S^2, R) \circledS_T L^{\uparrow}_{+}$, i.e., is the semi–direct product of the group of *globally* defined orientation–preserving conformal transformations of the unit two–sphere, which is isomorphic to the orthochronous homogeneous Lorentz group $L^{\uparrow}_{+}$, times the abelian normal subgroup $C(S^2, R)$ of so–called supertranslations (no relation to supersymmetry) ; $C^\infty(S^2, R)$ is the set of real–valued functions defined on the unit two–sphere. Supertranslations occur for a direct physical reason [39]. Imagine a family of observers in the far field zone ('near $\Im^+$') of a bounded source which emits gravitational radiation, at various angular coordinates (θ, ϕ) who synchronize their clocks. A gravitational wave passes and they examine their clocks again. They will find that they have become desynchronized, that is, they have passed from a common retarded time u to one of the form $u + \alpha(\theta, \phi)$, where $\alpha(\theta, \phi) \in C(S^2, R)$. In Special Relativity, there are no gravitational waves, this sort of desynchronization cannot occur, and therefore one has no need to introduce B.

Wigner's work for P, and the universal property of B for general relativity, make it reasonable to attempt to lay a similarly firm foundation for quantum gravity, by following through the analogue of Wigner's programme with B replacing the group P. Some years ago McCarthy constructed explicitly [40, 41, 42, 43] the IRs of B for exactly this purpose. He based his work on G.W.Mackey's pioneering work on group representations [8, 10]; in fact MacCarthy's construction necessitated an extension to the relevant infinite–dimensional case of Mackey's semi–direct product theory.

One striking difference between P and B is as follows. The 'little groups' for P (from which the IRs are induced) are compact for positive mass squared, but non-compact for zero or negative mass squared. While compact little groups always give discrete spins, the non-compact ones also give *continuous* spins. However, the little groups for B are *always* compact [44, 45, 46, 47], and this means that the IRs of B necessarily have only *discrete* spins. Since only discrete spins are observed in nature, this means that the presence of gravity, or General Relativity, via B, gives a possible explanation that continuous spins are never observed. Consequently, apart from its connection with gravitation, the BMS group seems to be suited to elementary particle physics [41] because the 'elementary particles' it defines have only discrete spins with a finite number of polarization states and because it leaves open the possibility of a coupling, which avoids the pitfall of O'Raifeartaigh's theorem [48], with an internal symmetry group. Evidently representations of the BMS group should be used in the context of an S-matrix theory.

Another striking difference between P and B is that, while the little groups of P are all of *infinite* order (they are all three dimensional connected Lie groups), some of the little groups of B are *finite*. These finite groups are precisely the (double covering groups of the) symmetry groups of the regular polygons or polyhedra in ordinary Euclidean 3-space. That is, certain IRs of B are induced from the complex linear IRs of the cyclic or dihedral groups, or symmetry groups of the tetrahedron, cube or icosahedron [42].

Much later Kronheimer classified completely [49, 50] (see also [51, 52]) the ALE gravitational instantons which solve the real nonlinear euclidean self-dual Einstein equations. He showed that the parametrization of their spaces (moduli spaces) intimately involves the same complex linear IRs of these same finite symmetry groups (and not just the groups themselves). However, his description only partially describes the moduli spaces since it still involves *constraint equations* which remain unsolved. Moreover, gravitational instantons arise for either complex or Euclidean space-times, and this made it important to find analogues of B appropriate to complex space–times, or space–times of different signatures.

The role of the IRs of B is, however, much less well understood than the role of Wigner's IRs of P. In order to make this role better understood, and to relate the group theoretical approach more closely to other approaches to quantum gravity, where complexified or euclidean versions of General Relativity are frequently considered, McCarthy constructed analogues of B for these versions of the theory, and a variety of further ones, either real in any signature, or complex and obtained 42 groups G [53]. Three of them are complex, whereas the remaining 39 are real. All these groups G have a semi–direct product structure :

$$G = C(A, R) ⓢ_T H, \tag{1}$$

where, $C(A, R)$ is an Abelian subgroup of G. A denotes in all cases a locally compact space. $C(A, R)$ are the supertranslations, real–valued functions defined on A. The representation 'T' of H on $C(A, R)$ which specifies in each case the semi–direct product is not relevant here. Among these 42 groups more relevant to the purposes of this note are the 'Spi' group B_S ('Spi' stands for spatial infinity; it rhymes with 'Scri' which denotes null infinity), the Euclidean BMS group EB, the complex BMS group $\mathcal{CB}$ and its two real sections : the original BMS group B and the

Ultrahyperbolic BMS group $B(2,2)$. These groups are given respectively by :

$$\begin{aligned} B_S &= C(S^2 \times I, R) \circledS_T SL(2,C), & (2)\\ EB &= C(S^3, R) \circledS_T (SU(2) \times SU(2)), & (3)\\ \mathcal{CB} &= C(S^2 \times S^2, C) \circledS_T (SL(2,C) \times SL(2,C)), & (4)\\ B &= C(S^2, R) \circledS_T SL(2,C), & (5)\\ B(2,2) &= C_e(S^1 \times S^1, R) \circledS_T (SL(2,R) \times SL(2,R)), & (6) \end{aligned}$$

where I and S^n denote respectively an open interval of R and the $n-$dim sphere. Ashtekar and Hansen [54], in examining both space−like and null infinity for real lorentzian space−times derived a BMS−like group based on the unit space−like hyperboloid. B_S is easily shown to be isomorphic to the group defined by Ashtekar and Hansen.

B is often interpreted via its classical action $B \times \Im^+ \to \Im^+$ on Penrose's future null infinity $\Im^+$. This classical action is vital for understanding, for example, the nature of classical linear and angular momentum for radiating space-times. Interestingly enough each of the generalizations G of B has an analagous classical action. In fact each group G has two actions, the first being a 'lifting' of the second 'projective' action. It is the *projective* actions of the groups G which generalize the Penrose action $B \times \Im^+ \to \Im^+$. The spaces on which the groups act are interpreted not as 'boundaries at infinity', but rather as 'blow-ups' at the origin of parts of punctured flat spaces [55, 56, 57].

Let us pause for a moment and revise the situation in $4-d$ General Relativity. The asymptotic structure of the gravitational field of isolated systems has been investigated in detail in two regimes; at large null separations from sources and at large space−like separations. At present there are two main approaches to the problem of the asymptotic structure of the gravitational field at large null separations; that of Bondi, Metzner, Van der Burgh [19, 20] and Sachs [21] and that of Penrose [30, 31] in the framework of his conformal technique. An important link between the two approaches was found by Tamburino and Winicour [58] who gave a construction of the Bondi−Sachs coordinates for metrics satisfying Penrose's assumptions and the vacuum Einstein equations.

In the spatial regime, on the other hand, the situation is less satisfactory as far as the geometrical results are concerned. The early work on the subject [59, 60, 61, 62] was based on slicing of spacetime into space and time: the boundary conditions were introduced on the initial data on a Cauchy surface rather than on the 4−geometry as a whole. Furthermore, the status of the asymptotic symmetry group at spatial infinity remained unclear because these frameworks are ill equipped to explore the spacetime supertranslations. Indeed, in the literature on these 3+1 frameworks, one can find conflicting statements even on what the asymptotic symmetry group is.

To overcome these problems, five manifestly covariant approaches were introduced in the late 1970s. In the first of these covariant approaches Geroch [63, 64, 65] geometrized Arnowitt, Deser and Misner's approach by using Penrose's conformal technique. He attached an additional 'point at infinity' to the 3−dim Cauchy surface. The resulting conformally completed manifold has the topology of a 3−sphere. To study further spatial infinity Geroch introduced a 3−dim manifold S_G (consisting in his formulation of all 'points' at spatial infinity) with an appropriate metric. In the second of these covariant approaches Ashtekar and Hansen [66, 67], unified spatial and null infinity. Their method is a four−dimensional formulation of Geroch's version [63, 64, 65] of the Arnowitt, Deser and Misner's approach. According to this method spatial infinity is brought metrically close by a conformal transformation of the physical metric and is represented by a single point i^0. This point is attached to the space−time and serves as its spatial boundary. This framework involves a conformal completion of the *entire* space-time, null infinity becoming a null cone with spatial infinity i^0 its vertex. Since spatial infinity is now naturally 'tied' to null infinity, this framework enables one to establish theorems relating the two regimes. This framework is used, for example, both to formulate and to prove the assertion that the ADM mass is the past limit of the future Bondi mass and the ADM 4−momentum at spatial infinity is indeed the past limit of the Bondi 4−momentum [68, 69]. However, precisely because all points at spatial infinity are now 'squeezed down' to a single point the smoothness of the completed manifold fails at this point. So, inevitably, one is forced to deal with complicated differentiable structures there. This circumstance is less satisfactory than that of null infinity $\Im^{\pm}$, which is formulated as a smooth boundary of space-time. To overcome these unwanted features Ashtekar and Hansen propose a suitable 'blowing up' of i^0. The result of this 'blowing up' is a 4−dim manifold called Spi constructed from various inextendible space−like curves 'regular' at i^0. Spi has the structure of a principal fibre bundle: The base space is the unit time−like hyperboloid and the structure group is the additive group of reals. But both $\mathcal{S}_G$ and Spi are somehow artificial and neither of them is a boundary surface of the space−time itself. Thus, e.g., it is meaningless to ask whether or not a curve of the physical space−time has an end point on $\mathcal{S}_G$ or Spi, or whether a tensor field defined on physical space−time has a limit on $\mathcal{S}_G$ or Spi.

There does exist a related but coordinate dependent treatment similar to that of Bondi et. al. [20] in which the awkward differentiability conditions are avoided. In this approach Beig and Schmidt [70, 71] postulated a Bondi−type expansion of the metric near spatial infinity and showed that Einstein's vacuum equations give rise to a hierarchy of linear equations for the coefficients in this expansion. They demonstrated that this hierarchy can be completely solved provided the initial data satisfy certain constraints.

In the remaining three covariant approaches spatial infinity now arises as a boundary of spacetime, rather than a single point. Therefore, these approaches are closer in spirit to the Penrose treatment of null infinity. In the third of these approaches Sommers [72] attached to an asymptotically flat space−time at spatial infinity a 3-dim boundary using projective rather than conformal completion. The boundary is assumed to be the unit timelike hyperboloid of the Minkowski metric and is constructed from Minkowski's space−time as the set of all 'end points' of (equivalence classes of) spacelike curves. Unfortunately however these investigations have remained rather incomplete. Thus, in [72] one finds a precise definition of asymptotic flatness only in the weak field limit. For full, non−linear General Relativity, one finds only guidelines as to how these boundary conditions may be modified. The weak field analysis is based on a projective completion of space−time. Detailed calculations show [73] that already in the case of Schwarzschild space−time, these projective techniques run into difficult problems and it is not clear if the programme outlined in [72] can in fact be completed. The fourth covariant approach developed in [74, 75, 76] is more complete in that it does specify the basic definitions in the full, non−linear case. However, almost none of the consequences of these definitions are then worked out: neither the asymptotic symmetries nor the structure of the asymptotic field equations is explored in any detail.

The previous work culminated in the the fifth covariant approach in which Ashtekar and Romano [73] completed the programme initiated in [72] and [74, 75, 76] by extracting the desirable features from the i^0 description [66] and the related coordinate−dependent treatment [70, 71]. In the new framework, spatial infinity arose as a boundary of spacetime. However, Ashtekar and Romano's framework was somewhat of a hybrid, in that it involved both the conformal and projective structure.

We conclude that at present there are two main approaches to the problem of the asymptotic structure of the gravitational field at large null separations, and three main approaches to the problem of the asymptotic structure of the gravitational field at large space−like separations; at large null separations the two main approaches are that of Bondi, Metzner, Van der Burgh [19, 20] and Sachs [21] and that of Penrose [30, 31] in the framework of his conformal technique; whereas at large space−like separations the four main approaches are that of Beig and Schmidt

[70, 71] which is similar to that of Bondi et al., that of Ashtekhar and Hansen [38] who employ Penrose's conformal technique, that of Sommers [72] and Persides [74, 75, 76] who use a projective technique, and finally that of Ashtekhar and Romano [73] who involve both the conformal and the projective structure. In the Penrose's treatment [30, 31] null infinity is described as smooth boundary of the unphysical space−time; similarly in Sommers',Persides', and, Ashtekhar and Romano's approach spatial infinity is defined as smooth boundary of the unphysical space−time. In both cases the unphysical space−time is conformally related to the original physical space−time.

In Penrose's treatment the asymptotic symmetry group B arises [77] as the group of diffeomorphisms preserving certain structures which are universal, i.e., which are common to all asymptotically flat space−times, *induced on null infinity from its embedding as a smooth boundary of the unphysical space−time; Penrose's strong conformal geometry is an example of such a universal structure.* Similarly, in Ashtekhar and Romano's approach the asymptotic symmetry group $B_{\mathcal{S}}$ arises [73] as the group of diffeomorphisms preserving certain universal structures induced on spatial infinity from its embedding as a smooth boundary of the unphysical space−time. In the null regime B was also derived by Bondi's approach [20], whereas, in the spatial regime $B_{\mathcal{S}}$ has not been derived by Bondi's approach so far [70, 71]. Ashtekhar and Romano showed [73] that $B_{\mathcal{S}} = C(S^2 \times I, R) \circledS_T SL(2, C)$, i.e., $B_{\mathcal{S}}$ has a structure similar to the structure of B. Indeed, $B_{\mathcal{S}}$ has a unique infinite−dimensional Abelian normal subgroup, the subgroup $C(S^2 \times I, R)$ of supertranslations. The quotient of $B_{\mathcal{S}}$ with this subgroup is just the Lorentz group $SL(2, C)$: $B_{\mathcal{S}}$ is the semi−direct product of the supertranslations $C(S^2 \times I, R)$ and the Lorentz group $SL(2, C)$. Finally $B_{\mathcal{S}}$ admits a preferred 4-*d normal* subgroup, the subgroup $\mathcal{T}$ of Spi translations.

The derivation of the aforementioned asymptotic symmetry groups in the spatial and in the null regime is much simpler in the group theoretic approach advocated in [53]. In this derivation no unnatural structures are invoked − no first and second order structures are employed and no 'blowing up' is needed [54], why one should stop at the second order structure, i.e., at the first derivative of the metric, and why one squeezes space−like infinity to a single point to 'blow it up' later? − and, moreover, the construction gives in a single stroke *all possible asymptotic symmetry groups in four dimensions both for real space−times in any signature and for complex space−times.* It is difficult to imagine that there is a simpler and more economical way to obtain these groups. The Einstein equations are not employed explicitly as it is done in the other approaches [54, 73].

Theories attempting to unify the forces often require a higher dimensional space−time. Still important and fundamental even in higher dimensional theories are the notion of an isolated system and the question arises if B and its 41 generalizations have higher dimensional analogs. In higher dimensions, there are only a few works [29, 78, 79, 80, 81, 82, 83] about asymptotic structure at spatial infinity or null infinity though recently the importance of higher−dimensional black holes is increasing in string theory and TeV gravity scenario [84, 85].

In higher dimensional space−times, the asymptotic structure of the gravitational field at spatial infinity has been studied in [82, 83]. In [82] Shiromizu and Tomizawa investigate asymptotic flatness in the spatial regime in higher dimensions following Ashtekar and Romano [73], whereas in [83] Tanabe, Tanahashi and Shiromizu define asymptotic flatness and investigate asymptotic structure at spatial infinity in higher dimensions, following Ashtekar and Hansen [66]. Much work still needs to be done since the structure of the asymptotic symmetry group is not stated clearly in [82, 83]. In [83] the asymptotic symmetry group is defined as the group of diffeomorphisms which preserve the universal structure at i^0. It is shown that the asymptotic symmetry group consists of transformations which have infinite translational directions. However the structure of the group is not given. Moreover in [83] it is shown that as in four dimensions, by imposing additional constraints on the behavior of the 'magnetic' part of the Weyl tensor the supertranslational ambiguity can be removed. Then, the asymptotic symmetry of the space–time reduces to the Poincare symmetry, which is a symmetry of background flat metric, and conserved quantities associated with this Poincare symmetry can be constructed. In [82] it is shown that in higher dimensions the geometric structure of spatial infinity is generally different from its structure in four dimensions: Spatial infinity', is an $(d-1)$–dimensional manifold in a general d–dimensional space–time. In four dimensions the spatial infinity $\mathcal{H}$ is homeomorphic to $S^2 \times I$, i.e., $\mathcal{H}$ is the 3–dim (timelike) hyperboloid of unit space–like directions [73]. The key result in [82] is that in higher dimensions there are many more varieties for the geometric structure of $\mathcal{H}$ due to the nontrivial $(d-1)$–dimensional Weyl tensor. In fact it appears [82] that this non triviality of the Weyl tensor makes the situation more complicated than one might have naively anticipated: The geometric structure of $\mathcal{H}$ depends on the number of dimensions $d > 4$ and also on the class of space–times considered. Therefore, the naive generalization of B_S to higher dimensions, i.e., of

$$B_S = C(S^2 \times I, R) \circledS_T SL(2, C)$$

to

$$B_S^d = C(S^{d-2} \times I, R) \circledS_T L_+^d,$$

where, L_+^d is the identity component of the Lorentz group in d dimensions, is not generally valid except in the very restricted class of static space–times [82]. Much work needs still to be done so that the structure of the asymptotic symmetry group is determined in the spatial regime and in higher directions for the different classes of space–times.

In higher dimensional space-times, the asymptotic structure at spatial infinity can be well-defined [82, 83]. On the other hand, the asymptotic structure at null infinity is not understood completely in higher dimensional space-times [78, 79, 80, 81]. Indeed, by using Penrose's conformal technique the definition of null infinity can be given only in even dimensions [78, 79, 80, 81]. This situation is to be contradistinguished with the situation at spatial infinity: by using Penrose's conformal technique the definition of spatial infinity can be given in *any* number of dimensions [83]. The difficulty in the definition of null infinity compared with spatial infinity is due to the presence of the gravitational waves at null infinity. At spatial infinity, since there are no gravitational waves, the asymptotic structure is 'stationary' and the total mass and total angular momentum are conserved, while the asymptotic structure at null infinity might be disturbed by gravitational waves. Hence, a stable definition of null infinity against gravitational waves is needed. Such a definition in even dimensions can be given with the employment of Penrose's conformal embedding method, but cannot be given in odd dimensional space–times [78, 79, 80] since the smoothness of Einstein equations at null infinity cannot be shown. This non–smoothness is related with the fact that in conformal embedding method the conformal factor $\Omega \sim 1/r$ is introduced and the behavior of gravitational waves near null infinity is of order $O(\Omega^{(d-2)/2})$ in d dimensional space–times. The problem comes from the half–integer power of Ω. In odd dimensions null infinity is defined not by using Penrose's conformal embedding method but by using the Bondi coordinates [81]. Neither in even dimensions [78, 79, 80] nor in odd dimensions [81] there is the well known infinite set of angle dependent translational symmetries (supertranslations). Furthermore, it appears [78, 79, 80, 81], that when $d > 4$ the asymptotic symmetry group in the null regime is the Poincare group in d dimensions. However the problem of the asymptotic symmetric group in the null regime in more than four dimensions has not been settled: in [29] a different definition of asymptotic flatness is given and this new definition results in a generalization of B in the null regime in higher dimensions which *does* entail supertranslations.

In the classical regime and in higher dimensions in order to compare the results of the aforementioned other approaches to asymptotic symmetries with the results of the group theoretic approach advocated here, and in order to make contact with other approaches to quantum gravity where complexified or euclidean versions of General Relativity are frequently considered, it is highly desirable to

(i) Construct the analogs of B in higher dimensions in complexified space–times as well as in real space–times of any signature, generalizing the work of McCarthy[53].

(ii) Search for and construct classical actions of the analogs of B in higher dimensions in analogy with the action of B, $B \times \Im^{+} \to \Im^{+}$ on Penrose's future null infinity.

(iii) Determine geometric structures for which the classical actions of (ii) are the automorphism groups in analogy with Penrose's 'strong conformal geometry' for which B is the automorphism group.

2. Wigner–Mackey's theory in a nutshell

Neither the 42 groups defined in [53] nor their (supersymmetric) generalizations in four and higher dimensional space–times are (locally) compact and there is no complete theory for such groups. Fortunately, we do not need to concern ourselves with the general case but only with the class of groups with the general structure

$$\mathcal{B} = C(A, R) \circledS_T H,$$

where, A and H have the same meaning as in (1), and,

(i) $C(A, R)$ is abelian normal subgroup of $\mathcal{B}$

(ii) T is an homomorphism

$$T : G \to Aut(C(A, R))$$

(iii) $C(A, R)$ and G are topological groups. In the product topology of $C(A, R) \times G$, $\mathcal{B}$ becomes a topological group. In fact in Wigner–Mackey's theory it is assumed that in the product topology of $C(A, R) \times G$, $\mathcal{B}$ becomes a *locally compact* topological group. This is *not* the case for B and its generalizations and some care is needed in the use of the results of Wigner–Mackey's theory.

To construct the IRs of these groups it would be sufficient to study just Wigner–Mackey's theory of induced representations of semi–direct products. We will present the bare essentials of this theory; our treatment will be purely functional analytic leaving aside the, more natural for induced representations, geometrical, fibre bundle dominated scheme [86, 87, 88].

2.1. The bare essentials

Let $\mathcal{A}$ denote $C(A, R)$. It is convenient to give a few definitions and state some simple facts:

(i) **Characters - The Dual Group** The IRs of $\mathcal{A}$ are one – dimensional and are called characters. They can be given the structure of an abelian group $\hat{\mathcal{A}}$, the dual group of $\mathcal{A}$, with group operation given by

$$(\chi_1 \chi_2)(\alpha) = \chi_1(\alpha)\chi_2(\alpha),$$

where, $\chi_1, \chi_2 \in \hat{\mathcal{A}}$, $\alpha \in \mathcal{A}$.

(ii) $\mathcal{A}$ **has a natural real vector space structure**

$$\begin{aligned} (\chi_1 + \chi_2)(\alpha) &:= \chi_1(\alpha) + \chi_2(\alpha) \\ (\lambda \circ \chi)(\alpha) &:= \lambda \chi(\alpha). \end{aligned}$$

(iii) **Dual Action** The action T of G on $\mathcal{A}$ induces a dual action $\hat{T}$ of G on $\hat{\mathcal{A}}$ defined by

$$(\hat{T}(g)\chi)(\alpha) := \chi(T(g^{-1})\alpha)$$

It is precisely this dual action which determines the structure of the IRs of $\mathcal{B}$

(iv) **Little Groups** L_χ For a *given* χ the largest subgroup L_χ of $\mathcal{B}$ which leaves χ fixed is called the little group of χ, i.e.,

$$L_\chi = \{g \in \mathcal{B} \mid \hat{T}(g)\chi = \chi\}$$

(v) **The Orbit of χ, denoted by $\mathcal{B}_\chi$, is defined by**

$$\mathcal{B}_\chi = \{g\chi | g \in \mathcal{B}\}$$

$\hat{\mathcal{A}}$ is partitioned by the orbits $\mathcal{B}_\chi$

2.1.1. Constructing the Hilbert space $\mathcal{H}_\nu$ in which the IRs of $\mathcal{B}$ are materialized

(i) **There is a natural bijection**

$$\mathcal{B}/L_\chi \quad \longleftrightarrow \quad \mathcal{B}\chi,$$
$$\text{given by,}$$
$$gL_\chi \quad \longleftrightarrow \quad g\chi,$$

where, $g \in \mathcal{B}$.

(ii) **The coset space $\mathcal{B}/L_\chi$ has a unique class of quasi-invariant measures for the $\mathcal{B}-$action [86]**; let ν be one of these

(iii) **Let U be a continuous irreducible of the little group L_χ on a Hilbert space $\mathcal{D}$**

(iv) **Let $\mathcal{H}_\nu$ be the space of functions $\psi : \mathcal{B} \to \mathcal{D}$ which satisfy the conditions**

(a) $\psi(gh) = U(h^{-1})\psi(g)$

(b) $\int_{\mathcal{B}\chi} < \psi(p), \psi(p) > \mathrm{d}\nu(p) < \infty$,

where, $h \in L_\chi, g \in \mathcal{B}$, and, where, the scalar product under the integral sign is that of $\mathcal{D}$.

(v) **Remark** The integrand is expressed as a function on $\mathcal{B}\chi \longleftrightarrow \mathcal{B}/L_\chi$ since, in view of 1, the integrand is constant on the cosets in $\mathcal{B}/L_\chi$

(vi) **$\mathcal{H}_\nu$ is turned into a Hilbert space by introducing the scalar product**

$$< \psi_1, \psi_2 >= \int_{\mathcal{B}\chi} < \psi_1(p), \psi_2(p) > \mathrm{d}\nu(p)$$

(vii) **Definition of the representation of $\mathcal{B}$ on $\mathcal{H}_\nu$**
Define an action of $\mathcal{B}$ on $\mathcal{H}_\nu$ by

$$(g_o\psi)(g) = \psi(g_o^{-1}g) \quad (7)$$
$$(\alpha\psi)(g) = [(\hat{T}(g)\chi)(\alpha)]\psi(g), \quad (8)$$

where, $g, g_o \in G$, and $\alpha \in \mathcal{A}$. This action gives a unitary representation of $\mathcal{B}$ on $\mathcal{H}_\nu$ which is continuous whenever U is.

(viii) **This is precisely the representation of $\mathcal{B}$ induced from χ and the irreducible representation U of the little group L_χ**

2.1.2. Mackey's Theorems

(i) **First Theorem :** Given the topological restrictions on $\mathcal{B} = \mathcal{A}Ⓢ_T G$ (separability and local compactness), *any* representation of $\mathcal{B}$, constructed by the method above , *is irreducible if the representation* U *of* L_χ *on* $\mathcal{D}$ *is irreducible.*

(ii) **Main Conclusion : An irreducible representation of $\mathcal{B}$ is obtained for each $\chi \in \hat{\mathcal{A}}$ and each irreducible representation U of L_χ**

(iii) **Second Theorem :** If $\mathcal{B} = \mathcal{A}Ⓢ_T G$ is a regular semi−direct product (i.e., $\hat{\mathcal{A}}$ contains a Borel subset which meets each orbit $G\chi$ in $\hat{\mathcal{A}}$ in just one point – **the G−action is not too pathological when the G−orbits can be enumerated in some way**) then **all** of its irreducible representations can be obtained in this way

3. Physics does not dictate the topology of $\hat{\mathcal{A}}$ and this gives rise to a host of possibilities

Sachs's original derivation [24] required that the supertranslations be twice diffrentiable, i.e., $\mathcal{A} = C^2(S^2, R)$, and Cantoni, in his investigation of representations [25, 26] , gave to this $\mathcal{A} = C^2(S^2, R)$ the pre−Hilbert topology determined by the area measure of S^2. McCarthy chose the same topology in his investigation [40, 42], but widened $\mathcal{A} = C^2(S^2, R)$ to $\mathcal{A} = L^2(S^2, R)$ i.e., to the square−integrable real−valued functions defined on the two sphere in order to simplify the treatment. The discreteness result of the spin parameterizing the ensuing representations is true with either the Cantoni or McCarthy choices.

The key observation here is that Sachs's original derivation [24] was superceded by that of Penrose [30, 31], who gave a precise and intrinsic derivation of B as that group of exact conformal motions of the future (or past) null boundary $\Im^{\pm}$ of (conformally compactified weakly asymptotically simple) space−times which preserve 'null angles'. Since *truly arbitrary* supertranslation functions describe symmetry transformations in Penrose's sense, *supertranslations need not have some minimum degree of smoothness.*

It is precisely this arbitrariness in the class of functions allowed in $\mathcal{A}$ which permits a wide range of choices of 'reasonable' topologies for $\mathcal{A}$. The catch is that the 'smoother' you require the functions in $\mathcal{A}$ to be, the 'rougher' you can allow the generalized functions in $\hat{\mathcal{A}}$ to be. This explains why little groups might change when the topology of $\mathcal{A}$ is modified: The induced representations are associated with the existence of invariant characters, i.e., of elements in $\hat{\mathcal{A}}$, the topological dual of $\mathcal{A}$ (supermomenta in B, generalizations of the Poincare momenta), which are left invariant by the action of some subgroup (little group) of H, $H = SL(2, C)$ in the case of B. Little groups are then connected, via $\hat{\mathcal{A}}$, to the topology of $\mathcal{A}$. Since, e.g., a refinement of the topology may broaden $\hat{\mathcal{A}}$, new invariant elements with associated little groups may come into existence. In fact, as we will explain, this is precisely what happens when $\mathcal{A}$ is endowed with the nuclear topology [44].

As mentioned already, Cantoni, in his investigation of representations [25, 26], gave to $\mathcal{A} = C^2(S^2, R)$ the pre−Hilbert topology determined by the area measure of S^2. Cantoni has shown [89] that if $SL(2, C)$ is given the usual topology, then $B_2 = C^2(S^2, R)Ⓢ_T SL(2, C)$ is a non−locally compact group in the product topology of $\mathcal{A} \times SL(2, C)$, $\mathcal{A}$ now being the group $C^2(S^2, R)$. McCarthy worked on two choices and proposed a third one:

3.1. First choice : B in the Hilbert topology

McCarthy in his first study of the representations of B [40, 42, 43, 46], chose the same topology with the one used by Cantoni [89], but widened $\mathcal{A} = C^2(S^2, R)$ to $\mathcal{A} = L^2(S^2, R)$ to simplify the treatment. $\mathcal{A}$ is endowed with the Hilbert topology as follows [40]:

- Introduce a scalar product into $\mathcal{A}$
$$< \alpha, \beta >= \int_{S^2} \alpha(x)\beta(x)\mathrm{d}\mu(x)$$
where, $x \in S^2$, and, $\alpha, \beta \in \mathcal{A}$.
- With this scalar product, $\mathcal{A} = L^2(S^2, R)$ becomes a real Hilbert space, and, in the induced metric topology, becomes an (Abelian) topological group.
- With a proof similar to Cantoni's, it can be shown that, when $\mathcal{A} = L^2(S^2, R)$ is endowed with the aforementioned topology, and $SL(2, C)$ is given the usual topology, then B is a non−locally compact group in the product topology of $\mathcal{A} \times SL(2, C)$.

The representations are determined by the action of $SL(2, C)$ on the topological dual of $\mathcal{A}$, $\hat{\mathcal{A}}$. It is a fundamental theorem, the Reisz−Frechet theorem, (the Hilbert space representation theorem), that $\mathcal{A} = L^2(S^2, R)$ and $\hat{\mathcal{A}}$ are isometrically isomorphic. It is precisely this isomorphism which simplifies the treatment when $\mathcal{A} = C^2(S^2, R)$ to $\mathcal{A} = L^2(S^2, R)$.

3.1.1. The more important results and open problems in the Hilbert topology

The four more important results in the Hilbert topology are:

- B irreducibles, i.e., B−elementary entities, describe 'elementary particles' [46].
- In the representation theory of B only compact little groups, hence discrete spins, arise.
- There is strong evidence which suggests that a class of IRs of B, those induced from non−connected discrete finite little groups, correspond to ALE gravitational instantons [53].
- Piard has proved [90, 91] that the inducing construction for B [40, 42] is exhaustive despite the fact that B is *not* locally compact in the Hilbert topology. This makes McCarthy's results even more important.

The more important open problem in the Hilbert topology is:

- Find the physical interpretation of the IRs of B, in particular of those induced from non−connected discrete finite little groups and establish and clarify the connection with the ALE gravitational instantons.

3.2. Second choice : B in the nuclear topology

McCarthy in his subsequent study of the representations of B [44, 92] chose $\mathcal{A} = C^\infty(S^2, R)$, i.e., chose $\mathcal{A}$ to be all the smooth real−valued functions on S^2 and not all the square−integrable functions on S^2. This choice by no means determines the topology uniquely. However, a natural choice for vector spaces whose elements are 'smooth' in some sense, e.g., for the set of smooth functions on a compact manifold, is the nuclear topology. The nuclear topology is determined by the following notion of convergence: a sequence of functions a_k on the sphere is said to converge to zero, if and only if, the functions a_k together with all their derivatives of all orders, converge uniformly to zero over the sphere. The expectation is that in a finer topology, the nuclear is finer than the Hilbert, more representations become continuous, and some have non−compact little groups, hence, possibly, continuous spins.

3.2.1. The more important results and open problems in the nuclear topology

The two more important results in the nuclear topology are:

- It appears [44] that the extra representations in the nuclear topology (which may have continuous spins) describe 'scattering' states, and the remaining ones are identified with bound states, corresponding to elementary particles.

- The particles appropriate to B have discrete spins irrespective of the choice of topology, Hilbert or nuclear.

The two more important open problems in the nuclear topology are:

- Some of the IRs of B in the nuclear topology are induced from non−compact infinite discrete subgroups of $SL(2, C)$ [44], which are far from being known!.
- It is not known if the inducing construction for B in the nuclear topology [44] is exhaustive, i.e., a proof corresponding to the one given by Piard [90, 91] for B in the Hilbert topology, is lacking for B in the nuclear topology.

3.3. Third choice: $\hat{\mathcal{A}}$ becomes enlarged to the space of real hyperfunctions $\mathcal{Y}(S^2, R)$ on the 2−sphere S^2

If $\mathcal{A}$ is taken to consist of real analytic $C^\omega(S^2, R)$ functions rather than $C^\infty(S^2, R)$ functions then, with an appropriately fine topology, the dual space $\hat{\mathcal{A}}$ becomes enlarged to the space of real hyperfunctions $\mathcal{Y}(S^2, R)$ on the 2−sphere S^2. The space of hyperfunctions is larger than the space of distributions on S^2. This new topology on $\mathcal{A} = C^\omega(S^2, R)$, and the associated hyperfunctional supermomenta, are probably more 'physical' than the nuclear topology and the associated distributional supermomenta [93]. It seems that the category of hyperfunctions is more appropriate than that of distributions for discussing S−matrix theory. Since the representation theory of B in nuclear topology strongly suggests [44] that the physical situations being analyzed are related to scattering problems, the present discussion of $\mathcal{A} = C^\omega(S^2, R)$ seems to imply that one should really consider hyperfunctional supermomenta, or more generally, hyperfunctional solutions to Einstein's equations.

Interestingly, quantum field theories in which fields are smeared by hyperfunctions show a non local behaviour and the density of states can have a non polynomial growth. This might in principle allow to recover bulk locality, in the exploration of the holographic principle in asymptotically flat spacetimes via the BMS group [94], although one should consider hyperfunctional solutions to the Einstein equations. More remarkably, if one assumes that the high energy behaviour of the density of states in the bulk is dominated by black holes, the exponential growth of states which suggests an intrinsic degree of non locality, might be explained by working with hyperfunctions.

4. Some more open problems

Some more open problems within this research program in the quantum regime are the following:

- One of the main objectives of the research program expounded here is to find the IRs of B and its generalizations [53] in four and higher dimensions in Hilbert spaces [95, 96, 97, 98, 99, 100]. These Hilbert space representations are related to elementary particles and quantum gravity via gravitational instantons. Thus, in the spirit of Klein's Erlangen programme, the geometries, symmetries and elementary quantum systems can all be studied from a single point of view.
- Prove the connection with instantons – ALE, complex and hyperbolic – in all dimensions. This should shed some light in the structure of the quantum gravity vacuum at least in the semi−classical regime.
- Explore if the holographic principle of t'Hooft and Susskind is realized in asymptotically−flat space-times. Two routes have been pursued: In one of them the holographic boundary is the future null infinity and the QFT at the boundary is 'B−invariant' [94], in the other, the symmetry algebra of asymptotically flat space−times at null infinity in 4 dimensions is taken to be the semi−direct sum of supertranslations with infinitesimal *local* conformal transformations [101] and not, as usually done, with the Lorentz algebra.

Try a third route: to put the holoscreen at spatial infinity, since then the analogy with AdS/CFT correspondence is better. This presupposes to find first the IRs of $B_{\mathcal{S}}$.

- Alternatively, find the IRs of B when $\hat{\mathcal{A}}$ becomes enlarged to the space of real hyperfunctions and explore if the holographic principle of t'Hooft and Susskind is realized in asymptotically flat space–times by putting the holographic screen at $\Im^+$ and by employing the IRs of B in this new topology, where the supermomenta are hyperfunctions, for the fields at $\Im^+$.
- B and its analogs result as asymptotic symmetry groups of Einstein–Maxwell isolated systems. It is highly desirable to find the corresponding groups in four and higher dimensions for Einstein–Maxwell–Yang-Mills isolated systems. This will reveal the structure of the quantum gravity–yang-mills vacuum and hopefully some conclusions will be drawn for the cosmological constant at least in the semi–classical regime.

5. References

[1] Wigner E 1939 *Ann. Math.* **40** 149
[2] Mackey G W 1949 *Proc. N. A. S.* **35** 537
[3] Mackey G W 1952 *Annals of Maths* **55** 101
[4] Mackey G W 1953 *Ann. of Math.* **58** 193
[5] Mackey G W 1957 *Trans. Amer. Math. Soc* **85** 134
[6] Mackey G W 1958 *Acta Mathematica* **99** 265
[7] Mackey G W 1963 *Bulletin of the American Mathematical Society* **69** 628
[8] Mackey G W 1955 *The theory of group representations* (Chicago notes)
[9] Mackey G W 1963 *The Mathematical Foundations of Quantum Mechanics* (New York: W.A.Benjamin, Inc.)
[10] Mackey G W 1978 *Unitary group representations in physics, probability and number theory* (London: Benjamin)
[11] Varadarajan V S 1968 *Geometry of Quantum Theory, vol. 1* (New York: D. Van–Nostrand Company, Inc.)
[12] Varadarajan V S 1970 *Geometry of Quantum Theory, vol. 2* (New York: Van–Nostrand Reinhold Company)
[13] Barut A O and Racza R 1977 *Theory of Group Representations and Applications* (Warsaw: Polish Scientific Publ.)
[14] Clifford A H 1937 *Ann. of Math.* **38** 533
[15] Mackey G W 1949 *Duke Mathematical Journal* **16** 313
[16] Stone M H 1930 *Proceedings of the National Academy of Sciences of the United States of America* **16** 172
[17] J. von Neuman 1931 *Mathematische Annalen* **104** 570
[18] Fleischhack C 2009 Representations of the Weyl Algebra in Quantum Geometry *Preprint* math–ph 0407006
[19] Bondi H 1960 *Nature* **186** 535
[20] Bondi H, Van der Burg M G J and Metzner A W K 1962 *Proc. R. Soc. London Ser.* A **269** 21
[21] Sachs R K 1962 *Proc. R. Soc. London Ser.* A **270** 103
[22] Trautman A 1958 *King's College notes on General Relativity, mimeographed notes (unpublished)*
[23] Chruciel P T, Jezierski J and MacCallum M A H 1998 *Phys. Rev.* D **58** 084001
[24] Sachs R K 1962 *Phys. Rev.* **128** 2851
[25] Cantoni V 1966 *JMP* **7** 1361
[26] Cantoni V 1967 *JMP* **8** 1700
[27] Komar A 1965 *Phys. Rev. Lett.* **15** 76
[28] Newman E T 1965 *Nature* **206** 812
[29] Awada M A, Gibbons G W and Shaw W T 1986 *Annals of Physics* **171** 52
[30] Penrose R 1963 *Phys. Rev. Lett.* **10** 66
[31] Penrose R 1965 *Proc. R. Soc. Lond.* A **284** 159
[32] Newman E T 1972 *JMP* **13** 1884
[33] Newman E T and Penrose R 1966 *JMP* **7** 863
[34] Schmidt B G 1974 *Commun. Math. Phys.* **36** 73
[35] Winicour J H 1968 *JMP* **9** 861
[36] Hawking S W and Ellis G F R 1973 *The large scale structure of space–time* (Cambridge: Cambridge University Press)
[37] Hansen R O, Newman E T, Penrose R and Tod K P 1978 *Proc. R. Soc. Lond.* A **363** 445
[38] Ashtekar A 1981 *Phys. Rev. Lett.* **46** 573
[39] Helfer A D 2007 *Gen. Relativ. Gravit.* **39** 2125
[40] McCarthy P J 1972 *Proc. R. Soc. Lond.* A **330** 517

[41] McCarthy P J 1972 *JMP* **13** 1837
[42] McCarthy P J 1973 *Proc. R. Soc. Lond.* A **333** 317
[43] McCarthy P J and Crampin M 1973 *Proc. R. Soc. Lond.* A **335** 301
[44] McCarthy P J 1975 *Proc. R. Soc. Lond.* A **343** 489
[45] McCarthy P J 1978 *Proc. R. Soc. Lond.* A **358** 141
[46] McCarthy P J and Crampin M 1974 *Phys. Rev. Lett.* **33** 547
[47] McCarthy P J 1972 *Phys. Rev. Lett.* **29** 817
[48] O'Raifeartaigh L 1965 *Phys. Rev. Lett.* **14** 575
[49] Kronheimer P B 1989 *J. Diff. Geom.* **29** 665
[50] Kronheimer P B 1989 *J. Diff. Geom.* **29** 685
[51] Kronheimer P B and Nakajima H 1990 *Math. Ann.* **288** 263
[52] Nakajima H 1990 *Invent. math.* **102** 267
[53] McCarthy P J 1992 *Phil. Trans. R. Soc. Lond.* A **338** 271
[54] Ashtekhar A and Hansen R O 1978 *JMP* **19** 1542
[55] McCarthy P J 1992 *Proc. R. Soc. Lond.* A **438** 585
[56] McCarthy P J 1993 *Proc. R. Soc. Lond.* A **441** 465
[57] McCarthy P J 1993 *Phys. Let.* A **174** 25
[58] Tamburino L A and Winicour J H 1966 *Phys. Rev.* **150** 1039
[59] Arnowitt R, Deser R and Misner C W 1960 *Phys. Rev.* **117** 1695
[60] Arnowitt R, Deser R and Misner C W 1960 *Phys. Rev.* **118** 1100
[61] Arnowitt R, Deser R and Misner C W 1961 *Phys. Rev.* **121** 1566
[62] Arnowitt R, Deser R and Misner C W 1961 *Phys. Rev.* **122** 997
[63] Geroch R 1970 *JMP* **11** 2580
[64] Geroch R 1972 *JMP* **13** 956
[65] Geroch R 1976 *Asymptotic Structure of Space−Time* ed P Esposito and L Witten New York: Plenum
[66] Ashtekar A and Hansen R 0 1978 *JMP* **19** 1542
[67] Ashtekar A 1980 *General Relativity and Gravitation vol 2, ed A Held* (New York: Plenum)
[68] Ashtekar A and Magnon-Ashtekar A 1979 *JMP* **20** 793
[69] Ashtekar A and Magnon-Ashtekar A 1979 *Phys. Rev. Let.* **43** 181
[70] Beig R and Schmidt B G 1982 *Comm. Math. Phys.* **87** 65
[71] Beig R 1984 *Proc. R. Soc. Lon.* **391** 295
[72] Sommers P 1978 *JMP* **19** 549
[73] Ashtekar A and Romano J D 1992 *Class. Quantum Grav.* **9** 1069
[74] Persides S 1979 *JMP* **20** 1731
[75] Persides S 1980 *JMP* **21** 135
[76] Persides S 1980 *JMP* **21** 142
[77] Penrose R 1972 Relativistic symmetry groups in *Group Theory in Non−Linear Problems* ed. by A. O. Barut (Netherlands: D. Reidel, Dordrecht)
[78] Hollands S and Wald R M 2004 *Class. Quant. Grav.* **21** 5139
[79] Hollands S and Ishibashi A 2005 *JMP* **46** 022503
[80] Ishibashi A 2008 *Class. Quant. Grav.* **25** 165004
[81] Tanabe K, Tanahashi N and Shiromizu T 2009 On asymptotic structure at null infinity in five dimensions *Preprint* gr−qc 0909042
[82] Shiromizu T and Tomizawa S 2004 *Phys. Rev.* **69** 104012
[83] Tanabe K, Tanahashi N and Shiromizu T 2009 *JMP* **50** 072502
[84] Randall L and Sundrum R 1999 *Phys. Rev. Lett.* **83** 3370
[85] Giddings S B and Thomas S 2002 *Phys. Rev.* D **65** 056010
[86] Simms D J 1968 *lie Groups and Quantum Mechanics. Lecture Notes in Mathematics. Vol. 52* (New York: Springer)
[87] DeWitt B S and Stora R eds. 1983 *Relativity, groups and topology II* (Les Houches: Elsevier Science Publisher B.V.)
[88] Sternberg S 1994 *Group theory and Physics* (Great Britain: Cambridge University Press)
[89] Cantoni V 1967 Ph.D.thesis, University of London, unpublished
[90] Piard A 1977 *Rep. Math. Phys.* **11** 259
[91] Piard A 1977 *Rep. Math. Phys.* **11** 279
[92] McCarthy P J and Craqmpin M 1976 *Proc. R. Soc. Lond.* A **351** 55
[93] McCarthy P J 1978 *Proc. R. Soc. Lond.* A **358** 495
[94] Dappiaggi C 2004 *Simplicial and asymptotical aspects of the holographic principle*, Ph.D. thesis, University of Pavia, *Preprint* gr−qc 0403072

[95] McCarthy P J and Melas E 2003 *Nucl. Phys. B* **653** 369
[96] Melas E 2004 *JMP* **45** 996
[97] Melas E 2006 *J. Phys. A : Math. Gen.* **39** 3341
[98] Melas E 2002 *Nucl. Phys. B. (Proc. Suppl.)* **104** 212
[99] Melas E 2006 *AIP Conference Proceedings* **861** 420
[100] Melas E 2006 *IOP Conference Series* **33** 260
[101] Barnich G and Troesssaert C Symmetries of asymptotically flat 4 dimensional spacetimes at null infinity revisited *Preprint* gr−qc 0909.2617

FRW models in the conformal frame of $f(R)$ gravity[1]

J Miritzis

Department of Marine Sciences, University of the Aegean, University Hill, Mytilene 81100, Greece

E-mail: imyr@aegean.gr

Abstract. We study the late time evolution of Friedmann-Robertson-Walker (FRW) models with a perfect fluid matter source and a scalar field arising in the conformal frame of $f(R)$ theories nonminimally coupled to matter. We prove using the approach of dynamical systems, that equilibria corresponding to non-negative local minima for V are asymptotically stable. We show that if γ, the parameter of the equation of state is larger than one, then there is a transfer of energy from the fluid to the scalar field and the latter eventually dominates. The results are valid for a large class of nonnegative potentials without any particular assumptions about the behavior of the potential at infinity.

1. Introduction

In this paper we study the late time evolution of initially expanding flat FRW models with a scalar field having an arbitrary bounded from below potential function $V(\phi)$. Since the nature of the scalar field supposed to cause accelerated expansion is unknown, it is important to investigate the general properties shared by all FRW models with a scalar field irrespectively of the particular choice of the potential. The scalar field is nonminimally coupled to ordinary matter described by a barotropic fluid with equation of state

$$p = (\gamma - 1)\rho, \quad 0 < \gamma \leq 2.$$

The motivation comes from higher order gravity (HOG) theories [1] which are conformaly equivalent to the Einstein field equations with a scalar field ϕ as an additional matter source.

2. Flat and negatively curved FRW with an arbitrary non-negative potential

For homogeneous and isotropic spacetimes the field equations, (see for example [2, 3]), reduce to the Friedmann equation,

$$H^2 + \frac{k}{a^2} = \frac{1}{3}\left(\rho + \frac{1}{2}\dot{\phi}^2 + V(\phi)\right), \tag{1}$$

and the Raychaudhuri equation,

$$\dot{H} = -\frac{1}{2}\dot{\phi}^2 - \frac{\gamma}{2}\rho + \frac{k}{a^2}, \tag{2}$$

[1] In collaboration with R. Giambò, University of Camerino, Italy.

while the equation of motion of the scalar field becomes

$$\ddot{\phi} + 3H\dot{\phi} + V'(\phi) = \frac{4-3\gamma}{\sqrt{6}}\rho. \tag{3}$$

The Bianchi identities yield the conservation equation,

$$\dot{\rho} + 3\gamma\rho H = -\frac{4-3\gamma}{\sqrt{6}}\rho\dot{\phi}. \tag{4}$$

We adopt the metric and curvature conventions of [4]. $a(t)$ is the scale factor, an overdot denotes differentiation with respect to time t, $H = \dot{a}/a$ and units have been chosen so that $c = 1 = 8\pi G$. Here $V(\phi)$ is the potential energy of the scalar field and $V' = dV/d\phi$.

In the following we consider initially expanding solutions of (2)-(4), i.e. $H(0) > 0$. For flat, $k = 0$, models the state vector of the system (2)-(4) is $\left(\phi, \dot{\phi}, \rho, H\right)$, i.e. we have a four-dimensional dynamical system subject to the constraint (1). Defining $y := \dot{\phi}$ and setting $(4-3\gamma)/\sqrt{6} =: \alpha$, we write the autonomous system as

$$\begin{aligned} \dot{\phi} &= y, \\ \dot{y} &= -3Hy - V'(\phi) + \alpha\rho, \\ \dot{\rho} &= -3\gamma\rho H - \alpha\rho y, \\ \dot{H} &= -\frac{1}{2}y^2 - \frac{\gamma}{2}\rho, \end{aligned} \tag{5}$$

subject to the constraint

$$3H^2 = \rho + \frac{1}{2}y^2 + V(\phi). \tag{6}$$

The function W defined by

$$W(\phi, y, \rho, H) = H^2 - \frac{1}{3}\left(\frac{1}{2}y^2 + V(\phi) + \rho\right), \tag{7}$$

satisfies

$$\dot{W} = -2HW, \tag{8}$$

and therefore, the hypersurface $\{W = 0\}$, representing flat, $k = 0$, cosmologies is invariant under the flow of (5). Denoting by

$$\epsilon = \frac{1}{2}y^2 + V(\phi),$$

the energy density of the scalar field, we see that

$$\dot{\epsilon} + \dot{\rho} = -3H(y^2 + \gamma\rho),$$

which implies that, for expanding models the total energy $\epsilon + \rho$ of the system decreases.

The equilibria of (5) are given by $(\phi = \phi_*, y = 0, \rho = 0, H = \pm\sqrt{V(\phi_*)/3})$ where $V'(\phi_*) = 0$. Regarding their stability for expanding cosmologies, we note the following facts. Critical points of V with negative critical value are not equilibria and they rather allow for recollapse of the model. Moreover, nondegenerate maximum points (with non negative critical value) for V are unstable, as can be easily seen by linearizing system (5) at the corresponding equilibria and verifying the existence of at least one eigenvalue with positive real part. It is interesting to study what happens near local minima of the potential with non negative critical value, and a stability result can be given when $k = 0, -1$.

Proposition. *Let ϕ_* a strict local minimum for $V(\phi)$, possibly degenerate, with nonnegative critical value. Then,* $\mathbf{p}_* = (\phi_*, y_* = 0, \rho_* = 0, H_* = \sqrt{\frac{V(\phi_*)}{3}})$ *is an asymptotically stable equilibrium point for expanding cosmologies in the open spatial topologies $k = 0$ and $k = -1$.*

Sketch of the proof. The proof consists in constructing a compact set Ω in $\mathbb{R}^4$ and showing that it is positively invariant. Applying LaSalle's invariance theorem to the functions W and $(\rho + \epsilon)$ in Ω, it is shown that every trajectory in Ω is such that $HW \to 0$ and $H(y^2 + \gamma\rho) \to 0$ as $t \to +\infty$, which means $y \to 0$, $\rho \to 0$, and $H^2 - \frac{1}{3}V(\phi) \to 0$. Since H is monotone and admits a limit, $V(\phi)$ also admits a limit, $V(\phi_*)$, thus the solution approaches the equilibrium point $\mathbf{p}_*$. For details, see [5]. □

Similar results were proved in [6] for separately conserved scalar field and perfect fluid.

3. Energy exchange

In the following we restrict ourselves to the flat, $k = 0$, case and study the energy transfer from the perfect fluid to the scalar field. We are interested in studying the late time behavior near the equilibrium point $(\phi = \phi_*, y = 0, \rho = 0, H = \sqrt{\frac{V(\phi_*)}{3}})$, which, by the previous Proposition is asymptotically stable. We suppose that the initial data in the basin of attraction of this equilibrium are such that the fluid is the dominant matter component, i.e.

$$\rho_0 > \epsilon_0,$$

and we investigate whether there is a time t_1 such that

$$\epsilon(t) > \rho(t), \qquad \forall t > t_1. \tag{9}$$

This question is relevant to the coincidence problem, that is, why dark energy and matter appear to have roughly the same energy density today (see [7] and references therein).

If $V(\phi_*) > 0$, the transition (9) does always happen. In fact, it easily follows that if the critical value of the potential is strictly positive, it behaves as an effective cosmological constant and the energy of the scalar field tends to this value whereas the energy of the fluid tends to zero. We conclude that the relevant case is when $V(\phi_*) = 0$ and hereafter we will focus on the case when ϕ_* is a nondegenerate minimum. Without loss of generality we suppose that $\phi_* = 0$ and therefore, the general form of the potential studied can be written in a neighborhood of $\phi = 0$ as

$$V(\phi) = \frac{1}{2}\omega^2\phi^2 + \mathcal{O}(\phi^3), \qquad \omega > 0. \tag{10}$$

Integrating (4), we get

$$\rho(t) = ce^{-\alpha\phi(t)}a(t)^{-3\gamma}. \tag{11}$$

From the Proposition in the previous Section we know *a-priori* that $\left(\phi, \dot{\phi}, \rho, H\right) \to (0, 0, 0, 0)$ as $t \to \infty$, hence we can write $\rho(t) \simeq ca(t)^{-3\gamma}$ as $t \to \infty$.

We now turn our attention to the equation of motion of the scalar field (3) with the potential (10), namely

$$\ddot{\phi} + 3H\dot{\phi} + \omega^2\phi + \mathcal{O}(\phi^2) = \alpha\rho. \tag{12}$$

This equation can be solved by the Kryloff-Bogoliuboff (KB) approximation [8]. We present an outline of the method for the convenience of readers with no previous knowledge of the KB approximation. Consider the differential equation

$$\ddot{\phi} + \eta f\left(\phi, \dot{\phi}\right) + \omega^2\phi = 0, \quad 0 < \eta \ll 1. \tag{13}$$

If $\eta = 0$, the solution can be written as

$$\phi(t) = \frac{1}{\omega} r \cos(-\omega t + \chi) \quad \text{and} \quad \dot{\phi}(t) = r \sin(-\omega t + \chi),$$

where r and χ are arbitrary constants. We are looking for a solution of (13) which resembles to the form of the simple harmonic oscillator, that is,

$$\phi(t) = \frac{1}{\omega} r(t) \cos(-\omega t + \chi(t)) \quad \text{and} \quad \dot{\phi}(t) = r(t) \sin(-\omega t + \chi(t)). \tag{14}$$

Setting $\theta(t) = -\omega t + \chi(t)$ and substituting (14) in (13) yields

$$\begin{aligned} \frac{dr}{dt} &= -\eta f\left(\frac{1}{\omega} r\cos\theta, r\sin\theta\right)\sin\theta, \\ \frac{d\chi}{dt} &= -\frac{\eta}{r} f\left(\frac{1}{\omega} r\cos\theta, r\sin\theta\right)\cos\theta. \end{aligned} \tag{15}$$

For η small, $r(t)$ and $\chi(t)$ are slowly varying functions of t. Consequently, in a time $T = 2\pi/\omega$, $r(t)$ and $\chi(t)$ have not changed appreciably, while $\theta(t) = -\omega t + \chi(t)$ will increase approximately by -2π. Therefore, we replace the rhs of (15) by their average values over a range of 2π of θ, i.e. the amplitude $r(t)$ is *regarded as a constant* in taking the average (this is the essence of the KB approximation). This leads to the equations

$$\begin{aligned} \frac{dr}{dt} &= -\frac{\eta}{2\pi} \int_0^{2\pi} f\left(\frac{1}{\omega} r\cos\theta, r\sin\theta\right)\sin\theta d\theta, \\ \frac{d\chi}{dt} &= -\frac{\eta}{2\pi r} \int_0^{2\pi} f\left(\frac{1}{\omega} r\cos\theta, r\sin\theta\right)\cos\theta d\theta. \end{aligned}$$

If we could apply the KB approximation to (12) with

$$\eta f\left(\phi, \dot{\phi}\right) = 3H\dot{\phi} + \mathcal{O}(\phi^2) - \alpha\rho, \tag{16}$$

then we should obtain for the amplitude of ϕ

$$\frac{dr}{dt} = -\frac{1}{2\pi} \int_0^{2\pi} \left(3H\omega r\cos\theta + \mathcal{O}(r^2\cos^2\theta) - \alpha c e^{-(\alpha/\omega) r\cos\theta} a^{-3\gamma}\right)\sin\theta d\theta. \tag{17}$$

The integral

$$\int_0^{2\pi} e^{-(\alpha/\omega) r\cos\theta} \sin\theta d\theta,$$

vanishes and since H is a slow varying function compared to the rapid oscillations of the scalar field, it can be considered as constant during one period, $T = 2\pi/\omega$. Therefore, (17) becomes

$$\frac{dr}{dt} = -\frac{3}{2} H r + \mathcal{O}(r^3), \tag{18}$$

since the integration of the quadratic terms vanish. Ignoring third-order terms, eq. (18) can be integrated to give $r(t) = r_0(t) a^{-3/2}$, where $r_0(t)$ is bounded and bounded away from zero uniformly as $t \to +\infty$.

Hence we find that for large t the amplitude of the scalar field varies as $a^{-3/2}$. Since the amplitude of $\dot{\phi}$ has by (14) the same time dependence as the amplitude of ϕ and in our approximation

$$\epsilon \cong \frac{1}{2}\dot{\phi}^2 + \frac{1}{2}\omega^2\phi^2,$$

it follows that

$$\epsilon \cong a^{-3}.$$

Comparing this result with the time dependence of the fluid density $\rho \cong a^{-3\gamma}$, we arrive at the following picture for the evolution of the universe.

If $\gamma < 1$ the energy density ρ eventually dominates over the energy density of the scalar field ϵ and this universe follows the classical Friedmannian evolution. For $\gamma > 1$, since ρ decreases faster than the energy density ϵ of the scalar field, ϵ eventually dominates over ρ.

A rigorous but long proof of these results is given in [5]. The main obstruction to apply the KB approximation is due to (16), indeed, the above argument applies if ρ/r goes to zero. Otherwise, the rhs in (15) is not infinitesimal and so $\chi(t)$ could in principle be comparable to ωt.

4. Summary

We analysed the late time evolution of flat and negatively curved expanding FRW models having a scalar field coupled to matter.

- Equilibria corresponding to non–negative local minima of V are asymptotically stable.
- In case the minimum is positive, say $V(\phi_*) > 0$, the energy density, ϵ, of the scalar field eventually rules over the energy density of the fluid, ρ, and the asymptotic state has an effective cosmological constant $V(\phi_*)$.
- In case the minimum is zero and nondegenerate, then ρ eventually dominates over ϵ if $\gamma < 1$ and ϵ dominates over ρ if $\gamma > 1$.
- These results could be interesting in investigations of cosmological scenarios in which the energy density of the scalar field mimics the background energy density. For viable dark energy models, it is necessary that the energy density of the scalar field remains insignificant during most of the history of the universe and emerges only at late times to account for the current acceleration of the universe.
- The above results can be rigorously proved only assuming that critical points are finite, and that $V(\phi)$ is non-negative as $\phi \to \pm\infty$. It must be remarked that the latter assumption does not enter in the study of the late time behavior around a critical point ϕ_*, because for that situation only the behavior of the potential near ϕ_* is important and no growth at infinity assumptions on V are actually needed.
- The cases studied of course do not cover all possible situations. In particular, one can take into account degenerate minima for the potential.
- For closed cosmologies further investigation is needed. We believe that a closed model cannot avoid recollapse, unless the minimum of the potential is strictly positive. In that case, the asymptotic state must be de Sitter space.

References

[1] Capozziello S and Francaviglia M 2008 *Gen. Rel. Grav.* **40** 357
[2] Bean R, Bernat Pogosian D L, Silvestri A and Trodden M 2007 *Phys. Rev.* D **75** 064020
[3] Magnano G, Sokołowski L M 1994 *Phys. Rev.* D **50** 5039
[4] Wainwright J and Ellis G F R 1997 *Dynamical Systems in Cosmology* (Cambridge University Press)

[5] Giambò R and Miritzis J 2010 *Class. Quantum Grav.* **27** 095003
[6] Miritzis J 2003 *Class. Quantum Grav.* **20** 2981
[7] Amendola L, Quartin M, Tsujikawa S and Waga I 2006 *Phys. Rev.* D **74** 023525
[8] Kryloff N and Bogoliuboff N 1943 *Introduction to Nonlinear Mechanics* (Princeton University Press)

Journal of Physics: Conference Series **283** (2011) 012025 doi:10.1088/1742-6596/283/1/012025

Genetic algorithms and the analysis of SnIa data

Savvas Nesseris

The Niels Bohr International Academy and DISCOVERY Center, The Niels Bohr Institute, Blegdamsvej 17, DK-2100

E-mail: nesseris@nbi.dk

Abstract. The Genetic Algorithm is a heuristic that can be used to produce model independent solutions to an optimization problem, thus making it ideal for use in cosmology and more specifically in the analysis of type Ia supernovae data. In this work we use the Genetic Algorithms (GA) in order to derive a null test on the spatially flat cosmological constant model ΛCDM. This is done in two steps: first, we apply the GA to the Constitution SNIa data in order to acquire a model independent reconstruction of the expansion history of the Universe $H(z)$ and second, we use the reconstructed $H(z)$ in conjunction with the Om statistic, which is constant only for the ΛCDM model, to derive our constraints. We find that while ΛCDM is consistent with the data at the 2σ level, some deviations from ΛCDM model at low redshifts can be accommodated.

1. Introduction

In the previous decade it was discovered that the Universe is undergoing an accelerated expansion [1], [2], [3]. This acceleration is usually attributed either to a cosmic fluid with negative pressure dubbed Dark Energy or to an IR modification of gravity. In order to identify the properties of Dark Energy or the structure of the IR modification of gravity it is necessary to know to a high precision the rate of the expansion of the Universe, parameterized as $H \equiv \frac{\dot{a}}{a}$ where $a = \frac{1}{1+z}$ is the scale factor and z is the redshift of the cosmological probe, as it is measured by the observations.

The behavior of the expansion of the Universe can be identified by studying two functions, the Equation of State (EoS) $w(z) \equiv \frac{P}{\rho}$ which can be rewritten as

$$w(z) = -1 + \frac{1}{3}(1+z)\frac{d\ln(\delta H(z)^2)}{dz}, \qquad (1)$$

where $\delta H(z)^2 = H(z)^2/H_0^2 - \Omega_{0\mathrm{m}}(1+z)^3$ accounts for all terms in the Friedmann equation not related to matter and the deceleration parameter $q(z) \equiv -\frac{\ddot{a}}{a\dot{a}^2}$ which can be rewritten as

$$q(z) = -1 + (1+z)\frac{d\ln(H(z))}{dz}, \qquad (2)$$

Obviously, the cosmological constant ($w(z) = -1$) corresponds to a constant dark energy density, while in general $w(z)$ can be time dependent. Also, an important parameter is the value of the deceleration parameter today, ie $q(z=0) \equiv q_0$, which for the cosmological constant model in GR it is $q_0 = -1 + 3\Omega_{0\mathrm{m}}/2$.

However, despite all the recent progress the origin of the accelerated expansion of the universe still remains unknown with many possibilities still remaining open, see for example [4]. The simplest choice that agrees well with the data is a positive cosmological constant which has to be small enough to have started dominating the universe at late times. As it was demonstrated by the Seven-Year WMAP data [5], the cosmological constant remains the best candidate and has the advantage of having only one free parameter related to the properties of the Dark Energy. Nonetheless, this model fails to explain why the cosmological constant is so small that it can only dominate the universe at late times, a problem known as the *coincidence problem* and there are a few cosmological observations which differ from its predictions [6],[7].

A very important complication in the investigation of the behavior of dark energy occurs due to the bias introduced by the parameterizations used. At the moment, there is a multitude of available phenomenological ansätze for the dark energy equation of state parameter w or dark energy density, each with its own merits and limitations (see [8] and references therein). The interpretation of the SNIa data has been shown to depend greatly on the type of parametrization used to perform a data fit [8], [9]. Choosing a priori a model for dark energy can thus adversely affect the validity of the fitting method and lead to compromised or misleading results.

The need to counteract this problem paved the way for the consideration of a complementary set of non-parametric reconstruction techniques [10], [11], [12], [13], [14] and model independent approaches ([15],[16], [9], [17], [18],[19]. These try to minimize the ambiguity due to a possibly biased assumption for w by fitting the original datasets without using any parameters related to some specific model. The result of these methods can then be interpreted in the context of a dark energy model of choice. Non-parametric reconstructions can thus corroborate parametric methods and provide more credibility. However, they too suffer from a different set of problems, mainly the need to resort to differentiation of noisy data, which can itself introduce great errors.

In this paper, we present a method that can be used as a model independent approach in testing the standard cosmological model. This is done by using the Genetic Algorithms (GA) technique, first used in the analysis of SNIa data in [20]. The GAs represent a method for non-parametric reconstruction of the dark energy equation of state parameter w, based on the notions of genetic algorithms and grammatical evolution. GAs are more useful and efficient than usual techniques when

- The parameter space is very large, too complex or not enough understood, as is the case with dark energy.
- Domain knowledge is scarce or expert knowledge is difficult to encode to narrow the search space.
- Traditional search methods give poor results or completely fail.

Naturally, therefore, they have been used with success in many fields where one of the above situations is encountered, like the computational science, engineering and economics. Recently, they have also been applied to study high energy physics [21],[22], [23] gravitational wave detection [24] and gravitational lensing [25]. Since the nature of Dark Energy still remains a mystery, this makes it for us an ideal candidate to use the GAs as a means to analyze the SNIa data and extract model independent constraints on the behavior of the Dark Energy.

This talk was delivered at the 14th conference in the series "Recent Developments in Gravity" (NEB-14) and is a brief overview of the use of the Genetic Algorithms to a model independent reconstruction of the expansion history of the Universe. In Section 2 a brief description of the Om statistic is given while in Section 3 we provide an overview of the general methodology of the GA paradigm and finally, we present our results in Section 4.

2. The Om statistic

The recently introduced *Om* diagnostic [26] (also look at Ref. [27]) enables us to distinguish ΛCDM from other dark energy models without directly involving the cosmic EOS. The *Om*

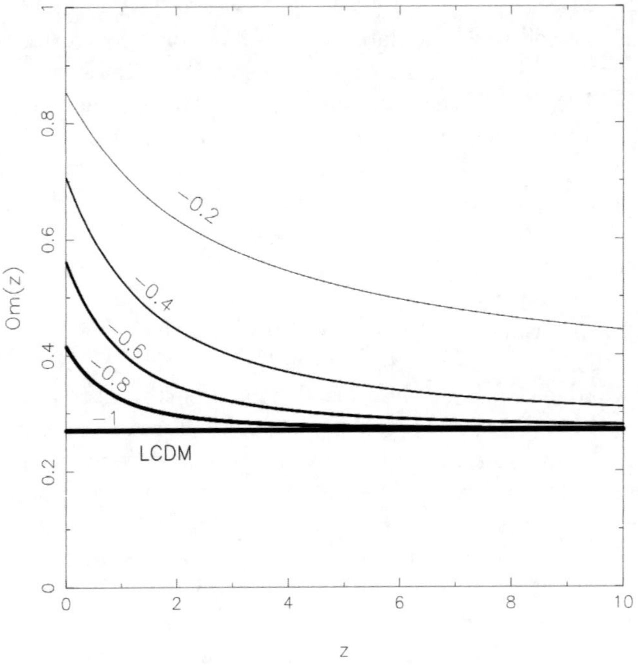

Figure 1. The Om diagnostic is shown as a function of redshift for dark energy models with $\Omega_{0m} = 0.27$ and $w = -1, -0.8, -0.6, -0.4, -0.2$ (bottom to top). For Phantom models (not shown) Om would have the opposite curvature.

diagnostic is defined as:

$$Om(x) \equiv \frac{h^2(x) - 1}{x^3 - 1}, \qquad x = 1 + z \ , \qquad h(x) = H(x)/H_0 \ . \tag{3}$$

For dark energy with a constant equation of state $w = const$,

$$h^2(x) = \Omega_{0m} x^3 + (1 - \Omega_{0m}) x^\alpha, \qquad \alpha = 3(1 + w) \tag{4}$$

(we assume that the universe is spatially flat for simplicity). Consequently,

$$Om(x) = \Omega_{0m} + (1 - \Omega_{0m}) \frac{x^\alpha - 1}{x^3 - 1} \ , \tag{5}$$

from where we find $Om(x) = \Omega_{0m}$ in ΛCDM, whereas $Om(x) > \Omega_{0m}$ in quintessence ($\alpha > 0$) while $Om(x) < \Omega_{0m}$ in phantom ($\alpha < 0$). We therefore conclude that: $Om(x) - \Omega_{0m} = 0$ if dark energy is a cosmological constant. Note that since $\Omega_\Lambda + \Omega_{0m} \simeq 1$ in ΛCDM, this model contains SCDM ($\Omega_{0m} = 1, \Omega_\Lambda = 0$) as an important limiting case. Consequently, the Om diagnostic cannot distinguish between large and small values of the cosmological constant unless the value of the matter density is independently known. In other words, the Om diagnostic provides us with a *null test* of the cosmological constant. This is a simple consequence of the fact that $h^2(x)$ plotted against x^3 results in a straight line for ΛCDM, whose slope is given by Ω_{0m}. For other dark energy models the line describing $Om(x)$ is curved, since the equality

$$\frac{d[h^2]}{d[x^3]} = constant \ , \tag{6}$$

(which always holds for ΛCDM for any $x = 1 + z$) is satisfied in quintessence/phantom type models only at redshifts significantly greater than unity, when the effects of dark energy on the expansion rate can safely be ignored. As a result the efficiency of the Om diagnostic improves at low redshifts ($z < 2$) precisely where there is likely to be an abundance of cosmological data in the coming years!

For a constant EoS: $1 + w \simeq [Om(z) - \Omega_{0m}](1 - \Omega_{0m})^{-1}$ at $z \ll 1$, consequently a larger $Om(z)$ is indicative of a larger w; while at high z, $Om(z) \to \Omega_{0m}$, as shown in Fig. 1.

On the other hand, for quintessence as well as phantom the line describing $Om(x)$ is curved, which helps distinguish these models from ΛCDM even if the value of the matter density is not accurately known – see Fig. 1.

In practice, the construction of Om requires a knowledge of the Hubble parameter, $h(z)$, which can be determined using a number of model independent approaches [15],[16], [9], [17], [18],[19].

3. Genetic algorithms

3.1. Overview

GAs were introduced as a computational analogy of adaptive systems. They are modeled loosely on the principles of the evolution via natural selection, employing a population of individuals that undergo selection in the presence of variation-inducing operators such as mutation and crossover. The encoding of the chromosomes is called the genome or genotype and is composed, in loose correspondence to actual DNA genomes, by a series of representative "genes". Depending on the problem, the genome can be a series (vector) of binary numbers, decimal integers, machine precision integers or reals or more complex data structures.

In order to evaluate the individual chromosomes a fitness function is used and reproductive success usually varies with fitness. The fitness function is a map between the gene sequence of the chromosomes (genotype) and a number of attributes (phenotype), directly related to the properties of a wanted solution. Often the fitness function is used to determine a "distance" of a candidate solution from the true one. Various distance measures can be used for this purpose (Euclidean distance, Manhattan, Mahalanobis etc.).

The algorithm begins with an initial population of candidate solutions, which is usually randomly generated. Although GA's are relatively insensitive to initial conditions, i.e. the population we start from is not very significant, but using some prescription for producing this seed generation can affect the speed of convergence. In each successive step, the fitness functions for the chromosomes of the population are evaluated and a number of genetic operators (mutation and crossover) are applied to produce the next generation. This process continues until some termination criteria is reached, e.g. obtain a solution with fitness greater than some predefined threshold or reach a maximum number of generations. The later is imposed as a condition to ensure that the algorithm terminates even if we cannot get the desired level of fitness.

The various steps of the algorithm can be summarized as follows:

(i) Randomly generate an initial population $M(0)$

(ii) Compute and save the fitness for each individual m in the current population $M(t)$.

(iii) Define selection probabilities $p(m)$ for each individual m in $M(t)$ so that $p(m)$ is proportional to the fitness.

(iv) Generate $M(t+1)$ by probabilistically selecting individuals from $M(t)$ to produce offspring via genetic operators (crossover and mutation).

(v) Repeat step 2 until a satisfying solution is obtained, or a maximum number of generations reached.

We should stress that the initial population $M(0)$ will only depend on the choice of the grammar and therefore it can only affect on how fast the algorithm will converge to the minimum.

Using the wrong grammar may result in the GA being trapped in a local minimum. Also, two important parameters that affect the transition from the population $M(t)$ to $M(t+1)$ are the selection rate and the mutation rate. The selection rate is typically of the order of 10% and it affects how many of the individuals, after they are ranked with respect to their fitness, will be allowed to produce offspring. The mutation rate is usually of the order of 5% and it expresses the probability that an arbitrary part of the genetic sequence will be changed from its previous state. This is done in order to maintain the genetic diversity from one generation of a population to the next.

The paradigm of GAs described above is usually the one applied to solving most of the problems presented to GAs. Though it might not find the best solution, more often than not, it would come up with a partially optimal solution. A more detailed overview and its application in cosmology can be found in [20].

The difference of the GA over the standard analysis of the data, ie having an a priori defined model with some free parameters, is that the obtained values of the best-fit parameters are model-dependent and in general models with more parameters tend to give better fits to the data. This is where the GA approach starts to depart from the ordinary parametric method. Our goal is to minimize a function, not using a candidate model function for the distance modulus and varying parameters, but through a stochastic process based on a GA evolution. This way, no prior knowledge of a dark energy model is needed to obtain a solution and our result will be completely parameter-free.

In simpler terms, the GA method does not require an a priori assumption for a DE model, but uses the data themselves to find this model. Also, it is parameter free as the end result does not have any free parameters that can be changed in order to fit the data. So, in this sense this method has far less bias any of the standard methods for the reconstruction of the expansion history of the Universe. This is the main reason for the use of the GAs in this paper.

3.2. General Methodology

We first outline the course of action we follow to apply the GA paradigm in the case of SNIa data. For the application of GA and grammatical evolution (GE) on the dataset, we use a modified version of the GDF [28] tool [1], which uses GE as a method to fit datasets of arbitrary size and dimensionality. This program uses the tournament selection method for crossover. GDF requires a set of input data (train set), an (optional) test sample and a grammar to be used for the generation of functional expressions. The output is an expression for the function which best fits the train set data.

Since the SNIa datapoints are given in terms of the distance modulus $\mu_{obs}(z_i)$, the fitness function for the GA we have chosen is equal to $-\chi^2_{SNIa}$, where

$$\chi^2_{SNIa} = \sum_{i=1}^{N} \frac{(\mu_{obs}(z_i) - \mu_{GA}(z_i))^2}{\sigma^2_{\mu\ i}}, \qquad (7)$$

where for the Constitution set $N = 397$ and $\sigma^2_{\mu\ i}$ are the errors due to flux uncertainties, intrinsic dispersion of SNIa absolute magnitude and peculiar velocity dispersion and $\mu_{GA}(z_i)$ is the *reduced distance modulus* obtained for each chromosome of the population by the GA.

An advantage of our method is the fact that neither the χ^2_{SNIa} nor the $\mu_{GA}(z)$ depend on any parameters. Also, as it can be seen from eq. (7), our method does not make an assumption of flatness or not in order to fit the data. Flatness (or non-flatness) only comes into play when one tries to find the luminosity distance $d_L(z)$ and the underlying dark energy model given the

[1] http://cpc.cs.qub.ac.uk/summaries/ADXC

best-fit form of $\mu_{GA}(z)$. So, in this aspect the result produced by the GA is independent of the assumption of flatness as well.

The GA evaluates (7) in each evolutionary step for every chromosome of the population. The one with the best fitness will consequently have the smallest χ^2_{SNIa} and will be the best candidate solution in its generation. Of all the steps in the execution of the algorithm, the evaluation of the fitness is the most expensive. GDF is appropriately modified to use (7) as the basis of its fitness calculation.

After the execution of the GA, we obtain an expression $\mu_{GA}(z)$ for the reduced distance modulus as the solution of best fitness and the corresponding χ^2_{SNIa}. Using this we can, through differentiation, obtain other functions or parameters of interest, such as the Om statistic or the deceleration parameter $q(z)$. For example, for a flat Universe the Hubble parameter will be given by

$$1/H(z) = \frac{d}{dz}\left(\frac{10^{\frac{\mu_{GA}(z)}{5}}}{1+z}\right) \tag{8}$$

and then the deceleration parameter can be found by Eq. (2).

A possible complication of this method is that it gives no direct way to estimate the errors for the derived parameters. One cannot expect to get an estimate by just running the algorithm many times and obtaining slightly different parameters. The GA usually tends to converge at the same solution for a given dataset, unless we change significantly the population size or the number of generations. A way to circumvent this problem is to use a bootstrap Monte Carlo simulation to produce synthetic datasets and rerun the algorithm on them. We can thus obtain a statistical sample of parameter values which will allow us to estimate the error.

We sketch the procedure we will follow [29]:

(i) The GA is applied on the original SNIa dataset with the chosen execution parameters and a solution for $\bar{\mu}_{GA}(z)$ is obtained.

(ii) Generate a number of synthetic datasets by drawing each time the same number of data points with replacement from the original set.

(iii) The GA is rerun for the synthetic datasets.

(iv) A new set of values for the Om statistic and the deceleration parameter $q(z)$ is generated.

(v) The 95% error of the desired parameter can be found by taking the 2.5 and the 97.5 percentiles of the bootstrap distribution [30].

Using the above steps, we can obtain error estimates for any desired parameter.

3.3. An example

In this Section we will briefly describe a simple example[2] of how the GA determines the best-fit. We will avoid describing the technicalities, like the binary representation of the solutions, and instead we will concentrate on how the generations evolve, how many and which individuals are chosen for the next generation etc.

As we mentioned earlier, the choice of the grammar is very important, however in order to keep our example as simple as possible we will assume that our grammar includes only basic functions like polynomials x, x^2 etc, the trigonometric functions $sin(x)$, $cos(x)$, the exponential e^x and the logarithm $ln(x)$.

The first step in the GA is setting up a random initial population $M(0)$ which can be any simple combination of these functions, eg $\mu_{GA,1}(z) = ln(z)$, $\mu_{GA,2}(z) = -1 + z + z^2$ and

[2] This is only a schematic description of how the GA works and this is done solely for the sake of explaining the basic mechanisms of the GA.

$\mu_{GA,3}(z) = sin(z)$. The number of candidate solutions (chromosomes) in the genetic population is usually a few hundreds and later on we will use the value of 500.

Next, the algorithm measures the fitness of each solution by calculating their χ^2, ie for our simple example $\chi_1^2 = 80579.9$, $\chi_2^2 = 292767.0$ and $\chi_3^2 = 412928.0$. The selection per se is done by implementing the "Tournament selection" method which involves running several "tournaments", sorting the population with respect to the fitness each individual and after that a fixed percentage, adjusted by the selection rate (see the previous subsection), of the population is chosen. As we mentioned, the selection rate is of the order of 10% of the total population, but lets assume that the two out of the three candidate solutions ($\mu_{GA,1}(z)$ and $\mu_{GA,2}(z)$) are chosen for the sake of simplicity.

The reproduction of these two solutions will be done by the crossover and mutation operations. The crossover will randomly combine parts of the "parent" solutions, for example in one such realization this may be schematically shown as

$$\begin{aligned} \mu_{GA,1}(z) \oplus \mu_{GA,2}(z) &\rightarrow (\bar{\mu}_{GA,1}(z), \bar{\mu}_{GA,2}(z), \bar{\mu}_{GA,3}(z)) \\ &= (ln(z^2), -1 + ln(z^2), -1 + ln(z)) \end{aligned}$$

After this is done, the GA will proceed to implement the mutation operation. The probability for mutation is adjusted by the mutation rate, which as we mentioned is typically of the order of 5%. In our example this may be a change in the power of some term or the change in the value of a coefficient. For example, for the candidate solution $\bar{\mu}_{GA,3}(z)$ this can be schematically shown as $\bar{\mu}_{GA,3}(z) = -1 + ln(z) \rightarrow -1 + ln(z^3)$, where the power of the z term was mutated from 1 to 3.

Finally, at the end of the first round we have the three candidate solutions

$$\begin{aligned} M(1) &= (\bar{\mu}_{GA,1}(z), \bar{\mu}_{GA,2}(z),\ \bar{\mu}_{GA,3}(z)) = \\ &\quad (ln(z^2), -1 + ln(z^2), -1 + ln(z^3)) \end{aligned}$$

At the beginning of the next round the fitness of each individual will again be determined, which for our population will be $(\chi_1^2, \chi_2^2, \chi_3^2) = (3909.2, 18435, 113327)$, and the selection and the other operations will proceed as before. It is easily seen that even after one generation the χ^2 of the candidate solutions has decreased dramatically and as we will see in the next section, it only takes a bit more than 100 generations to reach an acceptable χ^2 and about 500 to 1000 generations to start converging on the minimum.

After a predetermined number of generations, usually on the order of 10^3, has been reached or some other criteria have been met, the GA will finish, with the best candidate solution of the last generation being the best-fit to the data.

4. Results

We will now examine a number of different configurations, highlighted in Table 1. In order to keep our analysis as simple as possible we will use a Basic Grammar that includes polynomials like x, x^2 etc, the trigonometric functions $sin(x)$, $cos(x)$, the exponential e^x and the logarithm $ln(x)$ and a second grammar that includes the Basic Grammar and the Legendre polynomials. For the analysis we will use the Constitution SNIa dataset of [31] consisting of 397 SNIa. The steps followed for the usual minimization of (7) in terms of its parameters are described in detail in [32], [33],[34], [35].

In Fig. 2 we show the fitness (equal to $-\chi^2$) as a function of the number of generations. Clearly, all configurations seem to have converged very early or to be very close to converging to their minimum. An exception is Case 3 which even after 3000 generations still has not converged and we had to run the simulation for another 3000 generations to check its convergence. Since

Table 1. The different cases considered in the analysis. The Basic Grammar includes polynomials, trigonometric functions, exponentials and logarithms. For ΛCDM the best-fit corresponds to $\Omega_m = 0.289$.

Case 1	Basic grammar	$\chi^2_{min} = 465.19$
Case 2	Basic grammar + Legendre polyn.	$\chi^2_{min} = 477.87$
Case 3	Only polyn. in grammar (3000 gen.)	$\chi^2_{min} = 468.19$
Case 4	Only polyn. in grammar (6000 gen.)	$\chi^2_{min} = 462.56$
ΛCDM	$H(z)^2 = H_0^2(\Omega_m(1+z)^3 + 1 - \Omega_m)$	$\chi^2_{min} = 465.51$

we want to compare the configurations with equal standards, ie number of generations etc, we will treat this case (Case 4) separately.

As it can be seen from Table 1 and Fig. 3, most configurations agree well with the data and actually have a minimum χ^2 comparable or even better with respect to ΛCDM. An exception to this is Case 2, where the best fit differs from ΛCDM by a $\Delta\chi^2 = 12$ thus providing a much poorer fit. The reason for this is the fact that by increasing the grammar we effectively increase the landscape in which the algorithm has to search and therefore, we are making the fit poorer as the GA will take much much longer to converge.

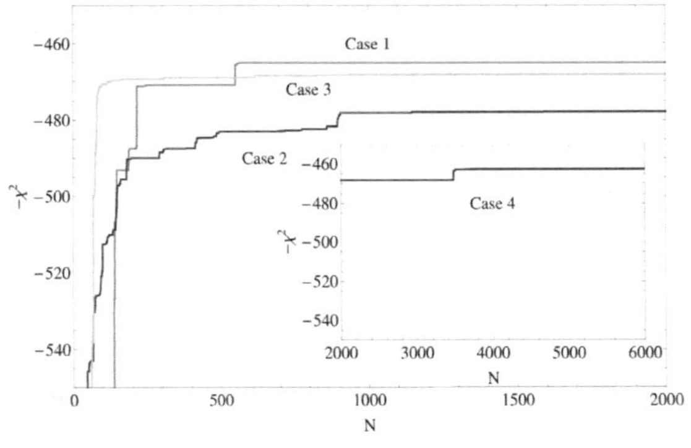

Figure 2. The fitness (equal to $-\chi^2$) as a function of the number of generations. The magenta line corresponds to case 1, the green line to case 2 and the blue line to case 3. The inset graphic shows case 4.

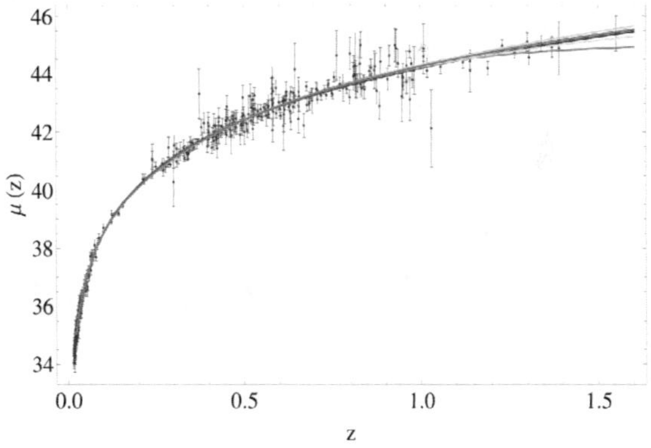

Figure 3. The SNIa distance modulus against the redshift. The red line represents the best-fit ΛCDM model with $\Omega_m = 0.289$ while the magenta, blue, green and cyan (best-fit) lines correspond to cases 1-4 respectively.

In Figs. 4 and 5 we show the corresponding results of each case for the Om statistic and the deceleration parameter $q(z)$ (our best-fit function corresponds to the cyan line). Due to the presence of trigonometric functions into the best fit and the fact that the Om statistic is derived from the Hubble parameter by differentiation, Case 1 exhibits strong oscillations. This is also apparent in the corresponding plot for the deceleration parameter for Case 2 (magenta line, Fig. 5).

As it can be seen in Fig. 1, a constant w (but different from $w = -1$) forces the Om statistic to be a monotonic function that is located completely either above or below the constant case

($Om = 0.289$). However, none of the curves in Fig. 4 are monotonic functions and all of them cross the constant Om case several times. Also, all cases remain reasonably close to the ΛCDM value of $\Omega_m = 0.289$ for intermediate redshifts but deviate from that value at high or low redshifts. Therefore, this could be the effect of a time-dependent equation of state $w(z)$ instead of a constant w. Comparing with Fig. 1, it is possible that this behaviour is due to an equation of state $w(z)$ more negative than -1.

This is also supported from Fig. 4 and Fig. 5 as at low redshifts all cases predict a much different form of acceleration than the best fit ΛCDM. In Fig. 4 and Fig. 5 our best fit results show an overall slowing down of the acceleration at the low redshifts (which is in concordance with results presented in [9] with a very rapid increase of acceleration at the very low redshifts). These are very similar with the results presented in Fig.2 of Ref. [19] where a smoothing method for the reconstruction of the expansion history of the universe is being used. It is obvious that when two very different model independent approaches reach to a similar result, we must be on the right track and very close to the best possible result. However, theoretical interpretation of these results is beyond the scope of this paper. In fact there might be very different mechanisms that result to a similar expansion history of the universe as we reconstructed in this paper (e.g see [36], [16], [37], [38]).

In Fig. 6 we show the Om statistic as a function of the redshift z. The black line corresponds to the best fit of case 4, while the gray-shaded area to the 2σ error region. The error region was calculated by implementing a bootstrap monte-carlo simulation as discussed in the previous section. Note that the overall slowing down of the acceleration at the low redshifts cannot be clearly seen from this figure as the errors are quite symmetrical around the best fit ΛCDM value. However, as we mentioned earlier this is clearly seen by the behavior of the deceleration parameter in Fig. 5. As it can be seen in Fig. 6, ΛCDM remains consistent with the data at the 2σ level but at the same time, some deviations from the standard ΛCDM model can be accommodated. Due to the large errorbars of the current datasets, many dark energy models are still consistent with them but we expect this to change in the near future with the advent of high quality SnIa observations.

In Fig. 7 we show a histogram of the bootstrap distribution found from the monte-carlo simulation that was used to create the error regions. Clearly, the distribution is quite symmetrical and centered, so we are confident that our error region in Fig. 6 was correctly reconstructed. Finally, we also tested the GA on a mock ΛCDM dataset and we found that it can reconstruct the luminosity distance curve to within 0.1%, so we are confident that is does not overfit the data, but instead it can successfully detect the underlying DE model.

5. Conclusions

We used the Om statistic and the GAs in order to derive a null test on the cosmological constant model ΛCDM. Our interest in the GAs stems from the fact that they represent a method for non-parametric reconstruction of the expansion history of the universe, based on the notions of genetic algorithms and grammatical evolution. These kinds of algorithms are more useful and efficient than usual techniques especially when the problem under examination is not well understood, as is the case with dark energy. Since the nature of dark energy still remains a mystery, this makes it for us an ideal candidate to use the GAs as a means to analyze the SNIa data and extract model independent constraints on the behavior of the Dark Energy. On the other hand, the Om diagnostic (Sahni et al. 2008) enables us to distinguish ΛCDM from other dark energy models without directly involving the cosmic EoS.

Our methodology is completely model independent and it can be summarized in two steps: first, we applied the GA to the Constitution SNIa data in order to acquire a model independent reconstruction of the expansion history of the Universe $H(z)$. After we reconstructed $H(z)$, the second step was to use it in conjunction with the Om statistic and derive our null test.

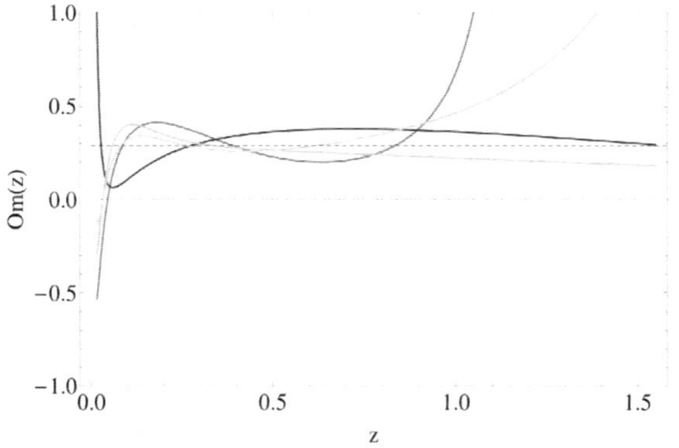

Figure 4. The Om statistic against the redshift. The black dashed line represents the ΛCDM value of $\Omega_m = 0.289$ while the magenta, blue, green and cyan (best-fit) lines correspond to cases 1-4 respectively.

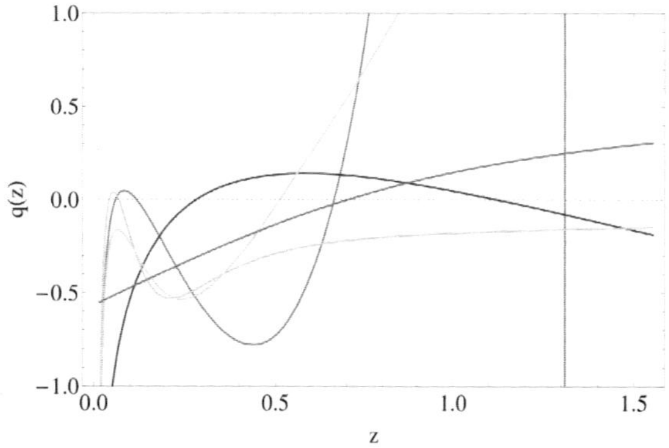

Figure 5. The deceleration parameter q against the redshift. The red line represents the best-fit ΛCDM model with $\Omega_m = 0.289$ while the magenta, blue, green and cyan (best-fit) lines correspond to cases 1-4 respectively.

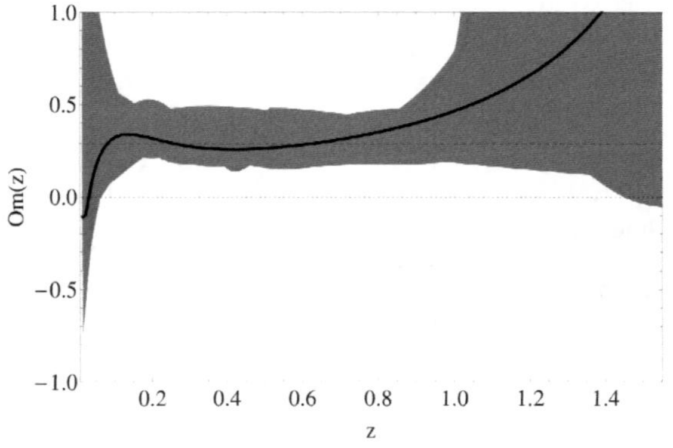

Figure 6. The Om statistic as a function of z. The black line corresponds to case 4 (the best-fit) and the gray-shaded area to the 2σ error.

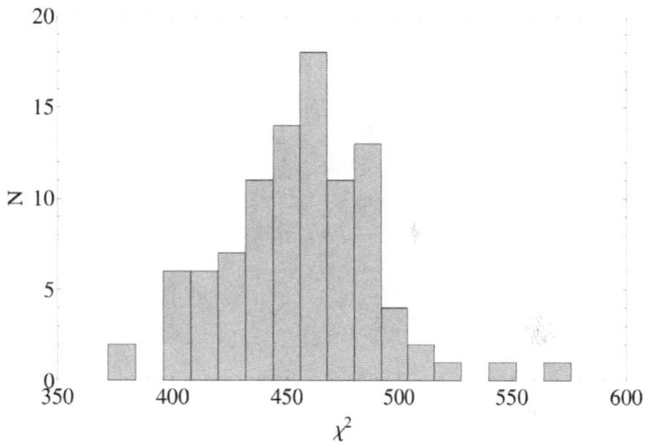

Figure 7. Histogram of the bootstrap distribution found from the monte-carlo simulation that was used to create the error regions of Fig. 6.

Our main result was that ΛCDM remains consistent with the data at the 2σ level, see Fig. 6, but at the same time, some deviations from the standard ΛCDM model can be accommodated and this is also in accordance with the data especially at low redshifts (see the behavior of the best fit in Fig. 6).

However, it should be mentioned that the slowing down at low redshifts for the Constitution set mentioned earlier can also be seen with the standard analysis with the CPL ansatz (see [39]) but only for the Constitution set. For the other SNIa datasets the reverse trend (speeding up the acceleration at low z) was observed. For this reason and due to the fact that the current data still have quite large errors we believe that the full potential of our method, while it is very promising, will only be realized in the near future when more high quality SNIa data will become available. When this happens, the error region at low z in Fig. 6 will be small enough for the Om statistic to completely discriminate between ΛCDM and the various Dark Energy models.

Acknowledgements

S.N. is supported by the Niels Bohr International Academy, the Danish Research Council under FNU Grant No. 272-08-0285 and the DISCOVERY center. The current work was presented on the 14th Conference on *Recent Developments in Gravity* (NEB14) and is based on Ref. [40].

References

[1] Riess A. G. et al. [Supernova Search Team Collaboration], 2004, ApJ, **607**, 665.
[2] Spergel D. N. et al. [WMAP Collaboration], 2007, ApJ.Sup., **170**, 377, [arXiv:astro-ph/0603449].
[3] Readhead A. C. S. et al., 2004, ApJ., **609**, 498, [arXiv: astro-ph/0402359].
[4] Perivolaropoulos L., 2006, AIP Conf. Proc. **848**, 698, [arXiv:astro-ph/0601014].
[5] Komatsu E. et al., 2010, arXiv:1001.4538.
[6] Perivolaropoulos L., 2008, arXiv:0811.4684 [astro-ph].
[7] Perivolaropoulos L., Shafieloo A., Phys. Rev. D **79**, 123502 (2009) [arXiv:0811.2802 [astro-ph]].
[8] Sahni V., Starobinsky A. , 2006, Int. J. Mod. Phys. D **15**, 2105, [arXiv:astro-ph/0610026].
[9] Shafieloo A., Alam U.,Sahni V., Starobinsky A. A., 2006, MNRAS, **366**, 1081, [arXiv:astro-ph/0505329].
[10] Daly R. A., Djorgovski S. G. , 2003, ApJ, **597**, 9, [arXiv:astro-ph/0305197].
[11] Wang Y., Mukherjee P., 2004, ApJ., **606**, 654, [arXiv:astro-ph/0312192].
[12] Saini T. D., 2003, MNRAS., **344**, 129 [arXiv:astro-ph/0302291].
[13] Wang Y., 2009, Phys. Rev. D **80**, 123525, [arXiv:0910.2492].
[14] Clarkson C., Zunckel C. , 2010, arXiv:1002.5004.
[15] Sahni V., Saini T. D., Starobinsky A. A., Alam U. , 2003, JETP Lett. **77**, 201 [Pisma Zh. Eksp. Teor. Fiz. 77, 249], [arXiv:astro-ph/0201498].
[16] Alam U. , Sahni V. , Starobinsky A. A. , 2003, JCAP, **0304**, 002
[17] Shafieloo A., 2007, MNRAS, **380**, 1573, [arXiv:astro-ph/0703034].
[18] Wang Y., Tegmark M., 2005, Phys. Rev. D **71**, 103513, [arXiv:astro-ph/0501351].
[19] Shafieloo A., Clarkson C., 2010, Phys. Rev. D **81**, 083537, [arXiv:0911.4858 [astro-ph.CO]].
[20] Bogdanos C., Nesseris S. , 2009, JCAP **0905**, 006, [arXiv:0903.2805 [astro-ph.CO]].
[21] Becks K. H. ,Hahn S., Hemker A. , 1994, Phys. Bl. **50**, 238;
[22] Allanach B. C. ,Grellscheid D., Quevedo F. , 2004, JHEP, **0407**, 069, [arXiv:hep-ph/0406277].
[23] Rojo J., Latorre J. I., 2004, JHEP **0401**, 055, [arXiv:hep-ph/0401047];
[24] Crowder J. ,Cornish N. J., Reddinger L. , 2006, Phys. Rev. D ,**73**, 063011, [arXiv:gr-qc/0601036].
[25] Brewer B. J. , Lewis G. F. , 2005, arXiv:astro-ph/0501202.
[26] Sahni V., Shafieloo A., Starobinsky A. A. , Phys. Rev. D **78**, 103502 (2008) [arXiv:0807.3548 [astro-ph]].
[27] Zunckel C., Clarkson C., 2008, Phys. Rev. Lett. **101**, 181301, [arXiv:0807.4304 [astro-ph]].
[28] Tsoulos I.G., Gavrilis D., Dermatas E., 2006, Computer Physics Communications **174**, 555-559.
[29] Press W. H. et. al., 1994, 'Numerical Recipes', Cambridge University Press.
[30] Efron B. , 1982, Society of Industrial and Applied Mathematics, CBMS-NSF Monographs **38**.
[31] Hicken M. et al., 2009, ApJ, **700**, 1097, [arXiv:0901.4804 [astro-ph.CO]].
[32] Nesseris S., Perivolaropoulos L., 2004, Phys. Rev. D **70**, 043531, [arXiv:astro-ph/0401556].
[33] Nesseris S., Perivolaropoulos L., 2005, Phys. Rev. D **72**, 123519, [arXiv:astro-ph/0511040].
[34] Lazkoz R., Nesseris S., Perivolaropoulos L. , 2005, JCAP **0511**, 010, [arXiv:astro-ph/0503230].
[35] Nesseris S., Perivolaropoulos L., 2007, JCAP 0701, 018 [arXiv:astro-ph/0610092].
[36] Csaki C. , Kaloper N. , Terning J. , 2002, Phys. Rev. Lett, **88**, 161302
[37] Kelly P. et al., 2010, ApJ., **715**, 743 (2009) [arXiv:0912.0929].
[38] Sullivan M. et al., Mon. Not. Roy. Astron. Soc. **406**, 782 (2010) [arXiv:1003.5119 [astro-ph.CO]].
[39] Sanchez J. C. B. , Nesseris S. , Perivolaropoulos L., 2009, JCAP **0911**, 029, [arXiv:0908.2636 [astro-ph.CO]].
[40] Nesseris S., Shafieloo A., Mon. Not. Roy. Astron. Soc. **408**, 1879 (2010) [arXiv:1004.0960 [astro-ph.CO]].

Journal of Physics: Conference Series **283** (2011) 012026 doi:10.1088/1742-6596/283/1/012026

Newton's law modifications due to a Sol manifold extra dimensional space

V K Oikonomou

Technological Education Institute of Serres,
62124 Serres, Greece

E-mail: voiko@physics.auth.gr

Abstract. The corrections to the gravitational potential due to a Sol extra dimensional compact manifold are studied. We compare the range of the corrections to the range of the 3-torus corrections. It is found that for small values of the radius of the extra dimensions the Sol manifold corrections are large compared to the 3-torus corrections. Also, Sol manifolds corrections can be larger, comparable or smaller compared to the 3-torus case, for larger compactification radii.

1. Introduction

In the last decade, the phenomenological implications of extra dimensional models [1] were extensively studied. Many of these studies are concentrated on the modifications caused on the gravitational potential due to extra dimensional manifolds [2, 3, 4, 5]. Generally the modifications are of Yukawa type (except in Randall-Sundrum warped extra dimensional models), differing only in their strength and range (for a very recent interesting application see [5]). We shall study the modification of the gravitational potential caused by a Sol extra dimensional manifold. Sol manifolds are compact and orientable three dimensional manifolds, denoted as M_A^3. We shall present the Sol manifolds geometric structure. Also we shall study the Laplace equation on such manifolds. Finally we compare the range of the corrections to the gravitational potential due to Sol manifolds, with the corresponding range caused by a T^3 extra dimensional manifold.

2. Sol Manifold Geometry and Topology

Sol manifolds are one of Thurston's 8 three dimensional geometries [6] (geometric structures). The geometrization conjecture is an approach to the geometry of three dimensional manifolds through general topological arguments. One starts with the fact that three dimensional manifolds are composed by 2-spheres or torii surfaces. According to the properties and the details of the topological gluing map of the above with R or S^1, the resulting manifolds acquire locally homogeneous metrics. Sol manifolds are described by one type of the 8 different homogeneous metrics.

Sol manifolds are obtained from $SL(2, Z)$ stiffennings of torus bundles over the circle. A theorem [7] states: given M a total bundle space of a T^2 bundle over S^1 with gluing map ϕ and let $A \in GL(2, Z)$ represent the automorphism of the fundamental group of the torus T^2 induced by the gluing map ϕ. Then the total bundle space admits a Sol geometric structure if A is hyperbolic, an E^3 structure if A is periodic and finally a Nil structure otherwise. A is hyperbolic and gives rise to an orientable manifold if $|TrA| > 2$ and we shall dwell on this choice. Thus Sol structure arises from the $SL(2, Z)$ stiffening of the mapping torus of a torus diffeomorphism ϕ [6].

Two different hyperbolic gluings with $A_1 \neq A_2$, give rise to different geometric Sol structures. A will be represented by,

$$A = \begin{pmatrix} 0 & 1 \\ -1 & n \end{pmatrix} \tag{1}$$

so the different Sol structures are classified by the integer n, with $|n| = |TrA| > 2$. Also only the positive eigenvalues of A will be considered.

Let us analyze more the Sol manifold structure. Consider the manifold $T^2 \times R$ described by the periodic coordinates (x, y) for the torus, defined modulo R (the radius of the compact dimensions) and $z \in (-\infty, \infty)$ be a coordinate of R. The combined action of the torus mapping through the hyperbolic gluing map A diffeomorphism (which we denote as $\bar{\Gamma}$) is :

$$\bar{\Gamma} : \begin{pmatrix} x \\ y \\ z \end{pmatrix} \rightarrow \begin{pmatrix} y \\ -x + ny \\ z + 2\pi R \end{pmatrix} \tag{2}$$

The Sol manifold, which is denoted as M_A^3 is the quotient of $T^2 \times R$ by the action of $\bar{\Gamma}$, that is $M_A^3 \equiv T^2 \times R/\bar{\Gamma}$. It is a total torus bundle over S^1, with T^2 the fiber, S^1 the base space and A the hyperbolic gluing map of the torus fibers. We consider Sol manifolds for which the eigenvalues λ of A are positive. For the case of (1) the characteristic polynomial of the matrix A reads:

$$\lambda^2 - TrA\lambda + 1 = 0 \tag{3}$$

or equivalently:

$$\lambda^2 - n\lambda + 1 = 0 \tag{4}$$

The solutions to equation (4) are λ and λ^{-1} with $\lambda + \lambda^{-1} = n = TrA$. Also the discriminant of A is:

$$D = (\lambda - \lambda^{-1})^2 \tag{5}$$

We use another coordinate system (u, v, z) on the Sol manifold. The (u, v) are linear coordinates of the torus fibres related to an eigenbasis of the hyperbolic map A. These coordinates correspond to a rotated torus lattice. The action of $\bar{\Gamma}$ in the new coordinate system reads:

$$\bar{\Gamma} : \begin{pmatrix} u \\ v \\ z \end{pmatrix} \rightarrow \begin{pmatrix} \lambda u \\ \lambda^{-1} v \\ z + 2\pi R \end{pmatrix} \tag{6}$$

The original lattice was orthogonal while the new torus lattice is not. If e_u and e_v are the basis of the lattice after the action of $\bar{\Gamma}$ identification, then $(e_u, e_v) = |e_u||e_v| \cos\theta$. Also the fiber coordinates T^2 are not periodic anymore. So in order two pairs (u_1, v_1) and (u_2, v_2) to define the same point on the torus lattice, the following must hold:

$$(u_1 - u_2, v_1 - v_2) = ke_u + me_v \tag{7}$$

with k, m integers and e_u, e_v defined previously.

2.1. Riemannian Sol group invariant metric on Sol manifolds

The Riemannian metric on Sol manifolds comes from the metric on the universal covering of M_A^3. The group invariant metric on the universal covering is a Sol group invariant metric. With the invariant metric on the universal covering of M_A^3 we can find the following class of metrics of Sol manifolds:

$$\mathrm{d}s^2 = Ee^{2z\ln\lambda}\mathrm{d}u^2 + 2F\mathrm{d}u\mathrm{d}v + Ge^{-2z\ln\lambda}\mathrm{d}v^2 + \mathrm{d}z^2 \tag{8}$$

We shall use the metric (8) in order to compute the Laplace-Beltrami operator on M_A^3.
Sol manifolds are hyperbolic manifolds with negative curvature. The hyperbolicity is a very interesting feature of Sol manifolds, regarding they are torus fibrations. Due to their hyperbolicity, the distribution of eigenvalues of the Laplacian is not Poisson. This might have effects on the gravitons mass splitting.

Newton's law and extra dimensions

Our purpose is to examine the corrections to the gravitational potential caused by an extra dimensional Sol manifold. The spacetime manifold is of the form $M^4 \times M_A^3$. Let us review the general technique to obtain these corrections. The presentation is based on reference [2, 8].

Consider a spacetime of the form $M^4 \times M^n$, with M^n an n-dimensional compact manifold and M^4 the four dimensional Minkowski spacetime and also a complete set of orthogonal harmonic functions on M^4, Ψ_m, satisfying the orthogonality condition:

$$\int_{M^n} \Psi_n(x)\Psi_m^*(x) = \delta_{n,m} \tag{9}$$

and the completeness relation:

$$\sum_m \Psi_m(x)\Psi_m^*(x') = \delta^{(n)}(x - x') \tag{10}$$

The functions Ψ_m are eigenfunctions of the n-dimensional Laplace-Beltrami operator Δ_n of the manifold M^n, with eigenvalues μ_m^2:

$$-\Delta_n\Psi_m = \mu_m^2\Psi_m \tag{11}$$

The gravitational potential V_{n+4} satisfies the Poisson equation in $n+3$ spatial dimensions, when the Newtonian limit is taken:

$$\Delta_{n+3}V_{n+4} = (n+1)\Omega_{n+2}G_{n+4}M\delta^{(n+3)}(x) \tag{12}$$

with M, the mass of the system, G_{n+4} the Newton constant in $n+4$ dimensions and

$$\Omega_{n+2} = \frac{2\pi^{\frac{n+3}{2}}}{\Gamma(\frac{n+3}{2})} \tag{13}$$

Equation (12) corresponds to the case of large compact radius limit and has the solution:

$$V_{n+4} = -\frac{G_{n+4}M}{r_n^{n+1}} \tag{14}$$

with r_n the radius of the $n+3$-dimensional space in spherical coordinates. Denoting by r the radius of the 3-dimensional space of our world, then $r_n = \sqrt{r^2 + x_1^2 + x_2^2 + ... + x_n^2}$. When the compact dimensions have small lengths, we find the harmonic expansion of V_{n+4} in terms of the eigenfunctions of the product space $M^4 \times M^n$, which reads:

$$V_{n+4} = \sum_m \Phi_m(r)\Psi_m(x) \tag{15}$$

with r denoting the coordinates of M^4 and x denoting the coordinates of M^n. Consequently, the Φ_m obey:

$$\Delta_3\Phi_m - \mu_m^2\Phi_m = (n+1)\Omega_{n+2}\Psi_m^*(0)G_{n+4}M\delta^3(x) \tag{16}$$

with solution:

$$\Phi_m(r) = -\frac{\Omega_n G_{n+4} M \Psi_m^*(0)}{2}\frac{e^{-|\mu_m|r}}{r} \tag{17}$$

The gravitational potential is written as:

$$V_{n+4} = -\frac{\Omega_n G_{n+4} M}{2r}\sum_m \Psi_m^*(0)\Psi_m(x)e^{-|\mu_m|r} \tag{18}$$

Since all point particles in the four dimensional spacetime have no dependence on the internal compact space M^n, we can take $x = 0$ in (18) to obtain the four dimensional gravitational potential:

$$V_4 = -\frac{G_4 M}{r}\sum_m \Psi_m^*(0)\Psi_m(0)e^{-|\mu_m|r} \tag{19}$$

which is valid for large values of r, compared to the lengths of the compact dimensions.

In the above general result of relation (19) we shall apply the eigenfunctions and eigenvalues of Sol manifolds.

3. Sol Manifold modification of Newton's Law

In order to compute the corrections to the gravitational potential we must solve equation (11) for the case of Sol manifold M_A^3. A much more elaborate analysis of this section can be found in [8, 9]. Using the (u, v, z) coordinates we introduced previously, the Laplace-Beltrami operator for the manifold M_A^3 is:

$$\Delta = Ee^{2z\ln\lambda}\frac{\partial^2}{\partial u^2} + 2F\frac{\partial^2}{\partial u\partial v} + Ge^{-2z\ln\lambda}\frac{\partial^2}{\partial v^2} + \frac{\partial^2}{\partial z^2} \tag{20}$$

which stems from the Riemannian metric (8). As usual $E = |e_u|^2$, $F = |e_v|^2$ and $G = |(e_u, e_v)|$, where e_u and e_v the basis of the T^2 lattice. Thus we must solve

$$-\Delta\psi = E\psi \tag{21}$$

A function $\Psi = e^{2\pi i(\gamma,w)}f(z)$ satisfies equation (21) if and only if $f(z)$ satisfies the modified Mathieu equation [8]:

$$\Big(-\frac{d^2}{dz^2} + |\nu(\gamma)|\cosh 2\mu(z+\alpha(\gamma))\Big)f(z) = \Lambda f(z) \tag{22}$$

with $\mu = \ln\lambda$, $\nu(\gamma) = 8\pi^2 CQ_{A^*}(\gamma)$ and $\alpha(\gamma) = \frac{\ln\left(\sqrt{\frac{E}{G}}\frac{|(\gamma,e_u)|}{|(\gamma,e_v)|}\right)}{2\ln\lambda}$. Also

$$C = \frac{1}{\sqrt{D}\mathrm{Vol}(M_A^3)\sin\theta} \tag{23}$$

The volume of the Sol manifold, $Vol(M_A^3)$, is equal to the volume of the total bundle space $T^2 \times S^1$, and is equal to $2\pi R^3$. D is the discriminant of the matrix A defined in (5) and $Q_{A^*}(\gamma)$ is the quadratic form (see [9]) corresponding to A^* acting to the dual lattice of T^2, with $Q_{A^*}(\gamma) = (\gamma, e_u)(\gamma, e_v)(\lambda - \lambda^{-1})$. The eigenvalues E and Λ are related as follows:

$$E = \Lambda + \nu(\gamma)\cos\theta \tag{24}$$

It is proved in reference [9] that the functions $\Psi_\gamma = e^{2\pi i(\gamma, w)} f_\gamma(z)$ form a complete orthogonal basis on the Sol manifold. The spectrum of the Laplace-Beltrami operator consists of two parts:

- The trivial part, with eigenvalues $E_k = \frac{4\pi^2 k^2}{R^2}$
- The non-trivial part with eigenvalues $E_{k,[\gamma]} = \Lambda_k(\nu[\gamma]) + \nu([\gamma])\cos\theta$ and eigenfunctions the solutions of (22).

with $k = 0, 1, 2, \ldots$ and R the radius of the universal covering space $T^2 \times S^1$. Thus we must compute the first eigenvalues and eigenfunctions. We shall use the most interesting case which is when $Vol(M_A^3)\sin\theta$ is large (or equivalently $\nu \to 0$). This term contains the compactification radius of the extra dimensions and the deformation of the lattice in terms of θ. When $Vol(M_A^3)\sin\theta$ becomes large, $\nu(\gamma)$ becomes small [9]. The eigenvalues of the Laplace-Beltrami operator are $E_k = \Lambda_k$, with Λ_k the eigenvalues of the modified Mathieu operator,

$$M = -\frac{d^2}{dz^2} + |\nu(\gamma)|\cosh 2\mu z \tag{25}$$

Let the order of the eigenvalues be $E_0 = 0 \leq E_1 \leq E_2 \leq ....$ We shall use the first two, since the corrections to the gravitational potential fall exponentially as the eigenvalues grow larger. Thus the first two are $E_0 = 0$ and E_1, which asymptotically read:

$$E_1 \sim \frac{(\ln\lambda)^2\pi^2}{(\ln C)^2} \tag{26}$$

with C:

$$C = \frac{1}{\sqrt{D}\mathrm{Vol}(M_A^3)\sin\theta} \tag{27}$$

The eigenfunctions corresponding to the Mathieu operator (25) are,

$$f_m(z) = \sum_{k=0}^{\infty} A_{2k}^{2m}\cosh[2kz] \tag{28}$$

with "m" counting the eigenvalues, $m = 0$ corresponds to E_0 e.t.c.

We substitute E_0 and E_1 in relation (19) with $|\mu_0| = \sqrt{E_0}$ and $|\mu_1| = \sqrt{E_1}$. Also we substitute the eigenfunctions of the Sol manifold, thus:

$$\Psi_m(0) = \sum_{k=0}^{\infty} A_{2k}^{2m} \tag{29}$$

The asymptotic behavior of the coefficients A_{2k}^{2m} for $\nu \to 0$ is really simple [?]. The terms of the form $A_m^{(m)}$ tend to 1, while terms of the form $A_m^{(k)}$, with $m \neq k$ tend to zero and the gravitational potential of relation (19) becomes:

$$V_4 = -\frac{G_4 M}{r}\sum_m \sum_{k=0}^{\infty} A_{2k}^{2m} A_{2k}^{2m} e^{-|\mu_m| r} \tag{30}$$

For the first two eigenvalues we have

$$V_4 = -\frac{G_4 M}{r}\left(A_0^0 A_0^0 e^{-|E_0|r} + A_2^2 A_2^2 e^{-|E_1|r}\right) \tag{31}$$

or (using the asymptotic behavior for the coefficients $A_m^{(m)}$):

$$V_4 = -\frac{G_4 M}{r}\left(e^{-\sqrt{E_0}r} + e^{-\sqrt{E_1}r}\right) \tag{32}$$

Since $E_0 = 0$ and using relations (26) and (27) we obtain finally:

$$V_4 \sim -\frac{G_4 M}{r}\left(1 + e^{-\left|\frac{(\ln\lambda)\pi}{(\ln C)}\right|r}\right) \tag{33}$$

It is clear that the Sol manifold correction to the Newton's law gravitational potential depends on the compactification radii of the extra dimensions R, the angle θ of the vectors e_u and e_v and on the eigenvalues of the hyperbolic gluing map A. In the next section we study the parameter space of the corrections found and we compare it with the T^3 manifold corrections.

4. Analysis of the parameter space and comparison with the T^3 corrections

We examine first the dependence of the Sol correction range $e^{-\mu_m r}$ on the parameters θ and n. In figure (1) we plot the dependence for the compactification radius value $R = 0.05\,mm$ (left) (smaller than the current experimental bound $r = 0.2mm$) and for $r = 0.1\,mm$, the values of R are $0.1\,mm$ (right) and $0.2\,mm$ (bottom) respectively. The Sol structure gives very

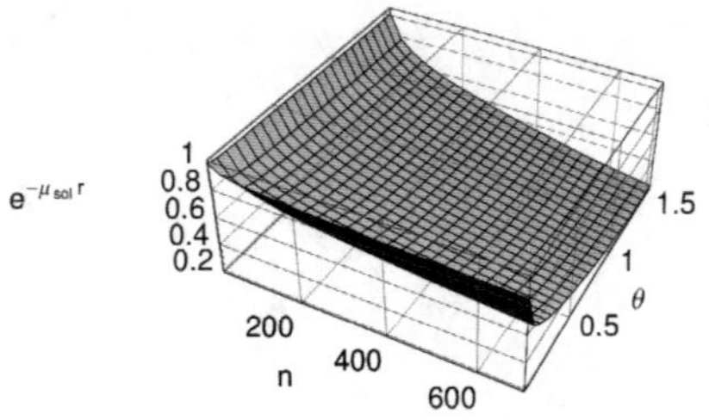

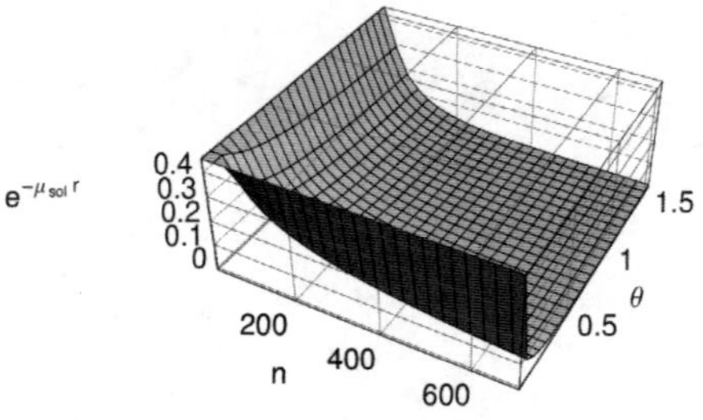

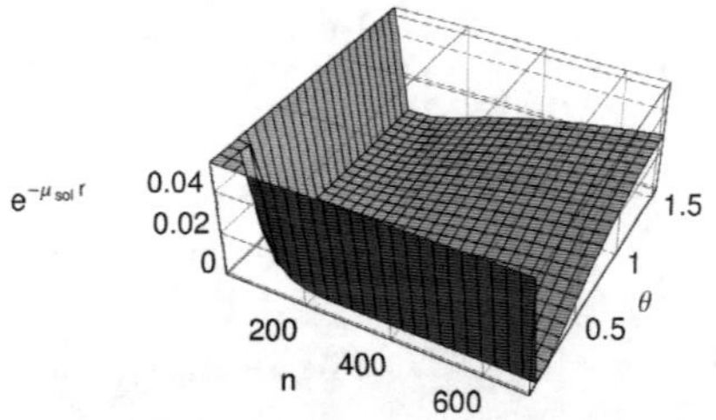

Figure 1. Dependence of the range $e^{-\mu_m r}$ of Sol corrections on θ and n, with $r = 0.1\,mm$, $R = 0.05\,mm$ (left), with $r = 0.1\,mm$, $R = 0.1\,mm$ (right), with $r = 0.1\,mm$, $R = 0.2\,mm$ (bottom).

large corrections to gravity if the compactification radius is very small (of order $R \sim 10^{-6}\,m$ and smaller). Also we shall check out the behavior of the range of the corrections around $R \sim 10^{-8}m$

which is the expected scale that three extra dimensional spaces should have. According to figure (1), we can see that the right and bottom figures are similar. Thus for large values of n (> 200) and for $\theta > 0.3$, the corrections are very small. Let us compare the range of Sol manifolds with the range of the T^3 torus corrections.

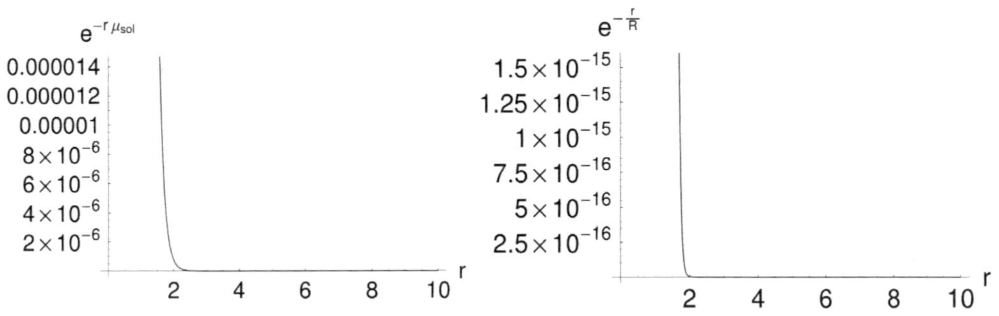

Figure 2. r-dependence of the range $e^{-\mu_m r}$ of Sol corrections with $R = 0.05\,mm$, $n = 250$ and $\theta = \frac{\pi}{3}$(left), r-dependence of the range of T^3 corrections for $R = 0.05\,mm$ (right)

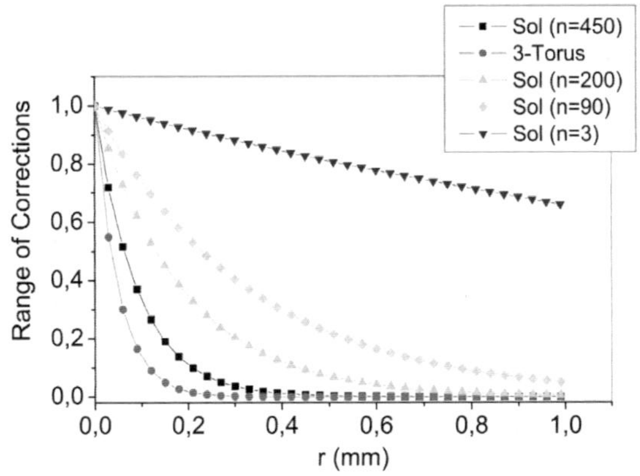

Figure 3. Comparison of the r-dependence of the range of T^3 and Sol corrections for $R = 0.2$ mm for various n, and $\theta = \pi/3$.

In figure (2) we plot the range of the T^3 corrections (left) and the corresponding dependence for the Sol manifold case (right) as a function of r with $R = 0.05mm$ and $\theta = \pi/3$ (left). As it can be seen from the figure (2), the term $e^{-\mu_m r}$ is very big compared to $e^{-r/R}$, for small n (~ 5). As n grows the term becomes smaller and smaller. As n grows, the range of Sol-corrections becomes comparable and after a value of n, smaller compared to the T^3 corrections.

The case with $R = 0.2\,$mm is very interesting. According to figure (3) the range of Sol corrections can vary significantly depending on the value that n takes. Compared to the corresponding 3-Torus range, Sol corrections can be much larger (for small n) and comparable (for $n > 450$) and even smaller.

Also Sol corrections can exist for compactification radii and at distances for which the corresponding T^3 torus range values are very large (and consequently not experimentally preferable).

5. Conclusions

We studied the corrections to the gravitational potential caused by a Sol manifold extra dimensional compact space. After investigating the parameter space we found that the range of Sol manifolds corrections can be similar to the T^3 torus results and can be very different compared to the T^3 results depending on the values of the parameters.

We left unanswered two issues, the graviton production and how do we distinguish Sol manifolds corrections from other corrections (the last due to the rich parameter space of Sol manifolds).

Acknowledgments

Acknowledgments V.O would like to thank Professor G. K. Leontaris for the hospitality at the University of Ioannina.

References

[1] Antoniadis I, Arkani-Hamed N, Dimopoulos S and Dvali G R 1998 *Phys. Lett.* B **436** 257
[2] A. Kehagias and K. Sfetsos 2000 *Phys. Lett.* B **472** 39
[3] Kehagias A and Russo J G 2000 *J. High Energy Phys.* **JHEP0007**(2000)027
[4] Floratos E G and Leontaris G K 1999 *Phys. Lett.* B **465** 95
[5] Floratos E G, Leontaris G K and Vlachos N D 2011 *Phys.Lett.* B **694** 410
[6] Thurston W P 1997 *Three-Dimensional Geometry and Topology* Vol.1, ed Silvio Levy (New Jersey: Princeton University Press) pp 109-208
Kodama H 1998 *Prog. Theor. Phys.* **99** 173
MacLachlan N W 1964 *Theory and Applications of Mathieu Functions*, (New York: Dover) pp 10-83
Hoyle C D,Kapner D J, Heckel B R, Adelberger E G, Gundlach J H, Schmidt U and Swanson H E 2004 *Phys. Rev.* D **70** 042004
Cullen S, Grojean C, Pilo L and Terning J, 1999 *Phys. Rev. Lett.* **83** 268
Kribs Graham D, Tasi Lectures on the Phenomenology of Extra Dimensions *Preprint* hep-ph/0605325
Cornish N J and Spergel D N 1999 *Preprint* math.DG/9906017
[7] Tanimoto M 2001 *Class. Quant. Grav.* **18** 479
[8] Oikonomou V K 2008 *Class. Quant. Grav.* **25** 195020
[9] Bolsinov A V, Dullin H R and Veselov A P 2006 *Commun. Math. Phys.* **264** 583

Tensor Perturbations in Starobinski's Inflation

D B Papadopoulos

Section of Astrophysics, Astronomy and Mechanics, Department of Physics
Aristotle University of Thessaloniki, Thessaloniki 54124, Greece

E-mail: papadop@astro.auth.gr

Abstract. In an almost de Sitter space-time, the stochastic semiclassical Einstein-Langevin equations with cosmological constant non-zero Λ have been written in the TT-gauge in a perturbative way (e.g. as zero-order and first order equations). Applying order reduction to zero and first order equations we found approximate solutions to those equations. To understand how large are the physical perturbations on different time scales, a two point correlation function of the intrinsic and induced fluctuations have been computed explicitly and their spectrum as well.

1. Introduction

A mathematically consistent and fairly well understood theory of gravity is the semiclassical one, which is based on the semiclassical Einstein equations.

The semiclassical Einstein equations are a generalization to the Einstein equations for the classical metric when the expectation value of the stress-momentum tensor of the quantum fields are source of curvature. Nevertheless, it has pointed out that this semiclassical theory may not be valid when the matter fields have important quantum stress-energy fluctuations. Because, in that case, the stress-energy fluctuations may have back-reaction effects on the space-time in the form of induced gravitational perturbations. Several authors have discussed this topic in cosmology and flat space-time and the necessity to extend the semiclassical theory to the effects of such fluctuations became obvious.

To address this problem Hu [1] proposed a theory, in the context of semiclassical cosmology, which leads to the influence functional formalism of Feynman and Vernon [2]. The theory is based on the observation that the semiclassical equations can be derived directly from the effective action of Feyman and Vernon [3], [4], [5].

An alternative approach have been introduced by Verdaguer [4]. He proposed a stochastic semi-classical theory of gravity as perturbative generalization of the semiclassical theory of gravity to describe the back reaction of the stress-tensor fluctuations. The resulting equations are the semiclassical Einstein-Lagevin equations which incorporate in a consistent way the stochastic tensor field as the source of linear perturbations to the semiclassical metric.

The theory has interesting applications in black hole physics and in cosmology, particularly in view of the problem of structure formation. Some examples have already been done [6].

The paper is organized as follows: In Sec.2., we present some aspects of the stochastic semi-classical gravity and write the Einstein-Langevin equations with non-zero cosmological constant in a perturbative way. In Sec.3., applying the so called order reduction to our higher than two derivatives Einstein-Langevin equations, we found approximately the scale factor $a(\tau)$ as

Journal of Physics: Conference Series **283** (2011) 012027 doi:10.1088/1742-6596/283/1/012027

solution to the zero-order equation and, later on, an approximate solution to the first-order inhomogeneous Einstein-Langevin equations with the aid of a retarded Green function adapting the so called Bunch-Davis initial conditions. A two point correlation function of the intrinsic and induced fluctuations have been computed explicitly and their spectrum as well. The conclusions are presented in Sect.4.

2. Gravitational waves in an almost de Sitter Universe-The equations

We assume that our universe is described by a homogeneous FRW space-time which is spatially flat with small metric perturbations

$$ds^2 \equiv \tilde{g}_{\mu\nu} dx^\mu dx^\nu \equiv a(\eta)^2 [\eta_{\mu\nu} + h_{\mu\nu}(x)] dx^\mu dx^\nu \tag{1}$$

where $\eta_{\mu\nu} = diag\{-1, 1, 1, 1\}$, $h_{\mu\nu}$ is a symmetric tensor representing small perturbations to the background flat metric, $a(\eta) = e^{\omega(\eta)}$ is the conformal scale factor and $\tilde{g}_{\mu\nu} = a(\eta)^2 [\eta_{\mu\nu} + h_{\mu\nu}(x)]$, is the physical metric. Notice that in general it is assumed that $\omega(x)$ is a scalar function independent of the metric and depends on the space time point only. In case of the spatially flat FRW it will be a function of the cosmological time t or conformal time η which is given by the expression $d\eta = dt/a(t)$.

We are interested in deriving the semiclassical corrections to Einstein's equations due to the quantum effect of the scalar field keeping the gravitational field classical. This may be achieved using the semiclassical Einstein's equations with cosmological constant Λ, since it is known [4] that quantum effects may be produced by the breaking of conformal flatness due to the coupling of the quantum field with the gravitational perturbations. These are:

$$\tilde{G}_{\mu\nu} \equiv \tilde{R}_{\mu\nu} - \frac{1}{2}\tilde{g}_{\mu\nu}\tilde{R} + \Lambda\tilde{g}_{\mu\nu} = 8\pi G_N < \tilde{T}_{\mu\nu} > \tag{2}$$

where tilde quantities refer to the physical metric $\tilde{g}_{\mu\nu}$,$\tilde{G}_{\mu\nu}$ is the Einstein's tensor, $\tilde{R}_{\mu\nu}$ and $\tilde{R}$ are the Ricci tensor and the Ricci scalar for the physical metric, respectively and $< \tilde{T}^{\mu\nu} >$ is the expectation value of the energy-momentum tensor. The $< \tilde{T}^{\mu\nu} >$, is written as $< \tilde{T}^{\mu\nu} > = < \tilde{T}^{\mu\nu}_{(0)} > + < \tilde{T}^{\mu\nu}_{(1)} >$ where $< \tilde{T}^{\mu\nu}_{(0)} >$ represents the zero order vacuum correction to the classical stress-tensor due to the spatially flat background and $< \tilde{T}^{\mu\nu}_{(1)} >$ is the first order vacuum expectation value of the stress-tensor. Subscripts (0) and (1) refer, respectively, to background and linear order metric perturbations.

Having the metric in the form (1), we modify the semiclassical Einstein equations (2) and write them in perturbative way. In this case the simplest equation to the first-order is derived by adding to the right of the Eq.(2) the stochastic tensor which characterizes the quantum fluctuations of the stress tensor.

The obtained stochastic equation up to the first order in $h_{\mu\nu}$, known as Einstein-Langevin equation, is

$$\{\tilde{G}^{(1)}_{\mu\nu}[g+h] + \Lambda e^{-2\omega} h_{\mu\nu}\} - 8\pi G_N < \tilde{T}^{(1)}_{\mu\nu}[g+h] >_{ren} = 8\pi G_N \xi_{\mu\nu}[g] \tag{3}$$

where the subscript "ren" means renormalized, the field $\xi_{\mu\nu}[g]$ is a Gaussian stochastic source defined by the following correlators (for details see Ref. [5],Sec.IV and references therein)

$$< \xi_{\mu\nu}(x) >_\xi = 0, \quad < \xi_{\mu\nu}(x)\xi_{\alpha\beta}(y) >_\xi = N_{\mu\nu\alpha\beta}(x, y) \tag{4}$$

$<>_\xi$ means statistical average [1], $N_{\mu\nu\alpha\beta}(x, y)$ is the noise kernel which describes the quantum fluctuations of the stress tensor operator

$$8N_{abcd}(x, y) \equiv < \{\hat{t}_{ab}(x), \hat{t}_{cd}(y)\} > [g](\Omega) \tag{5}$$

[1] In general, the two point correlation function of a stochastic tensor field ξ_{ab} must be a symmetric,in the sense $< \xi_{ab}(x)\xi_{cd} >_c = < \xi_{cd}(x)\xi_{ab} >_c$, and positive semi-definite real bi-tensor field

Journal of Physics: Conference Series **283** (2011) 012027 doi:10.1088/1742-6596/283/1/012027

$\{,\}$ means anticommutator and $\hat{t}_{ab}(x;\Omega) \equiv \hat{T}^R_{ab}(x;\Omega) - < \hat{T}^R_{ab}(x;\Omega) >$ and $\hat{T}^R_{ab}(x;\Omega)$ is renormalized and regularized operator depending on the regulator Ω.[2]

The Einstein-Langevin equation Eq.(3), which are gauge invariant, must be thought as a linear equation for the metric perturbations $h_{\mu\nu}$ which will behave as a stochastic field tensor.

Thus the homogeneous Eq.(3) give semiclassical equations [8]- [15]

$$\begin{aligned} &e^{6\omega}\{-[\tilde{G}^{\mu\nu}_{(0)} + \tilde{G}^{\mu\nu}_{(1)} + \Lambda e^{-2\omega}(\eta^{\mu\nu} - h^{\mu\nu})] - \frac{8\pi G_N k_1}{6}(\tilde{B}^{\mu\nu}_{(0)} + \tilde{B}^{\mu\nu}_{(1)}) \\ &+8\pi G_N k_3(\tilde{H}^{\mu\nu}_{(0)} + \tilde{H}^{\mu\nu}_{(1)})\} - 2(8\pi G_N k_3)\tilde{R}^{(0)}_{\alpha\beta}C^{\mu\alpha\nu\beta}_{(1)} \\ &+(8\pi G_N k_2)[-4(\omega C^{\mu\alpha\nu\beta}_{(1)})_{;\alpha\beta} + \int d^4y A^{\mu\nu}_{(1)}(y)H(x-y;\bar{\mu})] = 0 \end{aligned} \tag{6}$$

where k_1 is a numerical coefficient which has to be determined by experiment,k_2, k_3 are also numerical coefficients given by the relations [7]

$$k_3 = \frac{1}{2880\pi^2}(N_0 + 11N_{1/2} + 62N_1), \quad k_2 = \frac{3}{2880\pi^2}(N_0 + 6N_{1/2} + 12N_1) \tag{7}$$

with $N_0, N_{1/2}$ and N_1 quantum numbers with spins $0, 1/2$ and 1, respectively. Also, in Eq.(6) [8]-[16],

$$H(x-y;\bar{\mu}) = -\frac{1}{2}\int \frac{d^4p}{(2\pi)^4}e^{ip.(x-y)}[\ln\frac{|p^2|}{\bar{\mu}^2} + i\pi\theta(-p^2)sgn(-p^0)] \tag{8}$$

$G^{\mu\nu}(x)$ is the Einstein's tensor, $C^{\mu\alpha\nu\beta}(x)$ Weyl's tensor [17], $B^{\mu\nu}(x)$ and $A^{\mu\nu}(x)$ are the space time tensors given by the variation of $\int d^4xR^2(x)$ and $\int C^{\mu\alpha\nu\beta}(x)C_{\mu\alpha\nu\beta}(x)$ respectively with respect to the metric tensor $g_{\mu\nu}(x)$ and are expressed in terms of the metric tensor and its derivatives higher than two, while [4]

$$H^{\mu\nu}(x) = -R^{\mu\alpha}R^{\nu}_{\alpha} + \frac{2}{3}RR^{\mu\nu} + \frac{1}{2}g^{\mu\nu}R_{\alpha\beta}R^{\alpha\beta} - \frac{1}{4}g^{\mu\nu}R^2 \tag{9}$$

Their explicit expression are decomposed in the fashion $B^{\mu\nu}(x) \equiv B^{\mu\nu}_{(0)} + B^{\mu\nu}_{(1)}$, $A^{\mu\nu}(x) \equiv A^{\mu\nu}_{(0)} + A^{\mu\nu}_{(1)}$ and $H^{\mu\nu}(x) \equiv H^{\mu\nu}_{(0)} + H^{\mu\nu}_{(1)}$, where the bracket sub-indices are referred to zero (0) and (1) order perturbations.

We will focus only in perturbations up to the first order and we will study the correlation of the small perturbation $h_{\mu\nu}$ in an almost de Sitter universe (1). For our convenience we introduce the notations

$$H_0^2 = \frac{1}{8\pi G_N k_3}, \quad M^2 = \frac{1}{8\pi G_N k_1}, \quad H_1^2 = \frac{1}{8\pi G_N k_2} \tag{10}$$

Thus, starting with Eqs.(2) which do not include the stochastic source and adapting the so called TT-gauge which means that $h^{0\alpha} = 0$, the trace $h = h^{\alpha}_{\alpha}0$ and $h^{\mu\nu}_{,\nu} = 0$, we obtain [18]- [19]:

I. Zero-order equations:

$$\tilde{G}^{\mu\nu}_{(0)} + \Lambda e^{-2\omega}\eta^{\mu\nu} = 8\pi G_N < \tilde{T}^{\mu\nu}_{(0)} > \equiv \frac{1}{H_0^2}\tilde{H}^{\mu\nu}_{(0)} - \frac{1}{6M^2}\tilde{B}^{\mu\nu}_{(0)} \tag{11}$$

II. First-order equations:

$$\tilde{G}^{\mu\nu}_{(1)} - \Lambda e^{-2\omega}h^{\mu\nu}$$

[2] For further details see Ref. [10],Sec.II.

$$= 8\pi G_N < \tilde{T}^{\mu\nu}_{(1)} >\equiv \frac{1}{H_0^2}(\tilde{H}^{\mu\nu}_{(1)} - 2\tilde{R}^{(0)}_{\alpha\beta}\tilde{C}^{\mu\alpha\nu\beta}_{(1)}) - \frac{1}{6M^2}\tilde{B}^{\mu\nu}_{(1)}$$
$$+\frac{1}{H_1^2}e^{-6\omega}[-4(\omega C^{\mu\alpha\nu\beta}_{(1)})_{;\alpha\beta} + \int d^4y A^{\mu\nu}_{(1)}(y)H(x-y;\bar{\mu})] \tag{12}$$

Taking into account the cosmological constant Λ and introducing the Hubble function $H(t) = \dot{a}/a$ (dot stands for derivative in terms of the coordinate time t), Eq.(11) gives Friedman's equation

$$H^2 - \frac{\Lambda}{3} = \frac{H^4}{H_0^2} - \frac{1}{M^2}(2H\ddot{H} + 6H^2\dot{H} - \dot{H}^2) \tag{13}$$

Furthermore, in the TT-gauge, the first-order equations (12), become

$$-\frac{1}{2}\eta^{\sigma\tau}h^{\mu\nu}_{,\sigma\tau} + h^{\mu\nu}_{,0}\omega_{,0} + h^{\mu\nu}(2\omega_{,00} + \omega_{,0}) - \Lambda e^{2\omega}h^{\mu\nu}$$
$$= \frac{1}{H_0^2}e^{-2\omega}\{-\eta^{\sigma\tau}h^{\mu\nu}_{,\sigma\tau}(2\omega_{,00} - \omega^2_{,0}) + 2h^{\mu\nu}_{,0}\omega_{,0}\omega_{,00}$$
$$+h^{\mu\nu}\omega^2_{,0}(4\omega_{,00} - \omega^2_{,0}) - 2h^{\mu\nu}_{,00}(\omega_{,00} - \omega^2_{,0})]\}$$
$$-\frac{1}{M^2}e^{-2\omega}\{\eta^{\sigma\tau}h^{\mu\nu}_{,\sigma\tau}(\omega_{,00} + \omega^2_{,0}) - h^{\mu\nu}_{,0}(\omega_{,000} + 2\omega_{,00}\omega_{,0})$$
$$-\ h^{\mu\nu}[2\omega_{,0000} - 2\omega_{,000}\omega_{,0} - 12\omega^2_{,0}\omega_{,00} + \omega^2_{,00} + 3\omega^4_{,0})]\}$$
$$+\frac{1}{H_1^2}e^{-2\omega}\{[-2\omega_{,0}\eta^{\sigma\tau}h^{\mu\nu}_{,0\sigma\tau} + 2(h^{\mu\nu}_{,00} + \frac{1}{2}\eta^{\sigma\tau}h^{\mu\nu}_{,\sigma\tau})\omega_{,00}]$$
$$+\int d^4y A^{\mu\nu}_{(1)}(y)H(x-y;\bar{\mu})\} \tag{14}$$

Decomposing the perturbation $h_{\mu\nu}$ into modes of the form $h_{\mu\nu}(\eta,\vec{x}) = \frac{1}{(2\pi)^{3/2}}\int_{-\infty}^{\infty} d\vec{k} e_{\mu\nu}(\vec{k})g_k(\eta)e^{i\vec{k}\cdot\vec{x}}$, Eq.(14) simplifies considerably and the functions $g_k(\eta)$ satisfy the equation [15], [20] [3]

$$\frac{1}{(2\pi)^{3/2}}\int_{-\infty}^{\infty} d^3\vec{k}e^{i\vec{k}\cdot\vec{x}}e^{\mu\nu}(\vec{k})\{g_k''[1 + \frac{1}{3M^2}R + \frac{1}{3H_0^2}(2R^0_0 - R)]$$
$$+g_k'[\frac{2a'}{a}(1 + \frac{1}{3M^2}R - 2\frac{1}{3H_0^2}R^0_0) + \frac{1}{3M^2}R']$$
$$+g_k k^2[1 + \frac{1}{3M^2}R + \frac{1}{3H_0^2}(R - 6R^0_0)]$$
$$-2\frac{1}{H_1^2a^2}[\frac{2a'}{a}(g_k''' + k^2g_k') + (g_k'' - k^2g_k)(\frac{a'}{a})']\}$$
$$= \frac{2}{a^2H_1^2}\int d^4y A^{\mu\nu}_{(1)}(y)H(x-y;\bar{\mu}) \tag{15}$$

where prime stands for time derivatives in terms of the conformal time η, $R^0_0 = \frac{3}{a^4}(aa'' - a'^2)$ is the mixed 00−component of the Ricci tensor, $R = \frac{6}{a^2}\frac{a''}{a}$ is the scalar curvature, $R' = \frac{6a'''}{a^3} - \frac{18a''a'}{a^4}$, $g_k(\eta)$ is the amplitude of the gravitational wave (GW) and the last term, in the right hand side

[3] Note that in Eq.(14) the term with the cosmological constant Λ does not exist because we have applied the Eq.(11) into Eq.(14).

of the Eq.(15) is the so called non-local term with

$$H(x-y;\bar{\mu}) = \frac{1}{960\pi^2}\{\mathcal{P}f(\frac{1}{\pi}\theta(x^0-y^0)\frac{d}{d(x-y)^2}\delta((x-y)^2)$$
$$+(1-\gamma-\ln\bar{\mu})\delta^4(x-y)\}$$
$$= \frac{1}{960\pi^2}\lim_{\epsilon\to 0^+}\{\frac{1}{\pi}\theta(x^0-y^0)\theta(|\vec{x}-\vec{y}|-\epsilon)\frac{d}{d((x-y)^2)}\delta((x-y)^2)$$
$$+[1-\gamma-\ln(\epsilon\bar{\mu})]\delta^4(x-y)\} \tag{16}$$

$\bar{\mu}$ is a renormalization parameter and $\mathcal{P}f$ denotes the Hadamard finite part, which gives well defined expressions in the sense of distributions [21], [22].

At this stage we have to point out that in Eq.(15), if the right-hand side does not exist and the coefficient $\frac{1}{3M^2}$ is replaced by $-\frac{1}{3M^2}$, we obtain Vilenkin's equation corrected by a higher-order term proportional to $(1/H_1^2)$. In Eq.(15), if the terms proportional to $1/H_1^2)$ do not exist and we replace $\frac{1}{3M^2}$ with $-\frac{1}{3M^2}$, we recover the equation presented in Ref. [20].

3. The homogeneous Einstein-Langevin equations with $\Lambda \neq 0$

We start with the zero-order equation (13). Since $a = e^{\omega(\eta)}$, $d\eta = dt/a(t)$ and $H = \frac{\dot{a}}{a}$(dot stands for derivatives in terms of the coordinate time t) we find $\omega_{,0} = \frac{a'}{a} = Ha$ and Eq.(13) becomes

$$\omega_{,0}^2 - \frac{\Lambda}{3}e^{2\omega} = \frac{1}{H_0^2 e^{2\omega}}\omega_{,0}^4 + \frac{1}{M^2 e^{2\omega}}(\omega_{,00}^2 + 3\omega_{,0}^4 - 2\omega_{,0}\omega_{,000}) \tag{17}$$

If we consider that the background is driven by the cosmological constant $\Lambda \neq 0$ and the terms proportional to $1/H_0^2$ and $1/M^2$ which include higher derivatives, perturbations to the background, we may apply order reduction as follows:

We consider the equation

$$\omega_{,0}^2 - \frac{\Lambda}{3}e^{2\omega} = O(1/H_0^2, 1/M^2) \tag{18}$$

From the above equation we find that

$$\omega_{,0} = \pm\sqrt{\frac{\Lambda}{3}}e^{\omega} \tag{19}$$

Taking the second and third derivative of Eq.(19) and substituting them into Eq.(17) we find

$$\omega_{,0}^2 = \frac{\Lambda}{3}e^{2\omega} + \frac{1}{H_0^2}(\frac{\Lambda}{3})^2 e^{2\omega} \tag{20}$$

In the last equation setting $\omega_{,0} = \frac{a'}{a} = Ha$ we find

$$H = \frac{\dot{a}}{a} = \pm\sqrt{\frac{\Lambda}{3}}[1 + \frac{1}{H_0^2}(\frac{\Lambda}{3})]^{1/2} \tag{21}$$

which leads to the solution $a(t) = a_0 e^{Ht}$ if we assume that at $t = 0$ the universe has size $a(t=0) = a_0$. Eventually, the scale factor in terms of the conformal time reads

$$a(\eta) = -\frac{1}{H\eta}, \quad \forall\eta \in (-\infty, 0] \tag{22}$$

Since we have the explicit expressions for the scale factor, we will compute from Eq.(15) the function $g_k(\eta)$.

Thus we have

$$\frac{1}{(2\pi)^{3/2}}\int_{-\infty}^{\infty} d^3\vec{k}e^{i\vec{k}\cdot\vec{x}}e^{\mu\nu}(\vec{k})\{g_k''[1+\frac{1}{3M^2}R+\frac{1}{3H_0^2}(2R_0^0-R)]$$
$$+g_k'[\frac{2a'}{a}(1+\frac{1}{3M^2}R-2\frac{1}{3H_0^2}R_0^0)+\frac{1}{3M^2}R']$$
$$+g_k k^2[1+\frac{1}{3M^2}R+\frac{1}{3H_0^2}(R-6R_0^0)]$$
$$-\frac{2}{H_1^2a^2}[\frac{2a'}{a}(g_k'''+k^2g_k')+(g_k''-k^2g_k)(\frac{a'}{a})']\}$$
$$=\frac{2}{H_1^2a^2}\{\frac{1}{(2\pi)^{3/2}}\int_{-\infty}^{\infty} d^3\vec{k}e^{\mu\nu}(\vec{k})e^{i\vec{k}\cdot\vec{x}}N_k^{(nl)}\} \tag{23}$$

where

$$N_k^{(nl)} \equiv \{-\int_{-\infty}^{x^0} dy^0\frac{\sin k(x^0-y^0)}{k(x^0-y^0)}[g_k'''''(y^0)+2k^2g_k'''(y^0)+k^4g_k'(y^0))]$$
$$+[1-\gamma-\ln(\bar{\mu})][g_k''''+2k^2g_k''+k^4g_k]\} \tag{24}$$

To make further progress with Eq.(23) we will proceed as follows:

We will apply order reduction to the right-hand side of Eq.(23),e.g. to the non-local term, and to the left-hand side of the Eq.(23). Then, we will discuss the obtained results.

Eq.(23) has derivatives higher than two. This is well known in the semiclassical content and it is common in back reaction problems. The higher derivative terms are responsible for spurious solutions. However, approximate solution to the Eq.(23) may be found, applying the method of "order reduction" [9], [22], [23] of the differential equations(see Appendix A), since the terms in Eq.(23) multiplied by $\frac{1}{M^2},\frac{1}{H_0^2}$ and $\frac{1}{H_1^2}$ are corrections to the differential equation

$$g_k''+\frac{2a'}{a}g_k'+k^2g_k = O\left(\frac{1}{M^2},\frac{1}{H_0^2},\frac{1}{H_1^2}\right) \tag{25}$$

Taking higher order derivatives of Eq.(25) and applying them only in the non-local term in Eq.(23) we find that the non-local term vanishes e.g. the non-local term in a de Sitter space-time with the scale factor being $a=e^{\omega(\eta)}=-\frac{1}{H\eta}$ vanishes. Furthermore, applying once more order reduction to the left hand side of the Eq.(23) we find that the $g_k(\eta)$ satisfies the equation

$$g_k''-\frac{2}{\eta}(1-\nu_0)g_k'+k^2(1-2\nu_0)g_k=0 \tag{26}$$

where $\nu_0\equiv\frac{2H^2}{H_1^2}$ and prime stand for derivatives in terms of the conformal time η.

For further study of Eq.(26) we adapt the so called Bunch-Davies vacuum initial conditions, which are [24]- [27]

$$z_k(\eta_i)=\frac{1}{\sqrt{\omega_k}},\quad z_k(\eta_i)'=i\sqrt{\omega_k} \tag{27}$$

where $\eta=\eta_i$ is some initial time.

To compute the z_k and ω_k appearing in (27) we work as follows:

In Eq.(26) we set

$$g_k(\eta) = \eta^{(1-\nu_0)} z_k(\eta) \tag{28}$$

and find the equation

$$z_k''(\eta) + \omega_k^2 z_k(\eta) = 0 \tag{29}$$

where

$$\omega_k^2 = k^2(1-2\nu_0)[1 - \frac{(\nu_0-1)(\nu_0-2)}{(k\eta)^2(1-2\nu_0)}] \tag{30}$$

Eq.(29) admits the general solution of the form

$$z_k(\eta) = c_{1k} z_k^{(1)}(\eta) + c_{2k} z_k^{(2)}(\eta) \tag{31}$$

where

$$\begin{aligned} z_k^{(1)}(\eta) &= \sqrt{\eta} J_{-3/2+\nu_0}(\sqrt{(1-2\nu_0)}k\eta), \\ z_k^{(2)}(\eta) &= \sqrt{\eta} Y_{-3/2+\nu_0}(\sqrt{(1-2\nu_0)}k\eta) \end{aligned} \tag{32}$$

and $J_m(x)$ and $Y_m(x)$ are the Bessel functions of first and second order, respectively. We compute arbitrary constants c_{k1}, c_{k2} using the Bunch-Davies initial conditions in vacuum Eq.(31) and find [24]:

$$\begin{aligned} c_{1k} &= \frac{1}{W_i\sqrt{\omega_k}}[z_k^{(2)}(\eta_i)' - i\omega_k z_k^{(2)}(\eta_i)] \\ c_{2k} &= -\frac{1}{W_i\sqrt{\omega_k}}[z_k^{(1)}(\eta_i)' - i\omega_k z_k^{(1)}(\eta_i)] \end{aligned} \tag{33}$$

with W_i being the Wronskian of the solutions $z_k^{(1)}(\eta), z_k^{(2)}(\eta)$ evaluated at the initial time $\eta = \eta_i$.

Here we have to point out that $z_k(\eta, \eta_i)$ and its complex conjugate $z_k(\eta, \eta_i)^*$ are two linearly independent solutions of the Eq.(29) which are the same for all Fourier modes with the given magnitude of the wave vector $k = |\vec{k}|$. Their Wronskian $W[z_k, z_k^*]$ is different from zero but its derivative in terms of the conformal time η is zero which means that the Wronskian $W[z_k, z_k^*]$ is constant.

From Eqs.(31-33) we find the general solution of the Eq.(29) corresponding to the positive frequency k-mode to a graviton in the Bunch-Davies vacuum state in an almost de Sitter background with scale factor $a(\eta) = -\frac{1}{H\eta}$ which reads:

$$\begin{aligned} z_k(\eta, \eta_i) &= \frac{1}{W_i\sqrt{\omega_k}}\{[z_k^{(2)}(\eta_i)' z_k^{(1)}(\eta) - z_k^{(1)}(\eta_i)' z_k^{(2)}(\eta)] \\ &\quad -i\omega_k[z_k^{(2)}(\eta_i) z_k^{(1)}(\eta) - z_k^{(1)}(\eta_i) z_k^{(2)}(\eta)]\} \\ &= -\frac{1}{\sqrt{k}k\eta_i}(1-2\nu_0)^{-3/4}[1 - \frac{(1-\nu_0)(2-\nu_0)}{k^2\eta_i^2(1-2\nu_0)}]^{-1/4} \\ &\quad \times\{[\frac{1-\nu_0}{k\eta_i\sqrt{1-2\nu_0}} + i[1 - \frac{(1-\nu_0)(2-\nu_0)}{k^2\eta_i^2(1-2\nu_0)}]^{1/2}] \\ &\quad \times k\sqrt{1-2\nu_0}\sqrt{\eta\eta_i}[Y_{-3/2+\nu_0}(x_i)J_{-3/2+\nu_0}(x) - J_{-3/2+\nu_0}(x_i)Y_{-3/2+\nu_0}(x)] \\ &\quad +k\sqrt{1-2\nu_0}\sqrt{\eta\eta_i}[Y_{-1/2+\nu_0}(x_i)J_{-3/2+\nu_0}(x) - J_{-1/2+\nu_0}(x_i)Y_{-3/2+\nu_0}(x)]\} \\ &\quad \times[Y_{-1/2+\nu_0}(x_i)J_{-1/2+\nu_0}(x_i) - J_{-1/2+\nu_0}(x_i)Y_{-1/2+\nu_0}(x_i)]^{-1} \end{aligned} \tag{34}$$

where $x = k\eta\sqrt{1-2\nu_0}, x_i = k\eta_i\sqrt{1-2\nu_0}$, $\nu_0 = \frac{2H^2}{H_1^2}$ and $H = \sqrt{\Lambda/3}[1+\frac{\Lambda/3}{H_0^2}]^{1/2}$.

From Eq.(28) and (34) we obtain the solution of the Eq.(26).

For gravitational waves (GWs) with $k|\eta_i| \gg 1$, Eq.(34) gives:

$$z_k(\eta,\eta_i) = \frac{(1-2\nu_0)^{-1/4}}{\sqrt{k}}[1+\frac{i(1-\nu_0)(2-\nu_0)}{2k\eta}]e^{ik\sqrt{1-2\nu_0}(\eta-\eta_i)} \tag{35}$$

For $\nu_0 = 0$, Eq.(35) reduces to the equation:

$$z_k(\eta,\eta_i) = \frac{1}{\sqrt{k}}(1+\frac{i}{k\eta})e^{ik(\eta-\eta_i)} \tag{36}$$

which corresponds to Mukhanov's equation (8.122) [26].

3.1. A Green function to the inhomogeneous Einstein-Langevin equation with $\Lambda \neq 0$.

Our aim is to find the correlation function for the metric perturbations since they describe how large are the physical perturbations on different time scales. This can achieved using stochastic's gravity approaches which is different from the semiclassical gravity. Thus, starting with the Einstein-Langevin Eq.(3) we lead to an inhomogeneous Eq.(23).

In Eq.(23) we apply order reduction and considering the scale factor (22) we obtain an inhomogeneous Eq.(26) which includes a driving term due to the stochastic source which contributes to the metric.

Next, we will obtain the metric perturbations g_k in terms of the stochastic source.

To obtain the metric perturbations in terms of the stochastic source we need the retarded propagator for the g_k i.e. the required Green function to solve the inhomogeneous Eq.(26) with the appropriate boundary conditions. Thus, the problem now is to find a Green function to the inhomogeneous Eq.(26).

We consider the equation, [25]

$$G''_{(k)Ret}(\eta-\eta') - \frac{2}{\eta}(1-\nu_0)G'_{(k)Ret}(\eta-\eta') + k^2(1-2\nu_0)G_{(k)Ret}(\eta-\eta') = -\delta(\eta-\eta') \tag{37}$$

Since the Green function $G_{(k)Ret}(\eta-\eta')$ is proportional to the general solution of the Eq.(26), we find

$$G_{(k)Ret}(\eta,\eta') = -\left[\frac{a(\eta')}{a(\eta)}\right]^{(3/2-\nu_0)} \frac{\Theta(\eta-\eta')}{k\sqrt{1-2\nu_0}W(x')}$$
$$\times[J_{-3/2+\nu_0}(x)Y_{-3/2+\nu_0}(x') - Y_{-3/2+\nu_0}(x)J_{-3/2+\nu_0}(x')] \tag{38}$$

where $W(x') = Y_{-1/2+\nu_0}(x')J_{1/2+\nu_0}(x') - Y_{1/2+\nu_0}(x')J_{-1/2+\nu_0}(x')$ and $x' = k\eta'\sqrt{(1-2\nu_0)}$.

Eq.(38) is the retarded propagator appropriate to study correlation of perturbations using the Einstein-Langevin Eq.(3) in an expanding universe with scale factor (22) and non-zero cosmological constant. From Eq.(38) we have $G_{(k)Ret}(\eta,\eta') = 0$ for $\eta < \eta_i$ while for $\eta > \eta_i$ we have a non-zero $G_{(k)Ret}(\eta,\eta')$ due to the effect of the stochastic source after η_i.

Now we go back to the inhomogeneous Einstein-Langevin equation Eq.(3). Its final form in a perturbative way includes the zero-order Eq.(11) and the inhomogeneous first-order Eq.(12) which now reads [5]:

$$\tilde{G}^{\mu\nu}_{(1)} - \Lambda e^{-2\omega}h^{\mu\nu}$$
$$= 8\pi G_N < \tilde{T}^{\mu\nu}_{(1)} >\equiv \frac{1}{H_0^2}(\tilde{H}^{\mu\nu}_{(1)} - 2\tilde{R}^{(0)}_{\alpha\beta}\tilde{C}^{\mu\alpha\nu\beta}_{(1)}) - \frac{1}{6M^2}\tilde{B}^{\mu\nu}_{(1)}$$
$$+\frac{1}{H_1^2}e^{-6\omega}[-4(\omega C'^{\mu\alpha\nu\beta}_{(1)})_{;\alpha\beta} + \int d^4y A^{\mu\nu}_{(1)}(y)H(x-y;\bar{\mu})] + F^{\mu\nu}[\xi] \tag{39}$$

In Eq.(39), the tensor $F^{\mu\nu}(x)$ results from the functional variation of the stochastic term and its explicit expression is

$$F^{\mu\nu}(x) = -16\pi G_N e^{-6\omega}\partial_\alpha\partial_\beta\xi^{\mu\alpha\nu\beta}(x) \tag{40}$$

In Eq.(40), the tensor $F^{\mu\nu}(x)$ is symmetric and traceless, i.e. $F^{\mu\nu}(x) = F^{\nu\mu}(x)$ and $F^{\mu}_{\mu}(x) = 0$ which means that there is no stochastic correction to the trace anomaly. Also, in the same equation, the tensor $\xi^{\mu\alpha\nu\beta}$ plays the role of a stochastic source with the Gaussian probability given by Eq.(4.7) in Ref. [5]. This tensor has the symmetries of the Weyl tensor i.e. it has the symmetries of the Riemann tensor and vanishing trace in all its indices. This stochastic source is completely characterized by the relations

$$\begin{aligned}&< \xi_{\mu\nu\alpha\beta} >_\xi = 0,\\ &< \xi_{\mu\nu\alpha\beta}(x)\xi_{\rho\sigma\lambda\theta}(y) >_\xi = T_{\mu\nu\alpha\beta\rho\sigma\lambda\theta}N(x-y)\end{aligned} \tag{41}$$

where $N(x-y)$ is the noise kernel corresponding to the two-point correlation function of the stochastic source since the probability distribution is Gaussian and the tensor $T_{\mu\nu\alpha\beta\rho\sigma\lambda\theta}$ is the product of four metric tensors in such a combination that the right-hand side of the equation satisfies the Weyl symmetries of the two stochastic fields on the left-hand side. Its explicit expression is given in the Appendix of Ref. [5].

In Eq.(39), the stochastic term $F^{\mu\nu}$ will produce a stochastic contribution $h^{\mu\nu}_{(inh)}$ to the space time inhomogeneity

$$h^{\mu\nu}(\eta,\vec{x}) = h^{\mu\nu}_{(h)}(\eta,\vec{x}) + h^{\mu\nu}_{(inh)}(\eta,\vec{x}) \tag{42}$$

where $h^{\mu\nu}_{(h)}(\eta,\vec{x})$ is the solution to the homogeneous Einstein-Langevin Eq.(39) which is identical to the Eq.(6) and its explicit expression is

$$h^{\mu\nu}_{(h)}(\eta,\vec{x}) = \frac{1}{(2\pi)^{3/2}}\int_{-\infty}^{\infty} d^3\vec{k}e^{\mu\nu}(\vec{k})e^{i\vec{k}\cdot\vec{x}}\eta^{(1-\nu_0)}z_k(\eta) \tag{43}$$

with $z_k(\eta)$ given by Eq.(34).[4]

Furthermore, substituting Eq.(42) into eq.(39), taking into account that $h^{\mu\nu}_{(h)}(\eta,\vec{x})$ is the solution to the homogeneous Einstein-Langevin and repeating the computations of Sec.3., Eq.(39) becomes an inhomogeneous differential equation of the form

$$g''_{(inh)k} - \frac{2}{\eta}(1-\nu_0)g'_{(inh)k} + k^2(1-2\nu_0)g_{(inh)k} = 16\pi G_N F_k(\eta) \tag{44}$$

where we decomposed

$$\begin{aligned}h^{\mu\nu}_{(inh)}(\eta,\vec{x}) &= \frac{1}{(2\pi)^{3/2}}\int_{-\infty}^{\infty} d^3\vec{k}e^{\mu\nu}(\vec{k})e^{i\vec{k}\cdot\vec{x}}g_{(inh)k}(\eta),\\ F^{\mu\nu}(\eta,\vec{x}) &= \frac{1}{(2\pi)^{3/2}}\int_{-\infty}^{\infty} d^3\vec{k}e^{\mu\nu}(\vec{k})e^{i\vec{k}\cdot\vec{x}}F_k(\eta)\end{aligned} \tag{45}$$

and took into account the source term (40).

The left-hand side of Eq.(44) is of the same form as in Eq.(26) and because of its non-zero right-hand side, admits a Green function of the form (38) and thus the solution to the Eq.(44)

[4] To find the explicit expression of the $h^{\mu\nu}_{(inh)}(\eta,\vec{x})$, we assume that the $h^{\mu\nu}_{(inh)}(\eta,\vec{x})$ satisfy the TT-gauge and impose the boundary conditions $h^{\mu\nu}_{(inh)}(\eta=-\infty,\vec{x}) = 0$ and $\partial_\eta h^{\mu\nu}_{(inh)}(\eta=-\infty,\vec{x}) = 0$.

reads:[5]

$$h^{\mu\nu}_{(inh)}(\eta,\vec{x}) = \frac{1}{(2\pi)^{3/2}}\int_{-\infty}^{\infty} d^3\vec{k}e^{\mu\nu}(\vec{k})e^{i\vec{k}\cdot\vec{x}}g_{(inh)k}(\eta)$$
$$= \frac{1}{(2\pi)^{3/2}}\int_{-\infty}^{\infty} d^3\vec{k}e^{\mu\nu}(\vec{k})e^{i\vec{k}\cdot\vec{x}} \times \{\frac{16\pi G_N}{4\pi}\int^{\eta} G_{(ret)k}(\eta,\eta')F_k(\eta')d\eta'\} \tag{46}$$

To understand how large are the physical perturbations on different time scales, we will calculate the two-point correlation function of the metric perturbations which is given by

$$< h^{\mu\nu}(\eta_1;\vec{x})h^{\lambda\theta}(\eta_2;\vec{y}) > \; = \; < h^{\mu\nu}_{(h)}(\eta_1;\vec{x})h^{\lambda\theta}_{(h)}(\eta_2;\vec{y}) > + < h^{\mu\nu}_{(inh)}(\eta_1;\vec{x})h^{\lambda\theta}_{(inh)}(\eta_2;\vec{y}) > \tag{47}$$

In Eq.(47) there are two different contributions to the two-point correlation functions.

The first one,corresponding to the homogeneous Einstein-Langevin Eq.(26), is connected to the fluctuations of the initial state of the metric perturbations, we refer to them as intrinsic fluctuations. The second contribution,corresponding to the inhomogeneous Einstein-Langevin Eq.(3), is proportional to the noise kernel and thus connected with the fluctuations of the quantum fields, we will refer to them as induced fluctuations [28].

Now we will concentrate on the intrinsic fluctuations.

The intrinsic fluctuations are the expectation value of two-point correlations at equal times and are given by the relation [26]

$$< h^{\mu}_{(h)\nu}(\eta,\vec{x})h^{\nu}_{(h)\mu}(\eta,\vec{y}) >= \frac{8}{\pi^2 a^{2(1-\nu_0)}}\int_0^{\infty} k^3\frac{dk}{k}|z_k(\eta)|^2\frac{\sin kr}{kr} \tag{48}$$

where $r \equiv |\vec{x}-\vec{y}|$ and $z_k(\eta)$ is given by the Eq.(34).

However, the power spectrum of those fluctuations with comoving wave number k may be computed from the equation [24]

$$\delta^2(k,\eta) = \frac{8k^3}{\pi a^{2(1-\nu_0)}}|z_k(\eta)|^2 \tag{49}$$

where $z_k(\eta)$ is given by Eq.(34). For the Gws with $k|\eta_i| \gg 1$(e.g. in scales inside the horizon) we find:

$$\delta_h^2 = \frac{8H^2}{\pi}(1-2\nu_0)^{-1/2}k^2\eta^2[1+\frac{(1-\nu_0)^2(2-\nu_0)^2}{4k^2\eta^2}]\alpha^{2\nu_0} \tag{50}$$

where H is given by Eq.(21). Expressing Eq.(50) in terms of the physical wave number $k_{ph} = \frac{k}{a}$ we have

$$\delta_h^2 = \frac{8}{\pi}\frac{H^2}{\sqrt{(1-2\nu_0)}}[\frac{k_{ph}^2}{H^2}+\frac{(1-\nu_0)^2(2-\nu_0)^2}{4}]\alpha^{2\nu_0}$$
$$= \frac{8H^2}{\pi}\{1+\frac{k_{ph}^2}{H^2}+\nu_0[-2+\frac{k_{ph}^2}{H^2}+2\ln a(1+\frac{k_{ph}^2}{H^2})]+O(\nu_0^2)\} \tag{51}$$

where $\nu_0 = 2\frac{H^2}{H_1^2} \ll 1$, $\alpha = -\frac{1}{H\eta}$ and $H^2 = \frac{\Lambda}{3}[1+\frac{\Lambda}{3H_0^2}]$ (see Eq.(12)).

Setting $\nu_0 = 0$ in Eq.(51), we obtain Mukhanov's formula (8.123) in Ref. [24].

[5] Eq.(44) is linear in $g_{(inh)k}$ and resembles to the inhomogeneous equations (26). Thus its solution may be computed with the initial conditions (27).

Journal of Physics: Conference Series **283** (2011) 012027 doi:10.1088/1742-6596/283/1/012027

From Eq.(50) we compute the tensor spectra index([24])

$$\begin{aligned} n_T &\equiv \frac{d\ln\delta_h^2}{d\ln k} \\ &= [1+\frac{1}{k^2\eta^2}(1-3\nu_0+\frac{13}{4}\nu_0^2-\frac{3}{2}\nu_0^3+\frac{1}{4}\nu_0^4)]^{-1} \\ &\sim [1+\frac{H^2}{k_{ph}^2}(1-3\nu_0)]^{-1} \end{aligned} \tag{52}$$

To compute the two-point correlation function of the induced fluctuations at equal times we use Eq.(46) and find:

$$\begin{aligned} &< h^{\mu\nu}_{(inh)}(\eta,\vec{x})h^{\rho\lambda}_{(inh)}(\eta,\vec{y}) > \\ = &\frac{(8\pi G_N)^2}{(2\pi)^5} < \int_{-\infty}^{\infty} d^3\vec{k}e^{\mu\nu}(\vec{k})e^{i\vec{k}\cdot\vec{x}}\int^{\eta} G_{(k)Ret}(\eta,\eta')F_k(\eta')d\eta' \\ &\int_{-\infty}^{\infty} d^3\vec{k}'e^{\rho\lambda}(\vec{k}')e^{i\vec{k}'\cdot\vec{y}}\int^{\eta} G_{(k')Ret}(\eta,\eta'')F_{k'}(\eta'')d\eta'' > \\ = &\frac{(8\pi G_N)^2}{(2\pi)^8}\int_{-\infty}^{\infty} d^3\vec{k}d^3\vec{k}'e^{i\vec{k}\cdot\vec{x}}e^{i\vec{k}'\cdot\vec{y}}\int^{\eta} d\eta' d\eta''\{G_{(k)Ret}(\eta,\eta')G_{(k')Ret}(\eta,\eta'') \\ &\times\int d^3\vec{x}'d^3\vec{y}'e^{-i\vec{k}\cdot\vec{x}'}e^{-i\vec{k}'\cdot\vec{y}'} < F^{\mu\nu}(\eta',\vec{x}')F^{\rho\lambda}(\eta'',\vec{y}') >\} \end{aligned} \tag{53}$$

where we used the equation

$$e^{\mu\nu}(\vec{k})F_k(\eta') = \frac{1}{(2\pi)^{3/2}}\int d^3\vec{x}F^{\mu\nu}(\eta',\vec{x})e^{-i\vec{k}\cdot\vec{x}} \tag{54}$$

In Eq.(53),we perform some integrations and after some straightforward computations we find the two-point correlation function of the induced fluctuations at equal times

$$\begin{aligned} &< h^{\mu\nu}_{(inh)}(\eta,\vec{x})h^{\rho\lambda}_{(inh)}(\eta,\vec{y}) >= 6\alpha(8\pi G_N)^2\int_0^{\infty} k^3dk\frac{dk}{k}\frac{\sin(kr)}{kr}\{k_\alpha k_\beta k_\gamma k_\delta T^{\mu\alpha\nu\beta\rho\gamma\lambda\delta} \\ &\times\int^{\eta} d\eta' e^{-12\omega(\eta')}G_{(k)Ret}(\eta,\eta')G_{(-k)Ret}(\eta,\eta')\} \end{aligned} \tag{55}$$

4. Conclusions

In the context of stochastic semiclassical theory, we wrote the Einstein-Langevin equation with the cosmological constant Λ in an almost de Sitter universe which has been perturbed by the metric perturbations $h_{\mu\nu}$.

The equations have been written in a perturbative way, in the TT-gauge. Ignoring higher than the first order perturbation and keeping only first order corrections, we derived a zero order equation which gives the evolution of the scale factor and a first order equations which describes the evolution of the metric perturbation when $\Lambda \neq 0$.

In the case where $\Lambda \neq 0$, we have applied order reduction to the Eq.(13) and found that the zero order equation admits the solution $H = \sqrt{\Lambda/3}[1+\frac{1}{H_0^2}\frac{\Lambda}{3}]^{1/2}$ which shows that the Hubble parameter is shifted by the quantum terms (as it was expected) and the scale factor is given by the Eq.(22). The result corresponds to an expanding for ever universe.

Subsequently, we found explicit expression for the kernel $H(x-y;\bar{\mu})$ appearing on the right-hand side of the Eq.(15) and applying order reduction to the first order Eq.(23) we verified that, in a de Sitter space-time with a scale factor like Eq.(22), the non-local term vanishes and eventually the first order equation reduce to Eq.(26).

Eq.(26) is the homogeneous Einstein-Langevin equation with non-zero cosmological constant after order reduction. Its general solution has been computed uniquely in terms of Bessel functions of first and second kind adapting the Bunch-Davies initial conditions.

Since we are interested in finding the correlation function of the metric perturbations $h_{\mu\nu}$, we have considered the inhomogeneous Einstein-Langevin equation (3), which leads to the Eq.(44) e.g. the inhomogeneous Eq.(26), and computed its retarded Green function $G_{k(Ret)}(\eta, \eta')$(Eq.(38)).

As we have pointed out in the section 2, Eq.(3) is a stochastic equation which predicts that the gravitational field has stochastic fluctuations over the background space-time due to the stochastic term $F^{\mu\nu}$. The stochastic term, produces the stochastic contribution $h^{\mu\nu}_{(inh)}$ to the space time inhomogeneity in the fashion described by Eq.(42). Because of the form of Eq.(42), we found two different contributions to the two-point correlation function: One is due to the intrinsic fluctuations and is given by Eq.(48). For scales inside the horizon, the power spectrum of those fluctuations have been computed (see Eq.(50) and an approximate formula in terms of the physical wave number has been computed. The second contribution is due to the induced fluctuations, a closed form of which is given by Eq.(56). The explicit expression of this contribution is now under consideration.

Acknowledgments
This project is in collaboration with E. Verdaguer. The author expresses his gratitude to the head and staff of the Department of Fundamental Physics of the University of Barcelona, where part of the project was carried out, for their hospitality and also to the Dean of the Aristotle University of Thessaloniki for his support.

5. Appendix A

Order reduction: The method has been applied by several authors [22], [23] in theoretical physics.

The method has as follows. Modify the original equation so that the new one is second order and as accurate as the original one. Thus, if we have an equation of the form

$$\frac{d^n x}{dt^n} = P[\frac{d^k x}{dt^k}] + \tau Q[\frac{d^m x}{dt^m}] + O(\tau^2) \tag{56}$$

where $k < m, m > n$, differentiating Eq.(56) $m - n$ times, substituting $\frac{d^m}{dt^m}$ into Q to order zero, and repeating the process up until Q has lower derivative than n we obtain

$$\frac{d^n x}{dt^n} = P[\frac{d^k x}{dt^k}] + \tau Q[\frac{d^{(n-1)} x}{dt^{(n-1)}}] + O(\tau^2) \tag{57}$$

References

[1] Hu B L 1989 *Physica A* 1989 **158** 399
[2] Feynman R P and Vernon F L 1963 *Ann Phys* N Y **24** 118
[3] Calzetta E and Hu B L 1994 *Phys Rev D* **49**, 6636); Greiner C and Muller B 1997 *Phys Rev D* **55**, 1026
[4] Campos A and Verdaguer E 1994 *Phys Rev D* **49** 1861
[5] Campos A and Verdaguer E 1996 *Phys Rev D* **53** 1927 ; Campos A and Hu B L 1998 *Phys Rev D* **58** 125021

[6] Calzetta E Campos A and Verdaguer E 1997 *Phys Rev D* **56** 2163 ; Calzetta E and Verdaguer E 1998 *Phys Rev D* **59** 083513
[7] Hawking S W and Hertog T 2001 *Phys Rev D* **63** 083504-1
[8] Campos A Martin R and Verdaguer E 1995 *Phys Rev D* **52** 4319
[9] Campos A and Verdaguer E 1997 *Int Jour of Theor Phys* **36**, 2525 (1997)
[10] Martin R and Verdaguer E 1999 *Phys Rev D* **60** 084008
[11] Martin R and Verdaguer E 2000 *Phys Rev D* **61** 124024
[12] Perez-Nadal G Roura A and Verdaguer E 2008 *Class Quantum Grav* **25** 154013
[13] Roura A and Verdaguer E 2008 *Phys Rev D* **78** 064010
[14] Horowitz G T and Wald R M 1980 *Phys Rev D* **21** 1462
[15] Horowitz G T and Wald R M 1982 *Phys Rev D* **25** 3408
[16] Calzetta E and Hu B L 1987 *Phys Rev D* **35** 495
[17] Wald R M 1984 *General Relativity* (The University of Chicago Press Chicago and London)
[18] Starobinsky A A 1981 *Pis'ma Zh Eksp Teor Fiz* **34** 460 [1981 *JETP Lett* **34** 438]
[19] Starobisky A A 1980 *Phys Lett* **91B** 99
[20] Vilenkin A 1985 *Phys Rev D* **32** 2511
[21] Schwartz L 1966 *Theory des Distributions* (Paris:Hermann)
[22] Simon J Z 1991 *Phys Rev D* **43** 3308 ; Parker L and Simon J Z 1993 *Phys Rev D* **47** 1339
[23] Flanagan E E and Wald R M 1996 *Phys Rev D* **54** 6233
[24] Mukhanov V 2005 *Physical Foundations of COSMOLOGY* (Cambridge University Press)
[25] Weinberger H F 1986 *Equacionas diferencials en derivadas parciales* (Ed Reverte S A Barcelona Spain)
[26] Mukhanov V 2008 *INTRODUCTION TO QUANTUM EFFECTS IN GRAVITY* (Cambridge University Press)
[27] Olver F W J 1974 *Introduction to Asymptotics and Special Functions* pages 190-192 (Academic Press NY and London)
[28] Hu B L Roura A Verdaguer E 2004 *Phys Rev D* **70** 044002

Bulk Decay of $(4+n)$-Dimensional Simply Rotating Black Holes: Tensor-Type Gravitons

Nikolaos Pappas
Division of Theoretical Physics, Department of Physics,
University of Ioannina, Ioannina GR-45110, Greece

E-mail: npappas@cc.uoi.gr

Abstract. We study the emission in the bulk of tensor-type gravitons by a simply rotating $(4+n)$-dimensional black hole. The decoupling of the radial and angular part of the graviton field equation makes it possible to solve them analytically (in the limit of low-energy emitted particles and low-angular momentum of the black hole) and find the corresponding absorption probability. We also move to solve these equations numerically. The comparison between analytic and numerical results shows a very good agreement in low and intermediate energy regimes. Finally, the energy and angular momentum emission rates were calculated in order to explore their dependence on the number of additional spacelike dimensions of the spacetime background and the angular momentum of the black hole. Interesting conclusions about the significance of tensor-type gravitons as energy carriers in the context of Hawking radiation were reached.

1. Introduction

This work was motivated primarily by the exciting possibility that a mini-black hole could be created at the LHC should one of the theories, postulating the existence of additional spacelike dimensions in nature, is proven to be correct. All such scenaria, namely the Large Extra Dimensions [1] and Warped Extra Dimensions [2] ones, predict a new, lower than the four-dimensional, fundamental scale for gravity as low as one TeV. The most interesting phenomenological consequence of the new scale is the expectation that miniature black holes could be created during high-energy particle collisions at ground-based accelerators [3]. As these black holes would be higher-dimensional objects, their evaporation pattern by emission of Hawking radiation [4], will strongly depend both on the number and nature of the extra dimensions. Therefore the comparison between the theoretically expected behavior and the measurements performed by our detectors, will reveal valuable information about spacetime.

Because of this possibility, there has been a considerable amount of interest in the study of the radiation emission spectra from a higher-dimensional black hole in the literature in recent years (for some reviews, see [5, 6]). Rotating and non-rotating black holes have been exhaustively studied both analytically and numerically [7, 8, 9, 10, 11, 12, 13, 14, 15, 16, 17, 18, 19, 20, 21, 22, 23, 24, 25, 26, 27, 28, 29, 30, 31, 32]. These studies have shown that the brane channel is in most cases the dominant one, although at certain circumstances the bulk channel can be equally important at specific particle channels. Unfortunately, the question of the energy balance [33] between the 'bulk' and 'brane' channel during the Hawking radiation has not been answered yet.

Nevertheless, this balance is our principal guide to correctly evaluate signals correlated to the black hole evaporation should ever our detectors record any, in order to find decisive evidence about the existence of extra dimensions and their nature. For the aforementioned balance to be calculated one has to include the contribution of gravitons to the overall evaporation process.

Here we study the emission of Hawking radiation in the bulk in the form of tensor-type gravitational modes by a higher-dimensional black hole with at least three additional spacelike dimensions and one angular-momentum component using the perturbation equations derived in [34, 35]. After presenting the geometrical set-up and theoretical framework, we solve analytically the radial part of the graviton field equation (in the of low-energy emitted particles and low-angular momentum of the black hole limit). We repeat these calculations using numerical methods. Next, we perform a comparison of the analytical and numerical results to check the validity of the analytic, approximate method. The exact form of the energy and angular-momentum emission rates in the bulk from the simply rotating black hole in the form of tensor-type gravitons are also computed and studied. Finally, a comparison with the corresponding results for the emission of scalars is presented. We finish with our conclusions.

2. Theoretical Framework

We will focus on the case of a spacetime described by a line-element of the form

$$ds^2 = G_{MN}\, dz^M\, dz^N = g_{ab}\, dx^a\, dx^b + S^2(x)\, d\Omega_n^2, \tag{1}$$

where $\{a, b\} = (0, 1, 2, 3)$ and $d\Omega_n^2$ stands for the line-element of a n-dimensional unit sphere S^n. The above line-element is a special $(4+n)$-dimensional case of a more general class of gravitational backgrounds where the spacetime can be written as the warped product of a m-dimensional spacetime $\mathcal{N}$ and a n-dimensional space $\mathcal{K}$ of constant curvature [36, 37]. In these type of backgrounds it is possible to derive the perturbation equations for all types of gravitational modes in the case of a maximally-symmetric higher-dimensional black-hole background [38] where $m = 2$ and $\mathcal{K} = S^n$.

When the black hole rotates only in a single two-plane along the 4-dimensional spacetime, its line-element takes the well-known form [39]

$$\begin{aligned} ds^2 = &-\left(1 - \frac{\mu}{\Sigma\, r^{n-1}}\right) dt^2 - \frac{2a\mu \sin^2\theta}{\Sigma\, r^{n-1}}\, dt\, d\varphi + \frac{\Sigma}{\Delta}\, dr^2 + \Sigma\, d\theta^2 \\ &+ \left(r^2 + a^2 + \frac{a^2\mu \sin^2\theta}{\Sigma\, r^{n-1}}\right) \sin^2\theta\, d\varphi^2 + r^2 \cos^2\theta\, d\Omega_n^2, \end{aligned} \tag{2}$$

where

$$\Delta = r^2 + a^2 - \frac{\mu}{r^{n-1}}\,, \qquad \Sigma = r^2 + a^2 \cos^2\theta\,, \tag{3}$$

with $a_1 = a$ and $a_2 = a_3 = ... = a_N = 0$. Since colliding particles have a non-zero impact parameter only along the usual 3-space which results into a single angular momentum component once the black hole is created, this line-element is absolutely suitable for our study. Then, the black hole's mass M_{BH} and angular momentum J are related to the parameters a and μ as follows [39]

$$M_{BH} = \frac{(n+2)A_{n+2}}{16\pi G_D}\, \mu\,, \qquad J = \frac{2}{n+2}\, M_{BH}\, a\,, \tag{4}$$

with G_D being the $(4+n)$-dimensional Newton's constant, and A_{n+2} the area of a $(n+2)$-dimensional unit sphere given by $A_{n+2} = 2\pi^{(n+3)/2}/\Gamma[(n+3)/2]$.

The line-element (2) is described by Eq. (1) with $S(x) = r\cos\theta$. Tensor-type gravitational perturbations for the line-element (2) – which exist only for $n \geq 3$ – can be expanded in terms of a basis of transverse and traceless harmonic tensors $\mathbb{T}_{ij}^{(\ell,\alpha)}$ on the unit sphere S^n as follows [35, 40]

$$\delta G_{ij} = 2S^2(x) \sum_{l,\alpha} H_T^{(\ell,\alpha)}(x)\, \mathbb{T}_{ij}^{(\ell,\alpha)}(y)\,, \tag{5}$$

where $\{i,j\}$ refer to the y-coordinates along the sphere S^n, and $\mathbb{T}_{ij}^{(\ell,\alpha)}$ satisfy the eigenvalue equation

$$[\hat{\Delta} + \ell(\ell+n-1) - 2]\, \mathbb{T}_{ij}^{(\ell,\alpha)} = 0. \tag{6}$$

$\hat{\Delta}$ is the Laplace-Beltrami operator on S^n, $\ell = 2, 3, 4, \ldots$ labels the corresponding eigenvalues and α is used to distinguish harmonic tensors with the same eigenvalue.

With the use of the expansion (5), the (i,j)-component of the Einstein's equation in vacuum leads to the following second-order hyperbolic equation for the amplitude $H_T^{(\ell,\alpha)}(x)$ [40]

$$-\Box H_T - \frac{n}{r\cos\theta}\, g^{ab}\, \partial_a(r\cos\theta)\, \partial_b H_T + \frac{\ell(\ell+n-1)}{r^2\cos^2\theta}\, H_T = 0, \tag{7}$$

where $\Box$ is the d'Alembertian operator for the metric $g_{ab}(x)$ (for simplicity, we have omitted the labels $\{\ell,\alpha\}$). Using the ansatz

$$H_T(x) = e^{-i\omega t}\, e^{im\varphi}\, R(r)\, Q(\theta)\,, \tag{8}$$

the above partial differential equation reduces to a set of radial and angular equation, namely

$$\frac{1}{r^n}\, \partial_r\left(r^n \Delta\, \partial_r R\right) + \left(\frac{K^2}{\Delta} - \frac{\ell(\ell+n-1)a^2}{r^2} - \Lambda_{j\ell m}\right) R = 0\,, \tag{9}$$

$$\frac{1}{\sin\theta\cos^n\theta}\, \partial_\theta\left(\sin\theta\cos^n\theta\, \partial_\theta Q\right) + \left(\omega^2 a^2\cos^2\theta - \frac{m^2}{\sin^2\theta} - \frac{\ell(\ell+n-1)}{\cos^2\theta} + E_{j\ell m}\right) Q = 0 \tag{10}$$

where

$$K = (r^2 + a^2)\,\omega - am\,, \qquad \Lambda_{j\ell m} = E_{j\ell m} + a^2\omega^2 - 2am\omega\,, \tag{11}$$

with $E_{j\ell m}$ being the separation constant of the two equations, and j a new quantum number that labels the eigenvalues of the angular function $Q(\theta)$.

One needs to solve the above set of second-order ordinary differential equations (9)-(10) for R and Q. The solution of the radial equation will yield the expression for the corresponding absorption probability and the angular equation will provide us with the value of the separation constant $E_{j\ell m}$.

3. Analytic Solution

In spacetimes of the form of Eq. (1), the tensor-type gravitational perturbations are found to satisfy the same field equations that a massless scalar field obeys in the same background. Equations (9)-(10) are identical to the ones that follow from the scalar field equation

$$\frac{1}{\sqrt{-G}}\, \partial_M\left(\sqrt{-G}\, G^{MN} \partial_N \Phi\right) = 0\,, \tag{12}$$

using the expansion of Φ in terms of the hyperspherical harmonics $\mathbb{Y}^{(\ell,\alpha)}(y)$ on S^n

$$\Phi(x,y) = e^{-i\omega t} e^{im\varphi} R(r)\, Q(\theta)\, \mathbb{Y}^{(\ell,\alpha)}(y)\,. \tag{13}$$

Because of the different nature of the scalar and gravitational degrees of freedom, ℓ is constrained to values ≥ 2 in the case of gravitons whereas $\ell \geq 0$ in the scalar case [41].

The aim of the analytic approach is to find the asymptotic solutions near the horizon of the black hole ($r \simeq r_h$), and far away from it ($r \gg r_h$) and then match them in an intermediate zone, to create an analytical solution for $R(r)$ over the whole radial regime. The black hole's horizon radius r_h follows from the equation $\Delta(r_h) = 0$, and is given by the relation $r_h^{n+1} = \mu/(1 + a_*^2)$, where $a_* \equiv a/r_h$.

In the near-horizon regime ($r \simeq r_h$), Eq. (9) takes the form

$$f\,(1-f)\,\frac{d^2R}{df^2}+(1-D_*\,f)\,\frac{dR}{df}+\left[\frac{K_*^2}{A_*^2\,f(1-f)}-\frac{\left[\ell(\ell+n-1)a_*^2+\Lambda_{j\ell m}\right]\,(1+a_*^2)}{A_*^2\,(1-f)}\right]R=0\,, \quad (14)$$

with the use of the variable $f(r) = \Delta(r)/(r^2 + a^2)$, where

$$A_* = (n+1) + (n-1)a_*^2, \quad K_* = (1 + a_*^2)\omega_* - a_* m \quad (15)$$

$$\omega_* \equiv \omega r_h \quad \mathit{and} \quad D_* \equiv 1 - 4a_*^2/A_*^2. \quad (16)$$

Equation (14) can be brought to the form of a hypergeometric differential equation with general solution [42]

$$\begin{aligned} R_{NH}(f) = {} & A_-\, f^{\alpha}\,(1-f)^{\beta}\,F(a,b,c;f) \\ & + A_+\, f^{-\alpha}\,(1-f)^{\beta}\,F(a-c+1,b-c+1,2-c;f)\,, \end{aligned} \quad (17)$$

where the indices $(a,\, b,\, c)$ are defined as: $a \equiv \alpha + \beta + D_* - 1$, $b \equiv \alpha + \beta$, and $c \equiv 1 + 2\alpha$ and $A_\pm$ are integration constants. Parameters α and β are given by $\alpha_\pm = \pm iK_*/A_*$ and

$$\beta = \frac{1}{2}\left[(2-D_*) - \sqrt{(D_*-2)^2 - 4\left[\frac{K_*^2 - \left[\ell(\ell+n-1)a_*^2 + \Lambda_{j\ell m}\right](1+a_*^2)}{A_*^2}\right]}\,\right]. \quad (18)$$

Close to the horizon, we employ the convenient transformation of the radial variable $y = r_h(1 + a_*^2)\ln(f)/A_*$ and impose the boundary condition that no outgoing modes exist near the black hole's horizon. Then the near-horizon solution takes the form

$$R_{NH}(f) = A_-\, f^{\alpha}\,(1-f)^{\beta}\,F(a,b,c;f)\,. \quad (19)$$

In the far-field regime ($r \gg r_h$), Eq. (9) can easily be brought into the form of a Bessel differential equation by making the substitution $R(r) = r^{-\left(\frac{n+1}{2}\right)}\tilde{R}(r)$ and employing a new radial variable $z = \omega r$, with general solution

$$R_{FF}(r) = \frac{B_1}{r^{\frac{n+1}{2}}}\,J_\nu\,(\omega r) + \frac{B_2}{r^{\frac{n+1}{2}}}\,Y_\nu\,(\omega r)\,, \quad (20)$$

where J_ν and Y_ν are the Bessel functions of the first and second kind, respectively, and $\nu = \sqrt{E_{j\ell m} + a^2\omega^2 + \left(\frac{n+1}{2}\right)^2}$.

In order to match the two asymptotic solutions (19) and (20), they both need to be expanded for intermediate values of the radial variable. First we shift the hypergeometric function (19) so that its argument changes from f to $1-f$ by using a well-known relation [42, 30]. Then in the limit $r \gg r_h$ (or, equivalently, $f \to 1$) the near-horizon solution takes the 'stretched' form

$$R_{NH}(r) \simeq A_1\, r^{-(n+1)\,\beta} + A_2\, r^{(n+1)\,(\beta+D_*-2)}\,, \quad (21)$$

with A_1 and A_2 defined as

$$A_1 = A_- \left[(1+a_*^2)\, r_h^{n+1}\right]^{\beta} \frac{\Gamma(c)\Gamma(c-a-b)}{\Gamma(c-a)\Gamma(c-b)},$$

$$A_2 = A_- \left[(1+a_*^2)\, r_h^{n+1}\right]^{-(\beta+D_*-2)} \frac{\Gamma(c)\Gamma(a+b-c)}{\Gamma(a)\Gamma(b)}.$$

The far-field solution (20) in the limit of $r \to 0$ takes the polynomial form

$$R_{FF}(r) \simeq \frac{B_1 \left(\frac{\omega r}{2}\right)^{\nu}}{r^{\frac{n+1}{2}}\,\Gamma(\nu+1)} - \frac{B_2}{\pi\, r^{\frac{n+1}{2}}} \frac{\Gamma(\nu)}{\left(\frac{\omega r}{2}\right)^{\nu}}. \tag{22}$$

For the two stretched solutions to perfectly match, the power coefficients of r need to be the same. This is indeed the case in the limit of $a_* < 1$ and $\omega_* < 1$. Then, by ignoring terms of order $(\omega_*^2, a_*^2, a_*\omega_*)$ or higher, we find that $(n+1)\,\beta \simeq -j$, $(n+1)\,(\beta+D_*-2) \simeq -(j+n+1)$ and $\nu \simeq j + \frac{n+1}{2}$. Identifying the coefficients of the same powers of r, gives us the ratio

$$\frac{B_1}{B_2} = -\frac{(2/\omega r_h)^{2j+n+1}\, \nu\, \Gamma^2(\nu)\, \Gamma(\alpha+\beta+D_*-1)\, \Gamma(\alpha+\beta)\, \Gamma(2-2\beta-D_*)}{\pi\,(1+a_*^2)^{\frac{2j+n+1}{n+1}}\, \Gamma(2\beta+D_*-2)\, \Gamma(2+\alpha-\beta-D_*)\, \Gamma(1+\alpha-\beta)}, \tag{23}$$

that guarantees the existence of a smooth, analytic solution for the radial part of the tensor-type graviton wavefunction for all r (given that a_* and ω_* remain small).

Finally, we need to calculate the absorption probability $|\mathcal{A}_{j\ell m}|^2$ – or greybody factor, since it is the reason for the deviation of the black-hole spectrum from a pure black-body one. We derive it, by expanding the far-field solution (20) for $r \to \infty$, in which case we find

$$R_{FF}(r) \simeq \frac{1}{r^{\frac{n+2}{2}}\sqrt{2\pi\omega}} \left[(B_1+iB_2)\, e^{-i\left(\omega r - \frac{\pi}{2}\nu - \frac{\pi}{4}\right)} + (B_1-iB_2)\, e^{i\left(\omega r - \frac{\pi}{2}\nu - \frac{\pi}{4}\right)}\right]. \tag{24}$$

The greybody factor is then determined via the amplitudes of the outgoing and incoming spherical waves, as

$$|\mathcal{A}_{j\ell m}|^2 = 1 - |\mathcal{R}_{j\ell m}|^2 = 1 - \left|\frac{B_1-iB_2}{B_1+iB_2}\right|^2 = \frac{2i\,(B^*-B)}{BB^*+i\,(B^*-B)+1}, \tag{25}$$

where $B \equiv B_1/B_2$ is given by Eq. (23). The above result can be used to evaluate the absorption probability for the emission of tensor-type gravitons in the bulk, from a simply rotating black hole, in the low-energy and low-angular momentum limit.

4. Numerical Analysis

With the use of numerical analysis we manage to solve both the angular and radial equations for any value of the energy of the emitted particles and angular-momentum of the black hole. Consequently we were able to calculate the greybody factor and both energy and angular momentum emission spectrum for tensor-type gravitational modes in the bulk. Details concerning our method are extensively presented in our published article [43].

4.1. Absorption Probability

The results obtained analytically and numerically should agree in the low-energy and low-angular-momentum limit but are expected to deviate once we move outside these regimes. Fig. 1(a) depicts the results for the indicative modes $(j=2, \ell=2, m=0)$ and $(j=5, \ell=2, m=1)$:

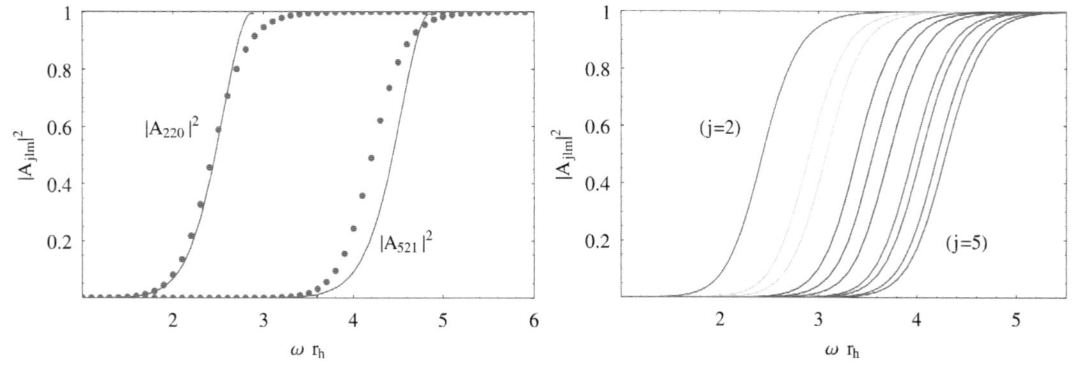

Figure 1. **(a)** Comparison between our analytical (solid lines) and numerical (data points) results for the greybody factor for the modes $(j = \ell = 2, m = 0)$ and $(j = 5, \ell = 2, m = 1)$, for $a_* = 0.5$ and $n = 3$. **(b)** Absorption probabilities for the sets of modes with $\ell = 2$ and $(j = 2, m = 0)$ (red), $(j = 3, m = -1, 1)$ (green), $(j = 4, m = -2, 0, 2)$ (blue), and $(j = 5, m = -3, -1, 1, 3)$ (magenda), for $a_* = 0.5$ and $n = 3$ [43].

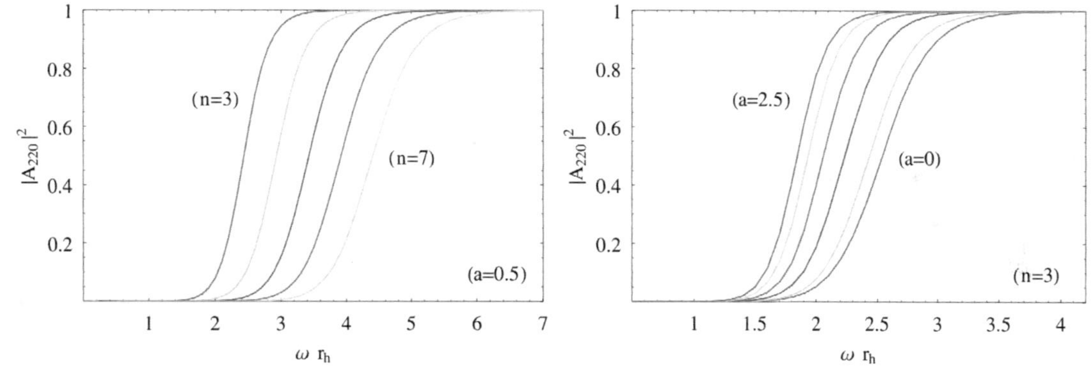

Figure 2. Absorption probabilities for the mode $(j = 2, \ell = 2, m = 0)$ as **(a)** a function of $n = 3, 4, 5, 6, 7$, for $a_* = 0.5$, and **(b)** a function of $a_* = 0, 0.5, 1, 1.5, 2, 2.5$, for $n = 3$ [43].

the analytical results are given by the solid lines and the numerical ones are presented as data points - both sets of results correspond to the case with $n = 3$ and $a = 0.5$ (in units of r_h). The agreement between the two curves is indeed very good in the low-energy and even intermediate-energy regime and, as expected, breaks down as we move towards the high-energy one. Furthermore, the agreement is better for the lowest modes and it worsens for higher modes for which the greybody factor raises to a significant value and approaches unity at an increasingly higher value of the energy parameter ωr_h.

In Fig. 1(b), we examine the behavior of the greybody factors for different tensor modes. The constraints $j \geq \ell + |m|$ and $\frac{j-(\ell+|m|)}{2} \in \{0, \mathbb{Z}^+\}$ [44] dictate that for each value of j, ℓ can take values in the range $[2, j]$ while, for given j and ℓ, m can take $j - \ell + 1$ values in total (j: total angular momentum, ℓ: angular-momentum along the compact space S^n, m: angular-momentum in the plane of rotation of the black hole). As either j or m increases, the corresponding greybody curve shifts to the right and to higher-energies. Note that we get a similar behavior if we vary also ℓ.

Finally, we investigate the dependence of the greybody factors on the spacetime parameters of the theory, namely the number of additional spacelike dimensions n and the angular-momentum parameter of the black hole a by studying the indicative mode ($j = 2, \ell = 2, m = 0$). Fig. 2(a) clearly shows that the greybody factors for the tensor-gravitational modes in the bulk decrease as the number of transverse-to-the-brane spacelike dimensions increases. In Fig. 2(b) the greybody factors are enhanced as the angular-momentum of the black hole increases. This behaviour is in total agreement with the one observed for bulk scalar fields [30, 32] propagating in the same background.

4.2. Energy and Angular Momentum Emission Rates

We now proceed to compute the differential emission rates of energy and angular momentum in the bulk in the form of tensor-type gravitons. These are given by the expressions

$$\frac{d^2E}{d\omega dt} = \frac{1}{2\pi} \sum_{j,\ell,m} \frac{\omega}{e^{\tilde{\omega}/T_H} - 1} N_{ST}^{\ell}(S^n) |\mathcal{A}_{j\ell m}|^2, \tag{26}$$

$$\frac{d^2J}{d\omega dt} = \frac{1}{2\pi} \sum_{j,\ell,m} \frac{m}{e^{\tilde{\omega}/T_H} - 1} N_{ST}^{\ell}(S^n) |\mathcal{A}_{j\ell m}|^2, \tag{27}$$

where $\tilde{\omega} \equiv \omega - \frac{am}{a^2+r_h^2}$ and T_H the temperature of the black hole given by

$$T_H = \frac{(n+1) + (n-1)a_*^2}{4\pi(1+a_*^2)r_h} . \tag{28}$$

N_{ST}^{ℓ} is the multiplicity of the 2nd-rank symmetric, traceless and divergence-free tensor harmonics T_{AB} that satisfy Eq. (6). This quantity first calculated by Rubin and Ordonez in [45], was reproduced in [43] via an alternative method and found to be

$$N_{\rm ST}^{\ell}(S^n) = \frac{(n+1)(n-2)(n+\ell)(l-1)(n+2\ell-1)(n+\ell-3)!}{2(\ell+1)!(n-1)!} , \tag{29}$$

for the ℓ-th eigenvalue.

Particular care was given to sum up to an appropriate maximum value of j which ensures that the derived emission rates are as close as possible to the real ones. In practise, we stopped the summation when the contribution to the total curve of the modes corresponding to the next higher value of j was negligibly small. Fig. 3 shows explicitly that as we sum up to $j_{max} = 5, 8, 10, 12$ and, finally, 15 the energy emission rate curve becomes wider and the slope of the tail decreases, whereas the low-energy behavior and the peak remains unchanged. In general, as either n or a_* increases, the number of modes that needed to be summed increases too. In the cases considered, we had to sum up to $j_{max} = 22$, i.e. $m = 20$ in order to obtain the optimum results.

We are mostly interested in exploring the dependence of the energy and angular momentum emission rates of the black hole on the spacetime parameters, namely n and a_*. In Figs. 4(a,b), we illustrate their dependence on the number of additional spacelike dimensions and the angular momentum of the black hole, respectively. As in the case of scalars [30, 32], the energy emission rate has a very strong dependence on n with an enhancement of almost two orders of magnitude as n changes from $n = 3$ to $n = 7$ and a more complex correlation with a_*. More specifically, the energy emission curve becomes taller and wider with the increase of n in the whole energy spectrum. On the other hand, Fig. 4(b) ($n = 3$) shows that, as a_* goes from 0 to 2.5, the emission rate of high-energy modes increases, whereas the one for low- and intermediate-energy modes decreases.

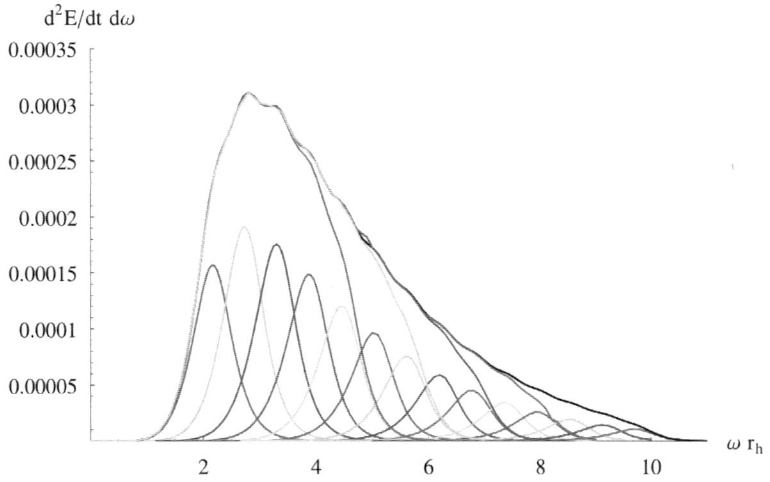

Figure 3. Energy emission of tensor-type gravitons for $n = 3$, $a_* = 1$. Upper curves, from left to right, correspond to the highest value of j considered in the sum: $j = 6$ (red), $j = 8$ (green), $j = 10$ (blue), $j = 12$ (magenta) and $j = 15$ (black) [43].

Table 1. Comparison between total energy emissivities in the bulk through scalars and tensor-gravitons

	Scalars	Tensor-Gravitons	ratio
$n = 3$	0.1646	0.0013	0.0018
$n = 4$	0.3808	0.0222	0.058
$n = 5$	0.7709	0.1853	0.24

Figs. 5(a,b) depict the dependence of the angular momentum emission rate on n and a_*. As either parameter gets larger, the angular momentum emission rate becomes more and more enhanced (even though the enhancement due to the increase of a_* is milder) and in all cases leads to the emission of larger number of modes in all energy regimes.

Finally, we move on to compare the significance of the emission of scalars and tensor-gravitons as independent channels, through which energy goes away from our brane into the bulk. We use the results found in [32] for the contribution of scalars to this procedure, in order to find the ratio of the total energy carried away by tensor-gravitons over the one carried away by scalars. Table 1 shows that scalar modes are the principal energy carriers in all cases. However, tensor-gravitons play an increasingly important role as the number of extra dimensions gets larger. Consequently, one has to carefully include their contribution as well to the overall process of energy being radiated away into the bulk by the black hole.

5. Conclusions

The potential creation of miniature black holes at the LHC and future high-energy experiments, means that we will also witness their evaporation through the emission of Hawking radiation. Of course, our detectors will record only the on-brane emitted degrees of freedom. Therefore, it is of absolute importance to have a clear knowledge of the balance between the amount of energy expected to go into the bulk and the one expected to rest on our brane, in order to evaluate any

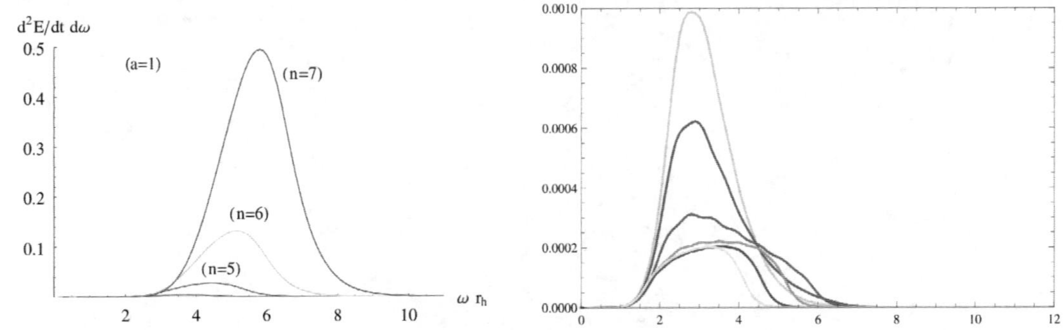

Figure 4. The energy emission rate for tensor-type gravitons in the bulk for **(a)** $a_* = 1$ and $n = 3, 4, 5, 6, 7$ (from bottom to top), and for **(b)** $n = 3$ and $a_* = 0, 0.5, 1, 1.5, 2, 2.5$ (from top to bottom) [43].

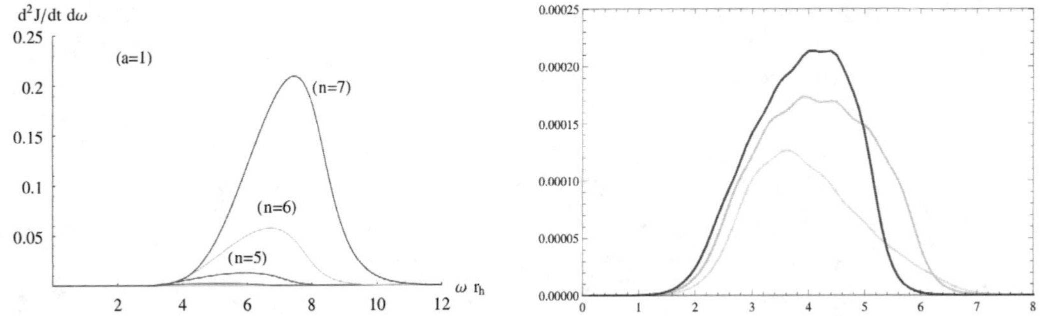

Figure 5. The angular momentum emission rate for tensor-type gravitons in the bulk for **(a)** $a_* = 1$ and $n = 3, 4, 5, 6, 7$ (from bottom to top), and for **(b)** $n = 3$ and $a_* = 0.5, 1, 1.5$ (from bottom to top) [43].

relevant data and find decisive evidence of the existence of extra dimensions and their nature as well. Previous studies [30, 32] have shown that the bulk emission of scalars, allowed to propagate in the whole spacetime, is subdominant compared to the on-brane restricted particles (scalars and all fermions and bosons of SM). However, the contribution of gravitons (scalar- vector- and tensor-type ones) has to be considered as well in order to calculate the actual brane-bulk channel balance. In this work we have shown that tensor-type gravitons do make the bulk channel less subdominant compared to the brane one (especially when the number of extra spacelike dimensions is large), even though the overall picture remains essentially unchanged. Hopefully, the contribution of scalar- and vector-type gravitons will also be calculated in the near future, to obtain full knowledge of the energy-loss-in-the-bulk procedure via of hawking radiation.

Acknowledgements
N.P. would like to thank Pangiota Kanti, Hideo Kodama, Roman A. Konoplya and Alexander Zhidenko for their collaboration. The author also acknowledges participation in the RTN networks UNIVERSENET-MRTN-CT-2006035863-1 and MRTN-CT-2004-503369.

References

[1] N. Arkani-Hamed, S. Dimopoulos and G. R. Dvali, *Phys. Lett.* B **429**, 263 (1998); *Phys. Rev.* D **59**, 086004 (1999); I. Antoniadis, N. Arkani-Hamed, S. Dimopoulos and G. R. Dvali, *Phys. Lett.* B **436**, 257 (1998).
[2] L. Randall and R. Sundrum, *Phys. Rev. Lett.* **83** (1999) 3370; *Phys. Rev. Lett.* **83** (1999) 4690.
[3] T. Banks and W. Fischler, hep-th/9906038;
R. Emparan, G. T. Horowitz and R. C. Myers, *Phys. Rev. Lett.* **85** (2000) 499;
D. M. Eardley and S. B. Giddings, *Phys. Rev.* **D 66**, 044011 (2002);
H. Yoshino and Y. Nambu, *Phys. Rev.* **D66**, 065004 (2002); *ibid.* **D 67**, 024009 (2003);
E. Kohlprath and G. Veneziano, *JHEP* **0206**, 057 (2002);
V. Cardoso, O. J. C. Dias and J. P. S. Lemos, *Phys. Rev.* D **67**, 064026 (2003);
E. Berti, M. Cavaglia and L. Gualtieri, *Phys. Rev.* **D 69**, 124011 (2004);
S. B. Giddings and V. S. Rychkov, *Phys. Rev.* D **70**, 104026 (2004);
H. Yoshino and V. S. Rychkov, *Phys. Rev.* D **71** (2005) 104028 ;
D. C. Dai, G. D. Starkman and D. Stojkovic, *Phys. Rev.* **73** (2006) 104037 ;
H. Yoshino and R. B. Mann, *Phys. Rev.* **74** (2006) 044003 ;
H. Yoshino, T. Shiromizu and M. Shibata, *Phys. Rev.* D **74**, 124022 (2006).
[4] S. W. Hawking, *Commun. Math. Phys.* **43**, 199 (1975).
[5] P. Kanti, *Int. J. Mod. Phys.* A **19**, 4899 (2004) [hep-ph/0402168]; *Lect. Notes Phys.* **769**, 387 (2009) [arXiv:0802.2218 [hep-th]]; arXiv:0903.2147 [hep-th].
[6] M. Cavaglia, *Int. J. Mod. Phys.* A **18**, 1843 (2003) [hep-ph/0210296];
G. L. Landsberg, *Eur. Phys. J.* C **33**, S927 (2004) [hep-ex/0310034];
K. Cheung, hep-ph/0409028;
S. Hossenfelder, hep-ph/0412265;
C. M. Harris, hep-ph/0502005;
A. S. Majumdar and N. Mukherjee, *Int. J. Mod. Phys.* D **14**, 1095 (2005) [astro-ph/0503473];
E. Winstanley, arXiv:0708.2656 [hep-th];
R. Emparan and H. S. Reall, Living Rev. Rel. **11**, 6 (2008). arXiv:0801.3471 [hep-th].
[7] E. Berti, V. Cardoso and A. O. Starinets, arXiv:0905.2975 [gr-qc]; R. A. Konoplya, Phys. Rev. D **68**, 124017 (2003) [arXiv:hep-th/0309030]; P. Kanti and R. A. Konoplya, Phys. Rev. D **73**, 044002 (2006) [arXiv:hep-th/0512257]; H. Ishihara, M. Kimura, R. A. Konoplya, K. Murata, J. Soda and A. Zhidenko, Phys. Rev. D **77**, 084019 (2008) [arXiv:0802.0655 [hep-th]]; E. Berti, K. D. Kokkotas and E. Papantonopoulos, Phys. Rev. D **68**, 064020 (2003) [arXiv:gr-qc/0306106].
[8] P. Kanti and J. March-Russell, *Phys. Rev.* D **66**, 024023 (2002); *Phys. Rev.* D **67**, 104019 (2003).
[9] C. M. Harris and P. Kanti, *JHEP* **0310** (2003) 014.
[10] A. Barrau, J. Grain and S. O. Alexeyev, *Phys. Lett.* B **584**, 114 (2004); J. Grain, A. Barrau and P. Kanti, *Phys. Rev.* D **72**, 104016 (2005); T. G. Rizzo, *Class. Quant. Grav.* **23**, 4263 (2006).
[11] E. l. Jung, S. H. Kim and D. K. Park, *Phys. Lett.* B **586** (2004) 390; *JHEP* **0409** (2004) 005; *Phys. Lett.* B **602** (2004) 105; *Phys. Lett.* B **614** (2005) 78; E. Jung and D. K. Park, *Nucl. Phys.* B **717** (2005) 272; hep-th/0506204.
[12] P. Kanti, J. Grain and A. Barrau, *Phys. Rev.* D **71** (2005) 104002.
[13] A. S. Cornell, W. Naylor and M. Sasaki, *JHEP* **0602** (2006) 012.
[14] D. K. Park, *Class. Quant. Grav.* **23** (2006) 4101.
[15] V. Cardoso, M. Cavaglia and L. Gualtieri, *Phys. Rev. Lett.* **96**, 071301 (2006) [Erratum-ibid. **96**, 219902 (2006)]; *JHEP* **0602**, 021 (2006).
[16] S. Creek, O. Efthimiou, P. Kanti and K. Tamvakis, *Phys. Lett.* B **635** (2006) 39; O. Efthimiou, hep-th/0609144.
[17] D. C. Dai, N. Kaloper, G. D. Starkman and D. Stojkovic, *Phys. Rev.* D **75**, 024043 (2007).
[18] V. P. Frolov and D. Stojkovic, *Phys. Rev.* D **67**, 084004 (2003).
[19] D. Ida, K. y. Oda and S. C. Park, *Phys. Rev.* D **67**, 064025 (2003) [Erratum-ibid. D **69**, 049901 (2004)].
[20] H. Nomura, S. Yoshida, M. Tanabe and K. i. Maeda, *Prog. Theor. Phys.* **114**, 707 (2005).
[21] C. M. Harris and P. Kanti, *Phys. Lett.* B **633**, 106 (2006).
[22] D. Ida, K. y. Oda and S. C. Park, *Phys. Rev.* D **71**, 124039 (2005); *Phys. Rev.* D **73**, 124022 (2006).
[23] E. Jung, S. Kim and D. K. Park, *Phys. Lett.* B **615**, 273 (2005); *Phys. Lett.* B **619**, 347 (2005);
[24] G. Duffy, C. Harris, P. Kanti and E. Winstanley, *JHEP* **0509**, 049 (2005).
[25] M. Casals, P. Kanti and E. Winstanley, *JHEP* **0602**, 051 (2006).
[26] M. Casals, S. R. Dolan, P. Kanti and E. Winstanley, *JHEP* **0703**, 019 (2007).
[27] S. Creek, O. Efthimiou, P. Kanti and K. Tamvakis, *Phys. Rev.* D **75** (2007) 084043; *Phys. Rev.* D **76** (2007) 104013.
[28] S. Chen, B. Wang, R. K. Su and W. Y. Hwang, *JHEP* **0803**, 019 (2008) [arXiv:0711.3599 [hep-th]].

[29] E. Jung and D. K. Park, *Nucl. Phys.* B **731**, 171 (2005). *Mod. Phys. Lett.* A **22**, 1635 (2007) [hep-th/0612043].
[30] S. Creek, O. Efthimiou, P. Kanti and K. Tamvakis, *Phys. Lett.* B **656**, 102 (2007) [arXiv: 0709.0241 [hep-th]].
[31] T. Kobayashi, M. Nozawa, Y. Takamizu, *Phys. Rev.* D **77**, 044022 (2008) [arXiv:0711.1395 [hep-th]].
[32] M. Casals, S. R. Dolan, P. Kanti and E. Winstanley, *JHEP* **0806**, 071 (2008) [arXiv:0801.4910 [hep-th]].
[33] R. Emparan, G. T. Horowitz and R. C. Myers, *Phys. Rev. Lett.* **85**, 499 (2000) [hep-th/0003118].
[34] H. K. Kunduri, J. Lucietti and H. S. Reall, *Phys. Rev.* **D74**, 084021 (2006) [hep-th/0606076].
[35] H. Kodama, Prog. Theor. Phys. Supple. 172, 11 (2008) [arXiv: 0711.4184]
[36] H. Kodama, A. Ishibashi and O. Seto, Phys. Rev. D **62**, 064022 (2000).
[37] H. Kodama, Lect. Notes Phys. **769**, 427 (2009).
[38] H. Kodama and A. Ishibashi, Prog. Theor. Phys. **110**, 701 (2003).
[39] R. C. Myers and M. J. Perry, *Annals Phys.* **172**, 304 (1986).
[40] H. Kodama, R. A. Konoplya and A. Zhidenko, Phys. Rev. D **79**, 044003 (2009) [arXiv:0812.0445 [hep-th]].
[41] D. Ida, Y. Uchida and Y. Morisawa, Phys. Rev. D **67**, 084019 (2003).
[42] M. Abramowitz and I. Stegun, *Handbook of Mathematical Functions* (Academic, New York, 1996).
[43] P. Kanti, H. Kodama, R. A. Konoplya, N. Pappas and A. Zhidenko, *Phys. Rev. D* **80**, 084016 (2009) [hep-ph/0906.3845];
[44] E. Berti, V. Cardoso and M. Casals, Phys. Rev. D **73** (2006) 024013 [Erratum-ibid. D **73** (2006) 109902];
[45] M.A. Rubin and C.R. Ordónez: *J. Math. Phys.* **25**, 2888 (1984).

Journal of Physics: Conference Series **283** (2011) 012029 doi:10.1088/1742-6596/283/1/012029

Dark Energy

David Polarski

LCC, CNRS, Université Montpellier 2, Montpellier, France

E-mail: David.Polarski@univ-montp2.fr

Abstract. A large number of observations suggest that our universe entered at low redshifts a stage with accelerated expansion rate. Many models, Dark Energy (DE) models, able to explain this departure from conventional cosmology have been proposed. These models are conceptually very different, either introducing some new component with sufficiently negative pressure, or modifying the gravitational interaction on cosmic scales. Some of these DE models are reviewed here. Future high precision observations probing both the background and the perturbations will single out viable models.

1. Introduction

There is growing *observational* evidence that our universe has entered a stage of accelerated expansion rate [1] at low redshifts. Allthough this constitutes a radical departure from conventional cosmology it is reassuring that a consistent picture of our universe seems to emerge from the various data of ever increasing accuracy. Though a wealth of models were proposed, the physical mechanism behind this unconventional expansion remains unclear. The basic problem is to explain why our universe went over from (standard) decelerated to accelerated expansion at a redshift $z \sim 0.5$.were extensively studied in the literature

This accelerated expansion rate could be produced by a new isotropic comoving perfect fluid with sufficiently negative pressure, coined Dark Energy. By extension, all models aiming to explain it are called Dark Energy (DE) models. The simplest candidate of this kind is a positive cosmological constant Λ whose pressure satisfies $p_\Lambda = -\rho_\Lambda$ or $w_\Lambda \equiv \frac{p_\Lambda}{\rho_\Lambda} = -1$. However it is well-known that its interpretation as the vacuum energy is problemetic because of its exceeding smallness (see e.g. [2] for recent comprehensive reviews). Observations seem consistent with a cosmological constant Λ so that ΛCDM has gradually become a fiducial DE model. One can also introduce a phenomenological smooth component with either constant or time varying equation of state and a sufficiently negative pressure. A prominent candidate with dynamical w_{DE} is a minimally coupled scalar field or Quintessence [3]. A further constraint arises if observations force us to consider DE models of the phantom type, i.e. satisfying $w_{DE} < -1$ [4] as allowed by observations on small redshifts $z < 0.5$ [5]. The most striking consequence in that case is that quintessence models are ruled out. However, the late-time acceleration could as well arise from a change in the laws of gravitation. Such models can typically produce DE of the phantom type. The growth of matter perturbations allow to distinguish these models from DE models inside General Relativity (GR). Probing the background *and* the inhomogeneities will establish whether this attractive possibility is consistent with observations, see e.g. [6]. We first review some basic properties of DE models.

2. Basics

When DE is a comoving isotropic perfect fluid in the framework of GR, its basic properties are easily derived. We consider a universe with metric given by

$$ds^2 = -dt^2 + a^2(t)\Big(\frac{dr^2}{1-\kappa r^2} + r^2\left(d\theta^2 + \sin^2\theta\, d\phi^2\right)\Big)\ , \tag{1}$$

with $\kappa = 0, 1, -1$ for a spatially flat, closed and open universe respectively. Let us write the corresponding Friedmann equations (we take c=1)

$$\left(\frac{\dot a}{a}\right)^2 = \frac{8\pi G}{3}\sum_i \rho_i - \frac{k}{a^2} \tag{2}$$

$$\frac{\ddot a}{a} = -\frac{4\pi G}{3}\sum_i (\rho_i + 3p_i)\ . \tag{3}$$

We introduce the important dimensionless quantities

$$\Omega_i = \frac{\rho_i}{\rho_{cr}} \qquad w_i \equiv \frac{p_i}{\rho_i} \qquad \Omega_k = -\frac{k}{a^2H^2} \qquad k = 0, \pm 1 \tag{4}$$

with

$$H^2 \equiv \frac{8\pi G}{3}\ \rho_{cr}\ . \tag{5}$$

The constraint

$$\sum_i \Omega_i + \Omega_k = \Omega_m + \Omega_r + \Omega_{DE} + \Omega_k = 1\ , \tag{6}$$

assuming three components, radiation (r), dustlike matter (m) and Dark Energy (DE) follows immediately. Each perfect fluid obeys the conservation equation

$$\dot\rho_i = -3H(\rho_i + p_i)\ , \tag{7}$$

where a dot stands for the derivative with respect to t. Introducing the deceleration parameter q with

$$q \equiv -\frac{\ddot a}{aH^2} = \frac{1}{2}\sum_{i\neq k}\Omega_i(1+3w_i)\ , \tag{8}$$

the following result is obtained at late times in a flat universe ($\Omega_k = 0$)

$$q \simeq \frac{1}{2}(1 + 3\ w_{DE}\ \Omega_{DE})\ . \tag{9}$$

Accelerated expansion will therefore occur provided

$$w_{DE} < -\frac{1}{3}\ \Omega_{DE}^{-1}\ . \tag{10}$$

For $\Omega_{DE,0} \approx 0.7$, accelerated expansion today requires $w_{DE,0} < 0.5$.

We did not specify in (9) or (10) whether w_{DE} evolves with time or not. It does generically as we shall see, while a cosmological constant Λ provides an important exception. Let us see now how a time-varying equation of state (EoS) parameter w_{DE} affects the background expansion and can be constrained with luminosity distances. Luminosity-distance data as function of redshift $d_L(z)$ (remember that we take $c = 1$) probe the universe background expansion

$$d_L(z) = (1+z)\ H_0^{-1}\ |\Omega_{k,0}|^{-\frac{1}{2}}\ \mathcal{S}\left(|\Omega_{k,0}|^{\frac{1}{2}}\ \int_0^z \frac{dz'}{h(z')}\right)\ , \tag{11}$$

where $\mathcal{S}(u) = \sin u$ for a closed universe, $\mathcal{S}(u) = \sinh u$ for an open universe while $\mathcal{S}$ is the identity for a flat universe. In eq.(11), $h(z)$ stands for the (dimensionless) reduced Hubble parameter $h(z) \equiv \frac{H(z)}{H_0}$, $H_0 \equiv H(z=0)$. Neglecting radiation at small z we have

$$h^2(z) = \Omega_{m,0}\ (1+z)^3 + \Omega_{DE,0}\ f(z) + \Omega_{k,0}\ (1+z)^2\ , \tag{12}$$

with all quantities defined in (4). The function $f(z)$ expresses the time evolution of Dark Energy

$$f(z) \equiv \frac{\rho_{DE}(z)}{\rho_{DE,0}}\ . \tag{13}$$

which is a priori unknown. It is directly related to the equation of state parameter $w_{DE}(z) = \frac{p_{DE}(z)}{\rho_{DE}(z)}$ through the energy conservation equation (7) for the DE component

$$f(z) = \exp\left[3\int_0^z dz'\ \frac{1+w_{DE}(z')}{1+z'}\right]\ . \tag{14}$$

Combining equations (12) and (14), we can also express directly $w_{DE}(z)$ using of $h^2(z)$ and the cosmological parameters $\Omega_{m,0}$ and $\Omega_{k,0}$. Note that this involves a second-order derivative with respect to z of $d_L(z)$. For a cosmological constant Λ, $f_\Lambda(z) = 1$ is recovered from (14) with $w_\Lambda = -1$: the energy density associated to Λ remains constant during the expansion.

A cosmological constant Λ, despite the theoretical problems that its exceeding smallness would inevitably pose, deserves special attention because of its simplicity. It is interesting that such a universe filled with Cold Dark Matter, a ΛCDM universe, is in good agreement with many large-scales observations. It should nevertheless be stressed that some observations on different cosmic scales could be challenging ΛCDM like the dark matter halo density profile which does not seem to be cuspy or large scales peculiar flows which are found to be much larger than expected in ΛCDM. On the other hand ΛCDM makes a very clean prediction: a constant $w_\Lambda = -1$ as well as a constant corresponding energy density. Therefore any tiny evolution of ρ_{DE} would rule out this model and most contenders have precisely such a behaviour.

A prominent example is provided by a time dependent homogeneous minimally coupled scalar field $\phi(t)$ called Quintessence. As is well-known such a field can be identified with a comoving perfect fluid with

$$\rho_\phi = \frac{1}{2}\dot{\phi}^2 + V(\phi) \tag{15}$$

$$p_\phi = \frac{1}{2}\dot{\phi}^2 - V(\phi)\ . \tag{16}$$

Hence the equation of state parameter w_ϕ is given by

$$w_\phi = \frac{\dot{\phi}^2 - 2V(\phi)}{\dot{\phi}^2 + 2V(\phi)}\ . \tag{17}$$

Clearly w_ϕ can be sufficiently negative if the kinetic energy is small enough compared to the potential energy. Actually this is also the mechanism which is so successful in implementing the inflationary phase in the very early universe. While w_ϕ is generically dynamical, it can never cross the boundary $w = -1$. So a time-varying equation of state parameter could in principle agree with such models provided there is no phantom phase. DE models inside GR, and in particular quintessence models, were extensively studied in the literature. We will consider next DE models based on modifications of gravity.

3. Modifying gravity ?

Rather than introducing a new component one can try a modification of the laws of gravity, an elegant solution to the DE puzzle. We note that a similar approach can be adopted for the dark matter problem. In both cases one replaces some hypothetical unknown kind of matter by an equivalent gravitational effect. A central problem in these attempts is to modify the laws of gravity in a way that complies with laboratory, solar-system and cosmic constraints. Cosmic tests go beyond the universe expansion, the growth of matter perturbations on cosmic scales play a crucial role.

3.1. Scalar-tensor models

Our first example of DE models outside General Relativity are scalar-tensor DE models [7]. Among other properties these models allow for a phantom phase [8]. In these models the existence of a new fundamental scalar field is also postulated.

In their most general form, scalar-tensor models have the following Lagrangian in the Jordan frame

$$L = \frac{1}{16\pi G_*}\Big(F(\Phi)\ R - Z(\Phi)\ g^{\mu\nu}\partial_\mu\Phi\partial_\nu\Phi - 2U(\Phi)\Big) + L_m(g_{\mu\nu})\ . \tag{18}$$

where L_m stands for the Lagrangian of any matter species. Here and below G_* is some bare gravitational constant whose precise value is found after the details of the theory are worked out. In this model , the Jordan frame is the physical frame. As there is no direct coupling of matter to the scalar field Φ, fermion masses are constant in this frame and atomic clocks measure the proper time t. In the above representation, the Brans-Dicke parameter is given by

$$\omega_{BD} = \frac{Z\ F}{(dF/d\Phi)^2} > -\frac{3}{2}\ , \tag{19}$$

where the last inequality comes from consistency of the theory. The quantity Z can be set equal to 1 or -1 by a redefinition of the field Φ. A positive Brans-Dicke parameter ω_{BD} corresponds to $Z = 1$. We consider in the rest of this section $\omega_{BD} > 0$ and we take $Z = 1$. The special case for which Z vanishes corresponds to the higher-derivative gravity theory of the type $f(R)$ that we will discuss later. ω_{BD} is very strongly constrained *today* by solar-system tests, viz.

$$\omega_{BD,0} > 4 \times 10^4\ . \tag{20}$$

We can define from (18) the effective Newton's constant G_N

$$G_N = G_*/F\ . \tag{21}$$

However its physical meaning will not necessarily be the same as that of G in General Relativity. Considering models where Φ is massless, the gravitational interaction potential between two test masses is given by

$$V = -G_{\rm eff}\ \frac{M_1\ M_2}{r}\ , \tag{22}$$

where

$$G_{\rm eff} = G_N\ \left(1 + \frac{1}{2\omega_{BD}+3}\right)\ . \tag{23}$$

We have from (20) the very important result

$$G_{\rm eff,0} \simeq G_{N,0}\ . \tag{24}$$

Hence these two quantities are almost identical while their numerical value must be very close to Newton's constant G. In order to make contact with the equations written down in the former section for models inside GR, we write down the modified Friedmann equations for a spatially flat universe

$$3FH^2 = 8\pi G_* \rho_m + \frac{\dot{\Phi}^2}{2} + U - 3H\dot{F} , \tag{25}$$

$$-2F\dot{H} = 8\pi G_* \rho_m + \dot{\Phi}^2 + \ddot{F} - H\dot{F} . \tag{26}$$

We *define* the DE energy density ρ_{DE} and pressure p_{DE} by writing these equations in the Einsteinian form [9]

$$3H^2 = 8\pi G_{N,0}(\rho_m + \rho_{DE}) \tag{27}$$

$$-2\dot{H} = 8\pi G_{N,0}(\rho_m + p_{DE} + \rho_{DE}) . \tag{28}$$

Taking into account (24), expressions for ρ_{DE} and p_{DE} are easily derived.

Defined in this way, ρ_{DE} and p_{DE} satisfy the usual conservation equation for perfect fluids (7) and w_{DE} has its usual physical meaning as in (14). The expression for the quantity $\rho_{DE} + p_{DE}$ reads

$$8\pi G_{N,0}(\rho_{DE} + p_{DE}) = \dot{\Phi}^2 + \ddot{F} - H\dot{F} + 2(F - F_0)\ \dot{H} < 0 . \tag{29}$$

We see from (29) that $w_{DE} < -1$ is impossible when F is constant the theory then basically reduces to quintessence. With a varying F however one can obtain a phantom phase.

3.2. $f(R)$ DE models

While there is an additional fundamental field Φ in scalar-tensor models, this is not the case in $f(R)$ modified gravity DE models . This a surprisingly rich class of models which has been intensively investigated in recent years. In these models the Ricci scalar R is replaced by some function $f(R)$ in the gravity Lagrangian

$$L = \frac{1}{16\pi G_*} f(R) + L_m . \tag{30}$$

To see how such a modification affects the background expansion, we write the corresponding modified Friedmann equations. When radiation can be safely neglected ($z \ll z_{eq}$), we obtain

$$3FH^2 = 8\pi G_* \rho_m + \frac{1}{2}(FR - f) - 3H\dot{F} , \tag{31}$$

$$-2F\dot{H} = 8\pi G_* \rho_m + \ddot{F} - H\dot{F} , \tag{32}$$

with

$$F \equiv \frac{df}{dR} . \tag{33}$$

We note the striking analogy with equations (25,26). It actually derives from the well-known fact that $f(R)$ models can be seen as particular cases of scalar-tensor theories. Of course one must not confuse $F(\Phi)$ and $F(R)$ this is however easily avoided from the context.

The first model of this kind was introduced by Starobinsky for the construction of the first internally consistent inflationary model [10]. In the context of DE models it was first suggested in [11]. In order to be a viable DE model, one requires a standard expansion history *followed* by an accelerated expansion stage at low redshifts. Surprisingly, this requirement turns out to be so constraining that simple Lagrangians of the type $R + \frac{\mu^2}{R}$ lead to an unviable cosmic expansion history [12]: they have no standard $a \sim t^{\frac{2}{3}}$ matter-dominated stage which is instead replaced by the behaviour $a \sim t^{\frac{1}{2}}$! Therefore, even before consideration of other gravitational constraints, these models are ruled out as viable DE models. Some viable $f(R)$ DE models do nevertheless exist and were recently proposed and extensively studied [13],[14],[15]. These viable models have the general form

$$f(R) = R - \lambda R_c \; f_1(x) \qquad\qquad x \equiv R/R_c \; , \tag{34}$$

where $R_c > 0$ defines some characteristic scale of the Ricci scalar, typically close to its cosmic value today, while $\lambda > 0$ is some free model parameter. In all viable models constructed in [13],[14],[15] $f_1(x)$ goes to one for $x \gg 1$. For illustration, let us give the model proposed recently in [13]

$$R - \lambda R_c \left(1 - \left(1 + \frac{R^2}{R_c^2} \right)^{-n} \right) \tag{35}$$

In the limit $R \to 0$, $f(R)$ reduces to R, in sharp contrast to the singular behaviour of models containing an inverse popwer of R and which, as stressed earlier do not yield a viable expansion history. It should be noted that cosmic constraints on DE coming from the late-time accelerated expansion rate typically probe the low curvature regime of these models, namely $R \sim R_c$. The high curvature regime $R \gg R_c$ in the (very) early universe or in compact relativistic objects, has subtleties of its own. In any case, in order to be viable DE models, modified gravity models must pass all gravitational tests in addition to purely cosmological tests.

Before considering how the growth of matter parturbations is affected in these models, let us return to the definition of the energy density and pressure of DE. For $f(R)$ models too we can define an energy density ρ_{DE} and pressure p_{DE} satisfying the usual conservation law as follows

$$3H^2 = 8\pi G_* \; (\rho_m + \rho_{DE}) \tag{36}$$

$$-2\dot{H} = 8\pi G_* \; (\rho_m + \rho_{DE} + p_{DE}) \; . \tag{37}$$

The requirement of a conserved energy density still leaves some arbitrariness – this is also true for scalar-tensor DE models – which is removed by requiring that dust besically couples to the expansion like in GR, otherwise we would not obtain a truly Einsteinian form of the Friedmann equations [16]. From eqs.(31),(32) one obtains straightforwardly the expressions

$$8\pi G_* \; \rho_{DE} = \frac{1}{2}(FR - f) - 3H\dot{F} + 3(1 - F) \; H^2 \tag{38}$$

$$8\pi G_* \; (\rho_{DE} + p_{DE}) = \ddot{F} - H\dot{F} - 2(1 - F) \; \dot{H} \tag{39}$$

This prescription (36,37) holds for $f(R)$ DE models considered above in which, as we will see be below, the numerical value of G_* is very close to G.

4. The growth of matter perturbations

The growth of matter perturbations provides a powerful constraint, particularly important in order to distinguish DE models in different theories of gravitation as we will show now. To explain why this is so, it will be sufficient to consider models where the growth of matter perturbations satisfies an equation of the type

$$\ddot{\delta}_m + 2H\dot{\delta}_m - 4\pi G_{\rm eff} \; \rho_m \; \delta_m = 0 \; , \tag{40}$$

valid on scales deep inside the Hubble radius. A model inside GR like ΛCDM would correspond to $G_{\rm eff} = G$. Of course, one can also have models for which the growth of perturbations satisfy still a different equation even inside GR. An equation of the type (40) is found for scalar-tensor DE models [8] with $G_{\rm eff}$ given by (23) when the dilaton field is essentially massless. The quantity $G_{\rm eff}$ encodes the modification of gravity in the weak-field regime due to the presence of the (essentially massless) dilaton in scalar-tensor DE models. It is useful to rewrite (40) in redshift space (we drop below the subscript m)

$$h^2\ \delta'' + \left(\frac{(h^2)'}{2} - \frac{h^2}{1+z}\right)\delta' = \frac{3}{2}(1+z)\frac{G_{\rm eff}(z)}{G}\ \Omega_{m,0}\ \delta\ . \tag{41}$$

Let us consider two models, one inside GR and one outside, that predict exactly the expansion history, in other words they yield the Hubble diagram $H(z)$. It is then clear that the perturbations data $\delta_m(z)$ yield a consistency test. If (41) is satisfied for $G_{\rm eff}(z) = G$, the model inside GR is consistent with the data. However if this is not the case, then the model outside GR with $G_{\rm eff}(z)$ satisfying (41) will be consistent with the data. Therefore, looking at the perturbations we can find whether a model is consistent with the background expansion. This explains why the growth of matter perturbations has become so important in the study of DE models. Another mathematical way to put it is that assuming some functional form $G_{\rm eff}(z)$, we can reconstruct $h(z)$. In particular we can check whether the data are consistent with eq.(40) inside GR i.e. for $G_{\rm eff}(z) = G$.

An equation of the same type is obtained for $f(R)$ DE models with

$$G_{\rm eff} = \frac{G_*}{F}\,\frac{1+4\frac{k^2F'}{a^2F}}{1+3\frac{k^2F'}{a^2F}}\ . \tag{42}$$

We can rewrite this expression in the suggestive way

$$G_{\rm eff} = \frac{G_*}{F}\,\frac{1+4\frac{k^2}{a^2\tilde{H}^2}}{1+3\frac{k^2}{a^2\tilde{H}^2}} = \frac{G_*}{F}\left(1+\frac{\frac{k^2}{a^2\tilde{H}^2}}{1+3\frac{k^2}{a^2\tilde{H}^2}}\right) \tag{43}$$

$$= \frac{G_*}{F}\left(1+\frac{1}{3}\,\frac{\frac{k^2}{a^2m^2}}{1+\frac{k^2}{a^2m^2}}\right) \tag{44}$$

with

$$\tilde{H}^2 = \frac{F}{F'} \equiv 3m^2\ . \tag{45}$$

In equation (41), $G_{\rm eff}$ depends on the cosmic curvature R

$$R = 6\left(2H^2 + \dot{H}\right)\ . \tag{46}$$

We see that in $f(R)$ models $G_{\rm eff}(z,k)$ has an explicit scale dependence in contrast to the scalar-tensor models considered above. To understand the physical meaning of (44) we note that Poisson's equation written in Fourier space with G replaced by the expression (44) for $G_{\rm eff}$ corresponds to a gravitational force whose potential in real space between two test masses is

$$V(r) = -\frac{G_*}{F}\,\frac{M_1\ M_2}{r}\left(1+\frac{1}{3}\ e^{-mr}\right)\ . \tag{47}$$

This corresponds to the modification of gravity in the weak-field regime as it is felt by the cosmological perturbations. In the high curvature regime, the quantity m corresponds to the scalaron mass introduced in [10]. The cosmic background value of the mass m can be large, inducing a finite range $\sim m^{-1}$ for the Yukawa type ("fifth-force") potential. In the high curvature regime $x \gg 1$, $F = 1$ and m is very large so that the usual Newtonian gravitation law is recovered. Hence at high redshifts Poisson's equation is unmodified and the growth of matter perturbations is as usual. However as the universe expands and R decreases, m decreases very steeply so that more and more scales will feel the modification of gravity as the universe expands and cosmic scales will feel it at low redshifts. The growth of matter perturbations is then modified, a crucial signature of these models. An additional modification occurs as F departs from 1. Laboratory experiments, like a Cavendish-type experiment, correspond to the high curvature regime so that no observable deviation from the usual Newtonian potential will manifest itself. It should be stressed that this mechanism easily allows to evade present laboratory constraints probing gravity on very small scales.

As a corollary, in this case it is the bare gravitational constant G_* whose numerical value must be very close to Newton's constant G. This fact was used when we defined ρ_{DE} and p_{DE} through (36,37). To summarize, the model is able through this subtle behaviour to pass laboratory tests of gravitation and at the same time to modify GR on cosmic scales, thereby affecting the growth of matter perturbations.

Matter perturbations can also be described using the "growth function" $f = \frac{d\ln\delta}{d\ln a} \equiv \frac{d\ln\delta}{dx}$. Equation (40) is then recast into

$$\frac{df}{dx} + f^2 + \frac{1}{2}\left(1 - \frac{d\ln\Omega_m}{dx}\right) f = \frac{3}{2}\frac{G_{\rm eff}}{G}\Omega_m \ . \qquad (48)$$

The quantity δ is easily recovered from f as follows

$$\delta(a) = \delta_i \ \exp\left[\int_{x_i}^{x} f(x')dx'\right] \ . \qquad (49)$$

Note that $f = p$ when $\delta \propto a^p$, in particular $f \to 1$ in ΛCDM for large z and $f = 1$ in an Einstein-de Sitter universe. From (44), there are two asymptotic regimes

$$G_{\rm eff} = \frac{4}{3}\frac{G_*}{F} \qquad k \gg am \ , \qquad (50)$$

$$= \frac{G_*}{F} \qquad k \ll am \ . \qquad (51)$$

Analytic solutions in the asymptotic regimes (50,51) can be found during the matter stage when $\Omega_m \approx 1$ and $F \approx 1$. For $C \equiv \frac{G_{\rm eff}}{G_*}\Omega_m =$ constant, we obtain (constant) growing mode solutions $f = p$ with (see e.g.[17])

$$p = \frac{1}{4}\left(-1 + \sqrt{1 + 24C}\right) \ . \qquad (52)$$

A similar modification in the growth of matter perturbations can occur for many models outside GR, but also for models inside GR. We have $C \approx 1$ for $k \ll a\tilde{H}$ and $C \approx \frac{4}{3}$ for $k \gg a\tilde{H}$, so that we obtain [13]

$$\delta_m \propto a \qquad k \ll am \qquad (53)$$

$$\delta_m \propto a^{\frac{\sqrt{33}-1}{4}} \qquad k \gg am \ . \qquad (54)$$

When subhorizon scales go from the first into the second regime, this results in a scale-dependent increase in the growth of matter perturbations. With the universe expansion, perturbations on ever increasing scales will feel the modification of gravity corresponding to the Yukawa-type potential. The quantity γ defined as

$$f = \Omega_m^{\gamma} \ . \qquad (55)$$

provides a powerful way to "parameterize" the growth of perturbations. For ΛCDM γ has a very distinctive behaviour: $\gamma_{\Lambda CDM} = \gamma_0 - 0.02z$ to exquisite accuracy [21], with $\gamma_0 \equiv \gamma(z=0) \approx 0.55$ only slightly dependent on $\Omega_{m,0}$. In the context of DE models this approach was revived as it could provide a powerful way to discriminate modified gravity models from ΛCDM [19, 18] and this could be probed on low redshifts with weak lensing data, see e.g. [20]. A quasi-constant γ holds only in particular cases. When $G_{\rm eff}$ is scale-independent but depends on z, and also for models inside GR satisfying (41), γ is a priori some unknown function of z and we should write $f = \Omega_m^{\gamma(z)}$. Starting from (48) we easily obtain the following equation

$$-(1+z)\ \ln\Omega_m\ \gamma' + \Omega_m^\gamma + \frac{1}{2}(1 + 3(2\gamma - 1)\ w_{\rm eff}) = \frac{3}{2}\ \frac{G_{\rm eff}(z)}{G}\ \Omega_m^{1-\gamma}\ , \tag{56}$$

where $w_{\rm eff} \equiv w_{DE}\ \Omega_{DE}$, $\Omega_{DE} = 1 - \Omega_m$ and $\gamma' \equiv \frac{d\gamma}{dz}$. From (56) the following equality is easily obtained [21]

$$\gamma_0' = \left[\ln\Omega_{m,0}^{-1}\right]^{-1}\ \left[-\Omega_{m,0}^{\gamma_0} - 3(\gamma_0 - \frac{1}{2})\ w_{\rm eff,0} + \frac{3}{2}\ \frac{G_{\rm eff,0}}{G}\Omega_{m,0}^{1-\gamma_0} - \frac{1}{2}\right]\ . \tag{57}$$

A model with constant γ should satisfy in particular $\gamma_0' \equiv \frac{d\gamma}{dz}(z=0) = 0$. For ΛCDM we have $\gamma_0 \equiv \gamma(z=0) \approx 0.55$ and $\gamma_0' \approx -0.015$. Substantial departure from these values would signal that we are not living in a ΛCDM universe. For DE models inside GR for which (40) holds, small slopes $|\gamma_0'| \le 0.02$ are obtained for smoothly varying $w_{DE}(z)$. Hence a large value for γ_0' and in general large variations of γ on low redshifts could be a hint for a DE model outside GR [21]. This was indeed shown to hold for some scalar-tensor DE models investigated in [17]. As we will see large slopes do also accur for some $f(R)$ DE models [16, 22].

For theories where the growth of perturbations $\delta_m(z,k)$ is scale dependent, where there is spatial dispersion in the growth of perturbations, we have a further generalization and we should write in full generality

$$\gamma = \gamma(z,k)\ . \tag{58}$$

As we have seen such a generalization applies to $f(R)$ DE models . The non trivial thing is to investigate the behaviour of γ today on very low redshifts where it could be directly probed. Equations equivalent to (56,57) where now $G_{\rm eff}$ depends also on k can be derived resulting in a constraint equation of the type [23]

$$C\left(\gamma_0(k),\ \gamma_0'(k),\ \Omega_{m,0},\ w_{\rm eff,0},\ \frac{G_{\rm eff,0}(k)}{G}\right) = 0\ . \tag{59}$$

For the scalar-tensor DE models considered earlier, $G_{\rm eff}$ depends only on Z and in addition $G_{\rm eff,0} = G$ to high accuracy, see eq.(24). Even in that case large slopes can be obtained due to the modification of gravity in the past resulting in a different value γ_0 today but of course no dispersion can occur. The constraint (59) could serve as a consistency check for $f(R)$ models when the relevant cosmic scales today are in the asymptotic regime (50). However this requires accurate knowledge of the background quantities $\Omega_{m,0}$,$w_{\rm eff,0}$ and H_0 [23].

The growth of matter perturbations and more specifically the behaviour of γ on very low redshifts was investigated in most of the viable DE models in [24]. Depending on the parameter values of the models different possibilities arise for the relevant scales

$$0.01\ h\,{\rm Mpc}^{-1} \le k \le 0.2\ h\,{\rm Mpc}^{-1}\ . \tag{60}$$

It is possible to distinguish three essentially different cases:
(i) γ_0 is close to the ΛCDM value $\gamma_0 = 0.55$ for all modes, small deviations of the order of a few percents at most can arise. In this case γ_0' is also small and there is no large variation of γ on low redshifts. This corresponds to the large-scales asymptotic regime $\lambda \gg \lambda_c$ where $\lambda_c \equiv \frac{2\pi}{m}$

gives the range of the "fifth-force" and defines the scale below which gravitational modifications are felt.
(ii) $\gamma_0 \approx 0.40$ for all modes with $0.40 \leq \gamma_0 \leq 0.43$. This corresponds to the small-scales asymptotic regime $\lambda \ll \lambda_c$. The slope γ'_0 is very large and γ has a large variation on low redshifts. This behaviour was first found in [16].
(iii) The values of γ_0 show a clear dispersion with $0.40 \leq \gamma_0 \leq 0.55$. In this case the growth is scale-dependent corresponding to scales which are in the intermediate regime between the two asymptotic regime defined above. One can have very large slopes γ'_0 and large variations of γ on low redshifts which are very different for different scales.

To find out accurately when the asymptotic regimes are reached, and the values of γ_0 in the intermediate regime, one has to resort to numerical calculations. We are led to the important conclusion that it is possible to distinguish $f(R)$ DE models from ΛCDM through the behaviour of the parameter γ on very low redshifts. This will presumably be probed by future observations, like weak lensing data. High precision observations probing both the background expansion and the growth of matter perturbations can potentially discriminate between modified gravity DE models and ΛCDM. Decisive progress about our knowledge of DE will come from these observations rather than from more or less appealing or well-motivated principles. In the meantime we have to explore in details all observationally viable DE models able to explain the accelerated expansion rate of the universe. We can expect that future observations will considerably restrict the number of potential candidates.

Acknowledgments
I would like to thank the organizers, and in particular L. Perivolaropoulos, for inviting me to this very enjoyable conference. It is a pleasure to thank my collaborators L. Amendola, B. Boisseau, G. Esposito-Farèse, R. Gannouji, B. Moraes, A. Ranquet, A. A. Starobinsky and S. Tsujikawa.

[1] Perlmutter S J et al 1997 Ap. J. **483** 565; Nature 1998 **391** 51; Riess A G, Filippenko A V, Challis P et al. 1998 Astron. J. **116** 1009; Perlmutter S J, Aldering G, Goldhaber G et al. 1999 Astroph. J. **517** 565; Astier P, Guy J, Regnault N et al. 2006 Astron. Astroph. **447**, 31; Riess A G et al. 2007 Astrophys. J. **659**, 98
[2] Sahni V, Starobinsky A A 2000 Int. J. Mod. Phys. D **9** 373; Padmanabhan T 2003 Phys. Rept.**380** 235; Sahni V, Starobinsky A A 2006 Int. J. Mod. Phys. D **15** 2105; Copeland E J, Sami M, Tsujikawa S 2006 Int. J. Mod. Phys. D **15** 1753; R. Durrer and R. Maartens arXiv:0811.4132
[3] Ratra B, Peebles P J E 1998 Phys. Rev. D **37** 3406; Wetterich C 1998 Nucl. Phys. B **302** 668
[4] Caldwell R 2002 Phys. Lett. B **545** 23; Dabrowski M P, Stachowiak T, Szydlowski M 2003; Phys. Rev. D **68** 103519; Bo Feng, Wang X L, Zhang X M 2005 Phys. Lett.B **607** 35; Hao Wei, Rong-Gen Cai, Ding-Fang Zeng 2005 Class. Quantum Grav. **22** 3189
[5] Alam U, Sahni V, Saini T D, Starobinsky A A 2004 Mon. Not. Roy. Astron. Soc. **354** 275
[6] Ishak M, Upadhye A, Spergel D 2006 Phys. Rev.D **74** 043513
[7] Esposito-Farèse G, Polarski D 2001 Phys. Rev. D **63** 063504; Bartolo N, Pietroni M 2000 Phys. Rev. D **61** 023518; Perrotta F, Baccigalupi C, Matarrese S 2000 Phys. Rev. D **61** 023507; Fujii Y, Maeda K 2003 *The scalar-tensor theory of gravitation* (Cambridge:Cambridge Univ. Press); Faraoni V 2004 *Cosmology in Scalar-Tensor Gravity* (Dordrecht:Kluwer Academic); Martin J, Schimd C, Uzan J-P 2006 Phys. Rev. Lett.**96** 061303; Nesseris S, Perivolaropoulos L 2007 JCAP **0701**, 018; Demianski M, Piedipalumbo E, Rubano C, Scudellaro P arXiv:0711.1043
[8] Boisseau B, Esposito-Farèse G, Polarski D, Starobinsky A A 2000 Phys. Rev. Lett. **85** 2236
[9] Gannouji R, Polarski D, Ranquet A, Starobinsky A A 2006 JCAP **0609** 016
[10] Starobinsky A A 1980 Phys. Lett. B **91** 99
[11] Capozziello S 2002 Int. J. Mod. Phys. D **11** 483
[12] Amendola L, Polarski D, Tsujikawa S 2007 Phys. Rev. Lett. **98** 131302; Amendola L, Gannouji R, Polarski D, Tsujikawa S 2007 Phys. Rev. D **75** 083504
[13] Starobinsky A A 2007 JETP Lett.**86** 157
[14] Hu W, Sawicki I 2007 Phys. Rev. D **76** 064004
[15] Appleby S A, Battye R A 2007 Phys. Lett. B **654** 7
[16] Gannouji R, Moraes B, Polarski D 2009 JCAP **0902** 034

[17] Gannouji R, Polarski D 2008 JCAP **0805** 018
[18] E. V. Linder 2005 Phys. Rev. D **72** 043529; D. Huterer and E. V. Linder 2007 Phys. Rev. D **75** 023519
[19] Wang L M, Steinhardt P J 1998 Astrophys. J. **508** 483
[20] Heavens A F, Kitching T D, Verde L 2007 Mon. Not. Roy. Astron. Soc. **380** 1029
[21] Polarski D, Gannouji R 2008 Phys. Lett. B **660** 439
[22] Motohashi H, Starobinsky A A, Yokoyama J 2010 Prog. Theor. Phys. **123** 887
[23] Gannouji R, Moraes B, Polarski D arXiv:0907.0393
[24] Tsujikawa S, Gannouji R, Moraes B, Polarski D 2009 Phys. Rev. D **80** 084044

Sgr A*: The Optimal Testbed of Strong-Field Gravity

Dimitrios Psaltis and Tim Johannsen

Astronomy and Physics Departments, The University of Arizona, 933 N. Cherry Ave, Tucson, AZ 85721, USA

E-mail: `dpsaltis@email.arizona.edu, timj@physics.arizona.edu`

Abstract. The black hole in the center of the Milky Way has been observed and modeled intensely during the last decades. It is also the prime target of a number of new experiments that aim to zoom into the vicinity of its horizon and reveal the inner working of its spacetime. In this review we discuss our current understanding of the gravitational field of Sgr A* and the prospects of testing the Kerr nature of its spacetime via imaging, astrometric, and timing observations.

1. Introduction

The Kerr spacetime of spinning black holes is one of the most unexpected predictions of Einstein's theory of General Relativity. The special role this spacetime plays in the theory of gravity is encapsulated in the no-hair theorem, which states that the Kerr metric is the only stationary, axisymmetric, asymptotically flat solution to the vacuum field equations that possesses a horizon but no closed timelike loops (Israel 1967, 1968; Carter 1971, 1973; Hawking 1972; Robinson 1975; Mazur 1982). Because of this theorem, it is expected that all astrophysical objects that have been identified as black-hole candidates are indeed described by the Kerr metric.

The Kerr nature of astrophysical black holes is a prediction that can be tested observationally. Doing so offers the unique opportunity of both rejecting alternative interpretations of their nature and of verifying General Relativity in the strong-field regime. There are at least three astrophysical settings we know of today, in which black holes appear to exist: in galactic X-ray binary systems, as ultraluminous X-ray sources, and in active galactic nuclei. Among all these black holes, the one in the center of our galaxy, Sgr A*, combines a large mass with a small distance from the sun (e.g., Ghez et al. 2008; Gillessen et al. 2009). It can also be probed by a variety of observations throughout the electromagnetic spectrum making it the optimal test bed of the Kerr metric.

Recent and anticipated advances in observations of Sgr A* throughout the electromagnetic spectrum have both secured our understanding of the basic properties of this black hole (e.g., Reid 2009), and opened new opportunities for devising tests of gravity theories. At the same time, there have been significant recent advances in the development of a theoretical framework with which observations of black holes can be used to test quantitatively the Kerr metric and search for violations of the no-hair theorem.

In this article, we review our current understanding of the properties of the black hole in the center of the Milky Way and discuss the prospect of testing the no-hair theorem with upcoming observations.

Journal of Physics: Conference Series **283** (2011) 012030 doi:10.1088/1742-6596/283/1/012030

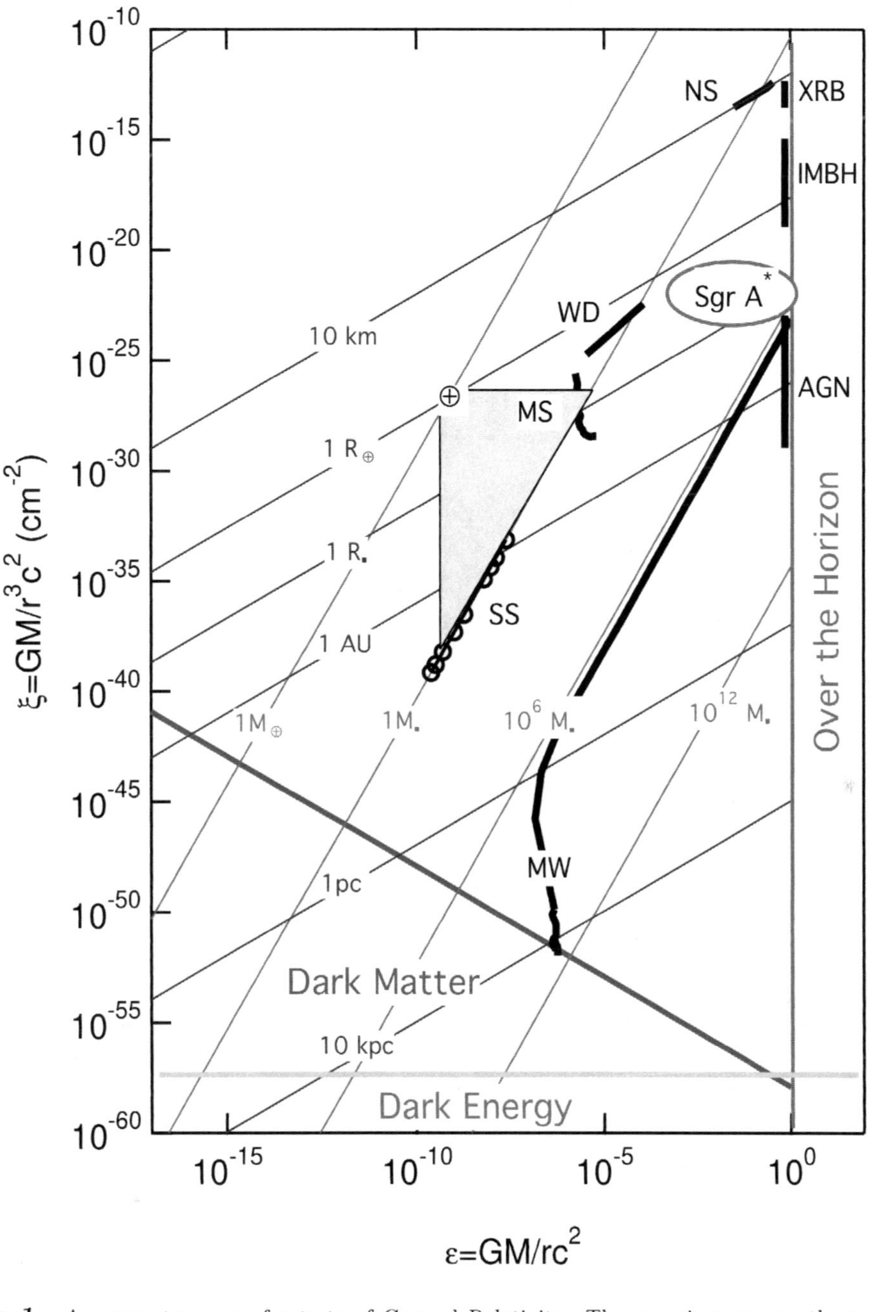

Figure 1. A parameter space for tests of General Relativity. The $x-$axis measures the potential and the $y-$axis the curvature of the field probed by different experiments. A number of astrophysical and cosmological objects are also shown. The black hole in the center of the Milky Way, Sgr A*, probes a region of the parameter space that has not been investigated with other current tests; the location of the latter is outlined by the inverted yellow triangle in the center of the figure (Psaltis 2008).

2. Sgr A*: The Optimal Black Hole for Tests of the No-Hair Theorem

The black hole in the center of the Milky Way is the optimal astrophysical object for testing our understanding of black hole spacetimes, for a number of reasons that we outline below.

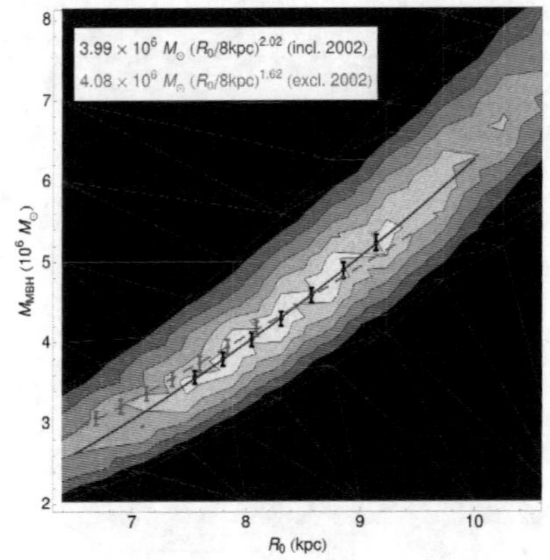

Figure 2. Confidence contours of the mass and distance to Sgr A*, based on the analysis of orbits of nearby stars (Gillessen et al. 2009).

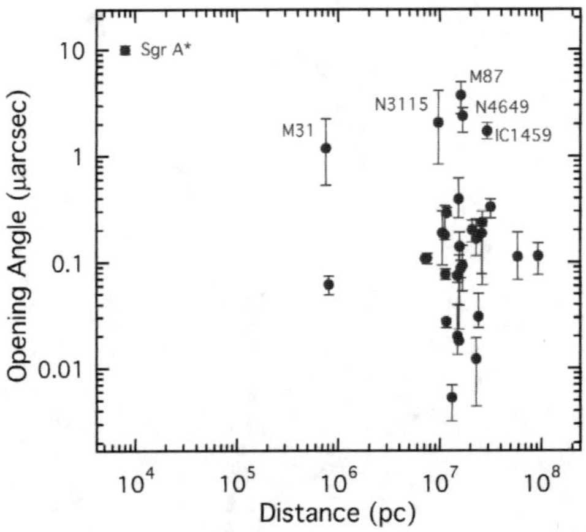

Figure 3. The opening angles, as viewed by an observer on Earth, of the horizons of a number of supermassive black holes in distant galaxies with a secure dynamical mass measurement (sample of Tremaine et al. 2002). The opening angle of the black hole horizon in the center of the Milky Way (Sgr A*) is also shown for comparison (after Psaltis 2008).

- **Sgr A* probes a new regime of gravitational field strengths.** General Relativity has been tested in a rather limited range of gravitational field strengths (Fig. 1). Phenomena in the vicinity of the black-hole horizon around Sgr A* probe gravitational fields with a potential and a curvature that are five orders of magnitude larger than the strongest fields probed by current tests of General Relativity.
- **Sgr A* has an accurately measured mass.** Fitting the trajectories of stars in the vicinity of Sgr A* has led to an accurate measurement of its mass. A recent careful analysis of the orbits by Gillessen et al. (2009; see also Ghez et al. 2008) resulted in a measurement of $M = (4.31 \pm 0.06 \pm 0.36) \times 10^6 M_\odot$ with the second uncertainty corresponding to the error in the estimate of the distance to the center of the Milky Way (see Fig. 2). This highly accurate mass measurement specifies one of the only two basic parameters of the black hole and sets uniquely the scale for any anticipated relativistic phenomena.
- **The mass measurement provides an independent distance measurement to Sgr A*.** Testing the Kerr spacetime via, e.g., analysis of the images of its inner accretion

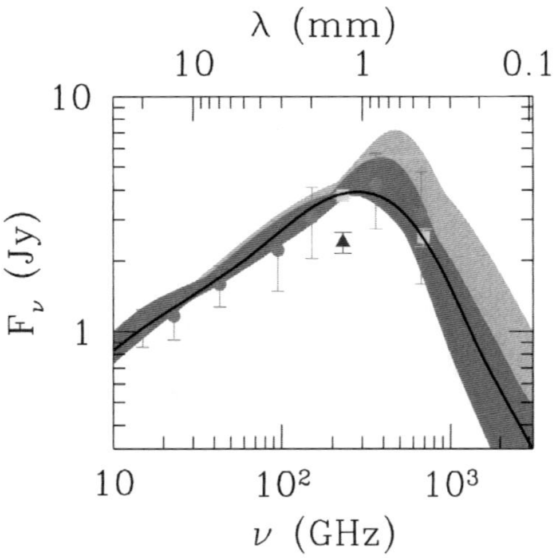

Figure 4. The long-wavelength spectrum of Sgr A*. The grey filled areas show the envelopes of semi-analytic models that best match the data (Broderick et al. 2009).

flow, requires an independent knowledge of the distance to the source in order to convert angular separations to physical lengths. Studies of the orbits of stars that approach Sgr A* lead to a measurement of the distance to the source of $R_0 = 8.33 \pm 0.35$ kpc (Gillessen et al. 2009).

- **Sgr A* is one of the only two black-holes with horizons that will be resolved with sub-mm VLBI in the very near future.** For a supermassive black hole in a distant galaxy, the opening angle of the horizon as viewed from the earth is

$$\theta = 20 \left(\frac{M}{10^9 M_\odot}\right)\left(\frac{1 \text{ Mpc}}{D}\right) \mu\text{arcsec} . \quad (1)$$

This is shown in Figure 3 for a number of supermassive black holes with secure mass determinations. Sgr A* is the black hole that combines the highest brightness with the largest angular size of the horizon.

- **At photon frequencies $> 5 \times 10^{11}$ Hz, the emission from Sgr A* is optically thin.** The long-wavelength spectrum of Sgr A* peaks at a frequency of $\simeq 5 \times 10^{11}$ Hz, suggesting that the emission changes from optically thick (probably synchrotron emission) to optically thin at a comparable frequency (see Fig. 4). As a result, observations at frequencies comparable to or higher than this transition frequency are expected to probe the region close to the black-hole horizon without significant obscuration. This fact is indeed confirmed by the measurement of the size of Sgr A* at this part of the spectrum (see below).
- **The size of the emitting region in Sgr A* at long wavelengths is comparable to the horizon scale.** Since the first measurements of the size of the source at 7 mm and at 1.4 mm (Krichbaum et al. 1998) demonstrated that the emitting region is only a few times larger than the radius of the horizon (see Figure 5), a number of observational investigations have aimed to probe deeper into the gravitational field of the black hole. Most recently, observations at 1.3 mm (Doeleman et al. 2008) revealed near-horizon sizes for the emitting region of Sgr A*.
- **The detailed structure of the inner accretion flow around Sgr A* will be imaged in the very near future.** Figure 6 shows two simulated images of Sgr A* at 345 GHz

Journal of Physics: Conference Series **283** (2011) 012030 doi:10.1088/1742-6596/283/1/012030

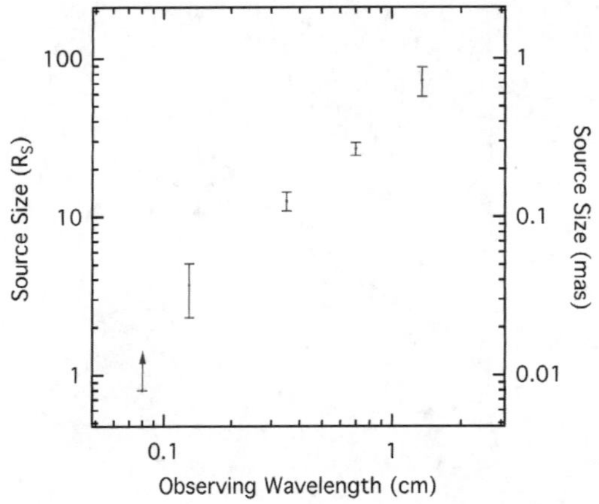

Figure 5. The major axis of the accretion flow around the black hole in the center of the Milky Way, as measured at different wavelengths, in units of the Schwarzschild radius (left axis) and in milliarcsec (right axis; adapted from Shen et al. 2005 and Doeleman et al. 2008; after Psaltis 2008). Even with current technology, the innermost radii of the accretion flow can be readily observed.

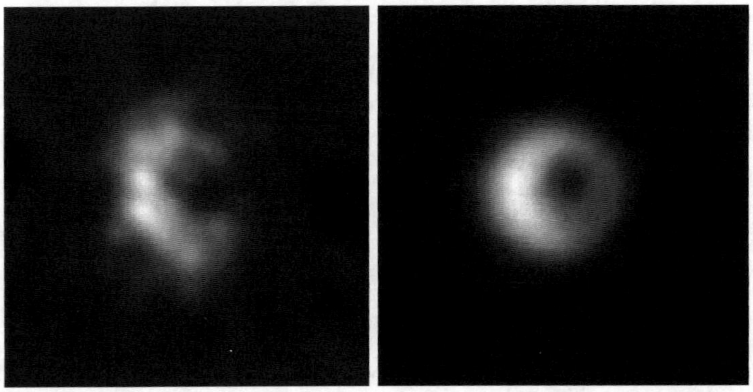

Figure 6. Simulated images of the inner accretion flow around Sgr A*, as they will be resolved with *(left)* a 7-telescope array in the near future (3-5 years) and *(right)* a 13-telescope array (Fish & Doeleman 2009).

as they will be resolved by a 7-telescope array in the near future (3-5 years) and by a 13-telescope array further in the future (Fish & Doeleman 2009). The characteristic asymmetry in the image caused by Lorentz boosting as well as the shadow of the black hole are clearly visible.

3. Parametrizing Deviations from the Kerr Metric

The Kerr metric, which describes the exterior spacetime of a spinning black hole, is not unique to General Relativity. Indeed, a large number of simple extensions to the general relativistic field equations are also satisfied by the Kerr solution (Psaltis et al. 2008). This is perhaps not surprising (see, e.g., Barausse & Sotiriou 2008), given the fact that the Kerr metric is derived by solving the vacuum field equation

$$R_{\mu\nu} = 0 \,, \tag{2}$$

where $R_{\mu\nu}$ is the Ricci tensor, and not the full Einstein field equation (the two are equivalent in vacuum). Equation (2), however, is also satisfied by the Minkowski metric of special relativity. As a result, the field equations of any theory that has special relativity as its vacuum limit should also be equivalent in vacuum to equation (2) and, therefore, be satisfied by the Kerr metric. This argument remains valid even when we allow for the possibility of a finite cosmological constant (see Psaltis et al. 2008).

Constructing a modification to General Relativity in which the black hole solutions are not described by the Kerr metric can be done, in principle, by violating one of the more fundamental assumptions in our understanding of gravity, such as the equivalence principle (see, e.g., Eling & Jacobson 2006). The only black-hole solution in a theory that obeys the equivalence principle that we are aware of is that of Yunes & Pretorius (2009) and Konno et al. (2009) in Chern-Simons gravity. In Chern-Simons gravity, a parity violating term is added to the Einstein-Hilbert action of General Relativity, with a coupling constant that is a dynamical field (see Alexander & Yunes 2009 for a review). This parity violating term is zero for a parity-symmetric metric and, therefore, the Minkowski and Schwarzschild spacetimes remain solutions to the field equations. However, the spacetime of a spinning black hole has a particular parity specified by its spin and, therefore, it is different than the Kerr metric.

Given the very limited range of available black-hole metrics in modified gravity theories that is known today, it is more fruitful to test the Kerr metric in a phenomenologial way, by adding parametrically terms to its various elements (see Psaltis 2009 for a discussion). This is equivalent to testing for violations of the no-hair theorem, even within General Relativity.

The no-hair theorem states that the only stationary, axisymmetric, asymptotically flat vacuum solution to the Einstein field equations that possesses a horizon and no time-like loops is the Kerr metric (we do not consider here the unlikely possibility that an astrophysical black hole will have a net charge). The Kerr metric is uniquely determined by only two parameters, the mass and the spin of the black hole. This allows us to define a formal test of the no-hair theorem, based on the work of Ryan (1995), in the following way (see also Collins & Hughes 2004; Glampedakis & Babak 2006; Vigeland & Hughes 2010; Vigeland 2010).

We can, in principle, expand the exterior metric of any compact object in multipoles and use observations to measure the coefficients of the expansion. Because of the no-hair theorem, only two of the multipole coefficients for the spacetime of a black hole are independent. The coefficient of the monopole is the mass M of the black hole and of the dipole is its spin a. All higher-order coefficients will depend on the first two, in the particular way dictated by the Kerr metric. Testing the no-hair theorem requires measuring at least the coefficient of the quadrupole q and verifying whether it satisfies the Kerr relation $q = -a^2$.

Four different approaches have been explored so far for introducing additional non-Kerr hair to the spacetimes of compact objects. Ryan (1995) studied a general expansion of stationary, axisymmetric, asymptotically flat spacetimes in Geroch-Hansen multipoles. Collins & Hughes (2004) as well as Vigeland & Hughes (2010) added Weyl-sector bumps to the Schwarzschild and Kerr spacetimes. Glampedakis & Babak (2006) modified the Kerr metric by adding an arbitrary quadrupole moment while ensuring that the metric to that order remains a solution to the vacuum field equation (2). Finally, Gair et al. (2008) explored in detail the parametric solution of Manko & Novikov (1992), which allows for deviations of all higher-order multipoles from their Kerr values. All these approaches were developed originally in order to test General Relativity with future observations of the gravitational waves generated during inspirals into supermassive black holes. However, they are also directly applicable to tests of gravity with Sgr A*.

In Johannsen & Psaltis (2010a, 2010b, 2010c), we focused on the parametric post-Kerr spacetime obtained by Glampedakis & Babak (2006) for two reasons. First, this approach uses a single parameter associated to the quadrupole moment of the spacetime to quantify potential deviations from the Kerr metric, making it the simplest and most concise possible avenue to testing the no-hair theorem. Second, the complete metric of Glampedakis & Babak (2006) remains a valid solution to the vacuum Einstein field equations, allowing us to perform

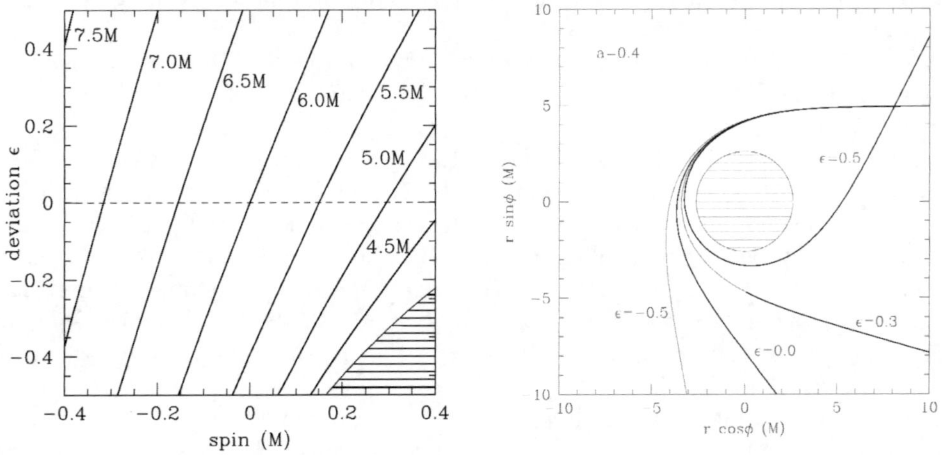

Figure 7. The dependence of *(Left)* the radius of the innermost stable circular orbit around a black hole and of *(Right)* the trajectories of photons on the black-hole spin a and on the parameter ϵ that measures the degree of violation of the no-hair theorem (Johannsen & Psaltis 2010a, 2010c).

a self-consistent test of the no hair theorem and of the black-hole identification of the compact object, within General Relativity.

The key drawback of all currently proposed parametric deviations from the Kerr metric is that they cannot be used in describing the exterior spacetimes of rapidly spinning black holes. Precisely because of the no-hair theorem, it is impossible to construct a spacetime that deviates from the Kerr metric, is a solution to the general relativistic vacuum field equations, and is regular everywhere outside the black-hole horizon; such a construction would prove that the no-hair theorem is violated. In all parametric deviations from the Kerr metric, the spacetimes become irregular at radii $\simeq 2M$ (see Gair et al. 2008; Johannsen & Psaltis 2010a). For moderate black-hole spins, radii comparable to $2M$ are much smaller than the radius of the photon orbit and of the innermost stable circular orbit. As a result, such radii can be artificially excised, without affecting the prediction of any observable. For rapidly spinning black holes, however, all characteristic radii become $\leq 2M$ and, therefore, the irregularities of the spacetimes preclude us from calculating various observables related, e.g., to the inner accretion flow around a black hole. We do not anticipate this to introduce significant problems for the case of Sgr A*, because preliminary attempts to model the VLBI images from the source show preference for small values of its spin ($a \leq 0.3$; Broderick et al. 2009, 2010).

4. Testing the No-Hair Theorem with Sgr A*

In Johannsen & Psaltis (2010a), we explored in detail the properties of the Glampedakis & Babak (2006) metric, addressing its potential in testing the no-hair theorem with astrophysical observations in the electromagnetic spectrum. Following their original analysis, we expressed the coefficient of the quadrupole multipole of the spacetime as

$$q = -(a^2 + \epsilon) \tag{3}$$

with the parameter ϵ measuring the degree of violation of the no-hair theorem. We then studied the trajectories of photons and particles in this spacetime and identified three important effects of the presence of a non-Kerr quadrupole.

Journal of Physics: Conference Series **283** (2011) 012030 doi:10.1088/1742-6596/283/1/012030

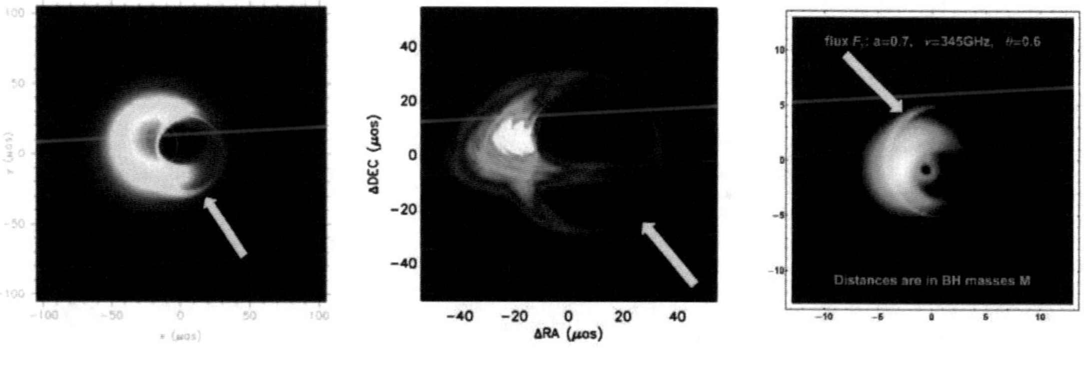

Figure 8. The presence of a bright photon ring surrounding the black-hole shadow is ubiquitous in all current simulations of images from radiatively inefficient accretion flows (from left: Moscibrodzka et al. 2009; Dexter et al. 2009; Shcherbakov & Penna 2010). The size, location, and shape of the photon ring is determined by the projected photon orbit at infinity and, as such, it depends only on the metric of the spacetime. The accretion flow is necessary only to provide the photons that will trace the photon ring for an observer at infinity.

First, the location of the innermost stable circular orbit (ISCO) is significantly altered, as shown in Figure 7 (left panel). The location of the maximum emission from the accretion flow around Sgr A* is expected to be very close to that of the ISCO (see, e.g., Krolik & Hawley 2002 and references therein) and, therefore, the brightness profile of the image from Sgr A* will depend on the value of the quadrupole. Second, the radius of the photon orbit is also significantly affected by the value of the quadrupole. The radius of the photon orbit determines the size of the shadow of the black hole (see Bardeen 1973; Falcke et al. 2000), and generates one of the most prominent features in the image of Sgr A* (i.e., the structure that we call the photon ring; see below). Finally, the detailed trajectories of photons that propagate close to the photon orbit depend also on the magnitude of the quadrupole (see Fig. 7, right panel) leading to non-trivial deformations of the calculated images.

Figure 7 also shows, that changing the quadrupole moment of a black-hole spacetime may even alter qualitatively the properties of particle and photon orbits in the vicinity of its horizon. In the Glampedakis & Babak (2006) metric, for each value of the spin, there exists a minimum quadrupole deviation beyond which all circular orbits are stable to radial perturbations (Johannsen & Psaltis 2010c; this results has also been discussed for the Manko & Novikov metric by Gair et al. 2008). For relatively large black-hole spins ($a > 0.4$), any reduction of the quadrupole moment of the spacetime from its Kerr value stabilizes all circular orbits against radial perturbations. In those spacetimes, equatorial particle orbits close to the black-hole horizon become instead unstable to vertical perturbations (see Gair et al. 2008; Johannsen & Psaltis 2010c) with implications for the inner accretion flows that have not been explored yet.

In Johannsen & Psaltis (2010b) we investigated the prospect of measuring the parameter ϵ for Sgr A* via imaging observations in the sub-mm. We identified in the simulated images of the accretion flow a bright ring (which we called the photon ring) that surrounds the black-hole horizon (see also, e.g., Beckwith & Done 2005). This ring is produced by photon rays that orbit the black hole multiple times at the radius of the photon orbit before escaping to infinity. Its presence is ubiquitous in all current simulations of images from radiatively inefficient accretion flows (see Fig. 8).

The location, size, and geometry of the photon ring corresponds to the projection of the photon orbit on the image plane of the distant observer. Because of this, its geometric

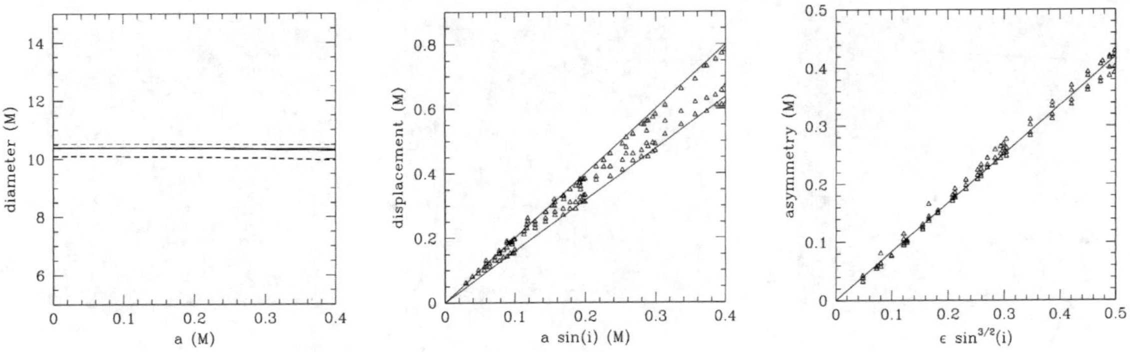

Figure 9. Three panels showing the *(Left)* diameter, *(Middle)* displacement, and *(Right)* asymmetry of the bright photon ring surrounding the black-hole shadow in the image of an accreting black hole, as a function of the parameters of the quasi-Kerr metric. The three depicted properties of the ring measure three different multipoles of the spacetime (modulo the sine of the observer's inclination $\sin i$) and can lead to a test of the no-hair theorem (Johannsen & Psaltis 2010b).

characteristics are independent of the particular accretion flow model assumed. The accretion flow is necessary only to provide the photons that will trace the photon ring for an observer at infinity. The diameter of the bright photon ring is $\sim 10.0 - 10.5\ M$, practically independent of the spin and the quadrupole moment of the black hole, providing a direct measure of the black-hole mass (Figure 9). The displacement of the ring from the center of mass of the system depends primarily on the spin of the black hole. Finally, the deviation of the shape of the photon ring from a circle is a direct measure of the violation of the no-hair theorem (modulo the observer's inclination). Indeed, photon rings for the Kerr metric remain practically circular for all but the fastest spins ($a \leq 0.9$), whereas even a small degree of violation of the no-hair theorem (as measured by the parameter ϵ) introduces a significant asymmetry to the photon ring (see Figure 9). Imaging the inner accretion flow and measuring the asymmetry of the photon ring offers, therefore, the possibility of a clean, quantitative test of the no-hair theorem.

5. Distinguishing Gravitational Effects from Astrophysical Complications

The biggest challenge in searching for violations of known physics with astrophysical measurements is ensuring that a particular measurement is not affected adversely by astrophysical complications. In the previous section, we identified a signature of gravitational effects that probes cleanly the metric of the black hole and is affected only marginally by other astrophysics, i.e., the bright photon ring that surrounds the shadow of the black hole. The location, size, and shape of the photon ring depends predominantly on the spin and quadrupole of the spacetime and only marginally on the properties of the underlying accretion flow. Even in this case, however, the narrow width of the ring (of order a few tens of M), as well as possible confusion with other emission from the accretion flow between the observer and the horizon may make such a measurement biased.

We can convert a compelling argument for possible violations of the no-hair theorem into a bullet proof result by using different observational probes for the spin and quadrupole moment of the same black hole and by testing whether all such probes agree quantitatively. Sgr A* is unique in that it potentially offers two additional, independent ways of testing the no-hair theorem.

Will (2008) showed that astrometric observations of the orbits of stars clustered very close to the black hole can be used to measure the quadrupole moment of the black hole via detection

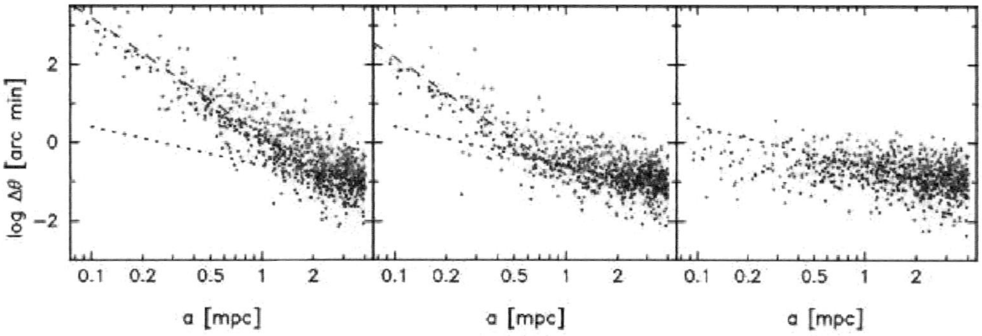

Figure 10. The cummulative change in the angle $\delta\theta$ between the initial and final orbital angular momenta of stars around Sgr A* with different orbital semi-major axes a. The dashed lines show the effect of relativistic frame-dragging while the dotted lines correspond to an approximate model of precession due to stellar perturbations. The three panels show the result for black hole spins *(left)* $a = 1$, *(middle)* $a = 0.1$, and *(right)* $a = 0$ (Merritt et al. 2010).

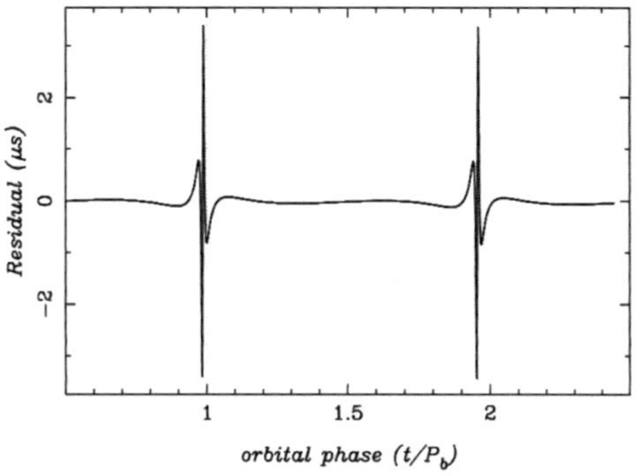

Figure 11. Typical residuals in the timing solution for a radio pulsar orbiting a black hole, when the effects of the quadrupole moment of the black hole spacetime have not been taken into account (Wex & Kopeikin 1999). The residuals shown here have been calculated for a $10^4 M_\odot$ black hole; they will be similar in shape but larger in magnitude for a pulsar orbiting Sgr A*.

of frame dragging. In a follow-up study, Merritt et al. (2010) showed that interactions between the stars of the cluster will mask frame dragging effects, unless the analysis is confined to stars within $\simeq$ 1000 Schwarzschild radii from the horizon (see Fig. 10). If such stars exist, future instruments (such as GRAVITY; Bartko et al. 2009) will be able to provide an independent measurement of the quadrupole moment of the black-hole spacetime.

The presence of a cluster of massive stars around Sgr A* makes it practically certain that a number of radio pulsars will also be orbiting the black hole. Current surveys put an upper limit of as many as 90 normal pulsars within the central parsec of the Galaxy (e.g., Macquart et al. 2010). Detecting such a pulsar in one of the current surveys will allow us to measure the quadrupole moment of the black hole by looking for particular spin-orbit coupling residuals in the timing solution for each pulsar (see Fig. 11; Wex & Kopeikin 1999).

The combination of observations of the images from the inner accretion flow around Sgr A*, of the relativistic precession of massive stars orbiting close to the black hole, and of the residuals in the timing solution of radio pulsars in similar orbits will allow for three, independent measurements of the spin and of the quadrupole moment of the black hole. Each observation will be performed at different wavelengths and with different telescopes. Moreover, each of the three observed phenomena will probe different regions of the black-hole spacetime and will be affected by different systematics. If all three measurements give consistent solutions for the black-hole spin and quadrupole moment, then this will lead to an unequivocal test of the no-hair theorem with an astrophysical black hole.

References

[1] Alexander S & Yunes N 2009 *Phys. Rep.* **480** 1
[2] Barausse E & Sotiriou T 2008 *PRL* **100** 099001
[3] Bardeen J 1973, *in Black Holes*, eds. DeWitt & DeWitt, 215
[4] Bartko H et al 2009 *New Astr. Rev.* **53** 301
[5] Beckwith K & Done C 2005 *MNRAS* **359** 1217
[6] Broderick A E et al. 2009 *ApJ* **697** 45
[7] Broderick, A E et al. 2010 arXiv:1011.2770
[8] Carter B 1971 *Phys. Rev. Lett.* **26** 331
[9] ———. 1973 in *Black Holes* eds. DeWitt and DeWitt
[10] Collins N A & Hughes S A 2004 *PRD* **69** 124022
[11] Dexter J Agol E & Fragile P C 2009 *ApJ* **703** L142
[12] Doeleman S et al. 2008 *Nature* **455** 78
[13] Dolence J C Gammie C F Mościbrodzka M & Leung P K 2009 *ApJS* **184** 387
[14] Eling C & Jacobson T 2006 *CQG* **23** 5643
[15] Fish, V L & Doeleman S S 2009 *IAU Symposium 261*, arXiv:0906.4040
[16] Falcke H Melia F & Agol E 2000 *ApJ* **528** L13
[17] Gair J R Li C & Mandel I 2008 *Phys. Rev. D* **77**, 024035
[18] Ghez A M et al 2008 *ApJ* **689** 1044
[19] Gillessen S et al 2009 *ApJ* **692** 1075
[20] Glampedakis K & Babak S 2006 *CQG* **23** 4167
[21] Hawking S W 1972 *Commun. Math. Phys.* **25** 152
[22] Israel W 1967 *Phys. Rev.* **164** 1776
[23] ———. 1968 *Commun. Math. Phys.* **8** 245
[24] Johannsen T & Psaltis D 2010a *ApJ* **716** 187
[25] ——— 2010b *ApJ* **718** 446
[26] ——— 2010c *ApJ* in press, arXiv:1010.1000
[27] Konno K Matsuyama T and Tanda S 2009 *Progr. Theoret. Phys.* **122** 561
[28] Krolik J H & Hawley J F 2002 *ApJ* **573** 754
[29] Krichbaum et al 1998 *A&A* **335** L106
[30] Lo K et al 1998 *ApJ* **508** L61
[31] Macquart J-P Kanekar N Frail D A & Ransom S M 2010 *ApJ* **715** 939
[32] Manko V S & Novikov I D 1992 *CQG* **9** 2477
[33] Mazur P O 1982 *J. Phys. A* **15** 3173
[34] Merritt D Alexander T Mikkola S & Will C M 2010 *PRD* **81** 062002
[35] Mościbrodzka M Gammie C F Dolence J C Shiokawa H & Leung P K 2009 *ApJ* **706** 497
[36] Psaltis D 2008 *Liv. Rev. Rel* **11** 9
[37] ———. 2009 *Journ. Phys. Conf. Ser.* **189** 012033, arXiv:0907.2746
[38] Psaltis D Perodin D Dienes K Mocioiu I 2008 *PRL* **100** 091101
[39] Reid M J 2009 *Int. J. Mod. Phys. D* **18** 889
[40] Robinson D C 1975 *Phys. Rev. Lett.* **34** 905
[41] Ryan F D 1995 *PRD* **52** 5707
[42] Shcherbakov R & Penna R 2010 *in The Galactic Center: a Window to the Nuclear Activity of Disk Galaxies* arXiv:0912.2106
[43] Shen et al 2005 *Nature* **438** 62
[44] Tremaine S et al 2002 *ApJ* **574** 740
[45] Vigeland S 2010 *PRD* **82** 104041
[46] Vigeland S & Hughes S 2010 *PRD* **81** 024030
[47] Wex N & Kopeikin S 1999 *ApJ* **514** 388
[48] Will C M 2008 *ApJ* **674** L25
[49] Yunes N 2010, *in Amaro-Seoane P & Schutz B* arXiv:1003.5553
[50] Yunes N & Pretorius F 2009 *PRD* **79** 084043

Cosmological consequences of the noncommutative spectral geometry as an approach to unification

Mairi Sakellariadou

Department of Physics, King's College, University of London, Strand WC2R 2LS, London, U.K.

E-mail: mairi.sakellariadou@kcl.ac.uk

Abstract. Noncommutative spectral geometry succeeds in explaining the physics of the Standard Model of electroweak and strong interactions in all its details as determined by experimental data. Moreover, by construction the theory lives at very high energy scales, offering a natural framework to address early universe cosmological issues. After introducing the main elements of noncommutative spectral geometry, I will summarise some of its cosmological consequences and discuss constraints on the gravitational sector of the theory.

1. Introduction

Approaching Planckian energies, gravity can no longer be considered as a classical theory, the quantum nature of space-time becomes apparent and geometry can no longer be described in terms of Riemannian geometry and General Relativity. At such energy scales, all forces should be unified so that all interactions correspond to one underlying symmetry. The nature of space-time would change in such a way, so that its low energy limit implies the diffeomorphism and internal gauge symmetries, which govern General Relativity and gauge groups on which the Standard Model is based, respectively. Thus, near Planckian energies one should search for a formulation of geometry within the quantum framework. Such an attempt has been realised within NonCommutative Geometry (NCG) [1, 2]. Even though this approach is still in terms of an effective theory, it has already led to very encouraging results and is now at a stage to be confronted with experimental and observational data.

Since all physical data are of a spectral nature, we will use the notion of a spectral triple, which is analogous to Fourier transform we are familiar with in commutative spaces. Distances are measured in units of wavelength of atomic spectra, while the notion of a real variable, taken as a function on a set X, will be replaced by a self-adjoint operator D in a Hilbert space. The space X will be described by the algebra $\mathcal{A}$ of coordinates, represented as operators in a fixed Hilbert space $\mathcal{H}$. Thus, the usual emphasis on the points of a geometric space is replaced by the spectrum of an operator and the geometry of a noncommutative space is determined in terms of the spectral triple $(\mathcal{A}, \mathcal{H}, D)$.

Noncommutative geometry spectral action, in its current version which is still at the classical level and it certainly considers almost commutative spaces, it nevertheless offers [3] an elegant geometric interpretation of the most successful phenomenological model of particle physics, namely the Standard Model (SM). Certainly if the Large Hadron Collider (LHC) supports evidence for physics beyond the SM, one should consider extensions of the effective theory we will adopt here. Closer to the Planck era one should consider the full, still unknown, theory, while at even higher energy scales the whole concept of geometry may lose its familiar meaning. Nevertheless, the effective theory we will consider here lives, by construction, at very high energies, offering an excellent framework to address open questions of early universe cosmology [4].

In what follows, I will briefly review the main elements of the NCG spectral action as an approach to unification and then discuss some of its cosmological consequences [4, 5, 6, 7, 8, 9].

2. NCG spectral action

In the context of noncommutative geometry, all information about a physical system is contained within the algebra of functions, represented as operators in a Hilbert space, while the action and metric properties are encoded in a generalised Dirac operator. The geometry is specified by a spectral triple $(\mathcal{A}, \mathcal{H}, D)$, defined by an algebra $\mathcal{A}$, a Hilbert space $\mathcal{H}$ and a generalised Dirac operator D.

Adopting the simplest generalisation beyond commutative spaces, we take the extension of our smooth four-dimensional manifold, $\mathcal{M}$, by taking its product with a discrete noncommuting manifold $\mathcal{F}$. Thus, $\mathcal{M}$ describes the geometry for the four-dimensional space-time, while the noncommutative space $\mathcal{F}$ specifies the internal geometry for the SM and is composed of just two points. Considering the Standard Model of electroweak and strong interactions as a phenomenological model, we will look for a geometry, such that the associated action functional leads to the SM with all specifications as determined by experimental data.

Within NCG, a space described by the algebra of real coordinates is represented by self-adjoint operators on a Hilbert space. Since real coordinates are represented by self-adjoint operators, all information about a space is encoded in the algebra of coordinates $\mathcal{A}$, which is the main input of the theory. Under the assumption that the algebra constructed in $\mathcal{M} \times \mathcal{F}$ is symplectic-unitary, it turns out that $\mathcal{A}$ must be of the form

$$\mathcal{A} = M_a(\mathbb{H}) \oplus M_k(\mathbb{C}) \ , \tag{1}$$

with $k = 2a$ and $\mathbb{H}$ denoting the algebra of quaternions. The choice $k = 4$ is the first value that produces the correct number of fermions, namely $k^2 = 16$, in each of the three generations [10]. The number of generations is a physical input. Certainly if at LHC new particles are discovered, one may be able to accommodate them by including a higher value for the even number k.

The operator D corresponds to the inverse of the Euclidean propagator of fermions, and is given by the Yukawa coupling matrix which encodes the masses of the elementary fermions and the Kobayashi–Maskawa mixing parameters. The fermions of the SM provide the Hilbert space $\mathcal{H}$ of a spectral triple for the algebra $\mathcal{A}$, while the bosons of the SM are obtained through inner fluctuations of the Dirac operator of the product $\mathcal{M} \times \mathcal{F}$ geometry.

The spectral action principle states that the bosonic part of the spectral functional S depends only on the spectrum of the Dirac operator and its asymptotic expression, and for large energy Λ is of the form $\mathrm{Tr}(f(D/\Lambda))$, with f being a cut-off function, whose choice plays only a small rôle. The physical Lagrangian has also a fermionic part, which has the simple linear form $(1/2)\langle J\psi, D\psi\rangle$, where J is the real structure on the spectral triple and ψ are spinors defined on the Hilbert space [11]. Applying the spectral action principle to the inner fluctuations of the product geometry $\mathcal{M}\times\mathcal{F}$, one recovers the Standard Model action coupled to Einstein and Weyl gravity plus higher order nonrenormalisable interactions suppressed by powers of the inverse of the mass scale of the theory [3].

To study the implications of this noncommutative approach coupled to gravity for the cosmological models of the early universe, one can concentrate just on the bosonic part of the action; the fermionic part is however crucial for the particle physics phenomenology of the model.

Using the heat kernel method, the bosonic part of the spectral action can be expanded in powers of the scale Λ in the form [3, 12, 13]

$$\mathrm{Tr}\left(f\left(\frac{D}{\Lambda}\right)\right) \sim \sum_{k\in\mathrm{DimSp}} f_k\Lambda^k \int\!\!\!\!\!- |D|^{-k} + f(0)\zeta_D(0) + \mathcal{O}(1)\ , \tag{2}$$

with the momenta f_k of the cut-off function f given by

$$f_k \equiv \int_0^\infty f(u)u^{k-1}\mathrm{d}u \ \ \text{for} \ \ k>0 \ \ , \ \ \text{and} \ \ f_0 \equiv f(0)\ ; \tag{3}$$

the noncommutative integration is defined in terms of residues of zeta functions, $\zeta_D(s) = \mathrm{Tr}(|D|^{-s})$ at poles of the zeta function, and the sum is over points in the *dimension spectrum* of the spectral triple.

In this way, one obtains a Lagrangian which contains in addition to the full SM Lagrangian, the Einstein-Hilbert action with a cosmological term, a topological term related to the Euler characteristic of the space-time manifold, a conformal Weyl term and a conformal coupling of the Higgs field to gravity. Writing the asymptotic expansion of the spectral action, a number of geometric parameters appear, which describe the possible choices of Dirac operators on the finite noncommutative space. These parameters correspond to the Yukawa parameters of the particle physics model and the Majorana terms for the right-handed neutrinos. The Yukawa parameters run with the Renormalisation Group Equations (RGE) of the particle physics model. Since running towards lower energies, implies that nonperturbative effects in the spectral action cannot be any longer neglected, any results based on the asymptotic expansion and on renormalisation group analysis can only hold for early universe cosmology. For later times, one should instead consider the full spectral action.

More precisely, the bosonic action in Euclidean signature reads [3]

$$\begin{aligned}\mathcal{S}^{\mathrm{E}} = \int\Big(&\frac{1}{2\kappa_0^2}R + \alpha_0 C_{\mu\nu\rho\sigma}C^{\mu\nu\rho\sigma} + \gamma_0 + \tau_0 R^\star R^\star + \frac{1}{4}G^i_{\mu\nu}G^{\mu\nu i} + \frac{1}{4}F^\alpha_{\mu\nu}F^{\mu\nu\alpha}\\ &+\frac{1}{4}B^{\mu\nu}B_{\mu\nu} + \frac{1}{2}|D_\mu\mathbf{H}|^2 - \mu_0^2|\mathbf{H}|^2 - \xi_0 R|\mathbf{H}|^2 + \lambda_0|\mathbf{H}|^4\Big)\sqrt{g}\ d^4x\ ,\end{aligned} \tag{4}$$

where

$$\begin{aligned}\kappa_0^2 &= \frac{12\pi^2}{96f_2\Lambda^2 - f_0\mathfrak{c}}\ ,\\ \alpha_0 &= -\frac{3f_0}{10\pi^2}\ ,\\ \gamma_0 &= \frac{1}{\pi^2}\left(48f_4\Lambda^4 - f_2\Lambda^2\mathfrak{c} + \frac{f_0}{4}\mathfrak{d}\right)\ ,\\ \tau_0 &= \frac{11f_0}{60\pi^2}\ ,\\ \mu_0^2 &= 2\Lambda^2\frac{f_2}{f_0} - \frac{\mathfrak{e}}{\mathfrak{a}}\ ,\\ \xi_0 &= \frac{1}{12}\ ,\\ \lambda_0 &= \frac{\pi^2\mathfrak{b}}{2f_0\mathfrak{a}^2}\ ,\end{aligned} \tag{5}$$

with $\mathfrak{a}, \mathfrak{b}, \mathfrak{c}, \mathfrak{d}, \mathfrak{e}$ given by [3]

$$\begin{aligned}
\mathfrak{a} &= \mathrm{Tr}\left(Y^\star_{(\uparrow 1)}Y_{(\uparrow 1)} + Y^\star_{(\downarrow 1)}Y_{(\downarrow 1)} + 3\left(Y^\star_{(\uparrow 3)}Y_{(\uparrow 3)} + Y^\star_{(\downarrow 3)}Y_{(\downarrow 3)}\right)\right) ,\\
\mathfrak{b} &= \mathrm{Tr}\left(\left(Y^\star_{(\uparrow 1)}Y_{(\uparrow 1)}\right)^2 + \left(Y^\star_{(\downarrow 1)}Y_{(\downarrow 1)}\right)^2 + 3\left(Y^\star_{(\uparrow 3)}Y_{(\uparrow 3)}\right)^2 + 3\left(Y^\star_{(\downarrow 3)}Y_{(\downarrow 3)}\right)^2\right) ,\\
\mathfrak{c} &= \mathrm{Tr}\left(Y^\star_R Y_R\right) ,\\
\mathfrak{d} &= \mathrm{Tr}\left(\left(Y^\star_R Y_R\right)^2\right),\\
\mathfrak{e} &= \mathrm{Tr}\left(Y^\star_R Y_R Y^\star_{(\uparrow 1)}Y_{(\uparrow 1)}\right) , \qquad (6)
\end{aligned}$$

with $Y_{(\downarrow 1)}, Y_{(\uparrow 1)}, Y_{(\downarrow 3)}, Y_{(\uparrow 3)}$ and Y_R being (3×3) matrices, with Y_R symmetric; the Y matrices are used to classify the action of the Dirac operator and give the fermion and lepton masses, as well as lepton mixing, in the asymptotic version of the spectral action. Note that $\mathbf{H}$ is a rescaling of the Higgs field, so that the kinetic terms are normalised. One should be cautious that the relations in Eq. (5) are only valid at unification scale Λ; it is incorrect to consider them as functions of the energy scale.

The noncommutative spectral geometry approach leads to various phenomenological consequences. Normalisation of the kinetic terms implies

$$\frac{g_3^2 f_0}{2\pi^2} = \frac{1}{4} \quad \text{and} \quad g_3^2 = g_2^2 = \frac{5}{3}g_1^2 ,$$

while

$$\sin^2\theta_{\mathrm{W}} = \frac{3}{8} ; \qquad (7)$$

a relation which holds also for SU(5) and SO(10), while assuming the *big desert* hypothesis, one can find the running of the three couplings $\alpha_i = g_i^2/(4\pi)$. One-loop RGE for the running of the gauge couplings and the Newton constant, shows that they do not meet exactly at one point, the error is though within just few percent. Therefore, the model in its simplified form, does not specify a unique unification energy, it however leads to the correct representations of the fermions with respect to the gauge group of the SM, the Higgs doublet appears as part of the inner fluctuations of the metric, and Spontaneous Symmetry Breaking mechanism arises naturally with the negative mass term without any tuning. Moreover, the see-saw mechanism is obtained, the 16 fundamental fermions are recovered, and a top quark mass of $M_{\mathrm{top}} \sim 179$ Gev is predicted. The mass of Higgs in zeroth order approximation of the spectral action is ~ 170GeV, which is strictly speaking ruled out by current experimental data. Nevertheless, the result depends on the value of gauge couplings at unification scale, which is certainly uncertain, while it was found neglecting the nonminimal coupling between the Higgs field and the Ricci curvature. It is however worth noticing that the NCG approach leads to the correct order of magnitude for the Higgs mass, a result which was not obvious *a priori*.

3. Cosmological consequences

To use the formalism of spectral triples in NCG, it is convenient to work with Euclidean rather than Lorentzian signature. Thus, the analysis of the cosmological consequences of the theory relies on a Wick rotation to imaginary time, into the Lorentzian signature.

The Lorentzian version of the gravitational part of the asymptotic formula for the bosonic sector of the NCG spectral action, including the coupling between the Higgs field and the Ricci curvature scalar, reads [3]

$$\mathcal{S}^{\mathrm{L}}_{\mathrm{grav}} = \int \left(\frac{1}{2\kappa_0^2}R + \alpha_0 C_{\mu\nu\rho\sigma}C^{\mu\nu\rho\sigma} + \tau_0 R^\star R^\star \xi_0 R|\mathbf{H}|^2\right)\sqrt{-g}\, d^4x , \qquad (8)$$

leading to the equations of motion [5]

$$R^{\mu\nu} - \frac{1}{2}g^{\mu\nu}R + \frac{1}{\beta^2}\delta_{cc}\left[2C^{\mu\lambda\nu\kappa}_{;\lambda;\kappa} + C^{\mu\lambda\nu\kappa}R_{\lambda\kappa}\right] = \kappa_0^2\delta_{cc}T^{\mu\nu}_{\rm matter} ,$$

with

$$\beta^2 \equiv -\frac{1}{4\kappa_0^2\alpha_0} \quad \text{and} \quad \delta_{cc} \equiv [1 - 2\kappa_0^2\xi_0\mathbf{H}^2]^{-1} . \tag{9}$$

In the low energy weak curvature regime, the nonminimal coupling between the background geometry and the Higgs field can be neglected, implying $\delta_{cc} = 1$. For a Friedmann-Lemaître-Robertson-Walker (FLRW) space-time, the Weyl tensor vanishes, hence the NCG corrections to the Einstein equation vanish [5], rending difficult to restrict β (or equivalently α_0, or f_0) via cosmology or solar-system tests. Imposing however a lower limit on β is very important, since it implies an upper limit to the moment f_0, corresponding to a restriction on the particle physics at unification. This has been achieved in Refs. [8, 9], by considering the energy lost to gravitational radiation by orbiting binaries.

Deriving the weak field limit of noncommutative spectral geometry, we have shown [8] that the production and dynamics of gravitational waves are significantly altered and, in particular, the graviton contains a massive mode that alters the energy lost to gravitational radiation, in systems with evolving quadrupole moment. Considering the rate of energy loss from a binary pair of masses m_1, m_2, in the far field limit, we have shown [8, 9] that the orbital frequency

$$\omega = |\rho|^{-3/2}\sqrt{G(m_1 + m_2)} , \tag{10}$$

where $|\rho|$ stands for the magnitude of their separation vector, has a critical value

$$2\omega_c = \beta c , \tag{11}$$

around which strong deviations from the familiar results of General Relativity are expected. This maximum frequency results from the natural length scale, given by β^{-1}, at which noncommutative geometry effects become dominant.

The form of the gravitational radiation from binary systems can be used to constrain β. There are several binary pulsars for which the rate of change of the orbital frequency has been well characterised, and the predictions of General Relativity agree with the data to high accuracy. Thus, one can restrict the parameter β by requiring that the magnitude of deviations from General Relativity be less than this uncertainty. Requiring that $\beta > 2\omega/c$, we have found [9]

$$\beta > 7.55 \times 10^{-13}\ \mathrm{m}^{-1} . \tag{12}$$

Since the strongest constraint comes from systems with high orbital frequencies, one expects that future observations of rapidly orbiting binaries, relatively close to the Earth, could improve it by many orders of magnitude.

Considering the background equations, the corrections to Einstein's equations can only be apparent at leading order for anisotropic models. Calculating the modified Friedmann equation for the Bianchi type-V model, we have shown [5] that the correction terms come in two types. The first one contains terms which are fourth order in time derivatives, hence for the slowly varying functions usually used in cosmology they can be neglected. The second one occurs at the same order as the standard Einstein-Hilbert terms, however, it vanishes for homogeneous versions of Bianchi type-V. Thus, although anisotropic cosmologies do contain corrections due to the additional NCG terms in the action, they are typically of higher order. Inhomogeneous models do

Journal of Physics: Conference Series **283** (2011) 012031 doi:10.1088/1742-6596/283/1/012031

contain correction terms that appear on the same footing as the ordinary (commutative) terms. In conclusion, the corrections to Einstein's equations can only be important for inhomogeneous and anisotropic space-times.

Certainly one cannot always neglect the coupling of the Higgs field to the curvature. Namely, as energies approach the Higgs scale, this nonminimal coupling can no longer be neglected, leading to corrections even for background cosmologies. To understand the effects of these corrections let us neglect the conformal term in Eq. (9), so that the equations of motion read [5]

$$R^{\mu\nu} - \frac{1}{2}g^{\mu\nu}R = \kappa_0^2 \left[\frac{1}{1-\kappa_0^2|\mathbf{H}|^2/6}\right] T^{\mu\nu}_{\rm matter} \ . \tag{13}$$

Thus, $|\mathbf{H}|$ plays the rôle of an effective gravitational constant [5].

The nonminimal coupling between the Higgs field and the Ricci curvature may turn out to be particularly useful in early universe cosmology [6, 7]. Such a coupling has been introduced *ad hoc* in the literature, in an attempt to drive inflation through the Higgs field. However, the value of the coupling constant between the scalar field and the background geometry should be dictated by the underlying theory.

In a FLRW metric, the Gravity-Higgs sector of the asymptotic expansion of the spectral action, in Lorentzian signature, reads

$$S^{\rm L}_{\rm GH} = \int \Big[\frac{1-2\kappa_0^2\xi_0 H^2}{2\kappa_0^2}R - \frac{1}{2}(\nabla H)^2 - V(H)\Big]\sqrt{-g}\; d^4x \ , \tag{14}$$

where

$$V(H) = \lambda_0 H^4 - \mu_0^2 H^2 \ , \tag{15}$$

with μ_0 and λ_0 subject to radiative corrections as functions of energy. For large enough values of the Higgs field, the renormalised value of these parameters must be calculated, while the running of the top Yukawa coupling and the gauge couplings must be evolved simultaneously.

At high energies the mass term is sub-dominant, and can be neglected. For each value of the top quark mass, there is a value of the Higgs mass where the effective potential is about to develop a metastable minimum at large values of the Higgs field and the Higgs potential is locally flattened [7]. Since the region where the potential is flat is narrow, the slow-roll must be very slow in order to provide a sufficiently long period of quasi-exponential expansion. Besides the slow-roll parameters, denoted by ϵ and η, which may be slow enough to allow sufficient number of e-folds, the amplitude of density perturbations $\Delta^2_{\mathcal{R}}$ in the Cosmic Microwave Background must be within the allowed experimental window. Inflation predicts that at horizon crossing (denoted by stars), the amplitude of density perturbations is related to the inflaton potential through

$$\left(\frac{V_*}{\epsilon_*}\right)^{\frac{1}{4}} = 2\sqrt{3\pi}\; m_{\rm Pl}\; \Delta^{\frac{1}{2}}_{\mathcal{R}} \ , \tag{16}$$

where $\epsilon_* \leq 1$. Its value, as measured by WMAP7 [14], requires

$$\left(\frac{V_*}{\epsilon_*}\right)^{\frac{1}{4}} = (2.75 \pm 0.30) \times 10^{-2}\; m_{\rm Pl} \ , \tag{17}$$

where $m_{\rm Pl}$ stands for the Planck mass.

Calculating [7] the renormalisation of the Higgs self-coupling up to two-loops, we have constructed an effective potential which fits the renormalisation group improved potential around the flat region. We have found [7] a very good analytic fit to the Higgs potential around the minimum of the potential:

$$V^{\rm eff} = \lambda_0^{\rm eff}(H)H^4 = [a\ln^2(b\kappa H) + c]H^4 \ , \tag{18}$$

where the parameters a, b are related to the low energy values of top quark mass $m_{\rm t}$ as [7]

$$\begin{aligned} a(m_{\rm t}) &= 4.04704\times 10^{-3} - 4.41909\times 10^{-5}\left(\frac{m_{\rm t}}{\rm GeV}\right) + 1.24732\times 10^{-7}\left(\frac{m_{\rm t}}{\rm GeV}\right)^2 \ , \\ b(m_{\rm t}) &= \exp\left[-0.979261\left(\frac{m_{\rm t}}{\rm GeV} - 172.051\right)\right] \ . \end{aligned} \tag{19}$$

The third parameter, c, encodes the appearance of an extremum and depends on the values for top quark mass and Higgs mass. An extremum occurs if and only if $c/a \leq 1/16$, the saturation of the bound corresponding to a perfectly flat region. It is convenient to write $c = [(1+\delta)/16]a$, where $\delta = 0$ saturates the bound below which a local minimum is formed.

This analysis was performed in the case of minimal coupling, so let us investigate the modifications introduced in the case of a small nonminimal coupling; within NCG the coupling is $\xi_0 = 1/12$. We have found [7] that the induced corrections to the potential imply that flatness does not occur at $\delta = 0$, but for fixed values of δ depending on the value of the top quark mass. Thus, for inflation to occur via the Higgs field, the top quark mass fixes the Higgs mass extremely accurately. Scanning carefully through the parameter space, we concluded [7] that sufficient e-folds are indeed generated provided a suitably tuned relationship between the top quark mass and the Higgs mass holds. However, while the Higgs potential can lead to the slow-roll conditions being satisfied once the running of the self-coupling at two-loops is included, the constraints imposed from the CMB data make the predictions of such a scenario incompatible with the measured value of the top quark mass [7]. Running of the gravitational constant and corrections by considering the more appropriate de Sitter, instead of a Minkowski, background we found [7] that do not improve substantially the realisation of a successful inflationary era.

4. Conclusions

Noncommutative spectral geometry is a beautiful mathematical construction with a rich phenomenological arena. In its present *simple* form, which still remains classical and refers only to almost commutative spaces, it offers an elegant explanation for the Standard Model of electroweak and strong interactions. The prediction for the top quark mass is in agreement with current experimental data, while the Higgs mass is within the correct order of magnitude; its precise value is excluded from the most recent experimental data but it is still remarkable how close it remains to the experimental allowed value besides the simplifications under which it was calculated.

Noncommutative spectral geometry lives by construction at very high energy scales, offering a natural set-up to study early universe cosmology. Late time astrophysics is a more difficult task due to technical issues at the current stage of the NCG spectral action [4]. Expecting further progress in computing exactly the spectral action in its nonperturbative form and performing the appropriate renormalisation group analysis, we expect that we will be able to tackle astrophysical issues in the near future.

Here, I have reviewed possible cosmological fingerprints of noncommutative spectral geometry, and proposed a mechanism to constrain physics at unification through the implications of production and dynamics of gravitational waves.

5. Acknowledgments

It is a pleasure to thank the organisers of the 14th Conference on Recent Developments in Gravity (NEB14) "$N\epsilon\omega\tau\epsilon\rho\epsilon\varsigma$ $E\xi\epsilon\lambda\iota\xi\epsilon\iota\varsigma$ $\sigma\tau\eta$ $B\alpha\rho\upsilon\tau\eta\tau\alpha$ 14" in Ioannina (Greece) for inviting me to present these results. This work is partially supported by the European Union through the Marie Curie Research and Training Network *UniverseNet* (MRTN-CT-2006-035863).

References

[1] A. Connes, *Noncommutative Geometry*, Academic Press, New York (1994).
[2] A. Connes and M. Marcolli, *Noncommutative Geometry, Quantum Fields and Motives*, Hindustan Book Agency, India (2008).
[3] A. H. Chamseddine, A. Connes and M. Marcolli, Adv. Theor. Math. Phys. **11**, 991 (2007) [arXiv:hep-th/0610241].
[4] M. Sakellariadou, [arXiv:1008.5348[hep-th]].
[5] W. Nelson and M. Sakellariadou, Phys. Rev. D **81**, 085038 (2010) [arXiv:0812.1657 [hep-th]].
[6] W. Nelson and M. Sakellariadou, Phys. Lett. B **680**, 263 (2009) [arXiv:0903.1520 [hep-th]].
[7] M. Buck, M. Fairbairn and M. Sakellariadou, Phys. Rev. D **82**, 043509 (2010) [arXiv:1005.1188 [hep-th]].
[8] W. Nelson, J. Ochoa and M. Sakellariadou, Phys. Rev. D (2010) (in press) [arXiv:1005.4276 [hep-th]].
[9] W. Nelson, J. Ochoa and M. Sakellariadou, Phys. Rev. Lett. **105**, 101602 (2010) [arXiv:1005.4279 [hep-th]].
[10] A. H. Chamseddine and A. Connes, Phys. Rev. Lett. **99**, 191601 (2007) [arXiv:0706.3690 [hep-th]].
[11] A. H. Chamseddine, A. Connes, Comm. Math. Phys. **186** (1997) 731-750 [arXiv:hep-th/9606001]
[12] A. H. Chamseddine and A. Connes, Phys. Rev. Lett. **77**, 4868 (1996).
[13] A. H. Chamseddine and A. Connes, Comm. Math. Phys. **186**, 731 (1977).
[14] D. Larson et al., [arXiv:1001.4758 [astro-ph]].

Affine gauge theory of gravity and its reduction to the Riemann-Cartan geometry

Rodrigo F. Sobreiro
UFF – Universidade Federal Fluminense, Instituto de Física, Campus da Praia Vermelha, Av. General Milton Tavares de Souza s/n, 24210-346, Niterói, RJ, Brasil.
E-mail: sobreiro@if.uff.br

Victor J. Vasquez Otoya
IFSEMG – Instituto Federal de Educação, Ciência e Tecnologia, Rua Bernardo Mascarenhas 1283, 36080-001, Juiz de Fora, MG, Brasil
E-mail: victor.vasquez@ifsudestemg.edu.br

Abstract. We discuss a possible framework for the construction of a quantum gravity theory where the principles of QFT and general relativity can coexist harmonically. Moreover, in order to fix the correct gauge group of the theory we study the most general one, the affine group and its natural reduction to the orthogonal group. The price we pay for simplifying the geometry is the presence of matter fields associated with the nonmetric degrees of freedom of the original setup. The result is independent of the starting theory.

1. Introduction

The description of gravity as a gauge theory based on the Lorentz group [1, 2] puts gravity at the same level as the other fundamental interactions. As extra features of the gauge approach we can point out the fact that spinors can minimally couple to gravity through torsion and there is no need to linearize the fields to define a local gauge symmetry. However, the standard Einstein-Hilbert action shows itself to be not renormalizable even in this approach. To face the renormalizability problem there were proposed several alternatives by extending the Einstein-Hilbert action to more general ones [3, 4, 5]. However, their quantum consistency still lacks, giving space to alternatives beyond the action level. One of them is to consider more general groups for the gauge symmetry that also allows the identification of the gauge space with the spacetime cotangent bundle, in particular we recall the so called metric-affine gravities [6, 7], based on the affine gauge group $A(d,\mathbb{R})$. This class of theories also requires generalized actions to account the quantum divergences and extra degrees of freedom. The present article is in fact devoted to the study of some geometrical properties of this class of theories and the development of a possible scenario for its quantization.

Gravity as geometry, where spacetime encodes the dynamical variables, is in fact incompatible with quantum field theory which considers coordinates as parameters. To avoid this ambiguity we start by improving the ideas developed in [8, 9, 10] in which gauge symmetry and spacetime are considered as independent from each other. This means that the gauge space is not identified with the cotangent bundle from the beginning. In fact, all quantum fields considered cannot be identified with gravity fields unless a dynamical mass parameter emerges in the theory. This property is ensured by considering fields that are of dimension 1 and thus cannot be identified with vierbeins nor premetrics. Once the mass parameters emerge, a rescaling is performed and those fields can be identified with geometric quantities of spacetime. Moreover, since at quantum level there is no relation between gauge space and spacetime, the spacetime is assumed to be Euclidean (a scenario that allows explicit computations for a QFT). This idea is indeed self consistent and can be applied to several gauge groups. In particular, the present work study the case of the affine group. However, due to the non-semisimplicity of the translational annex we are not able to write down an explicit action and thus we restrict ourselves to classical consequences of the model. On the other hand, we consider a general framework for the affine gauge theory that generates results that are independent of the specific dynamical equations. In fact, they depend exclusively on the topological features of a gauge theory and their description in terms of fibre bundles [11, 12, 13, 14, 15, 16, 17].

By considering the framework developed in [10] we are able to construct an affine gauge theory in a 4-dimensional spacetime in which the classical sector of the theory is a Riemann-Cartan geometrical gravity with extra matter fields. The path from a massless quantum gauge theory to a geometrical gravity is assumed to be mediated by mass parameters such as the Gribov parameter, always present in non-Abelian gauge theories with nontrivial topological sectors [12, 18, 19]. This main result is independent of the starting dynamics of the model, *i.e.*, it is valid for any affine gauge theory.

This work is organized as follows: In Section 2 we provide the fundamental ideas concerning the proposed quantum gravity model. At the same Section we also construct the gauge theory for an affine group and show how it is related to the standard metric-affine gravities [6, 7]. In Section 3 we show how the theory can in fact be contracted down to an orthogonal gauge theory with extra matter fields. Finally in Section 4 we display our conclusions.

2. Quantum gravity as an Euclidean gauge theory

A consistent gauge theory for gravity should encode both quantum and classical sectors. The classical regime is more or less well understood: It is a field theory whose dynamics determines the geometry of spacetime; the dynamical fields are the metric tensor and spacetime connection. To describe the quantum sector we propose that quantum gravity is a gauge theory in flat space, just like all other fundamental interactions. Thus, at high energies the quantum sector dominates and the model is a standard QFT with specific non-Abelian gauge symmetry embedded in flat space. Moreover, we introduce matter fields with dimension 1 which are translation algebra-valued[1,2]. The existence of a mass scale ensures the possibility of mapping the matter fields into coframes and then identify the effective dynamical variables with geometry and thus to relate the model with gravity. Let us discuss the present proposition. To do so we enunciate the postulates of the model and their consequences.

- **Flat spacetime**: *The spacetime of the fundamental theory, i.e., at very high energies, is a four dimensional Euclidean space* $\mathbb{R}^4$. This is the simplest choice to be made, so we made it. In fact, Euclidean spacetime is the proper scenario to make explicit computations in a QFT. Further, nonperturbative effects require Euclidean spacetime since Wick's theorem is not known to be valid outside perturbation theory.

[1] By algebra-valued we mean that the quantity takes values in the algebra of the group, in this case the group of translations, *i.e.*, the quantity is expanded into the generators of the group. As a consequence its components will carry the corresponding group index.

[2] This requirement restricts the possible gauge groups to be considered.

- **Gauge symmetry**: *The quantum sector is described by a gauge theory with symmetry group denoted by* $\mathfrak{G}$. At quantum level the gauge symmetry is totally independent of spacetime symmetries. Moreover, $\mathfrak{G}$ is required to enjoy any morphism with spacetime in the sense that some topological invariants of the spacetime Euclidean structure should coincide with those of the gauge group. This last requirement is necessary to identify the gauge symmetry with the spacetime structure. Furthermore, we also require that $\mathfrak{G}$ encodes as subgroup the group of translations $\mathbb{R}^4_g$ or at least that is related to it by forming a larger group. The index g establishes that the space is the gauge space and is not related with spacetime.
- **Matter**: *Massless matter fields algebra-valued in* $\mathbb{R}^4_g$ *carrying dimension 1 are introduced.* Those fields are required depending on the gauge group to be considered in order to avoid the intricacies that might emerge from groups that span non-flat gauge spaces.
- **Mass scales and geometry**: *A mass scale* m *should emerge dynamically from the quantum theory.* A mass scale is necessary in order to scale the UV and IR regimes. Once the energy decreases the mass scale dominates and throw the theory to a different behavior from the one at the UV scale. At this level, one can perform the rescaling of the matter fields and map them into a spacetime vierbein e^a field and premetric $\mathfrak{g}_{ab}$. Moreover, the gauge field should also be identified with the connection of a spacetime bundle (for instance the coframe bundle). Thus, the dynamics of the quantum theory are now dictating the spacetime structure at classical level. A metric tensor can then be defined

$$\mathfrak{g}_{\mu\nu} = \mathfrak{g}_{ab} e^a_\mu e^b_\nu \; . \tag{1}$$

  We remark that the most suitable candidate for the mass parameter is the so called Gribov parameter [18, 19, 20] associated with the improvement of the quantization of non-Abelian gauge theories.

The present program provides a self consistent way to define a quantum gravity theory in four dimensions. However, the formalization of the model is necessary in order to make explicit computations and predictions. The first step is to select the most suitable gauge group of the model. It turns out that the most general gauge group that can be fully mapped into a 4-dimensional spacetime is the 20 parameter affine group $A(4,\mathbb{R})$. The study of some properties of the respective gauge theory is the task of the rest of this work.

2.1. Algebraic structure of the affine group

The affine group is defined as $A(d,\mathbb{R}) \cong GL(d,\mathbb{R}) \ltimes \mathbb{R}^d$ where $GL(d,\mathbb{R})$ is the general linear group of $d \times d$ invertible real matrices and $\mathbb{R}^d$ stands for the group of translations. Generically, we have for the Lie algebra decomposition

$$[gl, gl] \subseteq gl \; , \quad [r, r] \subseteq \emptyset \; , \quad [r, gl] \subseteq r \; , \tag{2}$$

where gl is the algebra of the $GL(d,\mathbb{R})$ and r the algebra of the $\mathbb{R}^d$ translations. We remark that the affine group is not a semi-simple group due to the translational sector. Further, the algebra decomposition (2), implies that the coset space $\mathbb{R}^d \cong A(d,\mathbb{R})/GL(d,\mathbb{R})$ is a symmetric space and $GL(d,\mathbb{R})$ is a noncompact stability group of $A(d,\mathbb{R})$.

The general linear group can also be decomposed as $GL(d,\mathbb{R}) \cong K(d) \otimes O(d)$ where $K(d)$ is the space defined from the symmetric part of the algebra of the general linear group. Formally, $K(d)$ is the coset space $K(d) \cong GL(d,\mathbb{R})/O(d)$ and, differently from $\mathbb{R}^d$, it does not form a group. The orthogonal group $O(d)$ is a maximal compact subgroup of the affine group as well as for the general linear group. The algebra decomposition can be described through

$$[o, o] \subseteq o \; , \quad [k, k] \subseteq o \; , \quad [k, o] \subseteq k \; , \tag{3}$$

where o stands for the algebra of the orthogonal $O(d)$ group and k for the algebra of the coset space $K(d)$. From (3) one can infer that $K(d)$ forms a symmetric space.

The elements of the affine group are here represented by exponentials according to $\mathcal{U} = \mathrm{e}^{\zeta(x)} = \mathrm{e}^{\alpha^a{}_b(x)T^b{}_a + \xi^a(x)P_a}$ where we are considering the algebraic notation of the adjoint representation of the affine group. In fact, in this article, we are exclusively working in the adjoint representation. Thus, small Latin indices are vector indices of the d-dimensional differential manifold defined by the affine group, $a \in \{1, 2, \ldots, d\}$. The parameter ζ is a $A(d, \mathbb{R})$ algebra-valued quantity $\zeta = \zeta^A S_A$ where S_A represents the group generators in adjoint representation with $A \in \{1, 2, \ldots, d(d+1)\}$. At the same level, $\alpha^a{}_b$ and ξ^a are algebra-valued parameters associated with the group element $U \in GL(d, \mathbb{R})$ and $u \in \mathbb{R}^d$, respectively, while $T^b{}_a$ and P_a the respective generators. Further, it is a general property of Lie groups that the element $\mathcal{U}$ can always be decomposed as $\mathcal{U} = uU$. Furthermore, under decomposition (3) the elements U can be splitted according to $U = LC$, with $L = \mathrm{e}^{h^a{}_b(x)\Sigma^b{}_a} \in O(d)$ and $C = \mathrm{e}^{b^a{}_b(x)\lambda^b{}_a} \in K(d)$, where the notation speaks for itself.

2.2. *Affine gauge theory in Euclidean spacetime*

Following the construction of [10, 13, 16, 17, 21, 22] a gauge connection for the affine group, under the context of the previous section, arises from the principal bundle $A(4, \mathbb{R}) \equiv \{A_r(4, \mathbb{R}), \mathbb{R}^4\}$ where the index r stands for the rigid group and $\mathbb{R}^4$ the 4-dimensional Euclidean spacetime. The spacetime is then the base space while the affine group is the fiber as well as the structure group. A principal bundle of this form defines at each spacetime point a copy of the group, providing then a local character for the gauge group, *i.e.*, the principal bundle is indeed the local affine group. In this bundle a gauge connection $Y = \omega + E$ emerges, where $\omega = \omega^a{}_b T_a{}^b$ is the general linear connection and $E = E^a P_a$ the translational one. A translation in the principal bundle $A(4, \mathbb{R})$, when restricted to a single fibre ($x \in \mathbb{R}^d$ is fixed), defines a gauge transformation

$$Y \longmapsto \mathcal{U}^{-1}(\mathrm{d} + Y)\mathcal{U}\,, \tag{4}$$

where d is the space-time exterior derivative. At infinitesimal level the transformation (4) reduces to $Y \longmapsto Y + \nabla\zeta$ where the covariant derivative is defined as $\nabla = \mathrm{d} + Y$ as long as we are considering the adjoint representation. Also at infinitesimal level the transformation (4) decomposes as $\omega \longmapsto \omega + \mathcal{D}\alpha$ and $E \longmapsto E + \mathcal{D}\xi + E\alpha$, where the covariant derivative $\mathcal{D}$ is taken only with respect to the $GL(d, \mathbb{R})$ connection $\mathcal{D} = \mathrm{d} + \omega$.

From the definition of the covariant derivative one can easily compute the field strength two-form $\nabla^2 = \Theta = \Omega^a{}_b T^b{}_a + \Xi^a P_a$ where $\Omega^a{}_b$ is the $GL(d, \mathbb{R})$ curvature while Ξ^a the translational one,

$$\begin{aligned} \Omega^a{}_b &= \mathrm{d}\omega^a{}_b - \omega^a{}_c\omega^c{}_b\,, \\ \Xi^a &= \mathrm{d}E^a - \omega^a{}_b E^b = \nabla E^a = \mathcal{D}E^a\,. \end{aligned} \tag{5}$$

In despite of the fact that the field strength Θ is a covariant quantity under affine gauge transformations its components are not. In fact, at infinitesimal level it decomposes as $\Omega \longmapsto \Omega + \Omega\alpha$ and $\Xi \longmapsto \Xi + \Xi\alpha + \Omega\xi$.

The bundle $A(4, \mathbb{R})$ is very useful to define the correct gauge fields and their respective field strengths. However, as discussed in [10], it is a static bundle. To construct a dynamical theory one should migrate to an infinite dimensional principal bundle with the local gauge group $A(4, \mathbb{R})$ as fiber and the space of all independent gauge connections $\mathcal{Y}$ as base space [10, 14, 15, 16, 17]. By gauge independent connections we mean those that are not related to each other by a gauge transformation. Thus this principal bundle is denoted by $\mathbb{A} \equiv \{A(4, \mathbb{R}), \mathcal{Y}\}$. A typical fiber is the so called gauge orbit

$$Y^{\mathcal{U}} = \mathcal{U}^{-1}(\mathrm{d} + Y)\mathcal{U}\;\Big|\; Y \in \mathcal{Y}\,. \tag{6}$$

A problem to be faced is the fact that the gauge space spanned by the local group $A(4,\mathbb{R})$ is not flat. Thus, extra degrees of freedom have to be considered in the theory. Those are represented by the matter fields $\sigma = \sigma^a P_a$ and $g = g^{ab} P_a P_b$. The fields σ^a is a 1-form and possesses components with dimension 1. The field g^{ab} also carries dimension 1 whereas it is a 0-form field double algebra-valued[3] in $\mathbb{R}^4_g$. Being matter fields both transform covariantly under gauge transformations. See [10] for extra details.

The affine gauge theory here considered is then constructed with gauge fields (ω, E), matter fields (σ, g) and no mass parameters. A consistent simple gauge invariant action should then be considered through the gauge covariant quantities Θ, T and Q where the last two are the minimal couplings of the matter fields with the gauge fields

$$\begin{aligned} T^a &= \nabla\sigma^a = \mathcal{D}\sigma^a = \mathrm{d}\sigma^a - \omega^a{}_b\sigma^b \;, \\ Q^{ab} &= \nabla g^{ab} = \mathcal{D}g^{ab} = \mathrm{d}g^{ab} - \omega^a{}_c g^{cb} - \omega^b{}_c g^{ca} \;. \end{aligned} \tag{7}$$

It is worth mention that, in the presence of mass parameters, a relation between Ξ^a and T^a can be enforced if an identification between σ and E is assumed. However, such association requires the introduction of an auxiliary field φ^a in order to adjust the gauge field transformations between E^a and σ^a, see [7, 21, 10]. Anyhow, the affine group has a non-semisimple annex, the $\mathbb{R}^4_g$, that is characterized by the fact that there is no invariant Killing metric and thus disables a simple construction of a gauge invariant action [23]. Nevertheless, general linear algebra-valued quantities are allowed and can be used to define a theory on this sector (The most correct option is indeed to start from a $GL(4,\mathbb{R})$ gauge theory instead). We shall then skip the construction of an action and assume that an alternative method is successfully employed and study the consequences for the classical sector. Two remarks are in order: First that even in an affine gauge theory there are gauge invariant quantities and to construct them one should use a dual representation [10] avoiding then the use of the metric tensor of the gauge space spanned by the affine group. In here, those degrees are represented by σ and g and are not directly related with the geometry of any space as long as their components are carrying dimension 1. Second, the affine group has a compact subgroup and thus the respective gauge theory is plagued by Gribov ambiguities [12, 18, 19] implying that a mass parameter emerges in an improved quantization of the model, namely m. Notwithstanding, many other mass parameters might emerge as long as we consider non-Abelian gauge theories.

2.3. Rescales and metric-affine gravity

The idea behind a gravity gauge theory is that, after still unknown dynamical effects, the gauge space is identified space-time [4, 5, 7] and gravity as a geometric dynamical theory emerges. Considering, for instance, the Gribov parameter, one can rescale the fields σ and g together with the corresponding identification with vierbein and premetric of the cotangent spacetime bundle:

$$\begin{aligned} \langle\sigma^a\rangle &= m e^a \;, \\ \langle g^{ab}\rangle &= m \mathfrak{g}^{ab} \;, \end{aligned} \tag{8}$$

and similar relations for the dual variables [10]. As a consequence, expression (1) is fulfilled and a geometric theory emerges, being associated with a gravity theory. The rescaling and identifications (8) actually identify the gauge space with the cotangent spacetime bundle and thus say that the dynamics of the gauge theory dictates the geometry of the spacetime. In fact, the geometry that rises is the metric-affine geometry [6, 7] with no symmetry breaking to the

[3] By doble algebra-valued we mean that this quantity will carry two indices associated with the translation group. It is then expanded twice on the group generators.

general linear group. To see this one can simply take a quick look at the field strengths and minimal couplings through the map (8). The field strength Ω is then mapped into the curvature while Ξ is the translational curvature. Also, we have

$$T \longmapsto mC\ , \qquad Q \longmapsto mN\ , \tag{9}$$

where C is the Cartan Torsion 2-form and N is the 1-form nonmetricity.

3. Bundle contractions, Riemman-Cartan gravity and extra matter

An alternative path is to consider the topological properties of the bundles $A(4,\mathbb{R})$ and $\mathbb{A}$. In fact, as discussed in detail in [10], the affine group has two contractible pieces $\mathbb{R}^4_g$ and $K(4)$ that can be reduced to define simpler and smaller principal bundles.

3.1. Static bundle

The contractible sectors $\mathbb{R}^4_g$ and $K(4)$ are both symmetric coset spaces, acording to (2) and (3). As a consequence the $A(4,\mathbb{R})$ bundle can be contracted down to a smaller bundle [10, 11, 24]. Further, the connection on the reduced bundle is defined by the stability sector of the full connection [10, 11, 24]. In fact, the bundle contractions for the affine local group obey the following hierarchy

$$A(4,\mathbb{R}) \to GL(4,\mathbb{R}) \to O(4) \tag{10}$$

i.e., the affine gauge theory naturally reduces to the orthogonal one. The major consequence comes from the static nature of this bundle, the connection Y is also reduced to an orthogonal connection w implying that all nonorthogonal degrees of freedom decouples from the bundle. As discussed in [10, 24] this result implies that every connection on $A(d,\mathbb{R})$ imposes a connection on $O(d)$. Now, if the bundle $A(4,\mathbb{R})$ is identified with the coframe bundle then the resulting spacetime geometry is the Rieman-Cartan one [10, 24].

3.2. Dynamical bundle

The contraction (10) induces a contraction of the dynamical bundle $\mathbb{A}$. However, in that case there are two main differences: The fact that we consider all possible gauge configurations implies that the full affine connection cannot be simply reduced to the orthogonal one. To see this one should look at the gauge orbit (6) which is also the fiber of the bundle. It means that the connection Y is indeed part of the fundamental structure of the principal bundle and thus cannot simply decouple from it. The induced contraction on $\mathbb{A}$ can be performed by first contracting down the structure group $A(d,\mathbb{R})$ and then, by taking into account that all gauge connections and their equivalents were brought into the structure of the bundle $\mathbb{A}$ before the contraction and then analyze the respective effect on a typical fibre (6). Second, as we shall see soon, the fact that the subgroup $\mathbb{R}^4_g$ is contractible will induce a matter character to E that can be directly identified with the vierbein e. Thus, there is no need to introduce the matter field σ at all. This second statement is important because simplifies the original setting of fields[4] to $Y=(\omega,E)$ and g. This present approach slightly differs from the original developed in [10].

To perform the contraction we decompose the full affine connection according to $Y=w+q+E$ where q is the $K(d)$ sector of the connection, *i.e.*, $w=\omega^a{}_b\Sigma^b{}_a$ and $q=\omega^a{}_b\lambda^b{}_a$. Let us take then a generic fibre (6) and contract the group space according to $\mathcal{U}\longrightarrow L$. Thanks to the symmetric

[4] We remark that this procedure is equivalent to assume the identification of $\mathbb{A}$ with a metric-affine gravity in which the contraction is performed afterwards. In this case E and e are identified by extra relations.

nature of the coset spaces $\mathbb{R}^d$ and $K(d)$, the resulting decomposition is

$$\begin{aligned} w^L &= L^{-1}(\mathrm{d}+w)L\,, \\ q^L &= L^{-1}qL\,, \\ E^L &= L^{-1}E\,, \end{aligned} \tag{11}$$

i.e., there are no mixing between the coset sectors and the orthogonal one. Moreover, for the matter sector, the contraction implies in $g^L = L^{-1}g\left(L^{-1}\right)^T$. Thus, the non-orthogonal sector of Y abandon the geometric sector to migrate to the matter sector. Consequently, the metric-affine dynamical geometry described by $\mathbb{A}$ contracts down to a Riemann-Cartan one while the non-orthogonal components of the original connection are thrown into the matter sector. This effect is described by the contraction

$$\mathbb{A} \longrightarrow \mathbb{O} \oplus \mathcal{E} \oplus \mathcal{Q}\,, \tag{12}$$

where $\mathbb{O} \equiv \{O(4), \mathcal{L}\}$ is the dynamical principal bundle for an orthogonal theory. The space $\mathcal{L} \subset \mathcal{W}$ is the space of all independent orthogonal connections and $\mathcal{Q} = \mathcal{W}/\mathcal{L}$ is the space of the $K(4)$ connections. As an immediate consequence we can safely say that *If one starts with a dynamical metric-affine gauge theory of gravity it can end up in a Riemann-Cartan theory with two extra matter fields* [10]. In fact, the field strength decomposition yields

$$\begin{aligned} \Omega^a{}_b &= R^a{}_b + V^a_b\,, \\ \Xi^a &= \mathrm{D}E^a - q^a_bE^b\,, \end{aligned} \tag{13}$$

with $R^a{}_b = \mathrm{d}w^a{}_b - w^a{}_cw^c{}_b$ and $V^a_b = \mathrm{D}q^a{}_b - q^a{}_cq^c{}_b$ where $D = \mathcal{D}(w)$ is the orthogonal covariant derivative. Obviously, R is the orthogonal field strength while V is the orthogonal minimal coupling of q with a nonlinear mater interaction term. The same interpretation follows for E and its coupling Ξ. For the minimal coupling of g it is achieved

$$Q^{ab} = \mathrm{D}g^{ab} - q^a{}_cg^{cb} - q^b{}_cg^{ca}\,, \tag{14}$$

3.3. Rescales and Riemann-Cartan gravity

To perform an identification with an induced geometry of spacetime we perform the mapping between E and the vierbein together with g and the premetric. To do so we have to keep in mind that the group space is now flat with a deformed rigid dynamical premetric. Thus

$$\begin{aligned} \langle E^a\rangle &= me^a\,, \\ \langle g^{ab}\rangle &= m\delta^{ab} + \gamma^{ab}\,, \end{aligned} \tag{15}$$

where, again, m is the Gribov parameter associated with the nontriviality of the orthogonal sector of the $\mathbb{A}$ and γ^{ab} corresponds to the post-Riemannian sector of the original metric tensor of metric-affine gravity. Thus, δ^{ab} will work as the flat metric of cotangent space while γ^{ab} will be a generic covariant matter field, symmetric in the group indices. In fact, from the transformation of g under the affine group we obtain through contraction $\delta \longmapsto \delta$ and $\gamma \longmapsto L^{-1}\gamma\left(L^{-1}\right)^T$. The identification is then performed as

$$\begin{aligned} \mathfrak{g}_{\mu\nu} &= \delta_{ab}e^a \otimes e^b\,, \\ \mathfrak{g}^{\mu\nu} &= \delta^{ab}e_a \otimes e_b\,, \end{aligned} \tag{16}$$

ensuring the Riemannian character of the spacetime. The effect of this identification into the $\mathbb{R}^4_g$ curvature is that $\Xi^a = m(C^a - q^a_be^b)$ while for the minimal coupling Q it reduces to $Q^{ab} = \mathrm{D}\gamma^{ab} - 2mq^{ab} - q^a{}_c\gamma^{cb} - q^b{}_c\gamma^{ca}$. Obviously, Ξ is related to the Cartan torsion and a matter piece while Q is a pure matter sector.

4. Conclusions

In this work we have proposed a suitable scenario for the construction of a quantum theory of gravity by stating the independence between the internal gauge space and the cotangent bundle of spacetime. In particular we considered the affine gauge group $A(4, \mathbb{R})$ and the starting spacetime as an Euclidean 4-dimensional one. This independence is ensured by considering the premetric as a genuine dimension 1 matter field interacting with the affine gauge connection. Although we have not provided a specific consistent quantum theory, we have studied some geometrical properties of the formalism. An important point is that the present framework allow for the compatibility of the principles of general relativity and those of quantum physics by assuming that both sectors shall not overlap, *i.e.*, a consistent quantum model should be able to maintain the gauge internal space independent of the spacetime. Thus, dynamical effects associated to mass parameters, such as the Gribov parameter, are supposed to generate the identification between the cotangent space and gauge space. The theory is then automatically thrown in a geometric classical sector. We recall that this idea was already discussed in [8] under the Palatini formalism and in [10] already in the affine gauge model.

The main result of this article is the reduction of the metric-affine geometry to the Riemann-Cartan one. By considering the correct fibre bundle description of a gauge theory for the affine group, we have consistently contracted it to the orthogonal bundle. Thus, by decomposing the relevant fields into their orthogonal sector and the rest it is shown that we achieve a gauge theory for the orthogonal gauge group and extra matter fields. In fact, we have started with a theory with a gauge field Y and a matter field g and ended up in an orthogonal gauge theory with orthogonal spin-connection w, a vierbein field e and matter fields γ and q associated with the original nonmetric degrees of a metric-affine gravity. We remark that, differently from [10], we have improved the formalization of the ideas of the quantum sector by reducing the number of starting fields as well as defining those fields as dimension 1 matter fields, which ensures the impossibility of geometric identification unless a mass parameter emerges.

In summary, in this work we provide a few ideas that goes on the direction of a possible quantum framework for a gauge theory of gravity as well as the formal analysis for the reduction of metric-affine geometry to the Riemann-Cartan one, independently of the starting field equations, resulting on the migration of the original nonmetric degrees to the matter set of fields.

Acknowledgements

The authors express their gratitude to the Conselho Nacional de Desenvolvimento Científico e Tecnológico[5] (CNPq-Brazil) for financial support. RFS is also partially supported, and gratefull for, by the Fundação Carlos Chagas Filho de Amparo à Pesquisa do Estado do Rio de Janeiro[6] (FAPERJ) and the Pro-Reitoria de Pesquisa, Pós-Graduação e Inovação[7] of the Universidade Federal Fluminense (Proppi-UFF).

References

[1] R. Utiyama, Phys. Rev. **101**, 1597 (1956).
[2] T. W. B. Kibble, J. Math. Phys. **2**, 212 (1961).
[3] D. Lovelock, J. Math. Phys. **12**, 498 (1971).
[4] A. Mardones and J. Zanelli, Class. Quant. Grav. **8**, 1545 (1991).
[5] J. Zanelli, "Lecture notes on Chern-Simons (super-)gravities. Second edition (February 2008)," arXiv:hep-th/0502193.
[6] F. W. Hehl, G. D. Kerlick and P. Von Der Heyde, Phys. Lett. B **63**, 446 (1976).
[7] F. W. Hehl, J. D. McCrea, E. W. Mielke and Y. Ne'eman, Phys. Rept. **258**, 1 (1995) [arXiv:gr-qc/9402012].

[5] RFS is a level PQ-2 researcher under the program *Produtividade em Pesquisa*, 304924/2009-1.
[6] Under the program *Auxílio Instalação*, E-26/110.993/2009.
[7] Under the program *Jovens Pesquisadores 2009*, project 305.

[8] R. F. Sobreiro and V. J. Vasquez Otoya, Class. Quant. Grav. **24**, 4937 (2007) [arXiv:0705.2200 [hep-th]].
[9] R. F. Sobreiro and V. J. Vasquez Otoya, Braz. J. Phys. **40**, 370-374 (2010). [arXiv:0711.0020 [hep-th]].
[10] R. F. Sobreiro and V. J. Vasquez Otoya, J. Geom. Phys. **61**, 137 (2011) [arXiv:1003.1102 [hep-th]].
[11] S. Kobayashi and K. Nomizu, "Foundations of Differential Geometry, Vol. 1" *New York, USA: John Wiley & Sons (1963).*
[12] I. M. Singer, Commun. Math. Phys. **60**, 7 (1978).
[13] M. Daniel and C. M. Viallet, Rev. Mod. Phys. **52**, 175 (1980).
[14] P. Cotta-Ramusino and C. Reina, J. Geom. Phys. **1N3**, 121 (1985).
[15] G. Falqui and C. Reina, Commun. Math. Phys. **102**, 503 (1985).
[16] M. Nakahara, "Geometry, topology and physics," *Bristol, UK: Hilger (1990) 505 p. (Graduate student series in physics).*
[17] R. A. Bertlmann, "Anomalies in quantum field theory," *Oxford, UK: Clarendon (1996) 566 p. (International series of monographs on physics: 91).*
[18] V. N. Gribov, Nucl. Phys. B **139**, 1 (1978).
[19] R. F. Sobreiro and S. P. Sorella, arXiv:hep-th/0504095.
[20] D. Dudal, R. F. Sobreiro, S. P. Sorella and H. Verschelde, Phys. Rev. D **72**, 014016 (2005) [arXiv:hep-th/0502183].
[21] A. Trautman, Czech. J. Phys. B **29**, 107 (1979).
[22] A. Trautman, "Fiber Bundles, Gauge Fields, And Gravitation," *In *Held.A.(Ed.): General Relativity and Gravitation, Vol.1*, 287-308*
[23] R. Aldrovandi and J. G. Pereira, Phys. Rev. D **33**, 2788 (1986).
[24] B. McInnes, Class. Quant. Grav. **1**, 1 (1984).

Cosmologies with a time dependent vacuum

Joan Solà
HEP Group, Dept. Estructura i Constituents de la Matèria
and Institut de Ciències del Cosmos, Universitat de Barcelona,
Av. Diagonal 647, 08028 Barcelona, Catalonia, Spain

E-mail: sola@ecm.ub.es

Abstract. The idea that the cosmological term Λ should be a time dependent quantity in cosmology is a most natural one. It is difficult to conceive an expanding universe with a strictly constant vacuum energy density, $\rho_\Lambda = \Lambda/(8\pi G)$, namely one that has remained immutable since the origin of time. A smoothly evolving vacuum energy density $\rho_\Lambda = \rho_\Lambda(\xi(t))$ that inherits its time-dependence from cosmological functions $\xi = \xi(t)$, such as the Hubble rate $H(t)$ or the scale factor $a(t)$, is not only a qualitatively more plausible and intuitive idea, but is also suggested by fundamental physics, in particular by quantum field theory (QFT) in curved space-time. To implement this notion, is not strictly necessary to resort to ad hoc scalar fields, as usually done in the literature (e.g. in quintessence formulations and the like). A "running" Λ term can be expected on very similar grounds as one expects (and observes) the running of couplings and masses with a physical energy scale in QFT. Furthermore, the experimental evidence that the equation of state (EOS) of the dark energy (DE) could be evolving with time/redshift (including the possibility that it might currently behave phantom-like) suggests that a time-variable $\Lambda = \Lambda(t)$ term (possibly accompanied by a variable Newton's gravitational coupling too, $G = G(t)$) could account in a natural way for all these features. Remarkably enough, a class of these models (the "new cosmon") could even be the clue for solving the old cosmological constant problem, including the coincidence problem.

1. Introduction

The observed accelerated expansion of the universe [1, 2] is currently one of the central issues of observational and theoretical cosmology. It somehow came as a surprise to discover that the universe is not progressively slowing down its pace, but actually rocketing it up. The usual explanatory paradigm for this striking observation assumes the existence of the so-called *dark energy* (DE)– a mysterious cosmic component with negative pressure, which, in contrast to dark matter (DM), does not cluster (at least in a comparable way) and pervades all corners of the universe. An obvious candidate for the DE is a strictly constant cosmological term, Λ. Nevertheless, this option precludes the possibility of a time evolution of the DE (which might be appealing so as to face the so-called coincidence problem, namely the fact that the current matter density ρ_M is so close to the DE density $\rho_\Lambda = \Lambda/(8\pi G)$, or, equivalently, the fact that the acceleration epoch started so recently). This drawback could be cured by admitting that the CC can be a dynamical quantity: $\Lambda = \Lambda(t)$, an option which is perfectly admissible by the Cosmological Principle.

Actually, the mere possibility of a variable CC approach was explored long ago on purely phenomenological grounds – cf. [3, 4] – although many of the models are little motivated. At present, we know that the idea of a slowly running CC is in general compatible with the experimental data [5] and there are reasonable theoretical models grounded on fundamental physical principles that can support these ideas, meaning that there is no strict need to invoke ad hoc scalar fields to generate a time dependence for the DE – see [6] and references therein, see also [7, 8] for recent alternative approaches reinforcing this point of view. Still, the variable $\Lambda = \Lambda(t)$ models can be dealt with in terms of a self-conserved DE fluid in the manner of a scalar field (as we cannot know a priori the very nature of the DE, and hence we may be naturally tempted to keep on using this simple scalar field approach). However, although this description in terms of scalar fields (which may be called the "DE picture" of the original $\Lambda = \Lambda(t)$ model [9]) is perfectly possible, the corresponding effective equation of state (EOS) for the DE, $\omega = p_{\rm DE}/\rho_{\rm DE}$, will be a non-trivial function of time or of the redshift, $\omega = \omega(z)$, and in general it cannot be reduced to the standard forms proposed in the literature. In other words, the effect of having a variable cosmological term $\Lambda = \Lambda(t)$ cannot be described by the simple parameterizations usually employed for $\omega(z)$, characterized by two parameters (ω_0, ω_1) that become constrained by the observations [10]. The effective EOS of a variable vacuum model is in general more complicated than that. This is shown in detail, with specific examples, in reference [9]. In particular, the vacuum models that we are going to consider cannot be described with these simple parameterizations. Therefore, the $\Lambda = \Lambda(t)$ models must be studied on their own and constitute an independent class of DE models. It means that we should be more open minded on the form that the effective EOS can adopt, and we should not limit ourselves to some admittedly popular (albeit perhaps too simple) forms utilized too often in the literature.

Remarkably, the variable $\Lambda = \Lambda(t)$ approach can also be advantageously employed for solving (or highly palliating) the big or "old" CC problem [11], viz. the absolutely formidable task of trying to explain the value itself of the current vacuum energy density $\rho_\Lambda^0 \sim 10^{-47}$ GeV^4 on the face of its enormous input value $\rho_\Lambda^i \sim M_X^4$ (with $M_X \sim 10^{16}$ GeV) left over in the very early times, presumably just after the inflation phase induced by a Grand Unified Theory (GUT). This option has been recently investigated in [12, 13], and we will also review it here. It is based on the idea of "relaxation" of the effective vacuum energy density $\rho_\Lambda(t) = \Lambda(t)/(8\pi G)$. Relaxation offers indeed an entirely different perspective on the old CC problem; in particular, it provides the extremely appealing possibility to solve the CC problem *dynamically*, hence *without fine tuning* [13]. Moreover, the current value of the expansion rate H becomes determined by an approximate relation of the form $H_0 \sim 1/(\rho_\Lambda^i)^n$ $(n > 0)$. Interestingly enough, this relation tells us that H_0 is small just because the initial vacuum energy ρ_Λ^i was very large!

In the next sections, I summarize the general notion and properties of the models with time-dependent vacuum energy, and consider specific scenarios possessing very interesting features concerning the possible solution, or at least significant alleviation, of some of the most fundamental cosmological constant problems of modern times.

2. Two cosmological pictures: "CC picture" versus "DE picture"

Suppose that we are given a model where the vacuum energy density $\rho_\Lambda = \rho_\Lambda(\xi(t))$ is time evolving through its dependence on other cosmological function(s) of the cosmic time, e.g. $\xi(t) = \rho_M(t), H(t), a(t), ...$. We take this starting "picture" as the original (physical) representation of the DE, and we call it "the CC picture". In general, other fundamental "constants" can also be varying, in particular the gravitational coupling G. Therefore, in the CC picture we will have in general two functions, at least, of the cosmic time or the cosmological redshift, namely

$$\rho_\Lambda(z) = \rho_\Lambda(\rho_M(z), H(z), ...) , \qquad G(z) = G(\rho_M(z), H(z), ...) . \tag{1}$$

Of course other fundamental parameters could also be variable. For example, the fine structure constant has long been speculated as being potentially variable with the redshift. However, here we focus purely on the potential variability of the most genuine fundamental gravitational parameters of Einstein's equations, to wit: Λ and G.

For a cosmological model of the kind (1), the general Bianchi identity of the Einstein tensor leads to $\nabla^{\mu}\left[G(T_{\mu\nu}+g_{\mu\nu}\rho_{\Lambda})\right]=0$. In the FLRW metric, and modeling the universe as a perfect fluid, it implies

$$\frac{d}{dt}\left[G(\rho_M+\rho_\Lambda)\right]+3\,G\,H_{\rm CC}\,(\rho_M+p_M)=0\,. \tag{2}$$

This "mixed" conservation law connects the variation of ρ_Λ and G with that of the matter density ρ_M. As a result, the matter fluid in the CC picture is generally non-conserved and its evolution may be non-canonical, as we will see later on in an specific example. The general expression for the Hubble function in the CC picture, assuming a spatially flat FLRW universe, is

$$H^2_{\rm CC}(z)=\frac{8\pi G(z)}{3}(\rho_M(z)+\rho_\Lambda(z))\equiv H_0^2\left[\Omega_M^0\,f_M(z)(1+z)^{\alpha}+\Omega_\Lambda^0\,f_\Lambda(z)\right]\,, \tag{3}$$

where $\alpha=3(1+\omega_{\rm m})$, with $\omega_{\rm m}=0,1/3$ for the matter and radiation epochs respectively. Here f_M and f_Λ are certain functions of the redshift. They must satisfy $f_M(0)=1$ and $f_\Lambda(0)=1$, in accordance with the current cosmic sum rule $\Omega_M^0+\Omega_\Lambda^0=1$ for a flat universe.

On the other hand, the above cosmological dynamics in the CC picture can also be described in terms of an alternative fluid with covariantly self-conserved densities (ρ_{Ms},ρ_D), therefore satisfying $\dot\rho_{Ms}+\alpha\,H_{\rm DE}\,\rho_{Ms}=0$ and $\dot\rho_D+3\,H_{\rm DE}\,(1+\omega_{\rm eff})\rho_D=0$, where $\omega_{\rm eff}=p_D/\rho_D$ will be, in general, a function of the redshift: $\omega_{\rm eff}=\omega_{\rm eff}(z)$. Of course $\rho_{Ms}\neq\rho_M$ and $\rho_D\neq\rho_\Lambda$ in general. The Hubble rate in the new picture takes on the form:

$$H^2_{\rm DE}=\frac{8\pi G_0}{3}(\rho_{Ms}(z)+\rho_D(z))=H_0^2\left[\tilde\Omega_M^0\,(1+z)^{\alpha}+\tilde\Omega_D(z)\right]\,, \tag{4}$$

where

$$\tilde\Omega_D(z)=\tilde\Omega_D^0\,\exp\left\{3\int_0^z dz'\frac{1+\omega_{\rm eff}(z')}{1+z'}\right\}\,. \tag{5}$$

We have defined $\tilde\Omega_M^0=\rho_{Ms}^0/\rho_c^0$, with $\rho_c^0=3H_0^2/(8\pi G_0)$ the current value of the critical density. Notice that the parameter $\tilde\Omega_M^0$ need not coincide with Ω_M^0, as they are determined from two different parameterizations of the DE data. Similarly, we have defined $\tilde\Omega_D(z)=\rho_D(z)/\rho_c^0$ – and its current value $\tilde\Omega_D^0=\tilde\Omega_D(0)$. Notice also that $G_0=1/M_P^2$ (M_P being the Planck mass) is strictly constant, whereas G in eq. (3) is in general a variable function of the redshift: $G=G(z)$. The cosmological picture characterized by the *self-conserved* pair of energy densities (ρ_{Ms},ρ_D) at fixed G_0 will be called the "DE picture" [9].

Since the two pictures are assumed to be equivalent descriptions of the same cosmological evolution, their matching requires that the expansion history of the universe is the same in both pictures, i.e. that their Hubble functions are numerically equal:

$$H_{\rm DE}(z)=H_{\rm CC}(z)\,. \tag{6}$$

Recall that we take the original picture (1) as the physical one, whereas the DE picture plays the role of an effective description of the former in terms of the self-conserved DE component ρ_D at fixed G_0. Therefore, the function $\omega_{\rm eff}(z)$ defined above can be considered as the "effective EOS" (in the "DE picture") of the original model (1). A straightforward calculation, using the matching condition (6) and the general Bianchi identity (2), allows to determine the effective

EOS function $\omega_{\rm eff}(z)$ in terms of the fundamental parameters of the original theory (1), as follows [9]:

$$\omega_{\rm eff}(z) = -1 + \frac{1}{3}\frac{1+z}{\rho_D}\frac{d\rho_D}{dz} = -1 + \frac{\alpha}{3}\left(1 - \frac{\zeta_\Lambda(z)}{\rho_D(z)}\right), \tag{7}$$

where we have defined $\zeta_\Lambda(z) \equiv (G(z)/G_0)\,\rho_\Lambda(z)$, and

$$\rho_D(z) = (1+z)^\alpha \left[\rho_D(0) - \alpha \int_0^z \frac{dz'\,\zeta_\Lambda(z')}{(1+z')^{(\alpha+1)}}\right]. \tag{8}$$

In particular, for a model of the kind (1) with fixed $G = G_0$ in the matter epoch ($\alpha = 3$), the effective EOS simply reads $\omega_{\rm eff}(z) = -\rho_\Lambda(z)/\rho_D(z)$. The latter would shrink to just $\omega_{\rm eff} = -1$ only in case of a truly cosmological constant $\rho_\Lambda = const.$ because then ρ_D would coincide with it. With the help of the matching condition (6) of the two pictures, and the Bianchi identity, one can also show, with some effort, that

$$\frac{d\rho_D(z)}{dz} = \alpha\,(1+z)^{\alpha-1}\,\rho_c^0\,\left(\Omega_M^0\,f_M(z) - \tilde{\Omega}_M^0\right). \tag{9}$$

This is a remarkable relation, if we recall the conditions $f_i(0) = 1$ fulfilled by the functions f_M and f_Λ in eq. (3). Why is remarkable? As Ω_M^0 and $\tilde{\Omega}_M^0$ must be very close, and $f_M(z)$ is presumably monotonous, eq. (9) shows that a value $z = z^*$ *always* exists for which the *l.h.s.* of eq. (9) vanishes and, therefore, the effective EOS $\omega_{\rm eff}(z)$ crosses the phantom divide at that point: $\omega_{\rm eff}(z^*) = -1$. Whether the crossing is in the recent past or in the immediate future, it will depend on the details of the original model (1). Notice, however, that if $\tilde{\Omega}_M^0 = \Omega_M^0$, then the crossing will be exactly at $z^* = 0$, i.e. now. This shows that a generic model (1) with time varying vacuum energy, when described in the DE picture (i.e. as if it were a cosmological self-conserved DE fluid) generally leads to a crossing of the phantom divide, thus providing a possible natural explanation for the observational data – which still admit a tilt in the phantom domain [2].

3. Time dependent vacuum in the running ΛCDM model

In contradistinction to purely phenomenological models of time dependent Λ [3, 4], we adopt here the fundamental point of view that the Λ term is a running parameter in QFT in curved space-time [14]. As suggested in [6], we should have a corresponding renormalization group (RG) equation of the general form

$$(4\,\pi)^2\,\frac{d\rho_\Lambda}{d\ln\mu} = \sum_{n=1}^{\infty} A_n\,\mu^{2n}, \tag{10}$$

in which μ is an arbitrary scale associated to the RG running. Of course the full effective action is perfectly scale independent (i.e. RG invariant) [6], and the running of the vacuum energy should ultimately reflect the dependence of the leading quantum effects with respect to some physical cosmological quantity $\xi = \xi(t)$ associated with μ, hence $\rho_\Lambda = \rho_\Lambda(\xi)$. Despite it being a matter of debate, it has been argued by different methods [16, 17] that the physical scale ξ pointed by the μ-dependence is the Hubble rate H. Since $H = H(t)$ evolves with the cosmic time, the cosmological term inherits a time-dependence through its primary scale evolution with H. The coefficients A_n in eq. (10) are obtained after summing over the loop contributions of fields of different masses M_i and spins σ_i. The general behavior is $A_n \sim \sum M_i^{4-2n}$ [16, 17]. Therefore, for $\mu = H \ll M_i$, the series above is (for $n > 1$) an expansion involving powers of the small quantities $\mu = H$ and H/M_i. Given, however, that $A_1 \sim \sum M_i^2$, we see that the heaviest fields furnish the dominant contribution. This trait ("soft-decoupling") represents a generalization of the decoupling theorem in QFT. In addition, being $H_0 \sim 10^{-33}\,eV$, the condition $\mu \ll M_i$ is

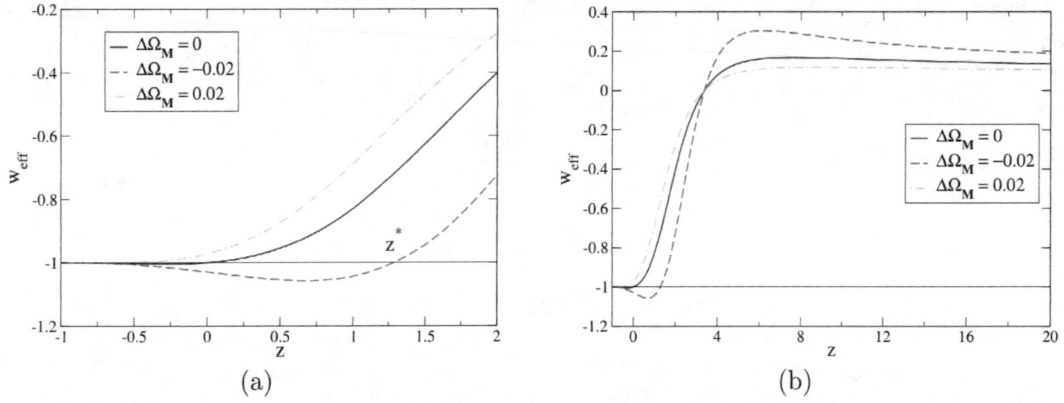

Figure 1. Time dependent vacuum model (12). **(a)** The effective EOS in the DE picture, given by eq. (15), as a function of z for various $\Delta\Omega_M$ at fixed $\nu = -\nu_0$, and $\Omega_M^0 = 0.3\,, \Omega_\Lambda^0 = 0.7$; **(b)** Extended z range of the plot (a). z^* is the crossing point of the phantom divide $\omega_{\rm eff} = -1$.

amply met for all known particles, and the series on the *r.h.s* of eq. (10) converges extremely fast at present. Also important is the fact that only even powers of $\mu = H$ are consistent with general covariance, see [6]. Finally, the $n = 0$ contribution is absent because it corresponds to terms $\propto M_i^4$ that give an extremely fast evolution. Actually, from the RG point of view they are already excluded because, as noted above, $\mu \ll M_i$ for all known masses. In practice only the first term $n = 1$ is needed, with M_i of the order of the highest mass available.

As the dominant masses M_i will be of order of a GUT mass scale M_X near the Planck mass M_P, it is convenient to introduce the ratio $\nu = \sigma/(12\pi)\,\left(M_X^2/M_P^2\right)$, in which $\sigma = \pm 1$ depending on whether bosons or fermions dominate in their loop contributions to (10). If the effective mass M_X of the heavy particles is just M_P, the parameter ν takes the value $\sigma\,\nu_0$, with $\nu_0 \equiv 1/(12\,\pi) \simeq 2.6 \times 10^{-2}$. In general we expect that ν will take values of this order or below. Under the very good approximation $n = 1$, and with $\mu = H$, eq. (10) abridges to

$$\frac{d\rho_\Lambda}{d\ln H} = \frac{3\,\nu}{4\,\pi}\,M_P^2\,H^2\,, \tag{11}$$

whose solution is [1]

$$\rho_\Lambda = c_0 + c_1\,H^2\,, \tag{12}$$

with

$$c_0 = \rho_\Lambda^0 - \frac{3\,\nu}{8\pi} M_P^2\,H_0^2\,, \quad c_1 = \frac{3\,\nu}{8\pi}\,M_P^2\,. \tag{13}$$

Solving the model in the original CC picture [9] leads to a non-standard evolution law for the matter density as a function of the redshift: $\rho_M(z) = \rho_M^0\,(1+z)^{3(1-\nu)}$, along with the following evolution of the vacuum energy density:

$$\rho_\Lambda(z) = \rho_\Lambda^0 + \frac{\nu\,\rho_M^0}{1-\nu}\,\left[(1+z)^{3(1-\nu)} - 1\right]\,. \tag{14}$$

For $\nu = 0$ we recover the ΛCDM results, of course. But for $\nu \neq 0$ we have a model in which the vacuum energy decays into matter (or vice versa) according to the generalized conservation law (2), with $G = const$ in this case. With this RG model of the cosmic evolution at hand, we may

[1] It is interesting to notice that this quadratic evolution law for the vacuum energy with the expansion rate has also been suggested recently by alternative QFT methods, see [15].

now use it to apply the procedure of section 2 for obtaining the effective properties of the model in the "DE picture" (in which matter and dark energy are separately conserved). Obviously, for this particular model, $\zeta_\Lambda(z) = \rho_\Lambda(z)$. The corresponding effective EOS cannot be described with standard parameterizations [10]. Indeed, from equation (7) – using (8) and (14) – we find:

$$\omega_{\rm eff}(z)\,|_{\Delta\Omega\neq 0} = -1 + (1-\nu)\,\frac{\Omega_M^0\,(1+z)^{3(1-\nu)} - \tilde{\Omega}_M^0\,(1+z)^3}{\Omega_M^0\,[(1+z)^{3(1-\nu)} - 1] - (1-\nu)\,[\tilde{\Omega}_M^0\,(1+z)^3 - 1]}\,. \tag{15}$$

Here $\Delta\Omega_M \equiv \Omega_M^0 - \tilde{\Omega}_M^0 \neq 0$ is the difference of the cosmological mass parameters in the two pictures (corresponding to two different fits of the same data). For $|\nu| \ll 1$ we may expand the previous result to first order in ν. Assuming for simplicity that $\Delta\Omega_M = 0$, we find

$$\omega_{\rm eff}(z) \simeq -1 - 3\,\nu\frac{\Omega_M^0}{\Omega_\Lambda^0}\,(1+z)^3\,\ln(1+z)\,. \tag{16}$$

This result reflects the essential qualitative features of the analysis presented in the previous section, where in this case $z^* = 0$. Indeed, for $\nu > 0$ eq. (16) shows that we get an (effective) phantom-like behavior ($\omega_{\rm eff} \lesssim -1$), whereas for $\nu < 0$ we have an (effective) quintessence behavior ($\omega_{\rm eff} \gtrsim -1$). In other words, this variable CC model can give rise to two types of very different qualitative behaviors by changing the sign of a single parameter. In Fig. 1 we show also more general cases where $z^* \gtrsim 0$, corresponding to $\Delta\Omega_M \neq 0$. In contrast to the previous situation, here the variable CC model may exhibit phantom behavior for $\nu < 0$ (if $\Delta\Omega_M < 0$), and manifests itself through the existence of a transition point z^* in our recent past – marked explicitly in the figure – as predicted by the general formalism of section 2.

Let us also examine another interesting situation. From eq. (2), we see that if $G = G(t)$ is variable and we assume that matter is conserved (thus no transfer of energy with the time dependent vacuum ρ_Λ), we have $d\zeta_\Lambda = -(\rho_M/G_0)\,dG = (\rho_M^0/G_0)(1+z)^\alpha\,dG$. From eq. (8) it is then not difficult to show that[2]

$$\frac{d\rho_D}{dz} = \alpha(1+z)^{\alpha-1}\,\frac{\rho_M^0}{G_0}\,[G(z) - G(z^*)]\,. \tag{17}$$

where z^* is the crossing point of the phantom divide, which satisfies $\rho_D(z^*) = \zeta_\Lambda(z^*)$. It follows that if G is asymptotically free – that is to say, if $G(z)$ decreases with redshift – we should observe quintessence behavior ($d\rho_D/dz > 0$) for $0 \leqslant z \leqslant z^*$; whereas if G is "IR free" (i.e. it increases with z, hence decreases with the expansion), we should then have phantom behavior ($d\rho_D/dz < 0$) in that interval. An interesting example of model of this kind appears if we take once more the evolution law (12) for the vacuum energy, but assume now that matter is conserved [16]. The basic cosmological equations can be formulated in the following compact way:

$$\begin{array}{ll} E^2(z) = g(z)\,\left[\Omega_M(z) + \Omega_\Lambda(z)\right] & \text{(Friedmann equation)} \\ (\Omega_M + \Omega_\Lambda)dg + g\,d\Omega_\Lambda = 0 & \text{(Bianchi identity)} \\ \Omega_\Lambda(z) = \Omega_\Lambda^0 + \nu\left[E^2(z) - 1\right] & \text{(running } \rho_\Lambda\text{)}\,, \end{array} \tag{18}$$

[2] One first shows that ρ_D in eq. (8) satisfies the differential equation $d\rho_D(z)/dz = \alpha\,(\rho_D(z) - \zeta_\Lambda(z))\,/(1+z)$. Next one notes that the previous equation is also solved by the alternative expression below:

$$\rho_D(z) = \zeta_\Lambda(z) - (1+z)^\alpha \int_{z^*}^{z} \frac{dz'}{(1+z')^\alpha}\frac{d\zeta_\Lambda(z')}{dz'} \quad\Rightarrow\quad \frac{d\rho_D(z)}{dz} = -\alpha\,(1+z)^{\alpha-1}\int_{z^*}^{z}\frac{dz'}{(1+z')^\alpha}\frac{d\zeta_\Lambda(z')}{dz'}\,.$$

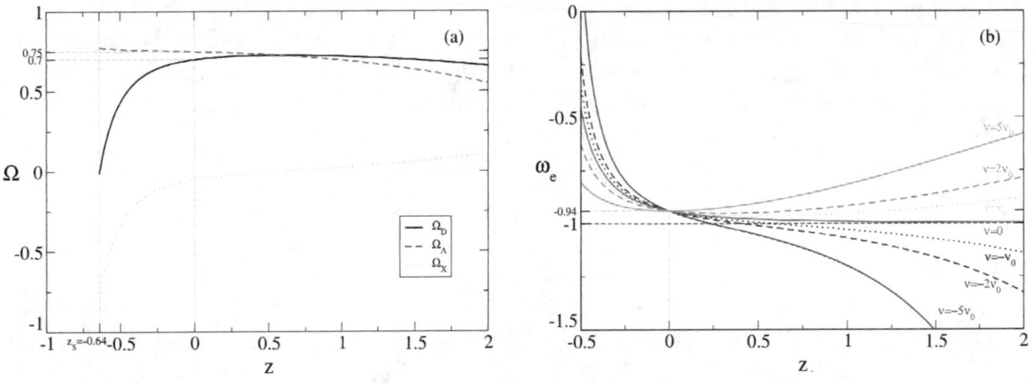

Figure 2. Features of the ΛXCDM model. **(a)** $\Omega(z)$ for the total DE and its individual components, $\Omega_X(z)$ and $\Omega_\Lambda(z)$, for $\Omega_M^0 = 0.3$, $\Omega_\Lambda^0 = 0.75$, $\omega_X = -1.85$, $\nu = -\nu_0$. The stopping of the expansion at $z = z_s$ is achieved here thanks to $\Omega_X(z)$ becoming eventually negative; **(b)** Behavior of the effective EOS, eq.(21), for the same ω_X and Ω_Λ^0 and different values of ν.

where $E(z) \equiv H(z)/H_0$ and $g(z) \equiv G(z)/G_0$. It is now straightforward to solve them for the gravitational coupling, which becomes a logarithmically running quantity of the Hubble rate:

$$G(H) = \frac{G_0}{1 + \nu \ln\left(H^2/H_0^2\right)}\,. \tag{19}$$

We see that G is asymptotically free (resp. IR free) for $\nu > 0$ (resp. $\nu < 0$). An alternative formulation of this running law for G and Λ, motivated by the quantum physics of inflation, was presented in [17].

The previous considerations show that we have plenty of theoretical and phenomenological motivations to support the possibility of observing running cosmological parameters, specially through the study of time-evolving vacuum cosmologies. Ultimately this issue must of course be backed up by observations, and in this sense it is very important to try to find all possible phenomenological indications suggesting such a vacuum dynamical picture. It is also remarkable that one could even find signs of the running Λ in the astrophysical context; for example, through the study of the implications that a variable Λ distributed (homogeneously or inhomogeneously) in a large domain could have in the process of local structure formation, see [18].

Time-evolving vacuum models inspired by QFT principles could do even more: they could provide a possible solution to the cosmic coincidence problem, and ultimately to the "old cosmological constant problem", namely the CC fine-tuning problem – the toughest cosmological conundrum of all times [11]. Possible avenues leading to a significant step in this direction are summarized in the next two sections.

4. The ΛXCDM model: a possible solution to the cosmic coincidence problem

The ΛXCDM model [19] makes a further step in our systematic approach to a cosmology with variable cosmological parameters. It is based on assuming that we have a composite DE energy density $\rho_D = \rho_X + \rho_\Lambda$. In the canonical case where $G = const$, it can be taken as covariantly conserved:

$$\dot{\rho}_D + 3(1+\omega_e)\,\rho_D\,H = \dot{\rho}_\Lambda + \dot{\rho}_X + 3(1+\omega_X)\,\rho_X\,H = 0\,, \tag{20}$$

and in these conditions matter is self-conserved too of course: $\dot{\rho}_M + \alpha\,H\,\rho_M = 0$. Here $\rho_\Lambda = \rho_\Lambda(t)$ is the energy density of the running Λ. Moreover, $\rho_X(t)$ is the energy density of a new dynamical entity X that we call the "cosmon". Its dynamics is completely determined by the above local conservation law (20) once the evolution for $\rho_\Lambda(t)$ is given. As emphasized in [19], this implies that in general it will be an effective entity, not a fundamental one (i.e. X is *not* supposed to be a scalar field ϕ) because the dynamics imposed by (20) will in general make it incompatible with the scalar potential provided for ϕ. The cosmon X, instead, embodies the behavior of terms in the effective action of QFT in curved space-time, and, far from being an elementary scalar field, it represents a convenient way to parametrize the effect of those terms in the cosmological evolution. The EOS of the cosmon is indicated by ω_X (assumed, for the moment, to be constant for simplicity). It should not be confused with the effective EOS of the total DE, which reads

$$\omega_e = \frac{p_D}{\rho_D} = \frac{-\rho_\Lambda + \omega_X\,\rho_X}{\rho_\Lambda + \rho_X} = -1 + (1+\omega_X)\,\frac{\rho_X}{\rho_D}\,. \tag{21}$$

Next we will assume that $\rho_\Lambda = \rho_\Lambda(H)$ is the same function of H as in the running ΛCDM model presented in section 3 – cf. eqs. (12) and (13). Notice that this does not mean that its redshift evolution is still given by eq. (14), as here the overall conservation law is no longer (2) but (20). After solving explicitly the details of this new cosmological model – which we refrain from reporting here [19] – the normalized energy densities $\Omega(z) = \rho(z)/\rho_c^0$ ($\rho = \rho_M, \rho_\Lambda, \rho_X, \rho_D$) with respect to the present critical density ρ_c^0 become determined as a function of the redshift (cf. Fig. 2a). We see that the cosmon density Ω_X can be negative, although this is not too surprising for an effective quantity that does not represent a fundamental field. On the other hand we observe that the total DE density $\Omega_D = \Omega_\Lambda + \Omega_X$ behaves properly, and its current value is the measured DE value, i.e. we have $\Omega_D^0 = 0.7$ with $\Omega_M^0 + \Omega_D^0 = 1$ (flat space). Furthermore, the EOS function $\omega_e = \omega_e(z)$ can display a rich variety of behaviors while being in agreement with the most recent data [2]. This is shown in Fig. 2b, where we see that, depending on the value of ν, the EOS can be quintessence-like, or mimic (in some cases almost exactly) a pure CC term ($\omega_e \simeq -1$), and it can even exhibit a mild transition from the phantom to the quintessence regime. From Fig. 2b we recognize that this model is a spectacular example of cosmology, in which despite of the fact that the vacuum energy $\rho_\Lambda = \rho_\Lambda(H)$ is a running quantity, the effective EOS of the composite DE can approach the standard ΛCDM model in an arbitrary way, and may display only small "quintessence-like" or "phantom-like" departures from it, thus mimicking a scalar field even in situations where such field could not behave canonically ($\omega_e \lesssim -1$).

The truly nice feature of a model of this kind, however, is that it can cure or highly alleviate the "cosmic coincidence problem". Recall that this problem is related to the behavior of the ratio $r(z) \equiv \rho_D(z)/\rho_M(z)$ between the dark energy density and the matter energy density. In the standard ΛCDM model, that ratio increases arbitrarily with the cosmic evolution because ρ_D is just the strictly constant vacuum energy density ρ_Λ^0, whereas $\rho_M \to 0$ with the expansion, so there is no apparent reason why we should find ourselves precisely in an epoch where $r = \mathcal{O}(1)$. Or, to put in other terms: there is no obvious reason why the epoch where the universe started to accelerate is so recent. This is an unexplained coincidence puzzle in the standard model. Not so in the ΛXCDM model, in which, thanks to the presence of the cosmon X, the function $r = r(z)$ presents a maximum at the resdhift [19]:

$$z_{\rm max} = \left[\frac{\Omega_\Lambda^0 - \nu}{\omega_X\,(\Omega_X^0 + \nu\,\Omega_M^0) - \epsilon\,(1-\Omega_\Lambda^0)}\right]^{\frac{1}{3\,(1+\omega_X-\epsilon)}} - 1\,, \tag{22}$$

where we have defined $\epsilon \equiv \nu(1+\omega_X)$. One can show that this point cannot be in our past [19], so it will generally be in the future (hence $-1 < z_{\rm max} < 0$). Beyond that maximum, there is a turning point $z = z_s < z_{\rm max}$ where the universe stops and reverses its evolution. At the stopping point, $\Omega_D(z_s) = -\Omega_M(z_s) < 0$. This is possible because we can have $\Omega_X < 0$, or $\Omega_\Lambda < 0$.

The following observation is now in order: notice, from eq. (20), that for $\nu = 0$ (which implies no running of ρ_Λ) the cosmon energy density is conserved; so, if in addition $\Omega_X^0 = 0$ (vanishing

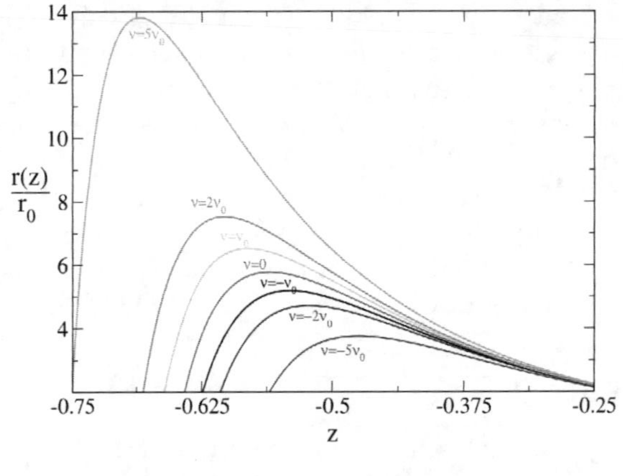

Figure 3. Behavior of the ratio $r = \rho_D/\rho_M$, normalized to its current value r_0, for $\omega_X = -1.85$, $\Omega_M^0 = 0.3, \Omega_\Lambda^0 = 0.75, \Omega_X^0 = -0.05$ (flat space) and different values of ν. Here $\Omega_X^0 < 0$ insures stopping and bouncing at a future point ($z = z_s < 0$). The maxima satisfy $r_{\rm max} < \mathcal{O}(10)$, which suggests a viable solution to the cosmic coincidence problem.

current cosmon density), we must then have $\Omega_X(z) = 0$ at all times. Therefore, in the limit $\nu = \Omega_X^0 = 0$ we exactly recover the standard ΛCDM model. And it is precisely in this limit that the formula (22) for the redshift location of the maximum of the ratio $r(z)$ ceases to make sense, as it is evident by simple inspection of that formula. In contrast, when we make allowance for the parameter space of the ΛXCDM model, the maximum does exist and moreover the ratio $r(z)$ stays bounded – typically $r < \mathcal{O}(10)$ – for the entire history of the universe. This is clearly seen in Fig. 3, for various values of the parameters. So, in fact, the ΛXCDM model can provide a nice solution to the cosmic coincidence problem without necessarily departing from the ΛCDM model in a significant way. Furthermore, the detailed analysis of the cosmic perturbations in the ΛXCDM model indicate excellent compatibility with the present observational data on structure formation [20].

5. The "new cosmon" model: towards solving the "old CC problem"

The "Relaxed Universe" is a special realization of the ΛXCDM model which points to a more ambitious aim, to wit: solving the "old CC problem" [11], i.e. understanding how the Universe could start up its thermal history with a huge value of the initial vacuum energy $\rho_\Lambda^i \sim M_X^4$ (e.g. triggered by a GUT with, say $M_X \sim 10^{16}$ GeV, presumably responsible for inflation) and then eventually ended up with its current tiny value $\rho_\Lambda^0 \sim 10^{-47}$ GeV$^4 \sim 10^{-110}\rho_\Lambda^i$! The history of this profound theoretical conundrum, which keeps in check the fundamental physics principles since more than forty years ago, traces back to Zeldovich's realization in the late sixties [21], that the vacuum fluctuations of the quantum fields should induce a very large value of the CC. Since then the problem stays with us, unbeaten and virtually invincible. The "traditional" fine-tuning solution [11] (i.e. the sheer adjustment by hand of the initial big CC value ρ_Λ^i by a carefully chosen finite "counterterm") is simply and fully inadmissible, especially at the quantum level – see [13] (particularly Appendix B of this reference) for a detailed account. A much more promising approach, instead, should be to find out a mechanism that neutralizes *dynamically* the presence of the initial vacuum energy, ρ_Λ^i, whatever it be its seed value in the early universe. We will call the cosmological model related to this relaxation mechanism, the "Relaxed Universe" [13][3].

[3] For other recent alternative approaches aiming to palliate the CC problem, see e.g. [22]

As we would like that the expansion of the universe be responsible for the triggering of the relaxation dynamics, it is reasonable to assume that such mechanism should be closely related to the expansion itself, or more precisely, to the interaction governing the expansion, i.e. *gravity*. A note of caution, however: although we aim at a modified form of gravity, we cannot just content ourselves with a "late time effect" (as usually done in the literature on modified gravity [23, 24, 25]). We rather must have a *persistent* effect, fully active throughout the *entire history* of the universe. Indeed, in order to solve the old CC problem we need that some automatic mechanism sets to work at a time prior to the onset of the radiation epoch – at which time it must cancel most of the value of the original vacuum energy present in the early universe– *and*, most important, we also need that this cancelation is maintained for the rest of the universe lifetime, leaving a small value today and maybe also in the future.

It is not easy to device such a mechanism, but some ideas on it are briefly explained here – see [12], and specially [13], for a detailed account; and [26, 27] for alternative forms. The effective action of the "Relaxed Universe" is proposed in the following way:

$$\mathcal{S} = \int d^4x\,\sqrt{|g|}\left[\frac{1}{16\pi G}R - \rho_\Lambda^i - \mathcal{F}(R,\mathcal{G}) + \mathcal{L}_\varphi(\text{matter fields})\right], \tag{23}$$

in which we have an (arbitrarily large) initial vacuum energy ρ_Λ^i, and we have included a functional (the "cosmon functional") $\mathcal{F} = \mathcal{F}(R,\mathcal{G})$ of the Ricci scalar R and the Gauss-Bonnet invariant $\mathcal{G} = R^2 - R_{\mu\nu}R^{\mu\nu} + R_{\mu\nu\rho\sigma}R^{\mu\nu\rho\sigma}$. The structure of this functional is convenient in order to avoid the presence of ghost degrees of freedom and also to dodge the Ostrogradski instability, i.e. the appearance of vacuum states of negative energy [28]. The corresponding (generalized) Einstein equations read

$$R^\mu_{\ \nu} - \frac{1}{2}\,\delta^\mu_\nu\,R = -8\pi G\,\left[\rho_\Lambda^i\,\delta^\mu_{\ \nu} + 2E^\mu_{\ \nu} + T^\mu_{\ \nu}\right], \tag{24}$$

where $T^\mu_{\ \nu}$ is the ordinary energy-momentum tensor of the matter fields, and $E^\mu_{\ \nu}$ is the new tensor in the field equations associated to the presence of the cosmon functional $\mathcal{F}$ in the action. We assume that matter is covariantly conserved, and therefore $E^\mu_{\ \nu}$ must also be conserved: $\nabla^\nu E^\mu_{\ \nu} = 0$. Clearly, our functional constitutes an extension of the Einstein-Hilbert's action with CC, in which we have added the $\mathcal{F}(R,\mathcal{G})$ part. The next point is to make an appropriate choice for the functional $\mathcal{F}$ such that it can have a bearing on the cosmological constant problem. We make the following Ansatz:

$$\mathcal{F}(R,\mathcal{G}) = \beta\,F(R,\mathcal{G}) = \frac{\beta}{B(R,\mathcal{G})}, \tag{25}$$

$$B(R,\mathcal{G}) = \frac{2}{3}R^2 + \frac{1}{2}\mathcal{G} + (y\,R)^3, \tag{26}$$

where $B(R,\mathcal{G})$ is a polynomial of the Ricci scalar and the Gauß-Bonnet invariant, and (β, y) are dimensionful parameters to be determined later. This model is in fact of ΛXCDM type, in the sense that the cosmon X here is related to the energy density $\rho_F = 2E^0_{\ 0}$ generated by $\mathcal{F}$ in the effective action, and so ρ_F plays the role of ρ_X in the ΛXCDM model. Explicitly, one finds

$$\rho_F = 2\,E^0_{\ 0} = \beta\left[F(R,\mathcal{G}) - 6(\dot{H} + H^2)F^R + 6H\dot{F}^R - 24H^2(\dot{H} + H^2)F^{\mathcal{G}} + 24H^3\dot{F}^{\mathcal{G}}\right]. \tag{27}$$

In our case, F has the form $F = 1/B$ as defined in (25), and $F^Y \equiv \partial F/\partial Y$ are the partial derivatives of $F = F(R,\mathcal{G})$ with respect to $Y = R, \mathcal{G}$. The parallelism with the ΛXCDM model becomes even more transparent if we notice that the mixed DE energy density defined in (20)

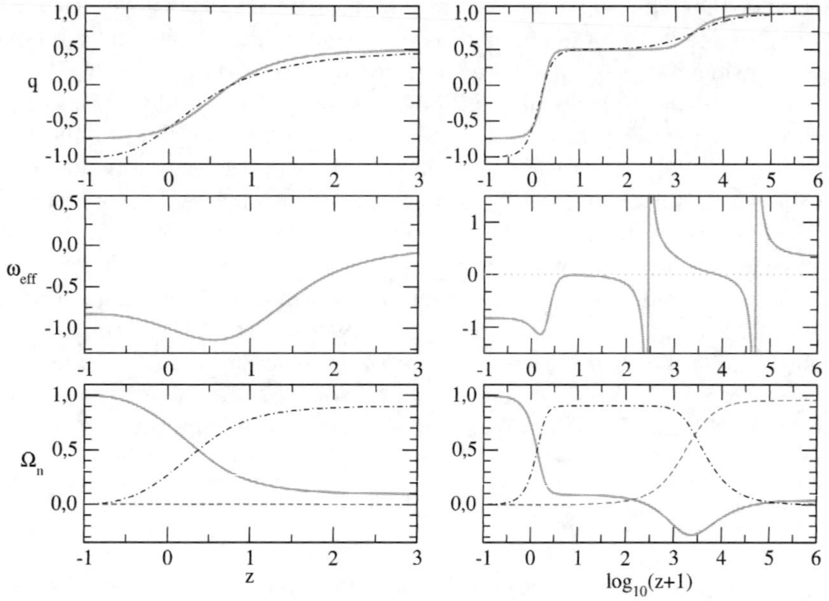

Figure 4. Deceleration parameter q, effective EOS $\omega_{\rm eff}$, and relative energy densities for each energy component (normalized with respect to the corresponding critical density) $\Omega_n(z) = \rho_n(z)/\rho_c(z)$ of dark energy $\rho_{\Lambda{\rm eff}}$ (orange thick curve), dark matter ρ_M (black dashed-dotted) and radiation ρ_r (red dashed) as functions of redshift z in the relaxation model (25) with $y = 0.7 \times 10^{-3}\, H_0^{-2/3}$, $\rho_\Lambda^i = -10^{60}$ GeV^4, $\Omega_M^0 = 0.27$, $\Omega_r^0 = 10^{-4}$, $q_0 \approx -0.6$, $\dot{q}_0 = -0.5\, H_0$. In the q-plot, the thick orange curve corresponds to the Relaxed Universe, and the black dashed-dotted curve to ΛCDM.

Incidentally, this value for M is the geometric mean of the two most extreme mass scales available in our universe below the Planck mass. It is rewarding to see that we do not obtain an extremely small mass scale of order $m_\phi \sim H_0 \sim 10^{-33}$ eV, as in the case of quintessence models.

From the above considerations, we see that the F-functional's main mission is to *dynamically* counterbalance the big initial ρ_Λ^i at *all epochs*, not just now; and in the last stage of this dynamical process the Hubble rate is driven to the very small value H_0 that we observe today. At the end of the day, what we get is not just a late time effect of the cosmological evolution (as in the traditional modified gravity models), but rather a truly perennial large scale influence on all periods of the cosmic history. One hope is that a relaxation mechanism of this sort should ultimate provide a dynamical solution to the tough fine tuning problem underlying the "old CC problem" [11]. For example, in the standard model of electroweak interactions such fine-tuning involves 55 decimal places, at least, and it must be retuned order by order in perturbation theory until reaching loop diagrams of order 20th! –see [13] for details. In the Relaxed Universe, in contrast, the tuning to all orders is automatic and performed dynamically by the universe's expansion. This is why such field-theoretical object associated to the functional F is called the "cosmon", as its primary mission is to compensate for the huge initial CC throughout the entire cosmological evolution. Clearly, it has the same aim as the cosmon scalar field first introduced in [29], except that here the cosmon has nothing to do with scalar fields, and moreover is *not* subdued by Weiberg's "no-go" theorem [11] – see [13] for a more detailed account.

While in the present version the cosmon is not a fundamental quintessence field [30], it can perfectly mimic quintessence-like behavior (cf. Fig. 4). The new cosmon, however, is a more complex object: it is a functional whose ultimate origin must be provided by a fundamental theory (say string theory or any other theory aiming at a final explanation of quantum gravity). In the meanwhile, what we have shown here is a practical working example, actually a class of such functionals which are able to truly adjust the vacuum energy dynamically, whatever it be its initial value, and without performing fine-tuning at any stage of the cosmological evolution. As this was the main aim of the original cosmon model [29], it is naturally suggested that the general class of these functionals is called the "new cosmon". For more details, see [13].

Let us also mention that the Relaxed Universe based on the cosmon functional (25) and extensions thereof (cf. Ref.[13]), provides also a possible natural solution to the cosmic coincidence problem (mentioned in section 4). This additional bonus is not too surprising if we recall that the Relaxed Universe is a particular implementation of the class of ΛXCDM models of the cosmic evolution [19]. We have already mentioned that, in the Relaxed Universe, there is a tracking behavior of matter/radiation and DE during the $q = 1$ (radiation) and $q = 1/2$ (matter epochs). However, when the matter epoch is left behind and the relaxation mechanism is enforced (dynamically) to pick a very small value of H (as we have seen above) in order to insure the condition $|\rho_\Lambda^i + \rho_F(H)| \ll \rho_\Lambda^i$ to hold, this crucial event decoupled once and forever the tracking behaviors of matter/radiation and DE, and since then the DE dominates the universe's expansion. Such breakdown of the tracking property at the end of the matter epoch, on the other hand, pushed the subdominant DE density curve upwards until crossing the decaying DM one at some point after the matter epoch and hence near our time. This feature, which can be very well appreciated in Fig. 4 (cf. third panel), encodes the very origin and the possible exciting explanation (within the Relaxed Universe) of the "coincidence $\Omega_\Lambda^0 \sim \Omega_M^0$" of the matter and DE densities in our late time neighborhood. Not only so: we can also understand now the smallness of both ρ_Λ^0 and H_0 in front of the huge initial vacuum energy density ρ_Λ^i. Indeed, from (29) we predict $H_0 \sim M^2/M_X \sim 10^{-42}$ GeV, for M not far from the ~meV scale, which is the mass scale itself of the current CC: $m_\Lambda \equiv \left(\rho_\Lambda^0\right)^{1/4} \sim 10^{-3}$ eV. We have thus arrived at a most natural scenario ("The Relaxed Universe") in which, at variance with the quintessence formulation of the DE, there is no need to severely fine-tune any parameter of the model, and moreover it is not required the existence of any extremely tiny mass scale.

6. Conclusions

We have shown that models with variable vacuum energy ρ_Λ (and may be also with variable G) generally lead to a non-trivial effective EOS when viewed from the point of view of a self-conserved DE fluid at fixed G. As a result, these models can effectively appear as quintessence, and even as phantom energy, without need of invoking the existence of fundamental quintessence and phantom fields. Combining a time-evolving ρ_Λ with a variable "cosmon" energy density ρ_X (in general *not* representing any scalar field), we obtain the class of ΛXCDM models, which are able to deal with the cosmic coincidence problem. Finally, one can provide a version of the ΛXCDM model in which the new cosmon stands for a self-adjusting functional, namely one which is able to counteract the effects of an arbitrarily large CC in the early universe, leaving a small and completely innocuous net value of the CC at any stage of the cosmological evolution, in particular the tiny value $\rho_\Lambda^0 \sim 10^{-47}$ GeV^4 that we observe at present. This particular cosmon functional is thus able to reduce the value of the CC *without fine-tuning*, and may be eligible as a prototype model ("The Relaxed Universe") for eventually solving the old CC problem.

Acknowledgments

I have been partially supported by DIUE/CUR Generalitat de Catalunya under project 2009SGR502 and by MEC and FEDER under project FPA2007-66665, and also by the Consolider-Ingenio 2010 program CPAN CSD2007-00042. It is my pleasure to thank F. Bauer, S. Basilakos, J. Fabris, J. Grande, A. Pelinson, M. Plionis, I.L. Shapiro and H. Štefančić for a nice and stimulating collaboration in some of the topics discussed here. I would also like to thank R. Woodard, B.F.L. Ward, M. Maggiore and N. Bilić for discussions. Last but not least, I am also very grateful to L. Perivolaropoulos and all the organizers of NEB 14, for the invitation to this excellent hellenic conference on Recent Developments in Gravity.

References

[1] R. Knop et al., Astrophys. J. **598** (2003) 102; A. Riess *et al.* Astrophys. J. **607** (2004) 665.
[2] D.N.Spergel, et al., Astrophys. J. Suplem., **170** (2007) 377; E. Komatsu et al. arXiv:1001.4538 [astro-ph].
[3] O. Bertolami, Nuovo Cimento, **93B**, 36, (1986); M. Ozer M. and O. Taha, Nucl. Phys., **B287**, 776, (1987); O. K. Freese K., et al., Nucl. Phys., **287**, 797, (1987); J. C. Carvalho, J. A. S. Lima and I. Waga, Phys. Rev. **D46**, 2404, (1992).
[4] J. M. Overduin and F. I. Cooperstock, Phys. Rev. D., **58**, 043506, (1998), and references therein.
[5] S. Basilakos, M. Plionis and S. Solà, Phys. Rev. **D 80**, (2009) 3511, arXiv:0907.4555 [astro-ph.CO]; J. Grande, J. Solà, J.C. Fabris, I.L. Shapiro, Class.Quant. Grav. **27** (2010) 105004, arXiv:1001.0259 [astro-ph.CO].
[6] I. L. Shapiro, J. Solà, Phys. Lett. **B682** (2009) 105, arXiv:0910.4925 [hep-th]; see also the reviews: I. L. Shapiro, J. Solà, J. Phys. **A40**(2007) 6583, and specially arXiv:0808.0315 for a detailed exposition of the QFT of the cosmological term, and references therein.
[7] B.F.L. Ward, PoS ICHEP2010 (2010) 477, arXiv:1012.2680 [gr-qc]; arXiv:1008.1046 [gr-qc]; Mod. Phys. Lett. **A25** (2010) 607, arXiv:0908.1764 [hep-ph], and arXiv:0910.0490,
[8] M. Hindmarsh, D. Litim, C. Rahmede, arXiv:1101.5401 [gr-qc].
[9] J. Solà, H. Štefančić, Phys. Lett. **B624** (2005) 147, astro-ph/0505133; Mod. Phys. Lett. **A21** (2006) 479, astro-ph/0507110; J. Phys. **A39** (2006) 6753, gr-qc/0601012.
[10] M. Chevallier and D. Polarski, Int. J. od Mod. Phys. **D10** (2001) 1939; E.V. Linder, Phys. Rev. Lett. **90** (2003) 091301; H. K. Jassal, J.S. Bagla, T. Padmanabhan, Phys. Rev. **D72** 103503, (2005); Mon. Not. Roy. Astron. Soc. Letters **356**, L11-L16, (2005).
[11] S.Weinberg, Rev.Mod.Phys., **61** (1989) 1; P.J.Peebles, B.Ratra, Rev.Mod.Phys., **75** (2003) 559; T.Padmanabhan, Phys.Rept. **380** (2003) 235; V.Sahni, A.Starobinsky, Int.J.Mod.Phys. **A9** (2000) 373.
[12] F. Bauer, J. Solà, H. Štefančić, Phys. Lett. **B688** (2010) 269, arXiv:0912.0677 [hep-th]; Phys. Lett. **B678** (2009) 427, arXiv:0902.2215 [hep-th]; F. Bauer, J.Phys. Conf. Ser. **259** (2010) 012083, arXiv:1010.1106 [hep-th]. For a comprehensive exposition, see [13].
[13] F. Bauer, J. Solà, H. Štefančić, JCAP **12** (2010) 029, arXiv:1006.3944 [hep-th].
[14] L.E. Parker and D.J. Toms, *Quantum Field Theory in Curved Spacetime: quantized fields and gravity* (Cambridge U. Press, 2009).
[15] M. Maggiore, arXiv:1004.1782 [astro-ph.CO]; N. Bilić, arXiv:1004.4984 [hep-th].
[16] I.L. Shapiro, J. Solà, H. Štefančić, JCAP **0501** (2005) 012, hep-ph/0410095.
[17] J. Solà, J. of Phys. **A41** (2008) 164066, arXiv:0710.4151 [hep-th].
[18] S. Basilakos, M. Plionis and J. Solà, *Phys. Rev.* **D82** (2010) 083512, arXiv:1005.5592 [astro-ph.CO].
[19] J. Grande, J. Solà and H. Štefančić, JCAP **08**, (2006) 011, gr-qc/0604057; Phys. Lett. **B645** (2007) 236, gr-qc/0609083.
[20] J. Grande, A. Pelinson, and J. Solà, Phys. Rev. **D79**, (2009) 043006, arXiv:0809.3462 [astro-ph]; J. Grande, R. Opher, A. Pelinson, and J. Solà, JCAP **0712** (2007) 007, arXiv:0709.2130 [gr-qc]; arXiv:0904.3293.
[21] Y. B. Zeldovich, *Cosmological constant and elementary particles*, Sov. Phys. JETP Lett **6** (1967) 3167.
[22] N.C. Tsamis, R.P. Woodard, Phys. Rev. **D81** (2010) 103509, arXiv:1001.4929 [gr-qc]; Phys. Rev. **D80** (2009)083512, arXiv:0904.2368 [gr-qc]; V. Branchina, D. Zappala, Gen. Rel. Grav **42** (2010) 141, arXiv:0705.2299 [hep-ph]; arXiv:1005.3657; J. Beltran Jiménez, A. L. Maroto, JCAP **0903** (2009) 016, arXiv:0811.0566 [astro-ph]; Phys.Lett. **B686** (2010) 175, arXiv:0903.4672 [astro-ph.CO].
[23] S. Capozziello, S. Carloni and A. Troisi, Recent Res. Dev. Astron. Astrophys. **1** (2003) 625 [arXiv:astro-ph/0303041].
[24] S. M. Carroll, V. Duvvuri, M. Trodden and M. S. Turner, Phys. Rev. D **70** (2004) 043528, astro-ph/0306438.
[25] S. Nojiri and S.D. Odintsov, eConf **C0602061** (2006) 06 [Int. J. Geom. Meth. Mod. Phys. **4** (2007) 115].
[26] F. Bauer, Class. Quant. Grav. , **27** (2010) 055001 [arXiv:0909.2237 [gr-qc]]; arXiv:1007.2546 [gr-qc].
[27] H. Štefančić, Phys. Lett. **B670** (2009) 246, arXiv:0807.3692 [gr-qc].
[28] R. Woodard, *Lect. Notes Phys.* 720 (2007) 403, astro-ph/0601672.
[29] R.D. Peccei, J. Solà and C. Wetterich, *Adjusting the Cosmological Constant Dynamically: Cosmons and a New Force Weaker Than Gravity*, Phys. Lett. **B195** (1987) 183.
[30] C. Wetterich, Nucl. Phys. **B302** (1988) 668; P.J.E. Peebles and B. Ratra, Astrophys. J. **325** (1988) L17; B. Ratra and P.J.E. Peebles, Phys. Rev. **D37** (1988) 3406; P.G. Ferreira and M. Joyce, Phys. Rev. **D58** (1998) 023503; R.R. Caldwell, R. Dave and P.J. Steinhardt, Phys. Rev. Lett. **80** (1998) 1582.

Hořava–Lifshitz gravity: a status report

Thomas P. Sotiriou

Department of Applied Mathematics and Theoretical Physics, Centre for Mathematical Sciences, University of Cambridge, Wilberforce Road, Cambridge, CB3 0WA, UK

E-mail: T.Sotiriou@damtp.cam.ac.uk

Abstract. This is intended to be a brief introduction and overview of Hořava-Lifshitz gravity. The motivation and all of the various versions of the theory (to date) are presented. The dynamics of the theory are discussed in some detail, with a focus on low energy viability and consistency, as these have been the issues that attracted most of the attention in the literature so far. Other properties of the theory and developments within its framework are also covered, such as: its relation to Einstein-aether theory, cosmology, and future perspectives.

1. Introduction

1.1. Gravity and renormalizability

It is a well known fact that general relativity is not a renormalizable theory. This means that, successful as it may be as a classical theory of gravity, it should be viewed as an effective theory that breaks down at some scale. Beyond that scale general relativity is not enough to describe the gravitational interaction or spacetime itself (depending on the perspective) and one cannot construct its quantum counterpart using conventional quantization techniques.

On the other hand, given that within this perspective general relativity is an effective theory, the Einstein-Hilbert action contains only the lowest order terms in a curvature expansion. A reasonable question is then whether a renormalizable theory can be obtained by including higher order curvature terms. Intuitively this could work, as such terms would modify the propagator of the graviton at high energies. Indeed, in 1977 Stelle showed that theories whose action includes invariants quadratic in the curvature are renormalizable [1]. However, unfortunately this comes at a very high price, as such theories contain ghost degrees of freedom and are, therefore, not unitary. This is because by adding higher order curvature invariants in the action, one adds higher order time derivatives.

One could attempt to modify the propagator by adding higher order spatial derivatives without adding higher order time derivatives. This could presumably lead to a theory with improved ultraviolet (UV) behaviour without having to face the dreadful consequences of having higher order time derivatives. Obviously, following such a prescription requires treating spatial and time derivatives on a different footing and as such is in clash with Lorentz invariance. On the other hand, since the modified behaviour of the propagator is strictly needed in the UV, it is conceivable that Lorentz invariance could be recovered in the infrared (IR), or at least that Lorentz violations in the IR could stay below current experimental constraints.

Hořava's proposal [2, 3], which has led to a spree of publications recently and is now commonly referred to as Hořava–Lifshitz gravity, can be seen as an attempt to put these heuristic arguments in a rigorous framework where one can test whether they are indeed valid when it comes to gravity theories. In what comes next, an overview of this framework is given and its major developments to date are briefly discussed. This is far from an exhaustive account of an already very extensive literature, but hopefully it covers the most critical topics and provides a concise introduction to the subject.[1]

1.2. Lorentz violations as a field theory regulator

What has been discussed above is essentially the idea of giving up Lorentz invariance in order to keep a quantum field theory finite. Clearly, this idea goes beyond gravity theories and it has been considered in the past for other fields, see for example Ref. [5]. In fact, it would be best to start from a simpler example, such as a scalar field. This section will closely follow Ref. [6].

Let us first of all point out that there is nothing wrong with using Lorentz violations as a field theory regulator from a perspective of logical consistency, as there is no reason why a theory should necessarily exhibit Lorentz symmetry in the far UV. On the other hand, Lorentz violations are severely constrained in a wide range of energies and especially in the IR. Therefore, as mentioned before, the real question is whether one can construct a field theory that exhibits Lorentz violations which in the far UV lead to renormalizability, but remain suppressed below experimental accuracy at lower energies. This is difficult to answer without concrete models and enough to motivate further investigation.

Consider the scalar field action

$$S_\phi = \int dt dx^d \left\{ \dot{\phi}^2 - \sum_{m=1}^{z} a_m \phi (-\Delta)^m \phi + \sum_{n=1}^{M} g_n \phi^n \right\}, \tag{1}$$

where $\dot{} \equiv \partial/\partial t$, $\Delta = \vec{\nabla}^2$ is the spatial Laplacian and z and M are positive integers still to be specified. A theory is said to be "power counting renormalizable" when all of its interaction terms scale like energy to some non-positive power, as in this case Feynman diagrams are expected to be convergent (or have at most a logarithmic divergence). To check whether this is true or not we first have to choose the engineering dimensions of space and time. The engineering dimensions

$$[dt] = [\kappa]^{-z}, \qquad [dx] = [\kappa]^{-1}, \tag{2}$$

where κ is a place holder symbol with dimensions of momentum, are compatible with the choice $a_z = 1$ and the requirement that the action has to be dimensionless. This choice of engineering dimensions, therefore, appears to be natural in the UV. The dimensions for the scalar field then turn out to be

$$[\phi] = [\kappa]^{(d-z)/2}. \tag{3}$$

It remains to check the scaling for the rest of the terms. Requiring that an interaction term scales like energy to some non-positive power is equivalent to the requirement that the coupling of this interaction scales like energy to some non-negative power (since the action is dimensionless). Also, we find it convenient to work in terms of momenta instead of energies. So, our theory will be power counting renormalizable if all couplings have non-negative momentum dimensions.

It is straightforward to verify that

$$[a_m] = [\kappa]^{2(z-m)}, \qquad [g_n] = [\kappa]^{d+z-n(d-z)/2}. \tag{4}$$

It is then obvious that a_m has non negative momentum dimension for all values of m. It is also easy to realize that, for $z \geq d$, g_n has non-negative momentum dimensions for all values of n. When $z < d$, g_n has non-negative momentum dimension only when $n \leq [2(d+z)]/(d-z)$. One

[1] See also Ref. [4] for an early review of the projectable Hořava–Lifshitz gravity.

can then easily recover well-known results. For example one can verify that the case $d = 3$, $z = 1$, $M = 4$ which corresponds to the usual relativistic ϕ^4 theory in $3+1$ dimensions is a power counting renormalizable theory. The $d = z = 1$ case, *i.e.* the relativistic scalar in $1+1$ dimensions, is power counting renormalizable for all values of M. In fact this case belongs to the class $z = d$ and is directly analogous to the case $d = z = 3$, which will also have the same property and is of interest to us. In fact one can argue that any theory with $z = d$ is actually finite with normal ordering, while theories with $z > d$ are finite even without normal ordering, see Ref. [6] for a more detailed discussion using the "superficial degree of divergence".

How would things go if instead of a scalar field we wanted to consider a graviton? The main difference lies in the graviton self-interaction vertices. In the case of the scalar field considered above, momenta did not enter these interactions. For a graviton, however, one would have to deal with self-interaction vertices which contain spatial derivatives [7]. This introduces some further complication, but it is not enough to spoil the power-counting renormalizability, as long as $z \geq d$, *i.e.* as long as the action contains operators with at least $2d$ spatial derivatives. The presence of self-interaction vertices which contain spatial derivatives is enough though to argue that, unlike the scalar theory consider above, its graviton analogue will not be finite (even with normal ordering) [7].

All of the arguments presented here support the statement that field theories that contain at least $2d$ spatial derivatives in $d+1$ dimensions are power counting renormalizable. Power counting renormalizability is a strong indication that a theory is indeed renormalizable. Little progress has been made so far in the case of gravity to check renormalizability beyond power counting for such theories.

2. From the Lifshitz scalar to non-relativistic gravity

We proceed to consider the exact structure of a gravity theory with the characteristics mentioned in the previous section. We will work in 3 dimensions from now on but generalizing to more than 3 dimensions is straightforward. The basic requirement is that the theory should have only two time derivatives but at least $2d$ spatial derivatives, that is 6 in our case. The fact that we want to have more spatial than time derivatives implies that we need to treat space and time on different footing. This is naturally achieved by working in the Arnowitt–Deser–Misner (ADM) decomposition of spacetime

$$ds^2 = -N^2 c^2 dt^2 + g_{ij}(dx^i + N^i dt)(dx^j + N^j dt). \tag{5}$$

One is essentially forced to pick a preferred foliation of spacetime in order to write down the action according to the prescription laid out in the previous section. Consequently, such an action cannot be invariant under standard diffeomorphisms, as general relativity is. It can, however, be invariant under a more restricted set: foliation preserving diffeomorphisms, *i.e.* space-independent time reparametrization together with time-dependent spatial diffeomorphisms

$$t \to \tilde{t}(t), \qquad x^i \to \tilde{x}^i(t, x^i). \tag{6}$$

This is the symmetry that the action will have to respect.

It is straightforward to realize that the only covariant quantity under spatial diffeomorphisms that contains a time derivative of the spatial metric is the extrinsic curvature

$$K_{ij} = \frac{1}{2N}\left\{\dot{g}_{ij} - \nabla_i N_j - \nabla_j N_i\right\}, \tag{7}$$

which also transforms like a scalar under time reparametrization. A dot denotes differentiation with respect to the time coordinate and ∇_i is the covariant derivative associated with the spatial metric g_{ij}. We want the theory to be second order in time derivatives, so we need to consider terms quadratic in the extrinsic curvature. Note that there are no invariants under the symmetry mentioned above that one can construct with time derivatives of the lapse N or the shift N_j, without including higher order time derivatives of g_{ij} as well. Therefore, the most general action one can write is

$$S = \frac{M_{\rm pl}^2}{2} \int d^3x dt N\sqrt{g} \left\{ K^{ij}K_{ij} - \lambda K^2 - V(g_{ij}, N) \right\} , \tag{8}$$

where $M_{\rm pl}$ is a constant which, with some foresight, we identify with the Planck mass, g is the determinant of the spatial metric g_{ij}, and λ is a dimensionless running coupling. V generically can depend on g_{ij} and N and their spatial derivatives. It does not contain time derivatives, neither does it depend on the shift N_i, as the symmetry of the theory does not allow one to construct any suitable invariants. Finally, power counting renormalizability requires V to contain terms which are at least sixth order in spatial derivatives. From now on we will restrict ourselves to theories that contain sixth order time derivatives but not higher, as these are the simplest ones which satisfy the power counting renormalizability requirement.

3. Potential, constraints and the various versions of the theory

Clearly there are numerous terms one could include in V. Different choices have led to different versions of the theory. In this section we list these versions, briefly presenting their basic characteristics and differences. We refrain from discussing or commenting upon their viability or consistency, as this, together with their phenomenology, will be the subject of the coming section.

3.1. Detailed balance

Hořava proposed a symmetry that V would have to satisfy, that would drastically reduce the number of invariants one should consider [3]. This symmetry is dubbed *detailed balance* and it is inspired by condensed matter systems. It sums up to the requirement that V should be derivable from a superpotential W:

$$V = E^{ij}\mathcal{G}_{ijkl}E^{kl}, \tag{9}$$

where

$$E^{ij} = \frac{1}{\sqrt{g}}\frac{\delta W}{\delta g_{ij}}, \tag{10}$$

and

$$\mathcal{G}^{ijkl} = \frac{1}{2}\left(g^{ik}g^{jl} + g^{il}g^{jk}\right) - \lambda g^{ij}g^{kl}. \tag{11}$$

The most general action one can write with V satisfying the conditions above is

$$S_{db} = \frac{M_{\rm pl}^2}{2} \int d^3x dt N\sqrt{g}\Bigg\{ K^{ij}K_{ij} - \lambda K^2 - \frac{\alpha^4}{M_{\rm pl}^4}C_{ij}C^{ij} + \frac{2\alpha^2\beta}{M_{\rm pl}^3}\frac{\epsilon^{ijk}}{\sqrt{g}}R_{il}\nabla_j R^l{}_k - \frac{\beta^2}{M_{\rm pl}^2}R^{ij}R_{ij}$$
$$+\frac{\beta^2}{4}\frac{1-4\lambda}{1-3\lambda}R^2 + \frac{\beta^2\zeta}{1-3\lambda}R - \frac{3\beta^2\zeta^2}{1-3\lambda}M_{\rm pl}^2\Bigg\}, \tag{12}$$

where ϵ^{ijk} is the Levi-Civita symbol,

$$C^{ij} = \frac{\epsilon^{ikl}}{\sqrt{g}}\nabla^k\left(R^j{}_l - \frac{1}{4}R\delta^j{}_l\right) \tag{13}$$

Journal of Physics: Conference Series **283** (2011) 012034 doi:10.1088/1742-6596/283/1/012034

is the Cotton tensor, which in 3 dimensions plays the role of the Weyl tensor, and α, β and ζ are dimensionless couplings. Notice that there are only 3 new couplings for a total of 6 terms in V.

The advantages of detailed balance is that it drastically reduces the number of terms one needs to consider and that it introduces a superpotential, which might simplify quantization. On the other hand, the action above contains a term that violates parity (the fifth order operator).[2] Additionally, as is obvious, the last term in the action, which plays the role of a cosmological constant, is restricted to have the wrong sign with respect to the one suggested by observations when $\lambda > 1/3$, as pointed out in Refs. [9, 10] (λ is supposed to be close to unity, if severe Lorentz violation are not to occur).[3] Note that there is nothing fundamental about detailed balance, so it is safe to consider it just a simplicity assumption.

3.2. Projectable Hořava–Lifshitz gravity

Another possible simplifying restriction proposed by Hořava in Ref. [3] is that of *projectability*, which is the assumption that the lapse is just a function of time. That is $N = N(t)$. Again there is no fundamental principle behind such an assumption. Hořava's main motivation for initially considering it was that under this assumption one has enough gauge freedom to set $N = 1$, as in general relativity. This is not possible without projectability, as the symmetries of the action allow only space independent time reparametrizations, as explained earlier.

In any case, projectability drastically reduces the number of invariants V can include even if detailed balance is not imposed, as all terms with spatial derivatives of N would now vanish. Let us see that in more detail, following the line of Refs. [9, 10]. Since N is now just a function of time there is no invariant one can built out of it without using time derivatives. So V will only depend on the metric and its spatial derivatives. Essentially, this means that the action should include all of the curvature invariants one can construct with g_{ij} with up to six spatial derivatives.

An important simplification comes from the fact that we are in 3 spatial dimensions. Therefore, the Weyl tensor vanishes identically and the Riemann tensor can be expressed in terms of the Ricci tensor. Taking Bianchi identities into account as well and ignoring surface terms, one can arrive at the conclusion that the most general action is

$$\begin{aligned} S_p \;=\; \frac{M_{\rm pl}^2}{2}\int d^3x dt N\sqrt{g}\Big\{ & K^{ij}K_{ij} - \lambda K^2 - g_0\, M_{\rm pl}^2 - g_1 R - g_2\, M_{\rm pl}^{-2}\, R^2 - g_3\, M_{\rm pl}^{-2}\, R_{ij}R^{ij} \\ & - g_4\, M_{\rm pl}^{-4}\, R^3 - g_5\, M_{\rm pl}^{-4}\, R(R_{ij}R^{ij}) - g_6\, M_{\rm pl}^{-4}\, R^i{}_j R^j{}_k R^k{}_i \\ & - g_7\, M_{\rm pl}^{-4}\, R\nabla^2 R - g_8\, M_{\rm pl}^{-4}\, \nabla_i R_{jk}\, \nabla^i R^{jk} \Big\}, \end{aligned} \qquad (14)$$

where the g_i are dimensionless couplings. As long as we have not coupled matter to the theory we are free to rescale the coordinate in order to set $g_1 = -1$, which is the value it has in general relativity. We then have 8 remaining couplings g_i which can be used to tune the scales suppressing the various operators. The following remarks are in order:

- We have suppressed parity violating terms.
- Enforcing projectability on top of detailed balance (without suppressing parity violating terms) would leave action (12) unaffected, apart from the fact that N would become just a function of time. As we will see later, this is quite an important difference.

[2] Parity violating terms, within or outside the framework of detailed balance, have been shown to lead to chiral primordial gravitational waves [8].

[3] This is a bare cosmological constant. Also, one could analytically continue the parameters as explained in Ref. [11], but in this case W would cease to be real.

- There are just 3 more operators in the most general projectable action than in the one with detailed balance. In this sense, detailed balance does not bring significant simplification once projectability has been assumed. For this reason, and for the drawbacks listed in the end of the previous section, we will not consider detailed balance further when assuming projectabiltity.
- g_0 controls the value of the cosmological constant, which, unlike the case with detailed balance, is not restricted.
- There are two types of Lorentz violating terms in the action. The ones in V, which are suppressed by some scale that can be determined by tuning the couplings g_2 to g_8, and one that comes from the kinetic part due to the fact that λ is not necessarily equal to 1. This will be a generic feature of all versions of the theory as we will see. Clearly, the term of the second kind is far more dangerous, as it introduces Lorentz violations at low energy scales.
- All couplings are running. One can, therefore, hope that λ will run to 1 in the IR (or sufficiently close to it for experimental constraints to be satisfied), the rest of the operators will be heavily suppressed, and the theory will effectively reduce to general relativity. Diffeomorphism invariance and Lorentz invariance would emerge as IR (approximate) symmetries. Whether or not this will indeed be the case requires a study of the renormalization group flow which is still pending.

3.3. Non-projectable Hořava-Lifshitz gravity

We now proceed to consider the case where neither detailed balance nor projectability are enforced. This version of the theory is called non-projectable. To avoid confusion it is worth pointing out explicitly, that the version with detailed balance can be a non-projectable version. However, in this section we wish to go beyond that. Once one abandons detailed balance, adding just some specific choice of extra terms is not really an option. Radiative corrections will generate all possible terms compatible with the symmetries of the theory and, thus, all such terms should be taken into account, much like the projectable case treated above. However, as first pointed out in Ref. [12], it is not only curvature invariants of g_{ij} that one can include in V in this case. One can also use the quantity

$$a_i = \partial_i \ln N\,, \tag{15}$$

as contractions of it with itself or curvature terms also lead to invariants. The lowest order invariant one can construct with a_i is $a^i a_i$, which comes at the same order as R. The action will then be of the form

$$S_{np} = \frac{M_{\rm pl}^2}{2}\int d^3x dt N\sqrt{g}\left\{K^{ij}K_{ij} - \lambda K^2 + \xi R + \eta\, a_i a^i + \frac{1}{M_A^2}L_4 + \frac{1}{M_B^4}L_6\right\}, \tag{16}$$

where L_4 and L_6 include all possible 4th and 6th order operators respectively that one can construct using a_i and g_{ij}. For example, R^2, $(a^i a_i)^2$, $\nabla^i\nabla_i R$ and $a_i a_j R^{ij}$ are some of the terms that are included in L_4.[4] As was the case before for g_1, ξ can now be set to unit by a coordinate rescaling when matter is not coupled to the theory. This version of the theory was first studied in Ref. [12] in order to resolve some inconsistencies in the dynamics of the non-projectable version with detailed balance, as we will discuss shortly. Although it is often referred to as an "extension" of Hořava-Lifshitz gravity, this terminology is probably unfortunate: this is Hořava-Lifshitz gravity, *i.e.* a theory constructed consistently according to the prescription described in Hořava's paper Ref. [3] and without any restriction, such as detailed balance or projectability.

The following points are worth stressing for the non-projectable version (apart from some of the remarks already made for the projectable case and apply here as well).

[4] $\nabla^i\nabla_i R$ and other similar terms are not surface terms, as the N in the 4-volume is space dependent.

- For simplicity the cosmological constant term has been suppressed, but it can be straightforwardly restored.
- The scales M_A and M_B suppressing the higher order operators have been left arbitrary. This is in analogy with having arbitrary dimensionless couplings g_i in the projectable case.
- The number of operators in the non-projectable case is an order of magnitude larger than that in the projectable case.
- If general relativity were to be recovered in the IR, not only λ would have to run to 1 but also η would have to run to 0.

3.4. Imposing further symmetries

All of the actions presented above, which correspond to different versions of Hořava–Lifshitz gravity, are invariant under foliation preserving diffeomorphisms only, and not the full set of diffeomorphisms. So, even though the action is in all cases quadratic in the time derivatives of g_{ij}, less symmetry generically implies that more degrees of freedom will be excited. We will see explicitly in the next section that this is indeed the case, and that this can, in most cases, have undesirable consequences for the consistency and viability of the theory.

A way to do away with the extra degrees of freedom is to attempt to nontrivially extend the gauge symmetry of the theory, so that it will have as many generators per spacetime point as general relativity. It was pointed out in Ref. [13] that this could be done if the action would be suitably modified to acquire an extra local $U(1)$ symmetry. More precisely in Ref. [13] a step by step construction of the action led to

$$\begin{aligned} S_{\text{extra}\,U(1)} &= \frac{M_{\rm pl}^2}{2}\int d^3x dt N\sqrt{g}\Big\{K^{ij}K_{ij} - K^2 - V(g_{ij}, N) \\ &\qquad + \nu\Theta^{ij}(2K_{ij} + \nabla_i\nabla_j\nu) - A(R - 2\Omega)/N\Big\}, \end{aligned} \tag{17}$$

where Ω is a constant, A acts as a Lagrange multiplier, ν is an auxiliary scalar field,

$$\Theta^{ij} \equiv R^{ij} - \frac{1}{2}g^{ij}R + \Omega g^{ij}, \tag{18}$$

and V should as usual contain 6th order operators and may or may not satisfy detailed balance. Note that A transforms as a spatial scalar and a time vector under foliation preserving diffeomorpisms, that is

$$A \to A + \dot{f}A + f\dot{A} + \xi^i\partial_i A, \tag{19}$$

where ξ^i is the generator. On the other hand, under a local $U(1)$ gauge transformation with generator φ, A and ν transform as

$$A \to A + \dot{\varphi} - N^i\nabla_i\varphi, \tag{20}$$

$$\nu \to \nu + \varphi. \tag{21}$$

See Ref. [13] for more details on constructing the action, as well as for discussions on anisotropic scaling, hamiltonian formulation, linearization, etc. It was argued there that in the IR this theory has very good chances to reduce to general relativity, as the high order operators in V would be suppressed (as in the previous versions of the theory), the extra fields in the action are just auxiliary fields, and the extra gauge symmetry leads to just the usual spin 2 graviton excitation. Moreover, it was claimed in Ref. [13] that the $U(1)$ symmetry forces the coefficient λ of K^2 in the action, which was a running coupling in all other versions of the theory, to be equal to 1 here, as in general relativity.

This last claim has been contested recently in Ref. [14] where it has been claimed that the extra $U(1)$ symmetry does not necessarily require $\lambda = 1$ and that even for arbitrary λ the theory propagates only a spin 2 graviton. If this is true, λ has to remain a running coupling and flow close to 1 in the IR for general relativistic phenomenology to be recovered according to Ref. [14]. So, at the moment, whether or not the value of λ is fixed to 1 can be considered a matter of debate.

Another potential difficulty with this version of Horava-Lifshitz gravity has to do with matter coupling. It is straightforward to notice that the Lagrange multiplier in action (17) forces R to be constant. Even though this is not necessarily a problem in vacuum, it cannot be generically the case once matter is coupled to the theory. This implies that matter would have to be coupled to A if the theory is to resemble general relativity at low energies, and that a suitable coupling that leads to standard phenomenology in the IR should exist. Whether this is indeed the case requires further investigation (some first steps where done in Ref. [14]). It is also worth pointing out that action (17) is susceptible to radiative corrections which could modify its form.

Since this last version of Horava-Lifshitz gravity has been proposed very recently and has not yet been studied extensively, we will not consider it further. Its cosmological aspects where recently studied in Ref. [15].

4. Projectable version: Dynamics, consistency and low energy behaviour

4.1. Degrees of freedom and propagators

We return now to the projectable version of the theory and take a closer look at its dynamics. We will consider the most general action in this version, action (14), *i.e.* we will not enforce detailed balance but our results will include it as a subcase. The field equations for this action have been derived in detail in Ref. [10]. We will refrain from rewriting them here due to space limitations but we will make the following remark: since $N = N(t)$, the variation with respect to N will not lead to a usual local super-Hamiltonian constraint, but to a global Hamiltonian constraint. That is, the constraint will come in the form of an integral over space. This will be a major difference with respect to the non-projectable case.

We now move straight to the linearization. If for simplicity one sets the cosmological constant term to 0, then flat space is a suitable background. After suitable gauge choices, see Ref. [10], one finds that the theory propagates a spin-2 mode, the usual graviton, that now satisfies a modified dispersion relation, *i.e.*

$$\ddot{\widetilde{H}}_{ij} = -\left[g_1\partial^2 + g_3 M_{\rm pl}^{-2}\partial^4 + g_8 M_{\rm pl}^{-4}\partial^6\right]\widetilde{H}_{ij}, \tag{22}$$

where $\widetilde{H}_{ij}$ is transverse and traceless. Notice that the couplings g_3 and g_8 control the scale that the lorentz violating terms become important in the dispersion relation. Recall also that in vacuum one has the freedom to rescale the coordinates and set $g_1 = -1$.

As mentioned earlier, less symmetry generically means more degrees of freedom, so since the theory is invariant under foliation preserving diffeomorphisms only, we expect to find extra excitations. This is indeed the case. There is an extra scalar degree of freedom, whose linearized dynamics are governed by the action [10]

$$S_2^p = -M_{\rm pl}^2\int d^3x dt\left[\frac{1}{c_h^2}\dot{h}^2 + h\partial^2 h + \frac{8g_2+3g_3}{M_{\rm pl}^2}h\partial^4 h - \frac{8g_7-3g_8}{M_{\rm pl}^4}h\partial^6 h\right], \tag{23}$$

where

$$c_h^2 = \frac{1-\lambda}{3\lambda-1}, \tag{24}$$

and we have set $g_1 = -1$ by rescaling the coordinates, in order to guarantee the stability of the spin 2 graviton. Given the overall minus sign in eq. (23), the scalar mode is a ghost whenever $1 > \lambda > 1/3$. On the other hand, the scalar is classically unstable at low energies whenever $\lambda > 1$ or $\lambda < 1/3$ [10]. It is conceivable that the instability can be cut off by the higher order derivatives. If we call $M_\star$ the scale where the higher order term would take over, then the time scale of the instability would be at least $1/|c_h|M_\star$. The relation between $M_\star$ and $M_{\rm pl}$ is clearly governed by the g_i coefficients in eq. (23). On the other hand, if the instability is not to develop within the lifetime of the universe then $|c_h|M_\star < H_0$, where H_0 is the Hubble parameter. [16]. Experiments rule out modifications of Newton's law at scales above $10\mu m$ which corresponds roughly to the constraint $M_\star \gtrsim 0.1$eV. In turn, this implies $|1-\lambda| \lesssim 10^{-61}$. This value is clearly very low and it seems unlikely that the renormalization group flow could drive λ to it.

In Refs. [17, 18] linearization around a de Sitter, instead of Minkowski, background was considered. As expected, the large scale modes exhibit better behaviour. However, qualitatively the result is the same overall: λ needs to be sufficiently close to 1 for the instability not to be a concern. Note that requiring λ to be very close to one is not only a potential fine tuning problem. $c_h \to 0$ as $\lambda \to 1$, which means that the low momentum phase velocity of the scalar mode will be much smaller than the speed of light for such values of λ. When a velocity of a certain mode is smaller than the speed of light in a medium, relativistic particles traveling in this medium decay via the Cherenkov process [19]. The fact that we observe high energy cosmic rays which need to travel a significant distance to reach us is evidence enough that such modes do not exist.

4.2. Strong coupling

In the previous section we only considered the linearized dynamics as described by the part of the perturbative action quadratic in h. However, it has been pointed out by several authors that such a perturbative treatment breaks down when λ approaches 1 and the scalar mode gets strongly coupled [20, 21, 16]. To see how this comes about we give below the cubic interactions of h,

$$S_3^p = M_{\rm pl}^2 \int dt d^3x \left\{ h(\partial h)^2 - \frac{2}{c_h^4}\dot{h}\partial_i h \frac{\partial^i}{\partial^2}\dot{h} + \frac{3}{2}\left[\frac{1}{c_h^4} h\left(\frac{\partial_i\partial_j}{\partial^2}\dot{h}\right)^2 - \frac{(2c_h^2+1)}{c_h^4} h\dot{h}^2\right]\right\}, \tag{25}$$

where we have neglected cubic interactions coming from higher order operators. The redefinitions $\hat{t} = |c_h|t$ and $\hat{h} = c_h^{-1/2} M_{\rm pl}\, h$ canonically normalize the lower operator part of the quadratic action (23) (first two terms). After these redefinitions the cubic action reads

$$S_3^p = \frac{1}{|c_h|^{3/2} M_{\rm pl}} \int d\hat{t} d^3x \left\{ c_h^2\, \hat{h}(\partial\hat{h})^2 - 2\hat{h}'\partial_i\hat{h}\frac{\partial^i}{\partial^2}\hat{h}' + \frac{3}{2}\left[\hat{h}\left(\frac{\partial_i\partial_j}{\partial^2}\hat{h}'\right)^2 - (2c_h^2+1)\hat{h}(\hat{h}')^2\right]\right\}, \tag{26}$$

where now $' = \partial/\partial_{\hat{t}}$. As is obvious from the equation, there are cubic interactions which are suppressed by the scale $|c_h|^{3/2}M_{\rm pl}$ with respect to the quadratic ones. Therefore, the theory gets strongly coupled at the scale $M_{sc} = |c_h|^{3/2}M_{\rm pl}$.

Given the constraints on λ from stability discussed previously, the strong coupling scale is phenomenologically unacceptably low, as we know that we can treat gravity perturbatively at low energies. There is also the issue of renormalizability: our arguments for power-counting renormalizability given above where based on the validity of the perturbative treatment at all energies. If there is strong coupling such arguments simply fail.

The only subtlety here is the fact that the strong coupling arguments presented so far neglect the higher order operators. However, for such operators to become important and invalidate our treatment, the strong coupling scale M_{sc} would have to be higher than the scale $M_\star$ that suppresses them. Given how low M_{sc} is here, this would imply that higher order operators modify the graviton dynamics at very low energies, which would be in conflict with current experimental evidence.

Since the analysis presented here is linear, one could ask whether including non-linear effects could resolve the strong coupling problem, by an analogue of the Vainstein mechanism in massive gravity [22], *i.e.* a non-perturbative restoration of the $\lambda \to 1$ limit. Recently, in Ref. [23] it has been claimed that this is indeed the case in spherically symmetric, static configurations. This matter deserves further attention.

5. Non-projectable version: Dynamics, consistency and low energy behaviour

5.1. Degrees of freedom and propagators

We now turn our attention to the non-projectable version of the theory. As before, we will not enforce detailed balance in order to obtain the most general results possible. However, we will discuss explicitly that case as well. The dynamics of the spin 2 graviton are essentially the same as in the projectable version so we will not consider them again in more detail. As in the projectable case, also here there is an extra scalar degree of freedom, and we will focus on that.

The low energy dynamics of this scalar graviton are governed by the action

$$S_2^{np} = -M_{\rm pl}^2 \int d^3x dt \left[\frac{1}{c_h^2}\dot{h}^2 + \frac{\eta - 2}{\eta} h\partial^2 h\right], \tag{27}$$

which one obtains by linearizing action (16) around flat space and considering only the lowest order operators. ξ has been set to 1 by a suitable coordinate rescaling. Note that c_h is defined in the same way as above and, thus, it is *not* the low momentum phase velocity of the scalar in the non-projectable version. The square of the latter is now

$$c_h'^2 = c_h^2 \frac{\eta - 2}{\eta}. \tag{28}$$

For h not to be a ghost one needs $c_h^2 < 1$, and for h to be classically stable one needs $c_h'^2 > 0$. These conditions are satisfied whenever [12]

$$\lambda > 1, \quad 0 < \eta < 2, \tag{29}$$

or

$$\lambda < 1/3, \quad 0 < \eta < 2. \tag{30}$$

In the second region in the parameter space λ is far from 1, which is the value it has in general relativity, so we will not consider this option further. The presence of the operator $a_i a^i$ in action (16), which is a lowest order operator and contributes to the quadratic action, has drastically improved the behaviour of the scalar graviton with respect to the projectable case we examined previously. In particular, there is now a region in parameter space, described by the constraints (29), for which h is neither a ghost nor classically unstable [12]. Even though we have neglected higher order operators here, the behaviour we have found would remain qualitatively the same if we had included them.

Let us examine what would happen if we had enforced detailed balance. The $a^i a_i$ term in the action would not be allowed which means $\eta = 0$. In this case, the coefficient of the second term in (27) blows up and the scalar field appears to freeze. Things, however, are actually more complicated, as a more detailed analysis shows [21]. When detailed balance is enforced, the scalar does propagate around backgrounds that are neither static nor homogeneous. In fact it appears to satisfy a first order differential equation, which constitutes a serious problem for the theory, see Ref. [21] for more details. Similar conclusions have been obtained by considering the hamiltonian formulation of the theory with detailed balance [24, 25]. In particular, the Hamiltonian constraint is not automatically preserved by time evolution (second class constraint) and the theory propagate 5/2 half degrees of freedom, the 1/2 corresponding to the scalar mode which satisfies a first order equation and the rest to the two graviton polarizations. Such a theory, though mathematically consistent, exhibits no time evolution and is, therefore, physically meaningless [25].

5.2. Strong coupling

We just saw that, judging from the quadratic action, the non-projectable version exhibits improved scalar dynamics with respect to the projectable version, as long as detailed balance is not enforced. However, we found earlier that the scalar mode in the projectable version gets strongly coupled at unacceptably low energies, so one has to check whether this is the case in the non-projectable version as well. To do so, we write down the cubic action for the scalar [26]:

$$\begin{aligned} S_3^{np} &= M_{\rm pl}^2 \int dt d^3x \left\{ \left(1 - \frac{4(1-\eta)}{\eta^2}\right) h(\partial h)^2 - \frac{2}{c_h^4}\dot{h}\partial_i h \frac{\partial^i}{\partial^2}\dot{h} \right. \\ &\left. + \left(\frac{3}{2}+\frac{1}{\eta}\right)\left[\frac{1}{c_h^4} h \left(\frac{\partial_i\partial_j}{\partial^2}\dot{h}\right)^2 - \frac{(2c_h^2+1)}{c_h^4} h\dot{h}^2\right]\right\}. \end{aligned} \tag{31}$$

Same as before for the projectable case, we canonically normalize the low energy quadratic action. This requires the redefinitions

$$t = \sqrt{\frac{\eta}{2-\eta}}\frac{\hat{t}}{|c_h|}, \qquad h = \left(\frac{\eta}{2-\eta}\right)^{1/4}\sqrt{|c_h|}\frac{\hat{h}}{M_{\rm pl}}. \tag{32}$$

Recall that we are only interested in the region of the parameter space where conditions (29) are satisfied. Under the redefinitions the cubic action reads

$$\begin{aligned} S_3^{np} &= \frac{(2-\eta)^2}{\eta^{1/2} c_h'^{3/2} M_{\rm pl}} \int d\hat{t} d^3x \left\{ c_h^2 \left(1 - \frac{8(1-\eta)}{(2-\eta)^2}\right) \hat{h}(\partial\hat{h})^2 - 2\hat{h}'\partial_i\hat{h}\frac{\partial^i}{\partial^2}\hat{h}' \right. \\ &\left. + \left(\frac{3}{2}+\frac{1}{\eta}\right)\left[\hat{h}\left(\frac{\partial_i\partial_j}{\partial^2}\hat{h}'\right)^2 - \left(\frac{2\,\eta\, c_h'^2}{2-\eta}+1\right)\hat{h}(\hat{h}')^2\right]\right\}, \end{aligned} \tag{33}$$

where we have expressed everywhere c_h in terms of the physical quantity c_h'. From eq. (33) one directly infers that the cubic interactions are suppressed with respect to the quadratic ones by various scales $f(|\lambda - 1|, \eta)M_{\rm pl}$, where f is an algebraic function whose functional form depends on which terms one considers. The scalar mode would become strongly coupled at the lowest of these scales.

Both $|\lambda-1|$ and η essentially measure deviations from Lorentz invariance which are present at arbitrarily low scales. They are to be small for the theory to avoid experimental constraints on Lorentz violations. It is rather straightforward to verify that in this case, also the scale that the theory gets strongly coupled has to be low. Instead of performing a general analysis leaving both $|\lambda - 1|$ and η small but arbitrary, we will focus on the most interesting case where $c_h' \sim 1$. This value of c_h' is the one dictated by Cherenkov radiation constraints [19], see discussion above. We then have $\eta \sim |\lambda-1|$ and the strong coupling scale turns out to be $M_{sc} \sim \sqrt{\eta}M_{\rm pl} \sim \sqrt{|\lambda-1|}M_{\rm pl}$ [26, 27].

Since $\eta, |\lambda - 1| \ll 1$ the strong coupling scale will be much lower than the Planck scale. Absence of preferred frame effects in the Solar system observations requires $\eta, |\lambda - 1| \lesssim 10^{-7}$

[28], which in turn implies that $M_{sc} \lesssim 10^{15}$GeV. Therefore, the strong coupling scale is too high to be phenomenologically accessible from gravitational experiments. In this sense, strong coupling in non-projectable Hořava–Lifshitz gravity would pose no threat if the theory were to be treated as an effective theory describing gravity at low energies. On the other hand, the mere presence of strong coupling even at relatively high energies, as in this case, is enough to seriously question the power counting renormalizability of the theory [26]. Indeed, as mentioned also earlier when considering the projectable case, renormalizability arguments were based on the assumption that a perturbative treatment can be used to arbitrarily high energies.

However, the strong coupling arguments given here neglect the role of the higher order operators in the action. As proposed in Ref. [29], a way to avoid strong coupling is to lower the scale that suppresses the higher order operators in action (16) below M_{sc}, *i.e.* impose that $M_A \sim M_B \sim M_\star$ and $M_\star < M_{sc}$. This clearly requires the introduction of a second scale, different from $M_{\rm pl}$, as well as a peculiar hierarchy of scales. Also, it does seem to involve some large dimensionless couplings as $M_{\rm pl} \gg M_\star$. However, as it has been argued in Refs. [29, 30] this hierarchy of scales is technically natural.

An important consequence of this effective constraint on $M_\star$ coming from the requirement to avoid strong coupling, first pointed out in Ref. [26], is that now $M_\star$ becomes bound from both above and below. Being the scale that suppresses higher order derivatives in the dispersion relations, one would expect that the larger $M_\star$ is, the better. Assuming that this same scale suppresses higher order derivatives in the matter dispersion relations (which would also be modified in a way similar to gravity), and in particular photons, places a strong lower bound on $M_\star$: The absence of a time delay in different frequency γ-rays coming from distant astrophysical sources implies that [31, 32] $M_\star \gtrsim 10^{11}$GeV. However, now one has also an upper bound if strong coupling is not to be a problem, and the combined set of constraints yields

$$10^{15}\text{GeV} \gtrsim M_\star \gtrsim 10^{11}\text{GeV}. \tag{34}$$

There seems to be a comfortable window for $M_\star$ within which non-projectable Hořava–Lifshitz gravity (without detailed balance or other restrictions) avoids strong coupling without exhibiting detectable Lorentz violations, at least with current experimental accuracy. Improving the accuracy with which we can constrain both preferred frame effects in the solar system and modifications in the dispersion relations could potentially close this window.

5.3. Relation to Einstein-Aether theory

Non-projectable Hořava–Lifshitz gravity is a theory with a preferred foliation, which will exhibit Lorentz invariance violations even at low energies. This should be clear by now by inspecting the action (16) and taking into account that, given the dynamics of the scalar discussed previously, one can no longer hope that λ and η will run to 1 and 0 respectively in the IR, in order for Lorentz symmetry to emerge. Note that the operators related to these couplings are of lowest order, *i.e.* dimension 2.

A model theory for gravity with preferred frame effects is Einstein-aether theory [33, 34]. It is the most general theory for a unit timelike vector field coupled to gravity (but not to matter), which is second order in derivatives. The most general action for Einstein-aether theory, up to total derivative terms and setting aside matter coupling, is

$$S_{æ} = \frac{1}{16\pi G_{æ}} \int \sqrt{-\bar{g}}\,(-\bar{R} + L_{æ})\, d^4x \tag{35}$$

where $\bar{g}_{\mu\nu}$ is the 4-dimensional metric, $\bar{g}$ its determinant, $\bar{R}$ the Ricci scalar of this metric, $\bar{\nabla}_\mu$ is the covariant derivative associated with $\bar{g}_{\mu\nu}$,

$$L_{æ} = -M^{\alpha\beta\mu\nu}\bar{\nabla}_\alpha u_\mu \bar{\nabla}_\beta u_\nu, \tag{36}$$

with $M^{\alpha\beta\mu\nu}$ defined as

$$M^{\alpha\beta\mu\nu} = c_1 \bar{g}^{\alpha\beta}\bar{g}^{\mu\nu} + c_2 \bar{g}^{\alpha\mu}\bar{g}^{\beta\nu} + c_3 \bar{g}^{\alpha\nu}\bar{g}^{\beta\mu} + c_4 u^\alpha u^\beta \bar{g}_{\mu\nu}. \tag{37}$$

Greek indices run from 0 to 3, the c_i are dimensionless coupling constants, and it is assumed that u_μ is constrained to be a unit vector, $g^{\mu\nu}u_\mu u_\nu = 1$. This constraint can be explicitly imposed by the use of a lagrange multiplier, or implicitly taken into account in the variation of the action by allowing only variations that respect it.

One may wonder whether there is any relation between Einstein-aether theory and the (naive) IR limit of non-projectable Hořava–Lifshitz gravity (that is, a theory described by action (16) without the higher order operators suppressed by M_A and M_B). It was shown in Ref. [35] that the latter is actually a limiting case of the former. Let us briefly go through the arguments presented there.

Starting from the action (35), impose the restriction that the aether be hypersurface orthogonal:

$$u_\alpha = \frac{\partial_\alpha T}{\sqrt{g^{\mu\nu}\partial_\mu T \partial_\nu T}}, \tag{38}$$

where the unit constraint on the aether has been taken into account. T will satisfy a dynamical equation. However, since it is a scalar, on shell its dynamical equation will be implied by the contracted Bianchi identity, provided that T is not a constant (this can be shown for a general scalar field coupled to gravity, provided that the theory is manifestly covariant and the other fields are assumed to satisfy their field equations). However, T cannot be a constant as it now defines the foliation. Thus, we are allowed to not explicitly impose the equation of motion for T, which in turn implies that T itself can be chosen as the time coordinate. We then have

$$u_\alpha = \delta_{\alpha T}(g^{TT})^{-1/2} = N\delta_{\alpha T}\,. \tag{39}$$

This choice selects a preferred foliation with N being the lapse function of this foliation. Substituting eq. (39) into the Einstein-aether theory action (35), one gets after some algebra

$$S = \frac{M_{\rm pl}^2}{2}\int dt d^3x\, N\sqrt{g}(K_{ij}K^{ij} - \lambda K^2 + \xi\,{}^{(3)}\!R + \eta\, a_i a^i), \tag{40}$$

with the following correspondence of parameters:

$$\frac{1}{8\pi M_{\rm pl}^2 G_{æ}} = \xi = \frac{1}{1-c_{13}}, \quad \lambda = \frac{1+c_2}{1-c_{13}}, \quad \eta = \frac{c_{14}}{1-c_{13}}, \tag{41}$$

where we use the convention $c_{ij} = c_i + c_j$.[5] Clearly, action (40) is the same as action (16) without the higher order operators. Therefore, the IR limit of non-projectable Hořava-Lifshitz gravity is equivalent to Einstein-aether theory with a hypersurface orthogonal aether.

This equivalence provides new insights to both theories. For Einstein-aether theory it implies that there may exist a UV completion, at least for some subclasses. For non-projectable Hořava–Lifshitz gravity it tells us that it can be viewed as a theory were a preferred foliation is dynamically defined by a scalar field. Note that one would expect to be able to supplement Einstein-aether gravity (with a hypersurface orthogonal aether) with suitable higher order operators and construct a covariant equivalent of the full non-projectable Hořava–Lifshitz gravity (see Ref. [36] for an early attempt).

Clearly, on account of the equivalence, some results obtained for Einstein-aether theory go through to the IR limit of Hořava–Lifshitz gravity and vice versa. A characteristic example

[5] Note that there is a typo in the formula relating ξ and c_{13} in eq. (10) of Ref. [35].

is spherically symmetric solutions, as the aether is always hypersurface orthogonal in spherical symmetry [35, 30]. For example, some spherically symmetric solutions found in Ref. [37] for Hořava–Lifshitz gravity had actually already been found earlier for Einstein-aether theory [38]. On the other hand, some important results do not go through from general Einstein-aether theory, as they hinge on the exact form of the field equations, and specifically on whether the vector is a gradient of a scalar or not. For instance, though qualitatively similar, quantitatively the PPN constraints on non-projectable Hořava–Lifshitz gravity cannot simply be inferred from those of Einstein-aether theory via a parameter mapping and have been obtain independently in Ref. [30].

6. Cosmology

As already mentioned, all versions of Hořava–Lifshitz gravity presented above are written in a preferred foliation, and, therefore, they do not constitute diffeomorphism invariant theories, but they are invariant under diffeomorphisms that preserve the foliation. This symmetry provides enough gauge freedom to choose

$$N = 1, \quad N^i = 0, \quad g_{ij} = a(t)^2 \delta_{ij}, \tag{42}$$

and bring the ADM line element (5) into the usual Friedmann–Lemaître–Robertson–Walker (FLRW) form

$$ds^2 = -dt^2 + a(t)^2 \left[\frac{dr^2}{1 - kr^2} + r^2 \left(d\theta^2 + \sin^2 \theta d\phi^2 \right) \right], \tag{43}$$

under the usual cosmological assumptions of homogeneity and isotropy. Here, $a(t)$ is the scale factor and $k = +1, 0, -1$ according to whether the universe is hyperspherical, spatially flat, or hyperbolic.

Clearly, in this setting N has no space dependence. Therefore, its spatial derivatives vanish. This will make the actions of the projectable and the non-projectable versions of the theory, eqs. (14) and (16) coincide (assuming that all operators allowed by the symmetries of the theory have been consistently taken into account). The field equations do not coincide completely, as it does make a difference whether N is assumed to be space-independent *a priori* or whether the latter is enforced as a gauge choice. The difference lies on the Hamiltonian constraint and the rest of the equations coincide. The former is global in the projectable theory and local in the non-projectable theory. So, when it comes to background cosmology, one can study both version of the theory at the same time, just taking into account the subtle issue of the Hamiltonian constraint. The dynamics of perturbation around the background will, of course, differ in the two versions, but we will not discuss perturbations here.

Cosmology in Hořava–Lifshitz gravity was first studied in Ref. [39, 10] (with detailed balance in Ref. [40]). A recent review of cosmology in the projectable version has recently been given in Ref. [23], and as explained above, most of the results summarized there are applicable also on the non-projectable version. Here we will only give a very brief overview of background cosmology and mention in passing some key advantages of Hořava–Lifshitz cosmology. We refer the reader to the relevant literature for more details.

We start by presenting the field equations in an FLRW background. The supermomentum constraint is trivially satisfied, so we are left with two equations, corresponding to the first and second Friedmann equations. In the projectable version, the Hamiltonian constraint yields

$$\int d^3x a^3 \left\{ \frac{3\lambda - 1}{2} \frac{\dot{a}^2}{a^2} - \frac{V(a)}{6} - \frac{8\pi G_N \rho}{3} \right\} = 0, \tag{44}$$

where $8\pi G_N \equiv M_{\rm pl}^{-2}$, $\rho \equiv -g^{-1/2}\delta S_M/\delta N$, S_M is the matter action, and

$$V(a) = g_0\, M_{\rm pl}^2 + \frac{6g_1 k}{a^2} + \frac{12(3g_2+g_3)k^2}{M_{\rm pl}^2\, a^4} + \frac{24(9g_4+3g_5+g_6)k}{M_{\rm pl}^4\, a^6}. \tag{45}$$

In the non-projectable case, the Hamiltonian constraint is local and one can do away with the integral:

$$\frac{3\lambda-1}{2}\frac{\dot a^2}{a^2} - \frac{V(a)}{6} = \frac{8\pi G_N \rho}{3}. \tag{46}$$

The dynamical equations yield in both cases

$$-\frac{3\lambda-1}{2}\frac{\ddot a}{a} \;=\; \frac{1}{2}\frac{3\lambda-1}{2}\frac{\dot a^2}{a^2} - \frac{1}{12a^2}\frac{d[V(a)\,a^3]}{da} + 4\pi G_N p, \tag{47}$$

where $p \equiv -g^{ij}(2/N\sqrt{g})\delta S_m/\delta g^{ij}$.

Eqs. (44) and (47) and Eqs. (46) and (47) are governing background cosmology in the projectable and non-projectable versions respectively. Notice the following: In the non-projectable version one can eliminate $\dot a$ from eq. (47) by using eq. (46) to get

$$-\frac{3\lambda-1}{2}\frac{\ddot a}{a} \;=\; -\frac{1}{12a}\frac{d[V(a)a^2]}{da} + \frac{4\pi G_N}{3}(\rho+3p). \tag{48}$$

Differentiating eq. (46) and subtracting from eq. (48) yields the standard conservation law

$$\dot\rho + 3\frac{\dot a}{a}(\rho+p) = 0. \tag{49}$$

This implies that the conservation law for matter is implied by the two (modified) Friedmann equations and, thus, imposing it separately is neither needed nor will it constrain the dynamics. This is in direct analogy to general relativity.

Things are quite different in the projectable case, given the global nature of the Hamiltonian constraint. It has been argued in Ref. [41] that a global constraint, such as eq. (44), is irrelevant for the local patch of the universe inside the Hubble horizon which the FLRW spacetime is supposed to approximate. Now, we could, instead of differentiating eq. (44), integrate eq. (47) and ignore eq. (44) completely. One then gets

$$\frac{3\lambda-1}{2}\frac{\dot a^2}{a^2} - \frac{V(a)}{6} = \frac{8\pi G_N}{3}\left(\rho + \frac{C(t)}{a^3}\right), \tag{50}$$

where the functional form of $C(t)$ depends on the conservation law matter satisfies. If we eliminate $\dot a$ from eq. (47) by using eq. (50) we get

$$-\frac{3\lambda-1}{2}\frac{\ddot a}{a} \;=\; -\frac{1}{12a}\frac{d[V(a)a^2]}{da} + \frac{4\pi G_N}{3}\left(\rho + \frac{C(t)}{a^3} + 3p\right). \tag{51}$$

Now, the conservation law for matter has become essential to the dynamics via the presence of $C(t)$. The terms containing $C(t)$ are the only difference between eqs. (50) and (51), which describe the background dynamics in the projectable case, and eqs. (46) and (48), which describe the background dynamics in the non-projectable case.[6]

[6] This analysis, leading to the same conclusions, can be performed also before imposing the FLRW ansatz and the relevant symmetries, see Ref. [41].

Now that we have the field equations in a suitable form for both cases we can proceed to discuss phenomenology. Suppose that matter is coupled in such a way that at low energies ρ and p have the usual meaning and satisfy the standard conservation law of eq. (49) (this is a strong assumption, see also next section). For the projectable case this implies that $C(t) = C_0$, where C_0 is a constant. The C_0 related terms then play the role of a dark matter component [41]. This dark matter component will not exist in the non-projectable case.

Setting aside the role of matter, the main difference between Hořava–Lifshitz gravity and general relativity in background cosmology is the presence of the last two terms in V. The first one is what is known as a dark radiation component, as it scales as a^{-4}. The second one is what one could call a stiff matter component, as it scales as a^{-6}. Note that the sign of the a^{-6} term depends on the sign of the k and that both terms vanish when $k = 0$. An interesting feature related to the presence of these terms is that they can lead to solutions with classical cosmological bounces, provided that their coefficients will have appropriate signs [10, 39].

One last important property of Hořava–Lifshitz cosmology which is worth mentioning before closing this brief overview, is that is seems to lead to a scale invariant spectrum of cosmological perturbations, without the need for inflation [42, 43]. This property is related to the anisotropic scaling between space and time that the theory exhibits at high energies, which is also the key ingredient for power counting renormalizability.

7. Conclusions and future perspectives

A brief overview of Hořava–Lifshitz gravity was given and special attention was paid in distinguishing between the various different versions. After that, the main focus of this review was on the dynamics and the viability of these versions. In summary, the projectable version has serious viability problems because it harbours a scalar mode which is either classically or quantum mechanically unstable (depending on the position in parameter space) and also exhibits strong coupling at low energies (whether or not the Vainshtein mechanism can alleviate the latter is not clear yet, see also above). The non-projectable version, on the other hand, seems to have similar (and other) problems if detailed balance is imposed. These problems, however, appear to be overcome once detailed balanced is abandoned and all operators allowed by the symmetries of the theory are consistently taken into account. Alleviating the strong coupling problem requires the introduction of a second scale in the theory which should be parametrically smaller than the Planck scale, and should be the scale at which higher order operators become important. The dynamics of the version exhibiting the extra $U(1)$ symmetry, which was very recently proposed, have not been adequately studied yet.

There are many open issues regarding Hořava–Lifshitz gravity, the two most important ones being the following. Firstly, in the introduction we argued that the theory is power-counting renormalizable, as is usually done in the literature. Even though this is a strong indication for UV completeness, renormalizability beyond power counting has not been explicitly shown (see Ref. [45] for some work in this direction). Additionally, the renormalization group flow of the various couplings has not been studied, which implies that, for the time being, one does not really know if the theory approaches GR in the IR ($\lambda \to 1$, $\eta \to 0$) or not. Secondly, the role of matter and its coupling to gravity have not been fully clarified yet. The matter action will have to include higher order spatial derivatives, which implies that there will be modifications in the dispersion relations of matter fields that can lead to serious constraints. Additionally, couplings between the matter and the scalar graviton could lead to violations of the equivalence principle. Even if initially omitted, such couplings would be generically generated by radiative corrections, unless some symmetry prohibits them (*e.g.* supersymmetry [44]). Hopefully, future work will shed some light onto these issues.

Acknowledgments
I am indebted to my collaborators Antonios Papazoglou, Matt Visser and Silke Weinfurtner for their hard work and invaluable input in our joint projects on Hořava–Lifshitz gravity. I would also like to thank Ted Jacobson, Ian Kimpton and Tony Padilla for numerous stimulating and enlightening discussions. Many thanks also to the participants of the Peyresq 2010 meeting. This work was supported by a Marie Curie Incoming International Fellowship.

References

[1] Stelle K S 1977 *Phys. Rev. D* **16** 953
[2] Horava P 2009 *JHEP* **0903** 020
[3] Horava P 2009 *Phys. Rev. D* **79** 084008
[4] Weinfurtner S, Sotiriou T P and Visser M 2010 *J. Phys. Conf. Ser.* **222** 012054, [arXiv:1002.0308 [gr-qc]].
[5] Anselmi D and Halat M 2007 *Phys. Rev. D* **76** 125001
[6] Visser M 2009 *Phys. Rev. D* **80** 025011
[7] Visser M 2009 arXiv:0912.4757 [hep-th]
[8] Takahashi T and Soda J 2009 *Phys. Rev. Lett.* **102** 231301
[9] Sotiriou T P, Visser M and Weinfurtner S 2009 *Phys. Rev. Lett.* **102** 251601
[10] Sotiriou T P, Visser M and Weinfurtner S 2009 *JHEP* **0910** 033
[11] Lu H, Mei J and Pope C N 2009 *Phys. Rev. Lett.* **103** 091301
[12] Blas D, Pujolas O and Sibiryakov S 2010 *Phys. Rev. Lett.* **104** 181302
[13] Horava P and Melby-Thompson C M 2010 *Phys. Rev. D* **82** 064027
[14] da Silva A M 2010 arXiv:1009.4885 [hep-th]
[15] Wang A and Wu Y 2010 arXiv:1009.2089 [hep-th]
[16] Koyama K and Arroja F 2010 *JHEP* **1003** 061 (2010)
[17] Huang Y, Wang A and Wu Q 2010 *Mod. Phys. Lett. A* **25** 2267
[18] Wang A and Wu Q 2010 arXiv:1009.0268 [hep-th]
[19] Elliott J W, Moore G D and Stoica H 2005 *JHEP* **0508** 066
[20] Charmousis C, Niz G, Padilla A and Saffin P M 2009 *JHEP* **0908** 070
[21] Blas D, Pujolas O and Sibiryakov S 2009 *JHEP* **0910** 029
[22] Vainshtein A I 1972 *Phys. Lett. B* **39** 393
[23] Mukohyama S 2010 arXiv:1007.5199 [hep-th]
[24] Li M and Pang Y 2009 *JHEP* **0908** 015
[25] Henneaux M, Kleinschmidt A and Gomez G L 2010 *Phys. Rev. D* **81** 064002
[26] Papazoglou A and Sotiriou T P 2010 *Phys. Lett. B* **685** 197
[27] Kimpton I and Padilla A 2010, *JHEP* **1007** 014
[28] Will C M 2005 *Living Rev. Rel.* **9** 3
[29] Blas D, Pujolas O and Sibiryakov S 2010 *Phys. Lett. B* **688** 350
[30] Blas D, Pujolas O and Sibiryakov S 2010 arXiv:1007.3503 [hep-th]
[31] Albert J *et al.* [MAGIC Collaboration and Other Contributors Collaboration] 2008 *Phys. Lett. B* **668** 253
[32] Ackermann M *et al.* [Fermi GBM/LAT Collaborations] 2009 *Nature* **462** 331
[33] Jacobson T and Mattingly D 2001 *Phys. Rev. D* **64** 024028
[34] Jacobson T 2007 *PoS* **QG-PH** 020
[35] Jacobson T 2010 *Phys. Rev. D* **81** 101502
[36] Germani C, Kehagias A and Sfetsos K 2009 *JHEP* **0909** 060
[37] Kiritsis E 2010 *Phys. Rev. D* **81** 044009
[38] Eling C and Jacobson T 2006 *Class. Quant. Grav.* **23** 5625
[39] Kiritsis E and Kofinas G 2009 *Nucl. Phys. B* **821** 467
[40] Calcagni G 2009 *JHEP* **0909** 112
[41] Mukohyama S 2009 *Phys. Rev. D* **80** 064005
[42] Mukohyama S 2009 *JCAP* **0906** 001
[43] Calcagni G 2010 *Phys. Rev. D* **81** 044006
[44] Groot Nibbelink S and Pospelov M 2005 *Phys. Rev. Lett.* **94** 081601
[45] Orlando D and Reffert S 2009 *Class. Quant. Grav.* **26** 155021

Conformal dynamical equivalence and applications[1]

N K Spyrou

Astronomy Department, Aristoteleion University of Thessaloniki, 541.24 Thessaloniki, Macedonia, Hellas (Greece)

Email: *spyrou@astro.auth.gr*

Abstract. The "Conformal Dynamical Equivalence" (CDE) approach is briefly reviewed, and some of its applications, at various astrophysical levels (Sun, Solar System, Stars, Galaxies, Clusters of Galaxies, Universe as a whole), are presented. According to the CDE approach, in both the Newtonian and general-relativistic theories of gravity, the isentropic hydrodynamic flows in the interior of a bounded gravitating perfect-fluid source are dynamically equivalent to geodesic motions in a virtual, fully defined fluid source. Equivalently, the equations of hydrodynamic motion in the former source are functionally similar to those of the geodesic motions in the latter, physically, fully defined source. The CDE approach is followed for the dynamical description of the motions in the fluid source. After an observational introduction, taking into account all the internal physical characteristics of the corresponding perfect-fluid source, and based on the property of the isentropic hydrodynamic flows (quite reasonable for an isolated physical system), we examine a number of issues, namely, (i) the classical Newtonian explanation of the celebrated Pioneer-Anomaly effect in the Solar System, (ii) the possibility of both the attractive gravity and the repulsive gravity in a non-quantum Newtonian framework, (iii) the evaluation of the masses - theoretical, dynamical, and missing - and of the linear dimensions of non-magnetized and magnetized large-scale cosmological structures, (iv) the explanation of the flat-rotation curves of disc galaxies, (v) possible formation mechanisms of winds and jets, and (vi) a brief presentation of a conventional approach - toy model to the dynamics of the Universe, characterized by the dominant collisional dark matter (with its subdominant luminous baryonic "contamination"), correctly interpreting the cosmological observational data without the need of the notions dark energy, cosmological constant, and universal accelerating expansion.

1. Introduction and Motivation

According to many current observational data, the realistic picture and morphology of an astrophysical-cosmological structure differs greatly from its corresponding optical picture. The *Solar*

[1] This is an extended version of the talk, under the same title, given by the author during the Conference **Recent Developments in Gravity (NEB) XIV** (Ioannina, Hellas, 8-11 June 2010), in which the main results of the research work, during approximately the last ten or so years, on the general subject **"Conformal Dynamical Equivalence"**, were presented. The members of the **Study Group,** participating in the above research, are the following: **M. Plionis** (1 and 2), **C. G. Tsagas** (1), **K. Kleidis** (1 and 3), **S. Basilakos** (4) , **K. Zagkouris** (1 and 5) and **N. K. Spyrou** (1).
1: Astronomy Department, Aristoteleion University of Thessaloniki, Thessaloniki, Hellas, 2: Institute of Astronomy and Astrophysics, National Observatory of Athens, Athens, Hellas, 3: Technological Education Institute of Serres, Serres, Hellas, 4: Research Center on Astronomy and Applied Mathematics, National Academy of Athens, Athens, Hellas, 5: Oxford University, Oxford, United Kingdom.

System, for example, viewed as a star accompanied by a number of nearby small objects and planets, differs from the one observed embedded in the solar wind, the outer boundaries of which extend up to the corresponding outer boundaries of the winds from nearby stars. Also, far beyond the *Kuiper Belt*, extending out of the orbits of the outer planets *Neptune* and *Pluto*, there is an enormous spherical array of icy worlds orbiting the Sun, called the *Oort Comet Cloud*, whose linear dimensions are of the order of 10^5 AU. This means that the Solar System extends up to half the distance of the *alpha-Proxima Centauri*, the star nearest to the Sun. Furthermore, according to far-ultraviolet spectroscopic observations of the properties of the highly-ionized oxygen, of the hot gas, and of the high-velocity clouds in the halo of the *Milky Way* galaxy and the *Local Group* [1] (see also [2] and references therein), there exists an extended, hot ($T > 10^6$ K), and sufficiently diffuse *Galactic Corona*, previously undetected by other means (e.g., X-rays), far beyond the nearby *halo* detected previously. This implies that the Galactic Corona, and hence the Milky Way galaxy itself, possibly extends out to the *Magellanic Clouds*. So, the expected linear dimensions of the Milky Way galaxy are at least 200 kpc, almost ten times larger than its optical linear dimensions (~30 kpc), and even larger on the basis of at least some of the high-velocity clouds at estimated distances as large as 850 kpc [3]. If such a result could be considered as typical, then the mutual (minimal) distance of the outer boundaries of the Milky Way and the nearby *Andromeda Galaxy*, at a distance of approximately 600 kpc from it, is of the order of their linear dimensions, approximately 200 kpc. Such a result for neighboring galaxies is generally believed to be valid for typical *clusters of galaxies*, in which the mutual distances of neighboring members are of the order of *a few hundred kpc*. Under such conditions, the linear dimensions of galaxies with haloes and coronas (in the range 0.1 kpc – 1 Mpc, depending on the type of the galaxies, e.g., elliptical, spiral, irregular) *is not necessarily* much smaller than their mutual distances. As a consequence, the dynamical description of the observed motions in a typical cosmological structure is not that of a system of gravitating point masses (namely, no tidal interactions), but rather of the form of hydrodynamic flows in a more or less continuous gravitating source.

Moreover, the structure of the interstellar medium itself in elliptical galaxies and of the intracluster medium in cluster of galaxies present special interest. Thus, on the one hand, the interstellar medium consists primarily of hot ($T > 10^6$ K) plasma; several molecular clouds with dimensions 20-50 pc and masses $\sim 10^6\ m_\odot$ in it, are partly associated with HII regions and embedded in a lower-density interstellar medium; the masses ($\sim 10^8 - 10^9\ m_\odot$) of the central dark objects in elliptical galaxies are only a few percent of the molecular-gas mass in the sub-kilo parsec region, as implied by the observational study of water-masering sources [4] - [7]. On the other hand, a cluster of galaxies can be treated [8] as an approximately isothermal sphere of hot ionized hydrogen (of dimensions a few Mpc, temperature, which can be as high as $\sim 10^8$ K, number density of electrons $\sim 10^{-4}\ cm^{-3}$, and number-density of molecular clouds $\sim 10^{-4}$ to $10^{-3}\ cm^{-3}$); the intracluster gas contains X-rays, fills the space between the galaxies, occupies much of the cluster's volume, and the X-ray luminosities fall in the range 10^{43} to 10^{45} erg. sec^{-1}; the mass of the intracluster medium exceeds the total mass of the luminous parts of the cluster's galaxies by a factor of several. (However, one must not forget that the surveys for the largest cosmological features so far cover only 0.001% of the observable universe!!!). In this spirit, it is interesting that, as a consequence of the improvement of the observing techniques, currently we witness in various ways (see, e.g., [9] – [12]; or even continuous citations-reports, prior the official publication) continuous observational detections of many new galaxies (e.g., small ones each orbiting another larger and brighter galaxy, or numerous, tidally-disrupted and densely-populated galaxies (ultra-compact dwarf; UCD galaxies) inhabiting at the central regions of huge clusters of galaxies, like the Fornax and the Virgo galaxy clusters), of intergalactic globular clusters (beyond the ~150 ones surrounding our *Milky Way* galaxy) and even stars [13], and also of hot filamentary networks connecting and surrounding the galaxies of a cluster of galaxies, seen in the ultraviolet but unseen in the optical, infrared and radio wavelengths [2], [3].

According to all the above, the morphology of galaxies is very different from the simple picture of spiral, elliptical, or, even irregular galaxies, in the sense that the galaxies (as well as the clusters of galaxies, and, possibly, the second-order clusters - super clusters - of galaxies) are almost spherically symmetric, very complex, practically continuous, and of much larger linear dimensions cosmological structures than previously assumed. This situation becomes even more interesting, considering, in particular, the shapes of *galactic dark-matter haloes* derived from *gravitational-lensing studies*, *N-body simulations*, *stellar tracers* (e.g., globular clusters), *HI studies and polar rings*, *studies of satellites* (e.g., various tails), as well as from *X-ray*, and *absorption lines* studies [14] (also, in the same volume, see, particularly, the contributions by Fillipi and Sepulveda, Ryden and Tinker, Zepf, Sparke, Johnston, Buote, Bowen). The observable Universe differs very much from the simple picture of a collection of galaxies (or higher-order cosmological structures), in which the mutual distances of the neighboring members are much larger than their linear dimensions. Consequently, the *constituents elements of the Universe and the Universe as a whole, can quite satisfactorily be treated dynamically as continuous gravitational systems and, more specifically, bounded, gravitating perfect-fluid sources, the physical-dynamical description of which is very well established in both the Newtonian and the general-relativistic levels.* So, we arrive at the very crucial result, that the motions of and in these constituents should be considered as hydrodynamic flows rather than geodesic motions. It is exactly this result, that enables us, through the *dynamical-equivalence approach,* to recast the geodesic motions, now taking into account the contribution to the observationally determined mass of the fluid source of all of its internal physical characteristics, not only its mass density, as sources of the observed motions.

Also, we recall [15], [16] that, according to the observations of the cosmic microwave background by the *Wilkinson Microwave Anisotropy Probe* (WMAP), the *Cosmos* is believed to be composed of *heavy elements (*0.03 %*), ghostly neutrinos* (0.3 %), *stars* (0.5 %), *free hydrogen and helium* (4%) (and so baryonic matter ~5%), *dark matter (22 %)*, and *dark energy (73 %).* Finally, we point out that the cosmological fluid is mostly considered as collisionless, and that, in the context of a cosmology dominated by a cosmological constant, it is not currently known what definitely this unknown is, and whether it is strictly constant, increasing, or decreasing with time.

In view of the above, we have applied the model of a bounded, continuous, gravitating perfect-fluid source of classical hydrodynamics or/and magneto hydrodynamics, in order to describe the isentropic motions in the large-scale cosmological structures, all their sources, and their consequences. This is in accordance with the approach of the *Conformal Dynamical Equivalence* (CDE) between the hydrodynamic flows and the geodesic motions, proposed and further developed and elaborated by the author and collaborators (for details and applications, see [17] – [38]). *Exactly, the results so far of the study group on this subject will be briefly presented here.*

The structure of this article is as follows: After an observational introduction in the beginning of the present *Section 1*, we recall in the next *Section 2* that, in general, the conservation of rest mass (baryon-number conservation) and the conservation of entropy are not mutually independent requirements of general relativity, a very important result frequently overlooked. Then, in *Section 3*, we first outline the dynamical equivalence in the case of the Newtonian theory of gravity. We prove that the isentropic flows in a bounded, gravitating perfect-fluid source are dynamically equivalent to geodesic motions in a *generalized scalar potential* (V). This generalized potential is explicitly expressed in terms of the standard gravitational potential (U, satisfying the standard gravitational *Poisson equation*), mass density, and, additionally, the internal physical characteristics of the gravitating source, namely, mass density, isotropic pressure and internal thermodynamic energy density. This enables us to define, through a Poisson-type equation for the generalized potential, the *generalized mass-energy density*, ρ_V (and the corresponding *generalized mass*, m_V) producing the generalized potential. As an astrophysical byproduct-application, we prove in *Section 3*, that, in the Newtonian theory of gravity for a static configuration, the mass-density and pressure functional laws cannot be chosen arbitrarily from each other, but their choice depends on each other's one. This can considerably limit the form of the (widely used in the literature, Plummer-type) mass-density law of an isentropic and isothermal equation of state.

In *Section 4*, it is proved that the generalized mass-energy density, ρ_V, can be positive/vanishing/negative, and that the generalized mass-energy, m_V, can, similarly, be positive/vanishing/negative. As a consequence, based on the form of the generalized Euler-Newton equations of motion, we prove that, in the Newtonian theory of gravity, a positive/vanishing/negative generalized mass-energy density, ρ_V, implies a spatially decreasing/non-changing/increasing acceleration, and, hence, dominance of the inwards acceleration (attractive gravity)/non changing acceleration/outwards acceleration (repulsive gravity), all of them stemming from the relative importance of the mass density and of the other internal physical characteristics of the fluid source.

In *Section 5*, based on the results of Section 4, we explore the discrimination of the regions of prevalence of the attractive gravity and repulsive gravity, leading to the determination of the so-called *inversion distance,* the latter's importance in determining-explaining the conventional linear dimensions of the source considered, as well as its temperature dependence.

In *Section 6*, we apply the results of the dynamical equivalence of *Section 3* to a typical inhomogeneous (with a *Plummer-type* mass-density configuration), and isothermal and isentropic perfect-fluid-cosmological structure. We explore the relative importance of the source's mass, m (theoretically evaluated through the Plummer-type density law) and its observationally determined mass, m_V (dynamical mass) and the latter's dependence on the mass density, ρ, (and mass, m) and on the fluid's internal physical characteristics. This relation enables us to prove that, under conditions generally met in the physical Universe, the mass m of the cosmological structure generally differs from, and actually it is *much larger* than the *observationally determined* mass m_V (dynamical mass). In this way, we *propose a possible, classical (namely, not quantum-mechanically oriented) partial explanation of the missing-mass problem.*

In *Section 7,* the distinction between the theoretical mass, m, and the observationally determined (dynamical) mass, m_V, (under the assumption and use of planar, equatorial, circular geodesics in disc galaxies), allow us to *propose a classical explanation of the flat-rotation curves problem.* Additionally, the dependence of the rotational velocity on the structure's (spatially constant) temperature (beyond its gravitational field) is critically examined.

In *Section 8*, we outline the use of the CDE approach in presenting a classical explanation of the celebrated *Pioneer Anomaly* Effect in the Solar System and, also, we provide the first analytical determination of the true linear dimensions of the Solar System (it extends up to almost half the distance to our closest star *α-Proxima Centaury)*, and the determination of the internal thermodynamic energy at the near and far regions of the Solar System.

In *Section 9*, based on the notions of the *repulsive gravity* and *inversion distance*, we outline the use of the CDE approach towards providing a possible theoretical physical explanation of the mechanism of formation of *winds* and *jets*, with possible applications to the phenomena of the expansion of a red-giant star and a supernova.

In *Section 10*, we outline the use of the CDE approach in the determination of the (cosmologically-dynamically important) masses of the globular clusters of galaxies, composed of both dark matter and baryonic matter, by relating theory and observational results.

In *Section 11*, we outline the generalization of the CDE approach in the case of a magnetized gravitating perfect-fluid source and the importance of the new magnetic contribution.

In *Section 12*, we outline the general-relativistic cosmological applications of the CDE , in the case of a collisional-dark-matter model. We prove that the extra energy, needed to compromise for the currently available cosmological data, can be compensated by the energy of the internal motions of the collisional cosmological fluid, that the post-recombination Universe remains *ever-decelerating,* and, also that the explicit form of the equation of state for the cosmological fluid can be determined. In other words, *in the collisional-dark-matter model, no dark energy and no acceleration of the universal expansion are necessary.* Also, we independently verify the particularly interesting facts, that, in the context of the standard, although, physically, less probable, collisionless-dark-matter model, the extra dark-energy component (to compromise the cosmic-microwave-radiation-background data) is, in fact, needed, the distant light sources appear to be dimmer than expected, and the expansion of the Universe keeps balancing between acceleration and deceleration, depending on the (fully and explicitly determined appropriate) value of the cosmological red-shift parameter. Therefore, these results of the collisionless-dark-matter model seem to be a *misinterpretation* of the cosmological observational data by an observer, who (although living in a Universe filled, mainly, with, collisional dark matter) *insists* in adopting the traditional (collisionless-dark matter) and less physical approach.

Finally, we conclude in *Section 13*.

2. Adiabaticity and Isentropicity versus Baryon Conservation in General Relativity

It is known that, in general relativity, the *field equations* for a gravitating perfect-fluid source

$$\mathcal{G}_{ik} = -\frac{8\pi G}{c^4} T_{ik} \tag{2.1}$$

are supplied by the *equations of motion*

$$T^{ik}{}_{;k} = 0 \tag{2.2}$$

In Eqs. (2.1) and (2.2), G is the gravitational constant and c the velocity of light in vacuum and, in the standard notation, $\mathcal{G}_{ik}$ is the *Einstein tensor,*

$$T_{ik} = (\varepsilon + p)u_i u_k - p g_{ik} \tag{2.3}$$

is the fluid's *energy-momentum tensor,* where g_{ik} are the covariant components of the *metric tensor,* u_i the covariant components of the *four-velocity*, a semi-colon denotes covariant derivative, and the *mass-energy density* ε is assumed (see, e.g., [39], pp. 81 – 84, 90 - 94) to split as

$$\varepsilon = \rho c^2 + \rho\Pi \tag{2.4}$$

with the proper quantities ρ, p and ρΠ being, respectively, *the rest-mass density*, *proper isotropic pressure*, and *internal specific-energy density* of the fluid source, the latter being the thermodynamic energy density that changes during the expansions and/or contractions of the fluid.

A direct consequence of Eqs. (2.2) and (2.4) is [40]

$$\varepsilon_{,i} u^i + (\varepsilon + p) u^i_{;i} = 0 \tag{2.5a}$$

or, equivalently,

$$\left[1 + \frac{1}{c^2}\left(\Pi + \frac{p}{\rho}\right)\right]\left(\rho u^i\right)_{;i} + \frac{1}{c^2}\left[\Pi_{,i} + p\left(\frac{1}{\rho}\right)_{,i}\right]\left(\rho u^i\right) = 0 \tag{2.5b}$$

where a comma denotes usual partial derivative with respect to the space-time coordinates.

Therefore, in general relativity, the *conservation of rest-mass (baryon-number conservation)* required by the equation

$$(\rho u^i)_{;i} = 0 \tag{2.6}$$

is compatible with the equations of motion if, and only if,

$$\left[\Pi_{,i} + p\left(\frac{1}{\rho}\right)_{,i}\right] u^i = 0 \tag{2.7}$$

This last equation is obviously satisfied under the assumption that the fluid's hydrodynamic flows are adiabatic, in other words, the net change, dQ, of the thermal content of a fluid's volume element, defined through the first classical axiom of thermodynamics

$$dQ = d\Pi + p d\left(\frac{1}{\rho}\right) \tag{2.8}$$

is always zero, namely,

$$dQ = Q_{,i}u^i = d\Pi + pd\left(\frac{1}{\rho}\right) = 0 \tag{2.9}$$

whence Eq. (2.7) is identically satisfied.

Proceeding further to *isentropic flows*, it might be helpful to recall here that in reaching the general result (2.9) we can also use (as, e.g., in [17]) the *equilibrium hydrodynamics hypothesis,* in the form of the constancy of the *entropy*, S, along the flow lines [41], [42], namely, in view of the second classical thermodynamic axiom,

$$dQ = TdS \tag{2.10}$$

T being the fluid distribution's absolute temperature,

$$\begin{aligned} 0 = dS &= S_{,i}u^i \\ &= \frac{1}{T}dQ = \frac{1}{T}\left[d\Pi + pd\left(\frac{1}{\rho}\right)\right] \\ &= \frac{1}{T}\left[\Pi_{,i} + p\left(\frac{1}{\rho}\right)_{,i}\right]u^i \end{aligned} \tag{2.11}$$

In fact, in relativistic astrophysics and cosmology, we usually assume S to be a *group invariant* [43], [44], whence

$$S_{,i}u^i = 0, \text{ for every } u^i \text{ and } T \tag{2.12}$$

or, equivalently,

$$\Pi_{,i} + p\left(\frac{1}{\rho}\right)_{,i} = 0 \tag{2.13}$$

direct consequence of which is the adiabaticity condition (2.9) (or (2.7)).

According to all the above, *the conservation of rest-mass* (Eq. 2.6) *and the conservation of entropy* (Eq. 2.13) *are not independent requirements in the framework of general relativity.* The reason of their independence in the Newtonian limit $\left("\frac{1}{c^2}" \rightarrow 0\right)$, is that Eq. (2.5b) reduces simply to *continuity equation*

$$\frac{\partial \rho}{\partial t} + \vec{\nabla}\cdot(\rho\vec{\upsilon}) = 0 \tag{2.14}$$

We emphasize that isentropicity (and, hence, also adiabaticity) of the hydrodynamic flows could be physically necessary and useful. Actually, then, beyond the constancy of the matter content (the number of baryons) of a finite fluid volume element (Eqs. (2.6) or (2.14)), also its thermodynamic content remains unchanged along the flow line, in accordance with such a usual assumption for an isolated physical system.

As emphasized by Chandrasekhar [40], the conservation of rest-mass (2.6), under the assumption of the absence of any dissipative mechanisms in the fluid source, should be considered as a fundamental physical law supplementing the field equations (2.1). If, according to [40], we furthermore assume that the only baryons present are protons and neutrons of practically equally rest masses, the conservation law (2.6) is equivalent to the equally fundamental law of the *conservation of the baryon-number* (practically Eq. (2.6), with ρ replaced by the *baryon-number density*). Consequently, if we assume that the rest-mass of a baryon remains constant, we conclude that the conservation of rest-mass (2.6) is actually expressing the *conservation of the baryonic mass*.

All these physical results concerning the adiabaticity or/and isentropicity of the hydrodynamic flows and its dependence on the conservation of rest-mass has long ago been emphasized by Chandrasekhar [40], but, in our opinion, since then, they, unfortunately, have been highly overlooked, especially in astrophysical applications. Here, we shall rely heavily on the adiabaticity of the hydrodynamic flows and, hence, on their dynamical equivalence to geodesic motions (see, e.g., [17]).

3. Newtonian Dynamical Equivalence of Hydrodynamic Flows and Geodesics

It has been suggested [17] that, in both the Newtonian and the general-relativistic theories of gravity, it is possible to give to the equations of the hydrodynamic flow motions in the interior of a bounded gravitating perfect-fluid source the *form* of the equations of geodesic motions.

In this *Section 3*, we limit ourselves to the Newtonian case (for the relativistic hydrodynamic case, see [17], [33], [37], [38] and for the magneto-hydrodynamic case, see [34], [35]; see also Sections 2 and 12 herewith). More precisely, in the case of the interior of a non-magnetized fluid, the *Newtonian* equations of the motion of a test particle (*geodesic motion* in the interior of the fluid) are

$$\frac{d\vec{\upsilon}}{dt} = \vec{\nabla} U \tag{3.1}$$

where the Newtonian gravitational potential U obeys the *Newtonian field equation* (Poisson's equation)

$$\nabla^2 U = -4\pi G \rho \tag{3.2}$$

with ρ being the gravitational source's mass-density function.

On the other hand, the *Euler's equations* for the hydrodynamic flow motion of a *fluid volume element* (in the interior of the non-magnetized fluid), are

$$\frac{d\vec{\upsilon}}{dt} = \vec{\nabla} U - \frac{1}{\rho}\vec{\nabla} p \tag{3.3}$$

supplied by the *continuity equation* (2.14).

These two kinds of motion in the interior of the fluid source differ from each other due to existence of the pressure-gradient term in Eq. (3.3). We remark that the continuity equation (2.14), as an expression of the conservation of the fluid's total mass, corresponds, physically, to the assumed constancy of the test-particle's mass.

Now we assume that the fluid-volume element's hydrodynamic flow is adiabatic and isentropic, whence, there is no exchange of heat quantities (Q) between the fluid-volume element and the rest of the fluid source. Then, according to Eq. (2.9),

$$\frac{dQ}{dt} = 0 = \frac{d\Pi}{dt} + p\frac{d}{dt}\left(\frac{1}{\rho}\right)$$

or, equivalently,

$$\frac{\partial \Pi}{\partial t} + p\frac{\partial}{\partial t}\left(\frac{1}{\rho}\right) + \vec{\upsilon}\cdot\left[\vec{\nabla}\Pi + p\vec{\nabla}\left(\frac{1}{\rho}\right)\right] = 0 \tag{3.4}$$

In analogy to Eq. (2.12), for dQ to be identically zero, Eq. (3.4) should be viewed as an identity for every value of the three velocity, $\vec{\upsilon}$, of the volume element, so that the following two equations

$$\vec{\nabla}\Pi + p\vec{\nabla}\left(\frac{1}{\rho}\right) = 0, \qquad \frac{\partial \Pi}{\partial t} + p\frac{\partial}{\partial t}\left(\frac{1}{\rho}\right) = 0 \tag{3.5a,b}$$

suffice to hold simultaneously. Notice that, for a time-independent source ($\frac{\partial}{\partial t}=0$), Eqs. (3.4) and (3.5a) are equivalent to each other.

In view of Eq. (3.5a), the Euler's equations of motion (3.3) are written in the form

$$\frac{d\vec{\upsilon}}{dt} = \vec{\nabla} V \tag{3.6}$$

where the *generalized (scalar) potential* V is defined as

$$V = U - \left(\Pi + \frac{p}{\rho} \right). \tag{3.7}$$

Eqs. (3.6) are of the *same functional form* as the geodesic equations of motion (3.1), namely, *the acceleration three-vector is equal to the gradient of a scalar potential*. Therefore, the adiabatic perfect-fluid hydrodynamic flows are *dynamically equivalent* to the geodesic motions in the generalized potential V. Interestingly enough, p and Π, beyond ρ, appear now as *sources* of geodesic motion, *in contrast to* the non-adiabatic-fluid case, in which only ρ (through U, Eq. (3.2)) produces geodesic motions.

Next we might ask: *What is the mass-energy density ρ_v, producing the generalized potential V?* Obviously, in view of the corresponding Newtonian "geodesic" equations (3.1) and (3.2), the density ρ_v can be defined through the *Poisson-type field equation*

$$\nabla^2 V = -4\pi G \rho_v \tag{3.8}$$

The generalized field equation (3.8), and, hence, the introduction of ρ_v is a natural consequence of the generalized equations of motion (3.6) (a property shared by the general-relativistic equations (2.1) and (2.2), but *not of classical Newtonian gravity*). In contrast to this, the standard Poisson equation (3.2), is not a consequence of the Euler's equations (3.3), but, independently of the latter, it, simply complements them.

It is straightforward to prove, with the aid of Eq. (3.7) along with Eqs. (3.2) and (3.5a), that

$$\rho_v = \rho + \rho_i \tag{3.9}$$

where

$$\rho_i = \frac{1}{4\pi G} \nabla^2 \left(\Pi + \frac{p}{\rho} \right) = \frac{1}{4\pi G} \left[\vec{\nabla} \cdot \left(\frac{1}{\rho} \vec{\nabla} p \right) \right] \tag{3.10}$$

will be called the *internal-mass density*, so that Eq. (3.8) is written in the form

$$\nabla^2 V = -4\pi G (\rho + \rho_i) \tag{3.11}$$

It is obvious that the mass m_v, which corresponds to the density ρ_v producing the generalized potential V, is

$$m_v = \int_v \rho_v d^3x \tag{3.12}$$

where *v* is the three-dimensional volume of the source considered. In view of Eqs. (3.9) and (3.12),

$$m_v = m + m_i \tag{3.13}$$

namely, the mass m_v differs from the *mass*

$$m = \int_v \rho d^3x \tag{3.14}$$

by the *internal mass*

$$m_i = \int_v \rho_i d^3x \tag{3.15}$$

Hence, the extra ingredient ρ_i to the generalized mass density, ρ_v, stemming from the source's internal physical characteristics, results in an extra mass, independent of the mass m of the source *(namely, the* volume integral *of ρ*), the so-called *internal mass (namely, the* volume integral *of* ρ_i), being also a part of the observationally determined mass (the *dynamical mass*; see Section 6, below) .

We note that, in general, the mass m_v is not constant. If we would wish to postulate the conservation of m_v

$$\frac{dm_v}{dt}=0 \Leftrightarrow m+m_i=const. \tag{3.16}$$

this would be equivalent to the *generalized continuity equation*

$$\frac{\partial \rho_v}{\partial t}+\vec{\nabla}\cdot\left(\rho_v \vec{\upsilon}\right)=0 \tag{3.17}$$

which, in turn, generalizes the classical continuity equation (2.14). The constancy of m + m_i permits m and m_i to change in time. For both to be constant in time, the corresponding two continuity equations must hold simultaneously and independently of each other at every moment. In any case, the possibly assumed necessity of Eqs. (3.16) and (3.17) would limit the character and properties of the physical source considered, probably, in a quite decisive way.

The *astrophysical and cosmological significance* of the above results (3.12 – 3.15) lies in the fact that (see *Sections 6* and *7,* below) the mass determined at any moment with the aid of the geodesic motions in the generalized potential V is not simply the mass m, determined with the aid of the geodesics in the field of the gravitational potential U, as it is generally believed, but the mass m_v, which is different from m. Moreover, it is obvious that V is not the standard gravitational potential U "modified by hand"; instead it is a natural generalization of U resulting simply from the assumption of the isentropicity of the hydrodynamic flows, and, so, *we are not dealing here with a modified-gravity approach.* Also, in the generalized potential V, the volume element moves as a point mass, but, now, *carrying along all the internal physical characteristics of the fluid source (not only its mass density).* We emphasize that both the above two remarks are of special importance in the context of both the Newtonian and the general-relativistic theories of gravity (see also *Section 12,* below).

In the special case of a static distribution $\left(\vec{\upsilon}=0,\ \frac{d\vec{\upsilon}}{dt}=0\right)$, from Eqs. (3.6) and (3.8) we obtain

$$\vec{\nabla}V=0,\ \ \nabla^2 V=0,\ \ \rho_v=0 \tag{3.18}$$

a situation similar to the motion in «vacuum», and also

$$\vec{\nabla}\cdot\left(\frac{1}{\rho}\vec{\nabla}p\right)=4\pi G\rho \tag{3.19}$$

The physical significance of Eq. (3.19) is best seen in the case of a spherically-symmetric fluid distribution with on *equation of state* of the form

$$p=f(\rho) \tag{3.20}$$

Then, Eq. (3.19) is written as

$$\left(\frac{1}{\rho}\frac{df}{d\rho}\right)\rho''+\frac{2}{r}\left(\frac{1}{\rho}\frac{df}{d\rho}\right)\rho'+\frac{d}{d\rho}\left(\frac{1}{\rho}\frac{df}{d\rho}\right)(\rho')^2=-4\pi G\rho \tag{3.21}$$

where a prime denotes total derivative with respect to the radial distance r.

Eq. (3.21) can be seen as either a consistency condition between Eqs. (3.22) and (3.20), for a given density distribution $\rho=\rho(r)$, or, as a differential equation to be solved for ρ(r). Therefore, for static and spherically-symmetric gravitating perfect-fluid sources, under adiabatic conditions, the functional form of an arbitrarily chosen equation of state f(ρ) depends on the functional form of the mass-density distribution law ρ(r). In other words, *the laws of the two parameters, mass density and pressure, cannot be chosen arbitrarily from each other, but their choice depends on each other's one.*

As an application we consider the *adiabatic and isothermal (adiabatic index γ = 1) equation of state* of the form (6.1) below and a *Plummer-type rest-mass density* of the form (6.2) below. In this case, Eq. (3.21) reduces to

$$\frac{\rho''}{\rho}+\frac{2}{r}\frac{\rho'}{\rho}-\left(\frac{\rho'}{\rho}\right)^2=-\frac{4\pi G}{\kappa}\rho \tag{3.22}$$

or, furthermore,

$$2n\left(1+\frac{r^2}{r_0^2}\right)^{-2}\frac{r^2}{r_0^4}-\frac{3n}{r_0^2}\left(1+\frac{r^2}{r_0^2}\right)^{-1}=-\frac{4\pi G}{\kappa}\rho_0\left(1+\frac{r^2}{r_0^2}\right)^{-\frac{n}{2}} \tag{3.23}$$

Applying Eq. (3.23) for r ~ 0, r = r_0 and r = R (the radius of the distribution), and comparing the values of κ in these three cases, we conclude that *necessarily*

$$n=2\frac{\ln 3}{\ln 2}=3.169 \tag{3.24α}$$

and, since n is, by assumption, an integer, finally

$$n=3 \qquad \text{(only!!!)} \tag{3.25β}$$

Such a relation between the adiabatic index γ (here 1) and n has already been derived independently and more generally by Kleidis and Spyrou [17], namely,

$$\gamma \le 1+\frac{1}{n} \tag{3.26}$$

and so n in Eq. (6.2) below is, practically, the adiabatic index γ (see also Section 5 of [17]).

4. Attractive Gravity and Repulsive Gravity in Newtonian Gravity

Now we remark that, contrary to the case of the equations of motion (3.1), the equations of motion (3.6) do not necessarily imply an inwards acceleration solely (namely, only attractive gravity), because

$$\vec{\nabla}\cdot\left(\frac{d\vec{\upsilon}}{dt}\right)=\vec{\nabla}\cdot(\vec{\nabla}V)=\nabla^2V=-4\pi G\rho_v \tag{4.1}$$

Therefore

$$\vec{\nabla}\cdot\left(\frac{d\vec{\upsilon}}{dt}\right)\gtreqless 0 \Leftrightarrow \rho_v \lesseqgtr 0 \tag{4.2}$$

or, equivalently,

Spatially decreasing $\frac{d\vec{\upsilon}}{dt}$ implies $\rho_v>0$ $(\nabla^2V=-4\pi G\rho_v<0)$ (4.3a)

Spatially non-changing $\frac{d\vec{\upsilon}}{dt}$ implies $\rho_v=0$ $(\nabla^2V=0)$ (4.3b)

Spatially increasing $\frac{d\vec{\upsilon}}{dt}$ implies $\rho_v<0$ $(\nabla^2V=-4\pi G\rho_v>0)$ (4.3c)

and vice versa. According to the above,

$\rho_v>0$ implies dominance of inwards acceleration (attracting gravity) (4.4a)

$\rho_v=0$ implies non changing (spatially) acceleration (4.4b)

$\rho_v<0$ implies dominance of outwards acceleration (antigravity, repulsive gravity) (4.4c)

The significance of the above results (4.1) – (4.4) is that, in contrast to the general belief, *they reveal the possibility of repulsive gravity in the framework of the Newtonian gravity.* Obviously, this possibility of repulsive gravity is a consequence of the CDE approach, according to which the source's internal physical characteristics appear to act a sources of gravitational geodesic motions.

Furthermore, it is rather straightforward from the definition (3.10) for ρ_i, that, in realistic cases, ρ_i cannot be a constant or vanishing (see Section 6, below) and, hence, ρ_v can, in principle, have any sign, $\rho_v \gtreqless 0$, as stated above.

5. The Inversion Distance and the CDE Approach

We recall that the generalized mass-density (3.9) producing the geodesic motions (3.6) and (3.7) can be either positive, or negative, or even vanish. This implies the possibility of a spatially-increasing, or spatially-decreasing, or even spatially-unchanging acceleration, depending on the distance from the centre of the source as compared to the so-called *inversion distance*.

Next, using Eqs. (3.2, for n=3), (3.9) and (3.10), we find

$$\frac{-\rho_i}{\rho} = \alpha \frac{2+z^2}{z} \tag{5.1}$$

where

$$z = \left(1 + \frac{r^2}{r_0^2}\right)^{\frac{1}{2}} \geq \sqrt{2}, \qquad (r \geq r_0) \tag{5.2}$$

and

$$\alpha = \frac{3kT}{\mu m_H} \frac{1}{4\pi G \rho_0 r_0^2} \tag{5.3}$$

Then, first we notice that, by standard theoretically-derived assumption, the radius of the innermost stable circular orbit (identified with r_0) around a central Schwarzschild black hole of mass M_c is three times the black hole's gravitational radius, R_s,

$$r_0 = 3R_s = \frac{6GM_c}{c^2}$$

If we approximate [28] M_c as

$$M_c = \frac{4}{3}\pi r_0^3 \rho(r = r_0) = \frac{4}{3}\pi r_0^3 \frac{\rho_0}{2^{\frac{3}{2}}} \tag{5.4a}$$

we find

$$4\pi G \rho_0 r_0^2 = \sqrt{2} c^2 , \tag{5.4b}$$

whence

$$\rho_0 = 6.439 \times 10^4 (M_c / M_\odot)^{-2} gr.cm^{-3} \tag{5.4c}$$

and

$$\alpha = \frac{3}{\sqrt{2}} \frac{kT}{\mu m_H c^2} \tag{5.5}$$

Although the choice (5.4a) seems arbitrary, the conditions (5.4a,b,c) can be compatible with the assumption of a black hole as a fictitious homogeneous dark object of mass M_c, radius r_0 and *mean* mass density equal to $\rho(r=r_0)$. Actually, in the case of a super-massive black hole of mass $M_c \sim 10^9$

$M_\odot$, we find $r_0 \sim 10^{15}$ cm, whence the condition (5.4b) implies $\rho_0 \sim 10^{-14}$ gr .cm^{-3}, namely, about ten orders of magnitude larger that the mass density of the Milky Way in the solar neighborhood ($\sim 10^{-24}$ gr.cm^{-3}).

Then, under the conditions assumed up to now, we notice that the condition

$$\frac{-\rho_i}{\rho} \gtreqless 1 \qquad (\text{equivalently, } 0 \gtreqless \rho + \rho_i = \rho_v) \qquad (5.6)$$

is equivalent to

$$\varphi(z) = z^2 - \frac{1}{\alpha} z + 2 \gtreqless 0 \qquad (5.7)$$

The discriminant

$$\Delta = \frac{1}{\alpha^2} - 8 \qquad (5.8)$$

of the binomial *φ(z)* satisfies

$$\Delta \gtreqless 0, \qquad \text{when} \qquad T \lesseqgtr T_{\lim} \qquad (5.9)$$

with (see, however, also Eqs. (5.3) - (5.5))

$$\frac{T_{\lim}}{\mu} = \frac{m_H c^2}{6k} \sim 1.817 \times 10^{12} K \qquad (5.10)$$

where the *threshold temperature* T_{tt} is defined via Eq. (7.4c). Therefore (see also Eqs. (4.21)),

For $T \geq T_{\lim}$ $(\Delta \leq 0)$, $\varphi(z) > 0$, $\rho + \rho_i < 0$ (5.11)

For $T < T_{\lim} (\Delta > 0)$, from the two solutions of the Eq. (5.7) only one, the largest $\left(z \geq \sqrt{2}\right)$, is acceptable, namely,

$$r = r_{inv};\ \frac{r_{inv}^2}{r_0^2} = \frac{1}{4}\left(\frac{1}{\alpha} + \left(\frac{1}{\alpha^2} - 8\right)^{\frac{1}{2}}\right)^2 - 1 \qquad (5.12)$$

such that

a) If $r < r_{inv}$, then $\rho + \rho_i > 0$ (inwards acceleration or attractive gravity dominates) (5.13a)

b) If $r = r_{inv}$, then $\rho + \rho_i = 0$ (spatially unchanging acceleration) (5.13b)

c) If $r > r_{inv}$, then $\rho + \rho_i < 0$ (outwards acceleration or repulsive gravity dominates) (5.13c)

In the special case $\Delta >> 0$, namely, $T << T_{\lim}$, which is believed to apply usually in almost all of the cosmological large-scale structures, Eqs. (5.12) and (5.2) reduce to, respectively,

$$\frac{r_{inv}}{r_0} = \frac{1}{\alpha} = \frac{\sqrt{2}}{3} \frac{\mu m_H c^2}{kT} \qquad (5.14)$$

with

$$z_{inv} \sim \frac{r_{inv}}{r_0} (= x_{inv},\ \text{for } r >> r_0) \qquad (5.15)$$

ASTRONOMICAL SYSTEM	$\mathbf{r_0}$	**Inversion Distance** $\mathbf{r_{inv}}$	***Numerical Values***
SOLAR SYSTEM	$R_\odot$ ($6.96X10^{10}$ cm)	$\frac{2.39\times10^8}{T_{(2)}/\mu}AU \Rightarrow$	2.39×10^7 AU, for $\frac{T}{\mu}\sim10^3$ K
GALAXY	$3R_s$ (Central Black Hole $M_c \sim 10^6 M_\odot$ $\sim 8.86x10^{11}$ cm)	$\frac{1.48}{T_{(6)}/\mu}pc \Rightarrow$	14.76 Kpc, for $\frac{T}{\mu}\sim10^2$ K (Conventional Outskirts of the Milky Way)
CLUSTER OF GALAXIES	$3R_s$ (Central Galaxy Black Hole $M_c \sim 10^{12} M_\odot$ $\sim 8.86x10^{17}$ cm)	$\frac{14.76}{T_{(8)}/\mu}Kpc \Rightarrow$	14.76 Mpc, for $\frac{T}{\mu}\sim10^5$ K (Conventional Dimensions of a Cluster of Galaxies)
SUPERCLUSTER OF GALAXIES	$3R_s$ (Central Galaxy Black Hole $M_c \sim 10^{12} M_\odot$ $\sim 8.86x10^{17}$ cm)	$\frac{1.48}{T_{(9)}/\mu}Kpc \Rightarrow$	147.56 Mpc, for $\frac{T}{\mu}\sim10^4$ K (Conventional Dimensions of a Super-cluster of Galaxies)

Table 1: The inversion distance for typical cosmological structures

Representative values of the invers*ion distance*, r_{inv}, based on Eqs. (5.14) and (5.15), are shown in *Table 1*, and are seen to be *comparable to the currently accepted linear dimensions of the corresponding structure.* This means that the *real dimensions of the structure are much larger,* than thought up to now.

We conclude that, for increasing r, smaller than the inversion distance r_{inv}, the gravitational repulsion of the negative internal mass is smaller than the gravitational attraction of the positive rest-mass. The matter accelerates inwards and eventually reaches an equilibrium situation due to the action of some kind of pressure (here, thermal pressure). At the inversion distance, r_{inv}, the gravitational repulsion and gravitational attraction cancel each other. Beyond the inversion distance an inversion occurs, namely, the gravitational repulsion dominates over the gravitational attraction, and matter accelerates outwards. Therefore, we conclude that, in general, *mass should exist also beyond the conventional limits of the structure.*

6. A Simplified Classical Treatment of the Problem of the Missing Mass

Today, the general cosmological belief is that the (dominant) dark matter coexists with its (subdominant, small) baryonic "contamination". The latter is tightly bounded to the former, and the two constitute the cosmological gravitating (perfect) fluid, in the sense that on top of a volume element of dark matter "sits" the baryonic (luminous) matter. So, in the spirit of CDE approach, the cosmological structures, e.g., the galaxies, should be treated as finite volume elements of the cosmological gravitating fluid in the context of a continuous Universe (in contrast to a collection of gravitating point masses (galaxies) at mutual distances large compared to their linear dimensions). In using the results of Section 3 (as in *Sections 10* and *12* below), we note that, generally, ρ is the total mass-energy density, including both the dark and the baryonic matter, and the equation of state $p(\rho) = \kappa\rho$, with κ being a positive constant, refers to the total pressure, namely, due to both the dark and baryonic masses. However, in this *Section* and for reasons of clarifying the physical importance of the results of *Sections 3-5,* we shall treat the source's matter density and pressure as simply of baryonic origin, and examine the relative importance of the (simply baryonic) mass-density, ρ, and the internal-mass density, ρ_i, in astronomical and cosmological structures, and its consequences.

Generally, all the physical parameters describing an astrophysical or cosmological source can be functions of both space and time. Here we shall limit ourselves to the case of physical parameters being explicitly functions of only the spatial coordinates. So, we assume that a gravitating fluid source (of *absolute temperature* T and *mean molecular weight* μ) is spherically-symmetric of radius R, and is characterized by an adiabatic and isothermal (adiabatic index $\gamma = 1$) equation of state, such that

$$p = \frac{kT}{\mu m_H}\rho \tag{6.1}$$

where k and m_H are, respectively, the *Boltzmann constant* and *rest-mass of the atomic hydrogen*, obviously compatible with the isentropicity conditions (3.5a,b). Furthermore, in choosing the baryonic-mass-density distribution law, we shall rely on the well-known observation-based "universal profile" of Navarro, Frenk and White [45], [46] and the Hernquist [47] profile (see also [48] – [50]) for the density distribution law of clusters of galaxies. So we shall adopt a special case of the above profiles, namely, the *Plummer-type density* described by

$$\rho(r) = \rho_0\left(1+\frac{r^2}{r_0^2}\right)^{-\frac{n}{2}} \quad (r_0 > 0,\ \rho_0 > 0,\ \text{n: a positive integer}) \tag{6.2}$$

Then using, additionally, the *Gauss- Stokes theorem,* from Eqs. (3.13) - (3.15) we find

$$m_i = -\frac{nkT}{G\mu m_H}\frac{R^3}{r_0^2\left(1+\dfrac{R^2}{r_0^2}\right)} < 0 \tag{6.3}$$

whence

$$m_i(R >> r_0) = -\frac{nkT}{G\mu m_H}R < 0 \tag{6.4}$$

Then, for n=3 and for every upper limit x_{max}, we find

$$\frac{-m_i}{m} = 1.946\times10^{-6}\frac{x_{max}}{\ln x_{max}}\frac{T_{(7)}}{\mu} \tag{6.5}$$

where

$$x = \frac{r}{r_0},\quad x_{max} = \frac{R}{r_0},\quad r_0 = 3R_s = 2.872\times10^{-5}M_{c(8)}pc \tag{6.6}$$

R_s, $T_{(7)}$ and $M_{c(8)}$ being, respectively, the *Schwarzschild radius* of the central dark object, the region's (constant) temperature in units 10^7 K and the mass-energy, M_c, of the central dark object in units of

10^8 $M_\odot$. Therefore, in this case, *the internal mass is definitely negative and, absolutely, it diverges with the source's linear dimensions.*

Under such conditions, Eq. (3.13) takes on the form

$$m = m_v + (-m_i) > m_v \tag{6.7}$$

Therefore, the baryonic mass m, determined with the aid of the geodesic motions in the gravitational potential U, is *larger* than the mass m_v, determined with the aid of the geodesic motions in the generalized gravitational potential V (for a proof of the above theorem in the relativistic case, see [33]). Obviously, it is the geodesic motions in the potential V that must be used rather than those in the gravitational potential U, because in the former (V) all the internal physical characteristics of the gravitating source are taken into account as sources of the observed motions. In other words, for simply physical reasons, as mass determined on the basis of geodesic motions, the mass m_v must be used, not the mass m. *But then the baryonic mass, m, of the (thermal) source is larger than the mass m_v determined observationally.*

We shall apply the above results to galactic and cosmological levels (for details see [17], [22] – [31], [37], [38]). Thus in the case of *masering galaxies* (e.g., NGC 4258 and NGC 1068) and *non-masering galaxies* (e.g., NGC 4261) we find that, depending on the linear dimensions of the *circumnuclear region* considered (ranging from the sub-parsec up to the kiloparsec), $-m_i$ is not always negligible compared to the mass M_c of the central dark object ($10^{-5} < -m_i / M_c < 10^{-2}$). On the other hand, $-m_i$ can be *comparable* to the total rest-mass, m, of the circumnuclear region ($10^{-3} < -m_i / m <$ 0.6; the upper bound 0.6 corresponding to the mass of the black hole believed to lurk in the Milky Way's nuclear region).

Specifically, in the case of the Milky Way, for the linear dimensions (R) of the region considered being in the range 0.1 pc and 100 kpc, the ratio $-m_i/m$ falls in the range (10^2 to 10^3) $T_{(7)}/\mu$. This means that, e.g., in the innermost region of dimensions 0.1 pc, this ratio can be *larger than* unity, provided that the temperature there is at least 10^8 K, and also that considering the broader region of dimensions 100 kpc, the same ratio can be at least 10^3 for temperature of the order 10^2 K.

It is very interesting that the above conditions can generally be met in the physical Universe. Thus, as mentioned in the *Section 1,* the interstellar medium in elliptical galaxies consists primarily of hot ($T > 10^6$ K) plasma; several molecular clouds with dimensions 20-50 pc and masses $\sim 10^6$ solar masses ($m_\odot$) are partly associated with HII regions and embedded in a lower-density interstellar medium; the masses ($\sim 10^8$ $m_\odot$) of the central dark objects in galaxies are only a few percent of the molecular-gas mass in the sub one kiloparsec region. *All the above data are in accordance with the modern aspect of a galaxy, different from that deduced on the basis of simply the galaxy's optical view.*

Analogously, in the case of typical clusters of galaxies with dimensions [11] of a few Mpc, the ratio $-m_i/m$ falls in the range (10^3 to 10^5) $T_{(7)}/\mu$ and so it can be larger than unity, provided that $T/\mu >$ 10^4 to 10^5 K. Again this is generally true, because, as mentioned in *Section 1*, a cluster of galaxies can be treated as an approximately isothermal sphere of hot ionized hydrogen (of dimensions a few Mpc, temperature $\sim 10^8$ K, number density of electrons $\sim 10^{-4}$ cm^{-3}, and number density of molecular clouds $\sim 10^{-4}$ to 10^{-3} cm^{-3}); the intracluster gas contains X-rays, fills the space between the galaxies, occupies much of the cluster's volume, and the X-ray luminosities fall in the range 10^{43} to 10^{45} erg. sec^{-1}; the mass of the intracluster medium exceeds the total mass of the luminous parts of the cluster's galaxies by a factor of several.

Finally, in the case of a second-order cluster (cluster of clusters or super-cluster) of galaxies, the ratio $-m_i/m$ is of the order of 10^3 $T_{(7)}/\mu$, and so it can exceed unity, for $T/\mu > 10^3$ K, a condition that again is met (notice, however, that current observations do not seem to support the idea of a spherically-symmetric super-cluster of galaxies; on this, see also, especially, [23], [28]).

r_{max}	$-m_i/m$	$-m_i/m \geq 1$
0.1 pc	$0.053\ T_{(7)}/\mu$	$\frac{T_{(7)}}{\mu} \geq 18.831$, $\frac{T}{\mu} \geq 1.883 \times 10^8 K$
1 pc	$0.449\ T_{(7)}/\mu$	$\frac{T_{(7)}}{\mu} \geq 2.223$, $\frac{T}{\mu} \geq 2.223 \times 10^7 K$
100 pc	$34.453\ T_{(7)}/\mu$	$\frac{T_{(7)}}{\mu} \geq 2.903 \times 10^{-2}$, $\frac{\mathrm{T}}{\mu} \geq 2.903 \times 10^5 K$
2 kpc	$597.979\ T_{(7)}/\mu$	$\frac{T_{(7)}}{\mu} \geq 1.672 \times 10^{-3}$, $\frac{\mathrm{T}}{\mu} \geq 1.672 \times 10^4 K$
15 kpc	$4.119 \times 10^3 T\ T_{(7)}/\mu$	$\frac{T_{(7)}}{\mu} \geq 2.428 \times 10^{-4}$, $\frac{\mathrm{T}}{\mu} \geq 2.428 \times 10^3 K$
30 kpc	$8.012 \times 10^3\ T_{(7)}/\mu$	$\frac{T_{(7)}}{\mu} \geq 1.248 \times 10^{-4}$, $\frac{\mathrm{T}}{\mu} \geq 1.248 \times 10^3 K$
100 kpc	$2.549 \times 10^4\ T_{(7)}/\mu$	$\frac{T_{(7)}}{\mu} \geq 3.922 \times 10^{-5}$, $\frac{T}{\mu} \geq 3.921 \times 10^2 K$
200 kpc	$4.969 \times 10^4\ T_{(7)}/\mu$	$\frac{T_{(7)}}{\mu} \geq 2.012 \times 10^{-5}$, $\frac{T}{\mu} \geq 2.012 \times 10^2 K$

Table 2: Relative importance $-m_i/m$, of the negative internal mass, m_i, and the positive rest mass m, for the Milky Way: $-\frac{m_i}{m} = 1.946 \times 10^{-6} \frac{x_{\max}}{\ln x_{\max}} \frac{T_{(7)}}{\mu}$, $M_{(8)} = 10^{-2}$, $r_0 = 2.872 \times 10^{-7} pc$

From all the above we conclude that, using the geodesic motions (inside and outside of) the cosmological structures for the observational determination of masses, in conjunction with the CDE approach, the extra (negative) mass m_i (which is meaningless in the context of the standard geodesic motion (3.1) of a test particle) appears in the observationally determined dynamical mass and, in many cases, it can largely exceed the mass of the gas of the (corresponding region of the) large-scale cosmological structure under consideration. *We consider this as an interesting contribution towards clarifying the concept of the missing mass.* The above results, for different values of M (and r_0), are shown analytically in *Tables 2 and 3.*

A typical cluster of galaxies: $-\frac{m_i}{m} = 1.946\times10^{-6}\frac{x_{\max}}{\ln x_{\max}}\frac{T_{(7)}}{\mu}$, $r_0 = 2.872\times10^{-5}M_{(8)}pc$

$M_{(8)}$	$r_{\max}$	$-m_i/m$	$-m_i/m\geq1$
10	1 Mpc $\frac{T_{(7)}}{\mu}\geq 3.582\times10^{-4}$,	$2.792 \times 10^3\ T_{(7)}/\mu$ $\frac{T}{\mu}\geq 3.5821\times10^3\ K$	
10^2	1 Mpc $\frac{T_{(7)}}{\mu}\geq 2.903\times10^{-2}$,	$34.453\ T_{(7)}/\mu$ $\frac{T}{\mu}\geq 2.903\times10^5\ K$	

A typical 2nd order cluster of galaxies : $-\frac{m_i}{m} = 1.946\times10^{-6}\frac{x_{\max}}{\ln x_{\max}}\frac{T_{(7)}}{\mu}$, $r_0 = 2.872\times10^{-5}M_{(8)}pc$

$M_{(8)}$	$r_{\max}$	$-m_i/m$	$-m_i/m\geq1$
10^2	100 Mpc $\frac{T_{(7)}}{\mu}\geq 3.582\times10^{-4}$,	$2.7917 \times 10^3\ T_{(7)}/\mu$ $\frac{T}{\mu}\geq 3.582\times10^3 K$	

Table 3: Relative importance $-m_i/m$, of the negative internal mass, m_i, and the positive rest mass m, for: (i) A typical cluster of galaxies and (ii) a typical 2nd order cluster of galaxies.

Finally, we stress that our method has been based on the assumption that only the baryonic density is considered, while the density of dark matter is ignored. However, our method can be generalized, in the spirit of *Sections* 10 and 12 below, so as to take into account both the above densities. This subject is currently under scrutiny.

7. Dynamical Masses and Flat Rotation Curves of Disc Galaxies

In this *Section*, we shall demonstrate that *the flat rotational curves of the disk galaxies* can naturally be explained, based on the generalized Euler's equations of motion (3.6), in which ρ is again simply the baryonic density. Thus, the case of a spherically-symmetric perfect-fluid source of generalized density $\rho_v(r)$, the velocity, $\upsilon_f(r)$, of an equatorial, circular geodesic orbital motion at a distance r from the center is

$$\upsilon_f^{\,2}(r) = \frac{Gm_v(r)}{r} = \frac{G}{r}\left[m(r)+m_i(r)\right] \tag{7.1a}$$

replacing the standard one (on the basis of the equations of motion (3.1))

$$\upsilon^2(r) = \frac{Gm(r)}{r}, \tag{7.1b}$$

where, we emphasize, that, by definition, the masses m(r) and $m_v(r)$ are the volume integrals of the corresponding densities ρ(r) and ρ_v(r), manifesting themselves in the corresponding Poisson equations, (3.2) and (3.8).

We recall that, for a homogeneous source of density ρ, Eq.(7.1b) reduces to

$$\upsilon(r) = \left(\frac{4\pi G\rho}{3}\right)^{\frac{1}{2}} r \tag{7.2}$$

Eq. (7.2), with the exception of the circumnuclear region, is in contradiction with observations. On the other hand, for ρ_v to be different than ρ, an isothermal gravitating perfect-fluid cannot be homogeneous. Moreover, as deduced from the *21 cm* radiation, in disk galaxies the HI clouds, at different distances from the center of the galaxy, all orbit at more or less the same speed. This can be explained, in the present context, however, if we assume a non-homogeneous, spherically-symmetric source and use Eq. (6.2), whence

$$m(r) = 4\pi\rho_0 r_0^3 \left[\frac{1}{\sqrt{2}} - \ln\left(1+\sqrt{2}\right) - \frac{\frac{r}{r_0}}{\left(\frac{r^2}{r_0^2}+1\right)^{\frac{1}{2}}} + \ln\left(\frac{r}{r_0} + \sqrt{\frac{r^2}{r_0^2}+1}\right) \right] \tag{7.3a}$$

$$\cong 4\pi\rho_0 r_0^3 \ln\frac{r}{r_0} \quad \text{for } r >> r_0 \tag{7.3b}$$

If, additionally, we assume the equation of state, Eq. (6.1), we find, in analogy to Eqs. (7.3) (for *n*=3),

$$m_i(r) = -\frac{3kT}{G\mu m_H} \frac{r^3}{r_0^2\left(1+\frac{r^2}{r_0^2}\right)} \tag{7.4}$$

Inserting Eqs. (7.3) and (7.1a) into Eq. (7.8a) and using also the condition (5.4a,b) we find

$$\frac{\upsilon_f^2(x)}{c^2} = \sqrt{2}\frac{1}{x}\ln\left(1+\sqrt{x^2+1)}\right) - \frac{\sqrt{2}}{\sqrt{x^2+1}} - \frac{1}{x}\left[\ln\left(1+\sqrt{2}\right) - \frac{1}{\sqrt{2}}\right] - \gamma\frac{x^2}{x^2+1} \tag{7.5}$$

where we have put

$$x = \frac{r}{r_0} \quad , \quad \gamma = \frac{3kT}{\mu m_H c^2} = 2.75\times10^{-13}\frac{T}{\mu} \tag{7.6}$$

The interest in Eq. (7.5) is, among others, in the explicit dependence of $\upsilon_f(x)$ on the distribution's temperature.

A more precise expression, however, for the velocity $\upsilon_f(x)$, to be used for comparison with observational data, should take into account the mass, M_c, of the central dark object, beyond of the density ρ_v. The velocity $\upsilon_{BH}(r)$, due to the mass M_c, of an equatorial, circular geodesic orbit at the distance, from the center is

$$\upsilon_{BH}^2 = \frac{GM_c}{r} = \frac{GM_c}{r_0 x} = \frac{GM_c}{3R_s x} = \frac{GM_c}{3\times\frac{2GM_c}{c^2}x} = \frac{c^2}{6x} \tag{7.7}$$

So, the total circular velocity, V_c, is

$$V_c = \left(\upsilon_f^2 + \upsilon_{BH}^2\right)^{1/2}$$

$$= c\left\{\begin{array}{l}\sqrt{2}\frac{1}{x}\ln\left(1+\sqrt{x^2+1}\right) - \frac{\sqrt{2}}{\sqrt{x^2+1}} \\ -\frac{1}{x}\left[\ln(1+\sqrt{2}) - \frac{1}{\sqrt{2}}\right] - \gamma\frac{x^2}{x^2+1} + \frac{1}{6x}\end{array}\right\}^{1/2} \tag{7.8a}$$

or

$$V_c\left(100\frac{km}{s}\right) = 2.989\times10^3\left\{\begin{array}{l}\sqrt{2}\frac{1}{x}\ln\left(1+\sqrt{x^2+1}\right) \\ -\frac{\sqrt{2}}{\sqrt{x^2+1}} - \frac{1}{x}\left[\ln\left(1+\sqrt{2}\right) - \frac{1}{\sqrt{2}}\right] - \gamma\frac{x^2}{x^2-1} + \frac{1}{6x}\end{array}\right\}^{1/2} \tag{7.8b}$$

or, equivalently,

$$V_c = 2.989\times10^5\left\{\sqrt{2}\frac{1}{x}\ln\left(1+\sqrt{x^2+1}\right) - \frac{\sqrt{2}}{\sqrt{x^2+1}} - \frac{1}{x}\left[\ln\left(1+\sqrt{2}\right) - \frac{1}{\sqrt{2}}\right] - \frac{3k_BT}{\mu m_H c^2}\frac{x^2}{x^2-1} + \frac{1}{6x}\right\}^{1/2} km/s \tag{7.8c}$$

In the form (7.8a) or (7.8b) or (7.8c), the velocity V_c is, generally, preferable than in the form (7.5), especially close to the central dark object. In *Figure 1*, the form of the velocity V_c is plotted as a function of x, for $T/\mu \sim 10^2$. The expected flat character of the rotation curve is prominent, resembling the case of the Milky Way ($V_c \cong 200\ km/s$). Similar results are presented in *Figure 2* (the group of curves in the middle), for several values of the temperature, namely, $T/\mu \sim 1.82 \times 10^6$ ($\gamma \sim 5.01\times10^{-7}$), $T/\mu \sim 1.45\times10^6$ ($\gamma \sim 3.99\times10^{-7}$), $T/\mu \sim 10^5$ ($\gamma \sim 2.75\times10^{-8}$), $T/\mu \sim 10^4$ ($\gamma \sim 2.75\times10^{-9}$) and, finally, $T/\mu \sim 10^3$ ($\gamma \sim 2.75\times10^{-10}$). The fourth case, in particular, is in good agreement with, e.g., the typical velocity-curve for a *Sc* spiral galaxy [9]. *The main difference, beyond the different scale, between the theoretically-derived velocity curve and the observed one, refers to the points close to the center, which are not considered in the theoretical treatment.*

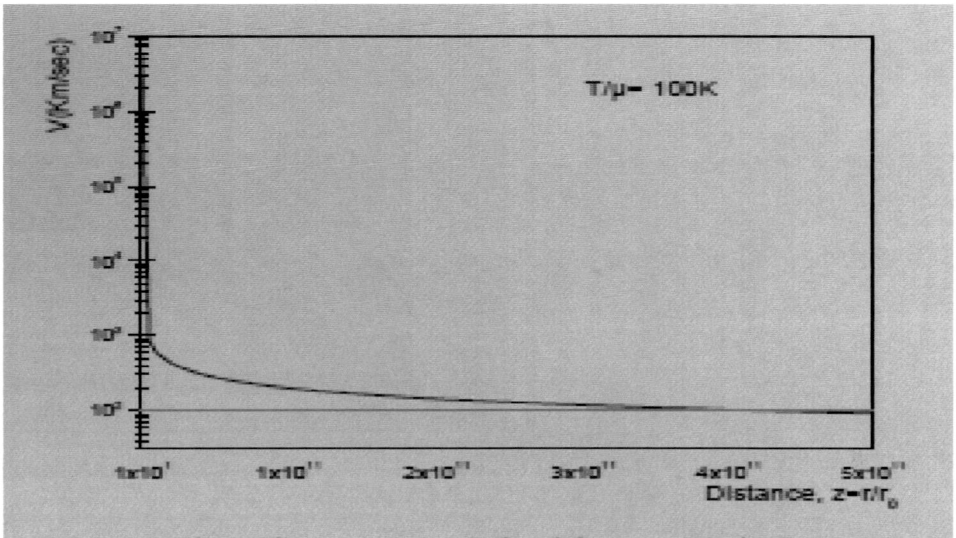

Fig. 1: The theoretically derived "flat" rotation curve in the Milky Way

It is remarkable that the "flat" part of the rotation curve correspond to the same range of values of x ($>10^{10} >> 1$) and, generally speaking, for such large values of x ($>10^{10} >> 1$), only a slight (but not vanishing) dependence of V_c on the temperature is noticed. Also, in our case, the theoretically derived flat rotation curve continues up to the distance of at least 200 kpc ($x \geq 10^{12}$). In the case of the Milky Way galaxy, such a distance (beyond the Magellanic Clouds!!!) defines the galaxy's real dimensions, as contrasted to its conventional radius ~15 kpc [1]. (For an independent method of determination of the true linear dimensions, applied in the case of the Solar System and based on the vanishing of the generalized acceleration $\vec{\nabla} V$ (Eq. (3.6)), see Section 8 below; and also Table 1 for the *inversion distance*). *We consider the results of Sections 6 and 7 as a partial classical solution to the dark-matter problem.* Obviously, as with *Section 6,* our method can be generalized, in the spirit of *Sections 10* and *12* below, so as to include both the dark-matter and the baryonic densities. This subject is currently under scrutiny.

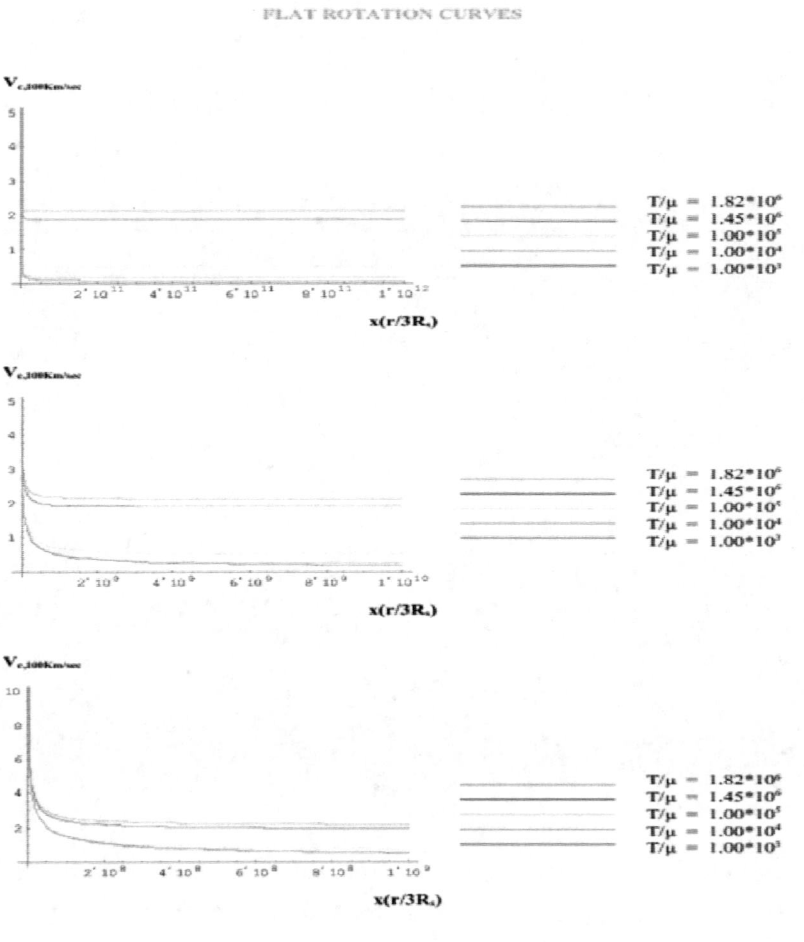

Figure 2: Flat rotation curves for several values of the temperature, in large-scale structures.

8. A Classical Treatment of the Pioneer-Anomaly Effect in the Solar System (with K. Zagkouris)

The *Pioneer 10* and *11* space probes have been launched in 1970s and are currently travelling outwards our Solar system. There has been a measured deviation in their (and other space probes in the Solar System) speeds, which shows a small *extra* inward acceleration towards the Sun, amounting in, approximately, the above space probe (and not only them), during their lifetime up to now, "loosing" a distance approximately equal to the distance of the Moon from the Earth. This extra acceleration has been given the name the *Pioneer Anomaly Effect*. Currently, no universally accepted explanation exists for this anomaly. In order to solve this mysterious acceleration, many theories have been proposed, most of which concentrate in observational errors, recording errors, new gravity sources in the Solar system and, finally, even new physics.

We studied the *Pioneer Anomaly Effect* in the framework of the CDE approach. In our study we use a model of the Solar System with the spherically-symmetric (or, equivalently, point-mass) Sun of mass M_c at its centre and an extra mass density (generally, both baryonic and dark matter) dispersed around it. The mass-density profile is described by a Plummer-type equation (see Eq. (6.2)). In the numerical applications, just for simplicity, we treat the extra, dispersed mass as a baryonic isothermal perfect fluid source (nevertheless, this may not be necessary, because the same steps can be followed for any mixture of dark matter and baryonic matter). For this system, the generalized Euler's hydrodynamic equations for the spherically-symmetric total acceleration of the isentopic motion are written in the form

$$\begin{aligned}\vec{\gamma}(r) &= -\frac{G}{r^2}\left[M_c + m_V(r)\right] \\ &= \vec{\gamma}_c + \vec{\gamma}_i \qquad (8.1) \\ &= -\frac{GM_c}{r^2}\hat{r} - \left[4\pi G\rho_0 r_0^2 \frac{\ln\left(r/r_0\right)}{r/r_0} - 3k\right]\frac{\hat{r}}{r}\end{aligned}$$

the two parts on the r.h.s. being due to, respectively, the central spherically-symmetric Sun and the extra mass (fluid) dispersed around it.

Furthermore, recalling that in the CDE approach the fluid volume element moves as a test particle, in accordance with the generalized laws (3.6) and (3.7), namely, but, now, *carrying along all the physical characteristics of the fluid source*, we identify the (Pioneer or other) space probe with the above volume element-test particle, moving as above.

From our analysis, applied to a baryonic and isothermal mass distribution, dispersed around the central Sun, we end up we some interesting results two of which are worth mentioning here. The first result is that we can actually evaluate an extra acceleration at the region of the Pioneer space probe and the numerical results are in agreement with the observed data, namely, the extra acceleration evaluated in our analysis is equal to $\gamma = -8.74311\ 10^{-8}\ \mathrm{cm\,s^{-2}}$, while the measured extra acceleration is $\gamma = -8.74 \pm 1.33\ 10^{-8}\ \mathrm{cm\,s^{-2}}$. *We consider the above result as a quite acceptable classical explanation of the celebrated Pioneer Anomaly Effect in the Solar System.*

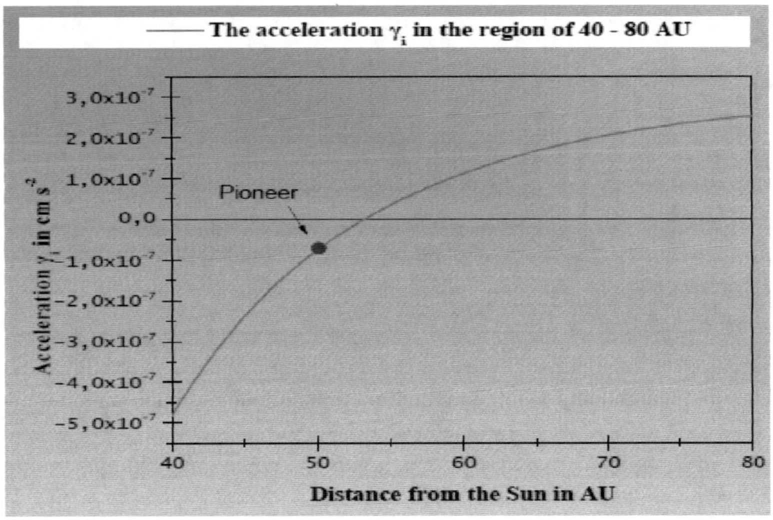

Figure 3: The extra acceleration γ_i as a function of the distance from the Sun.

The behaviour, with the distance from the Sun, of the evaluated extra acceleration is shown in *Figures 3* and *4*, and the corresponding result for the total acceleration is shown in *Figure 5*. The extra acceleration is an inward acceleration up to certain distance, after which it becomes an outward acceleration. In contrast, the total acceleration is always an inward acceleration vanishing at some far distance, R. This last result permits the first theoretical analytical determination of the *true linear dimensions of the Solar System,* R, as the solution of the algebraic equation (for r) $\gamma = 0$, for r = R.

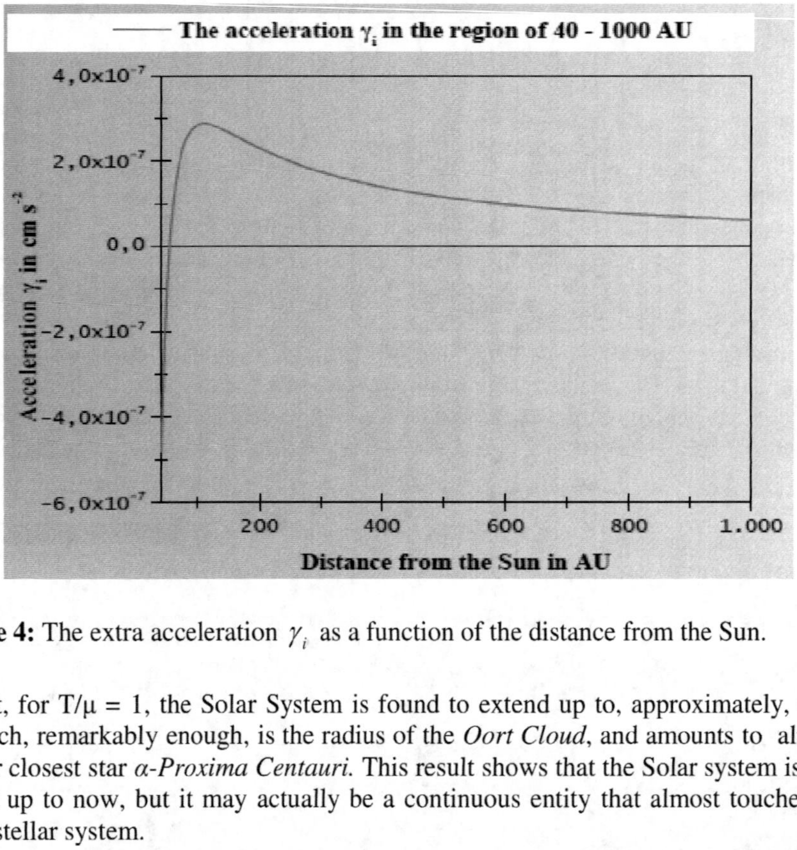

Figure 4: The extra acceleration γ_i as a function of the distance from the Sun.

In fact, for T/μ = 1, the Solar System is found to extend up to, approximately, the distance 100 kAU, which, remarkably enough, is the radius of the *Oort Cloud*, and amounts to almost half the distance to our closest star *α-Proxima Centauri.* This result shows that the Solar system is not as small as we thought up to now, but it may actually be a continuous entity that almost touches the nearest neighbouring stellar system.

Finally, we note that our model permits also the evaluation of thermodynamic parameters, like the internal thermodynamic energy (per unit mass), Π, whose values at the surface of the Sun is $\Pi(r_0) = 4.01\times10^9 cm^2 s^{-2}$ and at the Solar System's boundary is $\Pi(R) = 1.43\cdot10^8 cm^2 s^{-2}$. Of course, our model permits the analogous treatment of other space probes in the Solar System (e.g., *New Horizons).* For further details and results on the applications of the CDE approach to the Solar System, see [36].

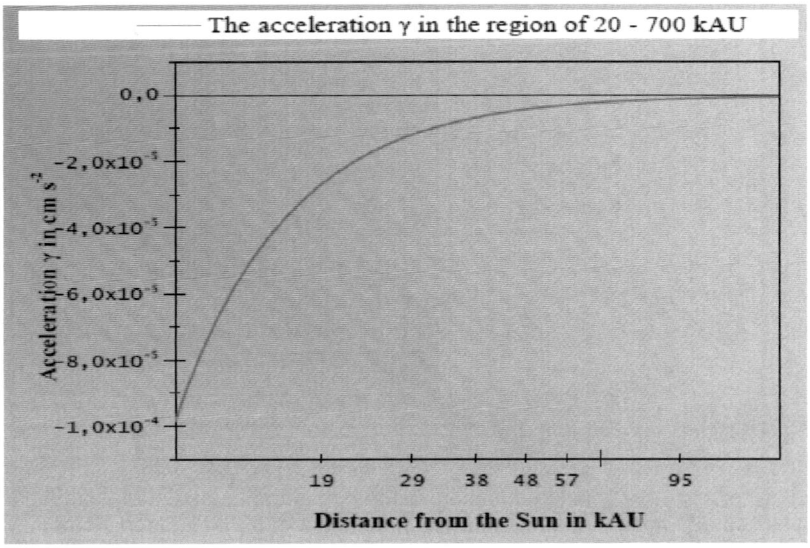

Figure 5: The total acceleration γ as a function of the distance from the Sun.

9. A possible theoretical explanation of the formation mechanism of winds, jets, supergiant stars, and supernovae

As described in Sections 4 and 5, the generalized gravity can be repulsive, provided that the generalized density ρ_V is negative. In this way, using the notion of the *inversion distance*, we can describe and justify the outwards acceleration necessary for the formation of winds and jets. Also, possible applications to the phenomena of the expansion of a red-giant star and a supernova are considered. This subject is, currently, under scrutiny.

10. Determination of the masses of clusters of galaxies (with M. Plionis and S. Basilakos)

Of obvious special cosmological importance and interest is the determination of the masses of the globular clusters of galaxies. Usually, the total gravitating mass (dark matter and baryons) of a typical cluster of galaxies is classically being treated using the hydrodynamic properties of the inter-cluster gas. More specifically (Plionis, *private communication*), for a cluster in hydrostatic equilibrium, from the combined use of the Poisson equation for the total gravitational potential (due to the total mass density, namely, dark matter and baryons), an equation of state for the gas (namely, the total pressure due only to the density of the gas – i.e., its baryonic mass), and an assumed Plummer-type gas-density profile (with, e.g., n = 3), all plugged into the Euler equation of hydrostatic motions, the hydrodynamic mass and the total mass-density are evaluated. Similar results are obtained for the same perfect-fluid cluster, using the idea of the CDE approach.

The latter method can be generalized for a non-static cluster of galaxies, so as to include the total mass-density and the total pressure of the non-static source, namely, considering the contributions to them of both the dark matter and the baryonic matter, and, additionally, using for the non-static source the notions of the spectral shift and the generalized dynamical mass. More precisely, assuming an equation of state in terms of the total density and total pressure,

$$p = w\rho c^2, \quad w > 0 \qquad (10.1)$$

we evaluate the generalized density ρ in terms of the mass density of the baryonic mass

$$\rho_V = \rho + \frac{wc^2}{4\pi G}\left[\vec{\nabla}\cdot\left(\frac{\vec{\nabla}\rho}{\rho}\right)\right], \qquad m_V = \int \rho_V d^3x \qquad (10.2a, b)$$

Then, according to an assumed total-mass-density profile, by its volume integration, we evaluate the generalized mass $m_v = \int_V \rho_v d^3x$. Independently, this same generalized (namely, the dynamical) mass, under the assumption of circular equatorial motions is directly related to the spectral shift, z,

$$z^2 = \frac{Gm_V(r)}{c^2 r} \tag{10.3}$$

immediately obtained from observations.

Therefore the whole game here, finally, is an interplay of the (assumed) constants appearing in the equation of state and the assumed total-mass-density profile. Next, using an observationally derived baryonic-gas-profile law (see also [46]), we can, by subtraction, evaluate the dark-mass-density profile and, then, by the latter's volume integration, evaluate the dark mass itself, and compare it with observational data, thus limiting the values of the constants appearing in the equation of state and the assumed total-mass-density profile. This subject is under scrutiny currently.

11. CDE in classical magneto hydrodynamics (with C. G. Tsagas)

The validity of the CDE approach in classical hydrodynamics (Section 3) has been extended into the realm of classical magneto hydrodynamics (MHD). This can be accomplished as follows. First, by a suitable redefinition of the pressure of the magnetized perfect-fluid source, so as to include also the *magnetic pressure,* the MHD equations of motion are put in the form of the Euler's hydrodynamic equations of motion. The above result is possible not for an arbitrary magnetic field B, but only for those characterized by a *"zero-curl Lorentz force".* Then, by a suitable redefinition of the internal thermodynamic energy of the magnetized fluid and of the condition of the adiabatic-isentropic motions, the above derived Euler's equations are put in the form of Newton's gravitational equations of motion. Specifically, the MHD equations of motion are written as

$$\frac{d\vec{\upsilon}}{dt} = \vec{\nabla} V, \quad \nabla^2 V = -4\pi G \rho_V, \quad \rho_V = \rho + \rho_{i(P)} + \rho_{i(B)} \tag{11.1a,b,c}$$

with the magnetically generalized mass-density (third equation above) containing also a magnetic contribution $\rho_{i(B)}$, which becomes important in the central region of a magnetized large- scale cosmological structure (for further details, see [34], [35]; see also [18], [20]).

12. Cosmology, collisional dark matter and CDE (with K. Kleidis)

This *Section 12* is a short presentation of the talk, at the same Conference, given by the first of the authors (K. Kleidis; *A Conventional Form of Dark Energy*; see the detailed program of the talk(s) at the official sight of the Conference (http://neb14.physics.uoi.gr/index.files/program.htm).

The current cosmological picture, known as the *concordance model* [51], includes two major unresolved issues: (i) According to the observational data on the temperature-variations of the cosmic microwave background, the Universe can be described, adequately, by a *spatially-flat* Robertson-Walker (RW) model (see, e.g., [52]) and, therefore, it must contain a considerably larger amount of energy, than the equivalent to the *total rest-mass density* of its matter content does. (ii) The cosmologically-distant indicators (*standard candles*) appear to be *dimmer* than expected [53], [54], something which has been accommodated by the assumption that, recently, the Universe entered into a phase of *accelerated expansion* [55] – [57].

In order to compromise the above mentioned observational results, as well as to deal with them into a unified theoretical framework, an extra, *dark energy* (DE) component, of *negative pressure*, has been introduced [58], occasionally referred to as *quintessence* [59]. However, there may be also another, *more conventional explanation*, which can be revealed, most appropriately, with the aid of the *conformal dynamical equivalence* technique [17], [27], [33], [35].

Nowadays, a lot of accumulated evidence suggests that, more than 85% (by mass) of the matter in the Universe consists of non-luminous and non-baryonic material [60]. Its name, *dark matter* (DM), reflects our ignorance on the exact nature of this constituent. Although we do not know for certain how the DM came to be formed, a sizeable relic abundance of *weakly interacting massive particles* (WIMPs) is generally expected to have been produced as a by-product of the Universe's hot youth [61]. Apart from their exact nature, the scientific community used to argue that, the WIMPs should be *collisionless* [62], [63]. However, many recent results from *high-energy-particle tracers*, such as the ATIC [64] and PAMELA [65], combined with those of the *Wilkinson microwave anisotropy probe* (WMAP) survey [66], have revealed an unusually-high *electron - positron* production in the Universe, much more than what is anticipated by *supernovae explosions* or *cosmic-ray collisions*. These results have led many scientists to argue that, among *the best candidate-sources of these high-energy events are the annihilations of WIMPs* (see, e.g., [67] – [76], for an extensive, though incomplete list), i.e., that the DM constituents can be *slightly collisional* [77] – [80]. If this is true, it could affect our perception on the nature of the DE (in connection, see, e.g., [81] – [85]).

Based on the perception that the DM constituents can be slightly collisional, we have assumed that, the matter-energy content which drives the evolution of the Universe (being modeled by a spatially-flat RW space-time) at every post-recombination epoch, is in the form of a *perfect fluid with positive pressure,* $p = w\rho c^2$, where ρ is the *rest-mass density* of the Universe matter content and w is a *dimensionless constant.* Now, together with all the other physical characteristics, *the energy of this fluid's internal motions,* Π, is (also) taken into account as a source of the universal gravitational field and the total energy-density of the Universe matter-energy content is written in the form (2.4). The dynamical evolution of this model is governed by the Friedmann equation with vanishing cosmological constant, which is written in the form

$$\left(\frac{H}{H_0}\right)^2 = \Omega_M \left(\frac{S}{S_0}\right)^3 \left[1 + \frac{\Pi_0}{c^2} + 3wln\left(\frac{S_0}{S}\right)\right]. \tag{12.1}$$

In Eq. (12.1), $H = \frac{S'}{S^2}$ is the *Hubble parameter,* $S(\eta)$ is the *scale factor* in terms of the *conformal time,* η, and a prime denotes differentiation with respect to η. Moreover, the constants H_0, S_0 and Π_0 are considered as denoting the corresponding present-time values, while $\Omega_M = \frac{\rho}{\rho_c} \approx 0.3$ is the *rest-mass density parameter*, which measures ρ in terms of the corresponding *critical* quantity, $\rho_c = \frac{3H_0^2}{8\pi G}$, with G being Newton's universal constant of gravitation. At the present epoch, we obtain

$$\Pi_0 = \left(\frac{1}{\Omega_M} - 1\right)c^2. \tag{12.2}$$

This result is quite interesting: It suggests that, in principle, the energy of the internal motions of the collisional-DM fluid can account for the (extra) DE, so that, at the present epoch, the *total-density parameter* is

$$\Omega = \frac{\varepsilon_0}{\varepsilon_c} = \frac{\rho_0 c^2}{\rho_c c^2} + \frac{\rho_0 \Pi_0}{\rho_c c^2} = 1, \tag{12.3}$$

where $\varepsilon_c = \rho_c c^2$ is the energy density equivalent to ρ_c. Accordingly, we have been able to determine the *"correct"* form of the scale factor, which (under the assumption that the DM is collisional) governs the evolution of the Universe, as follows

$$S(\eta) = S_0 \left(\frac{\eta}{\eta_0} \right)^{\frac{2}{1+3w\Omega_M}}, \tag{12.4}$$

which, for $w \neq 0$, is the natural generalization of the *Einstein-de Sitter (EdS) model*. Next, we have attempted to determine what is realized by someone who, although living in a collisional-DM model, insists in adopting the (traditional) collisionless-DM approach.

To do so, we have applied the *conformal dynamical equivalence* technique [17]. With the aid of this technique, we have found the (*conformal*) transformation, which relates the collisional-DM description of a cosmological model (the scale factor $S(\eta)$) to the corresponding collisionless-DM approach (i.e., to the scale factor of a pressureless Universe, $R(\eta)$). In terms of the cosmological redshift, the corresponding *conformal factor* is written in the form

$$f(z) = 1 + w\Omega_M [1 + 3ln(1+z)]. \tag{12.5}$$

With this *"tool"* at hand, we have explored the way that, a supporter of the collisionless-DM scenario interprets the observations carried out in a collisional-DM Universe. The result of the *"debate"* between collisional- and collisionless-DM approach is, definitely, in favor of the former. In particular, in the collisional-DM Universe there is a *characteristic value*, z_c, of the cosmological redshift, above which, the luminosity distance of the various light-emitting sources becomes larger than what is realized by an observer who treats the DM as a pressureless fluid. In other words, from the point of view of someone who (although living in a collisional-DM model) insists in adopting the (traditional) collisionless-DM approach, the cosmologically-distant indicators, located at $z > z_c$, seem to *lie farther* (i.e., they appear to be *dimmer*) than expected. Clearly, the similarity between the characteristic value z_c, and the (observationally-determined) *transition redshift*, z_t, that signals the onset of dimming of the SNe Ia standard candles [56], is more than obvious. On the other hand, after the thermodynamical content of a collisional-DM fluid is taken into account, the theoretical curve representing the *distance modulus*, $\mu(z)$ (green solid line), fits the *Hubble diagram* of an extended sample of SN Ia standard candles [86] to high accuracy (cf. Fig. 6), in contrast to the corresponding collisionless-DM quantity, $\tilde{\mu}(\tilde{z})$, being either appropriately (orange solid line) or falsely (dashed line) expressed in terms of the *truly measured quantity,* z. At the same time, as far as a supporter of the collisionless-DM scenario is concerned, the Universe appears to be *either accelerating or decelerating*, depending on the value of the cosmological redshift.

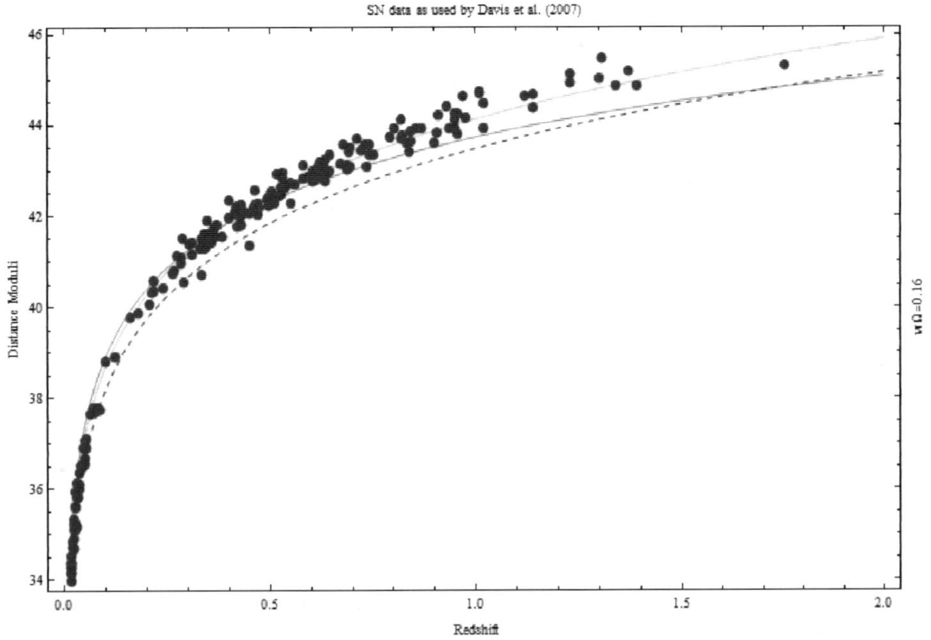

Figure 6: The theoretical curve, representing the distance modulus as a function of z in the collisional-DM model (green line), plotted against the corresponding collisionless-DM quantity (orange line), on the Hubble diagram of the SN Ia dataset used by Davis et al. [86].

At the same time, as far as a supporter of the collisionless-DM scenario is concerned, the Universe appears to be *either accelerating or decelerating*, depending on the value of the cosmological redshift. In this case, the quantity w, which, in the collisional-DM approach, parameterizes the various *isothermal flows*, plays also another (more interesting) role: As we have found, for $w \geq 0.24$, there exists a (theoretically-determined) *transition value*, $\tilde{z}_t$, of the (collisionless-DM-oriented) cosmological redshift, $\tilde{z}$, such that, for $\tilde{z} < \tilde{z}_t$, the (correspondingly-oriented) *deceleration parameter*, $\tilde{q}$, becomes *negative*, i.e., from the point of view of someone who adopts the (traditional) collisionless-DM approach, the Universe is *accelerating*. Accordingly, taking into account the observational result that, the transition redshift between accelerated and decelerated expansion is set at the value $z_t \approx 0.46$ of the *truly measured quantity* z, we have determined the exact value of w, for which the collisional-DM approach to the post-recombination Universe is compatible with observations, namely $w \approx 0.31$. This result implies that, compatibility of the collisional-DM approach with the observational data currently available, suggests that, the DM itself consists of *relativistic particles*, i.e., $w \approx \frac{1}{3} = 0.33$.

In conclusion, the assumption that the DM constituents can be *both collisional and relativistic*, could provide a reasonable and conventional explanation for some open aspects of modern Cosmology, including:

(i) The extra (dark) energy needed to flatten the Universe: It can be compensated by the energy of the internal motions of the collisional-DM fluid.

(ii) The observed dimming of the SNe Ia standard candles and the apparent accelerated expansion of the Universe: Both of them can be due to the misinterpretation of several cosmologically-relevant parameters by those observers who, although living in a collisional-DM Universe, insist in adopting the collisionless-DM approach.

Therefore, before inventing any new theory, it is useful to allow for a suitable use of the (so far) *neglected degrees of freedom* (hydrodynamic flows, pressure, energy of the internal motions, etc.). As we have shown, these internal physical characteristics can reveal their influence on several parameters of cosmological significance (scale factor, cosmological redshift, luminosity distance, Hubble and deceleration parameters) and yield a consistent alternative to the currently-accepted DE concept.

We consider our proposed cosmological model, with all its advantages and disadvantages, as a quite reasonable classical interpretation-explanation of the observational cosmological data and of the non-necessity of notions like dark energy, cosmological constant and accelerating universal expansion. This, in short, is expressed as: ***"Dark Matter: Yes! Dark Energy: No!"*** (for further details, see [37], while, for a slightly different approach, see [23]).

13. Discussion, Concluding Remarks, and Outlook

We have presented the theoretical concept of the "Conformal Dynamical Equivalence" approach, and used it for treating, in a unified and classically (namely, in a non-quantum) way, some interesting problems at various astrophysical levels (Sun, Solar System, Stars, Galaxies, Clusters of Galaxies, Universe as a whole).

The dynamical equivalence refers to the formal functional similarity between the adiabatic-isentropic hydrodynamic flows in a *physical* bounded gravitating perfect-fluid source and the geodesic motions in a *virtual,* fully defined one; and such problems are related to the celebrated *Pioneer-Anomaly Effect in the Solar System, missing mass, dynamical mass, flat-rotation curves of disc galaxies, masses of magnetized and/or non-magnetized large-scale cosmological structures, possible formation mechanisms of winds and jets, dark matter, repulsive/attractive gravity, non-necessity of the cosmological constant, the dark energy,* and the *accelerating expansion* of the Universe, as a whole.

The essential background idea at the heart of the classically (i.e., not quantum-mechanically oriented) unifying CDE approach is that, all the *internal physical characteristics* of the source-cosmological structure (beyond its rest-mass density) are taken into account as sources of geodesic motions. Treating the internal physical characteristics in this way was a known feature of the general-relativistic theory of gravity for geodesic motions, but not of the Newtonian one. Here, based on the CDE approach, we proved that, in the Newtonian theory of gravity, the internal physical characteristics can act as sources of geodesic motion, and examined some physical consequences of this fact. The interest in this fact stems from the methods and assumptions made in estimating the masses of astrophysical, galactic, and cosmological objects. In such mass determinations, observational techniques are used, which are based on the assumption of geodesic motions.

According to recent observational data, large amounts of hot gas exist in hydrodynamic (more generally, hydro-magnetic) flows, in, e.g., AGNs, or the extended haloes of galaxies, or in clusters and super clusters of galaxies and their haloes. In spite of this, however, the observational determination of e.g. the central masses and of the total masses of galaxies, and of clusters and super clusters of galaxies is based on the assumption of *purely geodesic motions,* namely, motions under the effect of gravity alone. However, obviously, in the presence of non-gravitational sources, due to, say, pressure gradients, viscosity or an inhomogeneous magnetic field, *the standard geodesic motion in a purely gravitation field is no longer sustainable and adequate, and the hydrodynamic (or, more generally, hydro-magnetic) description is more appropriate.* Therefore the emerging question is *how appropriate the standard mass measurements are*. In exactly this spirit we used the idea of the dynamical equivalence, and, for various cosmological structures, we took into account all the source's internal physical characteristics as sources of geodesic motions, we evaluated various forms of mass (*rest mass, baryonic mass, internal mass, total mass, dynamical mass*), and explored their relative importance. Thus, for the isothermal baryonic source, we proved that the baryonic mass is larger than the observationally determined (dynamical) mass, and that the (negative) internal mass can be (absolutely) larger than the baryonic mass, and, as such, it cannot be ignored. As a consequence, under these assumptions, there should a plenitude of baryonic matter in the cosmological structures. These results enabled (actually inspired) us to apply the same ideas in explaining the problems of the missing mass and of the flat-rotation curves, to explore the notion of repulsive gravity, but, also, to question the notions cosmological constant, dark energy, and accelerating expansion of the Universe.

As a result of the present treatment and in accordance with especially high-frequency electromagnetic observational data, for every cosmological structure, a new picture of a hot and quite extended structure emerges. This is determined by its temperature and chemical composition, and by the distinction between the actual linear dimensions of the structure and the (temperature-and chemical composition-dependent, and much smaller) *inversion distance* (expressing the structure's conventional linear dimensions). Interestingly, the region between the inversion distance and the actual (and larger) linear dimensions of the structure is the region of dominance of repulsive gravity over attractive gravity. Especially in the case of a super cluster of galaxies, the actual dimensions are much larger than its conventional dimensions. So, the interesting result is reached that, now, the super cluster appears as a hot and extended structure, resembling the typical picture of the whole cosmological Universe. Such a picture of a cosmological structure conforms with its more luminous, as denser, central regions, and its fainter, as more rarefied, outer regions (both assumed as having the same temperature and mean molecular weight). It is quite probable that this new picture excludes the existence of third (and higher)-order clusters of galaxies.

As for the general-relativistic cosmological applications of the CDE approach, we point out that, in the collisional-dark-matter model, the extra energy, needed to compromise for the currently available cosmological data, can be compensated by the energy of the internal motions of the collisional cosmological fluid, that the post-recombination Universe remains *ever-decelerating,* and, also that the explicit form of the equation of state for the cosmological fluid can be determined. In other words, *in the collisional-dark-matter model, no dark energy and no acceleration of the universal expansion are necessary.* The interest in these results is enhanced by proving that, in the context of the less probable collisionless-dark-matter model, the extra (dark)-energy component (to compromise the cosmic-background-microwave-radiation data) is, in fact, needed, the distant light sources appear to be dimmer than expected, and the expansion of the Universe keeps balancing between acceleration and deceleration, depending on the (fully and explicitly determined appropriate) value of the cosmological red-shift parameter. Therefore, these results of the collisionless-dark-matter model seem to be a misinterpretation of the cosmological observational data by an observer, who (although living in a Universe filled, mainly, with, collisional dark matter) insists in adopting the traditional (collisionless-dark matter) approach.

To the extent of our knowledge, the CDE approach is classical and novel, and it is characterized by simplicity of thought, inner consistency, conformity with observational data and generally acceptable theoretical predictions. Also it does not constitute any kind of modifications of gravity (Newtonian and relativistic), and it extends the realm of validity of the classical geodesic motions.

In spite of all its advantages above, in our treatment there exist many deficiencies. Thus, one can argue against the fundamental assumptions of the time-independence of the Newtonian theoretical framework used here and of the isentropicity of the hydrodynamic flow motions, although such assumptions are in line with the most usual assumption of an isolated astronomical system. Also, beyond the assumption of the (greatly simplifying) spherical symmetry, one can argue against the use of the Plummer-type mass-density profile, although such a density profile is a particular case of more general "universal" density profiles, which are supported by observations, and so they have been widely applied in the literature. We believe, that the choice of physically and mathematically reasonable density profiles will not affect our present conclusions considerably, *and, in any case, it will not affect our fundamental and conceptually important conclusion, that the theoretically evaluated mass is different from the observationally determined (dynamical) mass.* The same is true for the consequences of the isotropic, isothermal, and adiabatic equation of state assumed here, which enables one to evaluate the internal specific energy density in terms of the mass density, temperature and mean molecular weight of the distribution, and its temporal and spatial variations. Finally, although less probable and least justifiable, one can question even the necessity of the collisional dark matter itself. Such objections and disadvantages imply the need of more intense research work, both theoretical and observational, on these subjects. Therefore, for the future, it remains to be seen whether these novel ideas on the role and importance of the internal physical characteristics of the large-scale cosmological structures and on the new interpretation of basic cosmological observations will attract the interest of the academic society and, so, perhaps, it will be acceptable as new scientific knowledge concerning the past and future history and evolution of the Universe and its constituents, and their interpretation.

Acknowledgements
I wish to thank all my collaborators in this research work for their spirit of productive collaboration. Also, I wish to thank warmly the organizers of the Conference *Recent Developments in Gravity (NEB) XIV* (Ioannina, Hellas, 8-11 June 2010), Drs. L. Perivolaropoulos and P. Kanti, for the beautifully organized Conference and for their kind hospitality.

References

[1] Sembach K 2002 *Space Telescope Science Institute Newsletter* **19** 19
[2] Nicastro F *et al.* 2003 *Nature **421** 719*
[3] Braun R and Burton W B 2001 *A&A* **375** 219
[4] Hagiwara Y, Diamond P J, Nakai N and Kawabe R 2000 *A&A* **360** 49
[5] Hagiwara Y, Henkel C, Menten K M and Nakai N 2001 *A&A* **560** L37
[6] Hagiwara Y, Diamond P J and Miyoshi M 2002 *A&A* **383** 65
[7] Tarchi A, Henkel C, Peck A B and Menten K M 2002 *A&A* **385** 1049
[8] Felten J E, Gould R J, Stein W A and Woolf N J 1996 *ApJ* **146** 955
[9] Begeman K, Broeils A and Sanders R H 1991 *MNRAS* **249** 523
[10] Navarro J F 2002 in *The Shapes of Galaxies and their dark Halos,* Yale University Workshop, Ed. P Natarajan, World Scientific Publishing, pp. 114-118.
[11] Jones C, David L, Forman B, Markevich M, Murray S, van Speybroeck L and Vikhilinin A 2003, *Chandra News* **10** 1
[12] Simon J D, Robishaw T and Blitz L 2003 *"Dark Matter in Dwarf Galaxies: The First Dark Galaxy?"* in *Satellites and Tidal Streams,* ASP Conference Series, Eds. F Prada, D Martinez-Delgado and T Mahoney
[13] Alves J and Homeir N 2003 *ApJ* **589** L45
[14] Ellis R S 2002, in *The Shapes of Galaxies and their dark Halos,* Yale University Workshop, Ed. P Natarajan, World Scientific Publishing, pp. 254-262
[15] Livio M 2003 *Space Telescope Science Newsletter* **20** 16
[16] Page L *et al.* 2003 *ApJ Suppl. Series* **148** 39
[17] Kleidis K and Spyrou N K 2000 *Class. Quantum Grav.* **17** 2965
[18] Spyrou N K 1997a in *The Physics of Ionized Gases,* Proceedings of the 18th Summer School and International Symposium on the Physics of Ionized Gases (Kotor, Jugoslavia, 2-6 September 1996), Eds. B Vujicic, S Djurovic and I Puric, Institute of Physics, Novi Sad Univ. Yugoslavia, pp. 417-446
[19] Spyrou N K 1997b in *The Earth and the Universe,* Volume in Honorem L Mavrides, Eds. G Asteriadis, A Bandellas, M Contadakis, K Katsambalos, A Papademetriou and I Tziavos, Thessaloniki, Greece, pp. 277-291
[20] Spyrou N K 1997c *Facta Universitatis* **4** 7

[21] Spyrou N K 1999 in *Current Issues of Astronomical and Planetary Environmental Concern*, Proceedings of International Seminar (Thessaloniki, 6-7 April 1999), Ed. N K Spyrou, Astronomy Department, Aristoteleion University of Thessaloniki, Thessaloniki, Greece, pp.23-35

[22] Spyrou N K 2001 "Conformal Invariance and the Nature of Cosmological Structures", Invited Talk, Proceedings of the Conference on Applied Differential Geometry-General Relativity and the Workshop on Global Analysis, Differential Geometry and Lie Groups, Thessaloniki, 27 June-1July 2001, eds. G Tsagas, C Udriste and D B Papadopoulos, pp. 101-107.

[23] Spyrou N K 2002 *"On the determination of the masses of cosmological structures",* in *Modern Theoretical and Observational Cosmology,* Proceedings of the 2nd Hellenic Cosmology Meeting, Athens, 19-20 April 2001, Eds. M Plionis and S Cotsakis, Kluwer Academic Publishers, Dordrecht, **276,** pp.35-43

[24] Spyrou N K 2003 *"Conformal Dynamical Equivalence and the Cosmological Expansion of a Realistic Universe",* in Proceedings of the 10th Conference *Recent Developments in Gravity*, Eds. K Kokkotas and N Stergioulas, Kluwer Academic Publishers, Dordrecht, pp.90-96

[25] Spyrou N K 2004a, *"A classical treatment of the dark-matter and flat-rotation-curves problems,* in Proceedings of the 6th Hellenic Astronomical Conference (Penteli, Athens), September 2003, Ed. P G Laskarides (and the Edition Office of the University of Athens), Athens, pp. 229-234.

[26] Spyrou N K 2004b *"A classical treatment of the problems of dark energy, dark matter and accelerating expansion",* in Proceedings of the Conference *Recent Developments in Gravity (NEB) XI (Mytilini, Lesvos, Hellas) 2-6 June 2004, Eds. S Cotsakis and J Myritzis;* published as *Spyrou 2005.*

[27] Spyrou N K 2005 *JPhCS* **8** 122, http://www.iop.org/EJ/toc/1742- 55.1.6596/8/1

[28] Spyrou N K 2006 *"Large-Scale Cosmological Structures: A New Look",* in Proceedings of *the International "Workshop Cosmology and Gravitational Physics"*, Thessaloniki, December 15-16, 2005, Eds. N K Spyrou, N Stergioulas, and C G Tsagas, pp. 57-66

[29] Spyrou N K 2008 *"Negative mass and repulsive gravity in Newtonian theory and consequences,* in Proceedings of the Conference *Recent Developments in Gravity (NEB) XIII (Thessaloniki, Hellas) 4-6 June 2008 (in press)*

[30] Spyrou N K 2009 *"Negative mass and repulsive gravity in Newtonian theory and consequences,* in Proceedings of *the 2008 International Conference in honour of John Hadjidemetriou,* Litohoron, Greece, Eds. H Varvoglis and Z Knezevic, Aristoteleion University of Thessaloniki, Thessaloniki, Greece, and Astronomical Observatory, Belgrade, Serbia, pp.181-184

[31] Spyrou N K 2010 *"Conformal Dynamical Equivalence and Applications",* in Proceedings of *the* Conference *Recent Developments in Gravity (NEB) XIV* (Ioannina, Hellas) 8-12 June 2010*; see also Kleidis and Spyrou 2010a (in press)*

[32] Kleidis K and Spyrou N K 1999 *"On the Nature of Nuclear Galactic Masses"* in Proceedings of JENAM 1999, Samos, Greece.

[33] Spyrou N K and Tsagas C G 2004, *Class Quantum Grav.* **21** 2435

[34] Spyrou N K and Tsagas C G 2008 *MNRAS* **388** 187

[35] Spyrou N K and Tsagas C G 2010 *Cent. Eur. J. Phys* .**8** 989

[36] Spyrou N K and Zagkouris K 2010 "The *Solar System: A New Look"*, Aristoteleion University of Thessaloniki, Thessaloniki, Greece (Preprint)

[37] Kleidis K and Spyrou N K 2010a *"A Conventional Form of Dark Energy",* in Proceedings of *the* Conference *Recent Developments in Gravity (NEB) XIV* (Ioannina, Hellas) 8-12 June 2010

[38] Kleidis K and Spyrou N K 2010b *"Collisional Dark Matter: A conventional approach to the dark-energy concept* (Preprint)

[39] Fock V 1959 *"The theory of space, time and gravitation"* (London: Pergamon Press)

[40] Chandrasekhar S 1969 *ApJ* **158** 45
[41] Taub A 1959 *Arch. Ration. Mech. Anal.* **3** 312
[42] Misner C, Thorne K S and Wheeler J A 1973 *"Gravitation"* (San Francisco: Freeman)
[43] Anderson J 1967 *"Principles of Relativistic Physics"* (New York: Academic Press)
[44] Ryan M and Shepley L 1975 *"Homogeneous Relativistic Cosmologies"* (Cambridge: Cambridge University Press)
[45] Navarro J F, Frenk C S and White S D M 1995 *MNRAS* **275** 720
[46] Navarro J F, Frenk C S and White S D M 1997 *ApJ* **490** 493
[47] Hernquist L 1990 *ApJ* **356** 359
[48] Dehnen W 1995*MNRAS* **274** 919
[49] Syer D and White S D M 1998 *MNRAS* **293** 337
[50] Zhao H S 1996 *MNRAS* **278** 488
[51] Spergel D N et al. [WMAP Collaboration] 2003 *ApJ Suppl. Series* **148** 175
[52] de Bernardis P et al. [BOOMERanG Collaboration] 2000 *Nature* **404** 955
[53] Riess A G et al. [High-z Supernova Search Team] 1998 *Astron. J.* **116** 1009
[54] Perlmutter S et al. [Supernova Cosmology Project Group] 1999 *ApJ* **517** 565
[55] Riess A G et al. [High-z Supernova Search Team] 2001 *ApJ* **560** 49
[56] Riess A G et al. [High-z Supernova Search Team] 2004 *ApJ* **607** 665
[57] Tonry J L et al. [High-z Supernova Search Team] 2003 ApJ **594** 1
[58] Caldwell R, Dave R and Steinhardt P J 1998 *Phys. Rev. Lett.* **80** 1528
[59] Carroll S M 1998 *Phys. Rev. Lett.* **81** 3067
[60] Komatsu E et al. [WMAP Collaboration] 2009 *ApJ Suppl. Series* **180** 330
[61] Kolb E W and Turner M S 1990 *"The Early Universe"* (Menlo Park: Addison-Wesley)
[62] Olive K A 2003 *TASI Lectures on dark matter, arXiv: 0301505 [astro-ph]*
[63] Hooper D 2009 *TASI 2008 Lectures on dark matter, arXiv: 0901.4090 [hep-ph]*
[64] Chang J et al. [ATIC Collaboration] 2008 *Nature* **456** 362
[65] Adriani O et al. [PAMELA Collaboration] 2009 *Nature* **458** 607
[66] Hooper D, Finkbeiner D P and Dobler G 2007 *Phys. Rev.* **D 76** 083012
[67] Barger V, Keung W Y, Marfatia D and Shaughnessy G 2008, *Phys. Lett.* **B 672** 141
[68] Baushev A 2008, *"Dark matter annihilation at cosmological redshifts: Possible relic signal from weakly interacting massive particles annihilation", arXiv: 0806.3108 [astro-ph]*
[69] Bergstrom L, Bringmann T and Edsjo J 2008 *Phys. Rev.* **D 78** 103520
[70] Cholis I, Goodenough L, Hooper D, Simet M and Weiner N 2008a *"High-energy positrons from annihilating dark matter", arXiv: 0809.1683 [hep-ph]*
[71] Cholis I, Dobler G, Finkbeiner D P, Goodenough L and Weiner N 2008b, *"The case for a 700+ GeV WIMP: Cosmic-ray spectra from ATIC and PAMELA", arXiv: 0811.3641 [astro-ph]*
[72] Cirelli M and Strumia A 2008 *"Minimal dark-matter predictions and the PAMELA positron excess", arXiv: 0808.3867 [astro-ph]*
[73] Regis M and Ullio P 2008 *Phys. Rev.* **D 78** 3505
[74] Fornasa M, Pieri L, Bertone G and Branchini E 2009 *"Anisotropy probe of galactic and extra-galactic dark matter annihilations", arXiv: 0901.2921 [astro-ph]*
[75] Fox P J and Poppitz E 2009 *Phys. Rev.* **D 79** 083528
[76] Zurek K M 2009 *Phys. Rev.* **D 79** 115002
[77] Spergel D N and Steinhardt P J 2000 *Phys. Rev. Lett.* **84** 3760
[78] Arkani-Hamed N, Finkbeiner D P, Slatyer T R and Weiner N 2009 *Phys. Rev.* **D 79** 015014
[79] Cirelli M, Kadastik M, Raidal M and Strumia A 2009 *Nucl. Phys.* **B 813** 1
[80] Cohen T and Zurek K 2010 *Phys. Rev. Lett.* **104** 101301
[81] Zimdahl W, Schwarz D J, Balakin A B and Pavon D 2001 *Phys. Rev.* **D 64** 3501
[82] Balakin A B, Pavon D, Schwarz D J and Zimdahl W 2003 *New J. Phys.* **5** 85
[83] Lima J A S, Silva F E and Santos R C 2008 *Class. Quantum Grav.* **25** 205006
[84] Basilakos S and Plionis M 2009 *A&A* **507** 47

[85] Basilakos S and Plionis M 2010 *"Interactive dark matter as an alternative to dark energy"*, in Invisible Universe: Conference Proceedings, AIP Conference Proceedings **1241**, 721
[86] Davis T M et al. [ESSENCE Supernova Survey Team] 2007 *ApJ* **666** 71

Recent Developments in Gravity (NEB XIV) IOP Publishing
Journal of Physics: Conference Series **283** (2011) 012036 doi:10.1088/1742-6596/283/1/012036

Black hole spacetimes with self-gravitating, massive accretion tori

Nikolaos Stergioulas
Department of Physics, Aristotle University of Thessaloniki, Thessaloniki 54124, Greece
E-mail: niksterg@auth.gr

Abstract. We review the theory of stationary, axisymmetric spacetimes with black holes surrounded by a massive torus and present a numerical scheme for finding self-consistent solutions in a compactified grid that reaches to spatial infinity. Such numerical solutions have been utilized in studying nonaxisymmetric dynamical instabilities occurring in the torus.

1. Introduction

Thick relativistic accretion disks and tori around black holes (BHs) can form as transient structures in several astrophysical scenarios, including the core-collapse of massive stars [1, 2] and the merger of neutron star (NS) and NS-BH binaries [3, 4, 5]. Recent numerical simulations have demonstrated that the mass of the disk resulting from binary NS-NS or BH-NS mergers (which are candidates for the central engine of short GRBs), can be in the range of $\sim 0.01 - 0.2M_\odot$ [6, 7, 5, 4]. Early studies of the stability of accretion disks have revealed that they can be subject to several types of axisymmetric and/or non-axisymmetric instabilities in a number of circumstances [8, 9, 10, 11, 12, 13, 14], which can lead to highly variable and unstable accretion rates. It is thus important to create a comprehensive overall picture of the stability of accretion disks for a wide range of parameters.

We first review the theory of tori in a fixed spacetime, the so-called AJS disks [15]. This is then generalized to self-gravitating disks in quasi-isotropic coordinates, following [16]. We reformulate the equations in a compactified coordinate system, which allows for high accuracy and describe in detail a numerical scheme that leads to self-consistent solutions. First numerical results of this approach can be found in [17], where a detailed study of nonaxisymmetric instabilities or tori around black holes in a fully general-relativistic treatment is presented. A study of the parameter space of equilibrium solutions will appear elsewhere.

2. Disks in a stationary, axisymmetric spacetime

A stationary, axisymmetric spacetime is described by a metric of the general form

$$ds^2 = g_{tt}dt^2 + 2g_{t\phi}dtd\phi + g_{\phi\phi}d\phi^2 + g_{rr}dr^2 + g_{\theta\theta}d\theta^2, \quad (1)$$

(we assume a signature of $-,+,+,+$ and set $c = G = 1$). Excluding meridional circulation, for a disk to be stationary its 4-velocity must have the form

$$u^\alpha = (u^t, 0, 0, u^\phi). \quad (2)$$

The angular velocity measured by an observer at infinity is

$$\Omega = \frac{u^\phi}{u^t}. \tag{3}$$

Along each fluid trajectory, the existence of a Killing vector t^α describing time symmetry implies the conservation of $\mathcal{E} = -hu_t$ (energy per unit rest mass), while the existence of a Killing vector ϕ^α describing ϕ-symmetry implies the conservation of $j = hu_\phi$ (angular momentum per unit rest mass), where $h = (\epsilon + p)/\rho$ is the specific enthalpy, ρ is rest mass density, p is pressure and ϵ is energy density. When both t^α and ϕ^α exist, then also the ratio

$$\frac{j}{\mathcal{E}} = -\frac{u_\phi}{u_t} := l, \tag{4}$$

is conserved (the *angular momentum per unit energy*). One can eliminate u^t, u^ϕ to write

$$l = -\frac{g_{t\phi} + \Omega g_{\phi\phi}}{g_{tt} + \Omega g_{t\phi}}, \tag{5}$$

or

$$\Omega = -\frac{g_{t\phi} + l g_{tt}}{g_{\phi\phi} + l g_{t\phi}}. \tag{6}$$

From the normalization of the 4-velocity, $u^\alpha u_\alpha = -1$, one can obtain u_t as

$$u_t = -\left(-g^{tt} + 2lg^{t\phi} - l^2 g^{\phi\phi}\right)^{-1/2}, \tag{7}$$

$$= -\left[-\frac{g_{\phi\phi} + 2lg_{t\phi} + l^2 g_{tt}}{g_{tt}g_{\phi\phi} - (g_{t\phi})^2}\right]^{-1/2}, \tag{8}$$

where the components of the inverse metric were used

$$g^{tt} = \frac{1}{g_{tt} - (g_{t\phi})^2/g_{\phi\phi}}, \tag{9}$$

$$g^{\phi\phi} = \frac{1}{g_{\phi\phi} - (g_{t\phi})^2/g_{tt}}, \tag{10}$$

$$g^{t\phi} = -\frac{g_{t\phi}}{g_{tt}g_{\phi\phi} - (g_{t\phi})^2}. \tag{11}$$

From the same normalization it also follows that

$$u^t = \left(-g_{tt} - 2\Omega g_{t\phi} - \Omega^2 g_{\phi\phi}\right)^{-1/2}, \tag{12}$$

and

$$u^t u_t = -\frac{1}{1 - \Omega l}, \tag{13}$$

while (4), (13) imply

$$u^t u_\phi = \frac{l}{1 - \Omega l}. \tag{14}$$

Notice that the choice of sign when taking the square root in (7), (8) and (12) is made in agreement with the flat-space limit of u^α for the chosen signature of the metric.

The hydrostationary equilibrium equation can be written in the following equivalent forms

$$\frac{\nabla p}{\epsilon + p} = \nabla \ln u^t - u^t u_\phi \nabla \Omega, \tag{15}$$

$$= \nabla \ln u^t - \frac{l \nabla \Omega}{1 - \Omega l}, \tag{16}$$

$$= - \nabla \ln(-u_t) + \frac{\Omega \nabla l}{1 - \Omega l}, \tag{17}$$

and has a first integral only under specific conditions. For a barotropic equation of state of the form $p = p(\rho)$ one can define the log-enthalpy

$$H(p) := \int_0^p \frac{dp'}{\epsilon(p') + p'}. \tag{18}$$

Then, $dH/dp = (\epsilon + p)^{-1}$ and the l.h.s. of (15) becomes $(\epsilon + p)^{-1} \nabla p = \nabla H$. The r.h.s. of (15) can be written in the form of the gradient of a potential, only if $u^t u_\phi$ is a function of Ω, or equivalently, one can write the r.h.s. of (17) as

$$-\nabla \ln(-u_t) + \frac{\Omega \nabla l}{1 - \Omega l} := -\nabla W, \tag{19}$$

only if $\Omega = \Omega(l)$. In (19), $-\nabla W$ is the *effective gravity*, given by the gradient of the *effective potential*

$$W = \ln(-u_t) - \int^l \frac{\Omega}{1 - \Omega l} dl + \text{const.} \tag{20}$$

Notice that at zero pressure $H(0) = 0$, while W is only defined up to a constant, which can be set by requiring that W vanishes at infinity. Under these conditions, it follows that

$$\nabla H = -\nabla W, \tag{21}$$

and *the surfaces of constant pressure coincide with the equipotential surfaces.* At the location of maximum density $\nabla H = \nabla W = 0$.

The first integral of (17) is

$$H + \ln(-u_t) - \int^l \frac{\Omega}{1 - \Omega l} dl = \text{const.}, \tag{22}$$

or, equivalently, the first integral of (16) is

$$H - \ln(u^t) + \int^\Omega \frac{l}{1 - \Omega l} d\Omega = \text{const.}. \tag{23}$$

For a *homentropic flow* (uniform entropy throughout the fluid), the log-enthalpy becomes simply

$$H = \ln \left(\frac{h}{h_{\text{min}}} \right), \tag{24}$$

where $h_{\text{min}} = 1$ is the specific enthalpy at the limit of zero pressure[1]. At the disk surface, $h = h_{\text{min}}$ and $H = 0$.

[1] Notice that this differs from the Newtonian definition of specific enthalpy $h_{\text{Newt}} = e + p/\rho$, where e is the specific internal energy and the relation $h \to h_{\text{Newt}} + 1$ in the nonrelativistic, zero-pressure limit connects the two definitions.

Journal of Physics: Conference Series **283** (2011) 012036 doi:10.1088/1742-6596/283/1/012036

3. Constant specific angular momentum

A specific choice of the distribution of the specific angular momentum in the disk is to simply assume that $l = \text{const.}$, when (23) becomes

$$H - \ln[u^t(1 - l\Omega)] = \text{const.}, \tag{25}$$

or (22) becomes

$$H + \ln(-u_t) = \text{const.}. \tag{26}$$

If the disks terminates at some $r = r_{\text{in}}$, then the first integral becomes

$$H = \ln\left(\frac{u^t}{u^t_{\text{in}}}\right) + \ln\left(\frac{1 - l\Omega}{1 - l\Omega_{\text{in}}}\right), \tag{27}$$

$$= \ln\left(\frac{u_{t,\text{in}}}{u_t}\right), \tag{28}$$

and a similar relation holds when an outer termination radius $r = r_{\text{out}}$ is used. Notice that $\ln(-u_{t,\text{in}}) = \ln(-u_{t,\text{out}})$ and thus the the inner and outer radii of a disk are not independent, but are located on the same constant-pressure (or equipotential) surface. If the location $r = r_{\text{max}}$ of the density maximum ρ_{max} is used, then

$$H - H_{\text{max}} = \ln\left(\frac{u^t}{u^t_{\text{max}}}\right) + \ln\left(\frac{1 - l\Omega}{1 - l\Omega_{\text{max}}}\right), \tag{29}$$

$$= \ln\left(\frac{u_{t,\text{max}}}{u_t}\right), \tag{30}$$

where the subscripts in, out, max always indicate values at the corresponding radii.

For the homentropic, polytropic equation of state (EOS)

$$p = K\rho^{\Gamma}, \tag{31}$$

$$\epsilon = \rho + \frac{p}{\Gamma - 1}, \tag{32}$$

where K is the polytropic constant and Γ is the polytropic exponent, one obtains

$$H = \ln[1 + \Gamma/(\Gamma - 1)K\rho^{\Gamma-1}], \tag{33}$$

and from (28) one can solve algebraically for the density distribution, as an implicit function of Ω (for given EOS, l_0, r_{in} and stationary, axisymmetric spacetime):

$$\rho = \left\{\frac{\Gamma - 1}{K\Gamma}\left[\frac{u^t}{u^t_{\text{in}}}\left(\frac{1 - l\Omega}{1 - l\Omega_{\text{in}}}\right) - 1\right]\right\}^{\frac{1}{\Gamma-1}}, \tag{34}$$

$$= \left\{\frac{\Gamma - 1}{K\Gamma}\left[\frac{u_{t,\text{in}}}{u_t} - 1\right]\right\}^{\frac{1}{\Gamma-1}}. \tag{35}$$

The constant in the definition of the effective potential can be set by requiring that $W(r = \infty) = 0$. Then, the effective potential is

$$W = \ln\left(\frac{u_t}{u_{t,\infty}}\right). \tag{36}$$

4. AJS disks in a Schwarzschild background

In a spherically symmetric spacetime, the metric in Schwarzschild coordinates $(t, \mathfrak{r}, \theta, \phi)$ is

$$ds^2 = -\left(1 - \frac{2M}{\mathfrak{r}}\right) dt^2 + \left(1 - \frac{2M}{\mathfrak{r}}\right)^{-1} d\mathfrak{r}^2 + \mathfrak{r}^2 d\theta^2 + \mathfrak{r}^2 \sin^2\theta d\phi^2, \tag{37}$$

where M is the total mass-energy. In this metric

$$u_t = -\left[\left(1 - \frac{2M}{\mathfrak{r}}\right)^{-1} - \frac{l^2}{\mathfrak{r}^2 \sin^2\theta}\right]^{-1/2}, \tag{38}$$

and $u_{t,\infty} = -1$, so that the effective potential for $l =$ const. disks is

$$W = \ln(-u_t). \tag{39}$$

A disk of finite size extends between $\mathfrak{r} = \mathfrak{r}_{\rm in}$ and $\mathfrak{r} = \mathfrak{r}_{\rm out}$ in the equatorial plane. At the density maximum, $\nabla_{\mathfrak{r}} p = 0$ and the fluid elements there move as free test particles with Keplerian angular momentum, which restricts $\mathfrak{r}_{\rm max} > \mathfrak{r}_{\rm ms}$, where $\mathfrak{r}_{\rm ms} = 6M$ is the radius of the marginally stable circular orbit for test particles. The Keplerian angular momentum is $\Omega_{\rm K} = \sqrt{M/\mathfrak{r}^3}$ and the corresponding specific angular momentum is

$$l_{\rm K} = \frac{\sqrt{M\mathfrak{r}^3}}{\mathfrak{r} - 2M}, \tag{40}$$

which has a minimum at $\mathfrak{r} = \mathfrak{r}_{\rm ms}$ so that $\mathfrak{r}_{\rm max} > \mathfrak{r}_{\rm ms} \Rightarrow l > l_{\rm ms} = 3.674M$. For a given value of l, the largest of three roots of $l = l_{\rm K}$ corresponds to $\mathfrak{r} = \mathfrak{r}_{\rm max}$. The intermediate root, $\mathfrak{r} = \mathfrak{r}_{\rm cusp}$, corresponds to the existence of a cusp, where again $\nabla_{\mathfrak{r}} p = 0$ and fluid elements move as free test particles with Keplerian angular momentum. Thus, for given EOS and black-hole mass M and for a given value of l, there exists a one-parameter family of different finite-size disks with $\mathfrak{r}_{\rm cusp} < \mathfrak{r}_{\rm in} < \mathfrak{r}_{\rm max}$, which all have the same $\mathfrak{r}_{\rm max}$. For a chosen $\mathfrak{r}_{\rm in}$ in this range, one obtains the density distribution from (35). Members of this one parameter family differ in the maximum density, $\rho_{\rm max}$ and total mass.

Another factor that limits the parameter space of possible equilibrium configurations is the existence of the marginally bound orbit, which for nonrotating black holes is at $\mathfrak{r}_{\rm mb} = 4M$. For $l = l_{\rm mb} = 4M$, the location of the cusp is at $\mathfrak{r} = \mathfrak{r}_{\rm mb}$. At the same time, the location of $W = 0$ in the equatorial plane is relevant. For $l < l_{\rm mb}$, the effective potential W is always negative in the equatorial plane. For $l = l_{\rm mb}$, $W = 0$ at $\mathfrak{r}_{\rm mb} = 4M$ (an inflection point), while for $l > l_{\rm mb}$ there exists a region with $W > 0$. Thus, for any $l > l_{\rm mb}$, only disks without a cusp are possible, with the inner radius for this family of disks being limited by the location of $W = 0$.

Within each family of disks with same specific angular momentum l, the disk with the smallest $\mathfrak{r}_{\rm in}$ (i.e. the disk with $\mathfrak{r}_{\rm in} = \mathfrak{r}_{\rm cusp}$, for $l_{\rm ms} < l_{\rm mb}$, or the marginally stable disk for $l > l_{\rm mb}$) has the largest $\rho_{\rm max}$. Among all models with $\mathfrak{r}_{\rm in} = \mathfrak{r}_{\rm cusp}$, $\rho_{\rm max}$ *increases* with increasing l. For example, for $\Gamma = 4/3$ polytropes $\rho_{\rm max}$ reaches up to 1.341×10^{-6} (in units where $c = G = K = 1$) for the model with $\mathfrak{r}_{\rm in} = \mathfrak{r}_{\rm mb}$. For $l > l_{\rm mb}$ the largest $\rho_{\rm max}$ *decreases* with increasing l. Thus, among all finite size disks, the largest $\rho_{\rm max}$ is attained for the model with $\mathfrak{r}_{\rm in} = \mathfrak{r}_{\rm mb}$.

5. AJS disks in isotropic coordinates

The metric of the Schwarzschild spacetime in isotropic coordinates (t, r, θ, ϕ) is

$$ds^2 = -\left[\frac{1 - M/2r}{1 + M/2r}\right]^2 dt^2 + \left(1 + \frac{M}{2r}\right)^4 (dr^2 + r^2 d\theta^2 + r^2 \sin^2\theta d\phi^2). \tag{41}$$

The relation between the isotropic radial coordinate r and the Schwarzschild radial coordinate $\mathfrak{r}$ is

$$r = \frac{1}{2}\left[\mathfrak{r} - M + \sqrt{\mathfrak{r}^2 - 2M\mathfrak{r}}\right], \tag{42}$$

or

$$\mathfrak{r} = \frac{M^2}{4r} + M + r. \tag{43}$$

The Keplerian angular momentum is

$$l_{\mathrm{K}} = \frac{\sqrt{M}\,(1 + M/2r)^6\, r^2}{(1 - M/2r)^2\,(M + M^2/4r + r)^{3/2}}, \tag{44}$$

while

$$u_t = -\left[\left(\frac{M + 2r}{M - 2r}\right)^2 - \frac{16 l^2 r^2}{(M + 2r)^4 \sin^2\theta}\right]^{-1/2}. \tag{45}$$

The radius of the marginally stable orbit is at

$$r_{\mathrm{ms}} = \frac{1}{2}(5 + 2\sqrt{6}), \tag{46}$$

or, $r_{\mathrm{ms}} \simeq 4.950$. The radius of the marginally bound orbit is the location where $W = 0$ for $l = 4$, i.e. $r_{\mathrm{mb}} \simeq 2.914$.

6. Self-gravitating disks in quasi-isotropic coordinates

In quasi-isotropic coordinates, the metric of a stationary, axisymmetric spacetime is written as

$$ds^2 = -e^{2\nu}dt^2 + e^{2\alpha}(dr^2 + r^2 d\theta^2) + e^{2(\gamma-\nu)} r^2 \sin^2\theta (d\phi - \omega dt)^2, \tag{47}$$

where ν, γ, α and ω are metric functions that depend only on the coordinates r and θ. Because γ and ρ are divergent at the event horizon, we will use, instead, the functions

$$B := e^{\gamma}, \tag{48}$$
$$\lambda := e^{\nu}. \tag{49}$$

Then, the following conditions hold on the event horizon

$$B = 0, \tag{50}$$
$$\lambda = 0, \tag{51}$$
$$\omega = \omega_h, \tag{52}$$

where ω_h is the *constant* angular velocity of the horizon. Within the choice of metric (47), the coordinate r can be defined such that the horizon is a sphere of constant radius h_0 (see, e.g. Carter 1973). Along the rotation axis, the condition $\alpha = \gamma - \nu$ ensures local flatness.

In the angular direction we use the coordinate $\mu \equiv \cos\theta$ instead of θ, which distributes grid points more evenly near the rotation axis. In the radial direction, we use a compactified, dimensionless coordinate s, defined through

$$r := r_e \frac{s}{1 - s}, \tag{53}$$

where r_e is a suitably chosen radius (such as the outer radius of the disk). In this way, the infinite domain $r \to [0, +\infty)$ is mapped onto the finite domain $s \to [0, 1]$. In transforming the field equations to this new coordinate system, the following relations are useful

$$\frac{\partial}{\partial r} = \frac{(1-s)^2}{r_e}\frac{\partial}{\partial s}, \tag{54}$$

$$\frac{\partial}{\partial \theta} = -\sqrt{1-\mu^2}\frac{\partial}{\partial \mu}, \tag{55}$$

$$\vec{\nabla} a = \frac{(1-s)^2}{r_e}\frac{\partial a}{\partial s}\hat{r} - \frac{1-s}{r_e s}\sqrt{1-\mu^2}\frac{\partial a}{\partial \mu}\hat{\theta}, \tag{56}$$

$$\vec{\nabla} a \vec{\nabla} b = \frac{(1-s)^4}{r_e^2}\frac{\partial a}{\partial s}\frac{\partial b}{\partial s} - \frac{(1-s)^2}{r_e^2 s^2}(1-\mu^2)\frac{\partial a}{\partial \mu}\frac{\partial b}{\partial \mu}, \tag{57}$$

where a, b are some scalar functions.

For the 4-velocity, one can easily show that

$$u^t = \frac{e^{-\nu}}{\sqrt{1-v^2}}, \tag{58}$$

where

$$v = (\Omega - \omega)Be^{-2\nu} r \sin\theta, \tag{59}$$

is the 3-velocity, as measured by a zero-angular-momentum observer (ZAMO).

The field equations for the three metric functions λ, B and ω are of elliptic type

$$\nabla^2 \lambda = S_\lambda(s, \mu), \tag{60}$$

$$\left(\nabla^2 + \frac{(1-s)^3}{r_e^2 s}\frac{\partial}{\partial_s} - \frac{(1-s)^2}{r_e^2 s^2}\mu\frac{\partial}{\partial \mu}\right) B = S_B(s, \mu), \tag{61}$$

$$\left(\nabla^2 + \frac{2(1-s)^3}{r_e^2 s}\frac{\partial}{\partial_s} - \frac{2(1-s)^2}{r_e^2 s^2}\mu\frac{\partial}{\partial \mu}\right) \omega = S_\omega(s, \mu), \tag{62}$$

where ∇^2 is the flat-space Laplacian, and

$$\begin{aligned} S_\lambda :=& 4\pi\lambda e^{2\alpha}\left[(\epsilon + p)\frac{1+v^2}{1-v^2} + 2p\right] + \frac{1}{2}(-\mu^2)B^2\lambda^{-3}[s^2(1-s)^2(\omega_{,s})^2 + (1-\mu^2)(\omega_{,\mu})^2] \\ & - \frac{(1-s)^2}{r_e^2}\left[(1-s)^2(\gamma_{,s} - \nu_{,s})\lambda_{,s} - \frac{1-\mu^2}{s^2}(\gamma_{,\mu} - \nu_{,\mu})\lambda_{,\mu}\right], \end{aligned} \tag{63}$$

$$S_B := 16\pi p B e^{2\alpha}, \tag{64}$$

$$S_\omega := \frac{(1-s)^2}{r_e^2}\left[(1-s)^2(4\nu_{,s} - 3\gamma_{,s})\omega_{,s} + \frac{1-\mu^2}{s^2}(4\nu_{,\mu} - 3\gamma_{,\mu})\omega_{,\mu}\right] - 16\pi e^{2\alpha}(\epsilon + p)\frac{\Omega - \omega}{1-v^2}, \tag{65}$$

are source terms and *a comma in the subscript* denotes partial differentiation with respect to that coordinate.

The three elliptic-type equations are inverted using appropriate Green's functions

$$\lambda = 1 - \frac{h_0}{r} - \sum_{n=0}^{\infty} P_{2n}(\mu) \int_{h_0}^{\infty} dr' \int_0^1 d\mu' r'^2 f_{2n}^{(2)}(r,r') P_{2n}(\mu') S_\lambda(r',\mu'), \tag{66}$$

$$B = \left(1 - \frac{h_0^2}{r^2}\right) - \frac{2\pi}{r\sin\theta} \sum_{n=1}^{\infty} \frac{\sin(2n-1)\theta}{2n-1} \int_{h_0}^{\infty} dr' \int_0^1 d\mu' r'^2 f_{2n-1}^{(1)}(r,r') \sin(2n-1)\theta' S_B(r',\mu'), \tag{67}$$

$$\omega = \frac{\omega_h h_0^3}{r^3} - \frac{1}{r\sin\theta} \sum_{n=1}^{\infty} \frac{P_{2n-1}^1(\mu)}{2n(2n-1)} \int_{h_0}^{\infty} dr' \int_0^1 d\mu' r'^3 \sin\theta' f_{2n-1}^{(2)}(r,r') P_{2n-1}^1(\mu') S_\omega(r',\mu'), \tag{68}$$

where

$$f_n^{(1)}(r,r') := \begin{cases} \left(\frac{r'}{r}\right)^n - \frac{h_0^{2n}}{(rr')^n} & \frac{r'}{r} \le 1, \\ \left(\frac{r}{r'}\right)^n - \frac{h_0^{2n}}{(rr')^n} & \frac{r'}{r} > 1, \end{cases} \tag{69}$$

and

$$f_n^{(2)}(r,r') := \begin{cases} \frac{1}{r}\left(\frac{r'}{r}\right)^n - \frac{h_0^{2n+1}}{(rr')^{n+1}} & \frac{r'}{r} \le 1, \\ \frac{1}{r'}\left(\frac{r}{r'}\right)^n - \frac{h_0^{2n+1}}{(rr')^{n+1}} & \frac{r'}{r} > 1. \end{cases} \tag{70}$$

If we define the location of the horizon in terms of the coordinate s through

$$h_0 = r_e \frac{s_0}{1-s_0}, \tag{71}$$

then the metric function λ is given by

$$\lambda(s,\mu) = 1 - \frac{s_0(1-s)}{s(1-s_0)} - \sum_{n=0}^{\infty} P_{2n}(\mu) D_n^{\lambda(2)}(s), \tag{72}$$

where

$$D_n^{\lambda(2)}(s) := \int_{s_0}^1 ds' \tilde{f}_{2n}^{(2)}(s,s') D_n^{\lambda(1)}(s'), \tag{73}$$

$$D_n^{\lambda(1)}(s') := \int_0^1 d\mu' P_{2n}(\mu') \tilde{S}_\lambda(s',\mu'), \tag{74}$$

$$\tilde{f}_n^{(2)}(s,s') := \frac{r_e}{(1-s')^2} f_n^{(2)}(s,s'), \tag{75}$$

$$\tilde{S}_\lambda := r'^2 S_\lambda. \tag{76}$$

The metric function B is given by

$$B(s,\mu) = 1 - \left[\frac{s_0(1-s)}{s(1-s_0)}\right]^2 - 2\pi r_e^2 \sum_{n=1}^{\infty} \Pi_{2n-1}^1(\mu) D_n^{B(2)}(s), \tag{77}$$

where

$$D_n^{B(2)}(s) := \int_{s_0}^{1} ds' \tilde{f}_{2n-1}^{(1)}(s,s') D_n^{B(1)}(s'), \tag{78}$$

$$D_n^{B(1)}(s') := \int_0^1 d\mu' \sin(2n-1)\theta' S_B(s',\mu'), \tag{79}$$

$$\tilde{f}_n^{(1)}(s,s') := \frac{s'^2}{(1-s')^4}\left(\frac{1-s}{s}\right) f_n^{(1)}(s,s'), \tag{80}$$

$$\Pi_n^1(\theta) := \frac{\sin(n\theta)}{n\sin\theta}. \tag{81}$$

And, the metric function ω is given by

$$\omega(s,\mu) = \left[\frac{s_0(1-s)}{s(1-s_0)}\right]^3 \omega_h - r_e^2 \sum_{n=1}^{\infty} \Pi_n^2(\mu) D_n^{\omega(2)}(s), \tag{82}$$

where

$$D_n^{\omega(2)}(s) := \int_{s_0}^{1} ds' \tilde{f}_{2n-1}^{(3)}(s,s') D_n^{\omega(1)}(s'), \tag{83}$$

$$D_n^{\omega(1)}(s') := \int_0^1 d\mu' \sqrt{1-\mu'^2} P_{2n-1}^1(\mu') S_\omega(s',\mu'), \tag{84}$$

$$\tilde{f}_n^{(3)}(s,s') := \frac{s'^3(1-s)}{s(1-s')^3} \tilde{f}_n^{(2)}(s,s'), \tag{85}$$

$$\Pi_n^2(\mu) := \frac{P_{2n-1}^1(\mu)}{2n(2n-1)\sqrt{1-\mu^2}}. \tag{86}$$

In practice, we use the auxiliary variables

$$\bar{\rho} := \ln\frac{\lambda^2}{B}, \tag{87}$$

$$\hat{\nu} := \nu/r_e^2, \tag{88}$$

$$\hat{\gamma} := \gamma/r_e^2, \tag{89}$$

$$\hat{\omega} := r_e\omega. \tag{90}$$

The summations in (72), (77) and (82) are truncated to a maximum number of terms $n_{\max} = 10$. Both radial and angular integrations are performed with the 4th-order Simpson's rule.

The remaining metric function α is given in terms of the partial differential equation

$$\begin{aligned}\alpha_{,\mu} =& -\frac{1}{2}(\bar{\rho}_{,\mu} + \gamma_{,\mu}) - \{(1-\mu^2)[1+s(1-s)\gamma_{,s}]^2 + [-\mu + (1-\mu^2)\gamma_{,\mu}]^2\}^{-1} \\ &\times \Bigg[\frac{1}{2}\{s(1-s)[s(1-s)\gamma_{,s}]_{,s} + s^2(1-s)^2\gamma_{,s}^2 - [(1-\mu^2)\gamma_{,\mu}]_{,\mu} - \gamma_{,\mu}[-\mu + (1-\mu^2)\gamma_{,\mu}]\} \\ &\quad [-\mu + (1-\mu^2)\gamma_{,\mu}] + \frac{1}{4}[s^2(1-s)^2(\bar{\rho}_{,s} + \gamma_{,s})^2 - (1-\mu^2)(\bar{\rho}_{,\mu} + \gamma_{,\mu})^2] \\ &\quad [-\mu + (1-\mu^2)\gamma_{,\mu}] - s(1-s)(1-\mu^2)\Big[\frac{1}{2}(\bar{\rho}_{,s} + \gamma_{,s})(\bar{\rho}_{,\mu} + \gamma_{,\mu}) \\ &\quad + \gamma_{,s\mu} + \gamma_{,s}\gamma_{,\mu}\Big][1+s(1-s)\gamma_{,s}] + s(1-s)\mu\gamma_{,s}[1+s(1-s)\gamma_{,s}] \\ &\quad + \frac{1}{4}(1-\mu^2)e^{-2\bar{\rho}}\Big\{2\frac{s^3}{1-s}(1-\mu^2)\hat{\omega}_{,s}\hat{\omega}_{,\mu}[1+s(1-s)\gamma_{,s}] \\ &\quad - [s^4\hat{\omega}_{,s}^2 - \frac{s^2}{(1-s)^2}(1-\mu^2)\hat{\omega}_{,\mu}^2][-\mu + (1-\mu^2)\gamma_{,\mu}]\Big\}\Bigg]. \end{aligned} \tag{91}$$

This equation is integrated as an ODE along lines of constant s using Adams' multiple step method of local fourth-order accuracy.

In the chosen coordinate system, some problems appear in the source terms of the equations for λ and ω. Specifically, even though the term $(\gamma - \nu)_{,s}$ is regular at the horizon, its individual components, $\gamma_{,s}$ and $\nu_{,s}$ diverge (but their difference does not). In a numerical method, $\gamma_{,s}$ and $\nu_{,s}$ are evaluated separately and due to numerical finite-differencing errors it is not possible to exactly cancel the two terms at the horizon. One must therefore accept some approximation to keep the source S_λ regular. Similarly, problems appear in the second term of (63) and (65).

7. Numerical scheme

We use non-dimensional units of $c = G = M_{\rm BH}^{\rm AJS} = 1$ (the solution can then be scaled to any desired black hole mass). First, an AJS disk is computed as a trial configuration. A unique model is obtained by fixing the specific angular momentum l, the outer radius of the disk r_e (which also fixes the maximum enthalpy $H_{\rm max}$) and either of Γ, K or $\rho_{\rm max}$ (the latter three fluid properties are related to each other). Thus, for a chosen polytropic Γ, the solution space of AJS models is three-dimensional: the parameter l controls the rotation of the torus, r_e controls its distance from the black hole and its size and K (or $\rho_{\rm max}$) its mass content.

Once the AJS trial solution is obtained, one can then derive the location of the inner edge of the disk, $r_{\rm in}$, the location of the density maximum, $r_{\rm max}$, and the value of the density maximum, $\rho_{\rm max}$ (or K). The configuration is constructed so that the horizon of the black hole coincides with a grid point, s_0, in terms of the compactified coordinate s. The outer edge of the disk, r_e, also coincides with the grid point for which $s = 0.5$.

The AJS configuration is used as an initial guess for starting the self-consistent iteration between the field equations and the hydrostationary equilibrium equation. During this process, one can choose to keep some properties fixed, while others will necessarily vary. Since we always want the black hole horizon to coincide with a grid point, we choose to keep fixed the location of the horizon s_0 in terms of the compactified radial coordinate s. We also keep fixed the location of the outer radius at $s = 0.5$, so that the ratio h_0/r_e remains fixed (as do l and the EOS parameters K and Γ). Since r_e itself is allowed to vary during the iteration, we have to fix another quantity, in order to specify a model uniquely. This quantity can be $H_{\rm max}$ (or, equivalently, $\rho_{\rm max}$, since K is also fixed).

All other parameters of the BH-torus system (mass of the black hole, radius scale r_e, radius of horizon h_0, maximum density, mass of torus etc.) adjust themselves to self-consistent values during the iteration process. The iteration is carried out in units of $M_{\rm BH}^{\rm AJS} = 1$. After convergence, the results are rescaled to the newly computed mass of the black hole in the self-consistent solution.

The iteration proceeds as follows:

(i) A new value for r_e is obtained by solving the enthalpy equation (29) at $r = r_{\rm out}$, using (88) and (89)

$$r_e = \left\{ \frac{H_{\rm max} + 1/2 \ln \left[\left(\frac{1-v_{\rm max}^2}{1-v_{\rm out}^2} \right) \left(\frac{1-l\Omega_{\rm out}}{1-l\Omega_{\rm max}} \right)^2 \right]}{\hat{\nu}_{\rm out} - \hat{\nu}_{\rm max}} \right\}^{1/2}, \tag{92}$$

(ii) New angular velocity and 3-velocity distributions are obtained through (6) and (59)

$$\hat{\Omega} := \frac{\Omega}{r_e} = \left\{ \hat{\omega} + \frac{l}{r_e} \left[\frac{(1-s)^2 e^{r_e^2(4\hat{\nu}-2\hat{\gamma})}}{s^2(1-\mu^2)} - \hat{\omega}^2 \right] \right\} (1 - l\hat{\omega}/r_e)^{-1}, \tag{93}$$

$$v = (\hat{\Omega} - \hat{\omega}) e^{r_e^2(\hat{\gamma}-2\hat{\nu})} \frac{s}{1-s} \sqrt{1-\mu^2}. \tag{94}$$

(iii) New values $\hat{\Omega}_{\rm max}$ and $v_{\rm max}$ are obtained at the location of the density maximum $s_{\rm max}$.

(iv) A new enthalpy distribution is obtained through (29) and (58)

$$H = H_{\rm max} + r_e^2(\hat{\nu}_{\rm max} - \hat{\nu}) + \frac{1}{2} \ln \left[\frac{(1-v_{\rm max}^2)(1-l\hat{\Omega}/r_e)^2}{(1-v^2)(1-l\hat{\Omega}_{\rm max}/r_e)^2} \right]. \tag{95}$$

(v) New density, ρ, pressure, p, and energy density, ϵ, distributions are obtained through (33), (31) and (32).

(vi) New distributions of the four metric potentials λ, B, ω and α are obtained through (72), (77),(82), and (91).

(vii) The process is repeated from step (i) until the value for r_e converges with a desired relative accuracy (usually chosen as 10^{-6}).

8. Derived properties

The gravitational mass of the torus is

$$\begin{aligned} M_{\rm T} =& \int (-2T_t{}^t + T_i{}^i)\sqrt{-g} d^3x \\ =& 4\pi r_e^3 \int_0^1 \frac{s^2 ds}{(1-s)^4} \int_0^1 d\mu B e^{2\alpha} \left\{ \frac{\epsilon + p}{1-v^2} \left[1 + v^2 + \frac{2sv}{1-s} \sqrt{1-\mu^2} \hat{\omega} \frac{B}{\lambda^2} \right] + 2p \right\}. \end{aligned} \tag{96}$$

The rest mass of the torus is

$$\begin{aligned} M_0 =& \int \rho u^t \sqrt{-g} d^3x \\ =& 4\pi r_e^3 \int_0^1 \frac{s^2 ds}{(1-s)^4} \int_0^1 d\mu \frac{B}{\lambda} e^{2\alpha} \frac{\rho}{\sqrt{1-v^2}}. \end{aligned} \tag{97}$$

Physical units are obtained by multiplying with c^2L/G, where L is the fundamental length scale (set by $M_\odot = 1$). The internal energy of the torus is

$$\begin{aligned} U_{\rm T} &= \int (\epsilon - \rho) u^t \sqrt{-g} d^3x \\ &= 4\pi r_e^3 \int_0^1 \frac{s^2 ds}{(1-s)^4} \int_0^1 d\mu \frac{B}{\lambda} e^{2\alpha} \frac{\epsilon - \rho}{\sqrt{1-v^2}}, \end{aligned} \tag{98}$$

with physical units obtained by multiplying with c^4L/G. The angular momentum of the torus is

$$\begin{aligned} J_{\rm T} &= \int T^t{}_\phi \sqrt{-g} d^3x \\ &= 4\pi r_e^4 \int_0^1 \frac{s^3 ds}{(1-s)^5} \int_0^1 d\mu \sqrt{1-\mu^2} \frac{B^2 e^{2\alpha}}{\lambda^2} (\epsilon + p) \frac{v}{1-v^2}, \end{aligned} \tag{99}$$

with physical units obtained by multiplying with c^3L^2/G. The rotational kinetic energy of the torus is

$$\begin{aligned} T_{\rm T} &= \frac{1}{2} \int \Omega T^t{}_\phi \sqrt{-g} d^3x \\ &= 2\pi r_e^3 \int_0^1 \frac{s^3 ds}{(1-s)^5} \int_0^1 d\mu \sqrt{1-\mu^2} \frac{B^2 e^{2\alpha}}{\lambda^2} (\epsilon + p) \frac{v\hat{\Omega}}{1-v^2}, \end{aligned} \tag{100}$$

with physical units obtained by multiplying with c^4L/G.

The total mass of the spacetime can be obtained in two ways. From the asymptotic form of the metric potential λ, which becomes, to lowest order, $1 - M/r$ for $r \to \infty$, we find the ADM mass

$$M_{\rm ADM} = \lim_{r\to\infty} r^2 \frac{\partial \lambda}{\partial r} = \lim_{s\to 1} r_e s^2 \frac{\partial \lambda}{\partial s}. \tag{101}$$

On the other hand, the Komar mass, $M_{\rm K}$ is obtained from the integral

$$M_{\rm K} = \lim_{r\to\infty} \frac{1}{2} \int_0^\pi d\theta \sin\theta \left(r^2 B \frac{\partial \nu}{\partial r} - \frac{1}{2} \sin^2\theta r^4 B^3 e^{-4\nu} \omega \frac{\partial \omega}{\partial r} \right). \tag{102}$$

For stationary spacetimes, the total mass-energy is $M = M_{\rm ADM} = M_{\rm K}$ and since we use a compactified grid, we can evaluate M accurately from (101).

The Komar-charge of the black hole, $M_{\rm H}$, is equal to

$$M_{\rm H} = M - M_{\rm T}, \tag{103}$$

and does not coincide with the mass of the black hole, since it also includes part of the negative binding energy between the black hole and the torus. Our numerical results confirm that $M_{\rm H} < M - M_0$. The actual black hole mass should be larger than this difference, because of the negative binding energy. We compute the black hole mass from its circumference, i.e. from

$$M_{\rm BH} = \frac{C_{eq}}{4\pi}, \tag{104}$$

where

$$C_{eq} = \int_0^{2\pi} d\phi \sqrt{g_{\phi\phi}}|_{\theta=\pi/2} = 2\pi h_0 \frac{B}{\lambda}|_{\theta=\pi/2}, \tag{105}$$

is the proper equatorial circumference of the black hole. Since in our coordinate system both B and λ vanish on the horizon, we compute $M_{\rm BH}$ using a 3-point Lagrange extrapolation from near-by grid points. This procedure is highly accurate for vacuum black holes. Even for the most massive disks we examined, we estimate our extrapolation error to be less than 1%. Furthermore, our results confirm the finding of [18] that $M_{\rm BH}$ is larger than $M_{\rm H}$ by $\sim 0.1 M_0$.

The gravitational potential energy of the torus is

$$W_{\rm T} := M - M_{\rm BH} - M_0 - T_{\rm T} - U_{\rm T}, \tag{106}$$

and includes both self-energy and binding energy between the black hole and the torus.

The total angular momentum of the spacetime can be obtained from the Komar integral

$$J = \lim_{r\to\infty} -\frac{1}{8}\int_0^{\pi} d\theta \sin^3\theta r^4 B^3 e^{-4\nu}\frac{\partial\omega}{\partial r}, \tag{107}$$

which, for stationary spacetimes, agrees with the asymptotic expression

$$J = -\frac{1}{6}r^4\frac{\partial\omega}{\partial r} = -\frac{1}{6}r_e^3\frac{s^4}{(1-s)^2}\frac{\partial\omega}{\partial s}. \tag{108}$$

For the zero-angular-velocity black holes considered here, we find that the numerical determination of J and $J_{\rm T}$ agree to better than 1%. Part of this small difference is due to the fact that black holes with $\omega_h = 0$ have a small negative angular momentum in the presence of a disk.

Acknowledgments

I would like to thank Burkhard Zink, Oleg Korobkin and Marcus Ansorg for useful discussions and I am grateful to the University of Tuebingen for hospitality during an extended visit.

References

[1] Woosley S E 1993 *Astrophys. J.* **405** 273
[2] Woosley S E and Bloom J S 2006 *Annual Rev. Astron. Astrophys.* **44** 507
[3] Ruffert M, Janka H Th and Schäfer G 1996 *Astron. Astrophys.* **311** 532
[4] Duez M D 2010 *Class. Quant. Grav.* **27** 114002
[5] Rezzolla L, Baiotti L, Giacomazzo B, Link D and Font J A 2010 *Class. Quant. Grav.* **27** 114105
[6] Shibata M and Taniguchi K 2006 *Phys. Rev. D* **73** 064027
[7] Shibata M, Kyutoku K, Yamamoto T and Taniguchi K 2009 *Phys. Rev. D* **79** 044030
[8] Abramowicz M A, Calvani M and Nobili L 1983 *Nature* **302** 597
[9] Papaloizou J C B and Pringle J E 1984 *Mon. Not. R. Astron. Soc.* **208** 721
[10] Papaloizou J C B and Pringle J E 1985 *Mon. Not. R. Astron. Soc.* **213** 799
[11] Kojima Y 1986 *Prog. Theor. Phys.* **75** 251
[12] Woodward J W, Tohline J E and Hachisu I 1994 **420** 247
[13] Font J A and Daigne F 2002 *Mon. Not. R. Astron. Soc.* **334** 383
[14] Zanotti O, Rezzolla L and Font J A *Mon. Not. R. Astron. Soc.* **341** 832
[15] Abramowicz M, Jaroszynski M and Sikora M 1978 *Astron. Astrophys.* **63** 221
[16] Nishida S and Eriguchi Y 1994 *Astrophys. J.* **427** 429
[17] Korobkin O, Abdikamalov E B, Schnetter E, Stergioulas N and Zink B 2010 *Preprint* arXiv/1011.3010
[18] Shibata M 2007 *Phys. Rev. D* **76** 064035

Lensing in an elastically strained space-time

Angelo Tartaglia[1], **Ninfa Radicella**[2] **and Mauro Sereno**[1]

[1] Politecnico di Torino, DIFIS, Corso Duca degli Abruzzi 24, Torino, and INFN, Torino, Italy
[2] Universitat Autónoma de Barcelona, Departament de Física, Spain

E-mail: angelo.tartaglia@polito.it,ninfa.radicella@uab.cat,mauro.sereno@polito.it

Abstract. Describing curved space-time as a four-dimensional manifold strained by the presence of matter or of texture defects, an additional term in the Lagrangian of space-time has to be introduced besides the Ricci scalar, accounting for the strain. The additional term produces dark matter-like effects around any given body. These effects show up both in the angular speed of freely orbiting objects and in the gravitational lensing of light. These results are obtained and discussed while treating a spherically symmetric stationary space-time configuration.

1. Introduction

A problem hovering above the general relativity theory (GR) from the very beginning is the nature of space-time. It cannot be a simple mathematical artifact since it interacts with matter, so in a way or another it must have physical properties on its own.

Of course we know that space-time is a curved four-dimensional manifold endowed with Lorentzian signature, and we also know that space, at the cosmic scale, appears to be expanding with a typical symmetry: the Robertson-Walker (RW) symmetry. Furthermore some pieces of evidence tell us that the expansion is accelerated [1, 2]. This acceleration is not due to matter, so commonly it is attributed to "something else" which is dubbed in various ways according to different theories, but is mostly known as *dark energy* [3, 4, 5, 6, 7, 8, 9, 10, 11, 12, 13]. What is dark energy? There are as many answers as there are theories, but it is difficult to say that the situation is clear and that each interpretation of dark energy is physically well motivated.

A possible interpretation of the facts is given by the Strained State Theory (SST). It considers space-time as a four- dimensional deformable continuum with properties inferred by analogy and generalization from the corresponding ones of ordinary three-dimensional elastic continua. The SST theory is exposed in [14] and will be shortly reviewed in the next section. There are two relevant features with SST: a) the global symmetry of the universe is due to a defect in the texture of space-time working as defects do in material continua; b) the deformation of the manifold, due both to defects and to the presence of matter is expressed by a strain tensor which coincides with the non-trivial part of the metric tensor. To the strain tensor an *elastic* potential energy density corresponds, whose presence affects the global behavior of space-time and is responsible for the accelerated expansion. The theory has already been used to work out the luminosity/distance curve of type Ia supernovae with good results [14]; it is also consistent with the relative abundances of light elements in the universe (Big Bang Nucleosynthesis: BBN), the acoustic scale of the CMB and the primordial large scale structure (LSS) formation (these results are the object of another paper under consideration for publication). Here we shall

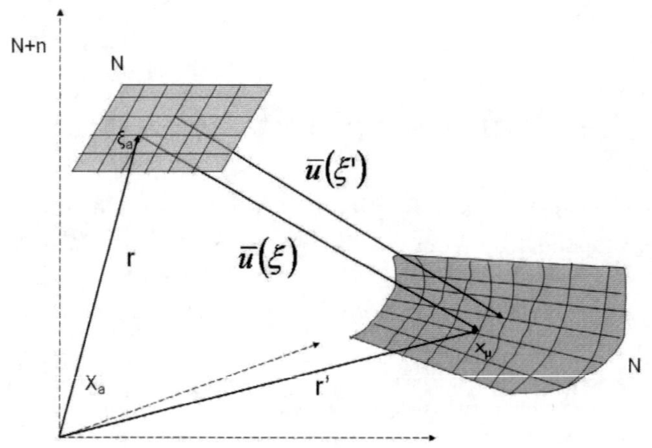

Figure 1. On the left we see an N-dimensional flat manifold (the reference manifold). On the right we have the natural N-dimensional manifold. The latter is curved. The curvature, in this case, is due to the presence of a defect which is indicated by the fact that a whole region of the flat manifold corresponds to a single point in the natural manifold. For the visualization everything is embedded in a flat $N+1$-dimensional manifold.

concentrate on the implications of a strained state in the vicinity of matter in a spherically symmetric space (typical Schwarzschild problem).

2. Review of the SST theory

As we have written in the introduction the core idea of the Strained State Theory is that the actual space-time manifold with its global and local curvature behaves as a four-dimensional elastic continuum, so that one may think that the natural situation is obtained introducing strain in an initially flat (Minkowski) manifold. The properties of the strained manifold are expressed in terms of two parameters, which are the Lamé coefficients of space-time, λ and μ. The details of the theory may be found in [14]; here we only recollect that the global RW symmetry is assumed to be induced by a cosmic defect and that the actual metric tensor is composed of two contributions:

$$g_{\mu\nu} = \eta_{\mu\nu} + 2\epsilon_{\mu\nu}, \tag{1}$$

where $\eta_{\mu\nu}$ is the Minkowski metric tensor and $\epsilon_{\mu\nu}$ is the strain tensor.

The situation is schematically reproduced in fig. 1, where the effect of a defect is also shown.

The strain is obtained comparing two corresponding line elements in the reference and in the natural manifold. Of course everything must be expressed in one single coordinates system. Most often it will be the one used for the natural frame. Again a picture may help in understanding what we mean. Fig. 2 shows the situation in a simplified three-dimensional embedding with a natural manifold endowed with axial symmetry.

According to the analogy with a deformed elastic material we expect the strain to be

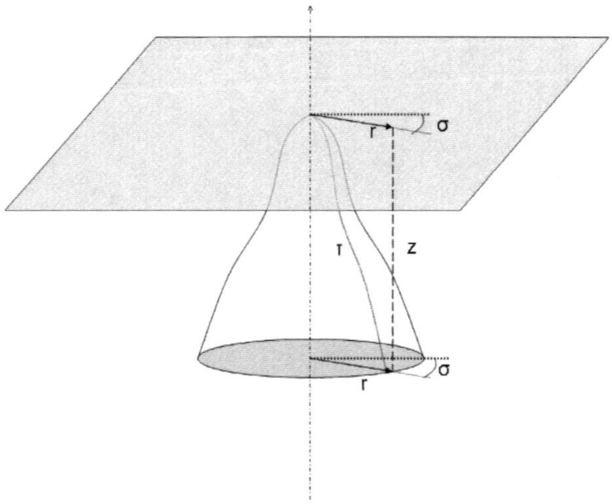

Figure 2. Bidimensional view of a strained axially symmetric manifold compared with its flat reference manifold. Corresponding line elements are different on the two manifolds, as well as the metric tensors. The strain tensor is proportional to the difference between the two metric tensors written with the same coordinates.

associated with an elastic deformation energy written as:

$$W_e = \frac{1}{2}\lambda\epsilon^2 + \mu\epsilon_{\alpha\beta}\epsilon^{\alpha\beta}. \tag{2}$$

Now $\epsilon = \epsilon^\alpha_\alpha$ is the trace of the strain tensor and λ and μ are the Lamé coefficients of space-time. Here the meaning of λ and μ, notwithstanding the four dimensions and the signature, is exactly the same as in the ordinary elasticity theory in three dimensions. We assume, for simplicity, that the "elasticity" of space-time is indeed linear.

The energy in (2) allows to write a new action integral including both space-time and matter/energy in the form

$$S = \int (R + \frac{1}{2}\lambda\epsilon^2 + \mu\epsilon_{\alpha\beta}\epsilon^{\alpha\beta} + L_{\text{matter}})\sqrt{-g}d^4x. \tag{3}$$

L_{matter} is the Lagrangian density of ordinary matter/energy and the rest has the usual meaning. The "elastic" potential energy term belongs to geometry, i.e. space-time, even though it looks like some matter contribution. Considered from the field theoretical viewpoint this new contribution implies a "mass" associated with the gravitational interaction; usually this is said as the graviton being massive, which fact has relevant consequences when studying the propagation of gravity and, in particular, of gravitational waves [12, 15, 16, 17]. In any case our conceptual framework is entirely classical so that, properly speaking, there are no gravitons, but rather it turns out that gravity has a finite range [18, 19].

As we have seen, everything depends on the strain tensor, which in turn depends on the way events on the reference manifold are associated to their corresponding events in the natural manifold. Actually there are infinite possible ways to get a given final situation starting from

Journal of Physics: Conference Series **283** (2011) 012037 doi:10.1088/1742-6596/283/1/012037

a flat initial one. This apparent freedom of choice has indeed a physical meaning since our manifolds are physical. So different choices correspond to different strains and the Hamilton principle permits to identify the least strain configuration. In practice what can be seen as a gauge freedom appears as a gauge function in the line element of the flat reference written using the coordinates of the curved natural manifold.

3. Specific symmetries

3.1. Robertson Walker symmetry

This case has been treated in [14]. Here we simply quote the result for the Hubble parameter of a universe containing matter (without further distinction) and radiation. It is:

$$H = \frac{\dot{a}}{a} = c\sqrt{\frac{B}{16}}\left\{3\left(1 - \frac{(1+z)^2}{a_0^2}\right)^2 + \frac{8\kappa}{3B}(1+z)^3\left[\rho_{m0} + \rho_{r0}(1+z)\right]\right\}^{1/2}. \tag{4}$$

Of course a is the scale factor of the universe; dot means derivative with respect to cosmic time; z is the cosmic redshift; ρ_{m0} is the present matter density and ρ_{r0} the present radiation energy density; $\kappa = 16\pi G/c^2$. The B parameter is interpretable as the bulk modulus of space-time; it is a function of the Lamé coefficients:

$$B = \frac{\mu}{4}\frac{2\lambda + \mu}{\lambda + 2\mu} \tag{5}$$

The best fit made on the luminosity/distance curve of type Ia supernovae tells us the order of magnitude of B, then of λ and μ. It is $B \simeq 10^{-52}\ m^{-2}$ [14]. The optimal values of the parameters from the fit are:

Table 1. Optimal values.

Parameter	Value
B	$(2.22 \pm 0.06) \times 10^{-52}\ m^{-2}$
ρ_{m0}	$(0.260 \pm 0.009) \times 10^{-26}\ kg \times m^{-3}$
B_{a0}^{-1}	$(0.011 \pm 0.006) \times 10^{52}\ m^2$

It is $B_{a0}^{-1} = 8\kappa\rho_{r0}a_0^4/9$.

4. The Schwarzschild symmetry

In order to treat the case of a spherically symmetric time independent configuration in space we start from the general form of a stationary spherically symmetric line element written using Schwarzschild coordinates:

$$ds^2 = fd\tau^2 - hdr^2 - r^2\left(d\theta^2 + \sin^2\theta d\phi^2\right) \tag{6}$$

where f and h are functions of r only.

This line element is confronted with the line element of a flat Minkowskian manifold, written with the same coordinates:

$$ds^2 = d\tau^2 - \left(\frac{dw}{dr}\right)^2 dr^2 - w^2\left(d\theta^2 + \sin^2\theta d\phi^2\right) \tag{7}$$

Now w is a gauge function depending on r, which is there because of the multiple ways in which we may establish a correspondence between pairs of events on the two manifolds (see fig. 1).

From the comparison between the two line elements we read out the components of the strain tensor of the natural manifold. It is:

$$\begin{array}{r}\varepsilon_{00}=\frac{f-1}{2}\\ \varepsilon_{rr}=-\frac{h-w'^2}{2}\\ \varepsilon_{\theta\theta}=-\frac{r^2-w^2}{2}\\ \varepsilon_{\phi\phi}=-\frac{r^2-w^2}{2}\sin^2\theta\end{array}\tag{8}$$

Primes denote derivatives with respect to r.

Once we have the strain tensor at hands, we may introduce it into the action integral (3), then applying the Hamilton principle we obtain the equations for the three unknown functions f, h and w. The explicit form of the equations is highly non-linear and rather formidable, as it can be seen hereafter:

$$\begin{array}{c}h-1+r\frac{h'}{h}-\frac{1}{16f^2h}\lambda r^2\left(2fh\frac{w^2}{r^2}-4fh-3h+fw'^2\right)\left(h-4fh+2fh\frac{w^2}{r^2}+fw'^2\right)\\ -\frac{1}{8f^2h}\mu r^2\left(2fh^2+4f^2h^2+2f^2h^2\frac{w^4}{r^4}-3h^2+f^2w'^4-4f^2h^2\frac{w^2}{r^2}-2f^2hw'^2\right)=0\end{array}\tag{9}$$

$$\begin{array}{c}h^2-h-\frac{1}{f}hrf'\\ -\frac{1}{16f^2}\lambda r^2\left(h-4fh+2fh\frac{w^2}{r^2}-3fw'^2\right)\left(h-4fh+2fh\frac{w^2}{r^2}+fw'^2\right)\\ -\frac{1}{8f^2}\mu r^2\left(h^2+4f^2h^2+2f^2h^2\frac{w^4}{r^4}-2fh^2-3f^2w'^4-4f^2h^2\frac{w^2}{r^2}+2f^2hw'^2\right)=0\end{array}\tag{10}$$

$$\begin{array}{c}\frac{\lambda}{2fh^2}w''\left(hr^2+3fr^2w'^2-4fhr^2+2fhw^2\right)\\ +\frac{\lambda}{h}ww'^2+\frac{\lambda r}{h^2}\left(\frac{f'}{4f}r-\frac{3h'}{4h}r+1\right)w'^3\\ +\lambda\frac{w'}{h}\left(\left(\frac{1}{2}w^2-r^2-\frac{1}{4f}r^2\right)\frac{f'}{f}+\left(r^2-\frac{1}{2}w^2-\frac{1}{4f}r^2\right)\frac{h'}{h}+\frac{1}{f}r-4r\right)\\ +\lambda w\left(4-\frac{2}{r^2}w^2-\frac{1}{f}\right)+\mu\frac{r^2}{h^2}w''\left(3w'^2-h\right)\\ +\frac{\mu}{h^2}\left(2r-\frac{3}{2h}r^2h'+\frac{1}{2f}r^2f'\right)w'^3\\ +\mu\frac{r}{h}\left(\frac{h'}{2h}r-2-\frac{f'}{2f}r\right)w'+2w\mu\left(1-\frac{w^2}{r^2}\right)=0\end{array}\tag{11}$$

4.1. Approximate solutions

The task of solving eqs. (9), (10), (11) is rather desperate, however we may look for approximated solutions under the hypothesis of weak enough effects. Actually from the numerical value for the Lamé coefficients inferred from subsection 3.1 we see that up to the scale of stellar systems or even galaxies, the dimensionless quantities λr^2 and μr^2 stay well below 1. By consequence we try solutions of the type:

$$\left\{\begin{array}{l}f\simeq f_0+f_1\\ h\simeq h_0+h_1\\ w\simeq w_0+w_1\end{array}\right.\tag{12}$$

assuming that the f_1, h_1 and w_1 functions are proportional to λr^2 and μr^2.

The explicit forms for f_0, f_1, etc. are then looked for recursively and perturbatively.

For the zero order of f and h we easily have the Schwarzschild solution:

$$h_0=\frac{1}{1-2\frac{m}{r}}\tag{13}$$

$$f_0 = \frac{1}{h_0} = 1 - 2\frac{m}{r} \tag{14}$$

Considering the fact that this solution is now referred to a deformable space-time, the central symmetry can be due to a matter distribution, in which case m is a mass, as well as to a linear space-time defect[1], in which case m is a parameter measuring the size and type of defect so that it can assume also negative values. Limiting our attention to the presence of an actual mass, we can also remark that in all physical situations of interest it is $m/r << 1$ even though at the scales of stellar systems or galaxies it remains many orders of magnitude bigger than λr^2 or μr^2.

Using (13) and (14) into eq. (11) we obtain the zero order solution for w in the form:

$$w_0 = \sum_{k=-1}^{\infty} a_k r^{-k} \tag{15}$$

It is $a_{-1} = 1$ and the other coefficients are all determined recursively. In practice the lowest order of w_0 is trivially r which corresponds to the natural frame coinciding with a flat Minkowski frame.

When we go back to (9) and (10), inserting the result we have found for w_0, we are enabled to calculate the first order approximation for f and h. Again we obtain a series solution whose elements, excepting the first, are all proportional to powers of m/r times powers of λr^2 and μr^2. The whole procedure is tedious but straightforward.

At the end we are left with the lowest order meaningful solution for the line element in the form:

$$\begin{aligned} ds^2 = &\left(1 - 2\tfrac{m}{r} + \tfrac{1}{8}\left(7\lambda + 2\mu\right) r^2\right) d\tau^2 - \left(\tfrac{1}{1-2\frac{m}{r}} + \tfrac{1}{4}\left(5\lambda + 2\mu\right) r^2\right) dr^2 \\ &-r^2\left(d\theta^2 + \sin^2\theta d\phi^2\right) \end{aligned} \tag{16}$$

5. Expected effects

Working on (16) we may deduce some effects of the strain that we would expect to see. For short let us put

$$\left\{ \begin{array}{l} \Psi = \frac{1}{8}\left(7\lambda + 2\mu\right) \\ \Lambda = \frac{1}{4}\left(5\lambda + 2\mu\right) \end{array} \right. \tag{17}$$

First we may identify various ranges of interest according to where the balance between $2m/r$ and Ψr^2 or Λr^2 is reached. Values are found in table 2:

Table 2. Typical ranges of the approximation in the text.

Case	Range in metres
Stars	$r \sim 10^{18}$
Galaxy	$r \sim 10^{22}$
Black hole at the galactic center	$r \sim 10^{20}$

[1] Linear in four dimensions: the spherical space symmetry is around a timelike straight worldline.

Out of the full study of the geodesics of line element (16) we may single an interesting case: the circular orbits of a test particle ($r = R$). The relevant quantity we can calculate is the angular velocity which is:

$$\frac{\omega^2}{c^2} = \frac{m}{R^3} + \Psi \qquad (18)$$

We see that the strain of space-time tends to increase the angular speed of a freely orbiting object by a constant amount with respect to the corresponding Keplerian orbit, provided $\Psi > 0$. This effect works in the same sense as the effect of dark matter. Of course we should keep in mind that, while going farther and farther the approximation we have made breaks down, because the dependence on the Lamé coefficients becomes much more complicated.

Another interesting phenomenology concerns the propagation of light. Let us consider the regions where the mass terms are non-negligible but comparable with the SST terms. The implicit equation of a light ray is:

$$\left(\frac{dr}{d\phi}\right)^2 = \frac{r^4}{b^2}\left(1 - \frac{b^2}{r^2} - \Psi r^2\right) - 2\frac{mr^3}{b^2}\left(1 - 2\frac{b^2}{r^2}\right) \qquad (19)$$

The b quantity is the geometric impact factor (the distance at which the light ray would pass from the central mass if the space-time were flat).

Again we see that the SST term acts in the same sense as the mass enhancing the curvature of the light ray.

6. Conclusion

We have applied the SST theory to the special and fundamental case of a space-time endowed with a spherical symmetry in space and not depending on time. The full treatment has necessarily been an approximated one, because of the complexity of the equations for the g_{00} and g_{rr} terms of the metric tensor and for the gauge function w. A perturbative treatment has been used with the small quantities λr^2 and μr^2. The condition is satisfied for scales up to $\sim 10^{20}$ m or so. Closer to the center of symmetry we may identify two other peculiar ranges. A near region, where the mass is dominating and the zero order solution for the metric coincides with the Schwarzschild solution. An intermediate region where the mass terms are comparable with the SST terms, so that a development in powers of m/r is in order. Studying the properties of the approximated metric tensor, we have found that the contribution from the strain of space-time produces a constant (at least in the intermediate region) increase of the angular velocity of a freely orbiting test particle. Furthermore we have seen that also the lensing effect about the center of symmetry is intensified. Both effects look like the ones commonly attributed to the dark matter. Quantitative evaluations do not allow for the moment to conclude that SST can completely eliminate the need for dark matter, however they go in the right direction.

The solution of the problem we have outlined in the present paper, though approximated, is a step forward in the determination of all implications of SST and follows a number of positive tests the theory has undergone, while being confronted with observation. Altogether the scenario which emerges is a physically motivated one and gives more and more confidence on the viability of SST as an interesting paradigm to describe the properties of our universe at large enough scales.

References

[1] S. Perlmutter *et al.*, *APJ*, 517:565–586, 1999.
[2] A. G. Riess *et al.*, *AJ*, 116:1009–1038, 1998.
[3] M. Kamionkowski, *ArXiv: 0706.2986*, 2007.

[4] A. Einstein, *Sitzungsberichte der Königlich Preußischen Akademie der Wissenschaften (Berlin)*, pages 142–152, 1917.
[5] A. Einstein, *Sitzungsberichte der Königlich Preußischen Akademie der Wissenschaften*, page 235, 1931.
[6] I. Zlatev, L. Wang, and P. J. Steinhardt, *Phys. Rev. Lett.*, 82:896–899, 1999.
[7] S. M. Carroll, *Phys. Rev. Lett.*, 81:3067–3070, 1998.
[8] R. R. Caldwell, *Phys. Lett. B*, 545:23–29, 2002.
[9] A. Vikman, *Phys. Rev. D*, 71(2):023515, 2005.
[10] Y.-F. Cai, H. Li, Y.-S. Piao, and X. Zhang, *Phys. Lett. B*, 646:141–144, 2007.
[11] M. Li, *Phys. Lett. B*, 603:1–5, 2004.
[12] N. Arkani-Hamed, H. Georgi, and M. D. Schwartz, *Ann. Phys.*, 305:96–118, 2003.
[13] C. Armendariz-Picon, V. Mukhanov, and P. J. Steinhardt, *Phys. Rev. Lett.*, 85:4438–4441, 2000.
[14] A. Tartaglia and N. Radicella, *Class. Q. Grav.*, 27(3):035001, 2010.
[15] M. Visser, *Gen. Rel. Grav.*, 30:1717–1728, 1998.
[16] T. Damour and I. I. Kogan, *Phys. Rev. D*, 66(10):104024–+, 2002.
[17] C. Deffayet, G. Dvali, and G. Gabadadze, *Phys. Rev. D*, 65(4):044023–+, 2002.
[18] D. G. Boulware and S. Deser, *Phys. Rev. D*, 6:3368–3382, 1972.
[19] S. V. Babak and L. P. Grishchuk, *Int. J. Mod. Phys. D*, 12:1905–1959, 2003.

Black holes, holography and thermodynamics of gauge theories

Nikolaos Tetradis

Department of Physics, University of Athens, Zographou 157 84, Greece

E-mail: ntetrad@phys.uoa.gr

Abstract. We review work published in [1, 2] on the thermodynamics of gauge theories on FLRW backgrounds through the AdS/CFT correspondence. Contrary to analogous studies of the so-called Bjorken flow, the backgrounds we consider are homogeneous and isotropic. This permits the exact determination of the gravitational dual, which is identified with the AdS-Schwarzschild geometry, expressed in coordinates that set the boundary metric in the FLRW form. We review the calculation of the stress-energy tensor of the dual CFT on the FLRW background, as well as the temperature and entropy of the CFT, which are related to the temperature and entropy of the black hole. We also derive the equation of cosmological evolution through the use of appropriate boundary conditions.

1. Introduction

An interesting application of the AdS/CFT correspondence [3] concerns the thermodynamics of strongly coupled gauge theories, such as the $\mathcal{N}=4$ super-Yang-Mills theory with $SU(N)$ gauge symmetry in the large-N limit, in four dimensions. The thermodynamics can be deduced from the properties of a Schwarzschild black hole in five-dimensional AdS space [4]. For a Minkowski boundary the CFT entropy is proportional to the volume of the black-hole event horizon, in agreement with the standard notions of black-hole entropy [5] and holography [6].

The AdS/CFT correspondence has also been employed for the derivation of properties of the quark-gluon plasma produced in heavy-ion collisions. This system can be modelled through the so-called Bjorken flow [7]. The construction of the space-time dual to this flow [8, 9] has led to the derivation of properties of the plasma, such as the ratio of shear viscosity to entropy density [10]. The holographic description of more general hydronamic regimes of strongly coupled gauge theories has also been studied [11]. The construction of gravity solutions dual to the Bjorken flow or its generalizations is difficult and has been achieved only through a perturbative expansion. The main reason is the anisotropic nature of the geometry. In general, one finds that the gravity duals are time-dependent. They also possess event and apparent horizons that can be distinct. It is not obvious which horizon is relevant for properties of the gauge theory, such as its entropy [12].

The above difficulties have motivated the study of properties of CFT on a time-dependent background, for which the gravity dual is exactly known [1, 2]. The theory is the standard $\mathcal{N}=4$ super Yang-Mills theory at non-zero temperature on a Friedmann-Lemaitre-Robertson-Walker (FLRW) background. The gravity dual is the AdS-Schwazschild geometry, but in a parametrization that sets the boundary metric in the FLRW form. The appropriate form of the metric was given in [1], where the stress-energy tensor of the dual CFT was also derived. The boundary metric can be made dynamical by including a boundary Einstein term and imposing mixed boundary conditions along the lines of [13]. This construction leads to the effective Friedmann equation governing the cosmological evolution [1].

The temperature and entropy of the CFT on a FLRW background were derived in [2]. The temperature can be deduced by requiring the smoothness of the five-dimensional metric. When the temperature and stress-energy tensor of the CFT are known, the entropy can be computed. It is not neccesary to assume that the entropy is proportional to the surface of the black-hole horizon. Moreover, between the distinct event and apparent horizons, we can identify the one that is relevant for the CFT on a time-dependent background.

2. Minkowski boundary

We consider an AdS-Schwarzschild black hole in five dimensions, which is a solution of the Einstein field equations with a negative cosmological constant ($\Lambda_5 = -6/l^2$),

$$R_{AB} = \frac{2}{l^2} g_{AB}\ , \tag{1}$$

where $A, B = 0, 1, 2, 3, 4$. The metric can be written in Schwarzschild coordinates as

$$ds^2 = -f(r)dt^2 + \frac{dr^2}{f(r)} + r^2 d\Omega_k^2\ , \qquad f(r) = r^2 + k - \frac{\mu}{r^2}\ , \tag{2}$$

where $k = +1, 0, 1$ for spherical, flat and hyperbolic horizons, respectively. We have set $l = 1$ for simplicity. The Hawking temperature can be computed through the elimination of the conical singularity in the Euclidean version of the metric [14]. The energy (or mass) of the black hole can be computed through the temperature dependence of the gravitational action on the AdS-Schwarzschild background, after subtraction of the action of pure AdS space with compactified time [4, 14]. The temperature and energy are, respectively,

$$T = \frac{2r_e^2 + k}{2\pi r_e}\ , \qquad E = \frac{3V_k}{16\pi G_5} r_e^2 (r_e^2 + k) = \frac{3\mu V_k}{16\pi G_5}\ , \tag{3}$$

where V_k is the volume of the three-dimensional space Σ_k spanned by Ω_k, r_e is the radius of the event horizon, satisfying $f(r_e) = 0$, and G_5 is the five-dimensional Newton's constant.

The case of a spherical horizon with $k = 1$ is important for the interpretation of the results reviewed in the following sections. For increasing $\mu \gg 1$ the temperature T increases as $\mu^{1/4}/\pi$. For decreasing $\mu \ll 1$, T again increases as $1/(2\pi\, \mu^{1/2})$. There are no black holes with T below a minimal value $\sqrt{2}/\pi$. For larger T, there are two black-hole solutions with the same temperature and different masses. The one with lower mass is known to be unstable [14]. The solution with the larger mass is dual to the high-temperature deconfined phase of the gauge theory [4].

In order to discuss the AdS/CFT correspondence, we must express the metric in terms of Fefferman-Graham coordinates [15]. We define z through $dz/z = -dr/\sqrt{f(r)}$, which gives (with an appropriate integration constant)

$$z^4 = \frac{16}{k^2 + 4\mu} \, \frac{r^2 + \frac{k}{2} - r\sqrt{f(r)}}{r^2 + \frac{k}{2} + r\sqrt{f(r)}}\ . \tag{4}$$

This equation may be inverted to give

$$r^2 = \frac{\alpha + \beta z^2 + \gamma z^4}{z^2}\ , \qquad \alpha = 1\ , \qquad \beta = -\frac{k}{2}\ , \qquad \gamma = \frac{k^2 + 4\mu}{16}\ . \tag{5}$$

The metric (2) reads

$$ds^2 = \frac{1}{z^2}\left[dz^2 - \frac{\left(1-\gamma z^4\right)^2}{1+\beta z^2+\gamma z^4}dt^2 + \left(1+\beta z^2+\gamma z^4\right)d\Omega_k^2\right] . \quad (6)$$

The coordinates (τ, z) are closely related to the isotropic coordinates that are often employed for the study of the Schwarzschild geometry. The isotropic coordinates do not span the full space. They cover the two regions of the Kruskal-Szekeres plane that are located outside the horizons. The same happens for the coordinates (τ, z) in the case of the AdS-Schwazschild geometry. As can be deduced easily from eq. (4), the coordinate z corresponds to values of r that satisfy $f(r) \geq 0$. On the other hand, eq. (5) indicates that the region (r_e, ∞) of r is covered twice by the coordinate z taking values in $(0, \infty)$. For fixed coordinate time t, the metric describes the Einstein-Rosen bridge connecting two asymptotically flat regions [16].

The Hawking temperature T of the black hole [14] can be determined from the metric (6). The partition function of a static thermalized system can be obtained if we switch to Euclidean time and assume that the time direction is periodic with period $1/T$. For arbitrary T, the resulting metric possesses a conical singularity at the location of the horizon $z_e = \gamma^{-1/4}$. This can be eliminated for a specific value of T, which is identified with the natural temperature of the black hole. Expanding z around z_e, we find that the conical singularity disappears for T given by

$$T = \frac{1}{\sqrt{2}\,\pi}\left(\frac{k^2+4\mu}{\left(k^2+4\mu\right)^{1/2}-k}\right)^{1/2}, \quad (7)$$

in agreement with the first of eqs. (3).

The energy and entropy of the black hole can be related to those of the dual CFT on the AdS boundary. For a metric in the form

$$ds^2 = \frac{1}{z^2}\left[dz^2 + g_{\mu\nu}dx^\mu dx^\nu\right], \qquad g_{\mu\nu} = g^{(0)}_{\mu\nu} + z^2 g^{(2)}_{\mu\nu} + z^4 g^{(4)}_{\mu\nu} + \ldots \quad (8)$$

the stress-energy tensor of the CFT is [17]

$$T^{(CFT)}_{\mu\nu} = \frac{1}{4\pi G_5}\left\{g^{(4)} - \frac{1}{2}g^{(2)}g^{(2)} + \frac{1}{4}\mathrm{Tr}\left[\mathrm{g}^{(2)}\right]\mathrm{g}^{(2)} - \frac{1}{8}\left(\left(\mathrm{Tr}\left[\mathrm{g}^{(2)}\right]\right)^2 - \mathrm{Tr}\left[\mathrm{g}^{(2)}\mathrm{g}^{(2)}\right]\right)\mathrm{g}^{(0)}\right\}_{\mu\nu}, \quad (9)$$

with indices lowered and raised by $g^{(0)}$ and its inverse. Applying this general expression to our metric (6), we obtain the energy density and pressure, respectively,

$$T^{(CFT)}_{tt} = 3T^{(CFT)}_{ii} = \frac{3\left(k^2+4\mu\right)}{64\pi G_5}, \quad (10)$$

on a four-dimensional space with Minkowski metric $g^{(0)}_{\mu\nu}$. The total energy $E = T^{(CFT)}_{tt}V_k$ is larger than the mass of the black hole (eq. (3)) by a constant (Casimir energy) in the case of a curved horizon ($k \neq 0$). The two quantities agree for flat horizons ($k = 0$).

In an intuitive derivation of the entropy, we consider an infinitesimal variation of the parameter μ of the metric (6). The volume V_k of the boundary is not affected. The variation of μ generates a variation of the internal energy E of the system that can be attributed to a change of its entropy S. Assuming that the process takes place sufficiently slowly, the variations of E and S obey $dE = TdS$. The integration with respect to μ gives

$$S = \frac{V_k}{4G_5}r_e^3. \quad (11)$$

Journal of Physics: Conference Series **283** (2011) 012038 doi:10.1088/1742-6596/283/1/012038

As expected, the entropy is proportional to the surface of the event horizon. It must be pointed out that the above determination of the entropy takes into account only the degrees of freedom that are directly related to the thermalized CFT. The variation of E with respect to μ eliminates the contribution from the Casimir energy.

3. FLRW boundary

We are interested in generalizing the previous discussion to the case of a time-dependent boundary. If we assume that the spatial part is maximally symmetric, the only allowed possibility is a boundary with the form of a FLRW spacetime

$$ds_0^2 = g^{(0)}_{\mu\nu} dx^\mu dx^\nu = -d\tau^2 + a^2(\tau) d\Omega_k^2 \ . \tag{12}$$

In order to apply holographic renormalization, we choose a different foliation away from the black hole, consisting of hypersurfaces whose metric is asymptotically of the form (12). To achieve this we need to make a change of coordinates $(t, r) \to (\tau, z)$ and bring the black-hole metric (2) to the form

$$ds^2 = \frac{1}{z^2}\left[dz^2 - \mathcal{N}^2(\tau, z)d\tau^2 + \mathcal{A}^2(\tau, z)d\Omega_k^2\right] \ , \tag{13}$$

where $\mathcal{N}, \mathcal{A} \geq 0$ and $\mathcal{N}(\tau, z) \to 1$, $\mathcal{A}(\tau, z) \to a(\tau)$ as we approach the boundary $z = 0$. Comparison with the static case (eq. (5)) suggests the ansatz

$$\mathcal{A}^2 = \alpha(\tau) + \beta(\tau) z^2 + \gamma(\tau) z^4 \ , \tag{14}$$

where $\alpha(\tau), \beta(\tau), \gamma(\tau)$ are functions to be determined. The solution of the Einstein equations gives [1]

$$\mathcal{N} = \frac{\dot{\mathcal{A}}}{\dot{a}} \ , \qquad \alpha = a^2 \ , \qquad \beta = -\frac{\dot{a}^2 + k}{2} \ , \qquad \gamma = \frac{(\dot{a}^2 + k)^2 + 4\mu}{16 a^2} \ , \tag{15}$$

where we fixed the integration constants by comparing with the static case (eq. (5)).

The derivation of the metric (13) has been achieved without reference to the static AdS-Schwarzschild metric of eq. (2). The two metrics can be shown to agree through a coordinate transformation. We do not need explicit expressions for the functions $t(\tau, z)$, $r(\tau, z)$, because we base our discussion on the metric in the (τ, z) coordinates (eqs. (13) - (15)). However, it must be emphasized that the coordinate transformation is singular at the location of the event horizon, where $f(r_e) = 0$.

The coordinates (τ, z), in terms of which the AdS-Schwazschild metric takes the time-dependent form (13), are a generalization of the isotropic coordinates we considered in the previous section. This is obvious, as eq. (13) reduces to eq. (6) for $a = 1, \dot{a} = \ddot{a} = 0$. The difference with the static case is that now the coordinate z spans a larger part of the Schwarzschild geometry. It can be shown [2] that

$$(r')^2 = \frac{\dot{a}^2 + f(r)}{z^2}. \tag{16}$$

We denote partial derivatives with respect to z by a prime, and partial derivatives with respect to τ by a dot. For a time-dependent boundary, $\partial r/\partial z$ does not vanish at the location of the event horizon r_e. Instead, for a given value of τ, it vanishes at

$$z_m^2(\tau) = \frac{4a^2(\tau)}{\left((\dot{a}^2 + k)^2 + 4\mu\right)^{1/2}}. \tag{17}$$

This value corresponds to some (in general time-dependent) r_m smaller than r_e. As a result, the new coordinate z spans a patch that reaches behind the event horizon. The region (r_m, ∞) of r is covered twice by the coordinate z taking values in $(0, \infty)$. Similarly to the static case discussed in section 2, for constant τ we have a Einstein-Rosen bridge connecting two asymptotically flat regions. The difference is that the minimal distance from the singularity is smaller than the horizon distance in this case. We emphasize the established fact that incoming photons, emanating from one of the flat regions, cannot cross the bridge and emerge in the other flat region. They always fall into the singularity [16].

A surface that will be important in the following is defined by the function $z_a(\tau)$ that satisfies $\mathcal{N}(\tau, z_a(\tau)) = 0$. For $\mu \neq 0$, $z_a(\tau)$ is given by the smallest root of

$$1 - \frac{\ddot{a}}{2a} z_a^2 + \frac{(\dot{a}^2 + k)(2a\ddot{a} - \dot{a}^2 - k) - 4\mu}{16a^4} z_a^4 = 0, \tag{18}$$

which is

$$z_a^2(\tau) = \frac{4a^2(\tau)}{a\ddot{a} + \left((\dot{a}^2 - a\ddot{a} + k)^2 + 4\mu\right)^{1/2}}. \tag{19}$$

Again, $z_a(\tau)$ corresponds to values r_a smaller than r_e.

An interesting example is obtained for $a(\tau) = \lambda\tau$. It is analogous to the Bjorken flow, but with isotropic expansion. We find constant values

$$r_m^2 = r_a^2 = \frac{1}{2}\left[-\tilde{k} + \left(\tilde{k}^2 + 4\mu\right)^{1/2}\right], \tag{20}$$

where $\tilde{k} = k + \lambda^2$. On the other hand, the event horizon has

$$r_e^2 = \frac{1}{2}\left[-k + \left(k^2 + 4\mu\right)^{1/2}\right]. \tag{21}$$

It can be checked that $r_m = r_a \leq r_e$. For $\lambda = 0$ all three surfaces defined by r_m, r_a and r_e coincide.

We would like to determine the horizons of the metric (13). An apparent horizon is defined as the boundary of the region of trapped surfaces. This boundary is a surface on which the expansion of outgoing null geodesics vanishes. The out/ingoing null geodesics obey $(dz(\tau)/d\tau)_\pm = \mp\mathcal{N}(\tau, z)$ and define a surface of areal radius $A(\tau, z(\tau))/z(\tau) = r(\tau)$. The growth of the volume of this surface is proportional to the total time derivative of r along the light path, i.e. to

$$\left(\frac{dr}{d\tau}\right)_\pm = \dot{r} + r'\left(\frac{dz}{d\tau}\right)_\pm = \mathcal{N}\left(\frac{\dot{a}}{z} \mp r'\right), \tag{22}$$

with r' given by eq. (16). The product $(dr/d\tau)_+(dr/d\tau)_- = -\mathcal{N}^2 f(r)/z^2$, which is expected to be invariant under relabelling of the scalars that define null hypersurfaces [9], vanishes on two surfaces. One of them is the event horizon, located at $r = r_e$, with $f(r_e) = 0$. The other is the surface parametrized by $z_a(\tau)$, for which $\mathcal{N} = 0$.

The outgoing geodesics have positive expansion at the location of the event horizon of the metric (2), where $f(r_e) = 0$ and $r' = -\dot{a}/z$. This surface is parametrized now by a function $z_e(\tau)$. On the other hand, the ingoing geodesics have zero expansion there. This behavior is identical to that encountered for the standard four-dimensional Schwarzschild metric in outgoing Eddington-Finkelstein coordinates. The outgoing null geodesics are well-behaved on the horizon, while the horizon is made up from world lines of ingoing photons [16]. It is worth noticing that the situation is reversed for contracting boundary metrics with $\dot{a} < 0$. In this case the event horizon is formed by outgoing photons, while the incoming ones are well-behaved on it. This is identical to what happens in the four-dimensional Schwarzschild geometry in ingoing Eddington-Finkelstein coordinates. Of course, there is nothing mysterious behind this behavior. It originates in the transformation from (t, r) to (τ, z), which is singular on the horizon. It would take an infinite time t for any geodesic to cross the horizon.

The expansion of all null geodesics also vanishes on the surface parametrized by $z_a(\tau)$, for which $\mathcal{N} = 0$. The nature of this surface is strongly linked to the evolving Einstein-Rosen bridge appearing in the system of coordinates we are employing [2]. The fact that no information can be retrieved from behind this surface means that it could define the boundary of trapped surfaces, and can be characterized as an apparent horizon. We mentioned earlier that the surface with $\mathcal{N} = 0$ is located behind the event horizon. This is consistent with the expectation that an apparent horizon can exist only behind an event horizon [18, 19]. The fact that the apparent horizon is located at a smaller areal distance than r_e has another implication: If the entropy in a dynamical situation is related to the area of the apparent horizon, then it is smaller than the maximal entropy of a black hole, given by the area of the event horizon.

We emphasize that we do not derive the entropy of the CFT by identifying it with the area of one of the horizons we discussed. Our approach is to compute the entropy directly from the energy density and the temperature. Only then we examine whether the derived entropy is proportional to a given area.

4. Temperature

A thermalized system fluctuates at the microscopic level with a characteristic time scale of order $1/T$, where T is its temperature. For strongly coupled theories, like the ones we are considering, this scale determines the interaction rates that help maintain the system thermalized. At the macroscopic level, the system, including its temperature, may evolve with a different, much longer, characteristic time scale. A typical example is the Universe at early epochs, during which several of its energy components were thermalized. For the problem we are considering, a Hawking temperature T can be assigned to the AdS-Schwarzschild solution with a time-dependent boundary in the limit that the variation of the scale factor $a(\tau)$ is negligible at time intervals of order $1/T$. This requires $T \gg \dot{a}/a$.

The calculation of the temperature in section 2 for the static case can be repeated by approximating $a(\tau)$ and its time derivatives as constant. A conical singularity appears again, but now it is located at the position of the apparent horizon $z_a(\tau)$. Eliminating the singularity in Euclidean space around the apparent horizon for $\mu \neq 0$ gives

$$T(\tau) = \frac{1}{2\pi}\left|\frac{-4a^3\ddot{a}z_a + \left[\left(\dot{a}^2 + k\right)\left(2a\ddot{a} - \dot{a}^2 - k\right) - 4\mu\right]z_a^3}{4a^3\mathcal{A}_a}\right| = \frac{1}{2\pi}\left|\frac{4a - \ddot{a}z_a^2}{\mathcal{A}_a z_a}\right|, \qquad (23)$$

where $\mathcal{A}_a = \mathcal{A}(\tau, z_a)$ and we have used eq. (18). For constant $a = 1$ the apparent and event horizons coincide, and eq. (23) reproduces the Hawking temperature in the static case, given by eq. (7). For time-dependent a the evolution of the apparent horizon and the temperature depend on the nature of the expansion.

The simplest case is analogous to the Bjorken flow, but for an isotropic expansion. The scale factor is $a(\tau) = \lambda\tau$ with λ constant. The apparent horizon is defined by $z_a^2 = 4a^2/\left(\tilde{k}^2 + 4\mu\right)^{1/2}$, where $\tilde{k} = k + \lambda^2$, or equivalently by eq. (20).

The temperature is given by

$$T = \frac{1}{\sqrt{2}\pi a}\left(\frac{\tilde{k}^2 + 4\mu}{\left(\tilde{k}^2 + 4\mu\right)^{1/2} - \tilde{k}}\right)^{1/2}. \qquad (24)$$

Its time evolution is equivalent to a redshift by the scale factor $a(\tau)$. However, the proportionality constant is not just the temperature in the static case, given by eq. (7). The two expressions differ by the change of the effective curvature $k \to \tilde{k} = k + \lambda^2$. This modification is natural as the total curvature of the boundary metric is proportional to $\tilde{k}$ for $a = \lambda\tau$. For sufficiently large λ we have $\tilde{k} > 0$ for any value of k. For this reason we expect behavior similar to that of a CFT on a sphere. A first consequence is that the temperature diverges for $\lambda^4 \gg \mu$. This is completely analogous to the divergence of the temperature for a static background with $k = 1$ and $\mu \to 0$.

The influence of the curvature resulting from the expansion on the temperature is an important physical effect. The expansion does not result in a simple redshift of the temperature, but induces additional corrections. These are reflected on the entropy as well.

For a general $a(\tau)$ the evolution is more complicated. Of particular interest are the cosmological spaces with $a = \tau^\nu$ and constant ν for large τ. We concentrate on the case $k = 0$. The cases with $k = \pm 1$ can be studied along similar lines. For $0 < \nu < 1$ the expansion is decelerating and for $\tau \to \infty$ we always have $\dot{a}^4 \ll \mu$. When this happens, the curvature of the boundary geometry becomes negligible compared to the scale set by the thermal energy of the CFT. In the same limit the apparent horizon approaches the event horizon. The temperature approaches that of the static case, redshifted by the scale factor.

For $\nu > 1$ the expansion is accelerating and at late times we have $\dot{a}^4 \gg \mu$. The apparent horizon deviates strongly from the event horizon and r_a eventually approaches zero. For example, for $\nu = 2$ (constant $\ddot{a}$) the leading behavior is $r_a = 2\sqrt{\mu}/(3\tau)$. The product Ta diverges asymptotically for $\tau \to \infty$. We discussed the origin of this behavior in the case $a = \lambda\tau$ with $\lambda \to \infty$: The regime $\dot{a}^4 \gg \mu$ is equivalent to the $\mu \to 0$ limit for the static case with $k = 1$. For $\nu > 1$ the solution always approaches this regime at late times, which leads to the asymptotic divergence of Ta. For $1 < \nu < 3/2$ the temperature falls with a power of the scale factor smaller than 1. For $\nu = 3/2$ we have asymptotically a constant temperature $T = 9/(8\pi\sqrt{\mu})$. Finally, for $\nu > 3/2$ the temperature grows with time and diverges asymptotically.

In summary, apart from the rescaling by a, there are two qualitatively different types of evolution. For $\nu < 1$ the CFT corresponds to a black hole with a mass that grows relative to the scale of the curvature induced by the expansion. For $\nu > 1$ the effective mass of the black hole seems to diminish and eventually vanish for $\tau \to \infty$. It will be clearer in the following that the two quantities that characterize the different types of evolution are the Casimir and the thermal energy of the CFT. For $\nu < 1$ the Casimir energy becomes negligible at late times, while for $\nu > 1$ it dominates over the thermal energy.

One should keep in mind that the small-mass AdS-Schwazschild solution in the static case is unstable [14]. The large-mass solution with the same temperature is dual to the deconfined phase of the CFT. It seems reasonable to interpret the black-hole configuration with an accelerating boundary as dual to a CFT in the deconfined phase on an accelerating FLRW background geometry. The fact that for such a configuration the black-hole mass diminishes with time relative to the curvature makes it likely that the corresponding CFT cannot exist in a stable phase. In the following we find more indications of this problematic behavior.

The case $a = \exp(H\tau)$, $k = 0$ corresponds to a de Sitter (dS) boundary. For $\mu \neq 0$ and large τ we find that the temperature quickly approaches the value $T = H/(\sqrt{2}\,\pi)$, which differs from the standard temperature of dS space by a factor $\sqrt{2}$. This means that, even though μ/a^4 becomes negligible relative to $(\dot{a}/a)^4 = H^4$, the temperature of the CFT does not approach the dS temperature. The discrepancy indicates that the configuration with $\mu \neq 0$ on a background with $a = \exp(H\tau)$ cannot evolve continuously to pure dS space. This conclusion is consistent with our earlier argument that a deconfined CFT on an accelerating background is not in a stable phase.

A crucial observation is that the limits $\mu \to 0$ and $\tau \to \infty$ are not equivalent for a dS boundary. Further insight can be obtained if we set $a = \exp(H\tau)$, $k = 0$, $\mu = 0$ directly in the metric (13). In this way we find $\mathcal{N}(\tau, z) = 1 - H^2 z^2/4$. For the determination of the temperature we can follow the same procedure as before. Despite the absence of a black hole, a conical singularity still exists at $z_a = 2/H$ for periodic Euclidean time. The location of the singularity is τ-independent for a dS boundary. As a result, we need not make any assumptions about the relative size of T and H. The singularity can be eliminated for an appropriate value of the temperature. We find $T = H/(2\pi)$. It is remarkable that the correct temperature of four-dimensional dS space can be reproduced through five-dimensional AdS space, when the coordinate frame defined by eq. (13) is used.

5. Stress-energy tensor and cosmological evolution on the boundary

The stress-energy tensor of the dual CFT on the cosmological boundary (12) is determined via holographic renormalization (eq. (9)). We obtain the energy density and pressure [1], respectively,

$$\langle (T^{(CFT)})_{\tau\tau} \rangle = \frac{3}{64\pi G_5} \frac{(\dot{a}^2 + k)^2 + 4\mu}{a^4} \quad (25)$$

$$\langle (T^{(CFT)})^i_i \rangle = \frac{(\dot{a}^2 + k)^2 + 4\mu - 4a\ddot{a}(\dot{a}^2 + k)}{64\pi G_5 a^4} , \quad (26)$$

where no summation over i is implied. We deduce the expected conformal anomaly

$$g^{(0)\mu\nu} \langle T^{(CFT)}_{\mu\nu} \rangle = -\frac{3\ddot{a}(\dot{a}^2 + k)}{16\pi G_5 a^3} . \quad (27)$$

The above results can also be derived directly from the standard expressions for the energy and pressure of a gauge-theory plasma in Minkowski space through an entirely four-dimensional calculation. To this end, we observe that the Robertson-Walker metric (12) is conformally equivalent to the flat Minkowski metric. The vacuum expectation values (25), (26) are calculated in the conformal vacuum. We may instead calculate the VEV in the Minkowski vacuum, in which case we obtain the static plasma result. The two VEVs are related through [20]

$$\langle T^{(CFT)}_{\mu\nu} \rangle \Big|_{RW} = \frac{1}{a^4} \langle T^{(CFT)}_{\mu\nu} \rangle \Big|_{Minkowski} + \mathbf{a} H^{(1)}_{\mu\nu} + \mathbf{b} H^{(3)}_{\mu\nu} , \quad (28)$$

with

$$H^{(1)}_{\mu\nu} = 2(\nabla_\nu \nabla_\mu - g^{(0)}_{\mu\nu} \nabla^2)\mathcal{R} - \frac{1}{2} g^{(0)}_{\mu\nu} \mathcal{R}^2 + 2\mathcal{R}\mathcal{R}_{\mu\nu}$$

$$H^{(3)}_{\mu\nu} = \frac{1}{12}\mathcal{R}^2 \delta^\nu_\mu - \mathcal{R}^{\rho\sigma}\mathcal{R}_{\rho\mu\sigma}{}^{\nu} . \quad (29)$$

The curvature tensor $\mathcal{R}^\mu{}_{\nu\rho\sigma}$ is derived from the Robertson-Walker metric $g^{(0)}_{\mu\nu}$ of eq. (12). The coefficients $\mathbf{a}$ and $\mathbf{b}$ are related to the conformal anomaly and depend on the field content of the CFT. For a theory with n_B spin-0 bosons, n_F fermions and n_V vector fields, we have

$$\mathbf{a} = -\frac{n_B + 3n_F - 18n_V}{1080(4\pi)^2} , \quad \mathbf{b} = \frac{n_B + \frac{11}{2}n_F + 62n_V}{180(4\pi)^2} . \quad (30)$$

For the $\mathcal{N} = 4$ $SU(N)$ super-Yang-Mills theory in the large-N limit, we have $n_B = 6N^2$, $n_F = 4N^2$ and $n_V = N^2$. Therefore

$$\mathbf{a} = 0 , \quad \mathbf{b} = \frac{N^2}{32\pi^2} . \quad (31)$$

After some algebra, eq. (28) is seen to agree with (25), (26), with $G_5 \sim N^{-2}$.

In the standard formulation of the AdS/CFT correspondence, one does not consider the boundary metric as being dynamical. Instead, one takes the form of $a(\tau)$ as an input and investigates the properties of the CFT on such a background. The boundary geometry can be made dynamical if one introduces an Einstein term for the boundary metric, with a four-dimensional Newton's constant G_4, and employs mixed boundary conditions [13].

The Einstein equations (1) are obtained by varying the bulk action $I_{\mathcal{M}}$. This is a five-dimensional Einstein-Hilbert action on $\mathcal{M}$ with a cosmological term. It also includes a Gibbons-Hawking boundary term, as well as boundary counterterms needed to render the system finite. In addition to the black hole solution in the bulk, the variation of the action yields a boundary term

$$\delta I_{\mathcal{M}} = \frac{1}{2}\int_{\partial\mathcal{M}} d^4x \sqrt{-\det g^{(0)}}\, T^{(CFT)}_{\mu\nu}\delta g^{(0)\mu\nu} , \tag{32}$$

where $g^{(0)}_{\mu\nu}$ $(\mu, \nu = 0, 1, 2, 3)$ is the boundary metric and $T^{(CFT)}_{\mu\nu}$ the stress-energy tensor of the dual conformal field theory (CFT) on the AdS boundary $\partial\mathcal{M}$.

If one adopts Dirichlet boundary conditions that fix $g^{(0)}_{\mu\nu}$, the additional term (32) vanishes. If we are interested in keeping $g^{(0)}_{\mu\nu}$ dynamical, we can follow [13] and adopt mixed boundary conditions instead. To define them, we introduce a boundary action consisting of two terms,

$$I_{\partial\mathcal{M}} = I^{(EH)}_{\partial\mathcal{M}} + I^{(matter)}_{\partial\mathcal{M}} . \tag{33}$$

The first term is the Einstein-Hilbert action in four dimensions

$$I^{(EH)}_{\partial\mathcal{M}} = -\frac{1}{16\pi G_4}\int_{\partial\mathcal{M}} d^4x \sqrt{-\det g^{(0)}}(\mathcal{R} - 2\Lambda_4) , \tag{34}$$

where G_4 is Newton's constant in the four-dimensional boundary and $\mathcal{R}$ the four-dimensional Ricci scalar (constructed from $g^{(0)}_{\mu\nu}$). The second term is an unspecified action for matter fields,

$$I^{(matter)}_{\partial\mathcal{M}} = \int_{\partial\mathcal{M}} d^4x \sqrt{-\det g^{(0)}}\mathcal{L}^{(matter)} . \tag{35}$$

The matter fields reside on the boundary and have no bulk duals.

To the variation of the bulk action (32) we must now add the variation of the new boundary action $\delta I_{\partial\mathcal{M}}$, given by (33), under a change in the boundary metric. Demanding that the sum vanish

$$\delta I_{\mathcal{M}} + \delta I_{\partial\mathcal{M}} = 0 \tag{36}$$

leads to two possibilities: (a) Dirichlet boundary conditions, i.e. fixed $g^{(0)}_{\mu\nu}$, or (b) mixed boundary conditions:

$$\mathcal{R}_{\mu\nu} - \frac{1}{2}g^{(0)}_{\mu\nu}\mathcal{R} - \Lambda_4 g^{(0)}_{\mu\nu} = 8\pi G_4\left(T^{(CFT)}_{\mu\nu} + T^{(matter)}_{\mu\nu}\right) . \tag{37}$$

The latter are nothing but the Einstein field equations in four dimensions. Moreover, the variation of the boundary matter action (35) under a change in the matter fields leads to the standard four-dimensional matter field equations.

In [13] it was shown that a general form of $I_{\partial\mathcal{M}}[g^{(0)}]$ leads to a sensible theory with normalizable metric fluctuations. Matter fields were not considered. However, the framework may be extended to include boundary matter fields. If one integrates over them in the path integral, an effective action is obtained as a functional of the boundary metric $g^{(0)}$. For this effective action, the discussion in [13] is applicable.

For a FLRW boundary, the boundary conditions (37) yield the equation of cosmological evolution [1, 21]

$$\left(\frac{\dot{a}}{a}\right)^2+\frac{k}{a^2}-\frac{\Lambda_4}{3}=\frac{8\pi G_4}{3}\left\{\frac{1}{16\pi G_5}\left[\frac{\left(\dot{a}^2+k\right)^2}{a^4}+\frac{4\mu}{a^4}\right]+\rho\right\}\,, \tag{38}$$

where $\rho = T_{00}^{(matter)}$ is the energy density of four-dimensional "ordinary" matter (without a bulk dual). This equation has the expected form, reflecting the conformal anomaly and the presence of a radiative energy component whose energy density scales $\sim a^{-4}$. The Casimir energy density is $\sim (\dot{a}^2+k)^2/a^4$. For $\dot{a}^4 \gtrsim \mu$ it becomes comparable to or dominates over the thermal energy $\sim \mu/a^4$ of the CFT.

6. Entropy

We would like to derive the entropy of the CFT in analogy to the static case. For this we follow a generalization of the intuitive approach of section 2. The entropy now depends on the parameter μ and the time τ through the scale factor $a(\tau)$ and its derivatives. We consider an infinitesimal adiabatic variation of μ that takes place within a time interval that is sufficiently small for the evolution of $a(\tau)$ to be negligible. In contrast to the determination of the temperature, the required time for the variation can be made arbitrarily small by sending $d\mu \to 0$. The fundamental relation $dE + pdV = TdS$ can be employed for the determination of the entropy. The volume a^3V_k of the boundary remains constant, while the temperature is given by eq. (23). The location of the apparent horizon satisfies

$$dz_a = -\frac{z_a^5}{4a^3(4a-\ddot{a}z_a^2)}d\mu = -\frac{z_a^4}{8\pi a^3\mathcal{A}_a}\frac{d\mu}{T} = -\frac{2G_5}{3V_ka^2}\frac{z_a^4}{\mathcal{A}_a}dS, \tag{39}$$

where we have employed eqs. (18), (23), (25) and $dE/T = dS$. Making use of eq. (18), we can express $\mathcal{A}_a$ as a function of z_a only, with implicit dependence on μ. It is given by

$$\mathcal{A}_a(z_a) = \left(2a^2 - \frac{\dot{a}^2+k+a\ddot{a}}{2}z_a^2 + \frac{(\dot{a}^2+k)\ddot{a}}{8a}z_a^4\right)^{1/2}. \tag{40}$$

We can now integrate eq. (39) in order to determine the CFT entropy S as a function of the location of the apparent horizon z_a. We obtain

$$S = \frac{V_k}{4G_5}\left(\frac{\mathcal{A}_a}{z_a}\right)^3 - \frac{3V_k}{32\,G_5}\frac{(\dot{a}^2+k)\,\ddot{a}}{a}\int^{z_a}\mathcal{A}_a(z)dz + F(a,\dot{a},\ddot{a}), \tag{41}$$

with $A_a(z)$ defined in eq. (40). The expression can be checked through simple differentiation. We emphasize again that τ is kept constant during these manipulations.

The constant of integration in the above expression is an arbitrary function of $a(\tau)$ and its derivatives. It appears because we have determined the CFT entropy by considering its variation with respect to μ. The form of $F(a,\dot{a},\ddot{a})$ can be fixed if we define the value of the entropy for a specific value of μ. We postulate that the CFT entropy vanishes for $\mu \to 0$. This is consistent with the fact that the thermal energy of the CFT (after the Casimir energy has been subtracted from $\langle (T^{(CFT)})_{\tau\tau}\rangle$ in eq. (25)) vanishes in this limit.

The first term in eq. (41) describes a contribution that is proportional to the area of the apparent horizon. This is time dependent for a general cosmological background. The integral in the second term can be expressed in terms of incomplete elliptic integrals. Unfortunately, the resulting expression is not very illuminating. For this reason, it is more instructive to consider specific cases, as we did for the temperature.

The expression for the entropy becomes transparent for $a(\tau) = \lambda\tau$. In this case we have $\ddot{a} = 0$, and eq. (41) becomes

$$S = \frac{V_k}{4G_5}\left(\frac{\mathcal{A}_a}{z_a}\right)^3 = \frac{V_k}{4G_5} r_a^3, \tag{42}$$

where the areal distance of the apparent horizon r_a is constant, given by eq. (20). We have set $F = 0$ in order to obtain consistency with eq. (11) for $\lambda = 0$. The total entropy is proportional to the constant area of the apparent horizon. The entropy density $s = S/(a^3 V_k)$ scales as a^{-3}. This particular example is the one closest to the Bjorken flow. The scale factor is linear in time, even though the expansion is isotropic. We find strong similarity with the conclusions of [12]. There is an apparent horizon that is distinct from the event horizon. It is the former that is relevant for the definition of the entropy of the dual CFT. In our example we were able to calculate the entropy and show that it is equal to the area of the apparent horizon, instead of postulating this fact.

A class of boundary geometries with relevance for cosmology are the spaces with $a = \tau^\nu$ and constant ν for large τ. We concentrate on the case $k = 0$. The cases with $k = \pm 1$ can be treated along similar lines. The resulting entropy displays different behavior depending on whether $\nu < 1$ or $\nu > 1$.

For $\nu < 1$ the expansion is decelerating ($\ddot{a} < 0$). The total entropy is proportional to the area of the apparent horizon. It grows with time and asymptotically approaches its maximum value, proportional to the area of the event horizon. The entropy density scales with an additional factor a^{-3} relative to the total entropy. We have seen that the temperature has similar behavior: It receives corrections related to the curvature of the boundary at early times. These become negligible at late times and the temperature becomes equal to the temperature of the static case, given by eq. (7), redshifted by a factor of a.

For $\nu > 1$ the expansion is accelerating ($\ddot{a} > 0$). The total entropy falls off with time and asymptotically approaches zero for $\tau \to \infty$. To leading order, the entropy is again proportional to the apparent horizon, even though the proportionality constant depends on ν. For an accelerating boundary, the areal distance of the apparent horizon asymptotically shrinks to zero for large τ.

This behavior is consistent with the evolution of the temperature that we discussed earlier. For $\nu < 1$, the CFT corresponds to a black hole with a growing effective mass that asymptotically reaches a constant value equal to μ. The entropy also becomes constant asymptotically. For $\nu > 1$, the effective mass of the black hole decreases and vanishes asymptotically, along with its entropy.

For the interpretation of the above results, one should keep in mind that the small-mass AdS-Schwazschild solution in the static case is unstable [14]. This fact and the conclusion that the total entropy decreases with time indicate strongly that there is a pathology in the case of the CFT on an accelerating background. On a static background, the large-mass AdS black hole is dual to the deconfined phase of the CFT [4]. It seems reasonable to speculate that the configuration with decreasing entropy on an accelerating background is again a thermalized CFT in the deconfined phase. Our findings indicate that this phase is probably unphysical and cannot exist in quasi-thermal equilibrium. A possible interpretation is that it is unstable towards the confined phase on the same background.

The above explain why taking the limit $\mu \to 0$ does not reproduce the correct temperature in the case of a dS boundary. The state corresponding to $\mu \neq 0$ in this background is not connected continuously to the pure dS space.

It would be very interesting to calculate the intrinsic entropy of the dS space-time along similar lines. However, the boundary geometry is introduced by hand in the standard holographic approach that we have followed in our derivation. This means that we have no dynamical information on the energy density and pressure that generate the boundary geometry. Even though the dS temperature can be computed from the five-dimensional geometry, we do not have sufficient information for the calculation of the entropy.

7. Summary

We have reviewed the calculation of the temperature and entropy of CFT on a FLRW background performed in ref. [2] through use of the AdS/CFT correspondence. Instead of conjecturing that the entropy is proportional to the area of a horizon of the relevant geometry, we computed it using the energy density and temperature. The relevant gravitational configuration is the five-dimensional AdS-Schwazschild solution, expressed in appropriate Fefferman-Graham coordinates, so that the boundary metric is of the FLRW type with a scale factor $a(\tau)$ [1]. The stress-energy tensor can be computed in a straightforward manner. It reflects correctly the Casimir energy and the conformal anomaly on this background.

The CFT temperature can be identified with the temperature of the black hole, as measured by an observer located near the boundary. The Schwarzschild radial coordinate r is $r = a(\tau)/z$ for a small Fefferman-Graham coordinate z. The five-velocity of an observer located at a fixed small distance z from the boundary is $(z/a, \dot{a}, \vec{0})$ in Schwarzschild coordinates. The boundary observer moves away from the black hole for an expanding FLRW metric. It is natural that he/she would see a temperature redshifted by the scale factor a that determines the velocity. Moreover, if the expansion velocity $\dot{a}$ is not constant, further corrections to the measured temperature are expected because of the Unruh effect [22].

The analysis of ref. [2] realizes all the above expectations and permits the determination of the corrections to the temperature. The black-hole temperature, even in the static case, has a dependence on the curvature of the boundary (which is proportional to that of the horizon for a given slicing). In the time-dependent situation, the total curvature of the boundary (or more precisely the Casimir energy), becomes the relevant quantity for the determination of the temperature. This definition of the temperature can reproduce correctly the temperature of dS space, if we employ an appropriate slicing of the five-dimensional AdS space and assume that the black-hole mass vanishes.

When the corrections to the temperature are taken into account, the total entropy cannot be identified with the area of the event horizon. Instead, a new surface appears, on which the expansion of outgoing null geodesics vanishes. The interpretation we suggested is that it represents a local apparent horizon. This is consistent with what has been found for the Bjorken-flow geometry [9, 11, 12].

The entropy of the CFT can be computed starting from the temperature and the energy density of the CFT, which are known. We found three distinct types of behavior.

- For a decelerating expansion of the FLRW background ($\ddot{a} < 0$) the total entropy at late times is proportional to the area of the apparent horizon, with the standard proportionality constant. The entropy is always smaller than the maximal entropy of the black hole, which is proportional to the area of the event horizon. The reason is that the apparent horizon is always located behind the event horizon, and therefore has a smaller area. As the rate of expansion decreases, the apparent horizon approaches the event horizon, while its area increases. Asymptotically, the two become identical. In this limit, the temperature and entropy density scale with simple powers of the scale factor a.
- For an accelerating expansion ($\ddot{a} > 0$) the CFT entropy decreases with time. This is a sign that the corresponding configuration is unphysical. Another indication is that the mass of the black hole relative to the boundary curvature decreases. Small-mass AdS black holes are known to be unstable [14]. One possible interpretation is that this configuration corresponds to a high-temperature CFT in the deconfined phase on an accelerating background. One may speculate that this phase cannot be realized in quasi-thermal equilibrium on an

accelerating background, because it is unstable, probably towards the confined phase on the same background.

- In the case with zero acceleration ($\ddot{a} = 0$) the boundary curvature is proportional to a constant factor $\sim (\dot{a}^2 + k)$. This results in a constant entropy, which is smaller than the area of the event horizon. In completely analogous fashion, the temperature has the simple scaling $\sim 1/a$, but the proportionality constant is different from the temperature of the static case. This case is the isotropic analogue of the Bjorken-flow geometry.

These results are consistent with the notion that the apparent horizon is the relevant entity for the determination of the entropy in time-dependent situations. The teleological character of the event horizon makes it appropriate only for static situations, in which there is no macroscopic time evolution.

We have also summarized the construction of ref. [1], where it is shown how the equations of cosmological evolution can be obtained in the context of the AdS/CFT correspondence. Two essential elements are necessary: a) the choice of appropriate boundary conditions, so that the boundary metric becomes dynamical, and b) the derivation of the gravity solution in terms of coordinates such that the boundary metric has the FLRW form. This procedure correctly reproduces the expected four-dimensional cosmological behavior.

References

[1] P. S. Apostolopoulos, G. Siopsis and N. Tetradis, *Phys. Rev. Lett.* **102** (2009) 151301 [arXiv:0809.3505 [hep-th]].
[2] N. Tetradis, *JHEP* **1003** (2010) 040 [arXiv:0905.2763 [hep-th]].
[3] J. M. Maldacena, *Adv. Theor. Math. Phys.* **2** (1998) 231 [*Int. J. Theor. Phys.* **38** (1999) 1113] [arXiv:hep-th/9711200]; S. S. Gubser, I. R. Klebanov and A. M. Polyakov, *Phys. Lett.* B **428** (1998) 105 [arXiv:hep-th/9802109]; E. Witten, *Adv. Theor. Math. Phys.* **2** (1998) 253 [arXiv:hep-th/9802150].
[4] E. Witten, *Adv. Theor. Math. Phys.* **2** (1998) 505 [arXiv:hep-th/9803131].
[5] J. D. Bekenstein, *Phys. Rev.* D **7** (1973) 2333; S. W. Hawking, *Commun. Math. Phys.* **43** (1975) 199 [*Erratum-ibid.* **46** (1976) 206].
[6] G. 't Hooft, arXiv:gr-qc/9310026; L. Susskind, *J. Math. Phys.* **36** (1995) 6377 [arXiv:hep-th/9409089].
[7] J. D. Bjorken, *Phys. Rev.* D **27** (1983) 140.
[8] R. A. Janik and R. B. Peschanski, *Phys. Rev.* D **73** (2006) 045013 [arXiv:hep-th/0512162]; *Phys. Rev.* D **74** (2006) 046007 [arXiv:hep-th/0606149].
[9] S. Kinoshita, S. Mukohyama, S. Nakamura and K. y. Oda, *Prog. Theor. Phys.* **121** (2009) 121 [arXiv:0807.3797 [hep-th]]; *Phys. Rev. Lett.* **102** (2009) 031601 [arXiv:0901.4834 [hep-th]].
[10] P. Kovtun, D. T. Son and A. O. Starinets, *Phys. Rev. Lett.* **94** (2005) 111601 [arXiv:hep-th/0405231].
[11] S. Bhattacharyya, V. E. Hubeny, S. Minwalla and M. Rangamani, *JHEP* **0802** (2008) 045 [arXiv:0712.2456 [hep-th]].
[12] P. Figueras, V. E. Hubeny, M. Rangamani and S. F. Ross, arXiv:0902.4696 [hep-th].
[13] G. Compere and D. Marolf, *Class. Quant. Grav.* **25** (2008) 195014 [arXiv:0805.1902 [hep-th]].
[14] S. W. Hawking and D. N. Page, *Commun. Math. Phys.* **87** (1983) 577.
[15] C. Fefferman and C. Robin Graham, *Conformal Invariants*, in *Elie Cartan et les Mathématiques d'aujourd'hui*, Astérisque, 1985, page 95.
[16] C. W. Misner, K. S. Thorne and J. A. Wheeler, *Gravitation*, W. H. Freeman and Company, 1973, chapter 31.
[17] S. de Haro, S. N. Solodukhin and K. Skenderis, *Commun. Math. Phys.* **217** (2001) 595 [arXiv:hep-th/0002230]; *Class. Quant. Grav.* **19** (2002) 5849 [arXiv:hep-th/0209067].
[18] S. W. Hawking and G. F. R. Ellis, *The Large scale structure of space-time*, Cambridge University Press, 1973.
[19] R. M. Wald, *General Relativity*, The University of Chicago Press, 1984.
[20] N. D. Birrell and P. C. W. Davis, *Quantum fields in curved space*, Cambridge Monographs on Mathematical Physics, Cambridge University Press, 1982.
[21] E. Kiritsis, *JCAP* **0510** (2005) 014 [arXiv:hep-th/0504219].
[22] W. G. Unruh, *Phys. Rev.* D **14** (1976) 870.

Filamentary structures of the cosmic web and the nonlinear Schrödinger type equation

E Tigrak[12], **R van de Weygaert**[1], **B J T Jones**[1]

[1] Kapteyn Astron. Inst., University of Groningen, PO Box 800, 9700 AV Groningen, The Netherlands
[2] Izmir Institute of Technology, Department of Mathematics, 35430 Urla-Izmir, Turkey

E-mail: `tigraks@astro.rug.nl`

Abstract. We show that the filamentary type structures of the cosmic web can be modeled as solitonic waves by solving the reaction diffusion system which is the hydrodynamical analogous of the nonlinear Schrödinger type equation. We find the analytical solution of this system by applying the Hirota direct method which produces the dissipative soliton solutions to formulate the dynamical evolution of the nonlinear structure formation.

1. Introduction

The large scale structure of the Universe is marked by prominent filamentary features embedded within a weblike network, the cosmic web. Extensive N-body simulations are used to model this complex and intricate dynamical structure. However we need to find a full analytical formalism to understand this complex structure. Analytical methods for studying the evolution of cosmological density perturbations which cause this intricate structure formation are classified into two broad classes: the Eulerian and the Zel'dovich approximations. While the Eulerian approximation provides an accurate description of the gravitational instability in the linear regime, the Zel'dovich approximation is an exact solution of the fluid equations as long as particle trajectories do not cross. When the trajectories cross, the velocity field becomes multi valued and this causes non-singularities in the density field. The adhesion theory has been proposed to avoid this velocity singularity problem in the Zel'dovich approximation. In the adhesion approximation, when shell crossing occurs, the particles are assumed to stick to each other by introducing an artificial viscosity term in the Burger's equation. In the special case, when the viscosity term tends to zero, the geometrical interpretation of the solution of the Burger's equation can be used to determine the skeleton of the large scale structure. In this limit structures formed in the adhesion model are infinitely thin and the adhesion approximation reduces to the Zel'dovich approximation outside of mass concentration.

An alternative method to these analytical approaches is based on the idea suggested by Spiegel. He showed the correspondence between the fluid structure equations and the nonlinear wave equations [1] and in 1993 Widrow and Kaiser were the first to apply the Schrödinger representation to the problem of the cosmological structure formation for cold dark matter (CDM) [2]. They developed an advanced nonlinear numerical model known as the Schrödinger Method to follow the nonlinear evolution of the dark matter field by offering an alternative particle mesh code. This new numerical model described the matter as a Schrödinger field obeying the coupled classical Schrödinger and Poisson equations. This code has been modified by Davis and Widrow [3]. Another extension of the Schrödinger Method has been done by Coles [4] in which he suggested that the nonlinear Schrödinger equation is a good candidate to model CDM but, in the same paper, he pointed out that this nonlinear equation is not easy to solve analytically. Coles and Spencer [5] demonstrated a wave mechanical approach to solve the caustic problem in the Zel'dovich approximation, when trajectories cross and this approach is similar to the adhesion approach except the pressure term. The pressure term in this new approach has the same effect as the viscosity term in the adhesion theory to avoid the singularities of the density field. By following this work, Short and Coles proposed a different approach to study the large scale structure formation based on a wave mechanical description of self gravitating CDM called free particle approximation [6]. They transformed the usual hydrodynamical equations of motion into the linear Schrödinger equation and they showed that the free particle approximation is useful into the mildly nonlinear regime and it has the same result as the adhesion approximation.

In this study, we derive fully analytical solution of the nonlinear wave equations (nonlinear Schrödinger type equation and the reaction diffusion equation) which are the fluid dynamical analogous of each other [7]. To do this, first we show that the cubic nonlinear Schrödinger type equation satisfies the cosmological fluid dynamical equations and by following the idea suggested by Coles and Spencer [5], we assume this fluid has the gas pressure that is related to the density via $p = K\chi^2$, where K is a positive constant. Then we derive a complete analytical solution of this nonlinear fluid by applying the Hirota direct method which produces the special soliton waves called dissipatons. To do this, we obtain the nonlinear Schrödinger equation from the fluid dynamical equation by applying the Madelung transformation and then we transform the nonlinear Schrödinger type equation into the reaction diffusion system which produces the soliton solutions by applying the Hirota direct method [8] to find the exact solution of the nonlinear wave equations. It means that the filamentary type structures of the cosmic web in the Einstein de Sitter (EdS) Universe can be modeled in a complete analytical way.

2. Fluid dynamical equations in the Einstein de Sitter Universe $\Omega(t) = 1$

To find the analytical solution of the nonlinear wave equation, we take our starting point as the scaled expressions of the Newtonian equations for a self gravitating perfect fluid in the comoving coordinate system. The continuity, the Bernoulli and the Poisson equations

$$\frac{\partial \chi}{\partial D} + \vec{\nabla}_x \cdot \chi \vec{\nabla}_x \phi_v = 0 \tag{1}$$

$$\frac{\partial v'}{\partial D} + v'\vec{\nabla}_x v' = -\vec{\nabla}_x \mathcal{V} - \frac{1}{\chi}\vec{\nabla}_x p \tag{2}$$

$$\nabla_x^2 \phi(x,t) = 4\pi G a^2 \rho_u \chi \tag{3}$$

where G is the gravitational constant and $a(t)$ is the scale factor. We express the equations of motion in terms of the density excess δ or density contrast $\chi(\vec{x},t)$, the peculiar velocity $\vec{v}(\vec{x},t)$ and the comoving gravitational potential $\phi(\vec{x},t)$. The density contrast χ is defined by

$$\chi(\vec{x},t) = 1 + \delta(\vec{x},t) = \frac{\rho(\vec{x},t)}{\rho_u(t)} \tag{4}$$

where $\rho_u(t)$ is the universal cosmic background density and D is the growth factor. In order to describe the evolution against the background of an expanding background, it is sensible to describe the evolution in terms of the expansion factor $a(t)$ or, even more convenient and appropriate, in terms of the linear density growth factor $D(t)$ (See [5]). In the case of an Einstein-de Sitter Universe, we have

$$D(t) = a(t) \tag{5}$$

The other important term in the Bernoulli equation (1) the effective potential $\mathcal{V}$ includes contributions from the velocity potential ϕ_v and the gravity potential ϕ,

$$\mathcal{V} = \frac{3\Omega}{2f^2 D}\left(\phi_v + \theta\right) \tag{6}$$

with the scaled gravity potential θ defined as

$$\theta \equiv \frac{2\phi}{3\Omega a^2 D H^2}. \tag{7}$$

We also introduce the comoving velocity potential ϕ_v for the scaled peculiar velocity $\vec{v}'$

$$\vec{v}' = \vec{\nabla}_x \phi_v \tag{8}$$

3. Relation between the nonlinear Schrödinger and fluid dynamical equations

The fluid equations (1) and (2) can be obtained by decomposing the nonlinear Schrödinger type equation into its real and imaginary parts by using a special transformation called Madelung transformation proposed by [9] in which the dark matter can be presented as complex scalar field. The form of this transformation is given by

$$\psi(\vec{x}, t) = \sqrt{\chi}\exp\left\{i\frac{\phi_v}{\nu}\right\} \tag{9}$$

where ν is an adjustable parameter with dimension equal to that of the comoving velocity potential ϕ_v. The scaled density function χ in (9) satisfies the relation

$$\chi = \psi\psi^* = |\psi|^2, \;\; |\psi|^2 \geq 0 \tag{10}$$

Note that this equality is the analogous of the probability density of the quantum mechanics. Since we use the nonlinear theory, $|\psi|^2$ gives the mass density of its constituent microscopic particles at that point [10] and it is positive definite $|\psi|^2 \geq 0$ but not essentially unity. This gives us the right to name the nonlinear wave equation as the nonlinear Schrödinger type equation and it is given by the following equation

$$i\nu\frac{\partial\psi}{\partial a} = -\frac{1}{2}\nu^2\nabla_x^2\psi + \mathcal{V}\psi + \kappa^2|\psi|^2\psi + \mathcal{P}\psi \tag{11}$$

where $\mathcal{P}$ is defined as the regulation term given as

$$P = \frac{\nu^2}{2}\frac{\nabla_r^2|\psi|}{|\psi|} \tag{12}$$

The reason for substituting the regulation term P into the nonlinear Schrödinger type equation is to avoid the extra term called the quantum potential in the Euler equation. Equation (11) originally explains the dynamical evolution of the Bose-Einstein condensate in which large fraction of the bosons occupy the lowest quantum state of the external potential and all wave functions overlap each other where the quantum effect becomes apparent on a macroscopic scale. Here we will use this fact to model the macroscopic wave function of bosons ψ obeys the nonlinear Schrödinger equation. The nonlinear term $\kappa^2|\psi|^2$ in (11) shows that the pressure term resulting from the inter particle interactions and κ is interpreted as the nonlinearity parameter which is a constant value. This nonlinear parameter and the constant parameter of the equation of sate satisfies $\kappa = \sqrt{K}$.

Journal of Physics: Conference Series **283** (2011) 012039 doi:10.1088/1742-6596/283/1/012039

4. Reaction diffusion system in the EdS Universe

We transform the nonlinear Schrödinger type wave equation into an equivalent reaction diffusion wave equation which is the fluid analogous of equation (11) [7]. To obtain the reaction diffusion analogous of equation (11) we define two real functions

$$Q^{+}(x,a) \equiv \sqrt{\chi}e^{\frac{\phi_v}{\nu}}, \quad Q^{-}(x,a) \equiv \sqrt{\chi}e^{-\frac{\phi_v}{\nu}} \tag{13}$$

These real functions and the wave function of equation (11) satisfy the following relations

$$-Q^{+}Q^{-} = \chi = \psi\psi^{*} = |\psi|^2 \tag{14}$$

so fluid equations (1) and (2) generated by (11) can be written as the reaction diffusion system which is the analogous of (11)

$$\begin{aligned} \nu\frac{\partial Q^{(+)}}{\partial a} + \frac{\nu^2}{2}\nabla_x^2 Q^{(+)} - \kappa^2 Q^{(+)}Q^{(-)}Q^{(+)} - \mathcal{V}Q^{(+)} = 0 \\ -\nu\frac{\partial Q^{(-)}}{\partial a} + \frac{\nu^2}{2}\nabla_x^2 Q^{(-)} - \kappa^2 Q^{(-)}Q^{(+)}Q^{(-)} - \mathcal{V}Q^{(-)} = 0 \end{aligned} \tag{15}$$

Here we represent the decoupling reaction diffusion system. The second equation in system (15) is the time reversible of the first reaction diffusion equation and it is crucial for the existence of the Hamiltonian structure and integrable system [7, 11]. In this study we do not take into account the time reversible reaction diffusion equation as we do not use the negative time values in order to have physical consistency in the large scale structure. We consider the full nonlinear Schrödinger Poisson system in which the evolution of CDM can be fully described for the gravitational field in terms of the cosmological model. Jones suggested an analytical model for nonlinear clustering in which he showed that the baryonic matter driven by CDM potential can be modeled in terms of the heat equation [12] and we know that the heat equation is the particular case of the reaction diffusion equation. It means that we can model the evolution of baryonic matter by solving (15) whose solution automatically gives the solution of (11) thanks to relation (14). It is interesting to note that in the linear regime the effective potential $\mathcal{V}$ is equal to zero. In fact, the conclusion of $\mathcal{V} = 0$ stretches out much further into the quasi-linear regime, for as long the Zel'dovich formalism still describes the motion of matter elements in the Universe. Hence the first equation in system (15) is reduced to

$$\nu\frac{\partial Q^{(+)}}{\partial a} + \frac{\nu^2}{2}\nabla_x^2 Q^{(+)} - \kappa^2 Q^{(+)}Q^{(-)}Q^{(+)} = 0 \tag{16}$$

5. Hirota bilinear form

Our approach is to find the exact solution of (16) by following the method developed by Hirota [8]. When we apply the Hirota direct method to the reaction diffusion equation, the soliton solutions are obtained. The crucial step of the Hirota method is a change of variables which converts the equation of motion to an equation of Hirota bilinear type. The appropriate change of variables is not obvious; fortunately we are guided by the change of variables which is used for the nonlinear Schrödinger equation.

We analyze equation (16) and construct one- and two- soliton solutions. To obtain the solution, we applied the following steps: First we make the bilinearization by applying the suitable transformations of (16) which is

$$Q^{(+)} = \frac{1}{\sqrt{2}}\frac{\nu}{\kappa}\frac{g^{+}}{f}, \quad Q^{(-)} = \frac{1}{\sqrt{2}}\frac{\nu}{\kappa}\frac{g^{-}}{f} \tag{17}$$

where $g^{\pm}(x,a)$ and $f(x,a)$ are the new differentiable functions which help us to write the reaction diffusion equation as a combination of bilinear equation. After substituting the derivatives of the new functions $Q^{(+)}(x,a)$ in terms of the independent variables into equation (16)

$$\begin{aligned}\frac{\partial Q^{(+)}}{\partial a} &= \frac{1}{\sqrt{2}}\frac{\nu}{\kappa}\left[\frac{D_a(g^+.f)}{f^2}\right]\\ \nabla_x^2 Q^{(+)} &= \frac{1}{\sqrt{2}}\frac{\nu}{\kappa}\left[\frac{D_x^2(g^+.f)}{f^2}-\frac{g^+}{f}\frac{D_x^2(f.f)}{f^2}\right]\end{aligned}$$

where D is called the Hirota D-operator or the Hirota derivative which is defined as

$$\begin{aligned}D_x^n(g^{\pm}.f) &= \sum_{l=0}^{n}(-1)^l\left(\begin{array}{c}n\\ l\end{array}\right)g^{\pm}_{(n-l)x}f_{lx}\\ &= g^{\pm}_{nx}f - ng^{\pm}_{(n-1)x}f_x + ... + (-1)^n g f_{nx}\end{aligned} \tag{18}$$

where the subscripts indicate the partial differentiation and dot (.) is multiplication symbol. Hence equation (16) can be expressed as the following pair of the Hirota bilinear equations

$$\begin{aligned}\left(\nu D_a + \frac{\nu}{2}D_x^2\right)(g^+.f) &= 0\\ D_x^2(f.f) &= -g^+g^-\end{aligned} \tag{19}$$

6. Perturbation analysis of the soliton solutions

To find the solution of equation (16) in terms of one-, two- ...etc soliton waves the Hirota perturbation method has been applied. To do this, we use two different differential functions defined in transformation (17) g and f. Let us define them as

$$g^+ = \epsilon^0 g_0^+ + \epsilon^1 g_1^+ + \epsilon^2 g_2^+ + ... \tag{20}$$

and

$$f = \epsilon^0 f_0 + \epsilon^1 f_1 + \epsilon^2 f_2 + ... \tag{21}$$

where g_0^+ and f_0 are non zero constants and f_i, g_i^+, $i = 1,2,...N$ are exponential functions. We insert f and g into the Hirota bilinear form (19) of equation (16).

6.1. One soliton solution

First we can take $f = 1 + \epsilon^2 f_1$ and $g = \epsilon g_1^{\pm}$ where $f_i = g_i = 0$ for all $i \geq 2$ and they are exponential functions. For a nontrivial solution, f_0 and $g_0^{\pm}$ should not vanish at the same time. At first let us examine the Hirota perturbation on the bilinear system and make the coefficients identically zero. The coefficient of ϵ's gives us

$$\epsilon^0 \;:\; D_x^2(1.1) = 0 \tag{22}$$

$$\epsilon^1 \;:\; \left(\nu D_a + \frac{\nu^2}{2}D_x^2\right)(g_1^+.1) = 0 \tag{23}$$

$$\epsilon^2 \;:\; 2D_x^2(1.f_1) = -g_1^+g_1^- \tag{24}$$

$$\epsilon^3 \;:\; \left(\nu D_a + \frac{\nu^2}{2}D_x^2\right)(g_1^+.f_1) = 0 \tag{25}$$

$$\epsilon^4 \;:\; D_x^2(f_1.f_1) = 0 \tag{26}$$

Journal of Physics: Conference Series **283** (2011) 012039 doi:10.1088/1742-6596/283/1/012039

and then by the coefficient of ϵ^1, we can easily show that $g_1^{\pm} = \pm e^{\eta_1^{\pm}}$ satisfies the differential form and here we choose the form of the function η_1 as $\eta_1{}^{\pm} = p_1^{\pm}x \pm \Omega_1^{\pm}t + \eta_0^{\pm}$. Hence again using the same form we find the following relations between the parameters of the $\eta_1^{\pm}$ which is

$$\Omega_1^{\pm} = -\frac{\nu}{2}p^{\pm}{}_1^2 \tag{27}$$

then $\eta_1{}^{\pm} = p_1^{\pm}x \mp \frac{\nu}{2}p^{\pm}{}_1^2 t + \eta_{10}^{\pm}$ so we get the solution as

$$g_1^{\pm} = \pm e^{\eta_1^{\pm}}, \quad f = 1 + e^{\eta_1^{+}+\eta_1^{-}+\gamma} \tag{28}$$

where γ is defined in the following form

$$\gamma \equiv \ln \frac{1}{2}\frac{1}{\left(p_1^{+} + p_1^{-}\right)^2}$$

Then the solution of the RD system has the form

$$Q^{+}(x,a) = \frac{1}{\sqrt{2}}\frac{\nu}{\kappa}\frac{e^{\eta_1^{+}}}{1 + e^{\eta_1^{+}+\eta_1^{-}+\gamma}}, \quad Q^{-}(x,a) = -\frac{1}{\sqrt{2}}\frac{\nu}{\kappa}\frac{e^{\eta_1^{-}}}{1 + e^{\eta_1^{-}+\eta_1^{+}+\gamma}} \tag{29}$$

This solution admits the exponentially growing and decaying components known as dissipation. It is useful to note that adiabatic parameter κ defines two different families of solutions of system (15) depending on the sign of κ. For $\kappa > 0$, the solution of equation (16) is given by a bright soliton, corresponding to a local maximum of the scaled density distribution $\chi = |\psi|^2$. For $\kappa < 0$, the solution of reaction diffusion (16) is given by a dark soliton, corresponding to a local minimum in the scaled density distribution. Here we investigate the bright soliton solution to obtain the density function of the local maxima represented as filaments. Hence the scaled density function can be written as

$$-Q^{+}(x,a)Q^{-}(x,a) = \psi\psi^{*} = |\psi|^2 = \frac{1}{2}\frac{\nu^2}{\kappa^2}\frac{e^{\eta_1^{+}+\eta_1^{-}}}{\left(1 + e^{\eta_1^{+}+\eta_1^{-}+\gamma}\right)^2} \tag{30}$$

which is equal to the density contrast

$$\chi = \frac{\nu^2}{\kappa^2}\frac{k^2}{\cosh^2\left[kx + \frac{\nu}{2}vkt + \frac{\eta_0^{+}+\eta_0^{-}}{2} + \frac{1}{2}\gamma\right]} \tag{31}$$

where k is represented as wave number and v is the velocity of the dissipative soliton

$$k \equiv \frac{p_1^{+} + p_1^{-}}{2}, \quad v \equiv p_1^{-} - p_1^{+}$$

The scaled density function shows the perfect soliton wave shape presented as filamentary type structure. When we change the parameters p_1^{+} and p_1^{-} we can see the different type of solitonic wave type structures which bear a striking geometric resemblance to the filaments of the cosmic web. One soliton solution provides the information of the distribution of the scaled density in terms of the expanding scale factor $a(t)$, in other words, it shows the evolution of the scaled density function in the EdS Universe in Figure 6.1.

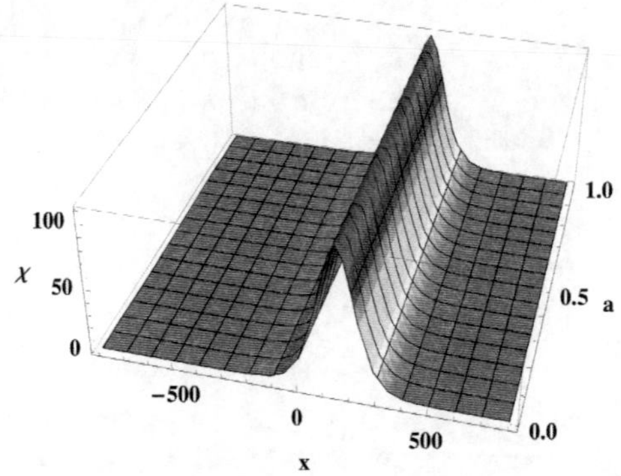

Figure 1. 3D plot of the scaled density function χ represented by one soliton solution of the reaction diffusion equation in terms of $\kappa = 1/\sqrt{2}$ with parameters $p_1^- = -0.1$ and $p_1^+ = 0.121$.

6.2. Two soliton solution

We give the analytical expression of two soliton solution of the RD system. Hence to find two-soliton solution, we take $f = 1 + \epsilon^2 f_1 + \epsilon^4 f_2$ and $g = \epsilon g_1 + \epsilon^3 g_2$ where $\eta_j^{\pm} = p_j^{\pm} x \mp \frac{\nu}{2} p_j^{\pm 2} t + \eta_{j0}^{\pm}$, $j = 1, 2$. Note that here $f_j = g_j = 0$ for all $j \geq 3$. At first we substitute f and g into the equation (19) and we try to make the coefficients of ϵ^m, $m = 0, ..., 8$ to vanish. By following the Hirota method we obtain the two soliton solution

$$\epsilon^0 \; : \quad D_x^2(1.1) = 0 \tag{32}$$

$$\epsilon^1 \; : \quad \left(\nu D_a + \frac{\nu^2}{2} D_x^2\right)(g_1^+.1) = 0 \tag{33}$$

$$\epsilon^2 \; : \quad 2D_x^2(1.f_1) = -g_1^+ g_1^- \tag{34}$$

$$\epsilon^3 \; : \quad \left(\nu D_a + \frac{\nu^2}{2} D_x^2\right)(g_1^+.f_1 + g_2^+) = 0 \tag{35}$$

$$\epsilon^4 \; : \quad D_x^2(2f_2 + f_1.f_1) = -(g_1^+ g_2^- + g_2^+ g_1^- \tag{36}$$

$$\epsilon^5 \; : \quad \left(\nu D_a + \frac{\nu^2}{2} D_x^2\right)(g_1^+.f_2 + g_2^+.f_1) = 0 \tag{37}$$

$$\epsilon^6 \; : \quad D_x^2(f_1 f_2 + f_2.f_1) = -g_2^+ g_2^- \tag{38}$$

$$\epsilon^7 \; : \quad \left(\nu D_a + \frac{\nu^2}{2} D_x^2\right)(g_2^+.f_2) = 0 \tag{39}$$

$$\epsilon^8 \; : \quad D_x^2(f_2 f_2) = 0 \tag{40}$$

By following the same way as in the one soliton solution, we obtain g

$$g^{\pm} = \pm[e^{\eta_1^{\pm}} + e^{\eta_2^{\pm}} + e^{\eta_1^+ + \eta_1^- + \eta_2^{\pm} + \alpha_1} + e^{\eta_1^{\pm} + \eta_2^+ + \eta_2^- + \alpha_2}] \tag{41}$$

where parameters α_1 and α_2 are defined as

$$\alpha_1 \equiv \ln \frac{1}{2} \frac{\left(p_2^+ - p_1^+\right)^2}{\left(p_1^+ + p_1^-\right)^2 \left(p_2^+ + p_1^-\right)^2}$$

$$\alpha_2 \equiv \ln \frac{1}{2} \frac{\left(p_2^+ - p_1^+\right)^2}{\left(p_1^+ + p_2^+\right)^2 \left(p_2^+ + p_2^-\right)^2} \tag{42}$$

and the general form of function f is

$$f = 1 + e^{\eta_1^+ + \eta_1^- + \gamma_{11}} + e^{\eta_2^+ + \eta_1^- + \gamma_{21}} + e^{\eta_1^+ + \eta_2^- + \gamma_{12}} + e^{\eta_2^+ + \eta_2^- + \gamma_{22}} + e^{\eta_1^+ + \eta_1^- + \eta_2^+ + \eta_2^- + \xi} \tag{43}$$

where the parameters are defined as

$$\begin{aligned}
\gamma_{11} &\equiv \ln \frac{1}{2} \frac{1}{\left(p_1^+ + p_1^-\right)^2} \\
\gamma_{12} &\equiv \ln \frac{1}{2} \frac{1}{\left(p_1^+ + p_2^-\right)^2} \\
\gamma_{21} &\equiv \ln \frac{1}{2} \frac{1}{\left(p_2^+ + p_1^-\right)^2} \\
\gamma_{22} &\equiv \ln \frac{1}{2} \frac{1}{\left(p_2^+ + p_2^-\right)^2} \\
\xi &\equiv \ln \frac{1}{4} \frac{\left|p_2^+ - p_1^+\right|^4}{\left(p_2^- + p_2^+\right)^2 \left(p_1^+ + p_1^-\right)^2 \left|p_1^+ + p_2^-\right|^4}
\end{aligned}$$

Then the scaled density function in terms of the two soliton solution (6.2) can be written as by applying (41) and (43) into the relation (14)

$$\chi = \frac{1}{2} \frac{\nu^2}{\kappa^2} \frac{g^+ g^-}{f^2} \tag{44}$$

7. Conclusion

In this study we show a methodology for dealing with the nonlinear Schrödinger type equation and the reaction diffusion equation which are derived from the cosmological fluid equations. This special methodology is called the Hirota direct method and by applying this method to the nonlinear Schrödinger or its hydrodynamical analogous the reaction diffusion equation, we obtain the analytical solution of these nonlinear systems. Due to the nature of the Hirota method which is based on the perturbation approximation, the solution of the reaction diffusion equation and the nonlinear Schrödinger type equation produce the dissipative soliton type waves called dissipatons. When we increase the order of the perturbations in the Hirota method from one-soliton solution to N- soliton solutions, these waves show striking similarity to the intricate structure of filamentary type features of the cosmic web in $2 + 1$ dimension. Apart from the filamentary structures represented as local maxima of the density field called bright solitons depending on the positive nonlinear term κ, we can construct the structures of the local minima called dark solitons and these dark solitons can be represented as empty negative density regions called voids around the filaments.

It is necessary to note that in this study we have focused on the detailed calculations of one soliton solution and added the graphics for two-soliton solutions because of the complexity of the calculations. We will give the technical details of the higher order solutions in the future paper.

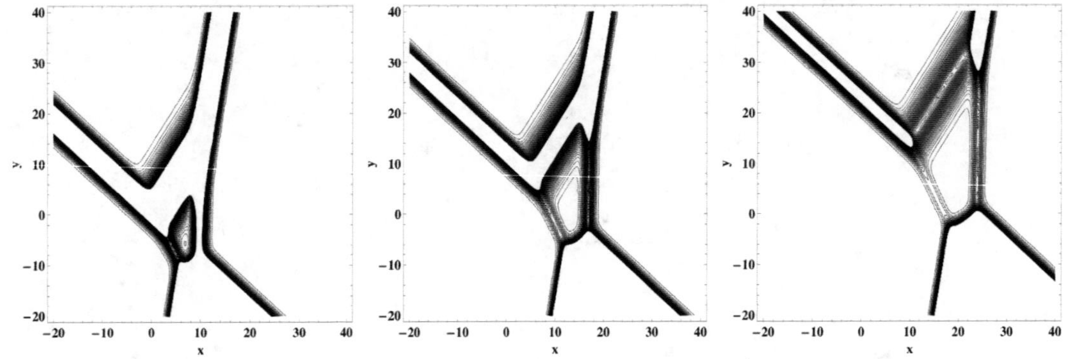

Figure 2. Contour plots of the scaled density or the density contrast function for two soliton solution of the reaction diffusion equation at the red shift values $z = 0.1, 0.05, 0$ in the phase space (from left to right). We can easily see that the scaled density of the filaments increase with respect to time evolution which means that the matter merges through the bridges into the filaments. In the late time steps, these matter bridges disappear because of the merging of the matter into the high density regions/lumps.

References

[1] Spiegel E A 1980 *Physica* **1D** 236
[2] Widrow L M and Kaiser N 1993 *Astrophys. J.* **416** L71
[3] Davis G and Widrow L M 1997 *Astrophys. J.* **485** 484
[4] Coles P 2002 *Mon. Not. R. Astron. Soc.* **330** 421
[5] Coles P and Spencer K 2003 *Mon. Not. R. Astron. Soc.* **342** 176
[6] Short C J and Coles P 2006 *Journal of Cosmological Astroparticle Physics* **12** 12
[7] Pashaev O K and Lee J H 1998 *Journal of Mathematical Physics* **39** 102
[8] Hirota R 1971 *Phys Rev. Letters* **27** 1192
[9] Madelung E 1926 *Zeits. F. Phys.* **40** 3325
[10] Johnston R, Lasenby A N and Hobson M P 2010 *Mon. Not. R. Astron. Soc.* **402** 2491
[11] Lee J H and Pashaev O K 2002 *Modern Physics Letters* A **17** 1601
[12] Jones B J T 1999 *Mon. Not. R. Astron. Soc.* **307** 376

Journal of Physics: Conference Series **283** (2011) 012040 doi:10.1088/1742-6596/283/1/012040

Apparent cosmic acceleration due to large-scale peculiar motions

Christos G Tsagas

Section of Astrophysics, Astronomy and Mechanics, Department of Physics
Aristotle University of Thessaloniki, Thessaloniki 54124, Greece

E-mail: tsagas@astro.auth.gr

Abstract. Observers drifting relative to the smooth Hubble flow have expansion rates different from that of the Universe itself. As a result, observers with small peculiar velocities in linearly perturbed Friedmann universes can experience accelerated expansion, while the universe is actually decelerating. The effect is local, but the affected scales can be large enough to give the (false) impression that the whole universe has recently entered an accelerating phase. Recent surveys reporting large-scale peculiar velocities substantially larger than previously anticipated add a certain degree of observational support to the aforementioned theoretical result.

1. Introduction

In idealised Friedmann-Robertson-Walker (FRW) cosmologies, comoving observers follow the universal expansion. In more realistic models, however, the Hubble flow is distorted and matter acquires 'peculiar' velocities. Thus, the dipole in the Cosmic Microwave Background (CMB) is interpreted as the result of our peculiar flow (at roughly 600 km/sec) relative to the cosmic rest-frame. This report considers the implications of such drift motions for the kinematics of the associated observers. First, we ask whether drifting observers and those following the Hubble expansion (in a dust-dominated FRW universe) can 'measure' substantially different deceleration parameters. Whether, in particular, it is theoretically possible for a peculiarly moving observer to 'experience' accelerated expansion while the universe is decelerating. We show that the answer to this question is positive. Accelerated expansion for a drifting observer does not necessarily imply the same for the universe itself. Then, we look for observational data that could turn this theoretical result into a realistic possibility. Based on the predictions of the ΛCDM model, this seems rather unlikely. However, based on recent independent surveys reporting bulk peculiar motions of up to 1000 km/sec on scales extending as far out as 1000 Mpc [1–5], the answer to the second question could be positive as well.[1]

2. Peculiar kinematics in perturbed FRW universes

The Microwave Background sets a preferred cosmological coordinate system, that of the smooth Hubble flow. This is the frame where the dipolar anisotropy of the CMB vanishes and with respect to which large-scale peculiar velocities can be defined and measured.

[1] The reader is also referred to [6] for an alternative/complementary discussion.

2.1. The expansion rates

Suppose that u_a is the reference 4-velocity of the CMB, relative to which the universe is an exact dust-dominated FRW model. In the perturbed spacetime, a typical observer in a galaxy like our Milky Way has

$$\tilde{u}_a = u_a + v_a\,, \tag{1}$$

with v_a representing their drift velocity.[2] Note that $u_a v^a = 0$ always and $v^2 = v_a v^a \ll 1$ in our case. The latter constraint reflects the almost-FRW nature of our model and guarantees that $\gamma = (1 - v^2)^{-1/2} \simeq 1$.

The mean kinematics of the tilded observers are determined by the volume scalar ($\tilde{\Theta} = \nabla^a \tilde{u}_a$) of their worldline congruence [7, 8]. Positive values for $\tilde{\Theta}$ imply that the mean separation between these observers increases and indicate expansion. Similarly, Θ (with $\Theta = \nabla^a u_a > 0$) monitors the expansion of the universe. To first order in v_a, we have[3]

$$\tilde{\Theta} = \Theta + \vartheta\,, \tag{2}$$

with $\vartheta = \tilde{\mathrm{D}}^a v_a$. This scalar measures the mean separation between neighbouring peculiar-flow lines. When ϑ is positive, the peculiarly moving observers expand faster than the universe (i.e. $\tilde{\Theta} > \Theta$). Here, we will always assume positive ϑ.

In multi-component systems each group of observers has its own time-direction. So, in our case, time can be measured relative to the CMB and along the tilded frame. The expansion rate in each time-direction is determined by the associated Raychaudhuri equation [7, 8]. In a dust-dominated FRW model with small drift velocities, the Raychaudhuri formulae in the CMB and the tilded frames are

$$\Theta' = -\frac{1}{3}\Theta^2 - \frac{1}{2}\rho \qquad \text{and} \qquad \dot{\tilde{\Theta}} = -\frac{1}{3}\tilde{\Theta}^2 - \frac{1}{2}\tilde{\rho} + \tilde{\mathrm{D}}^a \tilde{A}_a\,, \tag{3}$$

respectively. Here, primes indicate time-differentiation along u_a and overdots are time-derivatives relative to the $\tilde{u}_a$-field (i.e. $\Theta' = u^a \nabla_a \Theta$ and $\dot{\tilde{\Theta}} = \tilde{u}^a \nabla_a \tilde{\Theta}$). Also, ρ and $\tilde{\rho}$ are the matter densities in the CMB and the tilded frames respectively (with $\tilde{\rho} = \rho$ to linear order in v_a [10]). Finally, $\tilde{A}_a$ is the 4-acceleration in the tilded frame. This vector vanishes in the CMB frame (i.e. $A_a = 0$) but is nonzero in every other relatively moving coordinate system. In particular, $\tilde{A}_a = \dot{v}_a + (\Theta/3)v_a$, to linear order in v_a [10, 11]. The 4-acceleration term in Eq. (3b) is central to our analysis. Its presence means that expressions (3a) and (3b) are different, even when matter is pressureless dust and the peculiar velocities are small. In other words, drifting observers have expansion rates different from that of the actual universe simply because of their relative motion. For our purposes, this fact represents the most significant theoretical deviation from the conventional single-fluid studies.

2.2. The deceleration parameters

We have just shown that the expansion rate of a typical observer in a dust-dominated almost FRW universe differs from that of the Hubble flow. This means that the deceleration parameters measured in the two frames are different as well. One might then ask whether it is possible to experience accelerated expansion in one frame and decelerated in the other. Whether, in particular, the peculiarly moving observer could measure a negative deceleration parameter,

[2] The $\tilde{u}_a$-field is also timelike irrespective of the magnitude of the peculiar velocity. Each frame defines its own time direction and 3-space. The tensors $h_{ab} = g_{ab} + u_a u_b$ and $\tilde{h}_{ab} = g_{ab} + \tilde{u}_a \tilde{u}_b$, with g_{ab} representing the spacetime metric, project orthogonal to u_a and $\tilde{u}_a$ respectively. They also define the orthogonally projected covariant derivative operators, by means of $\mathrm{D}_a = h_a{}^b \nabla_b$ and $\tilde{\mathrm{D}}_a = \tilde{h}_a{}^b \nabla_b$ (∇_a is the standard covariant derivative) [7, 8].

[3] Recall that we have assumed non-relativistic peculiar velocities, which implies that we can drop terms of order v^2 and higher from Eq. (2) and the rest of our formulae. We also use geometrised units with $c = 1 = 8\pi G$.

while the universe is actually decelerating. Given that drift flows are expected to die away as we move on to successively larger scales, such an effect can only be local.

To investigate this possibility, we first need to write down the deceleration parameters as 'measured' in the u_a and $\tilde{u}_a$ frames. Expressed in terms of the corresponding volume scalars, these read

$$q = -\left(1 + \frac{3\Theta'}{\Theta^2}\right) \qquad \text{and} \qquad \tilde{q} = -\left(1 + \frac{3\dot{\tilde{\Theta}}}{\tilde{\Theta}^2}\right), \tag{4}$$

respectively. Substituting the above into Eqs. (3), the latter recast into

$$(1+q)\Theta^2 = \Theta^2 + \frac{3}{2}\rho \qquad \text{and} \qquad (1+\tilde{q})\tilde{\Theta}^2 = \tilde{\Theta}^2 + \frac{3}{2}\tilde{\rho} - 3\tilde{\mathrm{D}}^a\tilde{A}_a\,. \tag{5}$$

Although these already show that q and $\tilde{q}$ are generally different, it helps to relate the two parameters directly. First, recall that $\tilde{\rho} = \rho$ and $\tilde{A}_a = \dot{v}_a + (\Theta/3)v_a$. Then, using the definition $\vartheta = \tilde{\mathrm{D}}^a v_a$, expression (2) and the (linear in v_a) relation $\dot{\vartheta} = \tilde{\mathrm{D}}^a\dot{v}_a - \Theta\vartheta/3$ (see [11]), Eqs. (5) combine to give

$$1 + \tilde{q} = (1+q)\left(1 + \frac{\vartheta}{\Theta}\right)^{-2} - \frac{3\dot{\vartheta}}{\Theta^2}\left(1 + \frac{\vartheta}{\Theta}\right)^{-2}. \tag{6}$$

This clearly shows that q and $\tilde{q}$ generally differ. Moreover, as long as the right-hand side of (6) remains below unity, positive values for q do not a priori guarantee the same for $\tilde{q}$. It is therefore theoretically possible for the tilded observer to experience accelerated expansion in a decelerating universe.

At this point, it is worth noting that the condition $-1 < \tilde{q} < 0$ is equivalent to $-\tilde{\Theta}^2/3 < \dot{\tilde{\Theta}} < 0$ (see Eq. (4b)), which means that $\tilde{q}$ and $\dot{\tilde{\Theta}}$ can be simultaneously negative. Analogous relations also hold between q, Θ^2 and Θ'. One should therefore distinguish between accelerated expansion with simply $-1 < \tilde{q} < 0$ and that with $\dot{\tilde{\Theta}} > 0$. We may therefore view $-1 < \tilde{q} < 0$ and $\dot{\tilde{\Theta}} > 0$ (equivalently $\tilde{q} < -1$) as the conditions for 'weakly' and 'strongly' accelerated expansion respectively. It is also important to recognise that, as long as we only require $\tilde{q}$ to lie in the (-1,0) range, the 4-acceleration term in Eqs. (3b) and (5b) does not need to dominate the right-hand side of these expressions. This implies that peculiar motions can lead to weakly accelerated expansion within the limits of the linear (the almost-FRW) approximation. Noting that the supernovae results put the deceleration parameter close to -0.5 [12, 13], we will focus on the $-1 < \tilde{q} < 0$ case for the rest of this report.

3. Apparent acceleration in perturbed FRW universes

The main theoretical principle following from our discussion so far, is that measuring a negative deceleration parameter in a frame drifting relative to the CMB (like that of our Local Group for example) does not necessarily imply an accelerating universe. The question is whether current observations support such a scenario.

3.1. A scenario for apparent acceleration

Consider an expanding spatial region $\mathcal{A}$ – see Fig. 1, which largely complies with the FRW symmetries, but is still endowed with a weak, bulk peculiar velocity field that 'adds' to the background expansion (i.e. $\vartheta > 0$, with $\vartheta \ll \Theta$). Typical observers in $\mathcal{A}$ have peculiar velocities close to the bulk flow of the patch and their deceleration parameters obey Eq. (6). The simplest case corresponds to $3\dot{\vartheta}/\tilde{\Theta}^2 \simeq 0$, which occurs when ϑ varies very slowly with time (for example). Then, since $\vartheta/\Theta \ll 1$, a simple Taylor expansion reduces (6) to

$$1 + \tilde{q} = \left(1 + \frac{1}{2}\Omega\right)\left[1 - 2\left(\frac{\vartheta}{\Theta}\right)\right], \tag{7}$$

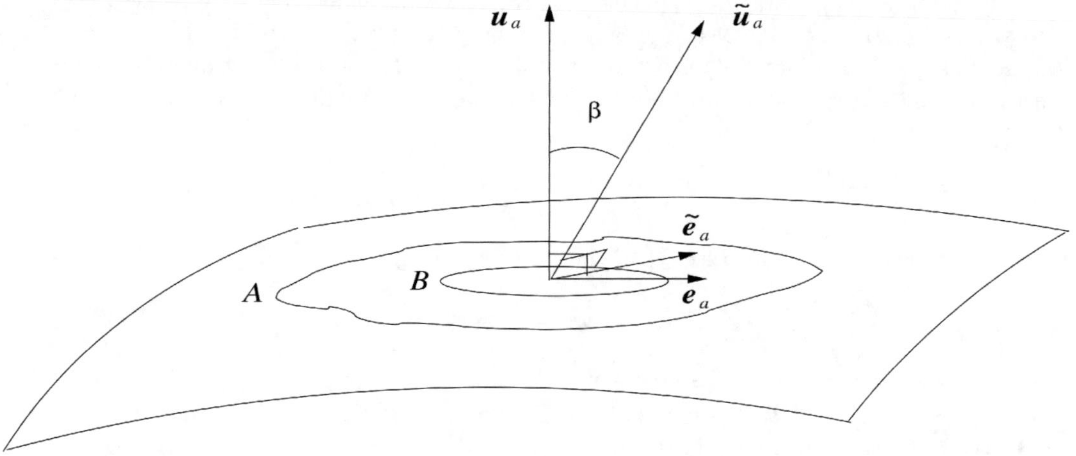

Figure 1. The patch (A) has positive $\vartheta = \tilde{\mathrm{D}}^a v_a$ and so expands faster than its surroundings (see Eq. (2)). Inside region (B) the right-hand side of expression (6) drops below unity and there the comoving observer 'measures' negative deceleration parameter.

given that $q = \Omega/2$ in dust-dominated FRW models. Note that $\Omega = 3\rho/\Theta^2$ (with ρ representing the matter density in the u_a-frame) may be seen as the effective density parameter of patch $\mathcal{A}$, rather than that of the universe itself. Following the above, when

$$\left(1 + \frac{1}{2}\Omega\right)\left[1 - 2\left(\frac{\vartheta}{\Theta}\right)\right] < 1 . \tag{8}$$

$\tilde{q}$ becomes negative, though q is still positive.

Let us now turn to the last term of Eq. (6). Qualitatively speaking, a positive $\dot{\vartheta}$ will assist the acceleration, relax the above given conditions and lead to lower values of $\tilde{q}$. So, here, we will assume that $\dot{\vartheta}$ is negative. We will also demand that $\dot{\vartheta}/\Theta' \simeq \vartheta/\Theta \ll 1$, to ensure that both ϑ and $\dot{\vartheta}$ are small perturbations relative to their background associates. The next step is to recast Raychaudhuri's formula (see Eq. (3a)) in the form

$$\Theta' = -\frac{1}{3}\Theta^2\left(1 + \frac{1}{2}\Omega\right) , \tag{9}$$

with $\Theta' < 0$. Solving the above for Θ^2, substituting into Eq. (6) and employing some straightforward algebra we arrive at

$$1 + \tilde{q} = \left(1 + \frac{1}{2}\Omega\right)\left(1 - \frac{\vartheta}{\Theta}\right) . \tag{10}$$

In this case, the necessary condition for accelerated expansion, in an otherwise decelerating universe, recasts into

$$\left(1 + \frac{1}{2}\Omega\right)\left(1 - \frac{\vartheta}{\Theta}\right) < 1 . \tag{11}$$

When either of conditions (8) or (11) is satisfied, around every typical observer in $\mathcal{A}$ there will be an essentially spherically symmetric region $\mathcal{B}$ (with $\mathcal{B} \subseteq \mathcal{A}$) where the expansion will be accelerated (see Fig. 1). In other words, the aforementioned observers will experience (locally) accelerated expansion in a globally decelerating universe. Whether this happens or not, as well as the scale of affected area (the size of section $\mathcal{B}$), depends mainly on the speed of the drift flow. So, the next question is whether observations allow for peculiar velocities large enough to trigger apparent acceleration.

3.2. Comparison to observations

Peculiar velocities are difficult to assess, since direct measurements only provide their radial component. One also needs to subtract the Hubble expansion from the data, which requires independent knowledge of the galaxy's distance. As a result, bulk peculiar velocities are estimated by means of statistical methods (e.g. see [14]).

The ΛCDM model predicts peculiar velocities no more than ~ 100 km/sec, on scales close to 100 Mpc, which die away as we move out to larger lengths. Recent independent surveys, however, have reported significantly larger drift velocities [1–5], putting the standard picture into doubt [15]. These surveys extend to lengths of $100h^{-1}$Mpc [4, 5], $300h^{-1}$ Mpc and $500h^{-1}$ Mpc [1–3], with h being the Hubble parameter in units of 100 km/sec Mpc. The results show bulk velocities as large as 500 km/sec [4, 5] and up to 1000 km/sec [1–3] on the corresponding scales. On smaller lengths (between 30 and 60 Mpc) the work of [16] suggests a (positive) variance in the local Hubble rate up to 10%. With the possible exception of the last survey, there is currently no way of knowing whether the reported bulk flows are of the desired type (i.e. with $\vartheta > 0$). Nevertheless, in the absence of better data, we will use these measurements to infer reasonable (order-of-magnitude) estimates for ϑ.[4] In particular, mainly for algebraic simplicity and illustration purposes, we will consider the intermediate value of 700 km/sec as a yardstick peculiar velocity. Note that this value is very close to the drift velocity of our Local Group.

Setting the Hubble parameter at 70 km/sec Mpc and extrapolating to 50 Mpc, 100 Mpc and 1000 Mpc, we find that ϑ/Θ is close to 1/5, 1/10 and 1/100 respectively. Then, following condition (8), the tilded observer will 'measure' negative deceleration parameter within a region of up to 50 Mpc (in an otherwise decelerating universe) if $0 < \Omega < 1.3$ in that region. This condition strengthens to $0 < \Omega < 0.5$ at 100 Mpc, while further out, near the 1000 Mpc mark for instance, $\tilde{q}$ will remain positive unless $0 < \Omega < 0.04$. Inserting these numbers into Eq. (7) we obtain a range of values for the deceleration parameter of the tilded observer on the corresponding scales. Thus, provided condition (8) is satisfied, $\tilde{q}$ varies within (-0.4, 0) on scales of 50 Mpc, between (-0.2, 0) when we move to 100 Mpc and within (-0.02, 0) near the 1000 Mpc threshold. So, in this example the size of accelerated region (i.e. that of patch (B) in Fig. 1) ranges between 50 Mpc and 1000 Mpc. Within these scales $\tilde{q}$ lies in the (-0.4, 0) range, taking its minimum value in small-scale regions of low density and approaching zero as we move on to larger lengths. These estimates are not far from those inferred by the supernovae data, which value the deceleration parameter close to -0.5 and put the transition to deceleration near $z = 0.5$ [12, 13]. The picture does not change much when we adopt the results of [16], the surveys of [4, 5], or those of [1–3]. Substituted into expressions (7) and (8), the former give $-0.2 < \tilde{q} < 0$ in regions of 50 Mpc when $\Omega < 0.5$ there. Similarly, close to 150 Mpc, the measurements of [4, 5] put $\tilde{q}$ in the range (-0.1, 0), provided $\Omega < 0.2$ there. Finally on lengths of 450 and 800 Mpc, the results of [1–3] suggest that $\tilde{q}$ varies the range (-0.07, 0) and (-0.04, 0) respectively, when $\O mega < 0.15$ and $\Omega < 0.07$ on the corresponding scales. Note that the same survey indicates bulk flows of 1500 km/sec on scales close to 150 Mpc. Inserted into Eqs. (7), (8) these values lead to $-0.3 < \tilde{q} < 0$ when $\Omega < 0.8$. One should keep in mind, however, that on relatively small scales the peculiar-velocity errorbars are large (see [1]).

Finally, substituting the previous values of ϑ/Θ into condition (11), we find that negative $\tilde{q}$'s on ~ 50 Mpc scales need $\Omega < 0.5$. Similarly, (11) translates into $\Omega < 0.2$ close to 100 Mpc and into $\Omega < 0.02$ near the 1000 Mpc mark, if $\tilde{q}$ is to become negative there. In this case, the accelerated patch extends from 50 Mpc to 1000 Mpc, with $\tilde{q}$ varying within (-0.2, 0). So, even with the last term of Eq. (6) accounted for (and in an unfavourable way), negative values for $\tilde{q}$ are still possible. Conventional almost-FRW kinematics can accommodate accelerated expansion.

[4] Using the approximate relation $\vartheta = \tilde{\mathrm{D}}_a v^a \simeq \partial v^a/\partial r^a \sim 3v/r$, where v is the magnitude of the bulk velocity and r the size of the associated region. Then, $\vartheta/\Theta \sim v/Hr$, with $\Theta = 3H$ and H representing the Hubble parameter.

4. Discussion

A decade ago, observations of high-redshift supernovae indicated that our universe was expanding at an accelerating pace [17, 18]. This conclusion was reached after applying the luminosity distances of the supernovae to the distance-redshift relation,

$$D_L = (1+z)H_0^{-1} \int_0^z e^{-\int_0^x (1+q)\mathrm{d}[\ln(1+y)]}\mathrm{d}x\,, \tag{12}$$

of an (exact) FRW model.[5] The results have repeatedly given negative values to q, indicating an accelerated expansion for our universe. In particular, the deceleration parameter was estimated close to -0.5. The same measurements also suggested that the accelerated phase was a relatively recent event, putting the transition from deceleration to acceleration around $z = 0.5$ [12]. Explaining the supernovae results has since dominated almost every aspect of contemporary cosmology. Dark energy, an unknown and elusive form of matter with negative gravitational mass, has so far been the most popular answer. Other suggestions include modifying General Relativity, introducing extra dimensions, or abandoning the Friedmann models altogether.

The reason behind such drastic proposals was that negative values for the deceleration parameter seemed impossible in conventional FRW (as well as in perturbed, almost-FRW) cosmologies. Here, however, we have shown that this is not the case. Accelerated expansion is theoretically possible in linearly perturbed FRW universes. Peculiar motions can locally mimic the kinematic effects of dark energy. Observers moving relative to the smooth Hubble flow can have local expansion rates appreciably different than that of the actual universe. Theoretically speaking, this reflects the fact that the Raychaudhuri equations in the two coordinate systems (that of the CMB and that of a drifting observer) are not the same. The difference is due to a 4-acceleration term, which vanishes in the CMB frame but takes nonzero values in any other relatively moving reference system. As a result, weakly accelerated expansion (i.e. with $-1 < \tilde{q} < 0$ – see § 2.2) is possible even when the drift velocities are small and matter is simple pressure-free dust, namely within the limits of the linear (almost-FRW) approximation. The effect is local, though the affected scales can be large enough to give the (false) impression that the whole universe has recently moved into an accelerated phase.

Extrapolating our drift velocity relative to the CMB frame, we found that peculiarly moving observers can measure negative deceleration parameter on scales between (roughly) 50 and 1000 Mpc, with $\tilde{q}$ varying in the range (-0.4, 0). Based on the surveys of [4, 5] and particularly those of [1–3], the deceleration parameter was confined within (-0.3, 0). These estimates are qualitatively in the right direction. Quantitatively, also, are not far from those inferred by the supernovae data [12, 13]. Finally, we should point out that our analysis has been based on the average peculiar kinematics without incorporating anisotropies. For instance, the symmetry of region $\mathcal{A}$ and the observers position in it can induce anisotropy in the spatial distribution of $\tilde{q}$. Generally, the higher the spherical symmetry of $\mathcal{A}$ and the closer the observer at the centre the better. These matters are less of an issue, however, when $\mathcal{A}$ is considerably larger than $\mathcal{B}$, namely as long as patch $\mathcal{B}$ lies well within region $\mathcal{A}$. The direction of the peculiar motion can also introduce an anisotropy in the $\tilde{q}$-distribution. This effect is maximised when the peculiar velocity maintains the same magnitude and direction throughout the integration period (i.e. from $z \simeq 0.5$ to the present – see expression (12)). In the opposite case, when the v_a-field has been sufficiently randomised, the anisotropy will be negligible. Estimating effects like these is currently impossible, however, as it requires detailed data on the distribution of peculiar velocities within regions of several hundred Mpc. Nevertheless, if peculiar velocities are responsible for the supernovae observations, their effect should be consistent with some degree of, most likely, dipolar anisotropy in the measured distribution of the deceleration

[5] To include the (linear) effects of peculiar motions into Eq. (12), we should replace q with $\tilde{q}$ by means of (7).

parameter. Interestingly, there have been recent reports in the literature suggesting that such kindof anisotropy may actually exist in the data [19].

Acknowledgments
The author wishes to thank Opher Lahav and Subir Sarkar for helpful comments/discussions and also for pointing out reference [19]. Special thanks also to the Conference organisers for their kind invitation and hospitality.

References

[1] Kashlinsky A, Atrio-Barandela F, Kocevski D and Ebeling H 2008 *Astrophys. J.* **686**, L49
[2] Kashlinsky A, Atrio-Barandela F, Kocevski D and Ebeling H 2009 *Astrophys. J.* **691** 1479
[3] Kashlinsky A, Atrio-Barandela F, Edeling H, Edge A and Kocevski D 2010 *Astrophys. J.* **712** L81
[4] Watkins R, Feldman H A and Hudson M J 2009 *Mon. Not. R. Astron. Soc.* **392** 743
[5] Feldman H A, Watkins R and Hudson M J 2010 *Mon. Not. R. Astron. Soc.* **407** 2328
[6] Tsagas C G 2010 *Mon. Not. R. Astron. Soc.* **405** 503
[7] Ellis G F R and van Elst H 1999 in *Theoretical and Observational Cosmology* Ed. M. Lachiéze-Ray (Kluwer, Dordrecht) 1
[8] Tsagas C G, Challinor C and Maartens R 2008 *Phys. Rep.* **465** 61
[9] KIng A R and Ellis G F R 1973 *Commun. Math. Phys.* **31** 209
[10] Maartens R 1998 *Phys. Rev. D* **58**, 124006
[11] Ellis G F R and Tsagas C G 2002 *Phys. Rev. D* **66** 124015
[12] Turner M S and Riess A G 2002 *Astrophys. J.* **569** 18
[13] Riess A G, et al 2004 *Astrophys. J.* **607** 665
[14] Trauss M A and Willick J A 1995 *Phys. Rep.* **261** 271
[15] Perivolaropoulos L 2008 preprint arXiv:0811.4684
[16] Li N and Schwarz D J 2008 *Phys. Rev. D* **78** 083531
[17] Riess A G, et al 1998 *Astrophys. J.* **116** 1009
[18] Perlmutter S, et al 1999 *Astrophys. J.* **517** 565
[19] Cooke R and Lynden-Bell D 2010 *Mon. Not. R. Astron. Soc.* **401** 1409

Recent Developments in Gravity (NEB XIV) IOP Publishing
Journal of Physics: Conference Series **283** (2011) 012041 doi:10.1088/1742-6596/283/1/012041

The Problem of Time in Quantum Cosmology: A Decoherent Histories View[1]

Theodosios Christodoulakis

University of Athens, Physics Department, Nuclear & Particle Physics Section
Panepistimiopolis 157-71, Ilissia, Athens, Greece

E-mail: tchris@cc.uoa.gr

Petros Wallden

[1] University of Athens, Physics Department, Nuclear & Particle Physics Section,
Panepistimiopolis 157-71, Ilissia Athens, Greece
[2] Technological Educational Institute of Chalkida, Department of Technological Applications, Psahna-34400, Greece

E-mail: petros.wallden@gmail.com, pwallden@phys.uoa.gr

Abstract.
The problem of time in quantum gravity arises due to the diffeomorphisms invariance of the theory and appears via the Hamiltonian constraint, in the canonical quantizations. There is a need for a description where one can ask some timeless questions that still encode some sense of temporality. The decoherent histories approach to quantum theory, already at the kinematical level admits an internal time. Several alternative proposals for resolving the problem of time via the decoherent histories, exist, and in this contribution we focus on one particular and examine how it manifests itself at some simple cosmological models.

1. The problem of time in quantum gravity

Quantum gravity is the synthesis of general relativity with quantum theory. In order to achieve this, several different approaches exist. While in this contribution we will focus on canonical approaches to QG, the problem of times exist in all approaches, manifesting itself in a different manner. The reason for this, is that it stems from the conceptual (and technical) different position that *time* has in quantum theory and general relativity[2]. In the former, time is a fixed external parameter as in Newtonian physics, while in the latter, time is on equal footing with space as manifested by the diffeomorphism invariance of the theory.

In quantum theory, time is not an observable, in the strict formal sense, as no self-adjoint operator can be constructed that corresponds to time. It solely, appears in the Schrodinger

[1] Based on talk given by P. Wallden at the NEB 14 conference
[2] Details on the nature and subtleties of the problem of time can be found in Refs. [1, 2]

equation as a parameter. In particular, any clock constructed from matter, does run backwards (even slightly) with respect to the abstract Newtonian time.

Time in general relativity is simply another coordinate, being in the same footing with the space dimensions. It is in general only locally defined. It is thus difficult to reconcile this with time in quantum theory that is based in the more-or-less Newtonian time.

Technically the problem arises due to the existence of diffeomorphisms invariance. The invariance of the theory appears as a gauge symmetry/constraint. Whenever such a constraint exist, the physical degrees of freedom do not coincide with the ones that appear at the kinematical level. For example in electromagnetism, the vector field A_μ, is not the true degree of freedom, but the equivalence classes of vector fields related by gauge transformations $A_\mu \equiv A_\mu + \partial_\mu \phi$. This is the case, because no physical property can depend on the arbitrary choice of gauge. When one is faced with the task to quantize a theory admitting such symmetries, has two options. (a) To impose the constraint classically, find the classical degrees of freedom, corresponding to the equivalent classes of gauge related configurations, and then quantize the resulting system with the usual prescription or (b) quantize the unconstrained system and impose the constraint on the quantum level (Dirac quantization). In particular if there is a classical constraint $\phi_i(x,p) = 0$ in quantum theory we require the operator version of this constraint (after suitable factor ordering choice) to annihilate any physical state

$$\hat{\phi}_i|\psi>= 0 \tag{1}$$

and furthermore that any observable $\hat{A}$ commutes with the constraints.

$$[\hat{A}, \hat{\phi}_i] = 0 \tag{2}$$

Therefore, while we start with a kinematical Hilbert space $\mathcal{H}_{kin}$ (unconstrained), after imposing the operator constraint Eq.(1) we end up with a subspace composed by the solutions of the constraint and this is the *physical* Hilbert space $\mathcal{H}_{phys}$.

Returning to general relativity, we see that the gauge symmetry in our case, is the invariance of the theory under diffeomorphisms Diff($\mathcal{M}$). In the canonical analysis of GR, we break space and time in 3+1, by selecting a foliation. The constraint then breaks into two parts:

(i) *Momentum constraint.* The theory is invariant under general diffeomorphisms in the 3-dimensional slice.

$$2iD_j \frac{\delta \Psi}{\delta h_{ij}} + H_i^{matter} \Psi = 0 \tag{3}$$

(ii) *Hamiltonian constraint.* The theory has as constraint, the Hamiltonian of the system, which reads

$$\hat{H} = -G_{ijkl} \frac{\delta^2}{\delta h_{ij} \delta h_{kl}} + h^{1/2} \left[-{}^3R(h) + 2\Lambda \right] + H^{matter} \tag{4}$$

This implies that the physical states solutions should obey the Wheeler-DeWitt equation

$$\hat{H}\Psi = 0 \tag{5}$$

and any observable should commute with the Hamiltonian/constraint

$$[\hat{H}, \hat{A}(t)] = 0 \tag{6}$$

However, the Hamiltonian, as we know from standard QT, corresponds to the generator of time translations. The time evolution of an observable $\hat{A}$ is given by

$$i\hbar\frac{d\hat{A}(t)}{dt} = [\hat{H}, \hat{A}(t)] \tag{7}$$

From Eqs. (6) and (7) we come to the profound conclusion that any observable should be independent of time. This is a general property of any theory that has weakly vanishing Hamiltonian[3]. These theories are called reparametrization invariant, since they are invariant under time reparametrizations.

There is a need therefore to construct a theory in which time plays no fundamental role. In such "timeless" theories, time should be an emergent property that appears due to coarse-grainning of the relative field configurations. In other words, all physical questions, should be translated to questions about the relative configurations of the universe and its material content. In this direction a lot of progress has been made, mainly in two approaches, the evolving constants (e.g. [3]) and the decoherent histories (e.g. [4, 5, 6, 7]). Here we will examine the latter.

2. The decoherent histories approach to quantum theory

It is an alternative formulation of quantum theory, specially designed to deal with closed systems (see for example [8]). The central objects in this approach, are histories of the system, and not one-moment-of-time propositions, and Hilbert space appears as a secondary object rather than as the starting point. The central mathematical objective of the approach is to assign probabilities to histories of a closed system. It can therefore easily deal with time-extended questions (as for example tunelling time, arrival time etc). It is an attempt to put time and space on equal footing, since we no longer deal with single time propositions (where time is obviously singled out)[4]. These considerations leads us to the conclusion that this formulation of quantum theory would be well suited to deal with the aforementioned problem of time.

If one wanted to calculate the probability, in standard "Copenhagen" QT, for a system being found in a sequence of positions in different times (a history), one would get:

$$P(\alpha_{t_1} \text{ at } t_1 \text{ and } \alpha_{t_2} \text{ at } t_2 \cdots \alpha_{t_n} \text{ at } t_n; \rho(t_0)) = \mathrm{Tr}(\alpha_{t_n}(t_n)\cdots\alpha_{t_1}(t_1)\rho(t_0)\alpha_{t_1}(t_1)\cdots\alpha_{t_n}(t_n)) \tag{8}$$

The $\alpha_{t_k}(t_k)$ are time evolved (Heisenberg picture) projection operators, and $\rho(t_0)$ is the initial state. We will call Eq.(8) *candidate* probability. However this is NOT the actual probability for this history to occur, for a given closed system. It assumes the existence of observers (and measuring apparatuses) at every moment t_k. If one attempted to use Eq.(8) as probability, he would fail, because the additivity of disjoint regions of the sample space, is not satisfied. The failure is caused due to interference. A well known example of the above, is the double slit, where the sum of the candidate probability (amplitude square in this trivial example) for crossing slit 1, and that of crossing slit 2, is not the candidate probability of crossing *any* slit. There are certain cases though, that we *can* assign the candidate probability to a closed

[3] The problem is more vital, when the Hamiltonian constraint is quadratic in all momenta, because in that case one cannot use any variable to "deparametrize" in order to recover time evolution.

[4] Other reasons to consider decoherent histories, involve emergence of classicality, decoherence etc, but these are not of interest for the topic of this contribution.

system. In particular, the decoherent histories approach, tells us when this is possible. The class operator is defined as

$$C_{\underline{\alpha}} = \alpha_{t_n}(t_n) \cdots \alpha_{t_1}(t_1) \tag{9}$$

The interference is measured by the decoherence functional, the key mathematical object in this approach. In the example of non-relativistic quantum mechanics we consider here, as a motivation, it corresponds to

$$\mathcal{D}(\underline{\alpha}, \underline{\alpha}') = \mathrm{Tr}(C_{\underline{\alpha}} \rho C^{\dagger}_{\underline{\alpha}'}) \tag{10}$$

Where $\underline{\alpha}$ is one history and $\underline{\alpha}'$ another, the decoherence functional simply measures the interference these two have[5]. We are interested in a set of histories $\{\underline{\alpha}_i\}$ that are *disjoint* (every pair of histories are orthogonal at at least one moment-of-time) and *exhaustive* (the set of histories span the full history space). Such a set is called *complete.* We can assign the probability given by Eq. (8) provided that we have a complete set of histories $\{\alpha_i\}$ that decoheres pairwise, i.e.

$$\mathcal{D}(\underline{\alpha}_i, \underline{\alpha}_j) = 0, \text{ for all } i \neq j \tag{11}$$

The probability we get for each history α_i to occur, is the candidate probability, that is equal to the diagonal parts of the decoherence functional

$$p(\underline{\alpha}_i) = \mathcal{D}(\underline{\alpha}_i, \underline{\alpha}_i) \tag{12}$$

One important issue is that there exist more than one complete set of histories (and possibly incompatible the one with the other) for a given quantum system. There is some interpretational ambiguity as to what is actually realized, or if there is a contextual reality (see for example Ref. [11, 12]).

One can generalize, all of the above, for the case of relativistic QT or even for quantum gravity. The important ingredient, one needs to consider in each of these cases, is (a) Construction of suitable Class Operators, that correspond to physical questions. (b) The use of the proper inner product, in order to be able to define probabilities. (c) The decoherence condition. Only sets of questions that decohere (obey Eq. (11)) can be used to get probabilities. The above steps will be taken below, for the case of timeless theories (with vanishing Hamiltonian). Particular (simple) quantum cosmological examples will conclude this contribution.

3. Decoherent histories and the problem of time

The first step is to consider classical (rather than quantum) timeless theories. Physical question that one can ask, which do not depend on time, could involve the full history/trajectory, of a classical object. For example:

Does a (classical) full trajectory cross a given region Δ of the configuration space?

This is indeed reparametrization invariant. However things get more complicated when we consider a quantum system. To deal with histories/trajectories, we immediately need to use the decoherent histories approach, and having the reparametrization invariance, adds further requirements, that would impose this invariance. In particular we need

[5] Further details can be found in e.g. Ref. [9] and its relation with the quantum measure in Ref. [10].

(i) An initial state obeying the constraint: $\hat{H}|\psi\rangle = 0$
(ii) Class operator (to be defined later): $[\hat{C}_\alpha, H] = 0$
(iii) We have to use the *induced* (or Rieffel) inner product, rather than the standard "Schrodinger" inner-product. It is an inner product defined on solutions of the constraint. The reader is referred to Ref. [13] for further details.
(iv) The decoherence condition needs to be satisfied in order to assign actual probabilities.

We need to find a class operator, that commutes with the Hamiltonian, gives (semi-classically) sensible results and corresponds to the question: "*What is the probability that the system crosses a region Δ of the configuration space, with no reference in time*". Here we will be following Ref. [5]. Since the classical reparametrization invariant object is the full trajectory, in our considerations we have the unphysical parameter time, running from $-\infty$ to $+\infty$. It is also clear, that the class operator for crossing the region Δ at any time, would be related to the Class operator of always remaining in the complement region $\bar{\Delta}$. The projections are related in this way $P_{\bar{\Delta}} = \mathbf{1} - P_\Delta$ and for notational simplicity we shall use $\bar{P}$ instead. The class operator for always remaining at $\bar{\Delta}$ is[6]

$$C_{\bar{\Delta}} = \prod_{t=-\infty}^{t=+\infty} \bar{P}(t) \tag{13}$$

One can see easily that this class operator commutes with the Hamiltonian,

$$[C_{\bar{\Delta}}, H] = 0 \tag{14}$$

while the crossing at some time class operator is given by

$$C_\Delta = \mathbf{1} - C_{\bar{\Delta}} \tag{15}$$

In particular one may rewrite the class operators with respect to the restricted propagator $g_r(t'', t')$ (see below for definitions and properties)

$$C_{\bar{\Delta}} = \lim_{t''\to\infty, t'\to-\infty} \exp(-iHt'') g_r(t'', t') \exp(iHt') \tag{16}$$

From this expression, one can see the resemblance with the arrival time problem in standard non-relativistic quantum mechanics (with different Hamiltonian though) as is examined in Refs. [6, 14, 15, 16][7].

The restricted propagator, seen above, is the normal quantum mechanical propagator, restricted to paths being only in a particular region ($\bar{\Delta}$ in our case). The path integral definition is

$$\begin{aligned} g_r(x, t \mid x_0, t_0) &= \int_{\bar{\Delta}} \mathcal{D}x \exp(iS[x(t)]) \\ &= \langle x | g_r(t, t_0) | x_0 \rangle \end{aligned} \tag{17}$$

[6] Some care is needed to define rigorously the continuous product of operators and taking the limits to infinity, but it can be done.
[7] Note that for periodic Hamiltonians (bounded systems), the above analysis changes slightly (see details in Ref. [5]).

One can generalize the above definition, using operators and taking the limit $\delta t \to 0$, $n \to \infty$ while keeping $\delta t \times n = (t - t_0)$:

$$\begin{aligned} g_r(t, t_0) &= \lim_{\delta t \to 0} \bar{P} e^{-iH(t_n - t_{n-1})} \bar{P} \dots \bar{P} e^{-iH(t_1 - t_0)} \bar{P} \\ &= \bar{P} \exp\left(-i(t - t_0)\bar{P} H \bar{P}\right) \bar{P} \end{aligned} \quad (18)$$

The restricted operator has the important property (Ref. [6])

$$g_r^\dagger(t, t_0) g_r(t, t_0) = \bar{P} \quad (19)$$

Using the latter, we can compute the candidate probability, provided that we have an initial state satisfying the constraint (solution of the Wheeler DeWitt equation). The probability for not crossing is

$$p_{\bar{\Delta}} = \langle \psi | C^\dagger_{\bar{\Delta}} C_{\bar{\Delta}} | \psi \rangle = \langle \psi | \bar{P} | \psi \rangle \quad (20)$$

And provided we have decoherence the probability for crossing becomes

$$p_\Delta = 1 - p_{\bar{\Delta}} = \langle \psi | P | \psi \rangle \quad (21)$$

Finally the decoherence condition is (requiring that the decoherence functional between the crossing and not-crossing class operators vanishes)

$$\lim_{t \to \infty, t_0 \to -\infty} e^{iE(t - t_0)} \langle \psi | g_r(t, t_0) | \psi \rangle = \langle \psi | \bar{P} | \psi \rangle \quad (22)$$

which becomes the very simple requirement, that *the wavefunction needs to vanish on the boundary of the region we consider*, i.e.:

$$\langle x | \psi \rangle = 0, \quad \forall x \in \partial \bar{\Delta} \quad (23)$$

At first sight, this seems to be very restrictive, since given an initial wavefunction (of a simple timeless system), there will be very few (if any) regions Δ that would satisfy the requirement. However, it is not too surprising, since with no external observer, and few or no matter degrees of freedom, one would not expect to have decoherence and be able to speak sharply about properties of the system. In realistic set-ups many non-trivial questions would exist. Below, we will see some very simple quantum cosmological models, that give non-trivial questions, and we will mention a possible comparison with the evolving constants approach to the problem of time.

4. Simple quantum cosmological examples

We will examine here the case of a Friedmann-Lemaitre-Robertson-Walker (FLRW) cosmology, i.e. a homogeneous and isotropic cosmology. In particular two specific examples will be considered. A $k = -1$ universe, with zero cosmological constant, (a) first empty and then (b) with a scalar field on it with potential $V(\Phi) = -\frac{\exp(2\Phi)}{2}$.

4.1. Empty FLRW

We have a universe with metric (in spherical polar coordinates)

$$(ds)^2 = -N^2(dt)^2 + \frac{\alpha^2}{1+r^2}(dr)^2 + r^2\alpha^2(d\theta)^2 + r^2\alpha^2(\sin\theta)^2(d\phi)^2 \tag{24}$$

Where $N(t)$ is the lapse function and $\alpha(t)$ is the scale factor. Following the usual prescription for quantization, we find the canonical Hamiltonian to be

$$H_c = N(\frac{\pi_\alpha^2}{4\alpha} - \alpha) \tag{25}$$

Where π_α is the conjugate momentum to the scale factor α. To get the Wheeler-DeWitt equation, we choose the operator ordering such that the kinetic part corresponds to the covariant Laplacian. We get the following equation

$$\Psi''(\alpha)/4\alpha - \Psi'(\alpha)/8\alpha^2 + \alpha\Psi(\alpha) = 0 \tag{26}$$

The general solutions are of the form

$$\Psi(\alpha) = C_1\alpha^{3/4}J_{-3/8}(\alpha^2) + C_2\alpha^{3/4}J_{3/8}(\alpha^2) \tag{27}$$

where the J's are Bessel functions of the first kind and C_1, C_2 are constants. We choose a particular solution as initial state of our universe

$$\Psi(\alpha) \propto \alpha^{3/4}J_{-\frac{3}{8}}(\alpha^2) \tag{28}$$

That looks like

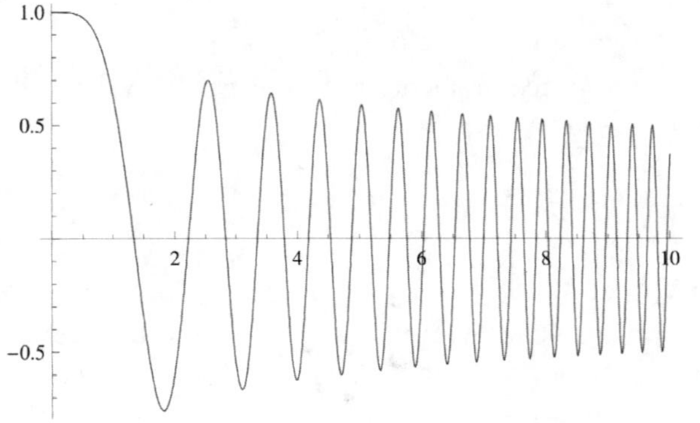

Following the prescription we analyzed in the previous sections, we know that the only questions that we can ask are those that satisfy the decoherence condition. They involve the zeros of the Bessel function. We could ask for example, which is the probability for never crossing the region $\alpha > 6$, in other words, "What is the probability that our universe (with initial condition of Eq.(28)) never reaches a scale factor α greater than 6". The probability is

$$p_c = \frac{\int_0^6 |\Psi(\alpha)|^2 d\alpha}{\int_0^\infty |\Psi(\alpha)|^2 d\alpha} \simeq 0.27 \tag{29}$$

If one wished to check, he would find that the probability for the universe to become at any time greater than $\alpha = 6$ turns out to be $p_{nc} \simeq 0.73$. This is expected, since here we have decoherence and the candidate probabilities are actual probabilities summing up to 1.

4.2. FLRW with an exponential scalar field

We now consider an FLRW universe with $k = -1$, and with an exponential scalar field of the form $V(\Phi) = -\frac{\exp(2\Phi)}{2}$. The canonical Hamiltonian is

$$H_c = N(-\frac{\pi_\alpha^2}{4\alpha} + \frac{\pi_\Phi^2}{4\alpha^3} + \alpha - \frac{\alpha^3}{2}\exp(2\Phi)) \tag{30}$$

which after the appropriate operator ordering gives the Wheeler-DeWitt equation

$$(2\alpha - \alpha^3 e^{2\Phi})\Psi(\alpha,\Phi) - \frac{1}{2\alpha}\frac{\partial^2\Psi(\alpha,\Phi)}{\partial\alpha^2} - \frac{1}{2\alpha^2}\frac{\partial\Psi(\alpha,\Phi)}{\partial\alpha} + \frac{1}{2\alpha^3}\frac{\partial^2\Psi(\alpha,\Phi)}{\partial\Phi^2} = 0 \tag{31}$$

The general solution of Eq.(31) is of the form

$$\Psi(\alpha,\Phi) = C_2 \exp\frac{\alpha^2 e^{-2\Phi}(-4C_1^2 - 4e^{4\Phi} + 6e^{6\Phi}\alpha^2)}{8C_1} \tag{32}$$

For our analysis we take one particular solution

$$\Psi(\alpha,\Phi) = \exp\frac{\alpha^2 e^{-2\Phi}(-4^2 - 4e^{4\Phi} + 6e^{6\Phi}\alpha^2)}{8} - \exp\frac{\alpha^2 e^{-2\Phi}(-4\cdot 1.5^2 - 4e^{4\Phi} + 6e^{6\Phi}\alpha^2)}{12} \tag{33}$$

Note that this solution, is not normalizable in the normal inner product, and we need to use the induced inner product (an inner product on solutions).The solution looks pictorially

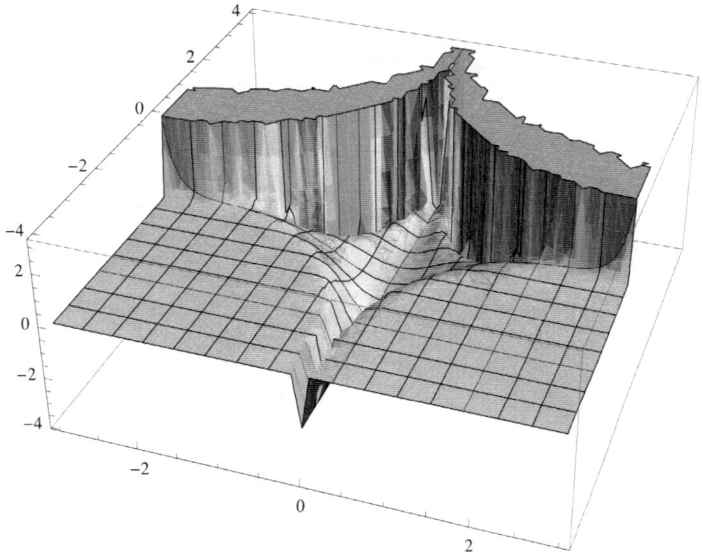

According to our scheme, we can ask a question, provided the wavefunction vanishes on the boundary of the region in question. In this example it vanishes at the following region[8].

(34)

[8] The closed form of this region is complicated

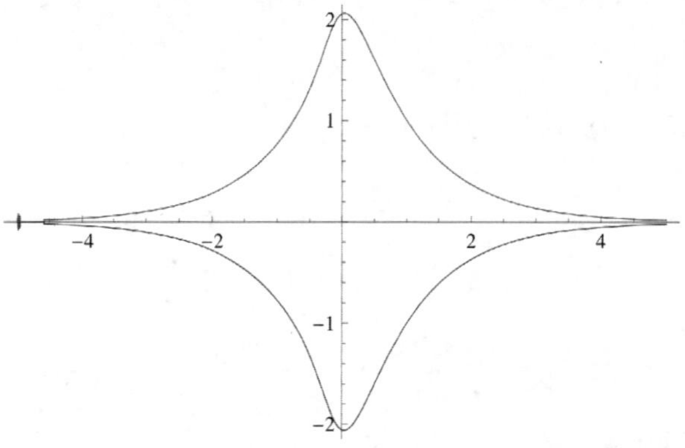

Since our solution vanishes at the above boundary, we have decoherence and we can assign the crossing probability, which however turns out to be zero $p_c = 0$. This happens even though the integral $|\Psi(\alpha, \Phi)|^2$ at the region is non-zero, because in the induced inner product it results to zero probability.

Other solutions (involving superpositions) or systems with more degrees of freedom (e.g. Bianchi Cosmologies, different matter content etc), will result to questions with non-trivial answers.

Here we should make a quick final note, on the relation of these results to similar questions in relational observables approach. One would classically evade the time dependence, by asking relational questions such as " which is the value(s) of α when $\Phi = 15$ or when $\Phi > 15$". This corresponds to an observable that projects at the range of Φ in question and since there is not explicit time dependence, classically it is reparametrization invariant. However, turning the above question to a quantum operator, it fails to commute with the Hamiltonian, and thus, is not a Dirac observable. A careful consideration, leads to the conclusion that the operator fails to commute with the Hamiltonian, due to some extra terms defined *only* on the boundary of the region considered. So while formally it is not an observable, if one was able to restrict attention to a single solution (the one realized for example) that vanishes on this boundary, then the operator would effectively commute with the Hamiltonian and be a well-defined observable. While this is not possible in the standard canonical quantization, it is allowed in the decoherent histories quantization, since the initial state is given and remains fixed in this approach. This is the advantage of using this approach, having however the major disadvantage that such questions can only give an answer, when we have decoherence. Interestingly the two conditions (commuting observable and decoherence) are both satisfied when the wavefunction vanishes at the boundary of the region considered.

5. Summary and Conclusions

We introduced the problem of time and the decoherent histories approach, in order to make a decoherent histories analysis of timeless quantum theories. We reviewed the class operator that results from this analysis and commutes with the Hamiltonian given in Ref. [5]. They gave a general enough set of physical questions of the type "Which is the probability that the system crosses a region in configuration space with no reference in time". However, following Ref. [6] we got a very restrictive but simple decoherence condition, namely "The initial state has to vanish on the boundary of the region considered". To see how this works in quantum cosmology, we took a couple of simple cosmological models (FLRW empty and with an exponential scalar field) and worked out the probabilities for crossing some regions that obeyed the decoherence condition we had obtained. Finally, some comments were made on how this result may compare with relational observables.

Acknowledgments
The authors thank Jonathan J. Halliwell for discussions in earlier stages of the work. PW was supported by I.K.Y. (State Scholarship Foundation).

References

[1] K. Kuchař, The problem of time in canonical quantization. in *Conceptual Problems of Quantum Gravity*, edited by A. Ashtekar and J. Stachel (Birkhäuser, Boston, 1991), K. Kuchař, in *Proceedings of the 4th Canadian Conference on General Relativity and Relativistic Astrophysics*, edited by G. Kunstatter, D. Vincent and J. Williams (World Scientific, Singapore, 1992).
[2] C.J. Isham, in *Integrable Systems, Quantum Groups, and Quantum Field Theories*, edited by L. A. Ibort and M. A. Rodrguez (Kluwer Academic Publishers, London, 1993).
[3] P. Hajicek, J. Math. Phys. **36**, 4612 (1995); M. Montesinos, C. Rovelli and T. Thiemann, Phys. Rev. D **60**, 044009 (1999); M. Montesinos, Gen. Relativ. Gravit. **33**, 1 (2001); R. Gambini and R.A. Porto, Phys. Rev. D **63**, 105014 (2001).
[4] K. Savvidou, J. Math. Phys. **40**, 5657 (1999); K. Savvidou, gr-qc/9912076; N. Savvidou, Braz. J. Phys. **35**, 307 (2005); C. Anastopoulos and N. Savvidou, Class. Quant. Grav. **22**, 1841 (2005); J.J. Halliwell and J. Thorwart, Phys. Rev. D **65**, 104009 (2002); J. Halliwell, in *The Future of Theoretical Physics and Cosmology*, edited by G.W. Gibbons, E.P.S. Shellard and S.J. Rankin (Cambridge University Press, 2003).
[5] J.J. Halliwell and P. Wallden, Phys. Rev. D **73**, 024011 (2006).
[6] P. Wallden, Int. J. Theor. Phys. **47**, 1512 (2008).
[7] J.J. Halliwell, Phys. Rev. D **80**, 124032 (2009).
[8] R.B. Griffiths, J. Stat. Phys., **36** 219, (1984); R. Omnès. J. Stat. Phys. **53**, 893 (1988); M. Gell-Mann and J. Hartle, in *Complexity, Entropy and the Physics of Information, SFI Studies in the Science of Complexity, Vol. VIII*, edited by W. Zurek, (Addison-Wesley, Reading, 1990); M. Gell-Mann and J. Hartle. Phys. Rev. D **47**, 3345 (1993).
[9] J.J. Halliwell, Annals N. Y. Acad. Sci. **755**, 726 (1995).
[10] R. D. Sorkin, Mod. Phys. Lett. A **9**, 3119 (1994); R. D. Sorkin, *Quantum Classical Correspondence: Proceedings of the 4th Drexel Symposium on Quantum Nonintegrability*, Philadelphia, September 8-11, 1994, pages 229-251, (International Press, Cambridge Mass. 1997) D.H. Feng and B-L Hu, editors. [arXiv:gr-qc/9507057].
[11] F. Dowker and A. Kent, Phys. Rev. Lett. **75**, 3038 (1995); F. Dowker and A. Kent, J. Statist. Phys. **82**, 1575 (1996).
[12] C.J. Isham, Int. J. Theor. Phys. **36**, 785 (1997).
[13] A. Ashtekar, J. Lewandowski, D. Marolf, J. Mourao and T. Thiemann, J. Math. Phys. **36**, 6456 (1995); A. Higuchi, Class. Quant. Grav. **8**, 1983 (1991). D. Giulini and D. Marolf, Class. Quant. Grav. **16**, 2489 (1999); Class. Quant. Grav. **16**, 2479 (1999). F. Embacher, Hadronic J. **21**, 337 (1998); N. Landsmann, J. Geom. Phys. **15**, 285 (1995); D. Marolf, gr-qc/0011112.
[14] J.J. Halliwell and E. Zafiris, Phys. Rev. D **57**, 3351 (1998).
[15] J. J. Halliwell. and J. M. Yearsley, Phys. Rev. **A** **79** 062101 (2009) .
[16] P. Wallden, J. Phys.: Conf. Ser. **67**, 012043 (2007).

AUTHOR INDEX

Institute of Physics
Dirac House, Temple Back
Bristol BS1 6BE UK

CURRAN ASSOCIATES INC
proceedings.com